# Classic Papers
# in Horticultural
# Science

# Classic Papers in Horticultural Science

Jules Janick, *editor*

*Sponsored by*
American Society for Horticultural Science

PRENTICE HALL, Englewood Cliffs, New Jersey 07632

**Library of Congress Cataloging-in-Publication Data**

Classic papers in horticultural science.

   1. Horticulture.  I. Janick, Jules, 1931–
II. American Society for Horticultural Science.
SB318.C53 1989     635     89-3537
ISBN 0-13-136847-8

Editorial/production supervision and
   interior design: **Kathryn Pavelec**
Manufacturing buyer: **Robert Anderson**
Page Layout: **Jennifer Maughan** and **Karen Noferi**

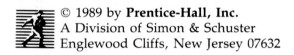
Prentice-Hall International (UK) Limited, *London*
Prentice-Hall of Australia Pty. Limited, *Sydney*
Prentice-Hall Canada Inc., *Toronto*
Prentice-Hall Hispanoamericana, S.A., *Mexico*
Prentice-Hall of India Private Limited, *New Delhi*
Prentice-Hall of Japan, Inc., *Tokyo*
Simon & Schuster Asia Pte. Ltd., *Singapore*
Editora Prentice-Hall do Brasil, Ltda., *Rio de Janeiro*

# Contents

## *part II*  Physiology and Culture

*part III*   **Breeding and Genetics**

# Preface

The origin of new information in horticulture derives from two traditions: empirical and experimental. The roots of empiricism stem from efforts of prehistoric farmers, Hellenic root diggers, medieval peasants, and gardeners everywhere to obtain practical solutions to problems of plant growing. The accumulated successes and improvements passed orally from parent to child, from artisan to apprentice, have become embedded in human consciousness via legend, craft secrets, and folk wisdom. This information is now stored in tales, almanacs, herbals, and histories and has become part of our common culture. More than practices and skills were involved as improved germplasm was selected and preserved via seed and graft from harvest to harvest and generation to generation. The sum total of these technologies makes up the traditional lore of horticulture.

The scientific tradition is not as old as the empirical but it is ancient, nevertheless. Its beginnings derive from attempts at systematic discovery of rational explanations for nature. Science, from the Latin "to know," is in reality a method for accumulating new information about our universe. The driving imperative is the desire to understand. If necessity is the mother of invention, curiosity is the mother of science. The scientific method involves experimentation, systematic rationality, inductive reasoning, and constant reformulation of hypotheses to incorporate new facts. When new explanations of natural phenomena are accepted, they must nevertheless be considered not as dogma but as tentative approaches to the truth and subject to change. The process is cumulative and science is alive only when it grows. When any society claims to know the complete truth such that further questioning is heresy, science dies.

One of the curious phenomena of science is that isolated pieces of information, even if of no immediate relevance or consequence, may have potential value. Thus the recording of experimentations in a growing body of knowledge is a prerequisite to the scientific process. The writing up of individual series of experiments and fitting the new information into the established order, or if necessary, creating a new framework, has become a specialized art form known as the scientific paper. Its theater is the scientific journal.

Horticultural science is concerned with all information relevant to the interaction of humans and the plants that serve them. Although the concerns are focused, they encompass a tremendous range of biological and physical phenomena. But because horticulture deals with the realm of the mundane—cabbages, apples, petunias—horticultural science was not always accepted as science, nor were horticultural scientists respected as scientists.

The nineteenth-century love–hate relationship between aristocratic botany, with its fascination of the irrelevant, and plebian horticulture, with its taint of the useful, is part of the resonance between the "pure" and "applied" sciences. Even the very term "pure" somehow suggests that the term "applied" is a synonym for "impure." This distinction, rather than being a remnant of Victorian snobbery, has been a persistent problem in biology and dates to the very origins of science and the nature of experimentation. Can one explain nature by studying "unnatural" events? Theophrastus neatly phrased the question 23 centuries ago: "Are we to study the nature of a plant in those that grow without human aid or in those growing under various forms of cultivation, and which of the two kinds of growth is natural?" (De Causis Plantarum I, 16.10) Theophrastus answers his own question under the heading *Cultivation Is Natural.* "For the nature of the plant is also fulfilled when that nature obtains through human art what it happens to lack, such as food of the right kind and in plentiful supply, and the removal of impediments and hindrances. . . ."

The answer was rephrased 2000 years later by William Shakespeare in *The Winter's Tale* when Polixenes justifies grafting (and cross pollination) to Perdita: ". . . this is an art which does mend nature, change it rather, but the art itself is nature."

The distinction between wild plants and cultivated plants continued to bedevil plant science. Even Gregor Mendel, the horticultural abbot, to justify his use of garden peas as a model system to study heredity, was forced to exclaim: "No one would seriously want to maintain that plant development in the wild and in the garden were governed by different laws."

This point was not settled in science and biologists and botanists continued to look down at agriculturists and horticulturists. Up until the twentieth century, horticulture was not felt to be a proper arena to deal with scientific questions. Thomas Andrew Knight in England and Liberty Hyde Bailey in the United States became the champions in the battle to treat horticultural science as science.

This compilation is an introduction to scientific papers in horticulture. Each paper chosen concerns some horticultural plant or phenomenon and has had an impact on the horticultural scientific tradition or the horticultural industry. The introductory essays have been contributed by various horticulturists, all of whom have added their special talents and perspectives. They have provided an account of each paper in its contemporary setting and in terms of its impact on the future course of horticulture. Each paper has a unique story to tell. Many of these papers and authors still engender controversy and passion. Together they underscore the diversity and richness of horticultural science.

JULES JANICK

# Contributors

Joseph Arditti
Department of Developmental
  and Cell Biology
University of California, Irvine
Irvine, California 92717

M. J. Bukovac
Department of Horticulture
Michigan State University
East Lansing, Michigan 48824

J. Scott Cameron
Southwestern Washington Research Unit
Washington State University
Vancouver, Washington 98665

Peter Carlson
Crop Genetics International
7170 Standard Drive
Hanover, Maryland 21076

Dermot P. Coyne
Department of Horticulture
University of Nebraska
Lincoln, Nebraska 68583

Frank G. Dennis, Jr.
Department of Horticulture
Michigan State University
East Lansing, Michigan 48824

Robert J. Downs
Southeastern Plant Environment Laboratories
North Carolina State University
Raleigh, North Carolina 27695

Louis J. Edgerton
Department of Pomology
Cornell University
Ithaca, New York 14853

O. J. Eigsti
American Seedless Watermelon Seed
  Corporation
Goshen, Indiana 46526

D. C. Elfving
Horticultural Research Institute of Ontario
Simcoe, Ontario, Canada

David A. Evans
DNA Plant Technology Corporation
2611 Branch Pike
Cinnaminson, New Jersey 08077

Miklos Faust
Fruit Laboratory
U.S. Department of Agriculture
Beltsville, Maryland 20705

J. Ray Frank
U.S. Department of Agriculture
Fort Detrick
Fredrick, Maryland 21701

W. H. Gabelman
Department of Horticulture
University of Wisconsin
Madison, Wisconsin 53706

Peter B. Goldsbrough
Department of Horticulture
Purdue University
West Lafayette, Indiana 47907

Jean H. Gould
Department of Soil and Crop Science
Texas A&M University
College Station, Texas 77843

Charles E. Hess
College of Agricultural and Environmental
    Sciences
University of California
Davis, California 95616

Jules Janick
Department of Horticulture
Purdue University
West Lafayette, Indiana 47907

A. D. Krikorian
Department of Biochemistry
State University of New York
Stony Brook, New York 11794

Robert W. Langhans
Department of Floriculture and Ornamental
    Horticulture
Cornell University
Ithaca, New York 14853

Cary A. Mitchell
Department of Horticulture
Purdue University
West Lafayette, Indiana 47907

Albert A. Piringer
Agricultural Research Service
U.S. Department of Agriculture
Beltsville, Maryland 20705

John T. A. Proctor
Department of Horticultural Science
Ontario Agricultural College
University of Guelph
Guelph, Ontario, Canada N1G 2W1

Lawrence Rappaport
Department of Vegetable Crops
University of California, Davis
Davis, California 95616

Edward J. Ryder
Agricultural Research Service
U.S. Department of Agriculture
Salinas, California 93905

Mikal E. Saltveit, Jr.
Department of Vegetable Crops
Mann Laboratory
University of California, Davis
Davis, California 95616

R. O. Sharples
Institute of Horticultural Research
East Malling
Maidstone, Kent ME19 6BJ England UK

Roberta H. Smith
Department of Soil and Crop Sciences
Texas A&M University
College Station, Texas 77843

Stephen C. Weller
Department of Horticulture
Purdue University
West Lafayette, Indiana 47907

Darlene Wilcox-Lee
Cornell University
Long Island Horticultural Research
    Laboratory
39 Sound Avenue
Riverhead, New York 11901

Sylvan H. Wittwer
Director Emeritus
Agricultural Experiment Station
Michigan State University
East Lansing, Michigan 48824

# Classic Papers
# in Horticultural
# Science

# part I

# Propagation and Tissue Culture

Theophrastus ca. 300 B.C. The modes of propagation in woody and herbaceous plants. Propagation in another tree: grafting. English translation of De causis plantarum by Benedict Einarson and George K. K. Link. 1976. Theophrastus: De causis plantarum. William Heinemann, London.

Theophrastus

The two botanical treatises of Theophrastus are the greatest treasury of botanical and horticultural information from antiquity and represent the culmination of a millenium of experience, observation, and science from Egypt and Mesopotamia. The first work, known by its Latin title, *Historia de Plantis* (Enquiry into Plants), and composed of nine books (chapters is a better term), is largely descriptive. The second treatise, *De Causis Plantarum* (Of Plants, an Explanation), is more philosophic, as indicated by the title. The works are lecture notes rather than textbooks and were presumably under continual revision. The earliest extant manuscript, *Codex Urbinas* of the Vatican Library, dates from the eleventh century and contains both treatises. Although there have been numerous editions, translations, and commentaries, the only published translation in English of *Historia de Plantis* is by (appropriately) Sir Arthur Hort (1916) and of *De Causis Plantarum* is by Benedict Einarson and George K. K. Link (1976). Both are available as volumes of the Loeb Classical Library.

Theophrastus (divine speaker), the son of a fuller (dry cleaner using clay), was born at Eresos, Lesbos, about 371 B.C. and lived to be 85. His original name, Tyrtamos, was changed by Aristotle because of his oratorical gifts. He came to Athens to study under Aristotle, 12 years his senior, at the Lyceum, a combination school, museum, and research center under the patrimony of Alexander. The Lyceum of Aristotle was to Plato's Academy as the land-grant colleges were to the established universities. Theophrastus inherited Aristotle's library, manuscripts, and gardens and succeeded him as head of the Lyceum, serving in that capacity for 35 years. His students (disciples) numbered over 2000. Diogenes Laertios lists 227 treatises attributable to Theophrastus, but the main works that have survived intact include his two botanical treatises and one on stones. Various fragments also exist on such diverse topics as odors, winds, and sweat. Theophrastus is also the author (perhaps editor) of a series of 30 character sketches of human weaknesses that have been considered a minor literary masterpiece for over 23 centuries. His influence was enormous. There is now a list of 1500 references to his work by ancient authors. A new collection of his work is in preparation (Project Theophrastus) in the series Rutgers Studies in Classical Humanities.

Theophrastus, called the father of botany, might better be called the father of horticultural science. He mentions over 300 species, practically all cultivated and most of them horticultural. The two treatises are a mixture of personal observations and reports of others, probably traveling students, but Theophrastus often specifically interjects his own belief. The striking feature is the systematic approach to information and the extensive compilation of botanical facts and plant lore. He is interested in the great problems of botany. How do plants grow and reproduce? How should plants be described and classified? What is the variation inherent in plant life, and what is its cause? What accounts for the geographic distribution in vegetation?

While interested in plants for their own sake, he was also concerned with the human applications of botany, in short, a horticulturist. Throughout, the work is infused with facts and the spirit is rational. He distinguishes between sexual and vegetative reproduction, dicotyledonous and monocotyledonous plants, and angiosperms and gymnosperms. On the horticultural level, he gives technical descriptions of graftage and cuttage and discusses the role of pollination of date palms and the caprification of fig.

Theophrastus's two treatises on plants are not scientific in the modern sense, for they represent descriptive rather than experimental science. Despite their clarity of thought and rational reasoning, one searches in vain for evidence on the mode of investigation or a process of verification. There are only facts and conclusions. There are brilliant insights, but also embarrassing credulity.

Theophrastus was centuries ahead of his time and little was added of substance until the Renaissance. Yet his influence in the medieval period was much less than that of Dioscorides of Anazarbos, a Roman physician of the first century and author of *Materia Medica*, who merely catalogued and described plants but proscribed medicinal uses, mostly fanciful. Thus the herbal tradition derived from Dioscorides, replacing the scientific spirit of Theophrastus, was responsible for the actual loss of botanical information. The significance of sexuality in plants, clearly defined by Theophrastus in his description of dates, seems to have been forgotten until its reintroduction in 1694 by Stephan Hales and Rudolph Jakob Camerarius. Modern botany and horticultural science are in one sense a return to the Theophrastian tradition.

Selections from Book I of *De Causis Plantarum*, translated as "The Modes of Propagation in Woody and Herbaceous Plants" and "Propagation in Another Tree: Grafting," are particularly rich in horticultural allusions. Although Theophrastus felt obliged to list spontaneous generation as a means of propagation, he is quick to note that most examples are better explained by imported seeds or very small, unnoticed seed. His discussion of grafting is very modern and with little editing would be appropriate for a modern manual. Theophrastus's discussion of the effects of scion selection and rootstocks forms the basis of modern fruit culture.

Jules Janick
Department of Horticulture
Purdue University
West Lafayette, Indiana 47907

## REFERENCES

EINARSON, BENEDICT, and GEORGE K. K. LINK (translators). 1976. Theophrastus: De causis plantarum. William Heinemann, London.

GREENE, EDWARD LEE. 1983. Landmarks of botanical history. (rev. ed.). Stanford University Press, Stanford. p. 128–211.

HORT, SIR ARTHUR (translator). 1916. Theophrastus: Enquiry into plants (2 vol.). G.P. Putnam, William Heinemann, London.

SARTON, GEORGE. 1953. A history of science: Ancient science through the golden age of Greece. Oxford University Press, Oxford. p. 547–561.

SINGER, CHARLES. 1921. Greek biology and its relation to the rise of modern biology. p. 1–101. In: C. Singer (ed.). Studies in the history and method of science. Oxford at the Clarendon Press, Oxford.

## The Modes of Propagation
### in Woody and Herbaceous Plants

4.1 We must study the modes of generation of these[c] in the light of the same considerations, laying it down that the mode most common to all is generation from seed. Still here too several modes occur, and we must distinguish the groups as they touch on the groups that have been discussed.

So with the mode under discussion,[d] generation

4.2 from the root. Some not only send up shoots from the roots spontaneously,[a] but are also propagated from these by growers, as bulbous plants and in general all with a thick and fleshy root.[b] But no such root should also have a watery fluid, as turnip and radish,[c] since these roots dry out easily and are too weak to survive. The root instead must either have several coats together with a certain viscosity (as in purse-tassel and squill), or else be quite succulent and with plenty of flesh (as in fresh sweet marjoram,[d] narcissus[e] and plants of the same kind). For only such roots as these can be planted and removed from their place. In these furthermore the duration of some is longer, of others shorter, the survival of some depending on the distinctive nature of each kind. Again, other roots send up a shoot when left in place (as roots of plants with an annual stem),[f] but the root is too dry to do so when taken up (since we must sup-

[a] That is, the root or bulb survives the winter in the ground and sends up a new plant in the next season.

[b] Cf. HP 7 2.1: "From the root are planted garlic, onion, purse-tassel, cuckoo-pint and in short such bulbous plants as resemble them." This refers to the practice of sowing seeds late in the season, then taking up the small bulbs and storing them through winter, planting them in spring and thus obtaining an earlier crop, with consequent advantages in marketing. The bulb is regarded as a root, and Theophrastus can thus call such planting "generation from the root," even though in onion the "lower roots" were removed (CP 1 4.5).

[c] Turnip and radish are always spoken of not as "planted" but as "sown" (HP 7 1.2; 7 1.7).

[d] Cf. HP 6 7.4: "Sweet marjoram grows in both ways, from a detached sucker and from seed." "Fresh" indicates that the root was left in the ground.

[e] Cf. HP 6 6.9 (of the fruit of narcissus): "This drops and produces a spontaneous sprouting; still it is also gathered and set in the ground. And the root is planted. It has a root that is fleshy, round and large."

[f] Cf. HP 7 2.1–2 and especially 7 2.1: "By root are planted garlic, onion, purse-tassel, cuckoo-pint and in general such bulbous plants. Such propagation is also possible in cases where the roots live for more than a year, although the shoots are annual;" 7 2.2: "Of those propagated from the root, the root is long-lived, although the plant itself may be annual . . ."

[c] The modes of generation of undershrubs and herbaceous plants are discussed in HP 6 6.6; 6 6.8–6 7.4; 7 2.1–9.
[d] CP 1 3.4–5.

pose the same explanation to apply to these roots as applies to certain others.[a]

Some can be planted and propagated both from a detached sucker and from their extremities. From a detached sucker grow cabbage[b] and rue,[c] and among the coronaries southernwood[d] for example, bergamot mint and tufted thyme; and some of the same—rue and southernwood with some coronaries—also grow from the other parts.[e] Indeed these last (at any rate), like ivy, have roots that come from their shoots and send them down at once, ivy being the plant which in general is best at living when cut off, both when it penetrates the trees themselves[f] and when it is stuck in the ground and covered with earth.

Of vegetables basil does this best, for it will even grow from cuttings taken from the upper parts,[g] in spite of being woody. It is why it does not readily dry out; this is why it not only survives for

4. 3

a long time, but also sprouts again when cut back.[a] Southernwood too is woody, but (like ivy) is protected by its close texture and pungency (for ivy too will grow from a cutting stuck into the ground).

These then are common to a large number of plants. We pass to forms of propagation that are both rare in occurrence and found in fewer plants.

In lily and rose[b] even the split stem grows and sends out shoots.[c] This is very similar to what happens with the olive[d] and other trees[e] that can grow from a cut piece of wood, and this is why these plants can come under the cause that was given[f] and preserve their generative fluid and heat. It is moreover reasonable that the wood should be cut up and the stem split: a start can be made more rapidly and easily from a smaller and open piece, whereas a large and closed piece is not so readily affected and thus does not sprout so easily. (This is why garlic when[g] planted is separated into its cloves[g] and the lower

4. 4

4. 5

[a] He does not specify the roots too dry to be treated in this way, since they in fact include most of the ones he has not mentioned.

[b] Cf. HP 7 2. 1: "Cabbage grows from a detached sucker, for one must include some of the root."

[c] For rue planted in a fig-tree cf. CP 5 6. 10.

[d] Cf. HP 6 7. 3: "Southernwood grows better from seed than from a root or a detached sucker . . ."

[e] Cf. HP 2 1. 3: "For . . . there are few plants that grow and are propagated more readily from the upper parts, as the vine from its twigs . . . and an occasional tree or undershrub of this description, as is held to be the case with rue, stock, bergamot mint, tufted thyme and calamint;" HP 7 2. 1: "From the shoots are planted rue, marjoram and basil . . ;"

[f] Cf. HP 3 18. 10: "Ivy . . constantly sends out roots from its shoots in the interval between the leaves, and with these roots it penetrates trees and walls . . . Hence, by removing and drawing to itself the fluid it causes the tree to wither; and if you sever the ivy below it is able to survive and live."

(This is due to incorrect observation. Ivy does not send its roots below the surface of the tree. It kills by shading the leaves. When the stem is severed it can survive for a while in a humid climate but eventually dies.)

[g] Cf. HP 7 2. 1: "From the shoots are grown rue, marjoram and basil; for basil too is propagated by cuttings when it has reached the height of a span or more, about half the shoot being cut off."

[a] Cf. HP 7 2. 4: "When the stems are broken off in practically all (sc. vegetables), the stem sprouts again . . . and most obviously . . . in basil, lettuce and cabbage;" CP 2 15. 6.

[b] Cf. HP 2 2. 1: "The rose and the lily are also generated when the stems are cut up, and so also dog's-tooth grass;" HP 6 6. 6: "The rose also grows from the seed . . ; nevertheless, since it then matures slowly, they cut up the stem and propagate it in this way."

[c] These shoots may develop from buds at the nodes or from adventitious buds which may develop from the callus which forms at the cut surface of the stem.

[d] Cf. CP 1 3. 3.

[e] Myrtle (cf. CP 1 3. 3; HP 2 1. 4) and wild olive (CP 1 3. 3).

[f] CP 1 3. 3: "they . . . preserve the vital starting-point by their close texture."

[g] Cf. HP 7 4. 11: "Garlic is planted . . divided into cloves."

The garlic of the market and kitchen is a cluster of bulbs. Each of these bulbs arises as a "side-growth" in the axils of the leaves of the parent bulb, which sends up a flowering stem. Each bulb when planted develops roots from the base and leaves from the sides of its very short stem. Buds develop in the axils of the leaves and grow into short stems with fleshy leaves, that is, into bulbs each of which is a clove. The apical end of the stem of the mother bulb grows into a floral stem. The scales which enclose each bulb and the cluster of bulbs are the bases of leaves of the parent bulb. The floral stems were used to braid the "garlic" into the chains of garlic which once festooned the vegetable markets.

roots [a] and outer scales of onions are removed, since these furnish everything that interferes with propagation, what is not alive (as the withered parts of trees) interfering with what is. Now in the olive and myrtle one must not peel off the coating when one cuts the pieces of wood,[b] since it seals off the piece and preserves the life; whereas here we are not removing coats that are alive or that determine growth. Indeed if here too we should remove this sort of coating, we are told that the pieces will not sprout.)

The most distinctive mode of generation is that 4. 6 from exudations,[c] as in the alexander, lily and a few others.[d] It is however not unreasonable, but accords with generation from split stems: all that is needed is that the generative starting-point [e] from this source as well should have been accumulated, since this kind of generation too is not without heat and fluid.

The explanation why sap and exudations [f] are not generative in all must be referred to the reasons

mentioned above [a] and earlier,[b] which explain why the stems too, and the roots too, are not generative.

The fact mentioned earlier [c] is also reasonable, that the lesser plants too have several modes of generation, in so far as it is easier to generate less perfect plants than the rest, and a smaller starting-point is needed.

The characters, then, and the causes of the modes of generation from the parts [d] are to be studied from this discussion. Indeed if anything has been omitted, it is not difficult to supply it [e] and perceive the explanation.

## (3) Spontaneous Generation

5. 1 Cases of spontaneous generation occur in the smaller plants (broadly speaking), especially in annuals and herbaceous plants. They nevertheless sometimes also occur in larger plants, either after spells of rain or when some other special condition has arisen in the air and the ground. For it is thus that silphium is said to have come up in Libya, when there had been a fall of rain described as "pitch-like" and thick, and the forest now existing there is said to have come from another such cause, not having existed before.[f]

---

[a] That is, the true roots, since the Greeks took the bulb to be a root: cf. HP 1 6. 8–9. The roots that are removed are the dried remnants of the roots of the parent bulb.

[b] Cf. HP 2 1. 4: "But both in this tree (sc. myrtle) and in the olive one must divide the pieces of wood into sizes not smaller than a span in length and not remove the bark."

[c] That is, from bulbils or bulblets. A bulblet is a small bulb formed above ground on some plants, as in the axils of the leaves of the common bulbiferous lily, and often in the flower clusters of leek and onion.

[d] Cf. HP 2 2. 1, 6 6. 8, 9 1. 4. The "others," mentioned also in HP 9 1. 4, are never specified.

[e] That is, a bud primordium must be formed. Cf. the discussion of "conflux" at CP 1 3. 5.

[f] Cf. CP 6 11. 16.

[a] CP 1 4. 4 (why the cut stems are not generative).

[b] CP 1 4. 1–2 (why the roots of lesser plants are not generative). For the need for adequate heat and fluid cf. CP 1 1. 3–4; 1 2. 3–4; 1 3. 1–5; 1 4. 2–3.

[c] CP 1 4. 1.

[d] The parts are seed, sucker (from root or stem), root, twig, branch, stem, pieces of wood, pieces of stem, and exudations. Cf. Aristotle, On Sophistical Refutations, chap. xxxiv (183 b 25–28); Nicomachean Ethics, i. 7 (1098 a 24–25); CP 6 15. 1.

[f] Cf. HP 3 1. 6: "... in some places they say that after a rain a special kind of forest has come up, as at Cyrene after a rain described as 'pitch-like' and thick; for in this way the forest near the town sprang up. And they say further that silphium, which did not exist before, made its appearance from such a cause."

Rainy spells [a] not only bring about certain cases of decomposition and alteration,[b] the water penetrating far and wide, but they can also feed what is formed and make it grow larger, while the sun warms and dries it, this [c] being also how most authorities account for the generation of animals as well.

## False Spontaneous Generation: (a) From Imported Seeds

And if the air too provides seeds which it carries down with the rain, as Anaxagoras [d] says, the rainy spells will be all the more prolific, since they would then produce an additional set of starting-points possessing supplies of food.[e] Rivers again and collections of water and streams bursting forth from the ground would do so too, importing from many sources

seeds both of trees and of woody plants (which is why rivers that shift their course make many regions wooded that were unwooded before).[a] These last forms of generation,[b] however, would not appear to be spontaneous, but a kind of propagation by sowing seeds (as it were) or setting pieces in the ground.[c]

## False Spontaneous Generation: (b) From Unnoticed Seeds

One might fancy that the generation of the fruit-less trees is rather a spontaneous one, since these trees are neither set in the ground nor produced from seed, and it is a necessary consequence that they are produced spontaneously if they are not produced in either of these ways.

But perhaps it is not true, at least of the larger plants, that they bear no seed, the truth being that we fail to observe all the cases of growth from seed, as we said in the History [d] of the willow and elm. Indeed among the smaller plants too we do not observe many cases of this among herbaceous plants, as we said [e] of thyme and others, whose seeds are not

[a] *Cf. HP* 3 1. 5 (of spontaneous generation, as presented by the natural philosophers): "But this kind of generation is somehow beyond the reach of our sense. There are other kinds that are admitted and evident to sense, as when a river overflows or makes a new bed . . . And again when there is a long spell of rainy weather, for here too plants sprout. It appears that the invasion of the rivers imports seeds and fruits . . .; and rainy weather does the same, for it deposits many kinds of seeds, and together with this it produces a certain decomposition of the earth and the water; indeed the mere mixture of water (sc. when there is no rain) with the Egyptian earth is held to produce a certain vegetation."

[b] *Cf.* note *c* on *CP* 1 1. 2.

[c] *Cf. HP* 3 1. 4 (of generation from seed, root, sucker, extremity): "We must suppose that these forms of generation belong to wild trees and also the spontaneous ones, of which the natural philosophers speak; as Anaxagoras says that the air has seeds of all and that these are carried down with the rain water and generate plants; Diogenes says plants are produced when the rain water decomposes and acquires a certain mixture with the earth, and Clidemus that plants are formed of the same components as animals . . . And certain others as well speak of the generation (sc. of plants)." *Cf.* also Anaximander, Fragments A 11 and A 30 (Diels-Kranz, *Die Fragmente der Vorsokratiker*, vol. i10, p. 84. 15–16, p. 88. 31) and Lucretius, v. 797–798.

[d] *Cf. HP* 3 1. 4, cited in note *c*.

[e] As in plants the seed, so in animals the egg and larva contain not only the starting-point, but food as well. *Cf.* Aristotle, *On the Generation of Animals*, iii. 11 (762 b 18–21): "Now the formation of plants that are generated spontaneously is from uniform substance: for they come from a certain portion of their source-substance, and one portion of it becomes the starting-point, the other the initial food for the plant that grows out."

[a] *Cf. HP* 3 1. 5, cited in note *a* (p. 34).

[b] By seeds (and "seed" can include any part from which a plant is propagated) imported by air and surface water.

[c] *Cf. HP* 3 1. 5: "It appears that the invasion of rivers imports seeds and fruits, and they say that irrigation ditches import the seeds of herbaceous plants; and rainy spells do the same, for they carry down many seeds . . .'

[a] *HP* 3 1. 2: "All (*sc.* wild trees) that have a seed and fruit, even if they grow from a root, grow also from these. So they say that even the trees considered to be fruitless generate (*sc.* from seed and fruit), as the elm and willow." Contrast Aristotle, *On the Generation of Animals*, i. 18 (726 a 6–7), cited in note *i* on *CP* 1 1. 2.

[e] *Cf. HP* 3 1. 3: "The occurrence (*sc.* with elm and willow, which appear to have no fruit, but yet generate themselves from seed) appears to be similar to what is found in certain undershrubs and herbaceous plants: they have no visible seed, some having a kind of down, some a flower (like thyme), and sprout from these;" *HP* 6 2. 3: "But no such seed (*sc.* evident to the eye) can be found in thyme. Instead it is mixed up somehow in the flower, since the flower is sown and the plant comes up from it;" *HP* 6 7. 1–2: "All the rest (*sc.* the undershrubs not grown especially for their flowers) flower and bear seed, but not all are held to do so because the fruit in some is not visible; indeed even the flower in some is hard to see. . . And yet some insist that they have no fruit . . . Nevertheless what we said first is truer, . . . and the nature of the wild congeners testifies to this . . ."

evident to the eye, but evident in their effect, since the plant is produced by sowing the flowers. Further in trees too some seeds are hard to see and small in size, as in the cypress. For here the seed is not the entire ball-shaped fruit, but the thin and unsubstantial bran-like flake produced within it. It is these that flutter away when the balls split open. This is why an experienced person is needed to gather it, by his ability to observe the proper season and recognize the true seed.

Here then is one point, propagation from unnoticed seed, and it applies to many trees, especially those that succeed each other without a break in wild forests and on mountains, since the succession could not easily be maintained if the trees were formed spontaneously.[a] Instead there are two alternatives: to come from a root or from seed.[b]

On the other hand woodcutters report that among trees of the self-same (and not just of a similar) kind a few individuals are fruitless.[a] Here it is likely that either the seed passes unnoticed or else that the tree becomes fruitless because it expends all its food on the other parts, as with vines that "get goatish",[b] and other trees[c] where this occurs. And when failure to bear is found in individuals of kinds that can or do bear fruit, what is to keep it from happening in whole kinds, which are maimed as it were in their capacity to engender fruit?

This however is to be taken as a mere opinion thrown in. We must examine the question more exactly and gather information about the cases of spontaneous generation. Broadly speaking it must occur when the earth is thoroughly warmed and the accumulated mixture[d] is qualitatively altered by the sun, which is what we observe when animals are spontaneously formed.[e]

[a] The so-called fruitless trees are all wild. Cf. HP 3 1. 2: '. . . since it is asserted that also the trees reputed fruitless generate seed, such as elm and willow. In proof is cited not only the fact that many of them grow separate from the roots of the parent, no matter where they are found, but certain localized occurrences are also taken into account, as at Pheneus in Arcadia, after the water broke out that had flooded the plain when the underground channels were blocked. Where willow had been growing near the lake, a willow grew up (they say) the next year in the part that was drained; and where elms had been growing, elms grew up, just as pines and silver-firs came up where pines and silver-firs had been growing before the lake was drained. This implies that the willows and elms were doing the same thing as the pines and silver-firs.' (Cf. HP 3 1. 2 [of wild trees]: '. . . excepting those that grow only from seed, as silver-fir, pine . . .')
[b] Cf. HP 3 1. 1 (of wild trees): "Now their modes of generation are of an uncomplicated sort: all come from a seed or a root."

[a] Cf. HP 3 3. 6–8.
[b] Cf. Aristotle, History of Animals, v. 14 (546 a 1–3): "He-goats when fat are less fertile, and from them vines are said to 'get goatish' when they fail to bear;" On the Generation of Animals, i. 18 (725 b 34–726 a 3): "Similar (sc. to failure of animals and plants to produce semen or seed because they expend their provision on bodily growth) is what happens to the vines that 'get goatish,' which get out of hand because of their feeding; thus he-goats when fat do less copulating, which is why they are previously made thin, and the vines are said to be goatish from what happens to them."
[c] Cf. CP 1 17. 10 (almond, fig, vine) and HP 2 7. 6–7 (vine, fig, almond, pear, sorb).
[d] Of earth and fresh water.
[e] Cf. Aristotle, On the Generation of Animals, ii. 6 (743 a 35–36) [of animals]: "To those produced spontaneously the movement and heat imparted by the season is the cause;" iii. 11 (762 a 9–12) [cited in note c on CP 1 1. 2].

## Propagation in Another Tree: Grafting

6. 1   It remains to discuss the cases where propagation occurs in other trees, namely in twig and bud-grafts.[a] What we have to say is simple and has (so to speak) been said already,[b] since the twig uses the stock as a cutting uses the earth.[c] So bud-grafting too is a kind of planting, and not a mere juxtaposition; here however it is evident that what produces both the sprout and the fruit is the generative fluid: the bud possesses this when it is fitted into the stock, and getting its food from the latter produces its own type of sprout.

6. 2   All grafts grow rapidly because their food has already been worked up; and this applies still more to the bud-grafts, for their food is the purest and just as it is already in the fruits[d] that are continuous with the stock. Like always coalesces readily with like,[e] and the bud is as it were of the same variety.[f]

It is also reasonable that grafts should best take hold when scion and stock have the same bark, for the change is smallest between trees of the same kind, and what occurs is as it were a mere shift in position.

For the impulse not only of the saps of the two but of the whole trees toward sprouting is then simultaneous, so that here, when graft and scion are like and have fruit with like responses, both circumstances make the rapidity of growth reasonable. In the rest the growth is more rapid as the difference in the kind of tree, the character of the sap and the seasons of development diminishes.

6. 3   The seasons of grafting are also reasonable, or rather perhaps are necessarily the ones that they are, when all further sprouting in general takes place: autumn, spring and the rising of the dog-star; for we must take a graft that feels the urge to sprout. The arguments in favour of each season are much like the arguments in favour of each as a time for planting.[a] Some persons recommend spring, the trees being still pregnant at the time of the vernal equinox, since the graft in that case will sprout at the time of the pregnancy,[b] and meanwhile the bark grows over the graft and encloses it. Others recommend the season at the rising of Arcturus, for the graft at once "takes root" (as it were)[c] and (as it were) "seals over," and once it has coalesced with the stock it puts forth its sprouts all at once at the coming of spring, having as it does a more powerful basis to start from.

[a] HP 2 1. 4 refers to the present chapter: " . . . for twig-grafting and bud-grafting are as it were mixtures or generations occurring in a different way, and of these we must speak later." Aristotle treats grafting with growth from seed and from a cutting as forms of generation: cf. On Youth and Age, Life and Death and Respiration, chap. iii (468 b 17–28).
[b] In the discussion of propagation from cuttings at CP 1 1. 3–1 3. 5.
[c] Cf. CP 2 14. 4.
[d] "Fruit" (καρπός) can be used of a fruiting shoot: cf. CP 1 12. 10.
[e] Cf. CP 5 5. 2.
[f] No great differentiation (as of flavour) has as yet occurred.

[a] At HP 2 5. 1 Theophrastus says of planting " the seasons when one should plant have been mentioned earlier;" the reference is perhaps to HP 2 4. 2, where he speaks of sowing vetch in spring and in autumn, but it is perhaps as likely that the passage has been lost. For spring and autumn as seasons for planting cf. CP 3 2. 6–3 3. 2 (for autumn cf. also CP 1 12. 2); for the dog days cf. CP 3 3. 3. For these as the seasons of further sprouting cf. CP 1 13. 3–7.
[b] And not wait until the subsequent year.
[c] Cf. the argument for autumn planting in CP 1 12. 2.

6. 4 The advice to graft buds on the smoothest and youngest axils is also reasonable. For here the buds best take hold because of the smoothness and youth of the axils, since what is young is full of life and sprouts well.

The stocks best fitted for bud-grafting, to put it in a word, are those with a certain stickiness in their fluid; further, those with bark that is soft and of the same kind and that have similar responses (which is why the best bud-grafting is on stocks close to the bud in nature and age). For the stickiness also establishes a hold;[a] and when the bark is soft and similar it favours the bud equally with the bud's own bark[b] and makes the change no great one.

6. 5 In the rest the time for grafting is short because of their rapid sprouting, but lasts longer for the olive, which keeps producing buds longer.[c] Further we are told that the new wood produced in spring stays tender and has a flow of fluid throughout the period, and the site of the graft remains moist all summer;

6. 6 and that with these advantages the graft grows better than that of any other tree; since some suppose that all this keeps the graft steeped in fluid for as long as four or even five months.[a]

Rain is harmful to a bud-graft, seeping in and decomposing it and killing it because of its weakness, and this is why it is considered safest to graft buds in the dog days, although nowadays some growers tie bark around the site to prevent rain from seeping in. For a twig-graft on the other hand rain is helpful if the graft is not naturally moist. This is why some growers plaster it with mud and others set a pot of water over it and let the water drip, in the belief that the wound is large enough for the scion to dry out quickly unless it gets fluid.

6. 7 We are rightly told (1) to keep the bud and bark from getting torn and (2) to trim the insert in such a way that no core wood is exposed at the site; for when the bark is torn[b] or the core exposed the scion dries out and perishes. This is why cultivators also first bandage the site with layers of lime bark and then plaster mud over it mixed with hair: to make the fluid remain and keep sun, rain and cold from doing

---

[a] As well as prevents drying: cf. CP 1 4.1.

[b] Since the bud is always young its bark is always soft.

[c] Columella (v. 11. 2, de arb. 26. 2) lets the time for grafting the olive last from the vernal equinox to the ides of April (April 13). The Geoponica at iii. 4. 3 give the month as April; at ix. 16. 3 they give the time from May 24 (?) to June 1. Palladius (v. 2. 3) says that the Greeks set the time as from March 25 to July 5.
The point of all this is that the shoots of the olive remain meristematic longer than those of other plants, and growth is not so rapid as in other plants but persists longer.

---

[a] That is, for the three months of summer and one or two months of spring, depending on when the graft is made.

[b] Especially if a break occurs in the continuity of the cambium of the stock.

any harm. So too after slitting the stock and giving the scion a wedge-like shape [a] they drive it in with a mallet to make the fit as tight as possible.

There must also be no excess of their own fluid in the scions. This is why in the case of the vine scions are cut two days before grafting, to allow the exudation that collects at the cut first to run off and save the scion from decomposition and mould. On the other hand scions of the pomegranate and fig and of trees drier than these are grafted at once.

6. 8

One must choose the proper seasons for grafting with both the country and the nature of the trees in view, since some combinations are too wet, others too dry. For thin soil spring is in fact [b] the better season; for what makes this combination appropriate is that thin soil contains but little fluid. For rich and muddy soil on the other hand the better season is autumn, since in spring there is far too much wetness to preserve the graft so long as bleeding still persists. Some set this autumnal season at thirty days.

6. 9

It is also reasonable that trees so grafted should bear finer fruit, especially when the scion is from a cultivated tree and the stock from a wild tree of the same bark, since the scion is better fed because the stock is strong (this is why it is recommended to plant wild olives first and later graft them with cultivated buds or twigs). For the grafts hold better to the stronger tree, and since this tree attracts more food they make it a finer producer. Indeed if one should reverse the procedure and graft wild scions on a cultivated stock, there would be a certain improvement in the wild crop but no fine fruit.

6. 10

Let this suffice for the discussion of planting in the sense of grafting.

[a] In cleft grafting.
[b] And not the dog days, as was believed (*CP* 1 6. 6).

 Lewis Knudson. 1922. Nonsymbiotic germination of orchid seeds. Botanical Gazette 73:1–25.

Lewis Knudson

Sometime between 1654 and 1701, Georgius Everhardus Rumphius described an orchid he called *Angraecum scriptum* (now known as *Grammatophyllum scriptum*). This description may include the first known references to orchid seeds. In describing the flower he wrote:

> . . . it remains fresh for a long time. . . . Finally they begin to wither, but do not drop, and . . . make the fruit . . . a six-edged capsule . . . inside entirely filled by a yellow, flaky flour. The ripe ones become dark grey and open up readily in six parts . . . and then the yellow flour is largely shed and is blown away on the wind; but whether this is endowed with a seed-virture, and settles to grow on other trees, is still unknown (de Wit 1977).

Orchid seeds are small, averaging about 1 mm in length and 0.8 mm in diameter, and are nearly invisible to the naked eye. Under natural conditions they germinate following penetration of their mycorrhizal symbiont by swelling and then forming a small body known as a protocorm, which produces leaves and, subsequently, roots.

Over the years orchids were thought to multiply by means of gemmae-like bodies and to produce seeds that were incapable of germination (for reviews, see Arditti 1967, 1979, 1984). These beliefs were replaced by reality in 1804 when the British botanist Richard Anthony Salisbury (1761–1829) observed and described orchid seedlings for the first time.

Tropical orchids started to arrive in Europe around the year 1700. *Brassavola nodosa* from Curaçao was in cultivation in Holland in 1698. *Bletia verecunda* from the Bahamas was brought to England in 1731 and flowered in 1732. *Vanilla* reached Great Britain before 1739 (Lawler 1984). The flow of species increased after that, but plants were propagated by division or not at all. Seed germination was unknown until 1822 or 1832 when seedlings of *Prescottia plantaginea* (which was sent to the Chiswick garden of the Horticultural Society of London in 1822 by John Forbes, a collector in Brazil) were ". . . raised abundantly. . . ." It is not clear how these seedlings were raised, but it is possible that seeds, either produced through self-pollination (this species tends to do that in Great Britain) or from a capsule present on a specimen collected in the wild, simply fell near the base of the plant and germinated. At approximately the same time efforts were made in France to germinate *Orchis* seeds by placing them on light soil and covering them with fine moss. In 1844, Josef Neumann, also in France, reported that he had pollinated *Calanthe veratrifolia*, obtained seeds, germinated them, and grew four seedlings which were expected to flower in 1845, but there are no reports regarding these seedlings and Neumann's statement may have been wishful thinking (Arditti 1984).

In the 1840s, seeds of *Epidendrum elongatum*, *Epidendrum crassifolium*, *Cattleya forbesii*, and *Phaius albus* were produced by David Moore (1807–1879), Director of the Glasnevin Botanical Gardens in Ireland; J. Cole, gardener to J. Willmore of Oldford near Birmingham, obtained seeds of *Phaius tankervilleae* and *Epidendrum elongatum*; and Robert Gallier, gardener to J. Tildesley of West Bromwich, Staffordshire, crossed *Dendrobium nobile* and *Dendrobium chrysanthum*. David Moore was the first to report his success in 1849, and his report in the *Gardeners' Chronicle* probably caused Cole and Gallier to report their experiences. Cole's plants seem to have survived, but Gallier's hybrid seedlings did not.

Around 1852, John Harris (1782–1855), a senior surgeon at the Devon and Exeter Hospital in Great Britain, suggested to John Dominy (1816–1891), orchid grower for the well-known English nursery firm of Veitch and Son, the possibility of orchid hybridization. Dominy followed the advice, and using the seed germination methods developed by Moore, Cole, and Gallier of scattering seed near the base of mature plants or on compost which supported orchids, produced the first man-made orchid hybrid, *Calanthe* Dominyi, in 1856. Using the same methods, he produced a second hybrid, the first in the genus *Cattleya* in 1859. A *Paphiopedilum* hybrid named after Harris flowered in 1869 and is still in existence. The development made orchid hybridization possible and the English, more than anyone else, produced many of the first hybrids in a number of popular genera. The seeds were always germinated through the method used by John Dominy. This method was symbiotic because it depended on mycorrhizal infection for germination. However, horticulturists and botanists alike seem to have been unaware of this, despite the fact that the fungus was observed and even drawn, but apparently not recognized, by S. Reisek as early as 1847. By 1800 it was clear that plants which harbored mycorrhizae benefited from them, and one of the first to appreciate this was the German botanist Albert Bernhardt Frank (1839–1900), who coined the term "mycorrhiza" around 1855. The role of mycorrhizae in the germination of orchid seeds was first recognized by the French botanist Noel Bernard (1874–1911), who described it in 1899.

Bernard's deduction was brilliant. He followed his initial observations by successful attempts to germinate symbiotically a number of European and tropical orchids. Bernard devoted much of his time to basic research and accomplished an amazing amount in the 10 years before his untimely death in 1911. He isolated a number of orchid endophytes, predicted that one day orchid establishments would have laboratories, and shortly before his death showed that penetration of the mycorrhizal fungus induced the formation of fungal inhibitors by orchid tissues. In other words, he discovered phytoalexins 30 years before K. O. Muller, working in Australia at the time, proposed the term and the concept, and 40 years prior to the isolation of the first orchid phytoalexin by Ernst Gaümann and his coworkers in Switzerland. His discovery was described in a posthumous paper written by his friends, mentor, and colleagues and published in 1911 (for a review, see Stoessl and Arditti 1984). Despite his great achievements, Bernard failed to develop an asymbiotic method for orchid seed germination.

After Bernard's death, Hans Bugeff (1883–1976), professor of botany at the University of Würzburg, Germany, became the pre-eminent expert on orchid mycorrhiza. He studied both the endophytes and the orchids, isolated and attempted to classify the fungi, worked on physiological aspects, and wrote five books and several articles (in German) and one review (in English) on the subject. Burgeff germinated many orchids symbiotically, but like Bernard, he failed to develop an asymbiotic method.

The asymbiotic method of orchid seed germination was developed by the American botanist Lewis Knudson (1884–1958), professor of botany at Cornell University. Lewis Knudson was born on October 15, 1884, in Milwaukee. He received a B.S. degree in agriculture from the University of Missouri (an honorary D.Sc. degree was awarded in 1942) and entered Cornell University as a graduate student, starting an association that lasted 51 years, until his death in 1958. His career began as an assistant in plant physiology under B. M. Duggar. Not long after that he was advanced to the rank of instructor and was awarded his Ph.D. degree in 1911.

Knudson was appointed assistant professor of plant physiology in 1911 and a year later he became acting head of the department on the resignation of Duggar. This is a meteoric rise, but those who knew Knudson claim that he was a master diplomat, a good administrator, a man who could make others want to work for him, and a hard taskmaster, all at the same time. In 1916 the plant physiology department was incorporated into the newly formed department of botany. Knudson was advanced to professor of botany and in 1941 became head of the department. He retired in 1952, but continued to visit the laboratory regularly and his last paper on orchids was published in 1957 in the *Botanical Gazette*. During his career Knudson served on the board of directors of the Bailey Hortorium, eventually becoming the director. In

1956, after his retirement, he was awarded the Gold Medal of the Federation of the Garden Clubs of New York State.

Knudson did not seem to have had a special interest in orchids at first. His interests were sugar utilization, the organic nutrition of angiosperms, tannic acid fermentation and the enzyme tannase, secretion of amylase by roots of maize and peas, some fungi, and axenic culture of plants (which he learned from J. K. Wilson), which must have led him to the papers of Bernard and Burgeff. He immediately perceived what the two European botanists missed: ". . . Bernard used a substance known as salep [to germinate orchid seeds] . . . it is rich in pentosans and starch . . . the influence of the fungus might be to digest some of the starch. . . ." On the basis of this he reasoned ". . . that germination of orchid seed might be obtained by the use of certain sugars." To test his assumptions, Knudson added sugar to Pfeffer's solution (which he probably designated as Solution A) and to a modification that he called Solution B.

Knudson's first paper on the subject was published in Spanish, probably as a result of a visit to Spain in 1920. On his return to the United States, Knudson published a much more detailed and classic paper (Knudson 1922) and continued to work on the subject after that (Knudson 1925, 1927, 1929, 1930) and eventually formulated an improved medium (Knudson 1946). This solution, known as Knudson C, is used universally for orchid seed germination at present. It was also the first solution to be used for *in vitro* clonal propagation (what would at present be called "micropropagation" or "mericloning") and is still employed for this purpose with a few orchids. One would have expected universal acclaim for Knudson's initial discovery, but this was not the case. His paper was greeted by bemused skepticism (Ramsbottom 1922a,b), attacks, and arguments (for reviews, see Arditti 1972, 1984). Captain John Ramsbottom, the great British mycologist, declaimed: ". . . an orchid seed without its fungus is like Hamlet without the Prince of Denmark."

Knudson's discovery of asymbiotic orchid seed germination had a great impact in Indonesia and Singapore. In the United States, Great Britain, and Europe, Knudson's discovery made large-scale orchid hybridization possible. Commercial and hobby growers as well as amateurs made thousands of new hybrids, many from intergeneric crosses.

Knudson's method reduced the cost of producing orchids and converted them from playthings of the rich to a plant that anyone could grow. Another effect of Knudson's procedure was to engender a need for mass rapid clonal propagation of the outstanding hybrids. It is therefore reasonable to state that Knudson was indirectly responsible for the invention of micropropagation, especially since the first successful attempts to culture orchid explants *in vitro* (Rotor 1949) were made on his medium C.

Joseph Arditti
Department of Developmental and Cell Biology
University of California, Irvine
Irvine, California 92717

# REFERENCES

ARDITTI, J. 1967. Factors affecting the germination of orchid seeds. Bot. Rev. 33:1–97.

ARDITTI, J. 1972. Professor Lewis Knudson and the asymbiotic germination of orchid seeds: The fiftieth anniversary. Am. Orchid Soc. Bull. 41:899–904.

ARDITTI, J. 1979. Aspects of the physiology of orchids. Adv. Bot. Res. 7:421–655.

ARDITTI, J. 1984. A history of orchid hybridization, seed germination and tissue culture. Bot. J. Linn. Soc. (London) 89:359–381.

DE WIT, H. C. D. 1977. Orchids in Rumphius' *Herbarium Amboinense*, p. 47–94. In: J. Arditti (ed.). Orchid biology: Reviews and perspectives. Cornell University Press, Ithaca, N.Y.

KNUDSON, L. 1922. Further observations on nonsymbiotic germination of orchid seeds. Bot. Gaz. 77:212–219.

KNUDSON, L. 1925. Physiological study of the symbiotic germination of orchid seeds. Bot. Gaz. 79:345–379.

KNUDSON, L. 1927. SYMBIOSIS AND ASYMBIOSIS RELATIVE TO ORCHIDS. NEW PHYTOL. 26:328–336.

KNUDSON, L. 1929. Physiological investigations on orchid seed germination. Proc. Int. Congr. Plant Sci. 2:1183–1189.

KNUDSON, L. 1930. Flower production by orchids grown nonsymbiotically. Bot. Gaz. 89:192–199.

KNUDSON, L. 1946. A new nutrient solution for germination of orchid seed. Am. Orchid Soc. Bull. 15:214–217.

LAWLER, L. J. 1984. Ethnobotany of the Orchidaceae. p. 27–149. In: J. Arditti (ed.) Orchid biology: Reviews and perspectives. Vol. III. Cornell University Press, Ithaca, NY.

RAMSBOTTOM, J. 1922a. Orchid mycorrhiza. Charlesworth & Co. Catalog, Hawards Heath, England.

RAMSBOTTOM, J. 1922b The germination of orchid seeds. Orchid Rev. 30:197–202.

ROTOR, G., JR. 1949. A method for vegetative propagation of *Phalaenopsis* species and hybrids. Am. Orchid Soc. Bull. 18:738–739.

STOESSL, A. AND J. ARDITTI. 1984. Orchid phytoalexins. p. 151–175. In: J. Arditti (ed.) Orchid biology: Reviews and perspectives. Vol. III. Cornell University Press, Ithaca. NY.

VOLUME LXXIII NUMBER 1

# THE
# BOTANICAL GAZETTE

*January 1922*

## NONSYMBIOTIC GERMINATION OF ORCHID SEEDS

LEWIS KNUDSON

(WITH THREE FIGURES)

The germination of orchid seeds for a long time has been recognized as difficult and generally uncertain of attainment. Practical orchid growers for years have attempted to find a method which will insure germination. They meet with success at times, but fail utterly on a second attempt with the same method. Moreover, two sowings made at the same time and under apparently identical conditions may result in germination in the one case and failure in the other. There are growers in England, France, and also the United States who, if one may believe reports, are consistently successful in germinating the seeds of the commercially important orchids. The grower, however, is naturally unwilling to part with the details of his method. From the scientific aspect it is doubtful whether he can explain the cause of his success. Generally speaking it may be stated that practical orchid growers have not yet solved the problem of producing orchid plants from seeds.

The difficulty of germinating seeds of orchids is due in part to inherent causes, but undoubtedly is due also to environmental factors. The extremely small size of the embryo renders it liable to death if it becomes desiccated. Generally the seeds are sown on a substratum rich in organic matter, such as sawdust, leaf mold, wood or bark, peat, sphagnum, or mixtures of the two last-named substances. These substances are favorable for the growth

of fungi and algae, and the embryos may be killed because of being covered by these organisms, or more likely by injurious substances produced by the decomposition of these organisms. The work of BURGEFF (4) and BERNARD (2) demonstrates that death may be due to pathogenic fungi, and the writer's experiments in transplanting seedlings from tubes to open pots demonstrate clearly this danger. In addition to these factors, attention must be given to preventing loss due to insect pests. As suggested, however, there are apparently inherent characteristics of the seeds which make for refractory germination. It is this which attracted the attention of BERNARD, who in a number of publications presented evidence tending to show that the germination of the seeds and the subsequent growth of the seedlings are dependent upon infection by certain strains of the fungus which generally is found living in the orchid root, and which BERNARD considered to be *Rhizoctonia*. BURGEFF came to substantially the same conclusions, maintaining that germination was possible only when the embryo became infected with the proper strain of the fungus, to which he gave the name *Orcheomyces*, without attempting to classify it.

BERNARD and BURGEFF both pointed out that infection of the embryo began at the suspensor end of the seed, and that in the case of *Cattleya* and related forms the primary infection occurred through the delicate suspensor. Growth occurred if only the lower portion of the embryo became infected, and if the infection continued beyond approximately the lower third of the embryo, then death of the embryo resulted. It was also observed in germinating embryos that the fungus disintegrated in the infected zone, forming clumps of disintegrated hyphal material in the cells similar to the clumps found in cells of the root. It was the opinion of BERNARD that the fungus was digested by the orchid embryo. The essential point to be noted, however, is that a delicate balance between the host and the fungus apparently must be maintained in order to insure germination and also to prevent death of the embryo.

Granting for the present that a symbiotic relationship exists between the fungus and the embryo, it is nevertheless true that failure of germination is more common than success, even when the fungus is provided. BERNARD's experiments reveal case after case in which the introduction of the fungus was followed by death of the seeds or failure to germinate. He states as follows:

The germination by inoculation is not obtained without certain difficulties. For five years I have sown seeds of diverse species of orchids in culture tubes, each of which contained 100 seeds, and these I have inoculated with *Rhizoctonia* obtained from the roots. Altogether, I have obtained a few hundreds of seedlings, but I underestimate when I place the number of seeds used in my experiments at 50,000. For the majority of the seeds, the association with the fungus that I have placed in their presence has been merely passive and without effect, or impossible or rapidly injurious to the embryos.

The explanation generally offered in these cases is that "activity" of the fungus was altered or the proper strain was not employed, so that the essentially delicate balance between the fungus and the embryo was not maintained.

In certain experiments BERNARD succeeded in germinating seeds of *Cattleya* and *Laelia* without the intervention of the fungus. This was accomplished by using a more concentrated solution of salep. Salep (KING, **6**) is the dry powder obtained by pulverizing tubers of certain orchids, and contains, principally, mucilage 48 per cent, starch 27 per cent, and proteins 5 per cent. It probably contains also some sugar as well as soluble mineral matter. The seedlings obtained in this way were in every respect normal and the germination was very regular. BERNARD suggests that some such method might be developed for practical purposes, since the results with the fungus are so unsatisfactory.

The increasing importance of orchid culture in America, the difficulties in and the restrictions on the importation of orchid plants, and the desirability of creating new hybrid forms, make particularly desirable a method for germinating the seeds. Certain data from the experiments of BERNARD and BURGEFF, indicating that soluble organic compounds might cause germination, and my own previous experiments (**7**) on the organic nutrition of plants, demonstrating that various sugars have a very favorable influence on growth, are indications that germination of orchid seeds might be obtained by the use of certain sugars. This proved to be true.

The results here reported describe a method for germinating the seeds under sterile conditions, the influence of certain sugars

on growth of the embryos, the influence of different concentrations of sugar on growth, the effects induced by certain plant extracts, the favorable influence of certain bacteria, and experiments on transplanting. In the discussion are treated critically the ideas expressed by BERNARD and BURGEFF with respect to the function of the fungus.

For a clearer understanding of the data that follow, it is desirable to trace briefly the mode of development in the germination of seeds of *Cattleya* and *Laelia*. For a detailed discussion, BERNARD's paper should be consulted. The embryo is somewhat oval-shaped, and is undifferentiated except that the cells at the basal region are large, while those at the apical region are smaller. This is the meristematic region. At the base is subtended a delicate suspensor. The embryo is inclosed within a transparent integument with an opening at the lower end through which the suspensor may protrude. The maximum length of the embryo of *Cattleya* or *Laelia* is about $250\mu$ and the width about $75\mu$.

Germination consists, first, in an enlargement of the embryo in a transverse direction until a small spherule stage is reached. Accompanying this development there is the formation of chlorophyll, generally more pronounced in the meristem region. The embryo when it ruptures the integument has a width of about $175\mu$ and a length of about $270\mu$. At the time of rupturing the integument, absorbing hairs begin to grow out from the epidermis. Subsequent development consists in a further enlargement of the embryo, other absorbing hairs begin to develop near the basal region, and there is attained a large spherule or top-shaped structure characterized by a marked depression at the upper surface. Following this there appears in the middle of the depression the first leaf point, which subsequently develops into the first leaf. During this period there is a continued increase in the diameter of the embryo, so that a disklike structure is formed which has been termed by BERNARD the protocorm. At the meristematic region a second and a third leaf may unfold, elongation may occur, and a distinct stem is apparent. The first root may arise either from the protocorm or from the stem below the second or the third leaf. The period required for these developments is generally from four to six months. Under greenhouse conditions this advanced stage has apparently been attained in some cases in a shorter time.

## Methods

Unless otherwise indicated, all cultures were made using agar slopes in culture tubes 180 mm.×18 mm. The nutrient solution used was either Pfeffer's or a modification referred to hereafter as solution B. The solutions were made up as follows:

SOLUTION B
$Ca(NO_3)_2$, 1 gm.
$K_2HPO_4$, 0.25 gm.
$MgSO_47H_2O$, 0.25 gm.
$Fe_2(PO_4)_3$, 0.05 gm.
$(NH_4)_2SO_4$, 0.50 gm.
Distilled $H_2O$, 1 l.

PFEFFER'S
$Ca(NO_3)_2$, 4 gm.
$K_2HPO_4$, 1 gm.
$MgSO_47H_2O$, 1 gm.
$KNO_3$, 1 gm.
$KCl$, 0.5 gm.
$FeCl_3$, 40 mgm.
Distilled $H_2O$, 5 l.

Solution B was used because BURGEFF stated that the orchid seeds utilized ammonium sulphate to better advantage than the nitrate salt. My own experience is not in accordance with this.

Generally 1.50 per cent agar was used, and all media and vessels were autoclaved at fifteen pounds pressure for thirty minutes. To prevent the lodging of spores and microorganisms on the cotton stopper of the culture tube, it was capped with a small vial which, fitting tightly over the cotton plug, inclosed the upper third of the tube. The use of the vial cap was essential because otherwise, under the moist greenhouse conditions, contamination resulted from spores growing down through the cotton plug or between the plug and the tube. By using the vial cap cultures remained pure even after a year in the greenhouse.

The cultures were all grown under aseptic conditions. For sterilizing the seeds, the calcium hypochlorite method of WILSON (13) was used. For this purpose 10 gm. of calcium hypochlorite was added to 140 cc. of distilled water. This was vigorously shaken for a few minutes and then filtered. The clear filtrate was used for sterilizing the seeds. The quantity of seeds desired was placed in a small test tube and the clear filtrate added. The tube was then shaken until each seed became moistened with the solution.

This was repeated several times, since the seeds generally float together in a mass at the surface of the liquid. The period of exposure was about fifteen minutes, although in some preliminary experiments with seeds of Cattleya and Laelia no injury was noted after a three hours' exposure. The seeds were transferred from the sterilizing solution, without any previous rinsing in water, by the use of a platinum needle. With the small loop used, it was possible to pick up about 100 seeds. These were scattered over the surface of the agar slope. The cultures were maintained in moist chambers in the greenhouse shaded by cheesecloth from direct sunlight, with the temperature between 15° and 35° C. In determining growth, the embryos were measured by means of an ocular micrometer. As was shown by both BERNARD and BURGEFF, the width of the embryo or the protocorm may be accepted as a good criterion of the degree of growth. Other data are included, such as percentage of germination, time of formation of first leaf, color, starch content, etc.

## Preliminary experiments

EXPERIMENT 1.—On December 7, 1918, seeds of Cattleya Schroederae×C. gigas were sterilized by treating them for two hours with the calcium hypochlorite solution. The seeds were sown on agar slopes. The medium used in one case was an extract of peat, made by autoclaving 300 gm. of bog peat, such as is used for potting orchids, with 1200 cc. of tap water. This was filtered and the clear brownish filtrate used. The other medium was made by autoclaving 400 gm. of dormant canna tubers with 600 cc. of water for thirty minutes. By January 7, 1918, the seeds on both were in the small spherule stage and were green. On April 10, four months after planting, the seeds on the canna medium had germinated, the seedlings having one and two leaves. On the peat agar medium the embryos were a little larger than on January 7, 1919, but not significantly different.

EXPERIMENT 2.—The media used were extracts of carrot and garden beet. The carrot extract was made by autoclaving 70 gm. of young carrots (root) with 75 cc. of tap water, and the garden beet extract was made by autoclaving 50 gm. of young beets (root) with 75 cc. of water. The extracts were filtered, and to the clear filtrate 1.25 per cent agar was added. Seeds of *Cattleya labiata* × *C. aurea* were sterilized and planted on February 14, 1919. On May 13 some of the seeds in each had germinated and the remainder were almost germinated, that is, they were just at the point of producing the first leaf.

EXPERIMENT 3.—The media used were Pfeffer's alone and Pfeffer's plus 1 per cent sucrose. Seeds of *Cattleya mossiae* were planted on January 14, 1919. On July 1 the seeds in the sucrose culture had germinated, one leaf showing. On the Pfeffer's alone the embryos were in a small green spherule stage, the diameter being about 250μ, while the diameter of the embryos on sucrose was about 1000μ.

EXPERIMENT 4.—Seeds of *Cattleya intermedia* × *C. Lawrenceana* were sown on July 18, 1919, on solution B plus 2 per cent glucose on the one hand and 2 per cent sucrose on the other. Owing to an absence from the University, the cultures were not examined until June 9, 1920. At that time, in both glucose and sucrose cultures, the seedlings were well developed, although the culture media had lost most of the water by evaporation. The seedlings had two or three leaves and one or two roots, some of the roots being 4 mm. in length.

## Influence of certain sugars and plant extracts on germination

The preliminary experiments show that germination of seeds of *Cattleya* and *Laelia* is possible without the aid of the fungus, provided soluble organic substances are present, particularly sugars. In all these cases the leaf point appeared only after three months, and yet under practical greenhouse conditions, when the seeds are sown merely on a compost of peat and sphagnum or other organic material, the leaf points may appear in a shorter time. For example, according to Mr. T. L. MEAD, of Oviedo, Florida, seeds of *Cattleya* have shown leaf points in as short a period as thirty-five days. Some of the media used by him were oak bark, magnolia bark, and a compost of decayed leaves and sphagnum. It is possible, of course, that under these practical conditions the fungus is a factor in the growth. Is it possible also that certain

of the products produced on decomposition of the organic substances, such as the auximones described by BOTTOMLEY (3) or the vitamine water-soluble B, are involved in the germination of orchid seeds? Of course other factors may be involved, such as the hydrogen ion concentration, mineral salts, or the rate of transpiration, particularly as influencing the organic composition of the plant.

That a full nutrient medium plus sugar is not capable of sustaining continued growth of higher plants was shown by KNUDSON and LINDSTROM (8) in their experiments with albino corn. The plants kept either in the light or in the dark and supplied with one of several different sugars all died after a month or two. These experiments, together with the work of BOTTOMLEY on auximones, the beneficial influence of vegetable extracts on the growth of fungi recently described by DUGGAR (5) and by WILLAMAN (11), and the beneficial influence of vegetable extracts on the growth of yeast as described by WILLIAMS (12) and by BACHMANN (1) suggest that more rapid germination and more vigorous plants could be obtained if a vegetable extract was added to the nutrient medium.

With no idea of determining what specific substances are involved in stimulating growth, but in the endeavor to develop a rapid and effective method for the germination of seeds of certain orchids, the experiments described in table I were made. The nutrient solution used was solution B.

The extracts used were prepared as follows. Potato extract: 200 gm. of new potato with the skin removed, with 300 cc. of distilled water; wheat extract: 200 gm. of air-dried soft wheat, with 300 cc. of distilled water; beet extract: 200 gm. of a red garden beet cut into small pieces, with 200 cc. of distilled water. Extraction was made by autoclaving for fifteen minutes at fifteen pounds pressure, and the extracts obtained by filtration. The yeast juice was obtained as follows: three four-liter flasks, each containing three liters of WILLIAMS' solutions, were inoculated with a cake of Fleischman's yeast, and after a week the yeast was filtered from the solutions, autolyzed at 37° for twenty-four hours, and then dried by suction and washing with ether. Seventy gm. of yeast was then steamed for ten minutes with 250 cc. of distilled water. The liquid was filtered and made up to a liter volume by the addition of distilled water.

All cultures were made in quadruplicate. The individual cultures of each series were strikingly uniform in growth, so that

## TABLE I

*Laelia-Cattleya* HYBRID NO. 1;* SEEDS PLANTED AUGUST 31, 1920; MEASUREMENTS MADE JANUARY 27, 1921

| CULTURE NO. | CULTURE SOLUTION | WIDTH OF EMBRYO IN MICRONS | | | PERCENTAGE OF GERMINATION | ORDER OF SUPERIORITY, MARCH 27 |
|---|---|---|---|---|---|---|
| | | Minimum | Maximum | Average | | |
| B 53... | Full nutrient 50 cc.+50 cc. beet extract | 242 | 582 | 407 | 0 | 5 |
| B 17... | Full nutrient 50 cc.+50 cc. potato extract | 339 | 630 | 459 | 0 | 5 |
| B 60... | Full nutrient 50 cc.+50 cc. wheat extract | 174 | 436 | 291 | 0 | 7 |
| B 39... | Full nutrient+2% glucose | 485 | 1261 | 970 | 20 | 4 |
| B 28... | Full nutrient solution alone | 194 | 242 | 213 | 0 | 8 |
| B 27... | Full nutrient+2% glucose | 485 | 1358 | 814 | 70 | 3 |
| B 34... | Full nutrient +2% glucose 90 cc.+10 cc. beet extract | 582 | 1358 | 979 | 90 | 1 |
| B 55... | Full nutrient +2% glucose 90 cc.+10 cc. wheat extract | 582 | 1552 | 1076 | 90 | 1 |
| B 66... | Full nutrient+2% fructose 98 cc.+2 cc. yeast extract | 776 | 1260 | 940 | 60 | 1 |
| B 48... | Full nutrient+2% fructose | 485 | 1358 | 902 | 80 | 2 |
| B 22... | Full nutrient+2% fructose 90 cc.+10 cc. beet extract | 485 | 1202 | 1008 | 90 | 1 |
| B 31... | Full nutrient+2% fructose 90 cc.+10 cc. potato extract | 582 | 1260 | 902 | 70 | 2 |
| B 5... | Full nutrient+2% fructose 90 cc.+10 cc. wheat extract | 582 | 1358 | 1047 | 80 | 1 |
| B 43... | Full nutrient+2% fructose 96 cc.+4 cc. yeast extract | 582 | 1746 | 970 | 60 | 2 |
| B 14... | Beet extract alone | 242 | 872 | 388 | 0 | 5 |
| B 69... | Potato extract alone | 339 | 679 | 594 | 0 | 5 |
| B 6... | Wheat extract alone | 242 | 436 | 358 | 0 | 6 |
| B 74... | Yeast extract alone | 87 | 339 | 194 | 0 | 7 |

* Composition: *L. Perrinii* Lindl. ¼; *C. labiata* Lindl. ¼; *C. amethystoglossa* Reichb. ¼; *C. intermedia* Grah. ¼.

for the first measurements only one culture for each series was taken, and forty individual measurements were made for each culture. The measurements given in table I were made on January 27, 1921. The order of superiority of the cultures on March 27 is recorded in the last column of the table. Similar data, not

included, were obtained in a like experiment with *Laelia-Cattleya* hybrid no. 2.

The degree of development represented by the numbers 1 to 8 is as follows: (1) dark green seedlings, most of these with two leaves and a few showing roots; (2) seedlings the same as no. 1, but light green; (3) seedlings light green, most of them with one leaf and a few showing two leaves; (4) seedlings light green, with only one leaf and that leaf short; (5) about 50 per cent of embryos showing leaf point; (6) embryos just showing a depression in meristem region; (7) advanced spherule stage; (8) smaller spherule stage.

The data in table I show that fructose is more favorable for growth of the embryos than glucose. This is apparent not only in the percentage showing leaves, but in the general appearance of the cultures. The embryos in the glucose cultures were whitish or yellowish in color. On the other hand, the fructose cultures were dark green. A more striking difference was noted on March 27, when in the glucose cultures the embryos were still yellowish and had shown no appreciable gain since January 27. The fructose cultures, on the other hand, had progressed and were still more markedly superior to the glucose cultures than on January 27. Fig. 1 shows the fructose culture and the nutrient solution culture minus sugar.

The addition of a plant extract to the glucose cultures has a marked effect on growth and chlorophyll development. In each case the percentage of germination is higher than with glucose alone, and the ranking of glucose-containing cultures on March 27 indicates that those with yeast or wheat extract rank with the best cultures, that with beet extract ranks in the third group, and the cultures with glucose alone fall in the fourth group. The addition of plant extract to the fructose-containing media is practically without any beneficial effect.

The loss or lack of development when glucose is supplied in the nutrient solution has been noted by MAZÉ and PERRIER (9) for corn, and by SERVETTAZ (10) in nutrition experiments with moss. In the case of orchid embryos the chlorophyll makes its appearance only when the leaf is developing, and then generally only in the leaf. Even then the leaves are only of a light green color. Is chlorosis due to a non-utilization of nitrogen or iron in the presence of glucose, or in the upbuilding of chlorophyll is glucose less favorable than fructose? Certain experiments now in progress may throw some light on this interesting point, and it is therefore desirable to await the results before speculating any further.

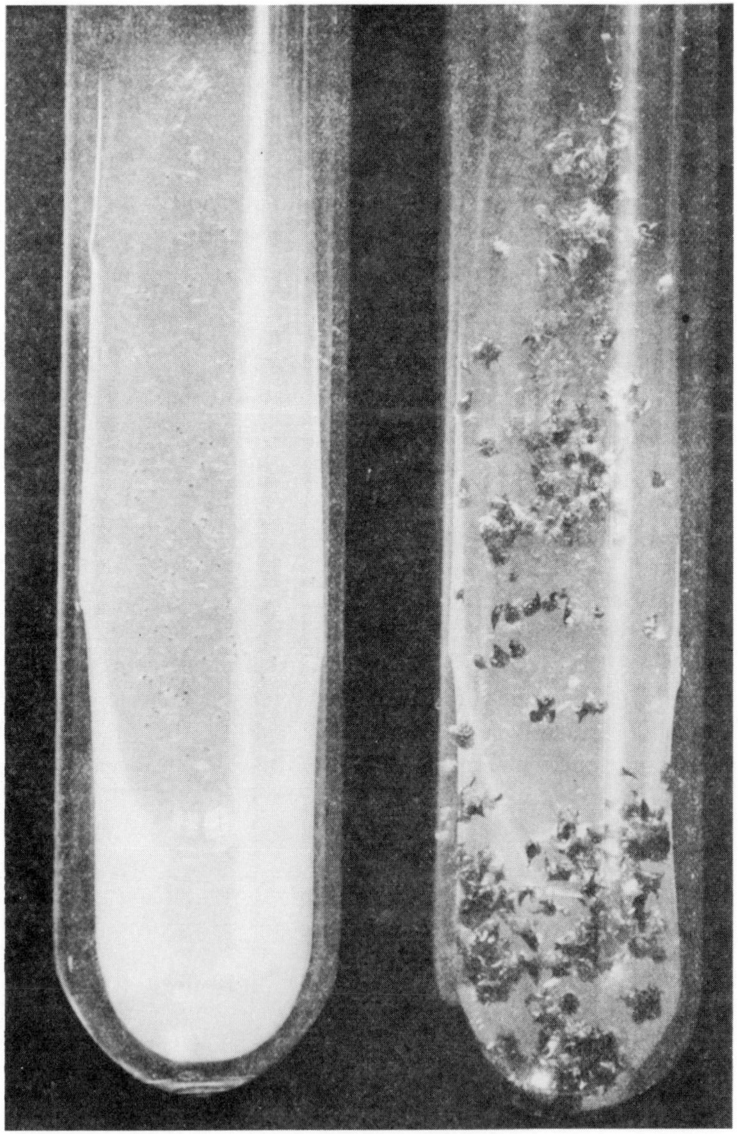

On January 27 the leaf point was not yet evident in any of the plant extract cultures. On March 10, however, in the wheat, beet, and potato extracts, embryos with leaf points were apparent, and a little later the same was noted for the yeast extract cultures. After several months more, seedlings with one and two leaves were to be noted in all these cultures.

Is germination on these extracts due to sugars or to other substances? Analyses made of the different extracts show that the potato, wheat, and yeast extracts had a sugar content of less than 0.025 per cent. The extracts diluted one-half with the nutrient solution, therefore, practically speaking had no sugar. The beet extract had a sugar content of 0.80 per cent, yet it did not permit any better germination than did the potato extract with merely a trace of sugar. As indicated previously, it should be borne in mind that the beet extract contains some substance injurious to the embryo. Furthermore, the stimulating effect of the plant extracts when added to the glucose solutions must be due to substances other than sugars. In the experiments on the influence of concentration of sugar, it will be noted that on the concentration of 0.05 per cent no germination has occurred even after four months.

## Influence of concentration of sugar

In view of the fact that germination is possible when sugar is supplied in the culture media, it seemed desirable to determine the concentration most favorable for growth. Accordingly several series of experiments were made, a number of which are here reported.

In the first experiment seeds of *Laelia-Cattleya* hybrid no. 2 were used. These were planted November 12, 1920, and notes were made December 16, 1920, and January 11, February 15, and March 15, 1921. Each of the average figures given represents the average

FIG. 1.—Solution B, embryos in small spherule stage; solution B+2 per cent fructose, seedling stage; ×2.

of thirty separate measurements. The data are given in table II. The slow growth is well shown. There is in general a corresponding increase with increase in concentration, but the increase in concentration beyond 0.80 per cent is without any significant effect. On February 15 the embryos of all the cultures were examined for starch. It was found only in those cultures with 0.80 per cent glucose or higher. This fact is evidence that the absorption of glucose at a concentration of 0.80 per cent is in excess of the utilization, and consequently a higher concentration

TABLE II

INFLUENCE OF CONCENTRATION OF GLUCOSE, *Laelia-Cattleya* HYBRID NO. 2;* SEEDS SOWN NOVEMBER 12

| CULTURE SOLUTION | AVERAGE WIDTH OF EMBRYOS IN MICRONS | | |
| --- | --- | --- | --- |
| | December 16 | January 11 | March 15 |
| Solution B................. | 126 | 145 | 174 |
| Solution B 0.05% glucose..... | 184 | 232 | 247 |
| Solution B 0.10% glucose..... | 200 | 252 | 339 |
| Solution B 0.20% glucose..... | 242 | 281 | 475 |
| Solution B 0.40% glucose..... | 310 | 291 | 455 |
| Solution B 0.80% glucose..... | 291 | 339 | 533 |
| Solution B 1.00% glucose..... | 320 | 417 | 543 |
| Solution B 2.00% glucose..... | 320 | 436 | 523 |

* Composition: *C. Trianaei* Reichb. ½; *C. Loddigesii* Lindl. ¼; *L. purpurata* Lindl. ¼.

should be without any increased beneficial effect. It should be borne in mind that glucose used with solution B is not particularly suited for the germination of orchid seedlings, since there is induced constantly in the embryos a distinct chlorosis. It is probable that higher concentrations of sucrose or fructose would permit of a more rapid germination.

The results of several other experiments on the influence of different concentrations of glucose on the germination of seeds of *Cattleya* are in agreement with these results, and need no repetition. In an experiment with seeds of *Epidendron*, germination was obtained with a concentration of 0.2 per cent glucose. In the cultures with less than 0.1 per cent glucose, not only was there a less development of the embryos, but a large percentage of the seeds never showed any initial swelling and development of chlorophyll.

The detailed data are given in table III. The figures given under average width represent the averages of forty individual measurements. Only seeds that had shown an initial increase in diameter and were green were included.

In this experiment another interesting observation was made. Just previous to the formation of the leaf point, the embryos were gorged with starch. With the formation of the leaf, however, there was a disappearance of the starch, it having been converted into

TABLE III

INFLUENCE OF CONCENTRATION OF GLUCOSE, *Epidendron tampense*×*E. inosmum*; SEEDS PLANTED DECEMBER 8, 1920; NOTES TAKEN MARCH 17, 1921

| CULTURE SOLUTION | CULTURE NO. | WIDTH OF EMBRYOS IN MICRONS | | | PERCENTAGE WITH LEAVES | REMARKS |
| --- | --- | --- | --- | --- | --- | --- |
| | | Minimum | Maximum | Average | | |
| Full nutrient............. | P 2 | 116 | 203 | 145 | .... | 95% no change |
| Full nutrient+0.05% glucose | P 72 | 116 | 291 | 178 | .... | 90% no change |
| Full nutrient+0.1% glucose | P 78 | 189 | 407 | 281 | .... | 80% no change |
| Full nutrient+0.2% glucose | P 81 | 339 | 630 | 465 | 10 | 30% no change |
| Full nutrient+0.4% glucose | P 86 | 291 | 582 | 397 | 10 | 40% no change |
| Full nutrient+0.8% glucose | P 91 | 194 | 582 | 446 | 20 | 50% no change |
| Full nutrient+1.00% glucose | P 99 | 339 | 582 | 446 | 10 | 30% no change |
| Full nutrient+2.00% glucose | P 104 | 291 | 630 | 446 | 10 | 25% no change |

sugar, as evidenced by the fact that some of the embryos still showed a slight presence of dextrins.

## Influence of microorganisms

Throughout the various experiments made a few cultures always became contaminated. Generally, if the contamination was a *Penicillium*, the embryos became covered by the mycelium and death resulted. Those embryos not covered often showed a marked increase in growth over the embryos in the corresponding uncontaminated cultures. The increase in growth may have been due to one or a combination of the following: an increase in the carbon dioxide content of the tube; a change in the chemical character of the nutrient medium, brought about either by secretion of organic substances from the fungus or by products produced on decomposition of the fungus; or changes in the sugar effected by extracellular enzyme action.

In an experiment with seeds of *Epidendron* growing on solution B plus 0.80 per cent glucose, one of the cultures became contaminated with a species of *Actinomyces*. The result was that when the embryos in the contaminated culture were dark green, with one and two leaves, the embryos in corresponding uncontaminated cultures were still white or yellowish and only one or two of them showing the leaf point. In the contaminated culture the embryo

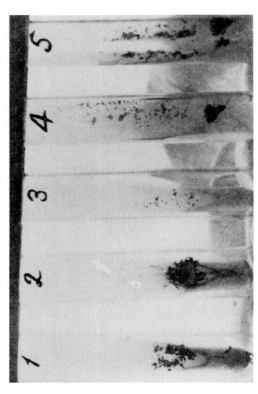

FIG. 2.—Culture no. 1, solution B+2 per cent glucose contaminated by *Actinomyces* sp. (corresponding check like no 3); culture no. 2, Pfeffer's +2 per cent sucrose inoculated with *Bacillus radicicola*; culture no. 3, same but not inoculated; culture no. 4, solution B+2 per cent glucose; culture no. 5, solution B+2 per cent glucose inoculated with *B. radicicola*; ×3.

had an average width of 975 $\mu$, while in the uncontaminated cultures the average width of the embryos was 600 $\mu$.

In view of these observations, and since BOTTOMLEY reported a beneficial effect on the growth of *Lemna* by the addition to the nutrient solution of an aqueous extract of *Azotobacter* or of *Bacillum radicicola*, an experiment was made to determine the influence of the latter organism on germination. Ten tubes were prepared with Pfeffer's solution plus 2 per cent sucrose; five were inoculated with *B. radicicola* from alfalfa, and five were left uninoculated. Seeds of *Epidendron* were sown on December 18, 1920. On March 5, 1921, 80 per cent of all the seeds in the inoculated tubes had

germinated and possessed one or two leaves. On the cultures not inoculated, the embryos lacked chlorophyll, and not one had produced even the leaf point. Most of the embryos exhibited the depression at the meristem region. Two months later some of the embryos, although still lacking chlorophyll, had formed the leaf point; while in the inoculated cultures the seedlings had two and three leaves and some had already produced roots (fig. 2).

The results appeared so unusual that the experiment was repeated with seeds of *Laelia-Cattleya* hybrid no. 3, using solution B and also Ashby's solution. The composition of the nutrient

TABLE IV

INFLUENCE OF *Bacillus radicicola* ON *Laelia-Cattleya* HYBRID NO. 3;* SEEDS SOWN MARCH 14, 1921; NOTES TAKEN AUGUST 10, 1921

| Nutrient medium | Average width in microns |
| --- | --- |
| Solution B | 271 |
| Solution B+0.1% glucose | 407 |
| Solution B+1.0% glucose | 814 |
| Solution B+inoculated | 194 |
| Solution B+0.1 glucose inoculated | 378 |
| Solution B+1.0% glucose inoculated | 834 |
| Ashby's solution | 194 |
| Ashby's solution+0.1% glucose | 446 |
| Ashby's solution+1.0% glucose | 698 |
| Ashby's solution+inoculated | 164 |
| Ashby's solution+0.1% glucose inoculated | 397 |
| Ashby's solution+1.0% glucose inoculated | 970 |

*Composition: *C. superba* Schomb. ‡; *C. Dormaniana* Reichb. ‡; *C. Warscewiczii* Reichb. ‡; *L. purpurata* Lindl. ‡.

solution and the width of embryos are given in table IV. The favorable influence of *Bacillus radicicola* was noted only in the cultures having 1 per cent glucose. On solution B+1 per cent glucose the diameters of the embryos averaged the same for both the inoculated and the uninoculated seeds, but there was a striking difference in the color and the number with leaves. In the uninoculated, 20 per cent of the embryos were showing the leaf point, but the embryos were whitish in color. In the inoculated, 50 per cent of the embryos showed leaves, the leaf development was greater, and the embryos were green.

On Ashby's medium plus 1 per cent glucose, the beneficial effect of the organism was more apparent. There was a marked increase in the width of the embryos. In the inoculated cultures all the embryos had produced one or two leaves and the embryos were dark green. In the uninoculated cultures only 25 per cent of the embryos had produced a leaf point and the embryos were chlorotic (fig. 2). On both solution B and Ashby's+1 per cent glucose, the influence of *Bacillus radicicola* was so strikingly beneficial that it was observable immediately.

In the cultures with 0.10 per cent glucose or no glucose, the influence of *B. radicicola* seemed to be injurious, for in all the inoculated cultures the average width of the embryos was less than in the uninoculated cultures.

The cause of this favorable influence of *Bacillus radicicola* on the growth of orchid embryos remains yet to be determined. Some experiments were also made in which the cultures were inoculated with *Azotobacter* sp. In every case, however, there was a marked retardation in growth.

**Transplanting experiments**

On July 12, seeds of *Laelia-Cattleya* hybrid no. 2 were sown on solution B plus 2 per cent fructose plus 1.5 per cent agar. As culture vessels, Erlenmeyer flasks of 150 cc. capacity were used, and 30 cc. of the medium was employed. On October 14 the embryos were just on the verge of producing the leaf point. They were then transferred to six Erlenmeyer flasks (D 18 to D 23) containing 50 cc. of nutrient media, as follows: Pfeffer's solution+2 per cent glucose+0.1 cc. carrot decoction.

On March 1, 1920, the seedlings in cultures D 18 to D 23 had two and three leaves with a pronounced protocorm, and from the protocorm one and two roots had grown out, the roots varying from 1 to 3 cm. in length; whereas in the corresponding tube cultures, four seedlings in a hundred had produced roots, and these roots were only 2 mm. and 3 mm. in length. On March 11, 1921, seedlings were transferred from cultures D 22 and D 23 to liter Erlenmeyer flasks containing 300 cc. of solution B plus or minus sugar. Eight cultures were made, four with 2.0 per cent sucrose

and four without sugar. On August 20, 1921, notes were made on these cultures. In the sucrose cultures the seedlings had made a marked development. The largest seedlings had four and five leaves, some of the leaves being 2 cm. in length; while the roots of these seedlings, two or three in number, were 2-5 cm. in length (fig. 3). In the cultures lacking sugar the growth was less striking, the leaves being 2-4 mm. and the roots 2-10 mm. in length.

Seedlings were transplanted at the same time from D 23 to a compost of peat and sphagnum in ordinary flower pots. These seedlings, on August 20, 1921, showed leaves 5-7 mm. and roots 1-2 cm. in length. These seedlings were better than those planted on solution B in the liter flasks, but not so good as the seedlings on solution B plus 2 per cent sucrose in the liter flasks.

That better growth is possible in 150 cc. to 500 cc. Erlenmeyer flasks than in culture tubes was demonstrated repeatedly during these experiments. Erlenmeyer flasks of from 150 to 500 cc. capacity, containing the culture media left over after supplying the tubes with the requisite amount, were generally planted with the seeds that remained after the tubes were sown. In practically every instance germination took place sooner in the flasks than in the tubes. The probable explanation is that in the tube cultures the inward diffusion of carbon dioxide is impeded to a certain extent by the cotton plug, and, the volume of air in the tube being small, the carbon

FIG. 3.—Seedlings one year old on solution B+2 per cent sucrose; ×⅔.

dioxide is soon exhausted. In the larger flasks, however, the volume of air, and therefore the volume of $CO_2$, is much greater, from six to twenty-five times as great, and furthermore, the area through which the $CO_2$ can diffuse is greater by virtue of the larger mouths of the flasks. That the diffusion of $CO_2$ is impeded by a cotton stopper was shown in a previous paper (KNUDSON 7).

From the practical standpoint this would seem to be a method for the propagation of orchid seeds. The seeds may be germinated in the small culture tubes or in larger containers, and when roots are produced they may be transplanted either to pots in the open or transferred to sterile culture in larger flasks. My efforts to develop the seedlings on peat sphagnum mixture in flower pots in the open resulted in failure on several different occasions, due in one case to the temperature running up to 40° C., which permitted a pathogenic fungus to destroy utterly the seedlings in about twenty-five pots. In another case, during an absence from the city the seedlings were destroyed by insects. Previous to the misfortunes which the seedlings experienced they had been growing for periods of three and four months and were making satisfactory development. Other experiments are now in progress on this phase of the question. Some tubes were sent to Mr. T. L. MEAD, of Oviedo, Florida. Some of these were transplanted four and five months ago, and according to a recent communication from Mr. MEAD, the seedlings transferred are continuing growth, and but few seedlings were lost as a result of transplanting. The results of certain experiments now in progress indicate that more rapid growth will be obtained if the culture seedlings are transferred to sterile media containing sugar and grown for a year or two under these conditions. This method, moreover, has the advantage that the seedlings are not exposed to the depredations of insects or the ravages of parasitic fungi. Furthermore, contamination of the cultures by *Penicillium* or *Aspergillus* is without any injurious effect, provided that at the time of transplanting the seedlings have roots.

## Discussion

What is the significance of these results in relation to the views advanced by BERNARD and BURGEFF, and quite generally believed

today, that for the germination of orchid seeds infection of the embryo by the appropriate fungus is essential? BERNARD believed that the action of the fungus was a physicochemical one, in that the fungus would cause an increase in a concentration of the cell sap, which increase in concentration would induce germination and the formation of a protocorm in somewhat the same way that the form of algae could change by increasing the concentration of the external solution. He points out that the fungus can invert sucrose and this may occur in the embryo.

The writer believes that the fungus may bring about germination in another way. As previously pointed out, in all his media BERNARD used a substance known as salep. This is a powder derived by grinding the dried tubers of certain species of orchids, and is rich in pentosans and starch, containing also about 5 per cent of organic nitrogenous substances. It probably contains some soluble organic and inorganic matter, judging from freezing-point determinations made by BERNARD. In view of the fact that organic matter is present, it is conceivable that the influence of the fungus might be to digest some of the starch, pentosans, and nitrogenous substances; which digestion products, together with secretions from the fungus or products produced on decomposition of the fungus, might be the cause of germination. In brief, it is conceivable that germination is induced not by any action of the fungus within the embryo, but by products produced externally on digestion or secreted by the fungus. Unfortunately I have not as yet succeeded in satisfactorily isolating the organism stated as necessary by BERNARD, nor has it been possible to purchase salep. Work is still in progress on this problem. There are, however, certain facts which support the idea that the action of the fungus is not necessarily internal. BERNARD does not give any analyses of the medium used, but he does give certain cryoscopic data. The medium generally used, made with salep, had a freezing-point depression of 0.01° C. Assuming that this depression ($\triangle$) is produced largely by hexose sugars, it would indicate at the outset of the experiment a concentration of hexose sugars equivalent to 0.1 per cent glucose. It is not possible, of course, to say that this depression is due entirely to hexose sugars; perhaps other sugars are present, as well as other soluble organic and inorganic substances. The significant fact is that at the outset some soluble organic substances are present.

In addition to the soluble substances present, which apparently are not sufficient in quantity nor suitable as regards quality to permit of germination, there are to be considered the insoluble organic substances, pentosans, starch, and organic nitrogenous substances. Digestion of starch by the fungus would augment the concentration of sugar, and digestion of the organic nitrogenous substances might produce certain products which would make possible the germination of the seeds.

In my experiments it is true that the sugar used generally was of a relatively high concentration, but in the case of *Epidendron*, germination was obtained on 0.2 per cent glucose, which sugar is not particularly favorable for growth. That other substances besides sugar exert a pronounced influence is shown by the experiments on the beneficial effects of adding certain plant extracts to the glucose-containing solutions. The fact that other substances besides sugars may be important in the germination is shown by the experiments in which germination was obtained on decoctions of yeast, wheat grains, or of potato. All of these extracts contained less than 0.02 per cent total sugar. The experiments on the influence of *Bacillus radicicola* also lend weight to the idea that certain extraneous products may markedly influence germination.

Burgeff, in certain of his experiments, used 2 per cent salep, but in other experiments he used starch, sucrose, or glucose. The explanations offered with respect to the function of the fungus in discussing Bernard's work may be used to account for the results obtained by Burgeff. There may appear to be rather more difficulty in explaining the function of the fungus in the cultures containing either glucose or sucrose. It will be necessary to discuss these in more detail.

In one experiment, seeds of *Cattleya* were sown in a tube containing a nutrient solution plus 0.33 per cent sucrose. After three months the embryos were 0.4–0.5 mm. in width. Then, according to Burgeff, they remained stationary. Cultures four months old, inoculated and maintained in the dark at 23° C., produced the first

leaf at the age of eight months. Burgeff does not state that the uninoculated cultures were maintained under the same condition, but presumably they were. Granting that the uninoculated culture did not produce leaves, it is possible to explain the germination on the basis of the inversion of sucrose, which would yield approximately concentrations of both glucose and fructose molecularly equivalent to the original sucrose concentration. In addition there is to be considered the possible influence of products secreted by the fungus or produced on decomposition of the fungus.

The favorable influence of saprophytic fungi and bacteria demonstrated by my experiments is paralleled by certain cultures of Burgeff. He transplanted four months' old seedlings to a mineral nutrient medium containing salep. Some of the cultures became contaminated with saprophytic organisms. The uncontaminated culture showed little growth, if any, after three months; while the culture contaminated by *Penicillium* made a marked growth, the leaves being 4 cm. in length. Another saprophytic fungus in another culture likewise caused a marked increase in growth, the leaves being about 9 mm. in length; and in a third culture contaminated by bacteria the seedlings were of similar character to those of the pure culture, but apparently darker green in color. All of these seedlings had produced vigorous roots. In the pure culture some of the seedlings had died. Burgeff considers that the more favorable growth in the tube with fungus contamination was due to the development of an acid reaction in the medium. There is a possibility that increased carbon dioxide content and other products produced by the organism are partly responsible. It should be stated that the seedlings originally transferred to these tubes had previously been infected by the essential fungus.

Another experiment of Burgeff lends weight, however, to the idea that the fungus is effective in inducing germination as a result of certain reactions brought about within the embryo. In this experiment the culture medium consisted of a weak nutrient solution plus $\frac{1}{20}$ per cent starch. Seeds of a *Laelia-Cattleya* cross were planted, and the root fungus from each of seventeen different orchids was tested for its ability to induce germination. In the cultures uninoculated the embryos attained a width of 0.45 mm. in four months. In certain cultures only the suspensor became infected. The diameter of the embryos in this case reached from 0.6 to 0.8 mm. in the same time. With infection still more advanced, but less than normal, a few seeds had leaves after seven months. Normally infected embryos produced leaves, and embryos that adhered to the wall of the tube likewise germinated. An infection more advanced than normal caused the same development as an infection slightly less than normal. When from one-half to two-thirds of the embryos became invaded, growth was less than in the embryos not infected, and in another case the seeds were killed outright.

As stated, certain facts from these experiments make it difficult to explain the action of the fungus as purely external. If so, why should the fungi behave so differently in inducing or retarding germination? Unfortunately BURGEFF gives no details so that one may judge whether or not these results could be duplicated on second trial. Another difficulty is an adequate explanation for the germination of seeds adhering to the inner surface of the culture tube. It is possible. of course, that decomposition products of the fungus growing on the surface of the tube may have been the cause. It is desirable to await experiments with the fungus before attempting to discuss these points further.

There are other phases of the problem presented by BERNARD, especially the loss by the fungus of its capacity to induce germination after prolonged culture in the laboratory. It is entirely possible that there has been no loss in the fungus, but that at the time of inoculating the culture the physiological state of the embryos was such as to resist or permit of infection. Those in which the infection was confined to the lower cell could still germinate despite the fungus. Those invaded to a greater extent would be killed. These and other experiments of BERNARD and BURGEFF suggest that one of the causes for the failure of germination is the parasitic character of the fungus. In other words, it is possible that the fungus, instead of being an aid in normal germination, is a factor in the death of the embryos and consequently in the failure of germination.

In conclusion, it may be stated that the evidence for the necessity of the fungus for germination has not yet been conclusively proved. The evidence is conclusive that under conditions of pure

culture employed by both BERNARD and BURGEFF germination of the seeds is dependent on the fungus. There is still considerable work to be done, however, before the validity of the fungus hypothesis can be proved or disproved.

## Summary

1. A method is given for sterilizing seeds of certain orchids and for growing them under sterile conditions.

2. Germination of seeds of *Laelia, Cattleya,* and related forms is possible without the aid of any fungus when certain sugars are supplied.

3. Fructose appears more favorable than glucose.

4. In the presence of glucose, chlorosis of the embryo generally results.

5. Germination is possible on certain plant extracts containing merely traces of sugar.

6. Embryos in sugar-containing cultures accumulate a considerable reserve of starch.

7. The concentration of glucose is important in the growth of the embryo.

8. *Bacillus radicicola* from alfalfa and certain other microorganisms on certain media have a favorable influence on the development of chlorophyll and germination.

9. Seedlings have been transplanted from tubes to large flasks and growth has continued.

10. The results thus far obtained indicate that the method is of value in the propagation of orchids from seeds.

11. The idea is advanced that the necessity of fungus infection for germination has not yet been proved.

12. One cause of failure of germination may be the pathogenic character of some of the endophytic fungi.

Mr. T. L. MEAD, of Oviedo, Florida, who has worked for many years on the practical problems of germinating orchid seeds, has supplied all the seeds used in these experiments. He has likewise supplied certain information on the practical difficulties and on various results obtained by him. For all of these favors and for constant interest I wish to express my thanks.

LABORATORY OF PLANT PHYSIOLOGY
CORNELL UNIVERSITY

## LITERATURE CITED

1. BACHMANN, F. M., Vitamine requirements of certain yeasts. Jour. Biol. Chem. **39**:235. 1919.

2. BERNARD, NOEL, L'évolution dans la symbiose, les Orchidées et leur champignons commensaux. Ann. Sci. Nat. Bot. **9**:1–196. 1909.

3. BOTTOMLEY, W. B., The effect of nitrogen-fixing organisms and nucleic acid derivatives on plant growth. Roy. Soc. London Proc. B **91**:83–95. 1919.

4. BURGEFF, HANS, Die Wurzelpilze der Orchideen, ihre Kultur und ihre Leben in der Pflanze. Jena. 1909.

5. DUGGAR, B. M., SEVERY, J. W., and SCHMITZ, H., Studies in the physiology of fungi. IV. The growth of certain fungi in plant decoctions. Ann. Mo. Bot. Gard. **4**:165–173. 1917.

6. KING, JOHN, King's American Dispensatory, re-written and enlarged by H. W. FELTER and J. W. LLOYD, 18th ed., 3d revision, **2**:1898 (p. 1699).

7. KNUDSON, L., Influence of certain carbohydrates on green plants. Cornell Univ. Agric. Exp. Sta. Memoir **9**:1–75. 1916.

8. KNUDSON, L., and LINDSTROM, E. W., Influence of sugars on the growth of albino plants. Amer. Jour. Bot. **6**:401–405. 1919.

9. MAZÉ, P., and PERRIER, A., Recherches sur l'assimilation de quelques substances ternaires par les végétaux à chlorophylle. Inst. Pasteur Ann. **18**:721–747. 1904.

10. SERVETTAZ, CAMILLE, Recherches expérimentales sur le développement et la nutrition des mousses en milieux stérilisés. Ann. Sci. Nat. Bot. IX. **17**:111–224. 1913.

11. WILLAMAN, J. J., The function of vitamines in the metabolism of *Sclerotinia cinerea*. Jour. Amer. Chem. Soc. **42**:549–585. 1920.

12. WILLIAMS, The vitamine requirements of yeasts. Jour. Biol. Chem. **38**:465–486. 1919.

13. WILSON, J. K., Calcium hypochlorite as a seed sterilizer. Amer. Jour. Bot. **2**:420–427. 1915.

L. C. Chadwick and D. C. Kiplinger, 1938. The effect of synthetic growth substances on the rooting and subsequent growth of ornamental plants. Proceedings of the American Society for Horticultural Science 36:809–816.

L. C. Chadwick                    D. C. Kiplinger

The formation of roots on cuttings has attracted the intellectual curiosity and pragmatic interests of plant physiologists and horticulturalists for years. The German plant physiologist, J. Sachs, postulated as early as 1879 that a green plant maintained in conditions suitable for growth synthesized in the leaves a specific root-forming substance which moved toward the root system. If a cutting were taken, the substance would accumulate above the cut surface and promote root formation. In the years that followed Sachs's observations, the concepts of root initiation became a muddle of vitalistic, nutritional, and hormonal hypotheses. In 1905, MacCallum suggested that the failure of an organ to develop was not due to the lack of conditions that favor growth, such as moisture or nutrition, or the lack of definite "formative substances," but to "some influence independent of all of these which an organ, acting perhaps along protoplasmic connections, is able to exert over other parts and so prevent their growth." MacCallum developed his hypothesis by sealing a vial containing water around the stem of an intact bean plant and observed that no roots would form. If the root system were severed, roots formed in the vial. If a notch were made in the stem of an intact bean plant below the vial, roots would develop in the vial on the notched side of the stem. MacCallum interpreted these results as an indication of a stimulus coming from the roots which prevented root formation and that the stimulus was interrupted by severing or notching the stem. However, the results may also be interpreted as the interruption of substances coming from the leaves that accumulated above the severed or notched area and stimulated root formation.

In 1917, Loeb provided a major contribution toward the hormonal concept of root initiation. His experiments were concerned with the influence of the leaf upon root formation and geotropic curvature in the stem of *Bryophyllum calycimum*. He demonstrated that the root-inducing substances accumulated on the lower side of a horizontal stem, and also observed that curvature of the horizontal stem occurred because of greater growth of the cortex on the lower side. Loeb found that leafless stems form roots more slowly, and that the curvature of the stem was correspondingly smaller. Loeb concluded that "in *Bryophyllum* the hypothetical geotropic hormone is associated (or identical) with the root forming hormone." In the same year, Loeb reported that the substances produced in the leaf inhibit shoot development, the inhibition was proportional to the mass of the leaf, and further, that the degree of root formation paralleled the degree of inhibition. Loeb's observations fit well three typical auxin responses: growth or cell elongation, bud inhibition, and root initiation. Loeb's investigations also provided the basis for the later conclusion that specific root forming substances and auxin are one and the same substance.

In 1924, van der Lek demonstrated the importance of buds in the development of roots in *Ribes nigrum*, *Salix*, *Populus*, and *Vitis vinifera*. He found that the presence of a bud was beneficial, even if the development of the bud was restricted by covering it with plaster. However, the intensity of root production could be correlated directly with the rate of bud development. A rapidly developing bud gave the greatest stimulation. If the bud were removed or a section of bark beneath the bud were removed, the stimulatory effect on rooting was lost. Van der Lek postulated that a hormone was formed in the buds and transported through the stem to the base of the cutting, where it stimulated rooting. Van der Lek also found that the stimulatory effect of the buds disappeared during the winter, and in fact the rooting response increased following bud removal. In later studies, van der Lek (1934) showed that the inhibition of rooting caused by dormant buds is proportional to the extent of dormancy in the buds.

In 1929, Went reported a "root promoting substance" which was extractable from leaves of *Acalypha*. He postulated that the substance was synthesized in the leaves and sprouting buds and was transported in the phloem. Went also demonstrated that the root-forming substances could be extracted from several plants, such as *Carica papaya* and germinating barley seeds. (It is interesting to note here that eighteenth-century European horticulturalists inserted grain seeds into the split base of a difficult-to-root cutting to stimulate root initiation.) Extracts from other plants applied to *Acalypha* cuttings have the same root-promoting effect as those obtained with extracts of *Acalypha* leaves. He concluded that the root-forming substance was not specific for each plant but was apparently similar in many plants.

In 1933, Bouillenne and Went introduced the term "rhizocaline" to describe the root-forming substance found in the cotyledons of *Impatiens balsamina*. Seedlings of *Impatiens* have five to seven initial roots. When the roots are cut off and bases of the hypocotyls (with cotyledons present) are placed in water, 18 to 30 adventitious roots are formed. If the cotyledons are cut off, no roots will form on the hypocotyls. When the decapitated hypocotyls are placed in glucose or sucrose solutions, "the hypocotyls increase their metabolism and a few (7–8) roots appear." Therefore, sugar was not the main limiting factor because it could not replace the stimulatory effect of the cotyledons. Bouillenne and Went described rhizocaline as extractable with water, thermostable, and active when freed from proteins by boiling and when diffused through an ultra filter. They also showed that an extract of rice polishings was active in increasing the formation of roots in *Impatiens* hypocotyls.

In 1934, Thimann and Went made a careful comparison of the properties of the root-forming hormone and of auxin and found that they were very similar. The following year Thimann and Koepfli (1935) found that synthetic indoleacetic acid was as active in the stimulation of root formation as any extract that had been tested. They concluded that since they had prepared the indoleacetic acid synthetically, the possibility of an active impurity was eliminated and therefore the root-inducing hormone must be identical with indoleacetic acid. Their results led Went and Thimann (1937) to conclude: "This provides final proof that one, at any rate, of the hormones causing root formation is identical with auxin. The names 'rhizocaline' and 'rhizogene,' in so far as they really refer to the action of auxin, can therefore be dropped."

The discovery that auxin was highly active in the formation of roots became a turning point in the investigation of root-promoting substances. Investigators turned their attention to the dramatic effects of auxin upon the initiation of roots in cuttings. Zimmerman and Wilcoxon (1935) tested eight growth substances on nine different plants, including marigold, tomato, sweet pea, windsor bean, and dahlia, and found that indolebutyric acid and $\alpha$-naphthaleneacetic acid were more effective than indoleacetic acid in stimulating root formation.

A young associate professor of horticulture at Ohio State University, Lewis Charles Chadwick (known as "Chad"), learned of Zimmerman and Wilcoxon's report from the Boyce Thompson Institute and wondered what the effects of the new synthetic growth substances would be on the rooting and subsequent growth of ornamental plants. Chad was interested in this area of research because his Ph.D. research at Cornell University involved a study of the influence of chemicals, the medium, and the position of the basal cut on the rooting of evergreen and deciduous cuttings. Chad joined the faculty at Ohio State in 1929 as an assistant professor. In 1937 he was joined in his research by Donald C. Kipplinger, a candidate for a Master's degree, known by his colleagues in the university and floricultural industry as "Kip." After completing his M.S. degree in 1938, Kip joined Ohio State as an assistant horticulturalist. He subsequently received his Ph.D. degree at Ohio State and gained a national reputation for his research in floriculture. Throughout their careers, Chad and Kip took leadership roles in translating theoretical information into practice for the nurseryman and florist. Chad was recognized for his contributions by many horticultural organizations, including being elected president of the International Plant Propagators' Society and being selected as the society's first recipient of the Award of Merit in 1957. Kip received the Vaughn Award from the American Society for Horticultural Science in 1947.

The paper reprinted here signaled a major advance in the application of technology to the field of commercial plant propagation. A broader range of plants could be propagated from cuttings more efficiently than had ever been possible before. Always the careful observers, so essential for outstanding research, Chad and Kip wrote in the summary of their paper: "In general, plants normally difficult to propagate by cuttings . . . were not benefited by treatment." They recognized the benefits and the limitations of the new growth substances and that additional research was necessary. The physiological basis of difficult-to-root cuttings and the failure of growth substances to overcome that difficulty is still not fully understood.

Charles E. Hess, Dean
College of Agricultural
  and Environmental Sciences
University of California, Davis
Davis, California 95616

## REFERENCES

Bouillenne, R., and F. W. Went. 1933. Recherches expérimentales sur la neoformati des racines dan les plantules et les boutures des plantes supérieures. Ann. Jard. Bot. Buitenzorg. 43:25–202.

Loeb, J. 1917a. Influence of the leaf upon root formation and geotropic curvature in the stem of *Bryophyllum calycinum* and the possibility of a hormone theory of these processes. Bot. Gaz. 63:25–50.

Loeb, J. 1917b. The chemical basis of regeneration and geotropism. Science 46:547–551.

Mac Callum, W. B. 1905. Regeneration in plants, I and II. Bot. Gaz. 40:97–120, 241–263.

Sachs, J. 1879. Stoff und Form der Pflanzenorgane, I. Arb. Bot. Inst. Wurtzburg 2:452–488.

Thimann, K. V., and J. B. Koepfli. 1935. Identity of the growth-promoting and root-forming substances of plants. Nature (London) 135:101–102.

Thimann, K. V., and F. W. Went. 1934. On the chemical nature of the root-forming hormones. Proc. K. Ned. Akad. Wet. 37:456–459.

Van der Lek, H. A. A. 1924. Over de wortelvorming van houtige stekken. Mededel. Landbhoogesch. Wageningen 28:1–230.

Van der Lek, H. A. A. 1934. Over den invloed der knoppen op de wortelvorming der stekken. Mededel. Landbhoogesch. Wageningen 38:3–95.

Went, F. W. 1929. On a substance causing root formation. Proc. K. Ned. Akad. Wet. 32:35–39.

Went, F. W., and K. V. Thimann. 1937. Phytohormones. Macmillan, New York.

Zimmerman, P. W., and F. Wilcoxon. 1935. Several chemical growth substances which cause initiation of roots and other responses in plants. Contrib. Boyce Thompson Inst. 7:209–229.

# The Effect of Synthetic Growth Substances on the Rooting and Subsequent Growth of Ornamental Plants

By L. C. Chadwick and D. C. Kiplinger, *Ohio State University, Columbus, Ohio*

SINCE the recent discovery by Zimmerman and Wilcoxon (8) of the effectiveness of indolebutyric and other acids on root initiation, the Floriculture Division of Ohio State University has been conducting experiments to determine the value of these synthetic growth substances to the commercial field of ornamental horticulture.

## Materials and Methods

The synthetic growth substance crystals were obtained from the Pennsylvania Chemical Company, Ambler, Pennsylvania. The dusts, Auxan and Rootone, were obtained from the Chemicals Limited, Montreal, Canada, and the American Chemical Paint Company, Ambler, Pennsylvania, respectively. Distilled water used in the preparation of the solutions was obtained from the Ohio State University chemical laboratory and had a pH of 5.8 as shown by tests on a Leeds and Northrup potentiometer. The plant materials were obtained from the campus of Ohio State University and from local greenhouses and nurseries.

The desired solutions were prepared by dilutions from a stock solution stored in the dark at room temperatures. The stock solution was prepared by dissolving 1 gram of indolebutyric acid in 125 cubic centimeters of 95 per cent ethyl alcohol and adding enough distilled water to bring the volume of the solution up to 250 cubic centimeters. The strength of this stock solution was 4 milligrams per 1 cubic centimeter. The diluted solutions were used only once, though preliminary tests showed they could be used several times.

Softwood cuttings of greenhouse and woody plant material were handled in the usual commercial manner. Leaf areas were kept at a maximum as better rooting responses were obtained than when leaf areas were materially reduced. This has also been reported by Zimmerman (7). After the cuttings were made, they were bunched, held with a rubber band, and placed in drinking glasses containing 100 cubic centimeters of the desired solution. About 1 to 1½ inches of the basal part of the cuttings were immersed. The glasses containing the cuttings were placed in the propagation house. All cuttings tested were soaked for 24 hours in the indolebutyric acid solutions. Soaking untreated cuttings in water resulted in rotting of the basal end, and since commercial methods were employed at all times, untreated cuttings were placed in the desired medium as soon as made.

When the cuttings were removed from the solutions, they were placed immediately into a greenhouse bench containing sand or a mixture of two parts sand and one part peat by volume, which had been previously steam sterilized. Cuttings were inserted in the media in the

accepted commercial manner as recommended by Laurie and Chadwick (3). All cuttings were rooted in a north lean-to which was heavily shaded with whitewash during the summer and was provided with electric cables thermostatically controlled at 74 degrees F for bottom heat. Air temperatures varied from 10 to 15 degrees lower in the cool months, and the relative humidity was between 60 to 80 per cent. The usual care was given the cuttings during the rooting period.

## Results

*Hardwood Cuttings:*—On November 25, 1937, 280 *Ligustrum vulgare* cuttings were tied in bundles of 10 and the basal 2 inches of each bundle was placed in a pot of moist German peat moss. The pot was placed in a closed case where the temperature was maintained between 80 degrees F and 84 degrees F and the relative humidity at 90 to 100 per cent. Ten cuttings were removed from the peat moss every day, and they were soaked 24 hours in a solution of indolebutyric acid at 5 milligrams per 100 cubic centimeters. Following this treatment the cuttings were inserted in sand in the closed case. In conjunction with these cuttings, 20 cuttings from similar Ligustrum plants were taken each day for 30 days, 10 of them being soaked in a 5 milligram per 100 cubic centimeters solution of indolebutyric acid, the other 10 receiving no treatment. These cuttings were also placed in sand in the closed case. On February 5, 1938, all cuttings were removed, and the results recorded.

In all cases of comparison of the various treatments, untreated cuttings rooted best. A comparison of the rooting percentages before and after the time of callusing (December 8, 1937) showed that the factor of callusing did not affect rooting. This is not in accord with Stoutemyer (6) who worked with locust cuttings taken in the spring. It is possible that 80 degrees F was too high for a storage temperature.

Hardwood cuttings of the Barbara Ecke poinsettia taken in January showed an increase in the rooting percentage over the period of time the cuttings were in the bench when treated with 1, 3, or 5 milligrams of indolebutyric acid.

*Softwood Cuttings, Greenhouse Plants:*—One claim for synthetic growth substances is their influence on the rooting percentage of cuttings. A number of greenhouse plants were tested by treatments with dusts and various concentrations of indolebutyric acid and results were compared with those of untreated cuttings. The treated cuttings rooted in greater percentages than untreated cuttings over the period of time the cuttings were in the bench. This is another way of stating that the time required to reach the normal rooting percentage was decreased. Dusts were slightly inferior to solutions in the percentage rooted but gave much better rooting percentages than untreated cuttings.

Skinner (5) has reported that the quality of the root system is improved when cuttings are treated with indolebutyric acid. Tests were conducted to determine the quality of root systems induced by the commercial dusts.

Dusts have one desirable advantage over solutions in that they can be used on cuttings of such plants as poinsettia, lantana, and so on, that are very subject to rot when soaked. The greatest disadvantage of dusts is the difficulty of increasing the concentration of growth substance. It can be partially accomplished by wetting the basal end of the cuttings before dusting as this materially increases the amount of dust that adheres.

In Table I are shown results tabulated from observations on ageratum, centaurea, feverfew, lantana, poinsettias, and scarlet plume cuttings taken in August and September, 1938. Since these plants exhibit similar rooting habits, the data is collective.

TABLE I—Comparison of the Quality Root Systems Produced on Untreated and Dusted Softwood Cuttings of Some Greenhouse Plants

| Treatment | Quality of Root System (Per Cent) | | | Total Rooted (Per Cent) |
|---|---|---|---|---|
| | Good | Poor | None | |
| Check | 27 | 44 | 29 | 71 |
| Auxan | 45 | 37 | 18 | 82 |
| Rootone | 60 | 30 | 10 | 90 |

It is readily seen that the use of dusts resulted in an increase in the percentage of cuttings possessing a good root system, although the results are not as striking as when solutions are used.

Due to inconvenience experienced in soaking the basal ends of cuttings, an effort was made to find some other method of treatment that could be used commercially. Cuttings of chrysanthemums were stuck in sand and sprayed once with indolebutyric acid at 10 milligrams per 100 cubic centimeters and three times with a 2 milligram per 100 cubic centimeter solution. Since preliminary tests showed that 100 per cent rooting may be obtained in 14 days by soaking the basal ends of chrysanthemums for 24 hours in indolebutyric acid at 0.25 milligram per 100 cubic centimeters, spraying was found not to be as effective in inducing roots as was the standard soaking method. Smaller rooting percentages were obtained and more growth substance was used when the cuttings were sprayed.

*Softwood Cuttings, Woody Plants:*—Similar studies have been made on softwood cuttings of over 100 woody ornamental plants. The results obtained substantiated those observed with cuttings of greenhouse material in that the time required to reach the normal rooting percentage was decreased and better quality root systems were obtained.

Treatment of cuttings of *Cornus florida rubra*, the Red Flowering Dogwood, for 24 hours in 1 milligram per 100 cubic centimeters solution of indolebutyric acid gave 90 per cent rooting in 6 weeks while untreated cuttings gave negative responses. Normally this plant is not propagated by cuttings, although it has been rooted successfully without the aid of growth substances. Tests with *Hydrangea petiolaris* showed that treatments with 1, 2, and 4 milligrams of indolebutyric acid for 24 hours gave negative responses after 5 months in the cutting bench.

Chadwick (2) has reported that roots of certain plants have a definite place of protrusion. To determine the effect of indolebutyric acid on this placement of roots, cuttings of seven of these plants were treated with 1 milligram per 100 cubic centimeters. Fifteen cuttings were used per treatment. Roots on cuttings of *Caragana arborescens* were distinctly polar. Treatment with indolebutyric acid did not change this position. *Clethra alnifolia* produced more roots in vertical lines below the buds when treated. Such a mass of roots occurred at the nodal region of Forsythia cuttings that some roots actually appeared to protrude from the internode. Roots appeared in the normal position with *Physocarpus amurense* and *Ribes odoratum*. Treatment of cuttings of *Pyracantha coccinea pauciflora* neither changed the position nor the number of roots that arose. All roots arose in the axil of the bud or thorn. Thirty per cent of the Ginkgo cuttings rooted at the second node from the base when treated, while untreated cuttings exhibited basal rooting only. Other tests showed that 1 milligram per 100 cubic centimeters was too strong for Ginkgo and hence the treated basal portion of the cutting was probably not the physiological base.

*Narrowleaf Evergreens:*—The distribution of roots on cuttings of narrowleaf evergreens was observed and results of tests on *Juniperus chinensis pfitzeriana* are shown in Table II.

TABLE II—Percentage of Root Distribution on *Juniperus chinensis pfitzeriana* Cuttings from the Base up in Centimeter Increments

| Treatment in Milligrams / 100 Cubic Centimeters | Centimeters from the Base | | | | | | | | | |
|---|---|---|---|---|---|---|---|---|---|---|
| | 1 | 2 | 3 | 4 | 5 | 6 | 7 | 8 | 9 | 10 |
| Check | 84 | 11 | 5 | — | — | — | — | | | |
| IB 1* | 79 | 16 | 4 | 1 | — | — | — | | | |
| IB 3 | 52 | 32 | 9 | 5 | 2 | — | — | | | |
| IB 5 | 60 | 16 | 10 | 9 | 4 | 1 | — | | | |
| IB 10 | 40 | 18 | 14 | 13 | 9 | 4 | 2 | | | |
| IB 15 | 29 | 27 | 18 | 14 | 7 | 4 | 1 | | | |

*IB 1—One milligram of indolebutyric acid per 100 cubic centimeters.

In general as the concentration passed the optimum, part or all of the soaked basal end became inactivated resulting in fewer roots being produced on this portion of the stem, but larger numbers further from the base of the cutting. This had been previously noticed on softwood cuttings of greenhouse and woody ornamental plants.

The effect of indolebutyric acid on the rooting percentage, number and length of roots was studied, and data are presented in Table III. With *Juniperus chinensis pfitzeriana* and *Thuja plicata atrovirens* the use of indolebutyric acid resulted in greater rooting percentages over the length of time the cuttings were in the bench and increased the average number of roots per cutting. There was no apparent relation between the number of roots induced and their length. Biale and Halma (1) have reported similar results. It is interesting to note that the use of high concentrations of growth substance when the plants had almost passed their rest period (January) resulted in a larger number of roots being produced than either in November or Decem-

TABLE III—THE EFFECT OF TREATMENT WITH INDOLEBUTYRIC ACID ON THE ROOTING OF CUTTINGS OF SOME NARROWLEAF EVERGREENS AFTER 3 MONTHS

| Date | Treatment (Milligrams per 100 Centimeters) | Average Number of Roots per Cutting | Average Total Length of Roots per Cutting (Centimeters) | Rooted (Per Cent) |
|---|---|---|---|---|
| *Juniperus chinensis pfitzeriana* | | | | |
| November 18 | Check | 1.22 | 18.64 | 10 |
| November 18 | IB 5 | 4.18 | 12.68 | 42 |
| November 18 | IB 10 | 3.92 | 12.95 | 51 |
| November 18 | IB 15 | 2.75 | 9.33 | 27 |
| December 19 | Check | 2.20 | 15.27 | 6 |
| December 19 | IB 5 | 4.93 | 17.11 | 19 |
| December 19 | IB 10 | 6.81 | 14.91 | 20 |
| December 19 | IB 15 | 5.70 | 17.33 | 13 |
| January 21 | Check | 1.44 | 16.15 | 11 |
| January 21 | IB 5 | 12.35 | 25.18 | 58 |
| January 21 | IB 10 | 19.67 | 18.32 | 45 |
| January 21 | IB 15 | 14.75 | 22.59 | 70 |
| February 24 | Check | 1.00 | 140.00 | 5 |
| February 24 | IB 1 | 7.00 | 47.97 | 65 |
| February 24 | IB 3 | 8.93 | 25.67 | 75 |
| February 24 | IB 5 | 6.50 | 62.91 | 70 |
| *Thuja plicata atrovirens* | | | | |
| November 21 | Check | 3.70 | 26.78 | 31 |
| November 21 | IB 5 | 9.42 | 21.65 | 65 |
| November 21 | IB 10 | 12.84 | 12.52 | 63 |
| November 21 | IB 15 | 15.47 | 13.43 | 48 |
| December 20 | Check | 15.75 | 16.47 | 13 |
| December 20 | IB 5 | 8.83 | 13.30 | 58 |
| December 20 | IB 10 | 14.09 | 19.49 | 55 |
| December 20 | IB 15 | 11.64 | 16.35 | 52 |
| January 20 | Check | 5.86 | 32.93 | 23 |
| January 20 | IB 5 | 6.84 | 34.04 | 55 |
| January 20 | IB 10 | 4.73 | 31.65 | 50 |
| January 20 | IB 15 | 13.63 | 33.83 | 63 |
| February 24 | Check | 3.00 | 31.39 | 30 |
| February 24 | IB 1 | 5.64 | 68.76 | 63 |
| February 24 | IB 3 | 4.29 | 51.42 | 70 |
| February 24 | IB 5 | 4.50 | 53.89 | 60 |
| *Taxus cuspidata* | | | | |
| November 20 | Check | 6.10 | 15.22 | 60 |
| November 20 | IB 5 | 5.69 | 14.57 | 44 |
| November 20 | IB 10 | 9.04 | 15.16 | 60 |
| November 20 | IB 15 | 9.04 | 13.99 | 69 |
| December 16 | Check | 3.63 | 12.33 | 45 |
| December 16 | IB 5 | 4.33 | 7.18 | 23 |
| December 16 | IB 10 | 5.72 | 7.02 | 45 |
| December 16 | IB 15 | 6.42 | 5.71 | 48 |

ber. Reducing the concentration of February cuttings resulted in a decrease in the number of roots induced but higher percentages of rooting. It follows that the time of taking the cuttings will result in variation of the response to growth substances. The length of the roots induced was independent of the concentration used, but the length increased as the season passed.

With *Taxus cuspidata* growth substances did not increase the percentage rooting in a 3 months period, nor was the number of roots induced significantly increased.

Treatment of cuttings of *Juniperus virginiana keteleeri* failed to give a response that was commercially significant. *Juniperus virginiana cannarti* failed to root or callus when treated with either indolebutyric or naphthaleneacetic acid. Untreated cuttings callused but did not root in a three months period.

*Broadleaf Evergreens:*—Tests were conducted with *Pyracantha coccinea pauciflora* using 5, 10, and 15 milligrams per 100 cubic centimeters of indolebutyric acid and 200, 700, and 1100 parts per million of nitrogen in the form of urea, alone and in all possible combinations. Five and 10 milligrams solutions increased the percentage rooting in a 2 months period when used alone or in any combination with urea. The 15 milligrams solution and concentrations of urea alone resulted in rotting of the basal ends. The use of indolebutyric acid on Pyracantha was of little value commercially since the increased percentage rooting was small and no better quality root system can be expected.

Cuttings of *Azalea kurume* were treated with commercial dusts and solutions of indolebutyric acid at 0.5, 1, and 2 milligrams per 100 cubic centimeters. Rootone gave 94 per cent rooting in 65 days, untreated cuttings rooting 61 per cent. Soaking the cuttings promoted rotting which reduced the percentage rooting.

*Miscellaneous Tests. Treatment of Rose Stocks:*—The purpose of this test was to determine the effect of indolebutyric acid on the development of new roots. It is generally thought the development of a good root system is associataed with greater percentages of successful grafts. Roots of rose stocks (*Rosa manetti*) were soaked 24 hours in 3 milligrams of indolebutyric acid per 100 cubic centimeters, and the stocks were potted in standard 2½ inch pots. Treatments and data recorded are shown in Table IV.

TABLE IV—THE EFFECT OF TREATING ROOTS OF *Rosa Manetti* (AMERICAN) STOCKS ON THE QUALITY OF THE ROOT SYSTEM PRODUCED

| Treatment | Quality of Root Systems (Per Cent) | | | | | | | | |
|---|---|---|---|---|---|---|---|---|---|
| | 5 Days in Greenhouse | | | 10 Days in Greenhouse | | | 15 Days in Greenhouse | | |
| | Good | Fair | Poor | Good | Fair | Poor | Good | Fair | Poor |
| Check | 33 | 52 | 14 | 32 | 60 | 7 | 45 | 51 | 4 |
| Indolebutyric acid 3 milligrams per 100 cubic centimeters | 20 | 55 | 21 | 24 | 65 | 10 | 31 | 62 | 6 |

In the case of each time period before placement in a grafting case more good root systems were produced on untreated stocks than on treated stocks. Also in each of the three time periods, fewer fair and poor root systems were produced by untreated stock plants. Records of successful grafts showed 65 per cent "take" on treated rootstocks and 75 per cent on untreated stocks.

*Root Cuttings:*—Pink *Bouvardia hybrida* was used to determine the effect of indolebutyric acid on root cuttings. Fifty cuttings were made for each of the four treatments shown in Table V, the treated cuttings being soaked 24 hours.

After 10 weeks no more growing points appeared, and final data were taken. The use of indolebutyric acid depressed the production of shoots from root cuttings. The greater the concentration, the fewer the number of growing points appeared. Inhibition of buds has been reported by Pearse (4).

TABLE V—The Effect of Various Concentrations of Indolebutyric Acid on the Shoot Production of Bouvardia Root Cuttings

| Treatment | Number Planted | Date Planted | Per Cent of Root Cuttings Forming Growing Points (Cumulative) Number of Weeks After Treatment | | | | | | | | | |
|---|---|---|---|---|---|---|---|---|---|---|---|---|
| | | | 1 | 2 | 3 | 4 | 5 | 6 | 7 | 8 | 9 | 10 |
| Check................ | 50 | December 24 | — | — | — | — | — | — | — | — | — | — |
| Indolebutyric acid 1 milligram per 100 cubic centimeters........ | 50 | December 24 | — | — | — | — | — | 24 | 50 | 94 | 94 | 94 |
| Indolebutyric acid 3 milligrams per 100 cubic centimeters........ | 50 | December 24 | — | — | — | — | — | 2 | 10 | 36 | 36 | 36 |
| Indolebutyric acid 5 milligrams per 100 cubic centimeters........ | 50 | Deeember 24 | — | — | — | — | — | — | — | 2 | 2 | 2 |

*Subsequent Growth After Rooting.*—Tests on pompon varieties of chrysanthemums showed that treatment of cuttings with growth substances did not result in any increase in yield over untreated cuttings. Growth observations on ageratum and poinsettias showed that if treated and untreated cuttings were removed from the propagation bench at the time the untreated cuttings had rooted, the ultimate growth was the same.

## Summary

Soaking the basal ends of cuttings gave more satisfactory results than spraying with solution of growth substances. Hardwood cuttings of *Ligustrum vulgare* did not respond favorably to indolebutyric acid applied either before or after callusing when stored at high temperatures. Treatment of softwood cuttings of greenhouse and woody ornamental plants and cutting of narrowleaf evergreens with synthetic growth substances decreased the time required to reach the normal rooting percentage. The final rooting percentage was not increased.

The quality of the root system produced on softwood cuttings by treatment with growth substances was superior to the root systems on untreated cuttings over the period of time the cuttings were in the bench. The use of growth substances in the majority of cases caused more roots to be produced over a larger stem area. Apparently no relation existed between the number of roots induced and their length. The external position of the roots on plants which exhibit specific rooting habits was not changed by applications of growth substances.

Treatment of roots of plants gave variable responses, in general causing a decrease in the root production. Plants normally propagated by softwood cuttings were rooted in less time than was usual for the species. In general plants normally difficult to propagate by cuttings, as for example, *Juniperus virginiana cannarti*, *Juniperus virginiana keteleeri*, and *Hydrangea petiolaris*, were not benefited by treatment.

Literature Cited

1. Biale, J. B., and Halma, F. F. The use of heteroauxin in rooting of subtropicals. *Proc. Amer. Soc. Hort. Sci.* 35: 443–447. 1938.
2. Chadwick, L. C. Studies in plant propagation. The influence of chemicals, of the medium, and the position of the basal cut, on the rooting of evergreen and deciduous cuttings. *Cornell Univ. Agr. Exp. Sta. Bul.* 571: 1–53. 1933.
3. Laurie, A., and Chadwick, L. C. Commercial Flower Forcing, p. 178–179. P. Blakistons' Sons and Co., Inc., Philadelphia. 1934.
4. Pearse, H. L. The effect of phenylacetic acid and of indolebutyric acid on the growth of tomato plants. *Jour. Pom. and Hort. Sci.* 14: 365–375. 1936–37.
5. Skinner, H. T. Rooting response of azaleas and other ericaceous plants to auxin treatments. *Proc. Amer. Soc. Hort. Sci.* 35: 830–838. 1938.
6. Stoutemyer, V. T. Root hardwood cuttings with acids. *Amer. Nurseryman* 68 (9): 3–5. 1938.
7. Zimmerman, P. W. Responses of plants to hormone-like substances. *Ohio Jour. of Sci.* 37: 333–348. 1937.
8. Zimmerman, P. W., and Wilcoxon, F. Several chemical growth substances which cause initiation of roots and other responses in plants. *Contrib. Boyce Thompson Inst.* 7: 209–229. 1935.

F. C. Steward, M. O. Mapes, and K. Mears. 1958. Growth and organized development of cultured cells. II. Organization in cultures grown from freely suspended cells. American Journal of Botany 45:705–708.

F. C. Steward

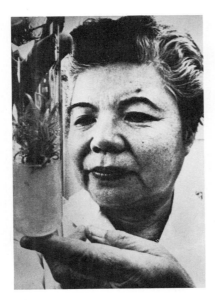

M. O. Mapes

K. Mears

It seems a platitude to state that all cells and organs of a sexually reproduced organism descend from the fertilized egg, that is, from a single cell. The cells that derive in turn from the zygote by mitosis are thought to perpetuate their genetic identity. Nevertheless, during development they also give rise to great biochemical and morphological diversity. Throughout the orderly course of development, the growing embryo and maturing plant body somehow limits the theoretical potential of each cell, thus restraining it from forming a new plant. Thus the paradox remains, since diversity and heterogeneity result from genetic uniformity.

Nowadays, when activity related to improvement of horticulturally and agriculturally important plants via the "new" biotechnologies such as cell and tissue culture and genetic engineering is approaching a frenetic pace, it seems difficult to contemplate a time when the expression of totipotency or morphogenetic competence of free cells grown in culture had not yet been demonstrated experimentally. Rereading the paper by Steward, Mapes, and Mears certainly helps to recapture some of that feeling. A noteworthy feature of that paper is that it is short and straightforward. This introduction is substantially longer than the paper itself, however, because it was the second of a series of three, originally published and later reprinted as a unit and intended to be read in sequence. As such, they are packed with background, theory, and analysis as well as information.

We now know that mature, living plant cells may be induced to proliferate, to multiply more or less indefinitely, and to release free cells and small units into their ambient medium. Via a number of pathways, including small proembryo-like globular masses, such cells can give rise to plantlets that can further develop, grow, and reproduce normally. This is especially true of carrot and other umbellifers, but there is an ever-lengthening list of other more or less responsive genera, species, and cultivars which can form plantlets from cells cultured *in vitro* (see Ammirato 1983).

The reporting of the experimental demonstration at Cornell University of totipotency or expression of morphogenetic competence in cells of the cultivated carrot (*Daucus carota* var. *sativa*) by F (rederick) C(ampion) Steward, his research technician (and later research associate) Marion Okimoto Mapes, and technician Kathryn Mears paved the way to the full recognition that, potentially, the living, normally diploid cells of the plant body retain intact, in their nuclei, the full information of the zygotic nucleus, and moreover, in their cytoplasm, the ability to make that information fully effective.

The work reported in this now classic paper demonstrated, unequivocally, the totipotency of somatic cells and brought about a new awareness of the nature of the controls that impinge on development. It also served as a focal point for new trends in higher plant research that continues vigorously to this day. Many investigators now view the carrot as providing *par excellence* the system with which to ascertain (1) the stimuli that release the otherwise suppressed totipotency of mature quiescent cells as they exist in the plant body, and (2) the factors that control the direction and pace of their subsequent development. Carrot, for better or worse, has in fact become a so-called "model system" for studying higher plant development from cells grown *in vitro* (see Terzi et al. 1985). Unfortunately, the relative ease with which one can nowadays manipulate the carrot system is encountered only infrequently in other, more economically important or even more genetically tractable species. Today especially, a prime objective still remains so to control development in cultured higher plant cells that single cells or their protoplasts growing in isolation will yield plantlets at a high level of efficiency and high level of genotypic fidelity. If this were now routinely feasible for a truly wide range of plants, it would not only greatly add to our knowledge of development but could provide the long-sought means for clonal multiplication or propagation of plants not currently possible in very large numbers from somatic cells (Krikorian 1982, Krikorian et al. 1986).

A number of important concepts in this overall field, and in this paper in particular, can be traced to the publications of the great German plant physiologist Gottlieb Haberlandt (1854–1945). Although experimentally unsupported in his concepts, Haberlandt was the first "to point to the possibility that . . . one could successfully cultivate artificial embryos from vegetative cells" (Haberlandt 1902, as translated by Krikorian and Berquam 1969, p. 84). His attempts to grow single isolated cells in simple salt solutions supplemented with a few organics were unsuccessful, but his later work on wound hormones set the stage for the discovery and isolation of plant cell division substances (Krikorian 1975, p. 75 et seq.). His studies on adventive embryony first published in 1921 demonstrated that select cells within the plant body could be stimulated to yield additional embryos and new plants (Haberlandt 1938). By pricking an ovary of the evening primrose, *Oenothera lamarckiana*, he initiated the formation of an additional embryo (more than likely of nucellar origin). Haberlandt credited the initial cell divisions to the wound hormones provided by injured cells, and the further embryonic development to embryo forming substances in the embryo sac.

To that early example, one can now add natural examples of embryo formation from cells additional to the zygote or fertilized egg which are not necessarily experimentally induced (Webber 1940). The term "adventive" or "adventitious embryony" is generally reserved for those somatic embryos that arise from the sporophytic cells of the nucellus or integuments. (Polyembryony in citrus was first noted by Antonie van Leeuwenhoek in 1719!) "Apomixis" is a broader term that should be applied to a range of asexual reproductive processes which substitute for, or supplement, the sexual process.

In his now classic book on plant embryology, Panchanan Maheshwari (1904–1966) characterized the cells that form adventitious embryos as "richly protoplasmic" and pointed out that they "actively divide to form small groups of cells, which eventually push their way into the embryo sac and grow further to form true embryos . . . sometimes a single cell may become a progenitor of an embryo, while on other occasions it is a small group of cells." According to Maheshwari (1950), the cell origins of adventitious embryos are comparable to sporophytic buds, and "although the adventive embryos have precisely the same germinal constitution as sporophytic buds, their developmental behavior is quite different. A sporophytic bud, whether terminal or axillary, directly proceeds to the formation of a stem, leaves, and flowers, while the adventive embryo recapitulates in a very striking manner the morphological features of true seedlings, viz. presence of cotyledons, radicle, plumule, epicotyl, and hypocotyl."

Haberlandt's early views on totipotency and prophecy on the production of embryos from somatic cells in culture went unfulfilled despite his impressive vision, his knowledge of adventive embryony and apomictic development, and considerable ingenuity in manipulating small tissue explants, as in the case of thin (a few cell layers thick) sections of potato tuber (see Krikorian 1975, p. 71 et seq.). Later forays by a number of investigators delving into aseptic culture work to study responses of isolated parts disclosed no apparent tendency to express totipotentiality, and some workers suggested that all living plant parts were not totipotent. William J. Robbins (1890–1979) and Walter Kotte are credited today for the first experiments in which excised roots of pea and maize were successfully maintained *in vitro*. Somewhat later, Philip R. White (1901–1968) investigated tomato roots and firmly established culture techniques for small tips and developed media which permitted their continuous and serial culture (White 1936). The prime purpose of those studies was, however, to demonstrate potentially unlimited growth and culturability.

Studies of regeneration from stem, leaf, and even root pieces had been carried out from time immemorial and formed the basis of vegetative multiplication so widely used in horticulture. Reference may be made to the work of Jacques Loeb (1859–1924) on regeneration (Loeb 1924), and to the still indispensable monograph on vegetative propagation from the standpoint of plant anatomy by J. H. Priestley and Charles F. Swingle (1929). But the first real tissue cultures (in contrast to organ cultures) were aimed primarily at initiating preparations demonstrably free of contaminating microorganisms and to assuring tissue survival by means of appropriate nutrition.

The strategies and procedures developed in France by Roger Jean Gautheret (1910–  ) and by Pierre Nobécourt (1895–1961) and in the United States by Philip R. White are the very backbone of much of tissue culture technology today (Gautheret 1959, White 1963). Even so, it is interesting to note that what the early workers sought to achieve is the very antithesis of what we strive for now. An ideal tissue culture in that period was defined as a "preparation of one or more isolated, somatic plant cells which grows and functions normally, *in vitro* without giving rise to an entire plant" (White 1936).

Hindsight now permits us to reflect on the constraints of the early concepts as they were promulgated by these pioneers. In addition to the unfortunate but conscious decision of avoiding the "complication" of differentiation and organized development from preexisting primordia, and *de novo* formation of new centers of growth by working predominantly with parenchymatous cell masses (callus), the early workers relied virtually exclusively on media made semisolid with agar. Also, they worked with rather large explants, very often chunks including cambium and numerous other tissue types. Regrettably, once growing cultures were initiated and were maintainable, few if any attempts were made to reinitiate fresh cultures. The large size of the primary explants made it virtually impossible to ascertain the role of endogenous or exogenous factors on the initiation of the cultures and development of any potential organized structures. Growth on agar medium was generally rather slow by present-day standards, and even when placed in liquid, inadequate aeration prevented rapid proliferation of callus and usually precluded smaller units from sloughing off. Concern with serial and "indefinite" subculture also worked against the early plant tissue culturists because variability in culture performance as to growth rate and diversity of cell and tissue types became exaggerated with time. Moreover, we now know that there is a general tendency for cultures to deteriorate in terms of their morphogenetic capacity with continued subculture.

A major turning point came when a few physiologists began to use tissue culture techniques not as a mere means of growing biological materials but as a tool to study problems. And at this point, radical changes in the design of tissue culture techniques occurred. F. C. Steward (1904–  ) was one of the innovators. His productive career as a plant physiologist and biochemist interested in the problems of salt uptake and growth had caused him to search for a controllable system that could be induced to grow *at will*. Tissue culture was the answer. His entry into plant tissue culture was unconventional and original, as was his earlier approach to the problems of salt uptake. Steward entered botany through chemistry and had been a graduate student at Leeds University in England of the highly original botanist Joseph H. Priestley (1883–1944). Priestley also had significant interests in horticulture and vegetative propagation, and perusal of Steward's work also shows that most of it was carried out on plants of horticultural or agricultural importance. His tissue culture work was to be no exception.

By growing tiny (2.5 mg) explants from thin (a couple of millimeters thick) slices of cultivated carrot root in a specially supplemented liquid culture medium in small T-tubes affixed to a rotating (1 rpm) apparatus, some of the limitations of the earlier culture techniques were overcome. In the first instance, the small explants derived from the region of the secondary phloem in the mature carrot root, far enough away from the cambium so that they would require specific and rather active stimuli to induce cell divisions. Any endogenous growth regulating substances or nutrients would be quickly depleted, thus rendering the carrot explant largely dependent on the nutrient medium provided. By rotating on a tumble tube apparatus which turned the culture vessels end over end on wheels mounted on a horizontal shaft, tissue explants were exposed alternately to liquid medium and to air. This provided ample and reasonably controlled aeration. Coconut water (commonly referred to in the usage of the day as coconut "milk"), the liquid endosperm of *Cocos nucifera*, was added as a growth supplement to the relatively simple basal medium comprised of macro- and microelements, sucrose, and vitamins. The whole system was grown in continuous light at a permissive temperature of around 70°F (21.1°C). But the most important feature of the entire system was that it could be induced to grow rapidly *at will* by the addition of coconut water, suggested by the work of J. van Overbeek, M. E. Conklin, and A. F. Blakeslee on development of heart-stage embryos of *Datura stramonium* (jimson weed) (see van Overbeek 1942). In 1948, Samuel Caplin and F. C. Steward first published the potentialities of the liquid endosperm of coconut in the induction of growth in otherwise-resting mature tissue. It was soon realized, however, that coconut water was only one of several fluids—for example, *Aesculus* (a horsechestnut) and black walnut liquid endosperm, the female gameto-phyte of *Ginkgo*, being nutritive fluids for immature embryos—that could produce the same effect. However, although the cell division factor was not specific to coconut water, it could not be replaced by any of a wide range of known substances. The initial idea was that a single substance was involved in eliciting the cell division response and was called *the* coconut milk factor (CMF). It did not take long, however, to establish the viewpoint that a number of substances which act in combination were involved—even to the extent of appreciating the balance between growth promotion and inhibition. This philosophy involving the chemi-cal basis of growth regulation in higher plants early found expression in the Steward aphorism that "no single substance unlocks the door to cell division" and is a view that has maintained itself even as it has become amplified in recent years (see Steward and Krikorian 1971).

The first appearance of free cells of carrot in a coconut water supplemented liquid medium was in 1955 in so-called "nipple flasks" (designed and made specifically to hold large amounts of tissue) contain-ing some 100 carrot root phloem explants. In 1954, W. H. Muir, A. C. Hildebrandt, and A. J. Riker of the University of Wisconsin had published the first successful culture of individual higher plant cells derived from agar-grown callus that had been disrupted in a liquid medium. Single cells so isolated could be induced to grow by placing them on a "nurse" tissue, provided in the form of a vigorously growing agar culture, from which it was separated by a piece of filter paper. This ingenious technique provided the milieu in which single cells could proliferate and form large calluses, but it was neither easy to follow their development, nor convenient for growing particularly large number of cells (see Muir et al. 1958). The nipple flasks were another matter. The nipple flasks full of carrot explants frequently appeared cloudy, and initially they were thought to be contaminated. Yet when examined carefully, the suspected contaminations were clearly not bacterial or fungal growth, and when aliquots of culture fluid were examined under the microscope, the cultures were full of floating plant cells. It further transpired that when these cells were separated either by fishing them out and depositing them on agar medium or if they were separated from larger tissue masses by straining through cheesecloth, the cells could be maintained through serial culture as cells which could, in turn, grow into larger masses.

These carrot cell cultures grown in this way represent for all practical purposes the first real batch cultures of higher plant cells. Other reports dealt with much smaller sample sizes and number of cells. Moreover, they were not nearly as predictable in their maintainability as carrot. Actively growing carrot suspensions are likened nowadays by some to yellow "pea soup" or "applesauce." Unless you have a preparation like these, you really do not have an actively growing cell or suspension culture.

In a series of three papers published in November 1958, observations on these cultures were presented by Steward and his various research technicians. In the first paper of the series, Joan Smith, a technician, is listed as junior author. Marion O[kimoto] Mapes (1913–1981) is second author. In the paper reprinted here, the junior author is Kathryn Mears, another laboratory technician, and Marion O. Mapes is the second author. (Both Mapes and Mears had Master's degrees.) The third paper, by Steward alone, is a more analytical and theoretical one, emphasizing the data of the previous two presentations.

In the first, a detailed survey is made of cell cultures and how they behave and grow. In the second, the observations are presented as to how cells in suspension can divide and give rise to essentially two kinds of organized forms. One involves the emanation of a root from a small polarized callus mass, which when placed on agar develops shoots and thence plantlets; or alternatively, a cluster of cells can polarize in their growth and yield a proembryo from which a nonzygotic or somatic embryo could develop. In either instance, the fact was appreciated that cells which had been maintained in suspension for long periods of time could be serially subcultured using relatively sparse inocula and, in time, these could undergo organized development in large numbers— eventually giving rise even to somatic embryos and, from these, plantlets.

F. C. Steward was clearly taken by the fact that cultured cells had the ability to regenerate plants. But what impressed him more was that it could occur from a subcultured suspension and by a series of events which in large measure seemed akin to a sort of zygotic embryogenesis. The significance for physiology, developmental genetics, and morphology was immediately obvious. From that point on, he directed much of the laboratory's tissue culture efforts to studying the phenomenon. Steward even wrote in a report to one of the sponsoring agencies that "Haberlandt's dream has thus been realized . . ." (see Krikorian 1975, p. 86).

The amazing fact is that most of the botanical community was not impressed by the finding. The animal biologists were quicker to appreciate what had been done, for they had labored under the constraining viewpoint that during development, genetic material was permanently lost, inactivated, or discarded. The carrot cell and somatic embryo work helped change all that. Because Steward had been a harsh critic of others, some tissue culturists were no less forgiving of him. Some botanists quibbled that a *single* cell in isolation had not been followed from beginning to end of the process. But Steward et al. never claimed to have taken *a* single cell and reared it to a plant. Reading the paper will show that it is, as is typical of anything F. C. Steward wrote, worded very carefully (Steward 1968). Others complained that the coconut water had not been shown to be absolutely necessary. Indeed, as so often happens in pioneer work, everything is not, at first, perfect or complete. Although we know that coconut water has a role to play when skillfully used, it is, indeed, not necessary for the regeneration of embryos and plantlets. Ralph Wetmore (1892–   ), a close friend of Steward, now professor emeritus, Harvard University, one of the few who fully recognized the work for what it was and even gave it "central billing" in his Sigma Xi lecture series, showed great insight when he said ". . . it is necessary, in our present knowledge of nutrition, to add the liquid endosperm of coconuts or some equivalent endosperm. It is significant that, in the culture of the embryos or parts of vascular plants other than the angiosperms, coconut milk proves unnecessary" (Wetmore 1959, p. 330). Nevertheless, Steward's laboratory continued to work on and extend the carrot somatic embryo story (see Steward et al. 1964). He lectured far and wide on "Carrots and Coconuts, Some Interpretations on Growth" and the like. One senior scientist even asked years later: "Is there anyone anywhere who has not witnessed a Steward "tour de force"? It was not until the late 1960s that a single cell of tobacco reared in isolation was shown by Vimla Vasil, a postdoctoral fellow in the botany department at the University of Wisconsin with A. C. Hildebrandt, to be capable of growing into a callus mass, which, in turn, could yield a shoot and root adventitiously, thus establishing a plantlet that could grow and mature (Vasil and Hildebrandt 1965). But this was different from what Steward, Mapes, and Mears did. For Steward totipotency came to mean that, at their best, cultured cells yielded somatic embryos.

The production of somatic or nonzygotic embryos à la Steward, which develop in ways very analogous to zygotic embryos—that is, *without* intervention of adventitious growth—is now being shown to be possible in an ever-increasing number of species. No one any more seriously doubts the totipotency of higher plant cells. Indeed, the regeneration of plants from single cells, be they protoplasts, components of pollen grains, or whatever, is now, of course, one of the essential aspects of many of the so-called genetic engineering, cloning, and plant improvement operations (see Krikorian 1982).

A. D. Krikorian
Department of Biochemistry
State University of New York
Stony Brook, New York 11794

# REFERENCES

AMMIRATO, P. V. 1983. Embryogenesis. p. 82–123. In: D. A. Evans et al. (eds.). Handbook of plant cell culture. Macmillan, New York.

GAUTHERET, R. J. 1959. La culture des tissus végétaux. Masson et Cie. Paris.

HABERLANDT, G. 1938. Über experimentelle Adventivembryonie. Sitzungsber. Preuss. Akad. Wissen. Phys.-Math. Kl. 24:243–248.

KRIKORIAN, A. D. 1975. Excerpts from the history of plant physiology and development. p. 9–97. In: P. J. Davies (ed.). Historical and current aspects of plant physiology. A symposium honoring F. C. Steward. New York State College of Agriculture and Life Sciences, Cornell University, Ithaca, N.Y.

KRIKORIAN, A. D. 1982. Cloning higher plants from aseptically cultured tissues and cells. Biol. Rev. 57:151–218.

KRIKORIAN, A. D., and D. L. BERQUAM. 1969. Plant cell and tissue cultures: The role of Haberlandt. Bot Rev. 35:58–88.

KRIKORIAN, A. D., R. P. KANN, S. A. O'CONNOR, and M. S. FITTER. 1986. Totipotent suspensions as a means of multiplication. p. 61–72. In: R. H. Zimmerman et al. (eds.). Tissue culture as a plant production system for horticultural crops. Martinus Nijhoff, Dordrecht, The Netherlands.

LEOB, J. 1924. Regeneration, from a physico-chemical viewpoint. Agricultural and Biological Publications. McGraw-Hill, New York.

MAHESHWARI, P. 1950. An introduction to the embryology of angiosperms. McGraw-Hill, New York.

MUIR, W. H., A. C. HILDEBRANDT, and A. J. RIKER. 1958. The preparation, isolation and growth in culture of single cells from higher plants. Am. J. Bot. 45:589–597.

PRIESTLEY, J. H., and C. F. SWINGLE. 1929. Vegetative propagation from the standpoint of plant anatomy. U. S. Dep. Agr. Tech. Bull. 151.

STEWARD, F. C. 1968. Growth and organization in plants. Addison-Wesley, Reading, Mass.

STEWARD, F. C., and A. D. KRIKORIAN. 1971. Plants, chemicals and growth. Academic Press, New York.

STEWARD, F. C., L. M. BLAKELY, A. E. KENT, and M. O. MAPES. 1964. Growth and organization in free cell cultures. Brookhaven Symp. Biol. 16:73–88.

TERZI, M., L. PITTO, and Z. R. SUNG (eds.). 1985. Proceedings of the International Workshop on Somatic Embryogenesis in Carrots, May 28–31. Presso La Lithografia, Pisa, Italy.

VAN OVERBEEK, J. 1942. Hormonal control of embryo and seedling. Cold Spring Harbor Symp. Quant. Biol. 10:126–134.

VASIL, V., and A. C. HILDEBRANDT. 1965. Differentiation of tobacco plants from single, isolated cells in micro cultures. Science 150:889–892.

WEBBER, J. M. 1940. Polyembryony. Bot. Rev. 6:575–598.

WETMORE, R. H. 1959. Morphogenesis in plants—a new approach. Am. Sci. 47:326—340.

WHITE, P. R. 1936. Plant tissue cultures. Bot. Rev. 2:419–437.

WHITE, P. R. 1963. The cultivation of animal and plant cells, 2nd ed. Ronald Press, New York.

# GROWTH AND ORGANIZED DEVELOPMENT OF CULTURED CELLS.

## II. Organization in Cultures Grown from Freely Suspended Cells[1]

F. C. STEWARD, MARION O. MAPES, AND KATHRYN MEARS

THE FIRST of this group of papers (Steward et al., 1958) described the various ways in which freely suspended cells from certain dicotyledonous plants grow and multiply to form a relatively unorganized multicellular mass. The growth in question occurs under prescribed and controlled nutritional and environmental conditions. The purpose of this paper is to show how the growth may be carried forward into the formation of roots and shoots and, in fact, the development of whole plants.

DIFFERENTIATION TO FORM ROOTS.—Tissue cultures grown directly from cambium-free explants of differentiated carrot root phloem in the basal medium supplemented with coconut milk do not normally form roots. On the very rare occasions when this has occurred, it may have been due to the presence of a preformed root initial in the explant. At any rate, root formation occurs extremely infrequently when the culture is started from freshly explanted phloem tissue. On the other hand, the minute tissue cultures formed by the development of cell aggregates from these freely suspended and frequently sub-cultured cells will develop roots with great ease. This may be due to the disappearance of some factor antagonistic to the formation of an organized root apical meristem that is present in the central core of tissue in the explant removed from the carrot root. Or, it may be said that the differentiated phloem tissue must first "de-differentiate"–whatever that may mean. Whatever explains the ease of root formation in these small aggregates, which are grown from cultured cells, the events that lead to root initials can be illustrated in fig. 1–5.

A root initial does not form until the cell aggregate or colony reaches such a size that the inner cells of the mass behave differently from the outer cells. A sign of this is the presence of a new type cell in the central region of the culture (fig. 1); this cell loses its contents, becoming somewhat lignified and forming a tracheid-like element of the kind so often observed in plant tissue cultures (Gautheret, 1956). A single xylem element is frequently followed by similar ones in close association. It is interesting that they form in a tissue which was wholly derived originally from the secondary phloem. Surrounding these nests of lignified elements, which lose their contents (fig. 2–4), there

develops a ring, or hollow spherical sheath, of cambium-like elements (c.f. Gautheret, 1956, fig. 8) which encloses a mass of cells; these cells now *have only limited access to the external medium*. This can be seen in fig. 3, 4.

It appears that so long as all the cells of the culture have free access to the external medium, they grow in a random and independent fashion by one or other of the methods already described (Steward et al., 1958). The ring of cambium-like tissue develops, however, as if in response to an injury or a wound, under the stimuli derived from the formation of these lignified elements. The cells in this cambium-like region are separated from the external medium, and they develop in a nutrient condition which is controlled by the closely packed zone of cells and the tissue which they enclose. In fact, the dividing cells lie along a gradient from the dead cells within, to the nutrient without, which contains the cell division stimuli. In this cambium- or pericycle-like region, a root apex forms and it subsequently grows out through the tissue mass into the surrounding medium (fig. 5). These events are not demonstrably different from the normal origin of lateral roots. When such root initials form, they grow apace and the more callus-like growth of the original colony of cells tends to be suppressed. When grown in this way, on cultures revolved around a horizontal axis, the roots emerge in all directions (fig. 6). Longitudinal sections of roots so formed show that they are normal and develop apparently normal protoxylem tissue.

FORMATION OF SHOOTS.—When culture flasks or tubes are inoculated by liquid suspensions which contain freely suspended cells or small cell aggregates, they form very large numbers of freely growing colonies, and many root initials may originate in the manner described. The conditions most conducive to root formation have not been investigated fully, though some early evidence seemed to indicate that in a culture which is prone to form roots this occurred more readily in the dark and more readily if the calcium content of the medium was reduced. Furthermore, it has been noted that certain active growth-promoting fractions, isolated from extracts of immature corn (Shantz and Steward, 1957) also tended to foster copious root formation. However, once a root initial has formed, it continues to grow, and thereafter the growth of the tissue culture is retarded, or suppressed. If, however, cultured cell aggregates with roots already developed are transplanted to nutrient agar in flasks, complete

[1] Received for publication June 1, 1958.

This work forms part of a program of research which has been supported by grants to one of us (F.C.S.) from the National Cancer Institute, National Institutes of Health, United States Department of Health, Education, and Welfare.

[The Journal for November (45:653–704) was issued November 22, 1958]
AMERICAN JOURNAL OF BOTANY, Vol. 45, No. 10, December 1958

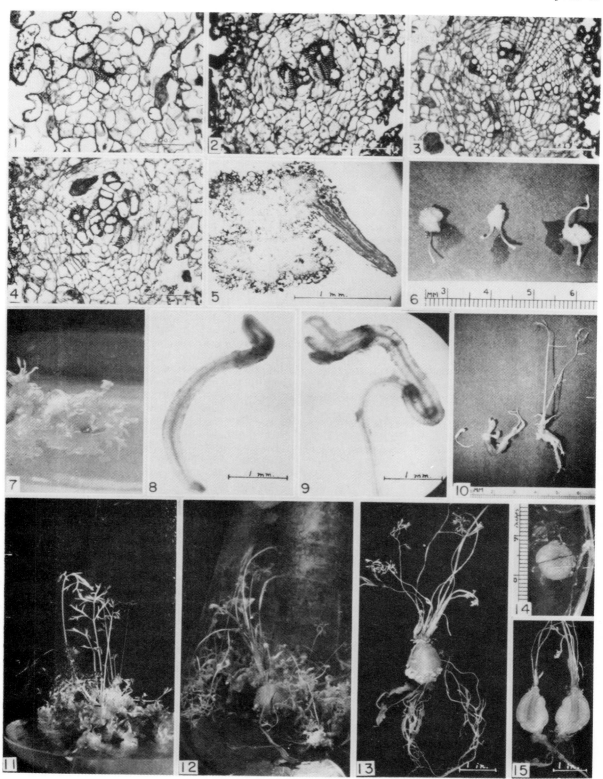

Fig. 1–15. Carrot.—Fig. 1–4. Stages in the formation of "nodules," or growth centers, which enclose "nests" of lignified elements. Fig. 1 shows the first lignified element and adjacent cell division; fig 3 shows complete ring of dividing tissue

with basal medium and coconut milk, the cultures can be indefinitely continued. Under these circumstances a copious growth of roots occurs. Indeed, if cultures remain in the normal rotating culture tubes in which explants are commonly grown, the roots may grow until they virtually fill the tube (Steward and Shantz, 1955, fig. 1d). If, however, tissue aggregates which have been reared from free cells are transplanted to agar media they may, as they grow, also form buds and shoots spontaneously.

The shoots which first emerge have leafy appendages which, like primitive leaves, tend to be entire (fig. 7), though later more typical, much dissected, carrot leaves develop (fig. 11, 12). Again, the stimuli which promote the development of buds and shoots have not yet been fully investigated. From the frequency with which this development occurs, it would appear that the irregular, randomly proliferated cultures that develop from the freely suspended cells do this spontaneously and with comparative ease when they are furnished with the basal medium plus coconut milk on a stationary, semisolid medium. No doubt the shoot is more dependent on regular orientation to gravity than is the root. The loosely proliferated culture, which originated from aggregates grown from free cells and later transfered to agar, presents the general appearance shown in fig. 7. From such a mass many minute plantlets can be dissected, such as those illustrated in fig. 8–10.

REGENERATED PLANTLETS: SIMILARITIES TO YOUNG EMBRYOS.—In the regenerated plantlets, the axis, with shoot and root apices, is completely established (fig. 8). The impression is inescapable that each of the many nodule-like growing centers (fig. 5) which develop in the cultured mass can first form roots; then, if transferred to a stationary agar medium, they will form shoots. Thus, the spherical masses of cultured cells, enclosed by their sheath of cambium-like initials (fig. 3, 4), really behave like a proembryo; they form both roots and shoots and, significantly enough, do this in a fluid (of another species) that normally nourishes immature embryos. Figures 7–12 show various stages in the organization of the shoot in cultures of this sort. Even more surprisingly, it may be observed that a secondarily thickened storage carrot root may grow in the tissue mass which had originated, in the manner described, from material that had been frequently sub-cultured and reduced to the free cell state (fig. 12, 13).

Thus the cycle of development is now complete,

because cells withdrawn from the phloem of the storage carrot root and which have passed through many transfers in which they were reduced to the single cellular state, have developed into cell aggregates, which have, in turn, differentiated to form roots and, when transplanted, have given rise also to shoots and to a secondarily thickened storage carrot root.

A curious, but suggestive, observation was made on the way the secondarily thickened carrot root grew in culture (fig. 13–15). The surface of the enlarging organ which was in contact with the agar medium containing coconut milk grew irregularly at its base, with green callous-like swellings (fig. 13). The side of the storage root which was away from the medium, and, therefore, not in direct contact with coconut milk, grew normally, forming a broad band of secondary phloem, rich in carotene, as seen in transverse section (fig. 14). This tissue appeared normal in every way, in contrast to the liquid-cultured, unorganized tissue which never achieves this rich, orange-red color or content of carotene. The only other abnormality was some green color in the xylem, which was seen in the longitudinal section (fig. 15); this is a not uncommon feature near the crown of normal carrot plants, especially when they have been grown in the light. The obvious suggestion is that direct contact with the stimuli to growth, which are in the coconut milk, leads to cell proliferation (fig. 13) but, when these stimuli are modified by the intervening tissue of the growing root, apparently normal secondary growth may occur (fig. 14, 15).

THE STIMULI TO DEVELOPMENT.—Much remains to be done to define the variables and the stimuli which regulate each of the definitive steps in the organization of cellular aggregates to form roots and shoots. It is suggestive that the effective nutrient conditions for this development also furnish all the nutrients and growth factors that normally nourish immature embryos. Thus, it is now clear that parenchyma cells, which are already far advanced toward maturity in the plant body, may return in culture, under appropriate nutrient conditions, to the dividing state and, as they do this, they can eventually recover the totipotency that was originally inherent in the egg. However, no single parenchyma cell can *directly* recapitulate the familiar facts of embryology, but, through the formation first of an unorganized tissue culture, which is in fact a colony of dividing cells, the necessary degree of organization is recaptured, first to form roots and then to form shoots.

enclosing "nests" of lignified elements; fig. 4 shows the beginning of organization in a nodule leading to root formation.—Fig. 5, 6. Emergence of roots from tissue cultures.—Fig. 7, 11, 12. Stages in the growth of carrot plants on agar.—Figs. 8, 9, 10. Stages in the development of young plantlets, which originate from individual nodules (fig. 8 shows very young stage in this development).—Fig. 13. Carrot plant with storage organ.—Fig. 14. Transverse section of cultured storage root shows callus-like growth in contact with the medium plus coconut milk and normal secondary phloem on side away from the medium.—Fig. 15. Longitudinal section of cultured storage root.

One may well ask why the cells of tissue explants, withdrawn from the phloem of a carrot root, will not as readily organize to form roots and shoots, even though they can be brought by coconut milk into a rapid state of cell division and may form very large callus-like masses if they are grown on an appropriate medium. The full explanation of the contrasted behavior of the cells which have passed through the freely suspended condition and the cells of the explants from which they were originally derived requires further investigation. Though the process may be described as "dedifferentiation," this term is hardly illuminating. It is, however, interesting to recall that in the early work of Van Overbeek et al. (1941), some young embryos were induced to grow in an unorganized proliferative fashion by the use of coconut milk. In the present investigation, already differentiated cells have been induced to grow again and eventually to produce structures that normally originate from the embryo. This surely means that the coconut milk contains the inherent stimuli and nutrients that make for active growth by cell division, but that the progress toward organized growth requires the growing system to acquire a measure of independence from the coconut milk stimuli which, if unregulated, lead to unorganized growth. This independence first occurs through the formation of an unbroken surface of dividing cambium-like cells, surrounding a region in which some cells of a tissue culture mature and die. Within this "walled-off" zone, some cells are confined and a root initial forms (fig. 7) and, thereafter, having formed roots, the culture may grow in response to the coconut milk and also form buds. It is difficult, therefore, to avoid the concept that the root initial must originate, under restraint, in a controlled environment, within a ring of cells which permit only *controlled access* to the coconut milk stimuli; buds, in turn, form only after this special nodule of cultured cells has responded to the culture medium in a polarized manner brought about by the presence of roots.

It will be a major landmark when the fertilized egg is removed from the environment of the embryo sac and the facts of early embryogeny can be recapitulated with the production of leaf, stem, and root under culture conditions in a synthetic medium. The present series of papers show that free-living, disassociated cells may be caused to grow in media containing coconut milk so that they achieve similar ends to those which result from the growth of the zygote, for they produce roots and shoots and even secondarily thickened storage organs. However, the freely suspended cells do not do this, as it were, directly, but only more deviously through the summation of the events which have here been described.

## SUMMARY

This paper describes the way in which organization may develop in a cultured mass of cells which originates by the growth of freely suspended cells that have been obtained in the manner previously described (Steward et al., 1958). For carrot tissue, it is demonstrated that normal roots may arise in the liquid medium and, once these have originated, the cultures will develop shoots when they are placed on a semisolid agar medium and are enabled to grow in a polarized way in stationary cultures exposed to air. The stages by which plantlets develop are illustrated, and eventually secondary thickening occurs around a newly formed nest of xylem cells, after which there may occur production of a carrot storage root. Thus the cycle from carrot root-phloem to free cell, to carrot root, to carrot plant has been completed. The essential stages in this reorganization involve: (1) The formation, within the cultured mass, of a spherical sheath of cambium-like cells which completely encloses a nest of lignified elements so typically observed in tissue cultures; (2) this spherical nodule, or growing center, plays a part equivalent to the pro-embryo in normal development, for it can give rise eventually to roots and subsequently to shoots; and (3) the formation of a shoot initial occurs at a point diametrically opposite to the emergence of the root apex, producing a simple, embryo-like structure reminiscent of that which occurs in normal development. The stages in the development of a plantlet into a completely organized plant with storage root are illustrated. It is emphasized that this orderly development becomes possible when cells are enclosed within, and limited by the restraints—probably more physiological than physical—of, this wall of cambium-like cells which effectively cuts off the internal cells from direct access to the coconut milk stimuli that cause random proliferation. Wherever the coconut milk has direct access, more random callus-like growth occurs.

DEPARTMENT OF BOTANY
CORNELL UNIVERSITY
ITHACA, NEW YORK

## LITERATURE CITED

GAUTHERET, R. J. 1956. Histogenesis in plant tissue cultures. Jour. Nat. Cancer Inst. 19: 555–573.

SHANTZ, E. M., AND F. C. STEWARD. 1957. The growth-stimulating substances in extracts of immature corn grain: a progress report. Proc. Amer. Soc. Plant Physiol., p. 8. A. I. B. S. meeting, Stanford, California.

STEWARD, F. C., MARION O. MAPES, AND JOAN SMITH. 1958. Growth and organized development of cultured cells. I. Growth and division of freely suspended cells. Amer. Jour. Bot. 45: 693–703.

———, AND E. M. SHANTZ. 1955. The chemical induction of growth in plant tissue cultures. I. Methods of tissue culture and the analysis of growth. *In* The chemistry and mode of action of plant growth substances. WAIN, R. L., AND F. WIGHTMAN (ed.). Butterworths. London.

VAN OVERBEEK, J., M. E. CONKLIN, AND S. F. BLAKESLEE. 1941. Factors in coconut milk essential for growth and development of very young *Datura* embryos. Science 94: 350, 351.

Georges M. Morel. 1960. Producing virus-free cymbidiums. American Orchid Society Bulletin 29:495–497.

Georges M. Morel

Gems occur where they are least expected. Georges M. Morel's 1960 paper is just such a gem. It is modest and unassuming in tone, published in a relatively specialized journal, and the important finding of Morel's work, "[that] each bud will give several plants so the stock of a rare or expensive variety can be increased," is not clearly stated until the concluding paragraph. Use of tissue culture for elimination of viruses in vegetatively propagated orchid cultivars was Morel's initial and primary goal. His observation that *in vitro* techniques can also be utilized to multiply desirable plant materials is mentioned almost as an afterthought.

There is no hint in Morel's publication that the techniques of *in vitro* plant multiplication which he mentions so casually would in just a few years attract the attention of many scientists, entrepreneurs, and businesspeople, and reorient the horticulture and nursery industries. However, Morel did recognize the importance of his initial observations with *in vitro* propagation and immediately set to work expanding and refining their application. These techniques provide the first commercialization of plant biotechnology and undergird an expanding worldwide industry.

This publication is typical of the understated style of Georges Morel. He was a modest, polite gentleman who was at the same time capable and practical. He rarely sought the limelight, remained dedicated in his laboratory experimentation, and let the results of his research speak for themselves. If the measure of a person is, indeed, not his or her actions but the consequences of these actions, Morel and his scientific legacy have transformed horticulture.

The use of *in vitro* manipulations for elimination of viruses began about 40 years ago in the United States when Wendell Stanley, of the Rockefeller Institute first began examining the biochemical components of tobacco mosaic virus. Sufficient amounts of this virus for Stanley's experimental purposes required large amounts of infected plant material. Stanley was acquainted with Philip White, also at the Rockefeller Institute, who was working with the first successful cultivation of plant tissues *in vitro*. Stanley suggested that White attempt to cultivate tobacco mosaic virus on isolated roots growing *in vitro*. White obtained virus multiplication in root tissues but noticed that the virus was eliminated if the explants used for subculture were small tips. Stanley subsequently demonstrated the reason for this unexpected observation: virus was not present in the meristematic cells of root tips. Later, French scientists demonstrated that meristematic cells of the shoot were virus free and that the virus titer of plant tissue decreased in proximity to the apical meristem. Georges Morel concluded that *in vitro* culture of shoots tips may be an effective means of eliminating viruses from vegetatively propagated plant varieties.

The credit for the initiation of apical meristem culture and propagation techniques goes to Morel, who was working at the National Agricultural Research Institute (INRA) in Versailles, France. In this 1960 paper, Morel first demonstrated the use of *in vitro* techniques for vegetative propagation. Prior to meristem culture of orchids reported in this paper, successful results had been obtained by Morel and his coworkers in removing viral pathogens from carnation, dahlia, and potato. Using meristem culture techniques cymbidium plants were generated freed of the systemic mosaic virus infection. In addition to obtaining virus-free clones, *in vitro* techniques developed by Morel also opened another important theoretical and commercially attractive avenue: that of plant multiplication and the production of large numbers of plantlets of valuable commercial orchids in a relatively short period of time. Other scientists had laid the groundwork for Morel's discovery. Lewis Knudson and others developed suitable culture media and Gavino Rodor, Jr. developed aseptic propagation; however, Morel invented *in vitro* proliferation.

Shoots of all angiosperms and gymnosperms grow by virtue of their apical or shoot meristems. The apical meristem is usually a dome of tissue located at the extreme tip of a shoot and measures about 0.1 mm in diameter and 0.25 to 0.3 mm in length. These tissues are comprised of cells that are indistinguishable but are commonly grouped into arbitrary zones such as tunica, corpus, and central mother cells. Apical meristems are first formed during embryo development and they remain, except for periods of dormancy, in an active state of division throughout the vegetative phase of the plant. The totipotency (i.e., the ability to regenerate an entire plant) of apical meristems forms the basis of meristem culture technique.

Commercial cultivars of orchids are genetically heterozygous, and the unique combinations of characters of a desirable cultivar are rarely transmitted through seeds. Cultivar integrity is maintained by vegetative propagation. However, vegetative propagation by division of entire plants is a slow, tedious process in orchids and risks disease infection, particularly from viruses.

Green bulbs containing an axillary meristem are the preferred source of explants of *in vitro* multiplication. Each bud is processed into an apical meristem varying in size from 0.5 to 1 mm. When cultured *in vitro*, each explant regenerates and gives rise to several protocorms or embryo-like bodies. Each protocorm can be divided into four to eight pieces and within one month in culture each piece develops into a full protocorm. These protocorms can be put through further divisions to increase their number of protocorms, or regenerated into entire plants.

Morel suggested that by using such tissue culture techniques, large-scale orchid plant production could be standardized. Production of large numbers of flowers of good quality and color and for the required period in the year could become feasible. Thus the application of tissue culture techniques could revolutionize orchid production.

The major stimulus for application of plant tissue culture techniques to the propagation of a large number or ornamental species may be attributed to the early work by Morel on the propagation of orchids and to the development and widespread use of new medium with high concentrations of mineral salts developed by Murashige and Skoog in 1962. Following success with rapid *in vitro* propagation of orchids, plant cell and tissue culture techniques were applied to other ornamental species. In many taxa, tissue culture propagation has been applied on a commercial scale.

As practiced today, *in vitro* multiplication of ornamental plants has several advantages over conventional methods of plant propagation.

1. The number of genetically identical plants recovered from a single stock plant is greatly increased. Through tissue culture techniques, a single stock plant may rapidly produce thousands or even millions of plants, depending on the capability of the culture system.
2. Disease-free plants may be obtained. Plants that have been propagated by tissue culture techniques are free of superficial bacteria and fungi, and virus-free plants may be obtained and maintained through meristem culture.
3. Stocks may be maintained and their growth controlled *in vitro* where greenhouse space for maintenance of plants is at a premium.

Morel's work was completed during an exciting period of botanical and horticultural experimentation. *In vitro* culture of plant cells and tissues is a technology that integrates basic findings in plant nutrition, physiology, genetics, developmental biology, and pathology. Morel clearly demonstrated that the laws which govern plant growth and development are the same in natural stands, in cultivated fields, in the greenhouse, and *in vitro*. He was one of the first *in vitro* horticulturalists and contributed to both investigation in and commercialization of his field.

Georges Morel was a botanist by inclination and family tradition, and a chemist by training. Tissue culture was for Morel a specialized form of chemically controlled gardening. His contributions to tissue culture were not limited to apical meristem culture but extended to experimental investigations with crown gall tumors and with plant protoplasts. Morel's death in 1973, at the early age of 57, cut short the remarkable career of a scientist who laid many of the foundations of modern plant biotechnology.

Peter Carlson
Crop Genetics International
7170 Standard Drive
Hanover, Maryland 21076

## REFERENCES

GRIESBACK, R. J. 1986. Orchid tissue culture. In: R. H. Zimmerman (ed.). Tissue culture as a plant production system for horticultural crops. Martinus Nojhoff, Dordrecht, The Netherlands.

KNUDSON, L. 1922. Nonsymbiotic germination of orchid seeds. Bot. Gaz. 73:1–25.

MORELL, G. 1974. Clonal multiplication of orchids. In: C. Withner (ed.). The orchids: Scientific studies. Wiley, New York.

ROTOR, G., JR. 1975. A method for vegetative propagation of *Phalaenopsis* species and hybrids. Am. Orchid Soc. Bull. 18:738–739.

SHEEHAN, T. J. 1983. Recent advances in botany, propagation, and physiology of orchids. Hort. Rev. 5:279–315.

# Producing Virus-Free Cymbidiums

GEORGES M. MOREL

CYMBIDIUM MOSAIC IS ONE of the most widespread of the orchid virus diseases. In recent years, many growers all around the world have become very much concerned about it.

Since the work of Jensen (1), the symptoms of the disease are well known and easy to recognize. It appears first on the new shoots as small elongate chlorotic spots. These spots enlarge quickly and become more defined. They show as elongated yellowish streaks scattered throughout the leaves, especially conspicuous by their transparency. On old leaves some of these areas become necrotic on the under side. Leaves with strong necrotic symptoms die earlier than normal ones.

The effect of the disease on different varieties is extremely variable. In some, like *Cymbidium Pauwelsii*, the mottle is so mild that the plant is almost a disease-free carrier. Growth of other plants is very much depressed, and they must be discarded. Some outstanding and expensive varieties, like the famous *C. Alexanderi* 'Westonbirt,' are now entirely contaminated with virus.

Jensen has shown that this virus is very stable and easy to transmit. According to him, the disease is mostly spread with knives and pruning shears during vegetative propagation of the plants or when the leaves or inflorescences of a healthy plant are cut immediately after cutting a diseased one. Sterilization of tools by alcohol or detergent is not entirely effective. It is also impossible to eliminate this very stable virus by inactivation with heat in the living plant without killing the plant first.

Like most plant viruses, Cymbidium mosaic is systemic. The virus nucleoprotein spreads through all the plant: root, bulbs, leaves and flowers. The embryos, however, remain virus free and all the plants obtained from seeds of a diseased mother plant are, accordingly, virus free. This is true also in many other diseases, tobacco mosaic, for example. But in this case the noninfected embryos are not much help; the cultivated hybrids are so heterozygous they never breed true.

But we have found (2) that in the case of most plant viruses another technique may be employed to secure virus-free plants. It has been shown by various authors that in a diseased plant the virus content of the meristematic parts is also very low. We have demonstrated that the growing point, the apical meristem, of the stem is virus free in most cases. We then devised a technique to isolate the meristem aseptically and cultivate it in a special environment so that it would develop into a full plant, which could then be propagated in the normal way.

This method, which was originally developed for potatoes, has since successfully been applied to various horticultural plants, such as dahlias, carnations and hyacinths. This method makes it possible to take up again the cultivation of varieties which in earlier periods had become worthless because of their contamination. In this paper we intend to describe how the technique has been applied to the Cymbidium mosaic.

Plants of different varieties, all issued from *Cymbidium eburneum, lowianum, insigne,* and *tracyanum,* such as *C. Talma, C. Doris,* or *C. Candeur,* were selected with very severe necrotic symptoms. The sap extracted from the leaves of all these plants gave a strong positive reaction with an antiserum obtained by the inoculation of the purified virus into a rabbit.

The leaves were removed from the pseudobulbs and all the buds separated, whether dormant or actively growing. Most of the leaves around the growing buds were taken off down to the size of about one-half inch. After careful washing with water the buds were sterilized by dipping in 75% ethyl alcohol for a few seconds, followed by immersion for half an hour in a solution of calcium hypochlorite at 80 grams per liter.

The buds were then dissected aseptically under a dissecting microscope by carefully removing the scales and leaf primordia to expose the apex. The dissection was carried out with needles and pieces of razor blade as well as tiny watchmaker's forceps.

The exposed apex showed up under the microscope as a tiny glowing ball. It was then excised by four cuts at right angles made with a piece of razor blade so it was about .1 mm thick. The apex was immediately planted on Knudson III medium in a small test tube. The test tubes were kept at a constant temperature of 22°C. with a 12-hour period of luminescent light. Under these conditions, the explants, colorless at first, became green and enlarged slowly, making a small flat bulblet looking exactly like the protocorm which develops from an embryo. Rhizoids were formed on the periphery and a small leaf showed up in the center (FIG. 1). From then on, the plant grew exactly like other seedlings, producing roots and leaves. Very often the protocorm-like body divided into a clump of four or five identical structures, each of them producing a new plant.

When the plants reached the size of about one cm. and had three to four leaves and several roots, they were cultivated in the greenhouse in very fine Osmunda fiber and watered occasionally with the mineral part of the Knudson solution. They all grew well. Some are now ten cm. high. None of them has shown any symptoms of the mosaic disease. All were tested with the antiserum of the Cymbidium mosaic and they appeared to be virus free.

FIGURE 1. Protocorm-like body obtained from an apical meristem of *Cymbidium Doris* after three months of culture. Note the rhizoids.

FIGURE 2. Young plant of *Cymbidium Talma* cultivated in vitro, 18 months after the excision.

In conclusion, we can state that it is relatively easy to free a Cymbidium plant from the mosaic virus. It will take about the same time as to raise a plant from seed. All the buds of a plant can be used for that purpose. Usually each bud will give several plants so the stock of a rare or expensive variety can be increased at the same time.

Experiments of the same kind are now being carried out with other orchids, such as Cattleya, Odontoglossum, and Miltonia, contaminated with different viruses. — *Institut National de la Recherche Agronomique, Versailles, France.*

BIBLIOGRAPHY

1. JENSEN, D. D. 1951. Mosaic black streak disease of Cymbidium orchids. *Phytopathology* **41**:401–414.
2. MOREL, G. and C. MARTIN. 1955. Guerison des plantes atteintes de maladies a virus; par culture de meristemes apicaux. *Report of XIV'ʰ Internatl. Hort. Cong.. Netherlands*, pp. 303–310.

T. Murashige and F. Skoog. 1962. A revised medium for rapid growth and bioassays with tobacco tissue cultures. Physiologia Plantarum 15:473–497.

T. Murashige

F. Skoog

Success in plant tissue culture can pivot on two critical factors: choice of an explant and culture medium. The choice of explant, with regard to the tissue of origin, developmental stage, and physiological state, must be determined by the goals of the project and the plasticity of the tissue in question. The selection and modification of an appropriate culture medium has been greatly simplified through the efforts of Philip R. White, K. V. Thimann, Folke Skoog, Carlos O. Miller, Toshio Murashige, and their students and colleagues. The inorganic salt formulation of Murashige and Skoog, published in 1962, represents one of the major achievements in the history of cell and tissue culture, facilitating widespread applications in horticulture, agriculture, and biology.

The first experiments involving isolated plant cells were reported by Gottlieb Haberlandt (1902) (Krikorian and Berquam 1969). These pioneering experiments using highly differentiated tissues of *Tradescantia*, *Erythronium*, and *Ornithogalum*, inspired plant, as well as animal, biologists to continue and intensify this approach. Haberlandt's failure to grow these isolated tissues were due, in part, to the choice of monocot plant species, which have been difficult to grow in culture until only recently. More important, due to limited understanding of plant nutrition at the turn of the century, Haberlandt had no knowledge of the special nutritional or hormonal requirements of tissues literally cut off from the metabolism of the complete plant. Haberlandt provided the isolated tissues with mineral nutrients; however, he believed that the isolated leaf tissue would continue to photosynthesize in light and supply the cells with the necessary organic requirements for survival and growth. He speculated that other growth factors had to be involved in his lack of success and that these may be present in the extracts of whole plants. This concept, that mineral nutrients and growth factors both played critical roles in maintaining plant cell growth, became fundamental to the successful development of media that could support the continued growth of plant tissues and cells.

A wider interest in the culture of isolated plant cells did not occur until many years later. In the 1920s and 1930s, reports began to appear in the literature regarding the culture of isolated plant cells, tissues, and organs. An important issue raised by Philip White (1934) was that isolated plant tissues would survive in culture, often for many months, but did not grow. White believed that many of the experiments in tissue culture were not conducted using an adequate nutrient medium and because of this, growth of the explant was ". . . dependent . . . on materials carried over in the original fragment." White argued the importance of completeness of the nutrient media and attempted to test his assumption that plant tissues provided with sufficient nutrients would not only survive, but grow in culture. White proved this proposition successfully by establishing a culture of tomato roots that not only survived, but grew indefinitely on a medium containing mineral salts, sucrose, and brewer's yeast. Although simple media had been developed capable of supporting limited growth of *Salix* cambial cells (Gautheret 1938) and carrot root tissue (Nobecourt 1938), a nutrient medium capable of supporting unlimited plant cell growth was unavailable in the 1930s.

In 1939, Philip White reported the successful establishment of cambial cultures of a *Nicotiana glauca* × *N. langsdorffii* hybrid on a medium containing inorganic salts, sucrose, and yeast extract. His success, however, was not due to the composition of the medium as he had thought, but to the utilization of a hybrid that had an unusual capacity to produce substantial wound callus on injured stem or leaf tissue. This particular tissue did not require exogenous growth regulators, and therefore, grew well on the simple medium. Because the growth substance autonomy of this unique tissue was not known, it was difficult to reproduce White's results with other species.

Folke Skoog, born in Fjärås, Sweden, completed his B.S. degree in 1932 and his Ph.D. degree in 1936 at the California Institute of Technology. His long and distinguished career includes appointments at Harvard, Johns Hopkins University, and the University of Wisconsin, where he is currently professor emeritus of botany. He is still interested in plant cell culture and plant hormones, especially cytokinin research. Skoog's career evolved with the developing field of the plant growth substances. While at Cal Tech he worked first with Herman Dolk and then with Kenneth Thimann and Frits Went. Also at Cal Tech at the time were J. Van Overbeek and James Bonner. This group formed a core of scientists interested in auxin, the first class of the plant growth regulating substances.

Thimann (1934) had shown that auxin is present in *Vicia faba* and *Raphanus*. Continuous production of auxin in these plants as well as

*Pisum* was light-dependent (Skoog 1944). In 1939, Philip White observed that indoleacetic acid, added to the culture medium, had no beneficial effects on the *Nicotiana glauca* × *N. langsdorffii* hybrid callus. He did not, at that time, believe that auxin was a bona fide plant growth factor, and in his report stressed that the *Nicotiana* hybrid cultures exhibited "growth without auxin." Nevertheless, at the same time he sent a culture to Skoog with a request that it be tested for auxin. This observation interested Skoog, who suspected that the hybrid tobacco tissue used by White actually produced ". . . auxin independently of photosynthesis and independent of light." Skoog extracted the tissues, which yielded "considerable amounts" of auxin (Skoog 1944). Thus the tissue used by White survived and grew in culture without added auxin.

Skoog and Tsui (1948) and Skoog and Miller (1957) using combinations of IAA and adenine on stem segments and combinations of auxin and kinetin on callus tissue, respectively, showed that root and shoot formation in *N. tabacum* cv. Wis. No. 38 could be regulated by the auxin/cytokinin ratio in the medium. Demonstration of this model was the culmination of many years of observation and experimentation on interaction between growth factors and nutrients by Skoog, who recognized the existence of ". . . a quantitative balance between parts, a degree of dominance, which is chemically controlled" (Skoog 1955). This body of work focused attention on the importance of the interaction of plant growth substances in the regulation of plant growth and morphogenesis. The application of this model of auxin and cytokinin interaction greatly facilitated the development of plant tissue culture.

Toshio Murashige was born in Kapoho, Hawaii (1930), received his B.S. degree from the University of Hawaii (1952), M.S. degree from Ohio State (1954), and Ph.D. degree from the University of Wisconsin (1958). Growing up in a large family in rural Hawaii, he worked when he was not occupied with school, first on his family's farm and later in the sugarcane industry. In school, and later in his job, he was recognized as an exceptional person and encouraged to continue his education, which he was able to do with scholarships from the sugarcane industry. The habit of hard work formed in these early years is reflected in the drive and meticulous care evident throughout his research, qualities that he has hoped to pass on to his students.

Murashige arrived in Skoog's laboratory in 1954, at which time a joint project with F. M. Strong's laboratory in the biochemistry department was well under way. Kinetin was obtained in crystalline form in 1954 and both kinetin and $N^6$-benzyladenine were synthesized in March 1955. Interest centered on the isolation and synthesis of additional cytokinins and their role in plant growth and development, especially their quantitative interaction with auxins and other factors in growth and organogenesis in plant tissue cultures. It had been shown (Wiggans 1951) that leaf extracts would give four- to five-fold increases in yield of callus grown on modified White's medium. Murashige's predoctoral work was an attempt to isolate organic factors other than cytokinins present in leaf extracts and which thus would interfere in bioassays of cytokinins (Murashige 1959). In a subsequent systematic analysis of biologically active components in leaf extract, Murashige found that more than half the increase in yield was due to inorganic constituents, mainly $K^+$, $NO_3^-$, and $NH_4^+$.

With cytokinins as well as auxins and vitamin $B_1$ present, the inorganic nutrients in the medium became suboptimal. To minimize interference from this source in quantitative cytokinin bioassays, it became necessary ". . . to develop a medium with such adequate supplies of all required mineral nutrients and common organic constituents that no appreciable change in growth rate or yield will result from the introduction of additional amounts in the range ordinarily expected to be present in materials to be assayed . . ." (Murashige and Skoog). In comparison with other inorganic salt formulations in use at the time, the resulting medium was relatively high in nitrate, potassium, and ammonium. Additional modifications in organic components were then made through extensive studies, especially of carbohydrates and vitamins (see Linsmaier and Skoog 1965), and also of amino acids and nucleic acid components. Results of some of Linsmaier's experiments on carbohydrates are included in the Murashige–Skoog article.

The Murashige–Skoog publication is a classic. The precise, careful groundwork laid by the Wisconsin group resulted in a basal salt formulation which is in widespread use, unchanged, 25 years later. Although a single medium has not been devised that will support the growth of all plants, the MS medium is the most widely used in plant tissue culture research. In a review concerning plant regeneration from cell cultures by Evans et al. (1981), over 70 percent of media formulations that resulted in successful regeneration of both monocot and dicot tissues used the MS basal salt formulation, and when new media were developed, the MS basal salt formulation served as the starting point. According to the Science Citation Index, the Murashige and Skoog publication was cited in over 2220 publications from 1984 to 1986 alone.

The Murashige–Skoog work demonstrates the humble origins of scientific breakthroughs. The investigation was not undertaken to develop a universal inorganic salt formulation for plant tissues in culture, nor was it one of the spectacular investigations involving cytokinins. The investigation was undertaken simply to optimize the tobacco callus bioassay system in use in Skoog's laboratory at that time, to facilitate the study of cytokinins.

Because of the great technical success of the formulation and its subsequent routine use throughout the world, this frequently cited publication is read infrequently. The background and purpose of the research described in the paper often comes as a pleasant surprise to many in the field who have used the media and then "discover" the publication many years later. Few articles in the plant sciences can come close to this work in impact and usefulness over the years.

A useful update, description of stock solution preparation, and comparison to other plant cell culture media can be found in Huang and Murashige (1976) and Gamborg et al. (1976). These include correction of the error that appeared in Table 6 of the original publication ($ZnSO_4 \cdot 4H_2O$ should have been $ZnSO_4 \cdot 7H_2O$).

Roberta H. Smith and Jean H. Gould
Department of Soil and Crop Sciences
Texas A&M University
College Station, Texas 77843

# REFERENCES

EVANS, D. A., W. R. SHARP, and C. E. FLICK. 1981. Plant regeneration from cell callus. Hort. Rev. 3:214–314.

GAMBORG, O. L., T. MURASHIGE, T. A. THORPE, and I. K. VASIL. 1976. Plant tissue culture media. In Vitro 12(7):473–478.

GAUTHERET, R. J. 1938. Sur le repiquage des cultures de tissu cambial de *Salix caprea*. C. R. Acad. Sci. 206:125.

HABERLANDT, G. 1902. Kulturversuche mit isolierten Pflanzenzellen. Sitzungsber. Akad. Wiss. Wien, Math.-Naturwiss. KI. III:69–92.

HUANG, L., and T. MURASHIGE. 1976. Plant tissue culture media: Major constituents, their preparation and some applications. TCA Man. 3(1):534–548.

KRIKORIAN, A. D., and D. L. BERQUAM. 1969. Plant cell and tissue cultures: The role of Haberlandt. Bot. Rev. 35:59–88.

LINSMAIER, E. M., and F. SKOOG. 1965. Organic growth factor requirements of tobacco tissue cultures. Physiol. Plant. 18:100–127.

MURASHIGE, T. 1959. Ph.D. dissertation. University of Wisconsin, Madison.

NOBECOURT, P. 1938. Sur les proliférations spontanées de fragments de tubercules de carotte et leur cultures sur milieu synthétique. Bull. Soc. Bot. Fr. 85:182–188.

SKOOG, F. 1944. Growth and organ formation in tobacco tissue cultures. Am. J. Bot. 31:19–24.

SKOOG, F. 1955. Growth factors, polarity and morphogenesis. Ann. Biol. 31:201–213.

SKOOG, F., and C. O. MILLER. 1957. Chemical regulation of growth and organ formation in plant tissue cultured *in vitro*. Symp. Soc. Expt. Biol. 11:118–131.

SKOOG, F., and C. TSUI. 1948. Chemical control of growth and bud formation in tobacco stem segments and callus cultured *in vitro*. Am. J. Bot. 35:784–787.

THIMANN, K.V. 1934. On the inhibition of bud development and other functions of growth substances in *Vicia faba*. Proc. Royal Soc. London 114:317–339.

WHITE, P. R. 1934. Potentially unlimited growth of excised tomato root tips in a liquid medium. Plant Physiol. 9:585–600.

WHITE, P. R. 1939. Potentially unlimited growth of excised plant callus in an artificial nutrient. Am. J. Bot. 26:59–64.

WIGGANS, S.C. 1951. Ph.D. thesis. University of Wisconsin, Madison.

Reprinted from

PHYSIOLOGIA PLANTARUM, VOL. 15. 1962

# A Revised Medium for Rapid Growth and Bio Assays with Tobacco Tissue Cultures

By

Toshio Murashige and Folke Skoog

Department of Botany, University of Wisconsin, Madison, 6, Wisconsin

(Received for publication April 1, 1962)

## Introduction

In experiments with tobacco tissue cultured on White's modified medium (basal medium in Tables 1 and 2) supplemented with kinetin and indoleacetic acid, a striking four- to five-fold increase in yield was obtained within a three to four week growth period on addition of an aqueous extract of tobacco leaves (Figures 1 and 2). Subsequently it was found that this promotion of growth was due mainly though not entirely to inorganic rather than organic constituents in the extract.

In the isolation of growth factors from plant tissues and other sources inorganic salts are frequently carried along with the organic fractions. When tissue cultures are used for bioassays, therefore, it is necessary to take into account increases in growth which may result from nutrient elements or other known constituents of the medium which may be present in the test materials. To minimize interference from contaminants of this type, an attempt has been made to develop a medium with such adequate supplies of all required mineral nutrients and common organic constituents that no appreciable change in growth rate or yield will result from the introduction of additional amounts in the range ordinarily expected to be present in materials to be assayed.

As a point of reference for this work some of the culture media in most common current use will be considered briefly. For ease of comparison their mineral compositions are listed in Tables 1 and 2. White's nutrient solution, designed originally for excised root cultures, was based on Uspenski and Uspenskaia's medium for algae and Trelease and Trelease's micronutrient solution. This medium also was employed successfully in the original cultivation of callus from the tobacco hybrid *Nicotiana glauca × N. langsdorffii*, and as further modified by White in 1943 and by others it has been used for the

31

474 TOSHIO MURASHIGE AND FOLKE SKOOG

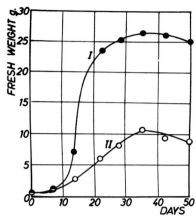

Figure 1.

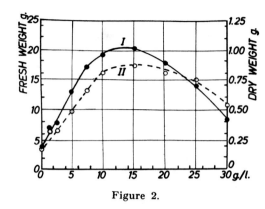

Figure 2.

Figure 1. *Effect of tobacco leaf extract on the growth of tobacco pith tissue* (T-2, 10 g/l. of medium: Expt. started Mar. 7, 1958). Curve I with, curve II without leaf extract added to the standard basal medium.

Figure 2. *Effect of concentration of leaf extract on the yield of tobacco callus tissue* (T-2; Growth period, Feb. 13 to Mar. 6, 1958). Curve I, fresh weight; curve II, dry weight.

cultivation of various tissues from numerous species (See Gautheret's 1959 compilation). Gautheret's medium (1939) was devised by combining a twice diluted Knop's macronutrient solution with a slightly modified Berthelot's micronutrient solution. Hildebrandt, Riker, and Duggar (1946) employed the triangulation technique to improve on White's medium for the cultivation of specific tissues. They reported two new formulae (included in Tables 1 and 2) which they designated as optimal for cultures of the above hybrid tobacco callus and for sunflower crown gall callus respectively. These media, according to Burkholder and Nickell (1949), were still unsuitable for cultivation of virus-induced tumor tissues. By employing the triangulation technique in initial trials and then by testing serial concentrations of each element independently of the others, these investigators devised still another formula based on the medium for sunflower tissue by Hildebrandt *et al.*, and which they reported as more suitable for Rumex tissue. Heller (1953) has made detailed studies of the mineral requirements in cultures of carrot and Virginia creeper. He first induced deficiencies of several of the elements by repeated transfers of the cultures to liquid media lacking the element in question and then reintroduced the element in serial concentrations to select satisfactory levels. For the most part the level of each element was varied only in combinations in which the respective levels of all other nutrients were kept unchanged. In comparative tests, Heller found his medium to give two to three times greater yields than White's or Gautheret's media which were employed as controls. The nutrient solution devised more recently by Nitsch and Nitsch (1956) for Jerusalem artichoke is based on the best estimate of available literature for cultures of this species and subsequent modifications

*Physiol. Plant., 15, 1962*

in accordance with results of testing variations in each element separately, but it lacks iron and other trace elements. It is now clear that none of the above media provide nearly the prerequisite amounts of some essential elements for the rapid growth rates and large yields that tobacco tissue is capable of attaining when various organic growth factors are also included in the medium.

## Materials and Methods

*Source and preparation of tissues.* Tobacco tissues, *Nicotiana tabacum,* var. Wisconsin 38, have been used exclusively in this study. Fresh pith was used for the bulk of the work, and continuously subcultured callus was used for the final confirmation tests. Stems were cut from ca. 1 m. tall tobacco plants grown in the greenhouse, the leaves were removed, the stems were swabbed with 95 % ethanol and cut into 5 to 7 cm. long cylinders. Only a 15 cm. region of the stem starting ca. 10 cm. from the tip was used. Cylinders of pith parenchyma were bored from near the center of the excised stems with a sterile No. 2 cork borer. The cylinders were extruded from the borer with a glass-rod, and sliced into discs approximately 2 mm. thick and weighing about 50 mg. each. The ca. half cm. end pieces in each cylinder were discarded as a precaution against contamination. Three discs, each placed with one of its flat surfaces in contact with the medium, were planted in each culture flask. As fresh pith was readily available, it was a convenient material to use in the large scale experiments. However, data obtained from one experiment to the next varied to some extent, possibly due to differences in the nutrient status of the plants from which the pith was excised. For this reason, improvements in yield at successive stages in the development of the nutrient solution and finally the suitability of the revised solution were confirmed in tests with tobacco callus. Firm, white callus from 4—5 weeks old stock cultures was cut into roughly rectangular pieces weighing 40—50 mg. each, and these were planted in groups of three into each culture flask. The stock callus, originally obtained from pith, had been subcultured 10 or more times on medium identical with the basal medium described below except that the kinetin level had been raised to 0.38 mg/l. and inositol had been omitted. The vigorous more compact tissue obtained by four-weekly subculturing on medium modified in this manner gave excellent growth and satisfactory reproducibility of data (within 10 %) from one experiment to another.

*Composition of the Basal Medium.* The basal medium which has served as control or reference medium throughout this study is a modified White's nutrient solution, which has been routinely employed in this laboratory and with two addenda, *myo*-inositol and Edamin (a pancreatic digest of lactalbumin furnished by Sheffield Chemical Co., Norwich, N.Y.). It had the following composition, in mg/l. of medium: (All salts were reagent grade unless noted to be otherwise.)

(a) Inorganic salts: $NH_4NO_3$, 400; KCl, 65; $KNO_3$, 80; $KH_2PO_4$, 12.5; $Ca(NO_3)_2 \cdot 4H_2O$, 144; $MgSO_4 \cdot 7H_2O$, 72; NaFe-EDTA, 25; $H_3BO_3$, 1.6; $MnSO_4 \cdot 4H_2O$, 6.5; $ZnSO_4 \cdot 7H_2O$, 2.7; and KI, 0.75.

(b) Organic substances: 3-Indoleacetic acid (IAA), 2.0; kinetin, 0.2 (for fresh pith) or 0.04 (for callus); thiamin · HCl, 0.1; nicotinic acid, 0.5; pyridoxine · HCl, 0.5; glycine (recrystallized), 2.0; *myo*-inositol, 100; Edamin, 1000; sucrose, 20,000; and Difco Bacto-agar, 10,000.

*Physiol. Plant., 15, 1962*

(c) pH adjustment: The pH of all media was adjusted to 5.7–5.8 with a few drops of either 1 N NaOH or 1 N HCl before the agar was added.

Fe was added as agricultural grade sodium-ferric-ethylenediaminetetraacetate obtained from Geigy Agricultural Chemicals, New York, N.Y. or $Na_2Fe$ EDTA prepared from $Na_2$-EDTA, MW 372.35 (Frederick Smith Chem. Corp. Columbus, Ohio) dissolved and heated in $H_2O$ with an equimolar amount of $FeSO_4 \cdot 7H_2O$.

As the $NH_4NO_3$ content of the basal medium far exceeded that of any other salt, the concentrations of the other elements were varied simply by adding them as either nitrate or ammonium salts and by adjusting the amount of $NH_4NO_3$ accordingly to attain the specified total inorganic nitrogen level. There was no evident effect on growth from possible changes in pH or in other conditions arising from this procedure.

In Tables 1 and 2 the inorganic constituents of the basal medium are listed in millimoles/l. (mM) or micromoles/l. (μM) of each element. To facilitate comparisons, the concentration of each element is also expressed in mg/l (values in parenthesis). These concentrations will be referred to as the 1 × levels, and in many experiments modifications will be made in multiples of these, such as 2 ×, 3 ×, etc.

*Preparation of tobacco leaf extract.* The leaf extracts, T-2 or T-3, used in this study are both from *Nicotiana tabacum* variety Wisconsin 38. T-2 is a water extract from 62.4 kg. of greenhouse grown plants and was used here only in the experiments represented by Figures 1 and 2. T-3 is an extract from field grown plants prepared as follow: Healthy green leaves picked from vigorous, nearly mature, ca 1 m. tall plants were packed into plastic bags, sealed, trucked to a cold storage plant, quickly frozen and kept indefinitely. The frozen leaves were then ground in a commercial type meat grinder and thawed. The juice was expressed with a hand press, heated to near 100°C, chilled to 5–10°C, centrifuged and concentrated under reduced pressure. The concentrate was stored in a frozen state in 1 l polyethylene bottles and

Table 1. *Macronutrient composition of some plant tissue culture media.*

| Medium | Element (Concentrations in millimoles per liter of medium | | | | | | | | |
|---|---|---|---|---|---|---|---|---|---|
| | N | K | Ca | Mg | S | P | Cl | Na | Fe |
| White (1943)[1] .......... | 3.2 | 1.7 | 1.2 | 3.0 | 4.4 | 0.14 | 0.9 | 3.0 | 0.013 |
| Gautheret[1] ............ | 5.5 | 2.2 | 2.1 | 0.5 | 0.5 | 0.9 | — | — | 0.125 |
| Hildebrandt et al.[1] (Tobacco) ............ | 4.2 | 1.7 | 1.7 | 0.7 | 6.4 | 0.24 | 0.9 | 11.7 | 0.143 |
| Hildebrandt et al.[1] (Sunflower) ............ | 8.4 | 3.3 | 3.4 | 2.9 | 3.6 | 1.0 | 1.8 | 1.7 | 0.018 |
| Burkholder & Nickell[1] .. | 8.0 | 12.0 | 6.0 | 2.0 | 1.0 | 8.0 | 10.0 | — | 0.009 |
| Heller[1] .............. | 7.1 | 10.0 | 0.51 | 1.0 | 1.0 | 0.9 | 11.0 | 8.0 | 0.004 |
| Nitsch & Nitsch[2] ....... | 19.8 | 39.9 | 0.23 | 1.8 | 1.0 | 1.8 | 0.5 | 1.8 | — |
| Basal Medium ........ | 12.05 | 1.76 | 0.61 | 0.29 | 0.32 | 0.092 | 0.87 | 0.1 | 0.053 |
| 1 × Level ........... | (169) | (68.8) | (24.4) | (7.10) | (10.6) | (2.85) | (30.9) | (2.3) | (2.94) |
| Revised .............. | 60.0 | 20.0 | 3.0 | 1.5 | 1.6 | 1.25 | 6.0 | 0.2 | 0.100 |
| Medium .............. | (840) | (782) | (121) | (36.8) | (52.3) | (39.0) | (212) | (4.6) | (5.57) |

[1] Data from Heller's compilation (1953). [2] Data from Nitsch and Nitsch (1956. Data in parenthesis are mg/l of medium.

Table 2. *Micronutrient compositions of plant tissue culture media.*

| Medium | Element (Concentrations in micromoles per liter of medium) | | | | | | | | | | |
|---|---|---|---|---|---|---|---|---|---|---|---|
|  | B | Mn | Zn | I | Cu | Mo | Co | Ni | Te | Be | Al |
| White (1943)[1] | 25 | 30 | 10 | 4.5 | — | — | — | — | — | — | — |
| Gautheret[1] | 0.4 | 4.5 | 0.2 | 1.5 | 0.1 | — | 0.1 | 1.0 | 1.0 | 0.3 | — |
| Hildebrandt et al. (Tobacco) | 6 | 20 | 2.2 | 8 | — | — | — | — | — | — | — |
| Hildebrandt et al. (Sunflower) | 50 | 20 | 1.0 | 2.2 | — | — | — | — | — | — | — |
| Burkholder & Nickell[1] | 10 | 2 | 4.6 | — | 1.6 | 1.0 | — | — | — | — | — |
| Heller[1] | 16 | 4.5 | 3.5 | 0.06 | 0.12 | — | — | 0.13 | — | — | 0.23 |
| Nitsch & Nitsch[2] | — | — | — | — | — | — | — | — | — | — | — |
| Basal Medium 1 × Level | 26 (0.28) | 29 (1.60) | 9.4 (0.62) | 4.5 (0.58) | — | — | — | — | — | — | — |
| Revised Medium | 100 (1.08) | 100 (5.50) | 30 (1.92) | 5.0 (0.64) | 0.10 (0.0064) | 1.0 (0.096) | 0.10 (0.006) | — | — | — | — |

[1] Data from Heller's compilation (1953). [2] Data from Nitsch and Nitsch (1956).

used as needed. T-3 represents 13.6 kg. of solids in 21.6 l of concentrate obtained from 465 kg. of fresh leaves.

*Culture conditions and growth measurements.* The tissues were grown routinely in 125 ml. erlenmeyer flasks with 50 ml. of medium. Fifteen pieces, three in each of five replicate flasks, were planted for each treatment. The cultures were kept on shelves under low intensity overhead florescent lights in a room at 26–28°C and about 85 % relative humidity.

A growth period of 4 weeks was selected because it extends past the end of the logarithmic phase of growth (see Figure 1), after which time the fresh and dry weights per flask were determined. Since in most cases the dry weights of the tissue were close to 5 % of the fresh weights, only the latter will be reported. However, certain trends in percentage dry weight of the tissue with change in composition of the medium will be considered. Replicate cultures had rather uniform yields, and the

standard error, $\pm \sqrt{\dfrac{\Sigma \Delta^2}{n(n-1)}}$, of the average final fresh weights in the 5–20 g range

was around $\pm 0.8$ g/flask.

*Testing procedure.* In most previous work on quantitative nutrient requirements either a single element was varied and the others kept constant, or three elements were varied at a time (the triangulation method). In the present study the requirement of a given element was established by varying its concentration in the presence of several levels of the remaining elements. Tests have been limited to 1 × and higher levels of the elements present in the basal medium. In addition Cu, Co and Mo have been tested.

# Results

Preliminary testing of the basal medium showed that doubling and quadrupling the levels of inorganic or organic constituents lead to increases in

Table 3. *Effect of increasing the concentration of the nutrients in the basal medium on the growth of tobacco callus tissue.* (Growth period, 11/15–12/18/58.)

| Change in composition of medium | Average final fresh weight g./flask. |
|---|---|
| None (Basal medium 1 ×) ..... | 5.6 |
| All constitutents .......... 2 × | 9.9 |
| „ „ .......... 4 × | 14.9 |
| All organic substances .... 2 × | 6.3 |
| „ „ „ .... 4 × | 6.4 |
| All inorganic salts ........ 2 × | 9.7 |
| „ „ „ ........ 4 × | 17.0 |

Table 4. *Effect of quadrupling the concentration of the inorganic elements of the basal medium all together and each separately on the growth of tobacco callus tissue.* (Growth period 12/17/58–1/19/59.)

| Change in composition of medium | Average final fresh weight g./flask. |
|---|---|
| None (Basal medium 1 ×) ..... | 7.1 |
| All elements .............. 4 × | 17.3 |
| N ...................... 4 × | 11.6 |
| K ...................... 4 × | 11.9 |
| P ...................... 4 × | 7.9 |
| Ca ..................... 4 × | 7.6 |
| Mg ..................... 4 × | 7.6 |
| S ...................... 4 × | 7.1 |
| Cl ..................... 4 × | 9.6 |
| Fe ..................... 4 × | 6.6 |
| B ...................... 4 × | 6.8 |
| Mn ..................... 4 × | 7.0 |
| Zn ..................... 4 × | 7.5 |
| I ...................... 4 × | 6.7 |

yield of tissue (Table 3) and that the principal, although not always the entire improvement was from the inorganic salts. When each element was increased separately to 4 times its level in the basal medium (4 ×), each major element except S gave appreciable improvement in yield (Table 4). N and K were especially effective, but increasing all the salts to the 4 × level definitely had a still greater effect than increasing any one element alone. Of the minor elements, excepting perhaps Cl, none was significantly better at the 4 × than at the 1 × level.

## The requirements for N, K, and P

In view of the above results further tests were carried out with possible combinations of N, K, and P raised to 2 ×, 4 × and 8 × levels. In addition these combinations of N, K and P were tested with the remaining elements increased to 2 ×, 4 × and 8 × levels. The greatest increase in tissue growth was obtained when N was increased to 4 ×, K to 8 ×, P to 8 ×, and all other elements were increased to 2× or 4×, *i.e.* under the more favorable conditions for growth the optimal levels of N, K and P would appear to be about 50, 15 and 1.0 m$M$ respectively.

*N.* The N requirement was examined in more detail by testing the growth of pith tissue on media containing varying amounts of $NH_4NO_3$ to give N levels of 12, 24, 48 or 96 m$M$ in combinations with 14.4 or 28.8 m$M$ K and 0.74 or 1.5 m$M$ P. All other inorganic constituents were kept at their 3× levels. The average final fresh weights of tissue obtained in one experiment are plotted against the N level in Figure 3. It may be noted that under the conditions of this experiment the increase in K for unknown reasons resulted in a marked increase in yield at the low N levels. In this experiment the high phosphate level (1.5 m$M$) depressed growth except perhaps in the case of high K and low N media. In experiments with other lots of plants, P levels up to 2.0 m$M$ were sometimes favorable, but generally P levels of 1.5 m$M$ or higher have tended to depress the yield of either pith or callus tissue.

The data show that 50 m$M$ is close to optimal for N irrespective of the K and P concentrations, and that the N content may be increased somewhat above this level with no appreciable effect on the yield. However, N levels of 80 m$M$ or higher consistently were found to depress the yield. Hence, a level of 60 m$M$ N has been selected as satisfactory.

*K.* The K requirement was determined with N kept at the 4× level (48 m$M$); P kept at three levels, 8×, 16× and 32× (0.74, 1.5 and 3.0 m$M$ respectively), and all other minerals kept at their 3× levels. As shown in Figure 4, the yields presented in curves 1 and 2 increased with the concentration of K up to 15 m$M$ and fell only slightly at 30 m$M$, the highest tested K level. In curve 3 the yields are low throughout, probably due to toxicity of the high P level (see above). A 20 m$M$ K level has been selected.

*P.* The P requirement was determined with N at the 4× level (48 m$M$) with K at three levels (7.2 14.4 and 28.8 m$M$) and with all other inorganic constituents at their 3× levels. The results are shown in Figure 5. It may be seen that the optimal requirement for P was strikingly dependent on the level of K, but with an ample supply of K it lies in the region between 0.7 and 1.5 m$M$. Because, as stated, high P levels were toxic to the pith explants in this particular lot of plants, the experiment was repeated in part with several other lots of plants. In these cases the K level was set at 20 m$M$ and as shown by typical results, curve 4 in Figure 5, the optimum P level was 1.5 m$M$. However, sometimes this level and frequently higher levels were growth inhibitory even though very healthy looking tissue might be produced, Because of these difficulties six additional experiments were done, three with pith and three with callus cultures. P levels were adjusted to 1.0, 1.25, 1.5, 1.0 and 3.0 m$M$. Other elements were as specified for the revised medium (Tables 1 and 2) except that Fe was used at both the 0.10 and 0.25 m$M$ levels. Of the organic constituents (Table 6 B) Edamin was omitted, and for some treatments sucrose was used at either the 2 % or 4 % level as well as at the specified 3 % level. Data from one experiment with pith, typical of the results obtained, are presented in curve 5 in Figure 5. It appeared that the P requirement may vary with the levels of Fe and sucrose as well as K. Nevertheless, the results showed conclusively that 1.25 m$M$ is close to the optimal P level;

TOSHIO MURASHIGE AND FOLKE SKOOG

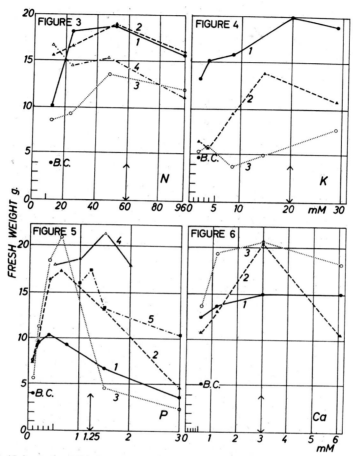

Figures 3 to 10 inclusive. *Fresh weight yields of excised tobacco pith cultures in response to increased concentrations of nutrient elements.*

Figure 3. *Effect of N.* (Growth period, May 5 to June 20, 1959). All elements at 3 × levels except K and P.
Curve 1: K 14.4, P 0.74 m*M*. Curve 2: K 28.8, P 0.74 m*M*. Curve 3: K 14.4, P 1.5 m*M*. Curve 4: K 18.8, P 1.5 m*M*. BC, basal control.

Figure 4. *Effect of K.* (Growth period, May 19 to June 16, 1959). N kept at 4 × (48.8 m*M*) and all other elements at 3 × levels except P.
Curves 1, 2 and 3, the level of P was 0.74, 1.5 and 3.0 m*M* respectievly. BC, basal control.

Figure 5. *Effect of P.* (Growth period, Curves 1–3, May 20 to June 17, 1959; Curve 4, 1960; Curve 5, July 19 to Sept. 1, 1962).
N kept at 4 × and all other elements at 3 × levels except K. Curves 1, 2, 3 and 4, the level of K was 7.2, 14.4, 28.8, and 20.0 m*M* respectively. Curve 5, see text. BC, basal control.

Figure 6. *Effect of Ca.* (Growth period, July 9 to Aug. 10, 1959) N kept at 55 m*M*, K at 20 m*M*, P at 1.0 m*M*, and micronutrients B, Fe, Mn and Zn at 3 × levels. Mg, S, and Cl varied. Each kept at its 1 ×, 2 ×, and 5 × levels in Curves 1, 2 and 3 respectively. BC, basal control.

*Physiol. Plant., 15, 1962*

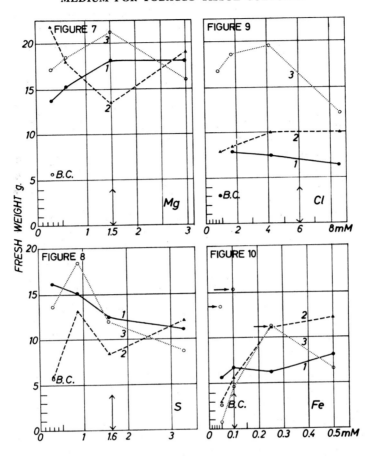

Figure 7. *Effect of Mg* (Growth period, July 9 to Aug. 10, 1959). N kept at 55 m*M*, K at 20 m*M*, P at 1.0 m*M*, and micronutrients B, Fe, Mn, and Zn at 3 × levels. Ca, S and Cl varied. Each kept at 1 ×, 2 × and 5 × levels in Curves 1, 2 and 3 respectively. BC, basal control.

Figure 8. *Effect of S* (Growth period, July 10 to Aug. 10, 1959). N kept at 55 m*M*, K at 20 m*M*, P at 1.0 m*M*, and micronutrients B, Fe, Mn, and Zn at their 3 × levels. Ca, Mg, and Cl varied. Each kept at 1 ×, 2 × and 5 × levels in curves 1, 2 and 3 respectively. BC, basal control.

Figure 9. *Effect of Cl* (Growth period, July 10 to Aug. 11, 1959). N kept at 55 m*M*, K at 20 m*M*, P at 1.0 m*M*, and micronutrients B, Fe, Mn, and Zn at their 3 × levels. Ca, Mg, and S varied. Each kept at its 1 ×, 2 × and 3 × levels in curves 1, 2 and 3 respectively. BC, basal control.

Figure 10. *Effect of Fe* (Growth period, July 15 to Aug. 12, 1959). Nutrients kept at the following levels: N 55, K 20, P 1.0, Ca 3.0, Mg 1.5, S 1.6 and Cl 6.0 m*M*. Micro-elements B, Mn, and Zn varied. Each kept at its 1 ×, 2 × and 5 × levels in curves 1, 2, and 3 respectively. Horizontal arrows represent yields attained in later experiments, see text. BC, basal control.

*Physiol. Plant., 15, 1962*

1.0 is on the low side, 1.5 on the high side, and 2 or more m$M$ definitely too high. For this reason, a 1.25 m$M$ P level has been selected.

### Requirements for Ca, Mg, S, and Cl

When the requirements for Ca, Mg, S, and Cl first were evaluated, the N, K, and P levels of the media were kept at 55, 20 and 1.0 m$M$ respectively; *i.e.*, close to the values selected above as satisfactory for each of these elements. Each of the four elements was tested at four levels (1 ×, 2 ×, 5 ×, and 10 ×) and in conjunction with three levels (1 ×, 2 ×, and 5 ×) of the other three elements. The micronutrients B, Fe, Mn and Zn were provided at their 3 × levels.

*Ca.* By comparison of the curves in Figure 6 the requirement for Ca would appear to be relatively higher with the 2 × than with the other levels of Mg, S and Cl. With these elements at the 5 × level the Ca concentration curve forms a broad plateau between 1.5 and 6 m$M$. A 3.0 m$M$ Ca level has been selected.

*Mg.* Results from one of the experiments with Mg are shown in Figure 7. Both with the 1 × and 5 × levels of Ca, S and Cl (curves 1 and 3) tissue yields increased as the Mg concentration was increased from 0.3 to 1.5 m$M$. The 3.0 m$M$ level of Mg was without further stimulating effect when the levels of Ca, S and Cl were 1 × and it was perhaps inhibitory when the levels of these elements were 5 ×. With 2 × levels of Ca, S, and Cl, the final fresh weights of the pith explants fell as the Mg concentration was raised from 0.3 to 1.5 m$M$ but was up again when the Mg concentration was 3.0 m$M$. Although in this case the lowest concentration of Mg resulted in the highest yield it should be noted that an imbalance existed as the cultures in this particular treatment became very necrotic. The 1.5 m$M$ level of Mg was selected.

*S.* As is shown by curve 1 in Figure 8, with 1 × levels of Ca, Mg and Cl, the lowest concentration of S (0,3 m$M$) was optimal and higher S concentrations caused a steady reduction in tissue yields. However, with the 2 × or 5 × levels of Ca, Mg and Cl, the 0.3 m$M$ S level was less than adequate, and a substantial increase in yield was obtained at the 0.66 m$M$ level. S concentrations higher than 0.66 m$M$ were inhibitory to the pith explants. In subsequent tests with callus, on the other hand, no toxicity was obtained even with S concentrations as high as 13 m$M$. On the basis of all results and even though it may be slightly higher than optimal for excised pith cultures, the 1.6 m$M$ level of S has been selected.

*Cl.* With Ca, Mg and S at their respective 1 × levels, increases in Cl concentration had no appreciable effect on growth of pith explants (Figure 9), whereas with Ca, Mg and S at their 2 × or 5 × levels, increases in the Cl level up to 4.4 m$M$ were stimulatory. A higher level of Cl (8.7 m$M$) gave neither a further increase nor inhibition of growth in the presence of 2 × levels of Ca, Mg and S, but it was far less stimulatory than the lower Cl levels in the presence of 5 × levels of Ca, Mg and S. With callus instead of pith, Cl concentrations as high as 12 m$M$ were without inhibitory effect. A Cl level of 6 m$M$ has been selected.

*Physiol. Plant., 15, 1962*

## *Micronutrient requirements*

Preliminary studies indicated that $3 \times$ to $5 \times$ levels of B, Fe, Mn and Zn were not inhibitory, and that in combinations with increased NaFe-EDTA levels they were sometimes stimulatory. In these experiments the other elements were supplied in the concentrations selected as optimal or very close to these levels.

*Fe.* Availability of Fe has long been recognized as a critical requirement for vigorous and prolonged growth of plant tissues. The use of chelating agents, citric or tartaric acid and more recently especially the use of EDTA has greatly improved the available supply of Fe and possibly some other micronutrients in culture media. In the present experiments Fe was supplied at first as a commercial NaFe-EDTA chelate which was tested in combination with $1 \times$, $2 \times$, and $5 \times$ levels of B, Mn and Zn. As shown in the curves in Figure 10, increases in the Fe chelate from 0.05 to 0.25 m$M$ resulted in increased yields of tissue especially in the presence of $2 \times$ and $5 \times$ levels of the other three trace elements. Increasing the NaFe-EDTA level to 0.5 m$M$ resulted in a definite decrease in yield when the other trace elements were raised to their $5 \times$ levels. A higher purity preparation of Na$_2$-Fe-EDTA was then used, which in a large number of tests with both callus and pith tissues consistently gave optimum yields at the 0.10 m$M$ level. Even 0.05 m$M$ Fe permitted excellent growth, and sometimes close to optimum yields, of both types of tissue, whereas 0.25 m$M$ Fe generally gave considerably lower yields. Representative yields at the 0.05, 0.10 and 0.25 m$M$ levels obtained in a series of three experiments with pith tissue are shown by horizontal arrows in figure 10. In view of these results the Fe content of the commercial Fe-chelate was verified by chemical analysis. Considering the complex interaction between the Fe-chelate and other elements in the nutrient medium various plausible reasons might account for the difference in the results obtained. The curves in figure 10 suggesting an optimal level of 0.25 m$M$ are included for the sake of consistency with the data for the other elements, even though on the basis of all results this level is too high and a 0.10 m$M$ of Na$_2$Fe-EDTA has been selected as optimal.

*B, Mn, and Zn.* Considering the likely presence of impurities, the B, Mn and Zn requirements were finally tested with callus grown on media made up with Difco purified agar and without addition of Edamin. Each element and all three in combination were tested in serial concentrations up to their $5 \times$ levels, and Na$_2$Fe-EDTA was kept at the 0.25 m$M$ Fe level. There was no inhibition from their presence, and perhaps a stimulation, at least when all three test elements were increased to their $5 \times$ levels (See Table 5). From these and other data the levels 0.10 m$M$ B, 0.10 m$M$ Mn, and 0.030 m$M$ Zn have been selected.

*Other Elements (Cu, Mo, Co, I, and Na).* Cu added in the range from 0.00003 to 0.03 m$M$ was without effect on the growth of the tobacco tissue, and so was Mo, 0.001 m$M$, the concentration employed by Torrey (1954) with pea cultures. Iodine, even when raised to the $4 \times$ level (0.02 m$M$), and Na tested at the 5 and 10 m$M$ levels, *i.e.*, in the range used by Heller (1953) and by Hildebrandt *et al.* (1946), similarly were without effect on yields.

In spite of the above negative results these elements have been included in

Table 5. *Effect of increased levels of B, Mn and Zn on the growth of tobacco callus tissue.* (Growth period 9/15–10/12/59.)

| Change in composition of medium | Average final fresh weight g./flask. |
|---|---|
| None (Basal medium 1 ×) . . . . . | 12.5 |
| B . . . . . . . . . . . . . . . . . . . . . 5 × | 11.6 |
| B and Mn . . . . . . . . . . . . . . . 5 × | 12.6 |
| B and Zn . . . . . . . . . . . . . . . 5 × | 12.6 |
| B, Mn and Zn . . . . . . . . . . . . 5 × | 14.3 |
| Mn . . . . . . . . . . . . . . . . . . . . 5 × | 13.7 |
| Mn and Zn . . . . . . . . . . . . . . 5 × | 12.2 |
| Zn . . . . . . . . . . . . . . . . . . . . 5 × | 14.0 |

the revised medium. Cu and Mo are added because they are known to be essential elements for plant growth. Similarly Co is included for reasons of its requirement in lower plants (Holm-Hansen *et al.* 1954) and the possible role of cobaltous ion in morphogenesis of higher plants (Miller 1954, Salisbury 1959). A test of $CoCl_2$ in eight serial concentrations between 0.0001 and 0.16 m$M$ and all other elements at the 1 × levels was included in an early experiment. No stimulatory effect of Co was obtained at any level but rather a toxic action at the two highest levels, 0.08 and 0.16. The selected levels are: Cu 0.0001, Mo 0.001 and Co 0.0001 m$M$. Iodine arbitrarily is retained at 0.005 m$M$. Na, ostensively nonessential but perhaps not entirely devoid of its often claimed stimulating effect on plant growth, being a part of the $FeSO_4$-$Na_2$ EDTA solution, therefore, is supplied at the 0.20 m$M$ level.

*pH.* The pH of the medium was adjusted to 5.7–5.8 with a few drops of 1 $N$ HCl or with NaOH or KOH either before or after the agar was added and had been heated a few minutes in the autoclave. The value was selected on the experience that the reaction of either more acid or more basic media tends to drift toward this region during the heat treatment for sterilization and subsequently with time even in uninoculated flasks. Preheating with the agar present avoids any appreciable change in pH during the subsequent autoclavation. The value 5.7 to 5.8 is suitable for maintaining all the salt in soluble form even with relatively high phosphate levels and low enough to permit rapid growth and differentiation of the tissue. The amounts of Cl, Na or K introduced in the process are not included in the stated levels, 6.0, 0.2 or 20 m$M$ respectively of these ions.

*Composition and growth effects of the revised medium.* The kinds and amounts of mineral salts finally adopted for the revised medium are listed in Table 6 A. Quantities have been adjusted upward, beyond actual needs in some cases to provide total concentrations of all elements in round numbers as shown in Tables 1 and 2.

The yield of tabacco tissue on the revised medium has been compared with those on the original basal medium (1 ×) and other media as shown in Table 7 and in Figure 11. Not only were the highest yields obtained consistently on the revised media but the pronounced tissue necrosis characteristic of old cultures on all other media was avoided as is also evident in Figure 11.

*Physiol. Plant., 15, 1962*

**Table 6.** *Composition of the revised medium.* pH adjusted to 5.7–5.8 with HCl, KOH, or NaOH (see text).

A. Mineral salts

| Major elements | | | Minor elements | | |
|---|---|---|---|---|---|
| Salts | mg/l. | m$M$ | Salts | mg/l. | $\mu M$ |
| $NH_4NO_3$ ............... | 1650 | N 41.2 | $H_3BO_3$ ............... | 6.2 | 100 |
| $KNO_3$ ................. | 1900 | 18.8 | $MnSO_4 \cdot 4H_2O$ .......... | 22.3 | 100 |
| $CaCl_2 \cdot 2H_2O$ ........... | 440 | 3.0 | $ZnSO_4 \cdot 4H_2O$ .......... | 8.6 | 30 |
| $MgSO_4 \cdot 7H_2O$ .......... | 370 | 1.5 | KI ................... | 0.83 | 5.0 |
| $KH_2PO_4$ ............... | 170 | 1.25 | $Na_2MoO_4 \cdot 2H_2O$ ........ | 0.25 | 1.0 |
| $Na_2$-EDTA ............. | 37.3[1] | Na 0.20 | $CuSO_4 \cdot 5H_2O$ .......... | 0.025 | 0.1 |
| $FeSO_4 \cdot 7H_2O$ .......... | 27.8[1] | Fe 0.10 | $CoCl_2 \cdot 6H_2O$ .......... | 0.025 | 0.1 |

[1] 5 ml/l of a stock solution containing 5.57 g $FeSO_4 \cdot 7H_2O$ and 7.45 g $Na_2$-EDTA per liter of $H_2O$.

B. Organic constituents

| | | | | |
|---|---|---|---|---|
| Sucrose ................. | 30 g/l. | | Agar ..................... | 10 g/l. |
| Edamin (optional) ........ | 1 g/l. | | *myo*-Inositol .............. | 100 mg/l. |
| Glycine ................. | 2.0 mg/l. | | Nicotinic acid ............. | 0.5 mg/l. |
| Indoleacetic acid[2] ....... | 1–30 mg/l. | | Pyridoxin·HCl ............. | 0.5 mg/l. |
| Kinetin[2] ................ | 0.04–10 mg/l. | | Thiamin·HCl .............. | 0.1 mg/l. |

[2] See text and figures 12 and 13.

**Table 7.** *Comparison of yields of tissue on different media*
A. Pith Tissue (Growth period 10/2–11/3/59).

| Medium | Fresh weight in g/flask | | Dry weight in g/flask | |
|---|---|---|---|---|
| | Without Edamin | With Edamin | Without Edamin | With Edamin |
| Revised.............. | 17.3 | 24.5 | 0.47 | 0.49 |
| Basal................. | 4.1 | 7.1 | 0.13 | 0.26 |
| Heller ............. | 5.0 | 17.9 | 0.19 | 0.43 |
| Nitsch ............. | 1.9 | 9.7 | 0.10 | 0.34 |
| Hildebrandt *et al.* | 5.3 | 9.2 | 0.20 | 0.37 |

B. Callus Tissue (Growth period 8/31–10/10/59).

| Medium | Fresh weight in g/flask | | Dry weight in g/flask | |
|---|---|---|---|---|
| | Without Edamin | With Edamin | Withouth Edamin | With Edamin |
| Revised.............. | 21.5 | 22.7 | 0.48 | 0.46 |
| Basal................. | 5.1 | 10.8 | 0.22 | 0.35 |
| Heller ........... .. | 3.6 | 16.2 | 0.24 | 0.38 |
| Nitsch and Nitsch | 3.0 | 17.6 | 0.28 | 0.48 |
| Hildebrandt *et al.* | 2.8 | 9.8 | 0.18 | 0.37 |

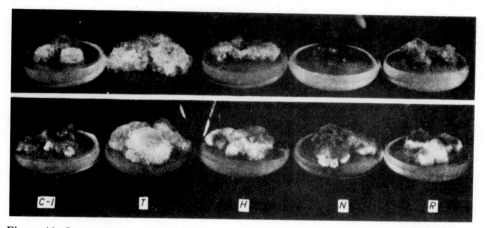

Figure 11. *Comparison of yields of tobacco pith tissue cultured on various media* (Growth period, Oct. 2 to Nov. 3, 1959). Upper row without, lower row with Edamin 1.0 g/l. Media from left to right: C-l, basal medium 1 ×, T, revised; H, Heller (1953), N, Nitsch and Nitsch (1956), and R, Hildebrandt *et al.* (1946).

## Organic Constituents

The required levels of organic constitutents in the basal medium have been tested only to a limited extent.

From the results in Table 3, no large increase in yield would be expected from increased concentrations of the specified organic ingredients, although with improved mineral supplies some beneficial effect might be obtained by modifications in these. The list of other potentially stimulatory organic substances and favorable combinations of these is of course inexhaustible.

*Vitamins.* The levels of the following vitamins have been retained unchanged from the basal medium: thiamin-HCl 0.1 mg/l, nicotinic acid 0.5 mg/l, pyridoxine · HCl 0.5 mg/l. On the basis of experience by others (Braun 1958, Steinhart *et al.* 1961, 1962) *myo*-inositol was tested and found to promote growth in cultures when other conditions would permit high yields. A level of 100 mg/l of *myo*-inositol has been selected.

*Nitrogenous compounds.* The combination of amino acids in yeast extract and also various simpler combinations are known to promote growth of tobacco callus on the basal medium (Sandstedt and Skoog 1960, La Motte 1960). Glycine has been retained without further testing at 2 : 0 mg/l. The casein hydrolysate, Edamin, was tested extensively. In the comparative tests with different media it substantially increased the yield on all except the revised medium (Table 7). The strikingly lower response on the latter medium probably is due mainly to the relatively high N content. The small increase which did result from addition of Edamin perhaps is due in part to unknown organic factors. An additional amount of $NH_4NO_3$ equivalent to the N content of the Edamin failed to replace it and instead lowered the yield. However, $NH_2$-N is not excluded by these results as being the principal active

*Physiol. Plant., 15, 1962*

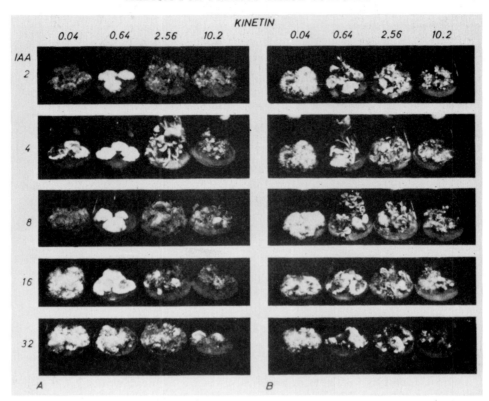

Figure 12. *Tobacco callus cultures grown on revised medium with increasing levels of IAA* (ordinate, mg/l.) *and kinetin* (abscissa, mg/l.). *A* without Edamin; growth period, Oct. 6 to Dec. 22, 1961. *B* with 1 mg/l. Edamin; growth period Oct. 11 to Dec. 22, 1961.

agent contributed by the hydrolysate, for it has been shown (Steinhart *et al.* 1961) that spruce tissue which is unable to survive on a medium with $NH_4NO_3$ as the sole N-source grows well on the same medium supplemented with either urea, arginine or other basic amino acid. The influence of different kinds of casein hydrolysate (kindly supplied by the Sheffield Chemical Co.) is quite variable, Edamin being among the more active preparations. The influence of Edamin itself seems to vary to some extent, increasing with age of the cultures and being even more striking with respect to the development of vigorous shoots and root systems than the increase in yield. As may be seen from comparisons of corresponding cultures under A and B in Figure 12, the presence of Edamin has resulted in a broader range of IAA and kinetin levels which permit vigorous organ development. It has shifted the levels and to a slight extent perhaps also the ratio of the two substances required for a particular developmental pattern to emerge. However, essentially no effects were obtained in the presence of Edamin that could not be obtained also in its absence by suitable modifications of the kinetin and IAA levels.

*Physiol. Plant., 15, 1962*

*Carbohydrates.* Although tobacco cultures are capable of utilizing various carbohydrates and organic acids to some extent as energy sources, sucrose is as favorable as any known material of this type. Concentrations of 2, 3, and 4 % are almost equally suitable for cultures with low yields. However, in several large series of comparative tests 3 % was definitely better than 2 %, and 4 % was often somewhat less effective than 3 % in cultures with moderate to high yields. For example in one experiment fresh weight yields of cultures with 2, 3 and 4 % sucrose were 14.3, 15.9 and 12.9 g/flask respectively and the corresponding dry weights were 0.44, 0.67, and 0.62 g/flask. Dry weights often were somewhat higher per fresh weight in cultures with 4 % sucrose than in the others, but on the average also the total dry weight per flask was less in these than in cultures with 3 % sucrose. The level of sucrose in the revised medium, therefore, has been set at 3 %.

Agar ordinarily is kept at 1 % to give a suitably moist but rigid medium. The concentration may need to be varied to some extent depending on the preparation and the salt content, pH, etc. of the medium. In the case of weakly growing tissue cultures agar content may be critical for survival, but for vigorously growing tobacco cultures it was unimportant. In liquid cultures, on the other hand, changes in composition of the medium definitely are required for vigorous bud development to occur (Loewenberg *et al.* unpublished.)

*Indoleacetic acid and kinetin.* The optimal level of IAA or kinetin will depend on the type of growth that is desired. It will vary with the endogenous contents of auxins and kinins and other factors influencing the sensitivity of a particular strain or clone of tissue to these substances.

The curves in figure 13 representing fresh weight yields in two experiments with different IAA-kinetin combinations clearly show that the optimal requirement for either one increased with the concentration of the other. The effect of a given combination, especially in the intermediate concentration range, varied from one experiment to the next as illustrated by the two examples in figure 13. Note that the response to 3 mg/l. IAA is much greater in one experiment than the response to 4 mg/l. in the other. Remarkable developmental patterns, differences in form and texture of the callus, extent of organ formation, number and size of organs, vigor and longevity of the cultures are associated with specific levels and ratios of IAA and kinetin. An impression of the range in size and form of the cultures may be gained from figure 12. Details of the strict quantitative aspects and other characteristics of the growth responses which have emerged as functions of the IAA-kinetin concentrations will be presented elsewhere. In general the following applied. For continuous growth of firm, healthy tobacco callus in sub cultures, 0.2 mg/l. of kinetin and 2.0 mg/l. of IAA have been used with excellent results for several years in our laboratory. However, 0.04 mg/l. of kinetin and 1.0 mg/l. of IAA has given much faster growing, loosely packed masses of cells which could be transferred and maintained as callus even though vigorous root formation ordinarily occurred in these cultures after 2 to 3 weeks. For long lasting cultures (2 to 6 months) higher kinetin levels were required, and for vigorously growing such cultures high levels of both kinetin (3 to 6 mg/l.) and IAA (up to 30 mg/l.) were needed. For induction of buds the IAA level

*Physiol. Plant., 15, 1962*

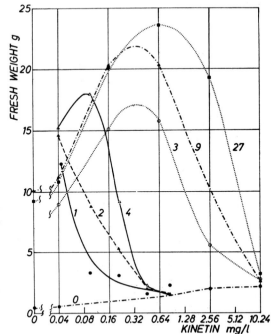

Figure 13. *Effects of increasing concentrations of IAA and kinetin on fresh weight yield of tobacco callus cultured on revised medium* (Growth period, 4 weeks, Aug. to Oct. 1960). Ordinate: average fresh weight per flask, abscissa: kinetin concentration. The numbers on the curves 0, 3, 9, 27 and 1, 2, 4 correspond to the concentration of IAA added in each case. Results are from two experiments represented by solid and broken lines respectively.

must be kept relatively low, between 1 and 5 mg/l., depending on the kinetin level.

Naphthaleneacetic and 2,4-dichlorophenoxyacetic acids have been tested as substitutes for IAA. In tissues with high IAA inactivation rates they were preferable if not essential for growth. Levels of 0.05 to 0.2 mg/l. have been satisfactory for the growth of callus. These substances are so much stronger or longer lasting auxins than IAA, that all but extremely low concentrations prevented bud formation.

*Re-test of the tobacco leaf extract.* In accordance with the objective to insure an adequate supply of minerals in tissue culture bioassays of organic growth promoting substances in complex extracts, the stimulating effect of tobacco leaf extract was now re-tested with the revised medium. The extract again promoted growth (Figure 15), but in contrast with the earlier tests on the basal medium, on the revised medium the ash of the extract was without significant effect. For example, in one experiment the fresh and dry weight yields on the revised medium alone were 13.6 and 0.43 g/flask respectively; with 3.75 g/l. T-3 extract they rose to 21.7 and 0.58 g/flask respectively; whereas the ash of this extract gave only 14.6 and 0.46 g/flask respectively.

The promotion of growth by the T-3 extract in this case, therefore, can be attributed to its content of organic substances rather than inorganic salts.

*Effect of volume of medium.* Further experiments revealed tissue yield to

32

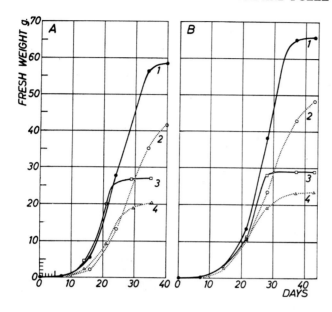

Figure 14. *Effects of the volume of revised medium and of the addition of tobacco leaf extract on fresh weight yield.*

A. Tobacco pith sections (planted Oct. 1, 1959).
B. Tobacco callus (subcultured Sept. 29, 1959).
Curves 1 and 2, 100 ml of medium in 250 ml Erlenmeyer flasks.
Curves 3 and 4, 50 ml of medium in 125 ml flasks.
Curves 1 and 3 with T-3 extract, 3.75 g solids/l of medium.
Curves 2 and 4 without T-3 extract.

be related not only to extract concentration, but also to the volume of medium used. It can be seen in Table 8, that at least at the higher extract levels, the yield was almost directly proportional to the volume of the medium. More important, yield differences between extract levels were pronounced in the 100 ml. volumes, much less in the 75 ml. volumes and only slight in the 50 ml. volumes of medium. Yields of tobacco tissue was increased from 30 to 60 g/flask by adding the optimum extract level and using 100 instead of 50 ml. of medium per flask. Figure 14 reveals, furthermore, that the tobacco extract enhanced the rate of growth of tissue, whereas increase in medium volume essentially prolonged the logarithmic growth period. It may be concluded that stoppage of growth is due to exhaustion of the entire medium, perhaps mainly the water (see figure 16). The greater final fresh weights of cultures with extract may in fact be due in part to the earlier absorption of most of the available water by the tissue and consequent lower loss by evaporation from the medium.

Although the fresh and dry weights increased in about the same manner, it is clear from the data in Table 8 as from other experiments that the dry weight continues to increase with the concentration of extract considerably beyond the level which is optimal for increase in volume of the tissue. The extent to which this additional increase in dry weight reflects biosynthesis of new cell materials or merely continued absorption of solids from the medium has not been determined.

*Attempts to replace the leaf extract by known substances.* Optimal fresh weight yields on 50 ml. volume of revised medium often have approached but rarely exceeded 25 g/flask, whereas with leaf extract added under otherwise the same conditions yields went over 30 g and once reached 35 g/flask.

*Physiol. Plant., 15, 1962*

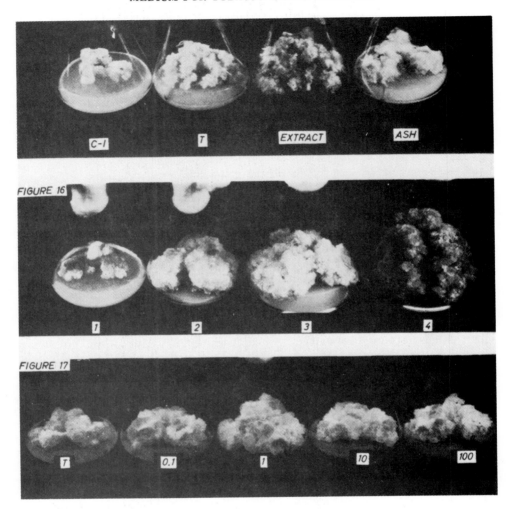

Figure 15. *Effects of tobacco leaf extract and its ash content on the growth of tobacco callus tissue* (Growth period, Sept. 3 to Oct. 2, 1959). From left to right: Basal control, revised medium alone, with T-3, 3.75 g/l, and with ash from the corresponding amount of T-3.

Figure 16. *Cultures illustrating growth and utilization of media of different composition and volume.* (Growth period, July 2 to 30, 1959.) Flask 1, basal medium 50 ml; flask 2, revised medium 50 ml; flask 3, revised medium 100 ml; and flask 4, revised medium with leaf extract 100 ml.

Figure 17. *Effect of concentration of gibberellic acid* (Growth period, Oct. 29 to Nov. 19, 1959). Numbers denote mg/l.

*Physiol. Plant., 15, 1962*

Table 8. *Effects of concentration of extract and volume of medium on tissue yield on revised medium* (Growth period, 7/2–7/30/59).

| Volume ml/flask | T-3 Extract (solids, g/l) | | | | |
|---|---|---|---|---|---|
| | 0 | 0.9 | 1.9 | 3.8 | 7.5 |
| | Fresh weight g/flask | | | | |
| 50 | 25.7 | 24.7 | 24.3 | 30.7 | 28.5 |
| 75 | 31.0 | 30.3 | 32.3 | 47.9 | 47.9 |
| 100 | 32.0 | 39.4 | 51.1 | 58.5 | 54.2 |
| | Dry weight g/flask | | | | |
| 50 | 0.65 | 0.67 | 0.70 | 0.77 | 0.86 |
| 75 | 0.90 | 0.82 | 0.87 | 1.17 | 1.36 |
| 100 | 0.93 | 1.08 | 1.31 | 1.41 | 1.55 |

With larger volumes of medium the differences in yield in the presence and absence of leaf extract were still more marked (see Table 8). About the same can be said for dry weight yields.

As a preliminary step in diagnosing the stimulatory effect of the extract various substances have been tested which are known to enhance the growth rate of tissue cultures.

"*Braun's supplements*" (cytidylic acid 200 mg/l., guanylic acid 200 mg/l., L-asparagine 500 mg/l. and L-glutamine 500 mg/l.: private communication) in some cases where the controls were relatively low, gave up to nearly 50 % increase in fresh weight but never more than between 25 and 30 g/flask. Furthermore, there were no corresponding increases in dry weight, so that apparently these addenda influenced mainly the rate of water uptake and cell expansion.

*Gibberellic acid* added to the basal medium (X) resulted in little or no growth stimulation. In contrast, with the revised medium cold-sterilized gibberellic acid markedly increased the growth rate. An example of the effect of increasing concentrations is shown in figure 17. In this case 1 mg/l. appears to be optimal, but generally perhaps higher levels may be preferable. As shown by the data in table 9 the gibberellin effect, except perhaps of very high levels, in these experiments was entirely on fresh weight. It was most

Table 9. *Influence of gibberellic acid on yield of tobacco callus tissue* (Growth period, 10/29–11/27/59).

| GA mg/l | Yield, g/flask | |
|---|---|---|
| | Fresh weight | Dry weight |
| 0 | 15.5 | 0.55 |
| 0.1 | 26.8 | 0.53 |
| 1.0 | 27.7 | 0.53 |
| 10 | 26.3 | 0.56 |
| 100 | 25.3 | 0.64 |

*Physiol. Plant., 15, 1962*

noticeable in the early stages of growth. In some cases where the tissues did not quickly exhaust the available medium and were kept for 8 weeks or longer the controls seemed to catch up with the treated cultures.

In pith cultures supplied with gibberellic acid, kinetin and IAA singly and in combinations it was clear that neither gibberellic acid nor kinetin alone, or in combination of the two, stimulated growth. In contrast, IAA alone of course gave rise to marked tissue expansion which effect of IAA was enhanced by gibberellic acid. IAA together with kinetin permitted continuous growth of the tissue and the rate of this callus development was speeded greatly by the further addition of gibberellic acid. Thus, on a suitable medium, the three substances in combination exerted growth stimulatory effects which readily could be distinguished each one from the others.

With the addition of all the above mentioned organic materials to the revised medium, rapid early growth rates comparable with those on medium with leaf extract added have been obtained, but so far the final yields, especially the dry weights, have been lower. In spite of this it is possible that the effect of the leaf extract may be accounted for in terms of its contents of gibberellins, purine derivatives and other known constituents. In fact, it now seems unlikely that its growth promoting activity is due to an unknown growth factor of major importance, although this possibility is not entirely excluded by the present results.

## Discussion

The use of tissue cultures as a tool in problems of inorganic nutrition of plants has been much advocated but so far for several obvious reasons has proved fruitless in practice. Their use for the detection and study of biologically active organic substances, natural and synthetic has been more purposeful. It is clear, however, that biological activity of any one substance not only varies with the dosage but depends greatly on the milieu in which it is placed. Conclusions as to the presence or function of a particular substance on the bases of growth responses must be drawn with caution. In bioassays for growth factors practically an all-or-none response is needed as evidence of a new active agent. Most claims for a new factor based solely on quantitative increases in yield or growth rate from the administration of unknown mixtures of materials have no validity and limited usefulness. Even in cases of unknown factors supplied in mixtures or in unknown amounts, very large differences between treatments and controls may be required for significant results. The medium specified in table 6 has been designed to minimize the response of the tobacco tissue employed to variations in its constituents and to inorganic salts, sugars, etc. commonly present it tissue extracts. A positive growth response, therefore, is likely to reflect the presence in the medium of growth promoting organic substances. This medium has provided for excellent growth of either tobacco callus or excised pith tissue under ordinary environmental conditions. Replicate cultures within a given treatment of an experiment were astonishingly alike in size and external appearance. The reproductibility of results from one experiment to the next also has been satisfactory, but some variability was observed in total yields and in morphogenic patterns in response to a given treatment. At least in part this variability can be ascribed to differences in environmental conditions and treatment of the stock cultures.

The reported optimal yields were obtained with the temperature of the culture room kept between 26 and 28°C., but the temperature was not carefully controlled nor was its influence systematically studied.

The cultures were kept on tiers of ca 1 meter wide shelves open only on one side, so that they were exposed to continuous weak and somewhat variable diffuse light from fluorescent tubes and Mazda lamps mounted on the ceiling. It has been shown earlier that growth and organ formation in response to IAA and kinetin treatments are obtained in darkness as well as in the light, but light is not without modifying effects. In general weak diffuse artificial light is employed. With increase in its intensity the tissue becomes more compact, the fresh weight tends to decrease and eventually also the dry weight is lowered. More uniform responses might be obtained by close regulation of the light intensity. It is likely also that more rapid and uniform growth might be obtained through suitable choices of light quality and especially of day length.

Variability in results also derives from disuniform planting material. Stock cultures of different age or grown on media with different levels of IAA or kinetin, for example, give sub-cultures which differ both in growth rate and yield. A standard pretreatment of stock cultures is needed but difficult to specifiy in advance, as it will depend on the type of growth desired and will vary with conditions and changes in the stock itself. Requirements of inorganic nutrients probably remain relatively fixed, but the need for organic factors may fluctuate widely. This is strikingly illustrated by the sporadic appearance in control cultures of tissue pieces which have the capacity to grow in the absence of IAA or kinetin and occassionally without either of these in the medium. That increased rates of biosynthesis is responsible for this behavior is well known in the case of auxin from the work of Gautheret (1959) and collaborators; and E. J. Fox in our laboratory recently has shown a relatively high kinin content in one tissue which grew without kinetin. Acquired capacity for increased rates of synthesis of both auxin and kinin is of course also known from Braun's studies of crown-gall in tobacco.

A significant case of acquired, graded capacity of tissue development is the report by Chouard and Aghion (1961) that stem segments from the apical region of a flowering tobacco plant produced buds with flowers *in vitro* whereas segments from the basal part of the stem produced only vegetative buds. Acquired differences in accumulative or biosynthetic capacities must be responsible for this difference in behavior. It is especially noteworthy that in this case the change in morphogenetic capacity is associated with normal ontogeny. It is not brought about by an exogenous agent in the same sense as is the case in tumor induction; nor is it an act that at present would find respectable status in the category of so called "biochemical mutations".

In the light of the above considerations it is to be expected that tissues even from the same plant may have different quantitative nutrient requirements for optimal growth *in vitro*, and that these may change with time and conditions. In fact it is surprising that cultures from all parts of the plant kingdom

*Physiol. Plant., 15, 1962*

apparently have rather similar requirements, and that often their growth is limited by one or a few of a small group of common growth factors, as for example by auxin or vitamin $B_1$.

The positive response of the tobacco tissue to gibberellic acid in the presence of the other growth factors on the revised but not on the basal medium is of special interest in showing the futility of attempts to classify tissues of various origin in terms of their requirements for a particular growth factor when cultured on a medium of unknown general adequacy for their growth.

During the five years this work has been in progress the medium in different stages of its revision has been used for cultivation of various species and strains in our and other laboratories. The results have varied from very good growth to no growth at all. In no case have yields approached those for the usual tobacco tissue. It seems that the medium must be modified to accommodate the needs of any given tissue. Furthermore, several passages may be required for the tissue to adapt to high nutrient levels.

The organic addenda in this connection may serve not merely to provide specific growth factors but also in part to buffer against excessive electrolyte activities. Complex formation by amino acids is mentioned as an example of this which might be important especially in early stages. In the course of its growth the tissue itself undoubtedly synthesizes and releases metabolites which not only react with specific constituents but may be said to modify the general physical and physiological properties of the medium.

The mechanisms which enable the tissue to derive nourishment for persistent growth, representing doubling in weight on the average every 2 days, for a 3 to 4 weeks period, are unknown and difficult to imagine. The completeness of the incorporation of water from the revised medium into tissue, especially in the presence of leaf extract, would appear to be a "new" and remarkable phenomenon. In cases of the highest yields, nearly 35 g. fresh weight on 50 ml. and 70 g. on 100 ml. of medium, the dry weights were ca. 0.9 and 1.5 g. respectively. Therefore, about 70 % of the water furnished originally was incorporated into living material and most of the remaining 30 % was lost by evaporation and transpiration during the four week growth period. As compared with this, high yielding dense suspensions of algae have been recorded to utilize about 10 to 15 % of the water available in the nutrient solution for the production of cells. It may be seen in figure 16 that very little medium remained in the 50 ml. revised nutrient solution and practically none in the 100 ml. revised medium with added leaf extract. The tissues must make intimate contact with the agar surface. In fact, breaks occur in the agar itself, and flakes adhere to the tissue as it "burrows" through the medium. Oxygen availability would appear to be no problem in these vigorously growing cultures. During the later stages of growth the osmotic forces required for transfer of water must be huge, *i.e.* of the same order as develop in root systems.

## Summary

A several fold increase in yield of excised pith or callus cultures of *Nicotiana tabacum,* variety Wisconsin 38 was obtained by addition of leaf extract to the standard modified White's nutrient medium.

*Physiol. Plant., 15, 1962*

In part the increase was due to inorganic constituents of the extract, especially N and K, which could be substituted for by either the ash of the extract or raised levels of N and K salts in the medium.

In part the increase was due to organic constituents of the extract which increased both the growth rate and final fresh weight of the cultures.

Similar increases in growth rate but not in final yield were obtained by adding gibberellic acid and Braun's supplements of purines and amino acids to a revised medium.

A revised medium has been developed (Table 6) with each element provided in sufficient quantity to insure that no increase in yield will result from the introduction of additional amounts in the range ordinarily to be expected in plant tissue extracts, etc.

The organic constituents have been retained unchanged except that sucrose has been raised to 3 per cent, *myo*-inositol has been added as a regular constituent and Edamin has been introduced as an optional constituent.

The revised medium is designed for use in bioassays of organic growth factors. It provides for rapid growth rate, increased response to organic growth factors and minimal interference from inorganic and common organic nutrients.

In the presence of plant extracts fresh weight yields of up to 35 g. on 50 ml. of medium have been obtained. Under these conditions water appears to be the limiting factor for growth of the tobacco tissues employed.

Aspects of general application, limitations and behavior of plant tissue cultures are discussed.

This work was supported in part by the Research Committee of the Graduate School, University of Wisconsin with funds from the Wis. Alumni Research Foundation and in part by grant G-5980 from the National Science Foundation. Edamin was kindly furnished by the Sheffield Chemical Co., Norwich, N.Y. The authors are grateful to Miss Joyce Klemm for technical assistance and especially to Dr. Elfriede Linsmaier who did the confirmatory experiments on P, Fe, and sucrose levels and on different combinations of IAA and kinetin.

Present address of T.M.: Department of Horticulture, Agr. Expt. Station, University of Hawaii, Honolulu, Hawaii.

## References

Braun, A. C.: A physiological basis for autonomous growth of the Crown-gall tumor cell. — Proc. National Acad. Sci. (Wash.) 44: 344. 1958.

Burkholder, P. R. & Nickell, L. G.: Atypical growth of plants I. Cultivation of virus tumor of Rumex on nutrient agar. — Bot. Gaz. 110: 426. 1949.

Chouard, P. & Aghion, D.: Modalités de la formation de bourgeons floraux sur des cultures de segments de tige de tabac. — Compt. rend. Acad. Sci. 252: 3864. 1961.

Gautheret, J. R.: Sur les possibilité de realiser la culture indefinie des tissus de tubercules de carotte. — Compt. rend. Acad. Sci. Paris 208: 118. 1939.

— La culture de tissus végétaux. — Masson et Cie, Paris. 1959.

Heller, R.: Recherches sur la nutrition minérale de tissus végétaux cultivées in vitro. — Ann. Sci. Nat. Bot. et Biol. Vég. 11th series 14: 1. 1953.

Hildebrandt, A. C., Riker, A. J. & Duggar, B. M.: The influence of the composition of the medium on the growth in vitro of excised tobacco and sunflower tissue cultures. — Amer. J. Bot. 33: 591. 1946.

Holm-Hansen, O., Gerloff, G. C. & Skoog, F.: Cobalt as an essential element for blue-green algae. — Physiol. Plant. 7: 665. 1954.

La Motte, C. E.: The effects of tyrosine and other amino acids on the formation of buds in tobacco callus. — Ph.D. Thesis. Univ. of Wisconsin. 1960.

Miller, C. O.: The influence of cobalt and sugar upon the elongation of etiolated pea stem segments. — Plant Physiol. 29: 79. 1954.

Nitsch, J. P. & Nitsch, C.: Auxin-dependent growth of excised Helianthus tuberosus tissue I. — Amer. J. Bot. 43: 831. 1956.

Salisbury, F. B.: Growth regulators and flowering II. — The cobaltous ion. — Plant Physiol. 34: 598. 1959.

Sandstedt, R. M. & Skoog, F.: Effect of amino acid components of yeast extract on the growth of tobacco tissue in vitro. — Physiol. Plant. 13: 250. 1960.

Steinhart, C. E., Standifer, L. C., Jr. & Skoog, F.: Nutrient requirements for in vitro growth of spruce tissue. — Amer. J. Bot. 48: 465. 1961.

Steinhart, C. E., Anderson, L. & Skoog, F.: Growth promoting effect of cyclitols in spruce tissues. — Plant Physiol. 37: 60. 1962.

Torrey, J. G.: The role of vitamins and micronutrient elements in the nutrition of the apical meristem of pea roots. — Plant Physiol. 29: 279. 1954.

White, P. R.: Nutrient deficiency studies and an improved inorganic nutrient medium for cultivation of excised tomato roots. — Growth 7: 53. 1943.

— Nutritional requirements of isolated plant tissues and organs. — Ann. Rev. Plant Physiol. 2: 231. 1951.

# part II

# Physiology and Culture

 Stephen Hales. 1727. Of vegetation (Experiment CXXII). Vegetable staticks. W. and J. Inny's and T. Woodward, London.

Stephen Hales

Stephen Hales was a child of the scientific revolution—a period when the primarily "biological view" of nature gave way to attempts to explain vital processes in chemical and physical terms. As a young man, Hales was profoundly influenced by the work of Newton, who left Cambridge to become Master of the Mint just as Hales completed his first degree. He undoubtedly mastered the main concepts of Newtonian science and demonstrated remarkable mechanical ingenuity. He sought to discover harmony in nature and in the functioning of plants and animals similar to that found in the celestial realm.

He was particularly interested in the circulatory fluids—blood in animals and sap in plants—quantitative growth of plants, and the role and chemistry of air. He was eager to study function in physical terms and to apply Newtonian science to the world of plant and animal biology. This was the basis of his "statical way [by number, weight, measure] of investigation." As a man of the church, he respected the work of the divine Architect, and felt, I'm sure, that his own work was consistent with and glorified the Creator. He writes in the introduction of his *Vegetable Staticks*:

> And since we are assured that the all-wise Creator has observed the most exact proportions, *of number, weight* and *measure,* in the make of all things; the most likely way therefore, to get any insight into the nature of those parts of the creation, which come within our observation, must in all reason be to number, weigh and measure.

Stephen Hales was born to a distinguished family in 1677 in the village of Bekesbourne, Kent, England. In 1696 he entered Bene't (now Corpus Christi) College, Cambridge, with the intent to become an Anglican priest. He studied the classics and attended lectures in mathematics and mechanics. After receiving his B.A. degree, he remained at Corpus Christi and obtained his M.A. degree and was made Fellow in 1703.

His interest in the natural sciences developed only after he became a Fellow and was stimulated by his association with a new undergraduate medical student, William Stukeley. Together, they explored the Cambridge area, collecting plants, making dissections and anatomical preparations, and venturing into chemistry and hydrostatics at Trinity College.

For reasons not clear, Hales left Cambridge in 1709 to become Perpetual Curate of a small church in Teddington, just a few miles up the Thames from London. After devoting the first few years completely to his parish, he began to use his spare time to conduct experiments on forces involved in blood circulation in animals, and was first to make direct measurements of blood pressure and calculations of cardiac output. His scientific efforts lapsed for a few years; nevertheless, his work came to the attention of the Royal Society and he was elected Fellow in 1718. Hales appeared to have a driving interest in exploring commonality between plants and animals, particularly the circulation of body fluids. He writes:

> . . . since in vegetables, their growth and preservation of their vegetative life is promoted and maintained, as in Animals, by the very plentiful and regular motion of their fluids, which are the vehicles ordained by nature, to carry proper nutriment to every part. . . .

His interest in the forces involved in sap flow in plants resurfaced when he accidentally discovered that a piece of bladder placed over a pruning cut of a bleeding stem became greatly extended. His reaction was:

> . . . if a long glass tube were fixed there in the same manner, as I had before done to the Arteries of several living Animals, I should thereby obtain the real ascending force of the Sap in that Stem. . . .

This was the beginning of a series of experiments that dramatically changed botanical research and laid the foundation for plant physiology. Over the next seven years, his science was a natural extension of his religious life. He critically measured transpiration (referred to as "perspiration") using sunflower plants, and showed that temperature affected water loss on both a plant and a leaf area basis. Having established that roots "imbibed" and leaves "perspired" water, he set out to measure velocity and force of sap movement in stems. Invariably seeing a "plan" in all life, he concluded that, mass for mass, a sunflower plant perspires 17 times more water than a "well sized man." He convincingly argued that capillary action alone could not account for the rise of sap in stems, but that leaves played an important role.

Further studies with apple, pear, cherry, plum, pea, and numerous other horticultural plants added significantly to our understanding of lateral movement of sap in stems and the effects of temperature, leaves, and fruits on transpiration. He discovered root pressure and measured it directly with mercury manometers. His concern for quantitative aspects of growth of plants led to measurements of growth rates and demonstration of unequal expansion of leaves and extension of internodes.

During his studies on sap movement in plants, he observed numerous air bubbles in the sap, and this led him into studies on analysis of air. Perhaps the most significant of his findings was that air was taken up not only by roots, but also by stems and leaves, especially at night. He viewed air as nourishing the plant and becoming a part of its substance. His concept of "fixed air" (carbon dioxide?) had a marked influence on subsequent investigators.

His studies on air led to development of two useful pieces of equipment: the pneumatic trough, which we have all used in high school chemistry to collect gases, and a so-called "pedestal apparatus." Here a glass cylinder is placed over a pedestal, standing in a vessel of water, so that the mouth of the cylinder is some distance underwater. Any change in air volume in the cylinder can be detected by rise or fall of the water level in the cylinder. Hales used this apparatus to conduct the experiment selected here as a classic—one of many described in papers read before the Royal Society and published in 1727 in his now famous book *Vegetable Staticks*.

Hales, in this study, demonstrated that spearmint plants held in the pedestal apparatus "looked florid for a month" and lived for about six weeks, during which time the water level in the cylinder had risen about 20 cubic inches. The account of this study is beautiful; was the rise in the water level caused by "imbibition" of air by the mint plants or by a temperature change over the course of the experiment? Regardless, he concludes ". . . therefore, do not depend on this Experiment . . .", and suggests repeating the study using a thermometer to measure temperature and an appropriate control—an identical apparatus with no mint plants. This simple account shows Hales's perceptiveness and recognition of the importance of utilizing appropriate controls.

Horticulture has gained much from the work of this most remarkable man. Of the many contributions, perhaps his most original and significant was the application of the "statical" method to the study of plants. He helped give birth to the discipline of plant physiology, which has since developed into a strong science and provided a foundation for the advancement of experimental horticulture. During his latter years at Teddington, he became acquainted with Princess Augusta, residing at Kew. With her help and interest, he began a botanical collection which developed into what is now known as the Royal Botanical Gardens.

He longed not only to understand the functioning of plants but also to apply his findings, as is evident from the following quotation from his conclusion in *Vegetable Staticks:*

> If therefore these Experiments and Observations give us any farther insight into the nature of plants, they will then doubtless be of some use in Agriculture and Gardening [Horticulture], either by serving to rectify some mistaken notions, or by helping farther to explain the reasons of many kinds of culture. . . .

His outstanding contributions to both church and science were recognized; Oxford conferred on him the degree of Doctor of Divinity, the French Academy of Sciences elected him a Fellow, and his countrymen erected a monument to his memory in Westminster Abbey. In America, the Carolina silver-bell tree was named *Halesia* in his honor, and the American Society of Plant Physiologists now recognizes outstanding contributions to plant physiology with the Stephen Hales Award.

Martin J. Bukovac
Department of Horticulture
Michigan State University
East Lansing, Michigan 48824

## REFERENCES

ALLAN, D. G. C. 1986. Stephen Hales DD FRS. Teddington, Parish Church, St. Mary with St. Alban, England. 12 pp.

COHEN, I. B. 1976. Stephen Hales. Sci. Am. 234(5):98–107.

GUERLAC, H. 1972. Stephen Hales. Dictionary Sci. Biograph. 6:35–48.

JAMES, P. J. 1985. Stephen Hales' statical way. Hist. Philos. Life Sci. 7(2):287–299.

WARREN, J. V. 1986. Stephen Hales revisited. ASPP Newsl. 13(5):5–9.

We find by the chymical analysis of vegetables, that their substance is composed of sulphur, volatile salt, water and earth; which principles are all endued with mutually attracting powers, and also of a large portion of air, which has a wonderful property of strongly attracting in a fixt state, or of repelling in an elastick state, with a power which is superior to vast compressing forces, and it is by the infinite combinations, action and re-action of these principles, that all the operations in animal and vegetable bodies are effected.

These active aereal particles are very serviceable in carrying on the work of vegetation to its perfection and maturity. Not only in helping by their elasticity to distend each ductile part, but also by enlivening and invigorating their sap, where mixing with the other mutually attracting principles they are by gentle heat and motion set at liberty to assimilate into the nourishment of the respective parts: "The soft and moist nourishment easily changing its texture by gentle heat and motion, which congregates homogeneal bodies, and separates heterogeneal ones." *Newton's Opticks*, qu. 31. The sum of the attracting power of these mutually acting and re-acting principles being, while in this nutritive state, superior to the sum of their repelling power, whereby the work of nutrition is gradually advanced by the nearer and nearer union of these principles, from a lesser to a greater degree of consistency, till they are advanced to that viscid ductile state, whence the several parts of vegetables are formed; and are at length firmly compacted into hard substances, by the flying off of the watry diluting vehicle; sooner or later, according to the different degrees of cohesion of these thus compacted principles.

But when the watry particles do again soak into and disunite them, and their repelling power is thereby become superior to their attracting power; then is the union of the parts of vegetables thereby so thoroughly dissolved, that this state of putrefaction does by a wise order of Providence

# CHAP. VII.
## Of Vegetation.

We are but too sensible, that our reasonings about the wonderful and intricate operations of nature are so full of uncertainty, that as the wise-man truly observes, *hardly do we guess aright at the things that are upon earth, and with labour do we find the things that are before us.* Wisdom Chap. ix. v. 16. And this observation we find sufficiently verified in vegetable nature, whose abundant productions, tho' they are most visible and obvious to us, yet are we much in the dark about the nature of them, because the texture of the vessels of plants is so intricate and fine, that we can trace but few of them, tho' assisted with the best microscopes. We have however good reason to be diligent in making farther and farther researches; for tho' we can never hope to come to the bottom and first principles of things, yet in so inexhaustible a subject, where every the smallest part of this wonderful fabrick is wrought in the most curious and beautiful manner, we need not doubt of having our inquiries rewarded, with some further pleasing discovery; but if this should not be the reward of our diligence, we are however sure of entertaining our minds after the most agreeable manner, by seeing in every thing, with surprising delight, such plain signatures of the wonderful hand of the divine architect, as must necessarily dispose and carry our thoughts to an act of adoration, the best and noblest employment and entertainment of the mind.

What I shall here say, will be chiefly founded on the following experiments; and on several of the preceding ones, without repeating what has already been occasionally observed on the subject of vegetation.

fit them to resuscitate again, in new vegetable productions; whereby the nutritive fund of nature can never be exhausted: Which being the same both in animals and vegetables, it is thereby admirably fitted by a little alteration of its texture to nourish either.

Now, tho' all the principles of vegetables are in their due proportion necessary to the production and perfection of them; yet we generally find greater proportions of Oil in the more elaborate and exalted parts of vegetables: And thus Seeds are found to abound with Oil, and consequently with sulphur and air, as we see by Exper. 56, 57, 58. which Seeds containing the rudiments of future vegetables, it was necessary that they should be well stored with principles that would both preserve the Seed from putrefaction, and also be very active in promoting germination and vegetation. Thus also by the grateful odours of flowers we are assured, that they are stored with a very subtile, highly sublimed Oil, which perfumes the ambient air, and the same may be observed from the high tastes of fruits.

And as Oil is an excellent preservative against the injuries of cold, so it is found to abound in the sap of the more northern trees; and it is this which in ever-greens keeps their leaves from falling.

But plants of a less durable texture, as they abound with Salt and Water, which is not so strongly attracting as sulphur and air, so are they less able to endure the cold; and as plants are observed to have a greater proportion of Salt and Water in them in the spring, than in the autumn, so are they more easily injured by cold in the spring, than in a more advanced age, when their quantity of oil is increased, with their greater maturity.

Whence we find that nature's chief business, in bringing the parts of a vegetable, especially its fruit and seed to maturity, is to combine together in a due proportion, the more active and noble principles of sulphur and air, that

chiefly constitute oil, which in its most refined state is never found without some degree of earth and salt in it.

And the more perfect this maturity is, the more firmly are these noble principles united. Thus Rhenish Wines, which grow in a more northern climate, are found to yield their Tartar, i. e. by Exper. 73. their incorporated air and sulphur in greater plenty, than the stronger Wines of hotter countries, in which these generous principles are more firmly united: And particularly in *Madera* Wine, they are fixt to such a degree, that that Wine requires a considerable degree of warmth, such as would soure many other Wines, to keep it in order, and give it a generous taste; and 'tis from the same reason, that small *French* Wines are found to yield more spirit in distillation, than strong *Spanish Wines*.

But when, on the other hand, the crude watry part of the nutriment bears too great a proportion to the more noble principles, either in a too luxuriant state of a plant, or when its roots are planted too deep, or it stands in too shady a position, or in a very cold and wet summer; then it is found, that either no fruit is produced, or if there be any, yet it continues in a crude watry state; and never comes to that degree of maturity, which a due proportion of the more noble principles would bring it to.

Thus we find in this, and every other part of this beautiful scene of things, when we attentively consider them, that the great Author of nature has admirably tempered the constituent principles of natural bodies, in such due proportions as might best fit them for the state and purposes they were intended for.

It is very plain from many of the foregoing Experiments and Observations, that the leaves are very serviceable in this work of vegetation, by being instrumental in bringing nourishment from the lower parts, within the reach of the attraction of the growing fruit; which like young animals is furnished with proper instruments to suck it thence. But the leaves seem also designed for many other noble and important services; for nature admirably adapts her instruments so as to be at the same time serviceable to many good purposes. Thus the leaves, in which are the main excretory ducts in vegetables, separate and carry off the redundant watry fluid, which by being long detained, would turn rancid and prejudicious to the plant, leaving the more nutritive parts to coalesce; part of which nourishment, we have good reason to think, is conveyed into vegetables thro' the leaves, which do plentifully imbibe the Dew and Rain, which contain Salt, Sulphur, &c. For the air is full of acid and sulphureous particles, which when they abound much, do by the action and re-action between them and the elastick air cause that sultry heat, which usually ends in lightning and thunder: And these new combinations of air, sulphur and acid spirit, which are constantly forming in the air, are doubtless very serviceable, in promoting the work of vegetation; when being imbibed by the leaves, they may not improbably be the materials out of which the more subtile and refined principles of vegetables are formed: For so fine a fluid as the air seems to be a more proper medium, wherein to prepare and combine the more exalted principles of vegetables, than the grosser watry fluid of the sap; and for the same reason, 'tis likely, that the most refined and active principles of animals are also prepared in the air, and thence conveyed thro' the lungs into the blood; and that there is plenty of these sulphureo-aereal particles in the leaves, is evident from the sulphureous exudations, which are found at the edges of leaves, which Bees are observed to make their waxen cells of, as well as of the dust of flowers: And that wax abounds with sulphur is plain from its burning freely, &c.

We may therefore reasonably conclude, that one great use of leaves is what has been long suspected by many, viz. to perform in some measure the same office for the support of the vegetable life, that the lungs of animals do, for the support of the animal life; Plants very probably drawing

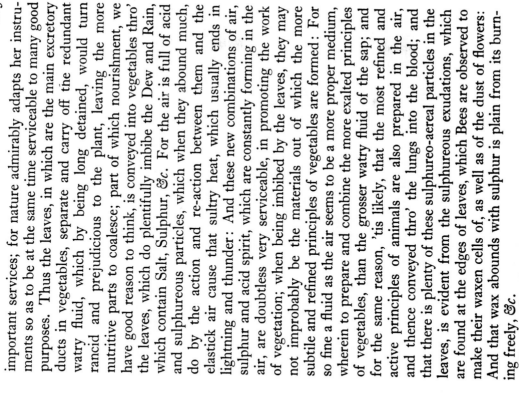

Pl. 19

Fig: 45.

Fig: 46.

S.G.

light convertible into one another? and may not bodies receive much of their activity from the particles of light, which enter their composition? The change of bodies into light, and of light into bodies, is very conformable to the course of nature, which seems delighted with transmutations. *Opt. qu.* 30."

## EXPERIMENT CXXII.

That the leaves and plants do imbibe elastick air, I have some reason to suspect from the following Experiment, *viz.* In *May* I set some well rooted plants of spear-mint in two glass cisterns full of water, which cisterns were set on pedestals, and had inverted chymical receivers put over them, as in (Fig. 35.) the water being drawn up to *a a*, half way their necks: In this inclosed moist state the plants looked pretty florid for a month, and made, as I think, some few weak lateral shoots, tho' they did not grow in height; they were not quite dead till after six weeks, when it was found that the water was risen in both glasses from *a a* towards *z z*, in bulk about 20 cubick inches: But as there was not so exact an account taken of the different temperature of the air, as to heat and cold, as there ought to have been, I am not certain, whether that rising of the water might not be owing to a greater coolness of the air at the six weeks end, than when they were first placed under the glasses; and therefore do not depend on this Experiment; but thought it proper to mention it, as well deserving to be repeated with greater accuracy, both with Mint, and other proper plants, by noting the temperature of the air on a *Thermometer,* hanging near the receivers, and observing after some time, whether the water *a a* be risen, notwithstanding the air be no cooler than when the Mint was first placed under the glass. And for greater certainty, it would be adviseable to suspend in the same manner another like receiver with no Mint, but only water in it, up to *a a.*

thro' their leaves some part of their nourishment from the air.

But as plants have not a dilating and contracting *Thorax,* their inspirations and expirations will not be so frequent as those of Animals, but depend wholly on the alternate changes from hot to cold, for inspiration, and *vice versâ* for expiration; and tis' not improbable, that plants of more rich and racy juices may imbibe and assimilate more of this aereal food into their constitutions, than others, which have more watry vapid juices. We may look upon the Vine as a good instance of this, which in Exper. 3. perspired less than the Apple-tree. For as it delights not in drawing much watry nourishment from the earth by its roots, so it must therefore necessarily be brought to a more strongly imbibing state at night, than other trees, which abound more with watry nourishment; and it will therefore consequently imbibe more from the air. And likely this may be the reason, why plants in hot countries abound more with fine aromatick principles, than the more northern plants, for they do undoubtedly imbibe more dew.

And if this conjecture be right, then it gives us a farther reason, why trees which abound with moisture, either from too shaded a position, or a too luxurious state are unfruitful, *viz.* because, being in these cases more replete with moisture, they cannot not imbibe so strongly from the air, as others do, that great blessing the dew of Heaven.

And as the most racy generous tastes of fruits, and the grateful odours of flowers, do not improbably arise from these refined aereal principles, so may the beautiful colours of flowers be owing in a good measure to the same original; for it is a known observation, that a dry soil contributes much more to their variegation than a strong moist one does.

And may not light also, by freely entring the expanded surfaces of leaves and flowers, contribute much to the ennobling the principles of vegetables? for Sir *Isaac Newton* puts it as a very probable query, "Are not gross bodies and

EXPERIMENT CXXIII.

In order to find out the manner of the growth of young shoots, I first prepared the following instrument, *viz.* I took a small stick $a$, (Fig. 40.) and at a quarter of an inch distance from each other, I run the points of five pins, 1, 2, 3, 4, 5, thro' the stick, so far as to stand $\frac{1}{4}$ of an inch from the stick, then bending down the great ends of the pins, I bound them all fast with waxed thread; I provided also some red lead mixed with oil.

In the spring, when the Vines had made short shoots, I dipped the points of the pins in the paint, and then pricked the young shoot of a Vine, (Fig. 41.) with the five points at once, from $t$ to $p$: I then took off the marking instrument, and placing the lowest point of it in the hole $p$, the uppermost mark, I again pricked fresh holes from $p$ to $l$, and then marked the two other points $i\ h$; thus the whole shoot was marked every $\frac{1}{4}$ inch, the red paint making every point remain visible.

(Fig. 42.) shews the true proportion of the same shoot, when it was full grown, the *September* following; where every corresponding point is noted with the same letter.

The distance from $t$ to $s$ was not enlarged above $\frac{1}{60}$ part of an inch; from $s$ to $q$, the $\frac{1}{26}$ of an inch; from $q$ to $p$, $\frac{3}{8}$; from $p$ to $o$, $\frac{3}{8}$; from $o$ to $n$, $\frac{6}{10}$; from $n$ to $m$ $\frac{9}{10}$ from $m$ to $l$, $1 + \frac{1}{10}$ of an inch; from $l$ to $i$, $1 + \frac{3}{10}$ inch nearly; and from $i$ to $h$ three inches.

In this Experiment we see that the first joint to $r$ extended very little; it being almost hardened, and come near to its full growth, when I marked it: The next joint, from $r$ to $n$, being younger, extended something more; and the third joynt from $n$ to $k$ extended from $\frac{3}{4}$ of an inch to $3 + \frac{1}{2}$ inches; but from $k$ to $h$, the very tender joint, which was but $\frac{1}{4}$ inch long, when I marked it, was when full grown three inches long.

We may observe, that nature in order to supply these young growing shoots with plenty of ductile matter is very

careful to furnish at small distances the young shoots of all sorts of trees, with many leaves throughout their whole length, which serve as so many joyntly acting powers placed at different stations, thereby to draw with more ease plenty of sap to the extending shoot.

The like provision has nature made in the Corn, Grass, Cane, and Reed kind; the leafy spires, which draw the nourishment to each joynt, being provided long before the stem shoots, which tender stem in its tender ductile state would most easily break and dry up too soon, so as to prevent its due growth, had not nature to prevent both these inconveniences provided strong *Thecas* or *Scabbards*, which both support and keep long in a supple ductile state the tender extending stem.

I marked in the same manner as the Vine, at the proper seasons, young *Honeysuckle* shoots, young *Asparagus*, and young *Sun-flowers*; and I found in them all a gradual scale of unequal extentions, those parts extending most which were tenderest. The white part of the *Asparagus*, which was under ground, extended very little in length, and accordingly we find the fibres of the white part very tough and stringy: But the greatest extension of the tender green part, which was about 4 inches above the ground when I marked it, separated the marks from a quarter of an inch, to twelve inches distance; the greatest distention of the *Sunflower* was from $\frac{1}{4}$ inch, to four inches distance.

From these Experiments, it is evident, that the growth of a young bud to a shoot consists in the gradual dilatation and distention of every part, they are extended to their full each other in the bud, as may plainly and distinctly be seen in the slit bud of the Vine and Fig-tree; but by this gradual distention of every part, they are extended to their full length. And we may easily conceive how the longitudinal capillary tubes still retain their hollowness, notwithstanding their being distended, from the like effect in melted glass tubes, which retain a hollowness, tho' drawn out to the finest thread.

The whole progress of the first joynt *r* is very short in comparison of the other joynts; because, at first setting out its leaves being very small, and the season then cooler than afterwards; 'tis probable, that but little sap is conveyed to it; and therefore it extending but slowly, its fibres are in the mean time grown tough and hard, before it can arrive to any considerable length. But as the season advances, and the leaves inlarge, greater plenty of nourishment being thereby conveyed, the second joynt grows longer than the first, and the 3d and 4th still on gradually longer than the preceding; these do therefore in equal times make greater advances than the former.

The wetter the season, the longer and larger shoots do vegetables usually make; because their soft ductile parts do then continue longer in a moist, tender state; but in a dry season the fibres sooner harden, and stop the further growth of the shoot; and this may probably be one reason why the two or three last joynts of every shoot are usually shorter than the middle joynts; *viz*, because they shooting out in the more advanced hot dry summer season, their fibres are soon hardened and dryed, and are withal checked in their growth by the cool autumnal nights: I had a vine shoot of one year's growth which was 14 feet long, and had 39 joynts, all pretty nearly of an equal length, except some of the first and last.

And for the same reason, Beans and many other plants, which stand where they are much shaded, being thereby kept continually moist, do grow to unusual heights, and are drawn up as they call it by the over shadowing Trees, their parts being kept long, soft and ductile: But this very moist shaded state is usually attended with sterility; very long joynts of vines are also observed to be unfruitful.

This Experiment, which shews the manner of the growth of shoots, confirms *Borelli's* opinion, who in his Book *De motu Animalium*, part second Chap. 13, supposes the tender

growing shoots to be distended like soft wax by the expansion of the moisture in the spongy pith; which dilating moisture, he with good reason concludes is hindered from returning back, while it expands, by the sponginess of the pith, without the help of valves. For 'tis very probable that the particles of water, which immediately adhere to, and are strongly imbibed into, and attracted by every fibre of the spongy pith, will suffer some degree of expansion before they can be detached by the sun's warmth from each attracting fibre, and consequently the mass of spongy fibres, of which the pith consists, must thereby be extended.

And that the pith may be the more serviceable for this purpose, nature has provided in most shoots a strong partition at every knot, which partitions serve not only as plinths, or abutments for the dilating pith to exert its force on, but also to prevent the rarified sap's too free retreat from the pith.

But a dilating spongy substance, by equally expanding it self every way, would not produce an oblong shoot, but rather a globose one, like an Apple; to prevent which inconvenience we may observe, that nature has provided several Diaphragms, besides those at each knot, which are placed at small distances across the pith; thereby preventing its too great lateral dilatation. These are very plain to be seen in Walnut-tree shoots; and the same we may observe in the pith of the branches of the sun-flower, and of several other plants; where tho' these Diaphragms are not to be distinguished while the pith is full and replete with moisture, yet when it drys up, they are often plain to be seen; and it is further observed, that where the pith consists of distinct vesicles, the fibres of those vesicles are often found to run horizontally, whereby they can the better resist the too great lateral dilatation of the shoot.

We may observe that nature makes use of the same artifice, in the growth of the feathers of Birds, which is very visible in the great pinion feathers of the wing, the smaller and upper part of which is extended by a spongy pith, but the lower and bigger quill part, by a series of large vesicles, which when replete with dilating moisture do extend the quill, but when the quill is full grown, the vesicles are always dry; in which state we may plainly observe every vesicle to be contracted at each end by a Diaphragm or Sphincter, whereby its too great lateral dilatation is prevented, but not its distention lengthwise.

And as this pith in the quill grows dry and useless after the quill is full grown, we may observe the same in the pith of trees, which is always succulent and full of moisture while the shoot is growing, by the expansion of which the tender ductile shoot is distended in every part, its fibres being at the same time kept supple by this moisture; but when each year's shoot is full grown, then the pith gradually drys up, and continues for the future dry and kiksey, its vesicles being ever after empty; nature always carefully providing for the succeeding year's growth by preserving a tender ductile part in the bud replete with succulent pith.

And as in vegetables, so doubtless in animals, the tender ductile bones of young animals are gradually increased in every part, that is not hardened and ossified; but since it was inconsistent with the motion of the joynts to have the ends of the bones soft and ductile as in vegetables; therefore nature makes a wonderful provision for this at the glutinous serrated joyning of the heads to the shanks of the bones; which joyning while it continues ductile the animal grows, but when it ossifies then the animal can no longer grow. As I was assured by the following Experiment, *viz.* I took a half-grown Chick, whose leg-bone was then two inches long, and with a sharp pointed Iron at half an inch distance I pierced two small holes through the middle of the scaly covering of the leg, and shin-bone; two months after I killed the Chick, and upon laying the bones bare, I found on it obscure remains of the two marks I had made at the same distance of half an inch: So that that part of the bone had not at all

distended lengthwise, since the time that I marked it: Notwithstanding the bone was in that time grown an inch more in length, which growth was mostly at the upper end of the bone, where a wonderful provision is made for its growth at the joining of its head to the shank, called by Anatomists Symphysis.

And as the bones grow in length and size; so must the membranous, the muscular, the nervous, the cartilaginous and vascular fibres of the animal body necessarily extend and expand, from the ductile nutriment which nature furnishes every part withal; in which respects animal bodies do as truly vegetate as do the growing vegetables. Whence it must needs be of the greatest consequence, that the growing animal be supplied with proper nourishment for that purpose, in order to form a strong athletick constitution: For when growing nature is deprived of proper materials for this purpose, then is she under a necessity of drawing out very slender threads of life, as is too often the case of young growing persons, who by indulging in spirituous liquors, or other excesses, do thereby greatly deprave the nutritive ductile matter, whence all the distending fibres of the body are supplied.

Since we are by these Experiments assured that the longitudinal fibres, the sap vessels of wood in its first year's growth, do thus distend in length by the extension of every part; and since nature in similar productions makes use of the same or nearly the same methods: These considerations make it not unreasonable to think, that the second and following years additional ringlets of wood are not formed by a meerly horizontal dilatation of the vessels; for it is not easy to conceive, how longitudinal fibres and tubular sapvessels should thus be formed; but rather by the shooting of the longitudinal fibres lengthways under the bark as young fibrous shoots of roots do, in the solid Earth. The observations on the manner of the growth of the ringlets of wood in Experiment 46 (Fig. 30.) do further confirm this.

I intended to have made father researches into this matter by proper Experiments, but have not yet found time for it.

But whether it be by an horizontal or longitudinal shooting, we may observe that nature has taken great care to keep the parts between the bark and wood always very supple with slimy moisture, from which ductile matter the woody fibres, vesicles and buds are formed.

Thus we see that nature, in order to the production and growth of all the parts of animals and vegetables, prepares her ductile matter: In doing of which she selects and combines particles of very different degrees of mutual attraction, curiously proportioning the mixture according to the many different purposes she designs it for; either for bony or more lax fibres of very different degrees in animals, or whether it be for the forming of woody or more soft fibres of various kinds in vegetables.

The great variety of which different substances in the same vegetable prove, that there are appropriate vessels for conveying very different sorts of nutriment. And in many vegetables some of those appropriate vessels are plainly to be seen replete either with milky, yellow, or red nutriment.

Dr. *Keill*, in his account of animal secretion, page 49, observes, that where nature intends to separate a viscid matter from the blood, she contrives very much to retard its motion, whereby the intestine motion of the blood being allayed, its particles can the better coalesce in order to form the viscid secretion. And Dr. *Grew*, before him, observed an instance of the same contrivance in vegetables where a secretion is intended, that is to compose a hard substance, *viz.* in the kernel or feed of hard stone fruits, which does not immediately adhere to, and grow from the upper part of the stone, which would be the shortest and nearest way to convey nourishment to it; but the single umbilical vessel, by which the kernel is nourished, fetches a compass round the concave of the stone, and then enters the kernel near its cone, by which artifice this vessel being much prolonged,

the motion of the sap is thereby retarded, and a viscid nutriment conveyed to the seed, which turns to hard substance.

The like artifice of nature we may observe in the long capillary fibrous vessels which lie between the green hull, and the hard shell of the Walnut, which are analogous to the fibrous Mace of Nutmegs, the ends of whose hairy fibres are inserted into the angles of the furrows in the Walnut-shell: Their use is therefore doubtless to carry in those long distinct vessels the very viscous matter which turns, when dry, to a hard shell; whereas were the shell immediately nourished from the soft pulpous hull that surrounds it, it would certainly be of the same soft constitution: The use of the hull being only to keep the shell in a soft ductile state till the Nut has done growing.

We may observe the like effect of a flower motion of the sap in Ever-greens, which perspiring little, their sap moves much more slowly than in more perspiring Trees; and is therefore much more viscid, whereby they are better enabled to outlive the winter's cold. It is observed that the sap of Ever-greens in hot Countries is not so viscous as the sap of more Northern Ever-greens, as the fir, &c. for the sap in hotter Countries must have a brisker motion, by means of its greater perspiration.

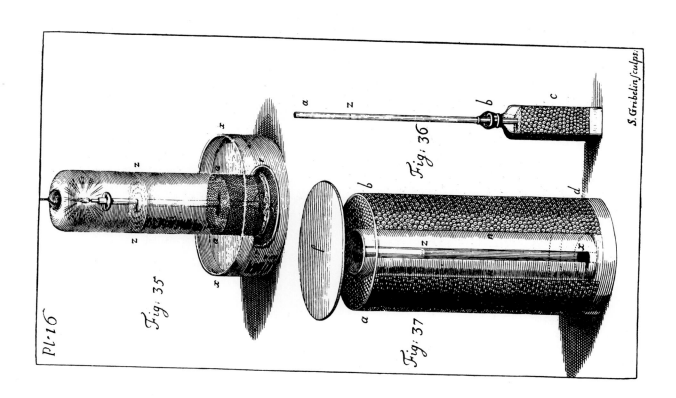

Pl. 16

Fig: 35

Fig: 36

Fig: 37

S. Gribelin sculps:

103

Thomas Andrew Knight. 1806. On the direction of the radicle and germen during the vegetation of seeds. Philosophic transactions of the Royal Society of London (part 1):99–108.

Thomas Andrew Knight

A short, descriptive scientific paper dealing with the subject of gravity and plant growth was read before the Horticultural Society of London in 1806 by Thomas Andrew Knight (1759–1838), a wealthy English country gentleman and amateur horticulturist. This obscure paper has since become the cornerstone of modern gravitational biology. Naturalists long had speculated that gravity was the stimulus for many plant orientation responses during the 140-year interval between recognition of Newton's theory of universal gravitation and the publication of Knight's paper, but this now classic paper provides the first experimental verification that gravity controls the directionality of root and stem growth by germinating seedlings.

A fine mind, a keen sense of observation developed from childhood, plus the leisure and resources available to the gentleman scientists of the eighteenth-century landed gentry, provided the opportunity for many of Knight's experimental contributions to plant science. Born into a well-to-do family, he settled on a farm and began experiments breeding cattle and raising new kinds of fruits and vegetables. At age 29 he inherited a castle and the management of a 10,000-acre estate, giving him even more resources with which to pursue his true passion—experimental horticulture and plant physiology. Knight conducted studies on stock/scion compatibility in grafting of fruit trees and the nature of fruit tree diseases. Although generally lacking quantitative data, Thomas Knight published many of his observations, and in his lifetime contributed nearly a hundred scientific papers on a wide range of topics, ranging from grafting to insect pollination vectors, diseases and pests of fruit trees, breeding of vegetables and fruits, as well as purely physiological studies with seedlings. His efforts as a plant breeder resulted in many new cultivars of apple, cherry, strawberry, red currant, plum, nectarine, pear, potato, pea, and cabbage. He later published an inheritance study on the garden pea and reported both dominance and segregation 42 years before Gregor Mendel's landmark paper entitled "Experiments on Plant Hybrids."

Knight was interested in developing improved culture methods to enhance earliness and yield, and his experiments spawned a large number of publications on plant response to temperature, fertilization, training, pruning, ringing, and graft compatibility. Although he also dabbled in studies of a purely physiological nature (especially involving the ascent and descent of sap, tropisms, and the nature of cambium), his principal objective was always utilitarian.

Recognition for his studies on grafting and inheritance gained Knight charter membership in the Horticultural Society of London (later the Royal Horticultural Society) in 1804. He became president of the society in 1811 and was to serve in that capacity for the remainder of his life. During the course of his association with the Royal Society, Knight was named a fellow of the society and received the Copley medal in recognition of his paper on geotropism, which is, perhaps, his most famous. Two years before his death at age 79, Thomas Knight was awarded the Royal Society's first Knightian medal, which bears his likeness and was founded in his honor. The medal is still given. A genus of the family Proteaceae also bears the name *Knightia* in honor of this early experimental horticulturist.

Thomas Andrew Knight's most famous contributions to basic plant science, summarized in his 1806 paper read to the Royal Society entitled "On the Direction of the Radicle and Germen During the Vegetation of Seeds," are delightful products of the powers of keen observation, clear thinking, and ingenuity. In the quaint setting of his country estate, Knight took advantage of a small rivulet (rill) flowing rapidly through his garden and, together with his gardener, constructed a small, vertical waterwheel that he subsequently used as a power source for his gravity experiments. Connected to the axis of the waterwheel was a smaller flywheel, which in turn was geared to a series of vertical and horizontal wheels of different sizes turning with various rotational velocities. He enclosed this entire "waterworks" in a box and protective grating for the stated purpose of protecting the fragile system of moving wheels against intrusion by any body that might impede their motion (squirrels?). It may be fortuitous that Knight so chose to enclose the system because the box must have provided a darkened, humidifying environment above the stream where seedlings could imbibe water, germinate, and grow, and where phototropic responses would not mask gravitropic responses. Thomas Knight did not include a drawing of his waterwheel system in the original article but the image it conjures up, including the rill in the English country garden, is rather charming. Unfortunately, his original notes and papers, which must have contained working drawings and data, have been lost.

By correctly imagining that continuous reorientation of germinated seedlings should nullify effects of the static gravitational field, Thomas Knight mentally invented the first plant clinostat (the simplest form of which would rotate a horizontal seedling about its long axis). Operationally, however, he invented the first plant centrifuge, by affixing bean seeds in various orientations about the circumference of a vertical wooden wheel. Following imbibition and germination on the centrifuge, Knight was delighted to observe that irrespective of the initial angle of radicle protrusion or germen (hypocotyl) orientation, shoot parts always grew inward toward the axis of the rotating wheel, while primary roots grew directly outward from the rim of the wheel. Given experimental data reported by Knight concerning wheel radius and rotational velocity, I calculated that his bean seedlings were centrifuged at about $3.5 \times g$ in those experiments. Knight astutely reasoned that during every rotation of the vertical wheel, each seedling briefly was exposed to more gravity at the perigee (lowest point) of rotation $[(3.5 + 1) \times g]$ than at the apogee (highest point) of rotation $[(3.5 - 1) \times g]$, and that his critics would use this fact to argue that gravity was not negated at all points 180° apart during rotation. Although a correct argument that he had not developed a clinostat, Knight still recognized the net $g$-force to be a small contribution to overall physical force imposed on the seedlings, and for the most part that the ever-changing direction of stronger centrifugal forces overwhelmed the effect of gravity in driving seedling orientation.

To finalize his argument on the gravity question, Knight germinated bean seeds around the rim of a wheel rotating horizontally at 250 rpm, in which case there was a constant, downward force of $1 \times g$ on the seedlings in addition to what I calculated to be $9.8 \times g$ horizontally. His quaint description in old English of how the vertical wheel splashed water up from the stream onto the seeds describes an effective aeroponic watering system! The fact that he mentioned placement of stationary seeds in various orientations about the inside of the enclosing box also indicates that Knight appreciated the value of having appropriate controls in experiments. Although he did not present tabulated data in this paper, he did mention angles of inclination, indicating that quantitative measurements were made. The reader also gets the impression that Thomas Knight appreciated the value of replication and verification in conducting experiments since he mentioned these practices several times in the article. Knight reported an 80° reorientation of bean hypocotyls (inward) and radicles (outward) from the vertical plane. The 10° declination from the (horizontal) direction of centrifugal force for each organ was correctly inferred by Knight to be a plant response to the downward 1-*g* vector. By slowing his horizontal centrifuge to 80 rpm, Knight obtained only a 45° deviation of stem and root growth from the vertical, in their respective directions. The horizontal centrifugal force at that lower speed of rotation would have to equal the downward gravitational force for the resultant vector to assume a 45° angle. Conversion of data from Knight's report revealed that rotating an 11-inch-diameter wooden wheel at 80 rpm would indeed generate approximately $1 \times g$ horizontally outward, thereby confirming theoretically what he found empirically.

Perhaps the only erroneous interpretations made by Knight in this classic article concern the physiological explanation of seedling gravitropic behavior, but even in this arena of plant science he was partially correct. Knight discerned that radicles grow from their tips but that hypocotyls elongate from a zone behind the tip. He was not aware that radicles also have discrete zones of cell division and cell elongation close to each other near the tip, but he can be forgiven that. He was hampered by operating more than a hundred years before the phytohormone era began, and before tropisms and differential growth were interpreted in terms of asymmetric redistributions of auxin and calcium. Knight based his findings on interconversion of matter between solid and fluid states, sap accumulation on the underside of a gravistimulated germen, more rapid subsequent growth on the underside, and eventual change in flexibility of "vessels and fibres" as the reoriented growth becomes fixed. Even his naive interpretations are consistent with modern concepts of hormone and ion asymmetry, differential growth, and changes in cell wall extensibility.

Like all scientists pioneering new frontiers of knowledge, Thomas Knight had his critics and detractors. For example, 25 years after publication of this now classic paper on gravity, his hypothesis that plant gravitropic responses have a physiological basis was challenged by M. Poiteau, a member of the Horticultural Society of Paris, who claimed to have proved that reorientation of a seedling on a centrifuge could be explained in strictly physical terms. Using an inanimate system of metal pieces shaped like gourd seeds attached by a pivot on the circumference of a spinning wheel, the Frenchman found the broader, heavier end of the metal piece to swing out during rotation, while the lighter, narrow end pointed inward. Poiteau therefore assumed that radicles, which orient away from the axis of rotation, must have greater specific gravity than hypocotyls, which grow toward the axis. Knight relished the defense of his science against such attacks. He dispatched this detractor with relative ease by demonstrating that excised hypocotyls of bean seedlings sank rapidly to the bottom of a vessel of water while radicles excised from the same seedlings floated, indicating that the specific gravities of these organs were just the opposite of what would be required for a strictly physical system responding to gravity. Knight also pointed out in a rebuttal read before the Royal Society (Knight 1831) that his seeds were fastened securely to the rim of the rotating wheel rather than by a pivot. Thus, plant organs reoriented independently of one another, not as a unit, as would rigid metal objects. Thomas Knight refuted such critics with eloquence and wit in his writings, providing exercise and recreation for his superior intellect.

It is gratifying that such an accomplished horticulturist as Thomas Andrew Knight also should be credited with a major contribution to fundamental plant physiology. He truly was a renaissance horticulturist.

Cary Mitchell
Department of Horticulture
Purdue University
West Lafayette, Indiana 47907

## REFERENCES

ANONYMOUS (G. BENTHAM, J. LINDLEY, and MRS. ACTON). 1841. A selection from the physiological and horticultural papers published in the Transaction of the Royal and Horticultural Societies by the late Thomas Andrew Knight to which is prefixed a sketch of his life. Longman, Orme, Brown, Green and Longmans, London.

BAGENAL, N. B. 1938. THOMAS ANDREW KNIGHT, 1759–1838. J. R. Hort. Soc. 63:319–324 (Fig. 79–84).

FLETCHER, H. R. 1969. The story of the Royal Horticultural Society, 1804–1968. Oxford University Press, Oxford.

KNIGHT, T. A. 1831. On the direction of the radicle and germen during the vegetation of seeds: A rebuttal to criticism of this work by M. Poiteau. J. Royal Institution of Great Britain. 2:80–82.

V. *On the Direction of the Radicle and Germen during the Vegetation of Seeds.* By Thomas Andrew Knight, *Esq. F. R. S. In a Letter to the Right Hon. Sir* Joseph Banks, *K. B. P. R. S.*

Read January 9, 1806.

MY DEAR SIR,

IT can scarcely have escaped the notice of the most inattentive observer of vegetation, that in whatever position a seed is placed to germinate, its radicle invariably makes an effort to descend towards the centre of the earth, whilst the elongated germen takes a precisely opposite direction; and it has been proved by Du Hamel* that if a seed, during its germination, be frequently inverted, the points both of the radicle and germen will return to the first direction. Some naturalists have supposed these opposite effects to be produced by gravitation; and it is not difficult to conceive that the same agent, by operating on bodies so differently organized as the radicle and germen of plants are, may occasion the one to descend and the other to ascend.

The hypothesis of these naturalists does not, however, appear to have been much strengthed by any facts they were able to adduce in support of it, nor much weakened by the arguments of their opponents; and therefore, as the phenomena

* *Physique des Arbres.*

O 2

observable during the conversion of a seed into a plant are amongst the most interesting that occur in vegetation, I commenced the experiments, an account of which I have now the honour to request you to lay before the Royal Society.

I conceived that if gravitation were the cause of the descent of the radicle, and of the ascent of the germen, it must act either by its immediate influence on the vegetable fibres and vessels during their formation, or on the motion and consequent distribution of the true sap afforded by the cotyledons: and as gravitation could produce these effects only whilst the seed remained at rest, and in the same position relative to the attraction of the earth, I imagined that its operation would become suspended by constant and rapid change of the position of the germinating seed, and that it might be counteracted by the agency of centrifugal force.

Having a strong rill of water passing through my garden, I constructed a small wheel similar to those used for grinding corn, adapting another wheel of a different construction, and formed of very slender pieces of wood, to the same axis. Round the circumference of the latter, which was eleven inches in diameter, numerous seeds of the garden bean, which had been soaked in water to produce their greatest degree of expansion, were bound, at short distances from each other. The radicles of these seeds were made to point in every direction, some towards the centre of the wheel, and others in the opposite direction; others as tangents to its curve, some pointing backwards, and others forwards, relative to its motion; and others pointing in opposite directions in lines parallel with the axis of the wheels. The whole was inclosed in a box, and secured by a lock, and a wire grate was placed to prevent the ingress of any body capable of impeding the motion of the wheels.

The water being then admitted, the wheels performed more than 150 revolutions in a minute; and the position of the seeds relative to the earth was of course as often perfectly inverted, within the same period of time; by which I conceive that the influence of gravitation must have been wholly suspended.

In a few days the seeds began to germinate, and as the truth of some of the opinions I had communicated to you, and of many others which I had long entertained, depended on the result of the experiment, I watched its progress with some anxiety, though not with much apprehension; and I had soon the pleasure to see that the radicles, in whatever direction they were protruded from the position of the seed, turned their points outwards from the circumference of the wheel, and in their subsequent growth receded nearly at right angles from its axis. The germens, on the contrary, took the opposite direction, and in a few days their points all met in the centre of the wheel. Three of these plants were suffered to remain on the wheel, and were secured to its spokes to prevent their being shaken off by its motion. The stems of these plants soon extended beyond the centre of the wheel: but the same cause, which first occasioned them to approach its axis, still operating, their points returned and met again at its centre.

The motion of the wheel being in this experiment vertical, the radicle and germen of every seed occupied, during a minute portion of time in each revolution, precisely the same position they would have assumed had the seeds vegetated at

rest; and as gravitation and centrifugal force also acted in lines parallel with the vertical motion and surface of the wheel, I conceived that some slight objections might be urged against the conclusions I felt inclined to draw. I therefore added to the machinery I have described another wheel, which moved horizontally over the vertical wheels; and to this, by means of multiplying wheels of different powers, I was enabled to give many different degrees of velocity. Round the circumference of the horizontal wheel, whose diameter was also eleven inches, seeds of the bean were bound as in the experiment, which I have already described, and it was then made to perform 250 revolutions in a minute. By the rapid motion of the water-wheel much water was thrown upwards on the horizontal wheel, part of which supplied the seeds upon it with moisture, and the remainder was dispersed, in a light and constant shower, over the seeds in the vertical wheel, and on others placed to vegetate at rest in different parts of the box.

Every seed on the horizontal wheel, though moving with great rapidity, necessarily retained the same position relative to the attraction of the earth; and therefore the operation of gravitation could not be suspended, though it might be counteracted, in a very considerable degree, by centrifugal force: and the difference, I had anticipated, between the effects of rapid vertical and horizontal motion soon became sufficiently obvious, The radicles pointed downwards about ten degrees below, and the germens as many degrees above, the horizontal line of the wheel's motion; centrifugal force having made both to deviate 80 degrees from the perpendicular direction each would have taken, had it vegetated at rest. Gradually diminishing the rapidity of the motion of the horizontal wheel, the radicles descended more perpendicularly, and the germens grew more upright; and when it did not perform more than 80 revolutions in a minute, the radicle pointed about 45 degrees below, and the germen as much above, the horizontal line, the one always receding from, and the other approaching to, the axis of the wheel.

I would not, however, be understood to assert that the velocity of 250, or of 80 horizontal revolutions in a minute will always give accurately the degrees of depression and elevation of the radicle and germen which I have mentioned; for the rapidity of the motion of my wheels was sometimes diminished by the collection of fibres of conferva against the wire grate; which obstructed in some degree the passage of the water: and the machinery, having been the workmanship of myself and my gardener, can not be supposed to have moved with all the regularity it might have done, had it been made by a professional mechanic. But I conceive myself to have fully proved that the radicles of germinating seeds are made to descend, and their germens to ascend, by some external cause, and not by any power inherent in vegetable life: and I see little reason to doubt that gravitation is the principal, if not the only agent employed, in this case, by nature. I shall therefore endeavour to point out the means by which I conceive the same agent may produce effects so diametrically opposite to each other.

The radicle of a germinating seed (as many naturalists have observed) is increased in length only by new parts successively added to its apex or point, and not at all by any

general extension of parts already formed: and the new matter which is thus successively added unquestionably descends in a fluid state from the cotyledons.* On this fluid, and on the vegetable fibres and vessels whilst soft and flexible, and whilst the matter which composes them is changing from a fluid to a solid state, gravitation, I conceive, would operate sufficiently to give an inclination downwards to the point of the radicle; and as the radicle has been proved to be obedient to centrifugal force, it can scarcely be contended that its direction would remain uninfluenced by gravitation.

I have stated that the radicle is increased in length only by parts successively added to its point: the germen, on the contrary, elongates by a general extension of its parts previously organized; and its vessels and fibres appear to extend themselves in proportion to the quantity of nutriment they receive. If the motion and consequent distribution of the true sap be influenced by gravitation, it follows, that when the germen at its first emission, or subsequently, deviates from a perpendicular direction, the sap must accumulate on its under side: and I have found in a great variety of experiments on the seeds of the horse chesnut, the bean, and other plants, when vegetating at rest, that the vessels and fibres on the under side of the germen invariably elongate much more rapidly than those on its upper side; and thence it follows that the point of the germen must always turn upwards. And it has been proved that a similar increase of growth takes place on the external side of the germen when the sap

* See Phil. Trans, of 1805.

is impelled there by centrifugal force, as it is attracted by gravitation to its under side, when the seed germinates at rest.

This increased elongation of the fibres and vessels of the under side is not confined to the germens, nor even to the annual shoots of trees, but occurs and produces the most extensive effects in the subsequent growth of their trunks and branches. The immediate effect of gravitation is certainly to occasion the further depression of every branch, which extends horizontally from the trunk of the tree; and, when a young tree inclines to either side, to increase that inclination: but it at the same time, attracts the sap to the under side, and thus occasions an increased longitudinal extension of the substance of the new wood on that side.* The depression of the lateral branch is thus prevented; and it is even enabled to raise itself above its natural level, when the branches above it are removed; and the young tree, by the same means, becomes more upright, in direct opposition to the immediate action of gravitation: nature, as usual, executing the most important operations by the most simple means.

I could adduce many more facts in support of the preceding deductions, but those I have stated, I conceive to be sufficiently conclusive. It has however been objected by Du HAMEL, (and the greatest deference is always due to his opinions,) that gravitation could have little influence on the direction of the germen, were it in the first instance protruded, or were it subsequently inverted, and made to

* This effect does not appear to be produced in what are called weeping trees; the cause of which I have endeavoured to point out in a former Memoir. Phil. Trans. 1804.

MDCCCVI.        P

point perpendicularly downwards. To enable myself to answer this objection, I made many experiments on seeds of the horse chesnut, and of the bean, in the box I have already described; and as the seeds there were suspended out of the earth, I could regularly watch the progress of every effort made by the radicle and germen to change their positions. The extremity of the radicle of the bean, when made to point perpendicularly upwards, generally formed a considerable curvature within three or four hours, when the weather was warm. The germen was more sluggish; but it rarely or never failed to change its direction in the course of twenty-four hours; and all my efforts to make it grow downwards, by slightly changing its direction, were invariably abortive.

Another, and apparently a more weighty, objection to the preceding hypothesis, (if applied to the subsequent growth and forms of trees,) arises from the facts that few of their branches rise perpendicularly upwards, and that their roots always spread horizontally; but this objection I think may be readily answered.

The luxuriant shoots of trees, which abound in sap, in whatever direction they are first protruded, almost uniformly turn upwards, and endeavour to acquire a perpendicular direction; and to this their points will immediately return, if they are bent downwards during any period of their growth; their curvature upwards being occasioned by an increased extension of the fibres and vessels of their under sides, as in the elongated germens of seeds. The more feeble and slender shoots of the same trees will, on the contrary, grow in almost every direction, probably because their fibres, being more dry, and their vessels less amply supplied with sap, they are less affected by gravitation. Their points, however, generally shew an inclination to turn upwards; but the operation of light, in this case, has been proved by BONNET* to be very considerable.

The radicle tapers rapidly, as it descends into the earth, and its lower part is much compressed by the greater solidity of the mould into which it penetrates. The true sap also continues to descend from the cotyledons and leaves, and occasions a continued increase of the growth of the upper parts of the radicle, and this growth is subsequently augmented by the effects of motion, when the germen has risen above the ground. The true sap is therefore necessarily obstructed in its descent; numerous lateral roots are generated, into which a portion of the descending sap enters. The substance of these roots, like that of the slender horizontal branches, is much less succulent than that of the radicle first emitted, and they are in consequence less obedient to gravitation: and therefore meeting less resistance from the superficial soil, than from that beneath it, they extend horizontally in every direction, growing with most rapidity, and producing the greatest number of ramifications, wherever they find most warmth, and a soil best adapted to nourish the tree. As these horizontal, or lateral, roots surround the base of the tree on every side, the true sap descending down its bark, enters almost exclusively into them, and the first perpendicular root, having executed its office of securing moisture to the plant, whilst young, is thus deprived of proper nutriment, and, ceasing almost wholly to grow, becomes of no importance to the tree. The tap root of the oak, about which so

* *Récherches sur l'Usage des Feuilles dans les Plantes,*

P 2

much has been written, will possibly be adduced as an exception; but having attentively examined at least 20,000 trees of this species, many of which had grown in some of the deepest and most favourable soils of England, and never having found a single tree possessing a tap root, I must be allowed to doubt that one ever existed.

As trees possess the power to turn the upper surfaces of their leaves, and the points of their shoots to the light, and their tendrils in any direction to attach themselves to contiguous objects, it may be suspected that their lateral roots are by some means directed to any soil in their vicinity which is best calculated to nourish the plant, to which they belong; and it is well known that much the greater part of the roots of an aquatic plant, which has grown in a dry soil, on the margin of a lake or river, have been found to point to the water; whilst those of another species of tree which thrives best in a dry soil, have been ascertained to take an opposite direction: but the result of some experiments I have made is not favourable to this hypothesis, and I am rather inclined to believe that the roots disperse themselves in every direction, and only become most numerous where they find most employment, and a soil best adapted to the species of plant. My experiments have not, however, been sufficiently varied, or numerous, to decide this question, which I propose to make the subject of future investigation.

I am, &c.

T. A. KNIGHT.

Elton, Nov. 22, 1805.

L. H. Bailey. 1891. Some preliminary studies on the influence of the electric arc lamp upon greenhouse plants. Cornell University Agricultural Experiment Station Bulletin 30.

L. H. Bailey

Liberty Hyde Bailey is remembered principally as a taxonomist. His contributions to plant classification include extensive work with the genera *Rubus* and *Carex*, among others, and with the palm family. However, Liberty Hyde Bailey should be remembered as far more than a plant collector and classifier. He was a man of many interests and talents, a man of perception and foresight. He was an extraordinarily successful scientist, teacher, administrator, poet, and philosopher, who profoundly influenced the direction of teaching, research, and extension in American horticulture.

Bailey was one of the fortunate few who developed and followed a logical plan, a schedule, for his life's work and interests. His plan was symmetrical. He allowed 25 years for training in his chosen, almost destined, field of horticulture, 25 years for public service, and 25 years for retirement, during which he would be free to pursue his own interests. He predicted the first 50 years of his life quite accurately. However, his years in retirement numbered over 40, and these years allotted to the pursuit of personal interests resulted in vast contributions to horticulture.

Bailey was born in 1858 in South Haven, Michigan. As the youngest child of a hardworking farm family, the young Bailey was largely dependent on his imagination and rural environment for his childhood entertainment. At an early age he became interested in collecting plants and animals, and speculation on the relationship between one organism and another.

In addition to the opportunity to observe nature both in the wild and under the cultivation of his father, a progressive fruit grower, the young Bailey was provided with intellectual stimulation. Since South Haven was a frontier town, Liberty Hyde Bailey Senior, although a conservative puritan, was nevertheless an open-minded man of independent thought and took pains to provide books as well as New York newspapers for his family. He allowed his precocious youngest son to read Darwin, whose ideas he considered incomprehensible, if not inflammatory, because Darwin seemed honest.

Two books that were to impress the young Bailey particularly were Darwin's *On the Origin of Species by Natural Selection* and Asa Grey's *Field, Forest and Garden Botany,* which introduced him to systematic botany and the theory of evolution. Bailey was excited to find the ideas in these books compatible with his own interest in nature. Further, it was illuminating for the young Bailey to realize that there was a systematic discipline for the habit of nature study, plant collection, and identification he had already developed as a country boy, roaming the fields and woods of nineteenth-century frontier Michigan.

In 1877, at the age of 19, Bailey left South Haven and began his secondary education at the Michigan Agricultural College. While at Michigan, Bailey worked with William Beal, a botanist with whom he had become acquainted through participation in the Michigan State Pomological Society. Beal was one of the first botanists to incorporate laboratory experience and student participation into his teaching of botany. These teaching methods, unorthodox for the time, were to become standard practice in later horticultural training. Due in large part to Beal's influence, Bailey began to understand the potential for using the skills of a naturalist and botanist, observation and classification, for horticultural purposes, and to become interested in plant breeding. In addition to his role as student and teaching assistant for Beal, Bailey also found time during his student years to organize and edit the student publication, *The College Speculum,* and to become involved in organizations such as the Natural History Society and student government, and to publish his first articles on identification of local flora in *The Botanical Gazette.* He also began his long-term involvement with classification of the genus *Rubus.* In short, the beginning of his academic career was also the beginning of the type, diversity, and quantity of activity that was to characterize Bailey's entire life.

Another important influence on the young Bailey during his years of professional training, was the botanist Asa Grey. After leaving Michigan

Agricultural College, Bailey went to Harvard to work for Grey, where he was responsible for sorting and classifying plant specimens received from Kew Gardens. Working with Grey, whose work had influenced Bailey since early youth, must have indeed been an honor, as well as a tremendous boost to his taxonomic skills and education. However, Grey's feelings about horticultural education were in distinct opposition to Beal's and those nascent in Bailey. Although Grey was enthusiastic about embracing new phases of botany–physiology and the use of horticultural species for scientific investigation, he held a very typical attitude of the period. Grey felt that research should be conducted in a laboratory or herbarium. Horticulture was considered an ornamental art, while botany was considered the science. The two disciplines were generally not considered compatible in the nineteenth century. Consequently, Grey was disappointed when Bailey returned to Michigan in 1885 to assume a newly established chair of horticulture and landscape gardening. So strong was the dichotomy between botany and horticulture that it was predicted that if Bailey returned to horticulture, he would forfeit his growing reputation as a scientist and disappear into professional obscurity.

Fortunately for the world of horticulture, that prediction could not have been more inaccurate. Bailey spent several years at Michigan working to cross the "garden fence," his term for the wall of prejudice that existed between botany and horticulture, between science and the art of growing plants. This prejudice limited training horticulturists as well as botanists, educating farmers, and conducting research relevant to production agriculture, and most certainly precluded the development of teaching, research, and extension as we have come to know it. A wide array of other projects also received the force of his energy: the aesthetics of the campus landscape, renovation of the campus orchards, continuation of his vegetable breeding work, and the publication of popular articles. Very importantly, he revised the agricultural curriculum, making horticulture come alive with student participation in laboratory courses. He also designed the first horticultural laboratory building in the country, which was subsequently approved and built by the college board of directors. As a professor at Michigan, Bailey also continued to publish his taxonomic works and, contrary to the predictions of those he left behind in Cambridge, established a reputation as a world authority on plant classification.

By 1888, the dynamic and prolific Bailey had a well-established reputation as a scientist and an innovator in the area of horticultural research and education. Consequently, he was recruited by Cornell to fill a new position: Professor of Practical and Experimental Horticulture. In this capacity, he continued his effort at establishing horticulture as a science by instituting research on practical problems and teaching science in horticulture, and publishing bulletins and books on a wide array of subjects. Extension activities were also becoming a large priority for Bailey at this time. He believed that not only the success of agriculture, but the quality and survival of life in rural communities, depended on extension education. He published many bulletins and books in order to extend information from the university to the people of the state. He also made countless trips around the state to give talks and establish a relationship of trust and credibility between the college and the rural population.

In 1903, Bailey temporarily left research behind as he became dean of the College of Agriculture, which was established as a separate state-supported college within the university in 1904. During the 10 years that Bailey served as dean, the college grew rapidly. The faculty increased from 11 to over 100, and student enrollment grew proportionately, from 100 to 1400. Funding was acquired for buildings and for the functions of teaching, extension, and research. The departments of pomology, vegetable crops, and floriculture were established, as well as the foundations for what would later become rural sociology.

By 1913, the road was well paved for American colleges of agriculture to fulfill what Bailey felt should be their three proper functions: teaching, the discovery of truth through research, and extending their work to all the population. In 1903, Bailey was a cofounder and first president of the first professional society for horticulturists, the (American) Society for Horticultural Science, which gave a large boost to the establishment of horticulture as a legitimate science. It must have seemed to Bailey that the garden fence had been effectively crossed. Amid protests from his colleagues, Bailey chose to retire from public service to return to his early interests of plant exploration and identification. He was adhering closely to the schedule he had defined for himself, having been in public service for 28 years, three years over his self-imposed 25-year limit.

From the time he retired in 1913 to his death in 1954, Bailey remained active in horticulture, publishing prolifically and collecting and classifying plants from all over the world, including the first extensive classification of the palms. The establishment of the Bailey Hortorium at Cornell also resulted from specimens collected by Bailey during his "retirement" years. Bailey died on Christmas Day in 1954, at the age of 96. His death ended nearly a century of horticultural achievements and in his pocket at the time of the injury ultimately responsible for his death were airplane tickets to Africa, the location for his next planned exploration and plant collection trip.

During his early years at Cornell Bailey did a great deal of applied horticultural research. In 1890 and 1891, he did a series of experiments to determine the feasibility of using electric lights in greenhouses and published a seminal bulletin. This paper on electro-horticulture is one relatively small work when considered among the voluminous output of Liberty Hyde Bailey, including such things as *Hortus* and the *Cyclopedia of Horticulture* and the publications on plant taxonomy. However, it is an important work, typical of Bailey's approach to horticulture.

Bailey was the first American to attempt research on the use of electric lights for horticultural production. Some preliminary work on the effects of electric lights on plants had been done in Europe as early as 1861, work with which Bailey presumably became familiar when he took a tour of European horticulture prior to assuming his position at Cornell in 1888. However, the early European works were more physiological in nature than those proposed by Bailey. The European research was concerned with the effect of artificial light on phototropism, chlorophyll production, and photosynthesis (the "decomposition of carbon dioxide and water"). Bailey was interested in the possibility of improving crop quality or yield by the use of supplemental lights. Thus Bailey's work was the first horticultural production-oriented investigation of the use of supplemental illumination.

The work was initiated in response to conflicting opinions held in the late nineteenth century on the effects on plants of what was then the new phenomenon of electric lights. Some assumed that electric lights would increase production in greenhouses, while others supposed that electric lights were injurious to plants. A separate issue from the practical questions of using lights in a greenhouse, Bailey felt, was the intellectual question that demanded attention: How would electric lights affect various crop plants under different conditions?

From his work with electric lights, Bailey concluded: "On the whole I feel that it will be possible some day to use the electric light in floricultural establishments to some pecuniary advantage." His conclusions were based on the fact that despite problems encountered in his greenhouse light project, Bailey observed a rapid maturation of plants grown under supplemental electric lights. He rightly concluded that practices which hastened crop maturity would ultimately prove to be commercially profitable. He also correctly noted that not all crops would benefit from additional light as certain species went to seed prematurely under these conditions. His observations were accurate. However, the cause of differences in maturity of species in response to light observed by Bailey, photoperiodism, would remain unknown for another 30 years. Similarly, the cause of the plant stunting and damage under his naked arc lamps could not be identified at the time as ultraviolet radiation damage.

It was not until several decades later, following the identification by Garner and Allard of the photoperiodic response, that the possibilities for supplemental illumination in greenhouses were further explored. With a nod of acknowledgment to Bailey and a few of his contemporaries who had also provided some preliminary information on the subject, Alex Laurie pursued the two ideas that were critical in the development of commercial floriculture: supplemental lights for enhancing crop growth, and photoperiodic responses in floricultural crops. Laurie, using three hours of supplemental lights, improved the growth of vegetatively propagated plants. Thus Laurie was able to restate Bailey's projection for the future in the present tense: "It is practically feasible and commercially possible to install additional lighting for profit." Further, Laurie determined the short-day responses of two important floricultural plants, chrysanthemums and poinsettias, responses that are critical to current commercial production of these crops.

It is now quite unremarkable for plants to be grown totally under artificial lights, although due to economic forces, this practice is usually limited to plants grown for research purposes. However, the commercial utilization of supplemental greenhouse light has grown from Bailey's belief that it would be possible to standard commercial practice. Use of efficient and cost-effective high-intensity lamps is a routine practice in rose and chrysanthemum production and in greenhouse vegetable production. Thus Bailey's belief that current technology or science could be utilized to improve horticultural practices has become commercial reality.

Darlene Wilcox-Lee
Cornell University
Long Island Horticultural Research Laboratory
39 Sound Avenue
Riverhead, New York 11901

## REFERENCES

DORF, PHILIP. 1956. Liberty Hyde Bailey. An informal biography. Cornell University Press, Ithaca, N.Y.

HANAN, J. J., W. D. HOLLEY, and L. L. GOLDBERG. 1978. Greenhouse management. Springer-Verlag, New York.

LARSON, O. F. 1958. Liberty Hyde Bailey's impact on rural life. Baileya 6:10–21.

LAURIE, A. 1930 Photoperiodism—practical application to greenhouse culture. Proc. Am. Soc. Hort. Sci. 27:319–322.

LAWRENCE, G. H. M. 1955. Liberty Hyde Bailey, 1858–1954. Baileya 3:27–40.

MOON, M. H. 1958. The botanical explorations of Liberty Hyde Bailey. 1. China. Baileya 6:1–9.

NELSON, K. S. 1980. Greenhouse management for flower and plant production. Interstate Printers and Publishers, Danville, Ill.

RODGERS, A. D., III. 1965. Liberty Hyde Bailey. A story of American plant sciences. Hafner Press, New York.

TUKEY, H. B. 1958. Liberty Hyde Bailey's impact on plant sciences. Baileya 6:58–68.

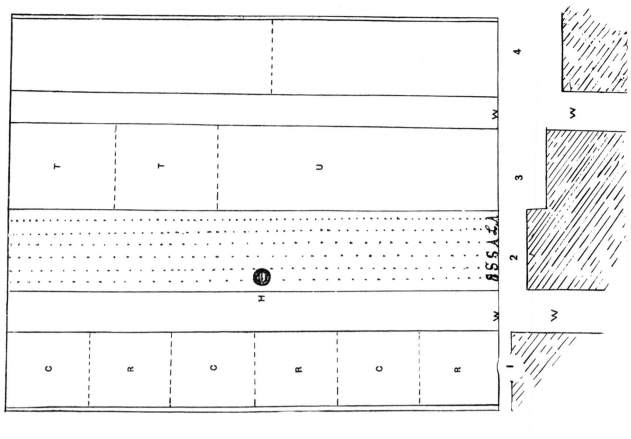

FIG. I.—*Plan of the Electric Light Compartment*

# SOME PRELIMINARY STUDIES OF THE IN-FLUENCE OF THE ELECTRIC ARC LIGHT UPON GREENHOUSE PLANTS.

## I. EXPERIMENTS AT CORNELL.

In the winter of 1889–90 we undertook experiments to determine what influence the ordinary street electric light exerts upon plants in greenhouses. Much has been said among gardeners concerning supposed retarding or accelerating influences of street lamps upon plants. Many have supposed that the electric light can be introduced profitably into greenhouses for the purpose of hastening growth. Still others have supposed the electric lights at exhibition halls to be injurious to plants, and have said that flowers fade quickly when placed near them. The whole subject of the relation of electric light to vegetation should be understood, and wholly aside from any thought of introducing the light into greenhouses, its influence upon plants, both under glass roofs and in the open, is a question which demands careful investigation.

For the purpose of our experiment, a forcing house 20x60 feet was set aside. The house is low, with a two thirds span and a very flat roof (22½°), and was designed for the growing of lettuce, radishes and cuttings. The house is ventilated entirely from the peak by small windows hinged at the ridge. It is heated by steam, the riser running overhead and the returns all lying under the benches. This house was divided into two nearly equal portions for our purpose by a tight board partition. One compartment was treated to ordinary conditions—sunlight by day and darkness by night—and the other had sunlight during the day and electric light during a part or whole of the night. In all the experiments the lamp was suspended from the peak of the house, the arc being 2½ feet above the soil of the bench over which it was placed. The arrangement of the benches is shown in Fig. 1. The lower portion of the figure shows a cross section of the benches and walks ; 1, 2, 3 and 4 are the benches, and w,

Plate I.

Fig. 9. *Bench of Lettuce in the Dark or Normal House.*

Fig. 10. *Bench of Lettuce in the Electric Light House.*

121

Plate II.

*Fig. 2. Normal Spinage Plant; Fig. 3 (to the right).*
*One of the same variety and age, from the light house.*

*Fig. 7. Six best Tubers of Scarlet Frame Radish from Dark and Light Houses.*

*Fig. 8. Six best Tubers of White Box Radish from Light and Dark Houses.*

w the walks. The outside benches, 1 and 4, are 4 ft. wide, and the middle ones 3½ ft. wide. Bench 3 is nearly two ft. lower than No. 2. The upper portion of the figure shows a surface diagram of the house. The lamp is at H; 1, 2, 3, 4 show the benches, and w, w the walks. The arc was 9¾ ft. in a direct line from the edge of the lowest bench, 4, and 3¼ ft. from the edge of the highest bench, 1.

During the first winter (January to April, 1890) we used a 10 ampere, 45 volt, Brush arc lamp of 2000 nominal candle power. This was run all night—from dusk until daylight—from January 23d to April 12th. At first the light was started at 4:30 in the afternoon and ran until 7:30 in the morning, but as the season advanced the run was shortened until in April it ran from seven o'clock till five. For the first six weeks the light was naked, but during the remainder of the time an ordinary white opal globe was used. The cropping of the compartment, which was exactly duplicated in the other compartment, is indicated on the plan, Fig. 1. R, R, R, indicate spaces devoted to radishes, and C, C, C to carrots. In these instances the rows ran crosswise the bench. The bench directly under the light, No. 2, was planted, in rows running lengthwise, to endive, E, spinage, S, S, cress, A, A, and lettuce, L. Bench No. 3 was planted mostly to peas, U, in transverse rows, and two cutting frames, T, T, were placed upon it. The lowest bench, No. 4, was planted to lettuce, but the first crop was nearly mature when the experiment began and it was not taken into consideration.

1. *Experiments with a naked light running all night.*—(1890).

The general effect of the light was to greatly hasten maturity, and the nearer the plants grew to the light the greater was the acceleration. This tendency was particularly marked in the leaf-plants—endive, spinage, cress and lettuce. The plants "ran to seed" before edible leaves were formed, and near the light the leaves were small and curled. This is well illustrated in spinage, Figs. 2 and 3, plate II. The cuts, which are made to the same scale, show Round Dutch spinage when seven weeks old. Fig. 2 shows an average plant from the dark or normal house, and Fig. 3 one from the light house within 7 ft. of the lamp.*

*All distances from the lamp are measured in a direct air line from the arc.

The electric light spinage matured and produced good seeds while that in the dark house was still making large and edible leaves with no indication of running to seed. An examination under the microscope of leaves of the plants illustrated in Figs. 2 and 3 showed that while there was apparently the same amount of starch in each, it was much more developed in the electric light specimen, the grains being larger and having more distinct markings and giving a better color test when treated with iodine.

Fig. 4, shows two representative plants of Prickly Seeded spinage at four weeks of age. The one at the left grew in the dark house, and it is making a normal, spreading growth. The other grew in the

FIG. 4.—*Spinage Plants at Four Weeks of Age.*

light house, and the plant is sending up a central stem preparatory to flowering, and the leaves are smaller than in the other. Landreth's Forcing lettuce, growing in a row nearly under the lamp (L, Fig. 1) behaved in a similar manner. For three feet either side of the lamp, most of the plants were killed outright soon after they came up, and the remaining ones in the entire row (35 plants) were seriously injured, the leaves curling and remaining very small. The plants increased in stature, vigor and size of leaves with increased distance from the lamp. Those nearest the lamp made most leaves early in their growth, and they maintained this advantage until about four weeks old, although the leaves were smaller. Five weeks after sowing, the average height of plants within four feet of the lamp was 1.2 in.; between four and five feet, 1.34 in.; between five and six ft., 1.8 in.; between six and seven ft., 2 in.; between seven and eight ft., 2.2 in. The average height of plants in the dark house at this time was 2¼ in., and the plants were much more vigorous and had larger and darker leaves. The increase in size was not uniform with increase in distance from the lamp. There were somewhat regular alternations of lower and higher plants, although there was a general progression in height. This is shown in the diagram, Fig. 5, in which each bar represents a plant one-sixteenth full size. The upper series represents those plants lying to

FIG. 5.—*Growth of Lettuce.*

the left of the light, and the lower one those lying to the right. This alternating elevation and depression is perhaps due to the concentric bands of varying intensity of light which fall from the arc and which are caused by the uneven burning of the carbons.

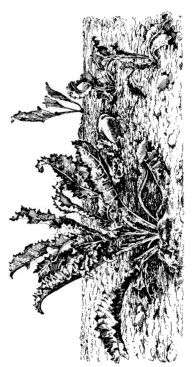

FIG. 6.—*Endive in Shadow and Full Light.*

Two varieties of cress (the French cresson *alénois* and *C.alénois frisé, Lepidium sativum*), which were grown almost directly under the light, behaved in the same way as the spinage and lettuce. For five feet either side of the light the plants died soon after coming up, and seven weeks after sowing all surviving plants in the light house, excepting a few which were shaded, were in bloom, and all were small and the leaves were curled. Those in the dark house at the same time were strong and vigorous, with good leaves and no blossoms.

Endive gave the same results. It chanced that for a time two rows of endive grew parallel to each other in the light house, but one stood in full light while the other was shaded by an iron post an inch and a half in diameter. The result is well shown in Fig. 6, page 89. The row to the left was shaded, and the other received the full light. There was great difference between the two rows. The following figures show the weights of the individual plants in both rows after having grown two months in the light, at the given distances :

| In full light. | | | In shade of post. | | |
|---|---|---|---|---|---|
| Plant. | Distance from lamp. | Weight. | Plant. | Distance from lamp. | Weight. |
| No. 1 | 4 ft. 4½ in. | 9 grains. | No. 1 a | 4 ft. 8 in. | 138 grains. |
| " 2 | 5 ft. | 16 " | " 2 a | 5 ft. 4½ in. | 57 " |
| " 3 | 5 ft. 9½ in. | 27 " | " 3 a | 6 ft. | 130 " |
| " 4 | 6 ft. 5½ in. | 26 " | " 4 a | 7 ft. 5 in. | 158 " |
| " 5 | 7 ft. 2½ in. | 72 " | " 5 a | 8 ft. 4 in. | 33 " |
| " 6 | 8 ft. | 67 " | " 6 a | 9 ft. 2 in. | 47 " |
| " 7 | 8 ft. 9 in. | 96 " | | | |
| " 8 | 9 ft. 6 in. | 84 " | | | |
| Average | | 49.6 grains. | | | 93.8 grains. |

An average plant in the dark house of the same age weighed 575 grains, and it was larger leaved and darker colored than those grown in the other compartment.

These figures show at once the damaging influence of the naked electric light upon plants near it. Not only were the plants in direct light smaller than those in shadow, but the weight and the vigor of plants in light increased rapidly as distance from the light became greater, while those in shade were heaviest near the light where the shadow was most dense.

Some of the most marked results in this first series of experiments were obtained with the radishes on the highest bench (No. 1, Fig. 1). The young radish plants were strongly attracted by the light, and in the morning they all leaned at an angle of from 60° to 45° towards the lamp. During the day they would straighten up, only to reach for the lamp again on the succeeding night. This was repeated until the roots began to swell and the plant became stiff. As the plants grew, the foliage became much curled, and the amount of this injury was in direct proportion to the nearness to the lamp. Those nearest the lamp—within three to six feet—were nearly dead at the expiration of six weeks, while those 14 ft. away showed little injury to the leaves. The following are figures of weights and sizes of radishes in the two houses :*

| Variety. | Average weight of entire plant. | | Average weight of top. | | Aver'ge weight of tubers. | | No. of marketable tubers. | |
|---|---|---|---|---|---|---|---|---|
| | Light house. | Dark house. | Light house. | Dark house. | Light house. | Dark house. | Light house. | Dark house. |
| Cardinal Globe.... | .14 oz. | .17 oz. | .08 oz | .08 oz. | .059 oz. | .089 oz. | 55 pct. | 62 pct. |
| Dreer Scar't Frame | .067 " | .14 " | .036 " | .057 " | .031 " | .083 " | 30 " | 74 " |
| White Box.... | .23 " | .43 " | .048 " | .21 " | .082 " | .22 " | 6 " | 75 " |
| | .27 | .50 | .16 | .23 | .11 | .27 | 18 " | 100 " |
| Average.... | .18 oz. | .31 oz. | .08: oz. | .14 oz. | .07 oz. | .16 oz. | 27 pct. | 78 pct. |

The table shows that the crops obtained in the dark or normal house were about twice greater than those in the light compartment. The entire plants and the tops were almost half lighter in the light house, and the tubers were more than half lighter, while the per cent. of tubers large enough for market, was as 9 in the light house to 26 in the dark house. And it should also be said that the average size of the tubers graded as marketable was less in the light house than in the other. This is well shown in Fig. 7, plate II, which represents the six best tubers of Dreer Scarlet Frame from the dark house upon the left, and the six best from the light compartment upon the right. A portion of the planta-tion of White Box radish in the light house — occupying the uppermost radish plot on bench No. 1 (see Fig. 1) — was protected by the fan-shaped shadow cast by an iron post an inch and a half

*The figures were made from the entire crop when the plants in the dark house gave the first good market picking, and this accounts for the small per cent. of marketable tubers in the table; but the plants in both houses were the same age.

in diameter. In the shadow the foliage was scarcely injured, while those leaves which projected into the light were curled. The following measures of these plants are instructive :

| Samples. | Average weight of entire plant. | Average weight of tops. | Average weight of tubers. | No. of marketable tubers. |
|---|---|---|---|---|
| Light house, in full light ........ | .11 oz. | .09 oz. | .02 oz. | 0 |
| " " shadow......... | .35 " | .21 " | .14 " | 45 pct. " |
| Dark house........ | .43 " | .21 " | .22 " | 75 " |

This variety stood near the light, ranging from four to nine feet away. The contrasts between the two houses were also very great. Fig. 8 shows upon the left the six best radishes from the light and upon the right the six best from the dark house.

A chemical analysis of samples of these radishes from the light house, in both full light and shadow, and the dark house, gave the following results :

| Samples. | Ash. | Potash K²O. | Chloro-phyll. | Total nitro-gen. | Albumi-noid ni-trogen. | Amide nitro-gen. | Albumi-noids. (Albu-minoid N. x 6.25.) |
|---|---|---|---|---|---|---|---|
| | Per ct. | Per ct. | Per ct. | Per ct. | Per ct. | Per ct. | Per ct. |
| Lighthouse, full light | 3.84 | 0.38 | 6.22 | 1.36 | 1.24 | .12 | 7.75 |
| " in shadow | 3.76 | 0.34 | 6.12 | 1.38 | 1.20 | .12 | 7.50 |
| Dark house........... | 3.26 | 0.15 | 5.02 | 1.34 | 1.01 | .33 | 6.31 |

These figures show that the plants under the electric light had reached a greater degree of maturity than those in the normal or dark house. The ash is more, potash more than double, and chlorophyll (including extracted gums) somewhat more. The total nitrogen is essentially the same in all samples, but it is noticeable that in the electric light plants more of the amide nitro-gen has been changed into other forms than in the other sample; and the electric light samples are richer in albuminoids. The light house plants from the shadow are much nearer to those in full light in composition than to those in the dark house.

Dwarf peas were grown in transverse rows upon bench No. 3, (U, Fig. 1). This bench is so low that No. 2 shaded about half of it. In the shaded portion the peas were larger and more pro-

125

ductive than those in full light, although the latter were further from the lamp. The average heights of plants were as follows : Light house, in full light, 4.8 inches. Light house, in shade, 5.3 inches. Dark house, 5.8 inches.

The plants in the light house, particularly those in direct light, blossomed about a week in advance of those in the dark house, and they gave earlier fruits, but the productiveness was less, being in the ratio of 4 in the light house to 7 in the dark house. The decrease in production was due largely to the fewer number of peas in each pod, for the number of fruitful pods produced in each case was as 7 in the light house to 9 in the dark house, and there were many seedless pods in the light house. In other words, the production of pods (or flowers) was about the same in both houses, but the plants in the light house produced only four-sevenths as many seeds as those in the dark compartment.

It is apparent from the above experiences that the plants in the electric light house were injured. The question at once arose whether the injury was due to the electric light itself or to continuous light during the whole twenty-four hours. To test this point, 10-inch pots were inverted over radish plants during the day and removed at night, so that the plants received no sunlight and about 12 hours of electric light. Rubber tubing was conducted underneath the pots and was connected with the hole in the bottoms in such a manner that perfect ventilation was secured and yet no light admitted. Seedling radishes, which had never received sunlight, made a slender and sickly growth, assuming a faint green color, but died in three or four weeks.

The experiment was now conducted upon a larger scale, and at a time when the hours of sunlight were about equal to the hours of electric light. A tight wooden frame was placed upon the soil of a bench at one end of the light house. This frame was provided with a tight cover which was kept on during the day and removed at night. Feb. 1st radish seeds were planted in this frame both in the soil and in pots. The plants appeared on the 5th, and for several days grew very rapidly, making a spindling and nearly colorless growth. They were shaded by the side of the box and received only diffused light. On the 11th, some of them measured 4 inches to the leaves and 6 inches in total

height. On the 12th they looked yellow and seemed to be failing. Feb. 21st they began to die, and they were all dead on the 28th. On the 8th of February radish seeds were sown in pots and these were elevated in the frame so that the young plants received direct light. The plants were up on the 11th. They behaved the same as those in diffused light, and all were dead March 3d. None of the radish plants in the entire experiment succeeded in making a third or true leaf.

Strong lettuce plants were set in the frame in diffused light, but all died in about two weeks. Beans were planted Feb. 3d in diffused light. The plants appeared on the 10th. On the 25th growth had ceased, and the stems were 10 to 12 inches high, slender and light colored. They were all dead March 13th. The second pair of true leaves appeared, but did not develop. Corn planted at the same time had grown 12 inches tall by Feb. 25th, and was dead Mar. 7th. Potato tubers planted Feb. 5th made a spindling growth and reached the top of the frame (somewhat over a foot) March 13th. The stems then fell over because of their weakness and grew about the bottom of the frame. April 3d the stems were nearly four feet in length, very spindling, and bore a few small and pale leaves. A small and healthy castor bean plant chanced to be growing on the bench when the frame was placed in position. This was covered by the frame Feb. 1st, and in two weeks, having received only electric light, it was dead.

Another series of tests was made by covering well established plants in the beds. A tight box 18 inches square and a foot high was placed over certain plants during the daytime, and was removed at night and placed over contiguous plants of the same kind. Thus one set of plants received only electric light and one only sunlight, and inasmuch as both were covered during half of the twenty-four hours, any error which might have arisen from the covering itself — as lack of ventilation and increased heat—was eliminated. Feb. 7th certain radishes in the light house which had been planted two weeks were covered. In eight days some of the plants which were covered during the day were dead and the remaining ones were very weak. At the same time, those which were covered during the night had made a better growth than they had before and better than contiguous plants which had not been covered. An examination of the leaves of the plants receiving only the electric light showed that they contained no starch and very little or no chlorophyll. Feb. 8th two lots of beans and radishes were planted in pots sunk to their brims in the soil between the radish rows in the light house. One lot was covered during the day and the other during the night, as above. Germination was the same in both lots. Feb. 25th, the daylight beans had made a stocky growth of 3½ to 4 inches, while the electric light lot had made a weak growth of 8 to 9 inches. Radishes behaved in a similar manner. March 3d the leaves of the electric light beans began to wither, and both beans and radishes were dead Mar. 10th. The daylight lots continued to grow thriftily.

On Feb. 15 two strong plants of German ivy (*Senecio scandens*), carnation and begonia were selected for a similar experiment. One plant of each was covered by day and the other by night. Feb. 28th the electric light ivy appeared as if dying, and Mar. 10th it was apparently dead. By Mar. 3d the electric light carnation was seen to be making an etiolated growth. A month later the new white growth had become 4 to 6 inches long. The electric light begonia began to drop its leaves Mar. 5th, and a month later all the leaves had fallen. Another begonia, which chanced to be standing in diffused electric light, did not lose its leaves. By the middle of March the ivy threw up a shoot from the root, but the sprout remained feeble. April 3d the boxes were removed entirely from both lots. The contrasts were most striking. The daylight plants were strong and dark colored, while the others were dead or nearly so. From that time until the close of the experiment for the season, Apr. 12th, all the plants had the same conditions,—daylight, and electric light at night. A week after the experiment had closed and the electric light had been stopped, the begonia began to send out new leaves, the new shoot on the ivy began to assume a natural color and habit, but the carnation plant was nearly dead and apparently past recovery.

Peas were also treated in a similar manner. A row of peas which had grown for a month in direct electric light and which was in ordinary health, was divided into two equal parts and covered. In 14 days a part of the electric light lot was dead, and the

others were rapidly dwindling, while the daylight plants had improved. At this time the boxes were permanently removed, but the remaining electric light plants did not recover, and soon died.

The above experiments show conclusively that within the range of an ordinary forcing-house the naked arc light running continuously through the night is injurious to some plants ; and in no case did we find it to be profitable. But the fact that the light hastens maturity or seed bearing suggests that a modified light may be useful under certain conditions. Our next step, therefore, was to temper the light by the use of a globe.

*2. Experiments with a protected light running all night.—(1890.)*

Early in March, 1890, an ordinary white opal globe was placed upon the lamp, and for five weeks experiments similar to those already described were conducted. The effect of the modified light was much less marked than that of the naked light. Spinage showed the same tendency to run to seed, but to a much less extent, and the plants were not affected by proximity to the lamp. Lettuce, however, was decidedly better in the electric light house. Radishes were thrifty in the light house, and the leaves did not curl, but they produced less than in the dark house, although the differences were much less marked than in the former experiments. These second series of experiments can scarcely be compared with the former ones, because of the greater amount of sunlight which the plants received in the lengthening days of spring. The figures obtained from radishes, however, may afford a practically accurate comparison because of their rapid growth. The following table should be compared with that on page 91 :

| VARIETY. | Average weight of entire plant. | | Average weight of top. | | Avg. weight of tubers. | | No. of marketable tubers. | |
|---|---|---|---|---|---|---|---|---|
| | Light house. | Dark house. | Light house. | Dark house. | Light house. | Dark house. | Light house. | Dark house. |
| Half-long Rose | .41 oz. | .41 oz. | .19 oz. | .16 oz. | .22 oz. | .25 oz. | 84 pct. | 97 pct. |
| Scarlet Globe | .26 " | .34 " | .12 " | .14 " | .14 " | .20 " | 76 " | 87 " |
| Prussian Turnip | .24 " | .26 " | .08 " | .08 " | .16 " | .18 " | 88 " | 95 " |
| | .28 " | .29 " | .11 " | .09 " | .17 " | .20 " | 100 " | 100 " |
| Blood-red | .27 " | .33 " | .09 " | .08 " | .18 " | .25 " | 90 " | 100 " |
| French Breakfast | .36 " | .40 " | .14 " | .15 " | .22 " | .25 " | 90 " | 75 " |
| Half-long Scarlet | .25 " | .30 " | .13 " | .11 " | .12 " | .19 " | 100 " | 100 " |
| | .22 " | .32 " | .08 " | .08 " | .14 " | .24 " | 85 " | 100 " |
| Average | .29 oz. | .33 oz. | .12 oz. | .11 oz. | .17 oz. | .22 oz. | 89 pct. | 94 pct. |

The loss due to the electric light averages from one to five per cent. in the different comparisons, while the loss occasioned by the naked light (see page 91), was from 45 to 65 per cent. It is noticeable, also, that while the tops or leaves were lighter under the naked light, they were heavier under the modified light than those of normal plants; and this is interesting in connection with the fact that lettuce did better under the modified light than in the dark house. The carrots (Short-horn Scarlet, C, C, Fig. 1) gave indifferent results. They did not appear to be affected greatly even by the naked light, even when growing directly opposite to it and but three or four feet away. Carrots require such a long period of growth that the first good picking was not obtained until the end of the experiment in April. The plants therefore grew under both the naked and protected lamps. In the following calculation the plat on the upper or east end (see Fig. 1) is omitted because it appeared to have been modified somewhat by the greater heat occasioned by the elbows in the steam pipes at that point:

| SAMPLES. (Carrot). | Average weight of entire plant. | Average weight of tops. | Average weight of tubers. | No. of marketable tubers. |
|---|---|---|---|---|
| Light house, center | .14 oz. | .08 oz. | .06 oz. | 51 per ct. |
| " west end | .16 " | .08 " | .08 " | 63 " |
| Dark house | .19 " | .11 " | .08 " | 64 " |

The figures show that the plants which grew directly in front of the lamp were but little inferior to those which stood 10 or 12 feet away, or even to those in the dark house. No other plant in our experiments has withstood the electric light so well.

Observations were made towards the close of the experiment upon the influence of the light on the blooming of plants, but the subjects were not numerous enough to enable us to draw definite conclusions. There was little difference in the length of time which flowers lasted in the two compartments. Flowers of primula and cineraria appeared to last longer by a day or so in the light house, while begonia and zonal geranium persisted rather longer in the dark house. These may have been incidental variations, and the subject was subsequently more fully investigated, as will appear in the sequel.

We had now found that the injurious effects of the electric lamp are lessened by the use of a thin globe, probably because glass cuts off the highly refrangible and invisible rays. The globe distributes the luminous rays more evenly, and probably absorbs a few of them, but the injury was lessened to an extent out of all proportion to the amount of modification in light; and we had found that lettuce and the leaves of radishes had grown larger under the light than in ordinary conditions. These facts indicate that there may be conditions under which the electric light can be made profitable to the gardener, and they suggested the experiments of the succeeding winter; and here the investigations for 1890 closed.

3. *Experiments with the naked light running a part of the night.* —(1891.)

From January 16th to May 1st, 1891, the experiment was conducted under new conditions. The arrangement of the house or compartments remained as before, but the lamp was connected with a street lighting system and the light ran but a few hours, and never on moonlight nights. In this test we used a 10 ampere 45 volt 2000 nominal candle power Westinghouse alternating current lamp. Radishes were grown upon bench No. 1 (see Fig. 1), peas upon No. 3 and lettuce upon the lowest one, No. 4. Upon bench No. 2, nearest the lamp, a variety of ornamental plants was grown, mostly tulips, verbenas, petunias, primulas, heli-

| Dates. | Light ran from | No of hours run. |
|---|---|---|
| Jan. 16 | 5 to 7 P. M. | 2 |
| " 17 | 5 to 6 | 1 |
| " 18 | 0 | 0 |
| " 19 | 5 to 7 | 2 |
| " 20 | 5 to 7 | 2 |
| " 21 | 0 | 0 |
| " 22 | 5 to 8 | 3 |
| " 23 | 5 to 7 | 2 |
| " 24 | 5 to 6 | 1 |
| " 25 | 0 | 0 |
| " 26 | 5 to 10 | 5 |
| " 27 | 5 to 10 | 5 |
| " 28 | 0 | 0 |
| " 29 | 5 to 11 | 6 |
| " 30 | 5 to 6 A. M. | 13 |
| " 31 | 5 to 11 | 6 |
| Feb. 1 | 5:30 to 11 | 5½ |
| " 2 | 5:30 to 11 | 5½ |
| " 3 | 5:30 to 11 | 5½ |
| " 4 | 5 to 11 | 6 |
| " 5 | 5 to 11 | 6 |
| " 6 | 5 to 11 | 6 |
| " 7 | 5:30 to 11 | 5½ |
| " 8 | 5:30 to 11 | 5½ |
| " 9 | 5 to 11 | 6 |
| " 10 | 5 to 11 | 6 |
| " 11 | 5 to 11 | 6 |
| " 12 | 5:15 to 11 | 5¾ |
| " 13 | 5 to 11 | 6 |
| " 14 | 5:15 to 6 | ¾ |
| " 15 | 0 | 0 |
| " 16 | 5:15 to 8 | 2¾ |
| " 17 | 5:15 to 8 | 2¾ |
| " 18 | 5:15 to 8:30 | 3¼ |
| " 19 | 6 to 8:30 | 2½ |
| " 20 | 5:30 to 8:30 | 3 |
| " 21 | 5:30 to 8 | 2½ |
| " 22 | 0 | 0 |
| " 23 | 6 to 7 | 1 |
| " 24 | 5:30 to 8 | 2½ |
| " 25 | 5:30 to 8:30 | 3 |
| " 26 | 5:30 to 8:30 | 3 |
| " 27 | 5:30 to 11 | 5½ |
| " 28 | 6 to 11 | 5 |
| Mar. 1 | 6 to 11 | 5 |
| " 2 | 6 to 11 | 5 |
| " 3 | 6 to 11 | 5 |
| " 4 | 6 to 11 | 5 |
| " 5 | 6 to 11 | 5 |
| " 6 | 6 to 5 A. M. | 11 |
| " 7 | 6 to 11 | 5 |
| " 8 | 6 to 11 | 5 |
| " 9 | 6 to 11 | 5 |
| " 10 | 6 to 11 | 5 |

| Dates. | Light ran from | No. of hours run. |
|---|---|---|
| Mar. 11 | 6 to 11 | 5 |
| " 12 | 6 to 11 | 5 |
| " 13 | 6 to 11 | 5 |
| " 14 | 7 to 11 | 4 |
| " 15 | 0 | 0 |
| " 16 | 0 | 0 |
| " 17 | 0 | 0 |
| " 18 | 0 | 0 |
| " 19 | 0 | 0 |
| " 20 | 0 | 0 |
| " 21 | 0 | 0 |
| " 22 | 0 | 0 |
| " 23 | 0 | 0 |
| " 24 | 0 | 0 |
| " 25 | 0 | 0 |
| " 26 | 7 to 8 | 1 |
| " 27 | 7 to 8 | 1 |
| " 28 | 7 to 11 | 4 |
| " 29 | 7 to 11 | 4 |
| " 30 | 7 to 11 | 4 |
| " 31 | 7 to 11 | 4 |
| Apr. 1 | 7 to 11 | 4 |
| " 2 | 7 to 11 | 4 |
| " 3 | 7 to 11 | 4 |
| " 4 | 7 to 11 | 4 |
| " 5 | 7 to 11 | 4 |
| " 6 | 7 to 11 | 4 |
| " 7 | 7 to 11 | 4 |
| " 8 | 7:30 to 11 | 3½ |
| " 9 | 7:30 to 11 | 3½ |
| " 10 | 7:30 to 11 | 3½ |
| " 11 | 7:30 to 11 | 3½ |
| " 12 | 7:30 to 11 | 3½ |
| " 13 | 0 | 0 |
| " 14 | 0 | 0 |
| " 15 | 0 | 0 |
| " 16 | 0 | 0 |
| " 17 | 0 | 0 |
| " 18 | 0 | 0 |
| " 19 | 0 | 0 |
| " 20 | 0 | 0 |
| " 21 | 0 | 0 |
| " 22 | 0 | 0 |
| " 23 | 0 | 0 |
| " 24 | 7:30 to 12 | 4½ |
| " 25 | 8 to 10:30 | 2½ |
| " 26 | 7:30 to 11 | 3½ |
| " 27 | 8 to 11 | 3 |
| " 28 | 8 to 11 | 3 |
| " 29 | 8 to 11 | 3 |
| " 30 | 8 to 11 | 3 |
| May 1 | 8 to 11 | 3 |
| Total | | ........308 hours. |

trope and coleus. These ornamentals were mostly named one-colored varieties, and they were grown to enable us to make observations concerning the influence of electric light upon color. In this experiment, the lighting of the compartments was reversed during the last month, in order to eliminate any error which might arise from any minor differences in the temperature or other conditions of the two portions of the house. In order to properly understand the significance of the observations which follow, the reader will need to consult the following record of the dates and hours of lighting :

Of radishes, the White Box and Cardinal Globe were grown. The foliage was noticeably larger in the electric light house, as it had been under the modified light (see page 97), but the tubers were practically the same in both houses, and the date of maturity was the same. Notwithstanding its greater size, the foliage in the light house showed some signs of curling.

American Wonder and Advance peas were grown, and in every case they were larger and more fruitful in the dark house. The electric light did not increase the size of leaves, as it did in the radishes. These results are similar to those obtained in 1890.

The lettuce, however, was greatly benefitted by the electric light. We had found that under the protected light (see page 96) the lettuce had made a better growth than in normal conditions, but now it showed still greater difference. Lettuce of two varieties—Landreth Forcing and Tennis Ball or Boston Market—was transplanted onto bench No. 4 when the light started. Both varieties were planted in each house, and the plants were all alike; and all the conditions in the two compartments were kept as nearly alike as possible. Three weeks after transplanting (Feb. 5), both varieties in the light house were fully 50 per cent. in advance of those in the dark house in size, and the color and other characters of the plants were fully as good. The plants had received at this time 70½ hours of electric light. Just a month later the first heads were sold from the light house, but it was six weeks later when the first heads were sold from the dark house. In other words, the electric light plants were two weeks ahead of the others. This gain had been purchased by 161¾ hours of electric light, worth at current prices of street lighting about $7.00.

This lettuce test was repeated and was watched very closely when the lamp was transferred to the compartment which had formerly been kept under normal conditions. The same results were obtained, and the differences in the two crops were so marked as to arrest the attention of every visitor. The electric light plants were in every way as good in quality as those grown in the dark house; in fact, the two could not be told apart except for their different sizes. Figs. 9 and 10, plate I, show representative portions of the crops as they appeared five weeks after being transplanted to permanent quarters. Fig. 9 is a view in the dark

house, and Fig. 10 in the light house. The variety in this case is Landreth Forcing. The history of the plants is as follows: The seeds were sown in flats Feb. 24th. Until March 17th they were grown under ordinary conditions, at which time they were set in their permanent positions in the two compartments. We began to pick lettuce from the light house April 30th, but the first of equal size from the dark house was obtained May 10th. The electric light plants were therefore upon the benches 44 days before the first heads were sold. During this time there were 20 nights in which the light did not run, and there had been but 84 hours of electric light, worth about $3.50. In order to compute the cost of growing lettuce by the aid of the electric light, it is necessary to know how far the influence of the light will extend. This we do not know; but the lamp exerted this influence throughout a house 20x30, and the results were as well marked in the most remote part as they were near the lamp.

The results obtained from lettuce suggest many questions, all of which must be answered by experiment. We need to know if there is any particular time in the life of the lettuce plant when the light has a predominating influence; if a mild light is as good as a strong one; if the failure of the light during the moonlight nights is a serious drawback; to what distance the influence of the light extends; if the same results can be obtained by hanging the lamp over the house, instead of inside it, and by that means lighting several houses at once; if other plants can be profitably forced by means of electric light. In all these directions, and many others, we are planning experiments for the coming years.

The influence of the light upon productiveness and color of flowers was found to vary with different species and different colors within the same species. Several named varieties of tulips gave interesting results. Careful observations were made upon Proserpine, light cherry color; Wourseman, maroon; Vander Neer, light cherry; Yellow Pottebakker, bright yellow; Belle Alliance, scarlet; and Cerise gris delin, cherry and white. Upon the 13th of February, when these came into full flower, it was found that in every case the colors were deeper and richer in the light house; but the colors lost their intensity after four or five days and were indistinguishable from those in the dark house.

The plants in the light compartment had longer stems and larger leaves than the others; and there was a greater number of floriferous plants in the light. The tulips were grown at a distance of 10 and 12 feet from the lamp.

Verbena flowers near the light were uniformly injured. February 26th, all plants within six feet of the light were stunted, the leaves were small and curled, and the flowers were short lived. The flowers were small, and those on the lower part of the clusters turned brown and died before those on the top opened. The buds within two or three feet of the lamp curled up and shrunk and became discolored before opening, but the discoloration did not extend to the inside of the flower until it had opened. Scarlet, dark red, blue and pink flowers within three feet of the light soon turned to a grayish-white, and this discoloration was noticeable to a

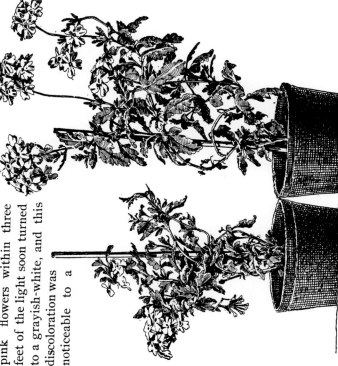

*Fig. 11.—Influence of Electric Light upon Verbenas.*

distance of six and seven feet. The plants bloomed somewhat earlier in the light house than in the other. Figs. 11 and 12 show the influence of the electric light upon verbenas. The left speci-

men in each instance stood four feet from the lamp, and the right hand specimen is from the dark house. The two light house specimens show the way in which different individuals are affected by the light. In Fig. 11—which shows a white-flowered variety—the electric light specimen is much dwarfed and weak

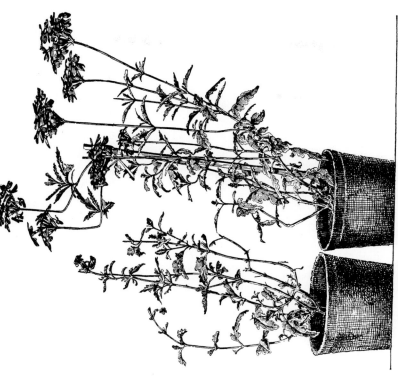

*Fig. 12.—Influence of Electric Light upon Verbenas.*

and it could not stand alone. It is nearly past flowering, while its companion is in good condition. Fig. 12 shows a red variety. The electric light plant in this case grew taller and was stronger, but it did not reach the stature of its neighbor from the dark house, and its flowers have all disappeared and the flower stems have broken off while the other is in its prime. At a distance of 9 and 10 feet from the light, verbenas grew tall and slender, even taller than in the dark house, and the difference in the time of flowering was very little.

A few fuchsias were grown in both houses. Those in the light house were about eight feet from the lamp, and they flowered three days earlier than the others. The colors were not changed.

Heliotropes of various named varieties standing nine and ten feet from the lamp did not appear to be affected in any way.

White ageratums stood at three feet from the lamp. The flowers soon turned brown and sere. Those in the dark house remained white three times as long.

Chinese primulas at seven feet from the light were not affected, but those four feet away, especially the lilacs, were changed in color. The lilac was bleached out to pure white wherever the light struck squarely upon the flowers, but any portion of the flower which chanced to be shaded by a leaf or another petal retained its color for a time, and then gradually became duller. The exact shape of the shadow could be distinguished during the day. It was noticed that the bleaching was most apparent when the lamp had not been running for a few days ; but this was simply because the loss of color took place within a few hours rather than slowly from the opening of the bud.

Petunias were much affected by the light. The plants were much taller and slenderer in the light, even at the farthest corners of the house, and they bloomed earlier and more profusely. Feb. 24th, when measurements were taken, it was found that the height of plants in the dark house was to height of those in the light house as 5 is to 6. Fig. 13 shows the difference well. The three front plants (A, B, C) grew 14 feet from the light, and the rear plants (D, E, F) grew in the dark house. White petunias were not changed in color by the light, but purple ones quickly became blue, especially near the lamp. The flowers near the light shrunk in size the first day or two and the texture became thin and flabby. During the moonlight nights of March, when the light was not running, many purple petunias opened near the lamp. The first night following, the lamp burned six hours, and the next morning nearly all the flowers five within feet were blue where the light had struck them squarely, and later they changed to a dirty white. Flowers which opened when the light was running every night were not so soon affected.

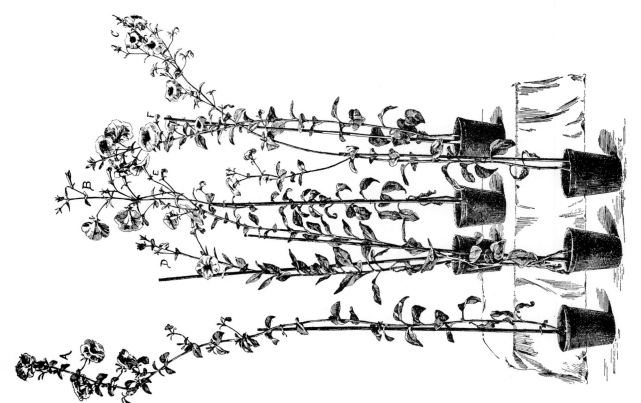

FIG. 13.—*Effect of Electric Light upon Petunias.*

SERIES A.  FROM JANUARY 16TH TO MARCH 31ST, 1891.

| PLANT. | Electric Light House. | | | | Normal or Dark House. | |
|---|---|---|---|---|---|---|
| | Color of flower. | Distance from light. Feet. | Hours of light upon the flower | Duration of flower. Days. | Color of flower. | Duration of flower. Days. |
| Cuphea hyssopifolia | Pink. | 5 | 70.5 | 14 | Pink | 10 |
| | " | " | 47. | 8 | " | 8 |
| | " | " | 46.5 | 8 | " | 9 |
| | " | " | 17. | 7 | " | 12 |
| | " | " | 35. | 6 | " | 7 |
| | " | " | 36.5 | 9 | " | 8 |
| | " | " | 24. | 8 | " | 10 |
| | | | | 8 57 Average | " | 7 |
| | | | | | " | 8 |
| | | | | | " | 7 |
| | | | | | " | 10 |
| | | | | | " | 8 |
| | | | | | " | 9 |
| | | | | | " | 8 |
| | | | | | | 8 64 Average |
| Heliotrope | White. | 8 | 45. | 16 | White | 16 |
| | " | " | 50.5 | 14 | " | 15 |
| | " | " | 55 | 16 | " | 22 |
| | " | " | 50 | 15 | " | 22 |
| | | | | 15.25 Average | " | 16 |
| | | | | | " | 17 |
| | | | | | " | 18 |
| | | | | | " | 19 |
| | | | | | | 18.12 Average |
| | Blue | 8 | 33 | 14 | Blue. | 15 |
| | " | " | 46.5 | 12 | " | 16 |
| | " | " | 36.5 | 13 | " | 12 |
| | " | " | 50 | 14 | " | 15 |
| | " | " | 39 | 14 | " | 14 |
| | " | " | 39 | 14 | " | 15 |
| | " | " | 29 | 13 | " | 13 |
| | " | " | 24 | 14 | " | 13 |
| | " | " | 14 | 14 | | 14.12 Average |
| | " | " | 14 | 14 | | |
| | | | | 13.6 Average | | |
| Ageratum | White | 3 | 2.15 | 10 | White | 25 |

Coleus plants of various colors were placed at different distances from the lamp March 31st. After two nights the plants within three feet of the lamp were much affected. Reds became yellow, browns turned to green, greens lost their brightness, and dark purple became glossy black. Whenever shadows of other leaves protected the foliage the color was unchanged, and the precise limits of the shadow, even to the teeth of the overhanging leaves, were visible for days afterwards. Plants five feet away were little affected at this time, and those 12 feet away were unchanged. Five days later, brown foliage seven feet from the lamp had become greenish in color. During April all coleuses 10 and 12 feet from the light were uninjured, but in January and February they gradually became duller in color at that distance, especially the browns, which appear to be particularly susceptible to the influence of the light. This difference between the results of April and January is undoubtedly due to the greater number of hours of electric light in the long nights of midwinter; and by a reference to the record on page 100 it will be seen that for eleven consecutive nights in April the light did not run.

Observations were made upon the duration of flowers of various colors and species in both houses, and they are tabulated below. By duration of the flower, as recorded in the third column of figures, is meant the total number of days which it remained "open," and held its shape. In many cases, especially near the lamp, the flowers became discolored soon after opening, but still remained open many days. The table also gives some interesting data concerning the duration of flowers as influenced by distance from the lamp. In the table, each entry represents an individual flower, which was ticketed and carefully watched.

### SERIES A. FROM JANUARY 16TH TO MARCH 31ST, 1891.—Continued.

| PLANT. | Electric Light House. | | | | Normal or Dark House. | |
|---|---|---|---|---|---|---|
| | Color of flower. | Distance from light. Feet. | Hours of light upon the flower. | Duration of flower. Days. | Color of flowers. | Duration of flower. Days. |
| Primula | White | 5 | 58 | 10 | White | 23 |
| | " | " | 56.5 | 12 | " | 21 |
| | " | " | 44.5 | 8 | " | 19 |
| | " | " | 53 | 14 | " | 21 |
| | " | " | 52 | 17 | " | 21 |
| | " | " | 36 | 8 | " | 17 |
| | " | " | 42 | 12 | " | 18 |
| | " | " | 27 | 9 | " | 14 |
| | " | " | 30.5 | 12 | " | 17 |
| | " | " | 32 | 10 | " | 13 |
| | " | " | 55 | 11.27 Average | " | 17 |
| | Lilac. | 5 | 39 | 11 | " | 25 |
| | " | " | 28 | 11 | " | 19 |
| | " | " | 24 | 10 | " | 16 |
| | " | " | 64 | 20 | " | 17 |
| | " | " | 44 | 10 | " | 15 |
| | " | " | 11 | 15 | " | 19 |
| | " | " | 55 | 11 | " | 13 |
| | " | " | 14 | 13 | " | 15 |
| | " | " | 56 | 16 | " | 15 |
| | " | " | 39 | 17 | 17.75 Average | |
| | " | " | 50 | 16 | Lilac. | 15 |
| | " | " | 39 | 17 | " | 17 |
| | " | " | 35 | 7 | " | 11 |
| | | | | 13.38 Average | " | 14 |
| | | | | | " | 17 |
| | | | | | " | 16 |
| | | | | | " | 18 |
| | | | | | " | 13 |
| | | | | | " | 18 |
| | | | | | " | 18 |
| | | | | | " | 14 |
| | | | | | " | 12 |
| | | | | | 15.25 Average | |

### SERIES A. FROM JANUARY 16TH TO MARCH 31ST, 1891.—Continued.

| PLANT. | Electric Light House. | | | | Normal or Dark House. | |
|---|---|---|---|---|---|---|
| | Color of flower. | Distance from light. Feet. | Hours of light upon the flower. | Duration of flower. Days. | Color of flower. | Duration of flower. Days. |
| Petunia | White | 5 | 30 | 7 | White. | 14 |
| | " | 7 | 34 | 13 | " | 11 |
| | " | 7 | 24 | 9 | " | 15 |
| | " | 7 | 19 | 10 | " | 11 |
| | " | 7 | 19 | 10 | " | 10 |
| | " | 8 | 50 | 11 | 12.2 Average | |
| | " | 10 | 14 | 4 | Purple | 16 |
| | " | 11 | 20 | 10 | " | 13 |
| | " | 11 | 39 | 9.33 Average | " | 11 |
| | Purple. | 5½ | 70.5 | 13 | " | 16 |
| | " | 5½ | 50.5 | 11 | " | 9 |
| | " | 6 | 27.5 | 12 | " | 15 |
| | " | 6 | 30 | 7 | " | 11 |
| | " | 7 | 39 | 10 | " | 12 |
| | " | 7 | 25 | 12 | " | 12 |
| | " | 8 | 27.5 | 8 | " | 15 |
| | " | 8 | 36 | 8 | " | 14 |
| | " | 8 | 19 | 11 | " | 12 |
| | " | 9 | 33 | 7 | " | 12 |
| | " | 12 | 14.5 | 8 | " | 15 |
| | " | 14 | 39 | 6 | " | 14 |
| | " | 15 | 24 | 11 | " | 10 |
| | | | 14 | 10 | " | 11 |
| | | | 55 | 11 | " | 13 |
| | | | | 9.81 Average | " | 11 |
| | | | | | " | 15 |
| | | | | | " | 16 |
| | | | | | " | 15 |
| | | | | | " | 14 |
| | | | | | " | 8 |
| | | | | | " | 10 |
| | | | | | " | 13 |
| | | | | | 12.77 Average | |

## SERIES B. FROM APRIL 1ST TO MAY 1ST, 1891.

| PLANT. | Electric Light House. | | | | Normal or Dark House. | |
|---|---|---|---|---|---|---|
| | Color of flower. | Distance from light. Feet. | Hours of light upon the flower. | Duration of flower. Days. | Color of flower. | Duration of flower. |
| Chrysanthemum | White. | 12 | 45 | 19 | White | 18 |
| | " | 12 | 33.5 | 21 | | |
| | | | | Average 20 | | |
| Fuchsia | — | 8 | 25.5 | 8 | — | 9 |
| | — | 8 | 17.5 | 8 | | |
| | | | | Average 8 | | |
| Heliotrope | White | 7 | 41.5 | 13 | White | 16 |
| | | | | | " | 19 |
| | | | | | " | 11 |
| | | | | | | Average 15.3 |
| | Blue | 4 | 49.5 | 14 | Blue | 14 |
| | " | 5 | 13.5 | 11 | " | 12 |
| | " | 8 | 49.5 | 13 | " | 11 |
| | " | 9 | 29.5 | 14 | " | 10 |
| | " | 10 | 14 | 11 | | Average 11.75 |
| | | | | Average 12.6 | | |
| Petunia | White | 4 | 7 | 9 | White | 14 |
| | " | 5 | 33.5 | 12 | " | 11 |
| | " | 6 | 14 | 9 | " | 14 |
| | " | 7 | 16.5 | 8 | " | 7 |
| | " | 9 | 46 | 11 | | Average 11.2 |
| | " | 10 | 37.5 | 10 | | |
| | " | 12 | 42.5 | 10 | | |
| | " | 13 | 13.5 | 7 | | |
| | | | 14 | 9 | | |
| | | | | Average 9.44 | | |
| | Purple | 1½ | Did not open | 13 | Purple | 13 |
| | " | 3 | 59.5 | 7 | " | 13 |
| | " | 3 | 25.5 | 8 | " | 7 |
| | " | 4 | 16.5 | 9 | " | 8 |
| | " | 4 | 34 | 9 | | Average 10.25 |
| | " | 7 | 14 | 10 | | |
| | " | 7 | 22.5 | 10 | | |
| | " | 10 | 37.5 | 15 | | |
| | " | 12 | 59.5 | 8 | | |
| | " | 13 | 25.5 | | | |
| | | | | 9.88 Average | | |

## SERIES A. FROM JANUARY 16TH to MARCH 31ST, 1891.—Continued.

| PLANT. | Electric Light House. | | | | Normal or Dark House. | |
|---|---|---|---|---|---|---|
| | Color of flower. | Distance from light. Feet. | Hours of light upon the flower. | Duration of flower. Days. | Color of flowers. | Duration of flower. Days. |
| Verbena | Blue. | 3 | 40 | 10 | Blue | 11 |
| | " | 5 | 61 | 11 | " | 9 |
| | " | 3 | 7 | 3 | " | 11 |
| | " | " | 3.5 | 2 | " | 6 |
| | " | " | 39 | 14 | " | 6 |
| | | | | Average 8 | " | 10 |
| | | | | | " | 10 |
| | | | | | " | 7 |
| | | | | | | 10 |
| | | | | | | 8.88 Average |
| | Red. | 1½ | 17 | 8 | Red | 7 |
| | " | 1½ | 17 | 6 | " | 8 |
| | " | 2 | 36 | 6 | " | 2 |
| | " | 3 | 12.5 | 4 | " | 11 |
| | " | 4 | 46 | 8 | " | 8 |
| | " | 3 | 23 | 4 | " | 5 |
| | " | " | 15.5 | 5 | " | 7 |
| | " | " | 7 | 4 | | 6.86 Average |
| | " | " | 24 | 7 | | |
| | " | " | 51 | 9 | | |
| | | | | 6.1 Average | | |

**SERIES B.   FROM APRIL 1ST TO MAY 1ST, 1891.—*Continued*.**

| PLANT. | Electric Light House. | | | | Normal or Dark House. | |
|---|---|---|---|---|---|---|
| | Color of flower. | Distance from light. Feet. | Hours of light upon the flower. | Duration of flower. Days. | Color of flower. | Duration of flower. |
| Primula............. | White | 3 | 49.5 | 14 | White | 14 |
| | " | 3 | 16.5 | 12 | " | 13 |
| | " | 4 | 23 | 18 | " | 13 |
| | " | 7 | 25.5 | 15 | " | 16 |
| | " | 7 | 25.5 | 19 | 14 Average | |
| | " | 8 | 7 | 13 | | |
| | " | 9 | 19.5 | 14 | | |
| | " | 10 | 37.5 | 12 | | |
| | | | | 14.62 Average | | |
| | Lilac | 4 | 49.5 | 14 | Lilac | 13 |
| | | | | | " | 17 |
| | | | | | " | 15 |
| | | | | | 15 Average | |
| Verbena............. | Blue | 4 | 10.5 | 8 | Blue | 15 |
| | " | 4 | 22.5 | 10 | " | 12 |
| | " | 5 | 31.5 | 10 | 13.5 Average | |
| | " | 5 | 10.5 | 9 | | |
| | " | 6 | 39.5 | 10 | | |
| | " | 6 | 7 | 9 | | |
| | " | 7 | 7 | 9 | | |
| | " | 7 | 0 | 7 | | |
| | " | 10 | 23.5 | 13 | | |
| | " | 11 | 7 | 11 | | |
| | | | | 9.7 Average | | |
| | Red | 2 | 7 | 7 | Red | 10 |
| | " | 2 | 10.5 | 11 | " | 9 |
| | " | 3 | 26 | 7 | " | 10 |
| | " | 4 | 25.5 | 14 | " | 5 |
| | " | 4 | 22.5 | 10 | " | 9 |
| | " | 6 | 19.5 | 9 | " | 11 |
| | | | | 9.66 Average | 9.14 Average | |

Perhaps the most noticeable feature of these figures is the lack of uniformity in duration under similar conditions. Neither the distance from the lamp nor the hours of light received by the flower appears to determine the duration. The longevity of the flower is probably determined more by the vigor and general condition of the plant than by the variations in the amount of light, although this subject is one which demands closer investigation. In all the observations recorded above, care was exercised to secure plants of like age and vigor. But if the figures of individual flowers give varying results, the averages in the first series (Jan. 16 to Mar. 31), when there were more hours of electric light, are uniformly larger in the dark house specimens. In the second series (April) the averages are not uniform. One is tempted to construct various comparisons from the tables, but the observations are so few that perhaps no other conclusion should be drawn at present but this,—that the duration of flowers is not so much influenced by the electric light at exhibition halls as is generally supposed.

Numerous auxanometer* readings were made upon representative petunia plants in both houses during 1890 and 1891. The following record of the growth of two petunia plants, alike in all respects, will be sufficient for the present purpose:

| DATE. 1891. | In electric light house. Plant 12 feet from the lamp. | | | In normal house. | | |
|---|---|---|---|---|---|---|
| | 8 p. m.—11 p. m. Electric light running. | 11 p. m.—8 a. m. Dark, until daylight. | 8 a. m.—8 p. m. Daylight. | 8 p. m.—11 p. m. Dark. | 11 p. m.—8 a. m. Dark, until daylight. | 8 a. m.—8 p. m. Daylight. |
| Apr. 27 .... | .25 in. | .0625 in. | .5 in. | .15625 in. | .0625 in. | .625 in. |
| " 29 .... | .00 | .3125 " | .125 in. | .00 | .125 " | .25 " |
| " 30 .... | .125 " | .125 " | .5 " | .00 | .375 " | .0625 in. |
| May 1 ... | .125 " | .375 " | .375 " | .125 | .25 " | .1875 " |
| | .50 | .875 in. | 1.5 " | .28125 in. | .8125 in. | 1.1250 in. |
| | Total growth 2.875 ins. | | | Total growth 2.21875 in. | | |

*An auxanometer is an instrument for measuring the growth of plants. The record is usually made continuously upon a revolving cylinder connected with a clock, so that the amount of growth can be determined for any hour of the day.

The average growth per hour is as follows:

| 8 P.M-11 P.M. | 11 P.M.-8 A.M. | 8 A.M.-8 P.M. | 8 P.M.-11 P.M | 11 P.M.-8 A.M. | 8 A.M.-8 P.M. |
|---|---|---|---|---|---|
| Electric light. | | | | | |
| .0416 in. | .0243 in. | .0312 in. | .0234 in. | .0225 in. | .0234 in. |

The greatest growth took place when the electric light was burning.

In all these experiments with ornamental plants, it was noticeable that the light exercised a very injurious effect within a radius of about six feet. Between six and eight feet the results were indifferent, and beyond that point there was usually a noticeable tendency towards a taller and straighter growth, and it seemed to us that at distances of a dozen feet or more the flowers were more intense in color, particularly when they first opened. There was usually a perceptible gain in earliness in the light house, also. On the whole, I feel that it will be possible some day to use the electric light in floricultural establishments to some pecuniary advantage. If we have not determined the distance to which the beneficial influence of the light extends, we have nevertheless defined with some degree of accuracy for a few species the limits of its injurious influence.

## II. EXPERIMENTS ELSEWHERE.

The first experiment to determine the influence of electric light upon vegetation was made by Hervé-Mangon in 1861.* This experiment showed that the electric light can cause the production of chlorophyll or the green color in plants, and also that the light can produce heliotropism, or the phenomenon of turning or bending towards the light.

In 1869 Prillieux† showed that the electric light, in common with other artificial lights, is capable of promoting assimilation, or the decomposition of carbon dioxide and water.

The next experiments appear to have been those of C. W. Siemens, in England, and P. P. Dehérain in France. These two, with our own, appear to be the only definite investigations of the subject upon what may be called a practical or horticultural scale.

*Compt. Rend. 53, 243.
†Compt. Rend. 69, 410.

The English experiments, although eminently practical, were conducted by an electrician, and the French were largely confined to physiological problems. It seemed proper that the third series of experiments should be approached from the particular standpoint of the gardener.

Dr. Siemens' experiments* may be divided into two series: in one series the lamp was placed inside the greenhouse, and in the other suspended over it. In both cases he observed marked effects upon vegetation in a short time. A great variety of plants was treated. The dynamo which Siemens' in his first experiment used "makes 1,000 revolutions a minute, it takes two horse-power to drive it, and develops a current of 25 to 27 webers of an intensity of 70 volts." "The light produced is equal to 1,400 candles measured photometrically." When the lamp was placed inside the house, plants within three or four feet of it suffered much, the leaves of melons and cucumbers "which were directly opposite the light turning up at the edges and presenting a scorched appearance." When these injured plants were removed to a distance of seven or eight feet they showed "signs of recovery, throwing out fresh leaves and pearls of moisture at their edges." In general, all plants which were exposed to normal conditions during the day and to six hours of electric light at night 'far surpassed the others in darkness of green and vigorous appearance generally." The flavor was fully as good in the electric light fruits as in the others. These results were supplemented by a larger experiment in the winter of 1880-81. In this case a lamp of 4,000 candle power was used, and it was placed inside a house of 2,318 cubic feet capacity. The light was run all night, and the arc was at first not protected by a globe. The "results were anything but satisfactory," the plants soon becoming withered. At this point a globe of clear glass was placed upon the lamp, and thereafter the most satisfactory results were obtained. Peas, raspberries, strawberries, grapes, melons and bananas fruited early and abundantly under continuous light,—solar light by day and electric light by night. The strawberries are said to have been "of excellent flavor and color," and the grapes of "stronger flavor than usual." The

*Proc. Royal Soc. xxx. 210 and 293. Rep. British A. A. S. 1881, 474. See also abstract in Nature xxi. 456 (Mch. 11, 1880), and an editorial in the same issue.

bananas were "pronounced by competent judges unsurpassed in flavor" and the melons were "remarkable for size and aromatic flavor." Wheat, barley and oats grew so rapidly that they fell to the ground of their own weight. The beneficial influence of the clear glass globe was therefore most marked. "The effect of interposing a mere sheet of thin glass between the plants and the source of electric light was most striking. On placing such a sheet of clear glass so as to intercept the rays of the electric light from a portion only of a plant—for instance, a tomato plant—it was most distinctly shown upon the leaves. The portion of the plant under the direct influence of the naked electric light, though a distance from it of nine or ten feet*, was shrivelled, whereas that portion under cover of the clear glass, continued to show a healthy appearance, and this line of demarkation was distinctly visible on individual leaves; not only the leaves but the young stems of the plants soon showed signs of destruction when exposed to the naked electric light, and these destructive influences were perceptible, though in a less marked degree, at a distance of twenty feet from the source of light."

In the other series of experiments, Siemens placed an electric lamp of 1400 candle power about seven feet above a sunken melon pit which was covered with glass. The light was modified by a clear glass globe. In the pit seeds and plants of mustard, carrots, turnips, beans, cucumbers and melons were placed. The light ran six hours each night and the plants had sunlight during the day. In all cases those plants "exposed to both sources of light showed a decided superiority in vigor over all the others, and the green of the leaf was of a dark rich hue." Heliotropism was observed in the young mustard plants. Electric light appeared to be about half as effective as daylight. A great difficulty experienced in this experiment was the film of moisture which condenses on greenhouse roofs at night and obstructs the passage of light. The light was at one time suspended over two parallel pits nearly four feet apart, and the effect was observed upon plants under the glass and in the uncovered space. In all cases the growth of the plants was hastened. Flowering was hastened in melons

*It is to be observed that the light used by Dr. Siemens in this case was 4,000 candle power, while ours was but 2,000.

and other plants under the glass. Strawberries which were just setting fruit, were put in one of the pits and part of them were kept dark at night while the others were exposed to the light. After fourteen days, the light having burned twelve nights, most of the fruits on the lighted plants "had attained to ripeness, and presented a rich coloring, while the fruit on those plants that had been exposed to daylight only had by this time scarcely begun to show even a sign of redness." He concludes that a lamp of 1400 candle power produced a maximum beneficial result on vegetation at a distance three metres (nearly ten feet) above the glass, but "the effect is nevertheless very marked upon plants at a greater distance."

At the close of his experiments Siemens was very sanguine that the electric light can be profitably employed in horticulture, and he used the term "electro horticulture" to designate this new application of electric energy. He anticipated that in the future "the horticulturist will have the means of making himself practically independent of solar light for producing a high quality of fruit at all seasons of the year." He had shown that growth can be hastened by the addition of electric light to daylight, that injury does not necessarily follow continuous light throughout the twenty-four hours, that electric light often deepens the green of leaves and the tints of flowers and sometimes intensifies flavors, and that it aids to produce good seeds; and he thought that the addition of the electric light enabled plants to bear a higher temperature in the greenhouses than they otherwise could. But whatever may be the value of the electric light to horticulture, the practical value of Siemens' experiments is still great. They have furnished data in several obscure relations of light to vegetation. Nature made the following comments upon this feature of the application of the electric light by Dr. Siemens: "But the scientific interest of its present application must rest mainly on the fact that the cycle of the transformation of energy engaged in plant life is now complete, and that, starting from the energy stored up in vegetable fuel, we can run through the changes from heat to electricity, and thence to light, which we now know we can store up in vegetable fuel again."

Dehérain's experiments* were conducted at the Exposition d' Électricité, Paris, in 1889. A small conservatory standing inside the Exposition building was divided into two compartments. One compartment was darkened and the glass painted white upon the inside; this received the electric light and all solar light was excluded. The other compartment was not changed. The amount of sunlight which the plants normally received in this conservatory within an exposition was not sufficient to maintain a healthy growth. A lamp of 2,000 nominal candle power was used. At first the naked electric light was used and it ran continuously. Barley in head and flax in flower were brought into the lighted compartment; also chrysanthemums, pelargoniums, roses, and a variety of ornamental plants. After seven days of continuous electric lighting most of the plants were seriously injured. All the pelargoniums lost their leaves, cannas were discolored, four o'clocks were tarnished, and bamboos were blackened. "But the most curious effect was produced upon the lilacs; all the parts of the leaves that had received the direct rays from the lamp were blackened, while those protected by the upper leaves preserved their beautiful green color, and the impression produced upon the epidermis by the electric rays had the clearness of a photographic plate." Similar effects were produced upon azaleas, deutzias and chrysanthemums. It was found that this discoloration did not extend beyond the first layer of palisade cells. Plants which received solar light by day and electric light at night only were injured in the same manner but in a less degree. The injury was most marked upon the jold leaves. The pelargoniums soon sent out new shoots and the young leaves resisted the action of the light much longer than did the mature ones. The flax continued to grow and the barley ripened. It was found that plants under the electric light alone were able to assimilate, but the action was very slow. As much assimilation took place in an hour on a bright summer day as in several days of electric light. At the expiration of two weeks the condition of the plants was so bad that a change was made, and thereafter a globe was used upon the lamp.

The experiment with modified light, by the use of a transparent glass globe, was conducted like the preceding. Sprouting seeds in electric light alone grew for a short time, then drooped and died,

*Ann. Agronom. vii. 551 (1881).

not being able to make true leaves. Sprouting maize turned black, but maize in full growth remained in apparently good condition though not growing, even for two months. New leaves appeared on roses and other plants, but growth was slow or none. Flowers did not appear, and seeds did not mature in previously formed fruits, except in the case of barley, which made good seeds. New growths appeared at the base of some plants and the petioles of pelargoniums became very much elongated. Many plants remained almost stationary throughout. Assimilation was more feeble than under the naked light. Plants which had been set out of doors during the day and brought into the electric light house at night, did not behave any better, if as well, than those left out of doors continuously. Dehérain's account is replete with interesting speculations upon the physiology of the plants under experiment. His general conclusions of the influence of electric light upon plants, are as follows:

"1. The electric light from lamps contains rays harmful to vegetation.

"2. The greater part of the injurious rays are modified by a transparent glass.

"3. The electric light contains enough rays to maintain full grown plants two and one-half months.

"4. The light is too weak to enable sprouting seeds to prosper or to bring adult plants to maturity."

Finally, observations* were made more recently upon the influence of the electric light upon plants in the winter palace at St. Petersburg. It was observed that in a single night ornamental plants turned yellow and then lost their leaves. Yet it is well known that incandescent lamps can be lodged in the corolla of a flower without injuring it.

RECAPITULATION.

It is impossible to draw many definite conclusions from the above researches. The many conflicting and indefinite results indicate that the problems vary widely under different conditions and with different plants. Yet there are a few points which are clear: the electric light promotes assimilation, it often hastens growth and maturity, it is capable of producing natural flavors

*Ann. Agronom. xiv. 281 (1888).

and colors in fruits, it often intensifies colors of flowers and sometimes increases the production of flowers. The experiments show that periods of darkness are not necessary to the growth and developments of plants. There is every reason, therefore, to suppose that the electric light can be profitably used in the growing of plants. It is only necessary to overcome the difficulties, the chief of which are the injurious influences upon plants near the light, the too rapid hastening of maturity in some species, and in short, the whole series of practical adjustments of conditions to individual circumstances. Thus far, to be sure, we have learned more of the injurious effects than of the beneficial ones, but this only means that we are acquiring definite facts concerning the whole influence of electric light upon vegetation; and in some cases, notably in our lettuce tests, the light has already been found to be a useful adjunct to forcing establishments.

The experiments suggest many physiological speculations upon which it is not the province of this bulletin to enter. Yet two or three of them may be mentioned. It is a common notion that plants need rest at night, but this is not true, in the sense in which animals need rest. Plants have simply adapted themselves to the conditions of alternating daylight and darkness, and during the day they assimilate or make their food and during the night, when, perforce, assimilation must cease, they use the food in growth. They simply practice an individual division of labor. There is no inherent reason why plants cannot grow in full light, and, in fact, it is well known that they do grow then, although the greater part of growth is usually performed at night. If light is continuous, they simply grow more or less continuously, as conditions require, as they do in the long days of the arctic regions, or as our plants did under continuous light. There is no such thing as a plant becoming worn out or tired out because of the stimulating influence of continuous light.

It would seem, therefore, that if the electric light enables plants to assimilate during the night and does not interfere with growth, it must produce plants of great size and marked precocity. But there are other conditions, not yet understood, which must be studied. Our radish plants, and many others, were earlier but smaller under the influence of the light. Observation and chemical examination showed that a greater degree of maturity had

been attained. Perhaps they assimilated too rapidly, perhaps the functions of the plant had been completed before it had had time to make its accustomed growth. Perhaps the highly refrangible and invisible rays from the electric lamp have something to do with it. In fact, this latter presumption probably accounts for much if not all of the injury resulting from the use of the naked light, for the effect of the interposition of a clear pane of glass is probably to absorb or obstruct these rays of high refrangibility. Good results which follow the use of a globe or a pane of glass show, on the other hand, that the injury to plants cannot result from any gases arising from the lamp itself, as has been supposed by some observers. In our own experiments, particularly with the Brush lamp, there was no perceptible odor from the gases of combustion; and it may also be said that commercial forcing-houses, like our own, are not tight enough to hold sufficient quantities of these gases to injure plants.

It is highly probable that there are certain times in the life of the plant when the electric light will prove to be particularly helpful. Many experiments show that injury follows its use at that critical time when the plantlet is losing its support from the seed and is beginning to shift for itself, and other experiments show that good results follow its later use. This latter point appears to be contradicted by Dehérain's results, but his experiments were not conducted under the best normal conditions.

On the whole, I am inclined towards Siemens' view that there is a future for electro-horticulture.

L. H. BAILEY.

E. J. Kraus and H. R. Kraybill. 1918. Vegetation and reproduction with special reference to the tomato. Oregon Agricultural Experiment Station Bulletin 149.

E. J. Kraus

H. R. Kraybill

This now famous bulletin describes experiments conducted at the University of Chicago in partial fulfillment of requirements for the Ph.D. degree. The work was done cooperatively by Ezra Jacob Kraus, a horticulturist, and Henry Reist Kraybill, a chemist, under the direction of William Crocker. Both authors had taken leaves of absence from their respective departments, Kraus from the department of horticulture at Oregon State College, Kraybill from the department of agricultural chemistry at Pennsylvania State College. E. J. Kraus, born in 1885 in Ingham County, Michigan, had earned his B.S. degree from Michigan Agricultural College in 1907, worked for two years as a scientific assistant for the U.S. Department of Agriculture (USDA), then accepted a position as professor of horticultural research at Oregon Agricultural College. A leave of absence permitted him to study for the Ph.D. degree at Chicago. H. R. Kraybill, born in Mt. Joy, Pennsylvania, in 1891, obtained his B.S. degree in agricultural and biological chemistry from Pennsylvania State College in 1913 and his M.S. degree from the University of Chicago in 1915. He was assistant chemist and subsequently an instructor in chemistry from 1913 to 1915, then held a teaching fellowship at the University of Chicago until 1917.

Kraus and Kraybill's study was undertaken at a time when environmental effects on plant growth and development were being interpreted in nutritional terms. Chief among the proponents of this approach was

George Klebs of the University of Heidelberg. His experiments with fungi, algae, and higher plants had demonstrated the importance of nutrition in reproduction (see Klebs 1910). When nutrients were abundant, the fungus *Saprolegnia* grew vegetatively; when the nutrient supply became depleted, sexual reproduction ensued; when placed in pure water, asexual zoospores were produced. Experiments with the houseleek (*Sempervivum*) demonstrated that the time of flowering could be hastened or delayed according to external conditions of temperature, light (including wavelength), and nutrition. Klebs concluded that the flowering of plants which were "ripe to flower" was inhibited by increasing the supply of inorganic salts and/or reducing carbon assimilation.

Klebs (1915) also suggested that periodicity of growth in tropical trees might have a nutritional basis. Kraus and Kraybill cite their supervising professor's criticisms of this suggestion, including his statement, ". . . there is need of a careful study of internal conditions of the plant, anatomical, chemical, and microchemical . . ." (Crocker 1916). Both students and professor were obviously concerned that scientists were attempting to interpret plant responses to environmental conditions without direct analysis of the tissues concerned. In their introduction, Kraus and Kraybill cite the need for more information as to the "internal changes, conditions, or compounds developed in response to the several external factors to which plants may be subjected. . . ." Once these were known they could be correlated and a determination made of "the external response to these internal conditions." This theme is repeated at several points in the introduction and general discussion.

The rationale for the research described is summarized in the foreword to the bulletin, written by C. I. Lewis, chief of the department of horticulture at Oregon State in 1918. He indicates that the bulletin is one of a series in connection with the pollination of pome fruits, and states that morphological and histological investigations had provided "little more than confirmation of the already well-recognized microscopic situations [thus] physiological and biochemical investigations must be made to establish a true basis for determination of the factors involved. . . ."

Thirty-two pages of the bulletin are devoted to a review of the information then available on fruit set and development, with emphasis on the effects of cultural practices, including soil management, nitrogenous fertilizers, and pruning. Although most of the discussion concerns fruit trees, tomato was chosen as an experimental plant for its ease of handling under greenhouse conditions, "clean-cut" fruit-setting responses, and ready availability. In the foreword, Lewis gives an additional reason, noting that "in its general responses in vegetation and fruit setting it accords very closely to . . . apple and pear trees. . . ." However, no evidence is provided by either Lewis or the authors to support this statement.

Kraus and Kraybill published the results of four major experiments (II, V, VI, and VII). Missing numerals suggest that at least three more were performed. Most of the tables (34 in all) contain data for one treatment only, making comparisons of treatment effects difficult. Comparative data for experiment VII are given in Figs. 19 through 22 for upper stems, lower stems, upper leaves, and lower leaves, respectively. From these data, the authors concluded that total nitrogen content was inversely correlated with total carbohydrates in the lower leaves. Independent statistical analysis (Cameron and Dennis 1986) confirmed significant, linear, negative correlations between the two components in all four figures. Mathematical ratios of carbohydrates to nitrogenous compounds were not calculated. Kraus (1925) objected to the use of such ratios on the grounds that they were unlikely to be related to physiological function. However, Cameron and Dennis (1986) later calculated ratios based on Kraus and Kraybill's data.

Given the volume of data, most reviewers have been content to cite the "four general conditions of the relation of nitrates, carbohydrates, and moisture within the plant itself, and the responses apparently correlated therewith . . ." (p. 6) or the Summary (pp. 85–87). The former appear to be based on the literature review; the latter is a list of 29 conclusions. The conclusions most often quoted state that fruitfulness is dependent on a balance between availability of nitrogen and opportunity for photosynthesis. If the supply of nitrogen is excessive, plants are too vigorous and therefore unfruitful. If nitrogen is insufficient, vigor is low and carbohydrates accumulate; these plants, too, are unfruitful. A moderate supply of nitrogen, on the other hand, allows adequate growth for fruit development.

Analysis of their data, based on the benefits of 70 years of hindsight, does not support many of the correlations for which they are most often cited (Cameron and Dennis 1986). Their data on flowering and fruiting are qualitative only and are insufficient to permit even a rough classification of response. The authors emphasized the effects of treatments on vegetative growth rather than reproduction. Subsequent work by one of Kraus's students, A. E. Murneek (1926), clearly demonstrated that tomato fruits are "sinks" for nitrogen and that their presence drastically alters the growth and appearance of the plant. This was unknown in 1917, but obviously complicated interpretation of the data obtained at that time.

Taken as a whole, the paper stands as a watershed between a period when physiological responses were explained by speculating on the effects of external conditions on biochemical processes and today's emphasis on attempting to determine what is actually happening within the plant. Kraus and Kraybill deserve the fame they have achieved, not so much for the data obtained, which have rarely if ever been cited, as for their willingness to attack a difficult problem head-on rather than merely speculating about it. Their work inspired many others, particularly those interested in the flowering of fruit trees and vegetables, to investigate the biochemical bases involved in these processes. These workers, like Kraus and Kraybill, believed that "It is quite essential to know what is present in the body of the plant itself, its extent and environment, as to know what nutrients are present in the soil" (p. 14).

Following completion of his graduate studies, Kraus returned to Oregon as dean of the College of Letters and Science. He served as professor of applied botany at the University of Wisconsin (1919–1927), then returned to the department of botany at Chicago, becoming chairman in 1934. From his retirement in 1948 until his death in 1960, he bred chrysanthemums and other ornamentals as a visiting professor of horticulture at Corvallis. Kraus maintained his interest in fruit development, particularly in the area of morphology, was a main figure in the discovery of the tumor-inducing and herbicidal activity of 2,4-D and related compounds, and was among the first to suggest that such chemicals might have commercial value as herbicides. Some of his students who became well known for their contributions to horticulture were J. H. Gourley, A. E. Murneek, F. P. Cullinan, and H. B. Tukey. C. M. Harrison, a former student, remembers Kraus, a confirmed bachelor, as a man of many interests. An excellent teacher, he not only could hold the interest of freshmen on the subject of plant morphology, but was also an authority on ants and oriental rugs!

A member of Phi Beta Kappa, Kraus served as president of the American Society for Horticultural Science (1927) and the American Society of Plant Physiologists (1928), and as vice-president of the American Association for the Advancement of Science (1930) and the American Society of Naturalists (1931). He edited the *Botanical Gazette* from 1928 until 1947. His memberships in the Entomological Society of Washington and the Association of Economic Entomology reflected his brief assignment with the USDA as an ant taxonomist. Among the many honors he received were honorary D.Sc. degrees from Oregon State College (1938) and Michigan State College (1949).

Kraybill served as assistant physiologist with the USDA (1917–1919), then professor and head of the department of agricultural chemistry at the University of New Hampshire (1919–1924). In 1924 he was hired by William Crocker, director of the newly opened Boyce Thompson Institute for Plant Research in Yonkers, New York, as the institute's first biochemist. In the period 1917–1926 he studied the effects of nutrition, shading, and ringing on flowering and fruit set of tomato, apple, and peach; compared fertilizers for tree fruits; and began working with virus diseases of tomato and cranberry. He moved to Purdue University in 1926, becoming state chemist and seed commissioner, responsible for the enforcement of state laws regulating the sale of fertilizers, as well as professor and later department head of agricultural chemistry. While at Purdue he continued work on tomato mosaic virus; introduced, together with S. F. Thornton, a new and much more efficient method for determining available potash in commercial fertilizers; studied tryptophan deficiency in mice; and in collaboration with R. B. Withrow, developed a new colorimeter. In 1941 he was named director of the department of scientific research, American Meat Institute, and subsequently (1947) director of research and education of the American Meat Institute Foundation in Chicago, becoming vice-president in 1955. Under his direction, institute scientists developed antioxidants that inhibited rancidity in animal fats; these are still used extensively in foods. Kraybill was elected a fellow of the American Association for the Advancement of Science, and served as president of the American Society of Plant Physiologists (1931), the Association of American Feed Control Officials (1932), and the Association of Official Agricultural Chemists (1938). In

1956—the year of his death—Purdue awarded him an honorary D.Sc. degree.

J. Scott Cameron
Southwestern Washington
   Research Unit
Washington State University
1919 N.E. 78th Street
Vancouver, WA 98665

Frank G. Dennis, Jr.
Department of Horticulture
322 Plant and Soil Sciences
   Building
Michigan State University
East Lansing, MI 48824-1325

## REFERENCES

CAMERON, J. S., and F. G. DENNIS, JR. 1986. The carbohydrate–nitrogen relationship and flowering/fruiting: Kraus and Kraybill revisited. HortScience 21:1099–1102.

CROCKER, W. 1916. Periodicity in tropical trees. Bot. Gaz. 62:244–246.

KLEBS, G. 1910. Alterations in the development and forms of plants as a result of environment. Proc. R. Soc. London Ser. B, 82:547–558.

KLEBS, G. 1915. Uber Wachstum and Ruhe tropischer Baumarten. Jahrb. Wiss. Bot. 56:734–792.

KRAUS, E. J. 1925. Soil nutrients in relation to vegetation and reproduction. Am. J. Bot. 12:510–516.

MURNEEK, A. E. 1926. Effects of correlation between vegetative and reproductive functions in the tomato (*Lycopersicon esculentum* Mill.). Plant Physiol. 1:3–56.

# Vegetation and Reproduction with Special Reference to the Tomato

(*Lycopersicum esculentum Mill.*)

By E. J. Kraus and H. R. Kraybill

## INTRODUCTION

The question of the differentiation of sexually reproductive parts, blooming, fruit setting, and fruit development has been a topic for investigation and speculation for many years. It has been approached in many different ways. Much has been learned; many facts remain unexplained and without correlation; not a few facts are still to be established. More recently the influences of self- and cross-pollination in various plants, particularly those of commercial importance, have been taken up for serious study. The whole subject is so vast that these studies must naturally concern themselves with special phases of the problem. It has been necessary to do much simple testing throughout a wide field and variety of plants under varying conditions. Morphological, anatomical, and histological investigations have been and still are necessary for the determination of the exact structures involved. Physiological studies must be extended and utilized in order to arrive at any final explanation of the conditions observed or the determination of their means of regulation. Not one of these types of study can be spared as a means of finally bringing the problem within the limits of practice.

More specifically the work with plants of commercial importance has dealt and must still deal with the determination of so-called affinities or compatibilities between plants in so far as fruit setting and seed development are concerned. This naturally has led to an investigation of the parts and processes concerned in fertilization, seed and fruit development, and their interrelation. While many of the results have simply furnished microscopic details of what was already well known macroscopically, yet some facts were added. There is still a wide opportunity for such work. Some insight into the mechanism and processes of abscission has been gained; much more is needed. The value of physiological studies can scarcely be over emphasized, but these of necessity must be so detailed and thorough, considering the multiplicity of factors involved, that at best individual investigations can cover only restricted fields.

Pending the more definite working out of details through any one or all of the foregoing methods, the very fertile field of established agricultural and horticultural practice is open for study. Whether such practices are good or bad from the commercial viewpoint, they furnish many suggestions that

may be correlated and interpreted in connection with the available results of controlled investigations. The material reported and the viewpoints expressed in this paper embody some of the results of such a study undertaken in connection with the fruit-setting problem, in so far as it concerns higher plants.

Four general conditions of the relation of nitrates, carbohydrates, and moisture within the plant itself, and the responses apparently correlated therewith, will be discussed. These are:

(1) Though there be present an abundance of moisture and mineral nutrients, including nitrates, yet without an available carbohydrate supply vegetation is weakened and the plants are non-fruitful;

(2) An abundance of moisture and mineral nutrients, especially nitrates, coupled with an available carbohydrate supply, makes for increased vegetation, barrenness, and sterility;

(3) A relative decrease of nitrates in proportion to the carbohydrates makes for an accumulation of the latter; and also for fruitfulness, fertility, and lessened vegetation.

(4) A further reduction of nitrates without inhibiting a possible increase of carbohydrates, makes for a suppression both of vegetation and fruitfulness.

This analysis is not intended, in any way, to convey the idea that only these compounds—carbohydrates, nitrates, and moisture—are concerned in vegetation and fruitfulness, but that the study in hand is principally concerned with them and the response resulting from an alteration of their relative proportions within the plant. It would be extremely difficult also to draw rigid lines between any particular class and the one next to it; since they intergrade insensibly one into another and yet, generally speaking, are recognizably distinct.

[A *General Discussion* section follows, emphasizing the similarities between fruits and vegetative organs, and the concepts of fruitfulness *vs.* sterility. A section of *Relations to Practice*, emphasizing apple, describes the effects of various treatments on flowering and fruit setting, especially in relation to the content of nitrates, carbohydrates and moisture in the plant. Emphasis is on apple. A *Historical* section follows dealing specifically with previous studies on chemical content *vs.* fruitfulness in both herbaceous and woody plants.]

## MATERIAL AND METHODS.
### Experimental.

In the present work with tomatoes, we were interested mainly in a comparative study of the internal conditions in plants which were setting fruit, and those which were not, particularly with reference to the presence of total nitrogen, nitrates, moisture, and carbohydrates, and the relations between them. Since our work was to deal largely with the conditions within the plant, we made little effort to determine the exact quantities of nutritive or other elements in the soil in which the plants were grown, beyond a knowledge that the supply was abundant or restricted in any particular case.

**Materials.** The tomato was selected as material for investigation because of the ease with which this plant can be handled and grown under green-house conditions, because it is clean cut in its fruit-setting responses to any set of conditions imposed, and because it is readily available in any quantity.

All the plants used, with the exception of one lot, were of the Lorillard variety and were raised from seed. The seeds were first sown in loam soil, the seedlings pricked off into individual two-inch pots containing rich potting soil and when they had made a height of some three or four inches, the most uniform plants were selected for transfer to any particular set of conditions which it was desirable to employ. For the most part the plants used were stocky, actively growing, and without external evidences of flower buds when they were selected for transplanting.

For the sand cultures ordinary white quartz sand was used. This was free from organic matter but was not subjected to any particular treatment or washing before being used. It contained considerable quantities of iron. Without added nutrients the plants did not grow in it. This sand was used as a medium in which to grow plants subjected to a very low supply of nitrates. It was not desired in any case completely to eliminate them from the soil. A thin layer of cotton batting was put in the bottom of each pot before filling it with sand so as to prevent the latter from being washed through when the plants were watered. After filling the pots a porous battery cup was sunk into each, allowing about one-half inch to protrude above the surface of the sand. A large, flat cork was fitted into the opening. The nutrient solutions were poured into these cups and allowed to seep out into the sand. In this way the taking up of the nutrient solution by the pot itself and its later appearance as an efflorescense on the rim and sides, was almost entirely avoided. The roots of the growing plants formed a solid mat about the cups.

The nutrient Knop's solution employed in connection with the sand cultures was made up as follows:

A.  Magnesium sulfate...................................2%
    Dibasic potassium phosphate...................2%
    Potassium nitrate...................................2%

B.  Calcium nitrate.......................................8%

For use, equal parts of A and B were diluted one to seven with tap water and then mixed. Some precipitate was always formed, but this did not in any way seem to interfere in the use of the porous cups. Every three weeks the cups were scraped out with a scalpel and the accumulated sand and precipitate, which is always but little, dug into the sand about the plant. In some cases this was not done; not the slightest difference was observed in the plants in the two cases.

When the plants were transferred to the pots containing sand, all the soil and adhering particles of organic matter were washed from the roots with great care. The plants were then placed in the sand about one-half inch deeper than they stood originally. This is desirable because a large part of the roots present at the time of transfer die back, but new ones are produced very quickly from the stem.

The particular conditions for any particular lot or series of plants are given in connection with the analysis of the same, but there are some general conditions which will apply to all of them. No special precautions were taken regarding the general green-house conditions. So far as possible a temperature suitable for growing tomatoes was maintained, the light conditions of course varied with the season and external conditions, but care was taken that all plants in any lot or series were uniformly subjected to the same general conditions. The plants were grown in ordinary ten-inch to twelve-inch earthenware pots. Every pot, however, was placed in a granite-ware basin, which served the double purpose, first, of preventing the roots which came through the bottom of the pots from coming into contact with any soil in the benches on which the plants stood, or the leeching of any material from such soil into the pots, and second, it was possible to maintain a uniform condition of moisture in the several pots. For the most part our effort was to maintain, in each basin, when the pot was in it, a depth of about one inch of water. Of course immediately after watering this level was often higher, and sometimes it was a trifle lower; but the soil was not permitted to become dry in any case, except in the experiment where the soil was intentionally allowed to dry out. Water from the Chicago Municipal supply was used throughout all the experiments and in making up the nutrient solutions.

[Plants were cut off just above the ground and separated into upper and lower stems and leaves. Samples of tissue (100 g) were stored in 95% ethanol for subsequent analysis. Ethanol extracts were analyzed for total nitrogen, nitrate nitrogen, free reducing substances, sucrose, and starch. Stem anatomy was also examined.

A total of 4 experiments are described, treatments consisting of growing plants in various media, including soil, fresh or rotted manure, or sand culture with or without nutrient solution. Data are expressed as percentage fresh or dry weight.]

## DISCUSSION.

Several points stand out clearly after a study of the foregoing data. It is particularly interesting to note that the interrelation of nitrogenous and carbohydrate substances in the leaves themselves is very variable in the several series and that these relations are frequently quite the reverse of those in the stems. This result might be anticipated perhaps from the general knowledge of photosynthesis. It will be better in any future studies of the reserves in leaves to collect samples after the plants have had a period in the dark as well as from plants which have had several hours of exposure to sunshine.

[The results of anatomical studies are next reviewed. The xylem/pith ratio was noticeably larger in the stems of non-vegetative plants, whereas the leaves and stems were larger in vegetative plants. Effects of treatments on plastid and chlorophyll content are also discussed.]

Throughout all of our experiments the plants grown with an abundant supply of available nitrogen were distinctly vegetative and non-fruitful. These plants as a whole were higher in total and nitrate-nitrogen and lower in free-reducing sugars, sucrose, and polysaccharides than were the distinctly non-vegetative plants. Within any given plant, especially those which were grown most vigorously and rapidly, the nitrate content was generally greater in that part of the stem which was the more vegetative. When the plants were not excessively vegetative, however, the total nitrogen was higher in the more vegetative portions, but the nitrate readings were greater in that portion of the plant where the starch content was also higher. It may be remarked that there was some disagreement between the quantitative chemical analyses and the microchemical analyses for nitrates; by the latter method the greatest quantity of nitrates was always indicated in the most actively growing portion of any given stem; whereas this relation was found to be variable according to the quantitative macrochemical methods. Just how to account for this or what the significance of it may be, must be left for future investigations. The general condition of an association of higher total nitrogen and nitrates with increased vegetation is in most instances valid, especially in the comparison of the stems as a whole in the various series.

There were some wide variations in the amounts of carbohydrate present in the different types of plants. The greatest fluctuation was in the amount of free-reducing substances. These were generally highest in the stems of the less vegetative plants, when considered as a whole, but within the stems themselves they were sometimes more, sometimes less in the more vegetative portions. Disaccharides and polysaccharides were far less variable in relation to any specific vegetative conditions of the stems, either as a whole or in any given portion of it, than were the free-reducing substances. Generally an increase in polysaccharides was closely associated with an increase in disaccharides, and both were almost uniformly greater in the less vegetative plants and in the less vegetative portions of any given stem. This association of starch content and condition of vegetation is clearly indicated in figure 18. Associated with greater polysaccharide content was a greater thickening of the walls of the xylem parenchyma cells, and in stems of equal age a far greater proportion of xylem to cortex and especially to pith. This is made clear in figure 14 and the diagrams of entire stems in figures 15 to 17. Starch was always found in the starch sheath or endodermis, even in the most actively vegetative stems, but was not found in any quantity in the pith cells. Frequently in the very vegetative plants, there was no starch storage in the bases of the stems. When nitrogen was limited, that is in those cases where the plants were less vegetative, starch storage was first noticeable in the pith close to the xylem; as more and more storage took place, all of the pith cells, the medullary rays and the wood parenchyma became filled with starch grains.

Our experiments indicate that sucrose is not the first sugar formed by synthesis but that it is present only in those instances where free-reducing substances are high and have been permitted to accumulate. The general situation seems to be a graded series from free-reducing substances through sucrose to polysaccharides. Our observations, therefore, are apparently not in close harmony with those of Parkin (35) on the Snowdrop, for he has stated that in that plant sucrose is the first sugar of synthesis.

The great fluctuation in the amount of free-reducing substances present in the various types of stems may be due to a variation in the extent of their utilization as well as their synthesis, dependent upon the presence of other substances in conjunction with which still other compounds are built from them. If this were the case, it might be expected that the quantities present at any given time or location would vary directly with the degree of such utilization. At least two alternatives are conceivable, and although neither of them can be proved from the work at hand nor from the various opinions as yet expressed by various workers, still they may be suggested. In the first place, if the simpler carbohydrates do serve as one of the building stones in the synthesis of amino-acids and proteins, or if the synthesis of the latter is conditioned by the available supply of carbohydrates, as well as a suitable nitrogen supply, it might be expected that the carbohydrates would be built over into these compounds more or less rapidly according to the amount of such suitable available nitrogen, and the presence of the other necessary conditions, whatever they may be. In the second place, if a suitable nitrogen supply were not utilized in the formation of nitrogen-containing compounds but accumulated as such, then there would be a possibility for their being built into the more complex forms such as disaccharides, polysaccharides, and the like. The tomato plant does not contain or store any considerable quantity of fat, hence estimations of it were not made in our experiments. Because of the close relationship between carbohydrates and fat synthesis, however, it would seem that there was at least a good possibility that similar relations may exist in fat-storing plants.

Our own experiments give indications that the foregoing ideas on the carbohydrate transformations may be correct, for with an abundance of available nitrates in the soil, the plants themselves are relatively high in total nitrogen and nitrate nitrogen, and relatively less in carbohydrates; but when there is a limitation of the nitrates, the carbohydrates, first the simple and then the more complex, accumulate rapidly, provided of course that other conditions for photosynthesis are not prevented. When available nitrogen is added to the soil in which such nitrogen-low, carbohydrate-high plants are growing, however, they very quickly increase in total nitrogen and nitrate-nitrogen content, and become actively vegetative. Associated with such a change is a decrease in the same complex carbohydrates. Microscopic examinations were made of the plants in series O before transferring to a soil abundant in available nitrogen and it was found that the cells of the pith, cortex, and medullary rays and even those of the xylem parenchyma were packed with starch grains. This was true for sections taken up to within one centimeter of the tip. Within three days following such transfer, the beginning of the disappearance of the starch grains from the center pith cells and cortical cells at the tips of the plants was very

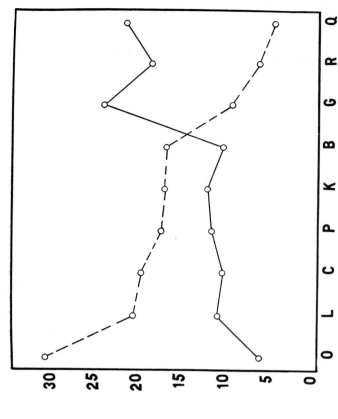

Fig. 19. Diagram to show the comparative quantitative relationships of the total carbohydrates (connected by broken line), and total nitrogen content x 7 (connected by solid line) arranged on the basis of the descending values for carbohydrates, in the upper stems of the several series.

noticeable. Successive examinations as growth progressed showed an active terminal elongation which contained no storage starch except in the starch sheath, an active development of secondary xylem in the older portion of the stem, and a very rapid, progressive, and finally complete disappearance of the starch from the pith and xylem parenchyma and also the cortical cells even down to the bases of the stems, where it was last to disappear. It may be added that some stems which were thus packed with starch were not given additional nitrates. They finally lost all their leaves except two or three at the very tip about one or two centimeters long. These stems remained alive for over seven months during which time there was a gradual disappearance of the starch in some of them until only traces in the medullary rays and pith could be demonstrated, while some of the others contained large starch reserves at death. Even after this long period a few of these old, yellowed, leafless, apparently dead stems put out new buds at a few of the nodes when calcium nitrate solution was applied to the sand in which they were growing. Every one of the plants which sprouted still contained carbohydrate reserves.

[A discussion follows on the regrowth of severed plants and effects on chemical constituents.]

The accompanying diagrams, figures 19 to 22, show the relation between the percentage of total nitrogen and the percentage of total carbohydrates (free-reducing substances plus sucrose plus polysaccharides) expressed as dextrose. It should be borne in mind that the free-reducing substances, sucrose, and polysaccharides are not absolute determinations, but that these terms are used with the significance given under the methods of determination in an earlier part of this paper. On the base line of the figure at equal distances apart are arranged the series of plants and on the vertical lines are arranged the percentages of total nitrogen multiplied by seven and of total carbohydrates, expressed on the dry weight. On account of the wide differences in composition of different parts of any plant grown under a given set of conditions, only similar portions are compared. With but few exceptions increased amounts of total nitrogen are associated with decreased amounts of total carbohydrates. This condition holds fairly uniformly throughout the plant with the exception of the lower leaves.

This relation between total nitrogen and carbohydrate storage may be due to any one or a combination of reasons, some of which are the following: (1) The presence of the nitrogenous compounds or nitrates may retard assimilation or the formation of the carbohydrates. (2) It may cause increased respiration of the carbohydrates. (3) It may aid in the utilization of the carbohydrates for the synthesis of organic nitrogenous substances. No definite, exact data on any one of these points are available. It is not worth while, therefore, to attempt conclusions concerning them, though a few suggestions may not be out

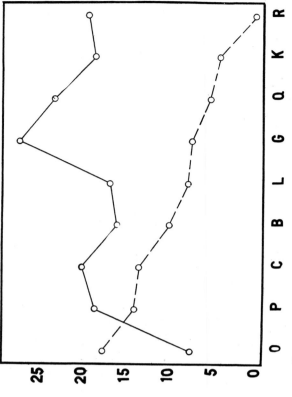

Fig. 21. Same as Fig. 19 except to show the relationships in the upper leaves.

Fig. 22. Same as Fig. 19 except to show the relationships in the lower leaves.

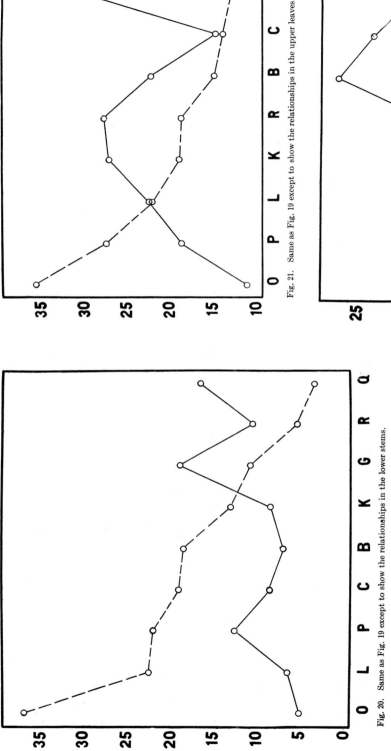

Fig. 20. Same as Fig. 19 except to show the relationships in the lower stems.

of place. The much greater leaf area developed by vegetative plants would seem to indicate the reverse of the first proposal, nor does the presence of increased amounts of carbohydrates in the non-vegetative plants of necessity indicate that they are therefore likewise synthesized in greater quantities. Evidence for or against the second point is not clear, but in keeping with the general findings of increased respiration accompanying more active growth there is a probability that more of the carbohydrates would be thus used in the vigorously vegetative plants. The third possibility has been previously suggested. The utilization of the carbohydrates in this manner as well as in the composition of portions of the walls of the new cells being formed and the thickening of others, probably affords the main reason why they are found as storage substances in relatively smaller quantities in the more actively growing stems.

In general, there is a close correlation between the amount of nitrate nitrogen, total nitrogen, and moisture. Among others, the several factors which follow might aid in accounting for this. (1) The nitrates may have a lyotropic effect in increasing the water-holding capacity of the plant. (2) Carbohydrates and dry matter, substances which have a relatively lower water-holding capacity, are greater where total nitrogen is less. (3) The nitrates may prevent the lignification and thickening of cell walls which have a relatively low water-holding capacity. (4) They may aid in rapid growth and the formation of new cells which have relatively thinner walls and a greater percentage of amphoteric substances whose water-holding capacity is relatively large. Then, too, the vacuoles are generally more numerous and larger in the decidedly vegetative tissues, and these may furnish more opportunity for the retention of water.

In the absence of conclusive evidence which might show that the lyotropic action of the nitrates is of significance, no definite conclusions can be drawn. The plants which constituted series O were grown in sand and the nitrates of the nutrient Knop's solution were eliminated and partly substituted by calcium chloride. Even with the presence of the chloride ion, which has a lyotropic effect somewhat similar to the nitrate ion, the plants were very low in moisture. Of course since no quantitative chlorine determinations were made there is no way of comparing the quantities within the plant, and also the presence of the calcium ion may overshadow the effect of the chloride ion. Microchemical tests indicated an abundance of chloride in all types of plants. The second and third points are self explanatory. There were, however, no specific experiments on the influence of vacuoles on moisture-holding capacity, but the effect of protein-like substances in this regard is fairly well established. Microscopic examinations showed a lesser increase in cellular thickenings in the vegetative stems.

From the investigations of others as well as our own, it has certainly been shown that blooming, pollination, or even fertilization do not necessarily assure actual fertility even in plants actually considered self-fertile, and it would appear that at least some cases of self- or even inter-sterility are due, not so much to a stable hereditary character as to the condition of the nutrition of the plant under investigation. Both heredity and nutrition must be taken into consideration in a study of this problem, and while it is possible profoundly to modify the expression of any particular plant dependent upon the conditions imposed, it may well be argued that such modifications still remain within hereditary limits. Just where such limits can be drawn certainly cannot, as yet, be determined off-hand, and much more than the average or so-called normal conditions must be investigated.

In general, the observed results and the analyses made in connection with the foregoing experiments tend to support our proposed classification of vegetative and reproductive tendencies insofar as they may be based on a relationship of the carbohydrate and nitrogenous compounds. Throughout the investigation, many questions naturally have suggested themselves; a few of them have been indicated. We hope that further and more extended investigations may be instituted and conducted not only to establish or deny the general hypotheses proposed, but to furnish accurate and reliable data on which to base interpretations of the more intimate processes and compounds concerned.

# SUMMARY

1. Plants grown with an abundant supply of available nitrogen and the opportunity for carbohydrate synthesis, are vigorously vegetative and unfruitful. Such plants are high in moisture, total nitrogen, nitrate nitrogen, and low in total dry matter, free-reducing substances, sucrose, and polysaccharides.

2. Plants grown with an abundant supply of nitrogen and then transferred and grown with a moderate supply of available nitrogen are less vegetative but fruitful. As compared with the vegetative plants, they are lower in moisture, total nitrogen, and nitrate nitrogen, and higher in total dry matter, free-reducing substances, sucrose, and polysaccharides.

3. Plants grown with an abundant supply of nitrogen and then transferred and grown with a very low supply of available nitrogen are very weakly vegetative and unfruitful. As compared with the vegetative plants, they are very much lower in moisture and total nitrogen and are lacking in nitrate nitrogen; they are much higher in total dry matter, free-reducing substances, sucrose, and polysaccharides.

4. When plants which have been grown with a large supply of available nitrogen and moisture are subjected to a reduced moisture supply just about the wilting point there is a decrease in vegetative activity. These plants compared with those which are vigorously vegetative, are lower in total nitrogen and nitrate nitrogen and higher in free reducing substances, sucrose, and polysaccharides.

5. Whatever the conditions under which a plant has been grown, considering the whole plant as a unit, increased total nitrogen and more particularly increased nitrate nitrogen are associated with increased moisture and decreased free-reducing substances, sucrose, polysaccharides, and total dry matter.

6. Fruitfulness is associated neither with highest nitrates nor highest carbohydrates but with a condition of balance between them.

7. There is a correlation between moisture content and nitrate nitrogen. This is probably due largely to the preponderance of non-carbohydrate materials to carbohydrates in the cases where nitrates are abundant.

8. In general, within the plant itself, in the stem from the top to bottom, there is a descending gradient of total notrogen and moisture, and an ascending gradient in total dry matter, polysaccharides and sucrose. The proportion of free-reducing substances to other carbohydrates, total nitrogen, and nitrate nitrogen is variable.

9. The great variations in the amount of carbohydrates in plants grown under different nutrient conditions and in different parts of the same plant indicate that in studying problems concerned with plant metabolism it is necessary to know the specific environment of the plant as a whole and of its several parts.

10. The conditions for the initiation of floral primordia and even blooming are probably different from those accompanying fruit setting. The greatest number of flowers are produced neither by conditions favoring highest vegetation nor by conditions markedly suppressing vegetation.

11. Lack of fruit development is not alone due to the lack of pollination or fertilization. The flowers may fall soon after pollination (markedly vegetative plants) or remain attached for many days without development of the fruit (markedly non-vegetative plants).

The effectiveness of the nitrogen value of leguminous cover crops is dependent upon the accompanying moisture supply.

23. Cultivation is largely effective in conserving moisture and in promoting the supply of available nitrogen. If in any given soil, moisture and available nitrogen are already present in quantities such that the plants growing upon it are largely vegetative, a decrease in cultivation will tend towards fruitfulness.

24. Non-leguminous companion crops or cover crops remove from the soil both available nitrogen and moisture. In regulating vegetation and fruitfulness by this means the relations of the available moisture, nitrogen, and carbohydrates largely determine the result.

25. Pruning is largely effective in promoting or retarding fruitfulness by its effects in balancing the carbohydrate supply within the plant, or the means for its manufacture, with the available moisture and nitrogen supply.

26. Girdling or ringing of the cortex or bark is effective through a modification of the carbohydrate-nitrate relationship. In practice the entire range of effects due to such a relationship may be expected from its application.

27. Fruit production is seemingly a specialized vegetative function usually more or less closely associated with the function of gametic reproduction. Parts concerned in reproduction range from but little-modified vegetative parts to those highly modified portions classified as fruits. The degree in which such modification is expressed, is dependent upon physiological changes within any specific plant, and may vary widely within the same variety or even the same individual.

28. At least some of the instances of sterility considered to be the result of physiological incompatibility may be due to the state or condition of nutrition of the plant itself.

29. Until more exact information is available, both environmental and hereditary factors must be considered in any attempted explanation of the reproductive or vegetative behavior of plants.

12. The tomato stem in cross section is made up of an epidermis from which arise glandular hairs, several layers of cortical cells, endodermis, a more or less interrupted layer of bast cells, the phloem with small patches of sieve cells, primary and secondary xylem, small patches of internal phloem and internal xylem separated from each other and the protoxylem of the outer bundles by pith cells, and lastly the pith.

13. Vigorously vegetative stems are much greater in diameter than those which are feebly vegetative. This is due to the greater number and size of the pith cells in the former and is accompanied by a marked proportional reduction in xylem. The collenchyma of the cortex is much less and the walls of the bast and internal xylem much thicker in feebly vegetative stems than in those which are vigorously vegetative.

14. Starch is present in the starch sheath of all stems. Starch storage in the stems begins first in the pith cells near the primary vascular bundles, then extends throughout the pith, xylem, and the cortical cells.

15. In vegetative stems there is a much greater number of chloroplasts. These are present even in the central cells of the pith. In stems very feebly vegetative there are no observable chloroplasts in the pith and their number and intensity of coloration is greatly reduced both in the cortex and in the leaves.

16. Stems without storage starch at the base when cut off close to the surface of the soil, fail to sprout but decay quickly, whereas those with large storage produce new shoots. Accompanying such growth there is a total or complete disappearance of the starch, depending upon the relative amount of growth made and the available nitrogen supply. If the latter is abundant vegetative extension is relatively great; if not, such extension soon ceases and starch is again stored in the new growth.

17. The available carbohydrates or the possibility for their manufacture or supply, constitute as much of a limiting factor in growth as the available nitrogen and moisture supply. When the opportunity for carbohydrate manufacture within the plant itself is greatly reduced or eliminated even though there is a relative abundance of moisture and available nitrogen, vegetation is decreased. But when there is a carbohydrate reserve within the tissues under the same conditions of nitrogen and moisture supply, growth is active. Very large proportional reserves of carbohydrates to moisture and nitrate supply, also accompany decreased vegetation.

18. Parts of the stems or cuttings of plants with a large amount of storage carbohydrates and particularly those parts where such storage is localized, when supplied with moisture or moist conditions, produce roots abundantly. This would be of particular interest in vegetative propagation.

19. Microchemical tests indicate very little difference in potassium content of individual cells whatever the condition of the plant.

20. Withholding moisture from plants grown under conditions of relative abundance of available nitrogen results in much the same condition of fruitfulness and carbohydrate storage as the limiting of the supply of available nitrogen itself.

21. Fertilizers containing available nitrogen or that which may be made available, are mainly effective in producing vegetative response. They may either increase or decrease fruitfulness, according to the relative available carbohydrate supply.

22. Irrigation or moisture supply is effective in increasing growth or fruitfulness only when accompanied by an available nitrogen supply and vice versa.

W. W. Garner and H. A. Allard. 1920. Effect of the relative length of day and night and other factors of the environment on growth and reproduction in plants. Journal of Agricultural Research 4(11):553–606.

W. W. Garner

H. A. Allard

One of the most fundamental discoveries in biology was first reported in 1920 by U.S. Department of Agriculture (USDA) plant physiologists W. W. Garner and H. A. Allard. This was photoperiodism, or dependence of plants, as well as animals, on the length of day and was to have a major influence on plant science and particularly on horticulture. Garner and Allard, working mainly at USDA's Arlington Experimental Farms in Virginia, continued reporting their research reports about photoperiodism into mid-1940. In midsummer (June 29) 1935, the U.S. Congress ordained the Bankhead–Jones Act, providing for expansion of in-depth research in agriculture. Under that act, in 1936, the USDA created a small group at Beltsville, Maryland, to look further into the nature of photoperiodism and its significance in agriculture. Because the photoperiodic response in plants partly regulates flowering, it was thought that progress might best result from attention to the morphological aspects involved in plants changing from vegetative to reproductive growth and to the underlying physiology. University of California, Davis, botanist H. A. Borthwick was recruited to lead the work. Of course, Borthwick knew and counseled with department peers Garner and Allard. In 1956, Borthwick, in his presidential address to the American Society of Plant Physiologists about "Light in Relation to Growth and Development," characterized Garner and Allard and their impact. Borthwick's unpublished address (presented in italics below) included the following verbatim remarks concerning these men and their work.

*Few papers in plant physiology have been more frequently cited and quoted than their 1920 paper in the Journal of Agricultural Research. This work opened up an entirely new phase of developmental physiology, all aspects of which have engaged plant physiologists in increasing numbers throughout the world ever since. Immediately upon its discovery and with almost no understanding of its mechanism, photoperiodism became a useful tool in the hands of plant breeders, plant pathologists, and other botanists. It particularly found quick and constantly increasing commercial application in floriculture, principally in the chrysanthemum industry but to some extent in production of other ornamental plants.*

*Garner and Allard started their first photoperiod experiment July 10, 1918, and their first paper bears the date of March 1, 1920. During that brief interval they not only repeated their original experiments with soybean and tobacco several times but also investigated more than a dozen and a half other plants—enough to convince them that the phenomenon they had discovered was one of widespread occurrence. In the years immediately following, they added large numbers of plants to the list of those known to be photoperiodically sensitive. During the period of their active investigation of photoperiodism, which extended from 1918 to 1945 when Dr. Garner retired, they published a score or more of papers on the subject.*

*Although the names of Garner and Allard have come to be synonymous with photoperiodism, each of these men had earlier distinguished himself in an entirely different field. Dr. Garner was an authority on the culture and physiology of the tobacco plant and Dr. Allard had become widely known for his fundamental discoveries on the insect transmission of plant viruses.*

*The diversity of interests and experience of the two men no doubt contributed greatly to their effectiveness as a research team. Dr. Garner gave emphasis to basic plant physiology, whereas Dr. Allard was presumably the one who stressed the ecological and plant distributional significance of the new discovery.*

*It was my privilege to have become acquainted with both of these men; somewhat better with Dr. Allard than with Dr. Garner. Dr. Garner's training was in chemistry, having been a student of Remsen at Johns Hopkins. In 1908 he was placed in charge of the Office of Tobacco and Plant Nutrition Investigations. In that unit he initiated basic studies on methods of tobacco curing, on the mineral nutrition of tobacco, on the occurrence of the alkaloid nicotine, and later on the details of the photoperiodic reaction. He was a capable administrator, who inspired loyalty and devotion on the part of his associates to the organization in which they worked and to him personally. Despite his administrative duties, he found time to participate in research himself and at the time of his retirement in 1945, he had earned world recognition as a leader in the physiology and culture of the tobacco plant and as a co-discoverer of photoperiodism.*

*Dr. Allard is a naturalist—a man of great curiosity and keen imagination concerning all kinds of biological phenomena of nature that surround us. We are* apt to think of him as primarily a botanist, but his biological interests range over the entire field of living organisms. He has been concerned from time to time and has published on such a diversity of subjects outside the fields of his official research as the musical notes of insects, the relation of light to the timing of bird song in the morning, the migration of birds, and the ratios of clockwise and counterclockwise spirality in the phyllotaxy of some wild plants. I am told that he currently has some turtles that he has trained to come at his call. He at one time became interested in crickets. He observed that for a particular species occurring in the eastern part of the United States the frequency of chirping was dependent on temperature. He accordingly plotted the chirping frequency of the crickets as a function of temperature and was thereafter able to make reasonably accurate estimates of temperature with his watch.

Garner and Allard's first paper appeared before the discovery of photoperiodism in animals, but they called attention to the possibility that animals might be found to exhibit it. Garner and Allard had both been interested in bird migration and had wondered about the regularity of timing of such events from season to season; so it is not surprising to find the following comment in their paper:

> As to animal life nothing definite can be said, but it may be found eventually that the animal organism is capable of responding to the stimulus of certain day lengths. It has occurred to the writers that possibly the migration of birds furnishes an interesting illustration of this response. Direct response to a stimulus of this character would seem to be more nearly in line with modern teachings of biology than are theories which make it necessary to assume the operation of instinct or volition in some form as explaining the phenomena in question.

Marcovitch, an entomologist in Tennessee, was trying to understand the cause of the very clearly defined seasonal sequence of different forms of the strawberry root louse at this time. He learned of Garner and Allard's results with plants and in them he saw a possible explanation for his own observations. He accordingly set up a photoperiod experiment with this organism in May 1922, and by September of that year he had demonstrated that the appearance of sexual forms in late autumn was regulated by photoperiod, not temperature, as was previously supposed. This piece of work, the inception of which must be attributed to Garner and Allard, was thus among the first to demonstrate the existence of the photoperiodic phenomenon in animals.

After a phenomenon such as control of flowering by daylength has been pointed out, one sometimes wonders how he could have missed seeing it himself. He wonders even more that all the previous generations of people interested in plants also failed to see it. In the case of photoperiodism here were, in fact, many near misses before Garner and Allard came along to find it. One notable example was that of L. H. Bailey, who in 1890 set up an experiment at Cornell University to study the potentialities of Edison's new electric light on plants. Among other plants he grew spinach in the greenhouse and during the nights subjected some plants to natural darkness and others to electric light. The lighted plants flowered and the unlighted ones made vigorous vegetative growth. Bailey reported his results in Bulletin No. 30, 1891, entitled "Some preliminary studies of the *influence of the electric arc lamp upon greenhouse plants." In this report he states:*

> It is impossible to draw many definite conclusions from the above researches. . . . Yet there are a few points which are clear: the electric light promotes assimilation, it often hastens growth and maturity, it is capable of producing natural flavors and colors in fruits, it often intensifies colors of flowers and sometimes increases the production of flowers. The experiments show that periods of darkness are not necessary to the growth and development of plants. There is every reason, therefore, to suppose that the electric light can be profitably used in the growing of plants.

*Bailey failed to appreciate the most important result of his experiment, so photoperiodism remained unknown for another 30 years.*

*When Garner and Allard's paper was submitted for publication, the manuscript was subjected to the editorial review of a group of scientists. Their duty was to pass on its scientific significance and acceptability. That it passed the test is obvious, but that it had rough going is perhaps not generally known. One man on the committee was unalterably opposed to it.*

*Dr. Allard once read me a few paragraphs from the memorandum written by that man. Never have I heard a more scathing and sarcastic attack on a piece of scientific work. The name of the man, who is no longer living, is of no importance here. I merely mention the incident because of its historic interest and because it illustrates so well that even though we might not have the vision to comprehend fully the significance of a new discovery, we should at least be tolerant of it. It might turn out to be important.*

*For a period of 10 or 15 years after the discovery of photoperiodism, work in the field was characterized largely by studies that described the photoperiodic requirements for flowering of a great many kinds of plants. By the late 30s, which was about the time the group with which I have been associated was organized, questions of the cause and mechanism of the reaction were beginning to receive increasing attention. The photoperiodic response, however, was still identified primarily with flowering despite the fact that Garner and Allard, in some of their first papers, had shown that various vegetative responses of plants were also controlled by photoperiod. The significance of the dark period to the response was beginning to be appreciated as was also the importance of the leaf as the perceptive organ. Grafting experiments had indicated that the ultimate cause of flowering in both long- and short-day plants, as well as in day-neutral ones, was the same. The term "florigen" had been invented and many workers were engaged in unsuccessful attempts to isolate it.*

About 1950, a pigment that controlled photoresponsive development of plants was identified and during the subsequent decade acquired the sobriquet phytochrome. Much has been learned of the structure and function of phytochrome during the 30 years since its initial physiological discovery. Much of this information may be found in an article by Shropshire and Mohr (1983). This photoregulatory pigment, now known to be a relatively large chromoprotein, has been purified from and studied in a large number of plant tissues. Current evidence suggests that phytochrome functions as a molecular dimer, having two identical chromoproteins per molecule. The photoreversibility and other photochemical properties of isolated phytochrome were found to be very similar to those predicted by Borthwick and Hendricks. Genes for the synthesis of phytochrome have been isolated and sequenced, providing the amino acid sequence of the protein. This information and that from other physical and immunochemical studies will soon provide knowledge of the three-dimensional structure of the molecule necessary for an understanding of its function.

Many additional characteristics of the behavior of phytochrome are now known and a few will be mentioned here. In etiolated tissues, the active, far red-absorbing form of phytochrome is rapidly destroyed and its presence greatly reduces the rate of phytochrome synthesis. As a result, the amount of phytochrome in light-grown green tissue is only a few percent of that in dark-grown seedlings. Immunochemical studies indicate that phytochrome in etiolated and green tissues may be distinctly different protein species. Phytochrome functions over a very broad range of light intensities and three types of responses in relation to light quantity have been distinguished: (1) induction/reversion responses at low photon fluences, such as characterized by Borthwick and Hendricks; (2) high irradiance responses, where under continuous irradiation the responses appear related to the rate of cycling in the photoreversible transformation of phytochrome; and (3) very low fluence responses, in which response sensitivities to light and to phytochrome photoconversion may be increased as much as 10,000-fold by temperature, hormonal, and other factors.

The primary action of phytochrome remains as yet unknown. The molecule possesses no known enzymatic activity. Current evidence suggests that it functions in association with a membrane, possibly that surrounding the cell. Here changes in electrical charge and in the transport of calcium appear to be closely linked to the action of phytochrome.

Summaries of recent advances in studies of the photoperiodic control of flowering are available in a book by Vince-Prue (1984). Although behavioral aspects of flowering processes and their complex interaction with light, temperature, and other factors have been characterized extensively, the processes and mechanisms of the control of flowering are still poorly understood. Progress in photoperiodism research has been limited in part by the universal difficulty of understanding the mechanism of the endogenous biological clock. Advances in membrane biology now provide an avenue for more rapid progress in this area. Extensive effort, as yet unrewarded, has been given to the identification of transmissible factors that appear necessary to mediate the control of flowering. Much present research emphasis is given to the photoperiodic regulation of the distribution of photosynthetic products throughout the plant. The possible direct involvement of such regulation of photosynthate partitioning in the photoperiodic control of flowering is one of the promising directions of current research.

Albert A. Piringer
Agricultural Research Service
U.S. Department of Agriculture
Beltsville, Maryland 20705

## REFERENCES

ALLARD, H. A., and W. W. GARNER. 1940. Further observations of the response of various species of plants to length of day. U.S. Dep. Agr. Tech. Bull. 727.

BAILEY, L. H. 1891. Some preliminary studies of the influence of the electric arc lamp upon greenhouse plants. Cornell Univ. Agr. Expt. Sta. Bull. 30.

BORTHWICK, H. 1972a. History of phytochrome. p. 4–23. In: K. Mitrakos and W. Shropshire, Jr. (eds.). Phytochrome. Academic Press, New York.

BORTHWICK, H. 1972b. The biological significance of phytochrome. p. 27–44. In: K. Mitrakos and W. Shropshire, Jr. (eds.). Phytochrome. Academic Press, New York.

CATHEY, H. M., and L. E. CAMPBELL. 1980. Lights and lighting systems for horticultural plants. Hort. Rev. 2:491–537.

MARCOVITCH, S. 1924. The migration of the Aphidiae and the appearance of the sexual forms as affected by the relative length of daily light exposure. J. Agr. Res. 27:513–522.

SALISBURY, FRANK B. 1982. Photoperiodism. Hort. Rev. 4:66–105.

SHROPSHIRE, W., JR., and H. MOHR. 1983. Photomorphogenesis. In: A. Pirson and M. H. Zimmermann (eds.). Encyclopedia of plant physiology. New series, Vol. 16. Springer-Verlag, Berlin.

VINCE-PRUE, D., B. THOMAS, and K. E. COCKSHULL. 1984. Light and the flowering process. Academic Press, New York.

# JOURNAL OF AGRICULTURAL RESEARCH

VOL. XVIII     WASHINGTON, D. C., March 1, 1920     NO. 11

## EFFECT OF THE RELATIVE LENGTH OF DAY AND NIGHT AND OTHER FACTORS OF THE ENVIRONMENT ON GROWTH AND REPRODUCTION IN PLANTS

By W. W. GARNER, *Physiologist in Charge*, and H. A. ALLARD, *Physiologist, Tobacco and Plant Nutrition Investigations, Bureau of Plant Industry, United States Department of Agriculture* [1]

### INTRODUCTION

The importance of the relationships existing between light and plant growth and development has been so long recognized and these relationships have been of so much interest to investigators that a very extensive literature on the subject has been developed. For present purposes it will not be necessary to attempt even a brief review of this literature, and only some of the leading features bearing upon the particular problems in hand need to be touched upon. For more extended discussions of the work in this field the monographs of MacDougal (*18*)[2] and Wiesner (*26*) may be consulted. Three primary factors enter into the action of light upon plants—namely, (1) the intensity of the light, (2) the quality, that is, the wave length of the radiation, and (3) the duration of the exposure. Most phases of these three factors have been more or less extensively investigated. In the present investigation we are concerned chiefly with the general growth and development of plants and the reproductive processes as affected by the daily duration of the light exposure.

As regards intensity, it seems to be pretty well established that there is an optimum for growth in each species and that for many species this optimum is less than the intensity of the full sunlight on a clear day. Within limits, reduction in light intensity tends to lengthen the main

[1] The authors desire to acknowledge their indebtedness to Prof. C. V. Piper, in charge of Forage Crop Investigations, Bureau of Plant Industry, for helpful suggestions, to Mr. W. J. Morse, of the Office of Forage Crop Investigations, Bureau of Plant Industry, for seed of certain varieties of soybeans and information as to the characteristics of these varieties, to Dr. D. N. Shoemaker, of the Office of Horticultural and Pomological Investigations, Bureau of Plant Industry, for similar assistance as to certain varieties of ordinary beans, and to Prof. H. H. Kimball, of the Weather Bureau, United States Department of Agriculture, for important data relating to the shading effects of nettings of different mesh used in these investigations.

[2] Reference is made by number (italic) to "Literature cited," p. 605-606.

Journal of Agricultural Research,
Washington, D. C.

Vol. XVIII, No. 11
Mar. 1, 1920
Key No. G-186

axis and branches and to increase the superficial area of the foliage of many species. Also, the thickness of the leaf lamina may be reduced, and there may be marked departures from the normal in internal structure, the tendency being toward a less compact structure. So far as is known, no important general relationships between differences in light intensity and reproductive processes have been experimentally demonstrated.

The comparative effects produced by different regions of the spectrum, including the ultra-violet, have been extensively investigated but with more or less conflicting results. The most extensive investigations on the subject, perhaps, have been made by Flammarion (8). It was found that there is abnormal elongation of the principal axis in several species under the influence of the red rays, while growth is markedly reduced under the green and especially under the blue rays. In some plants, however, such as corn, peas, and beans, growth is greatest in white light. Some plants blossomed considerably earlier in red light than in white. White light produced the greatest weight of dry matter. Leaves of Coleus developed decided differences in color patterns under differently colored lights. In subsequent work Flammarion has extended his studies to a large number of species.

The duration of the daily exposure to light needs to be considered in three separate phases—(1) continuous illumination throughout the 24-hour period, (2) continuous darkness throughout, leading to the phenomena of etiolation, and (3) illumination for any fractional portion of the 24-hour day. Under natural conditions continuous sunlight throughout the 24-hour period occurs, of course, only in very high latitudes. Schübeler (23) observed the behavior of several species transported from lower latitudes and grown in northern Scandinavia under continuous sunlight lasting for a period of two months. In the species under observation the vegetative period was shortened and the seeds produced were larger than the normal. It is stated, also, that there was an increased formation of aromatic and flavoring constituents. Another method of securing continuous illumination consists in the use of artificial light for illumination or in the supplementing of normal daylight with artificial light, though, of course, the quality and the intensity from the two sources will not ordinarily be the same. Using electric light alone, of an intensity one-third that of sunlight, Bonnier (6) observed a marked increase in chlorophyll formation which extended inwardly to unusual depths. He found also incomplete differentiation of the tissues, recalling, in this respect, the effects of continued darkness. In some instances the color of blossoms was deepened.

Etiolation, resulting from exposure to continuous darkness, has been the subject of much study. In this connection special mention should be made of the work of MacDougal (18) covering a very large number of species. This author also presents a comprehensive survey of previous work on the subject.  In most instances stems, and frequently leaves, exhibited negative geotropism in the absence of light.  In all species investigated etiolated tissues show a lesser degree of differentiation than the normal.  In this connection MacDougal points out that the differences exhibited between etiolated specimens and normal plants demonstrate the fact that growth, or increase in size, and development, or differentiation, are distinct processes capable of separation. For present purposes perhaps his most important observation is that in no plant investigated had the stamens and pistils attained functional maturity.

The effects of differences in the length of the daylight period, the subject of the present study, have not been so extensively investigated as most other phases of light action.  Obviously the problem may be approached in any one of four ways: by comparing the behavior of plants when propagated in different latitudes, by growing plants at different seasons of the year in the same latitude, by supplementing the daylight period with artificial light, and by preventing light from reaching the plant for a portion of the normal daylight period.  In the records of attempts to grow various plants in different parts of the world there are undoubtedly a great deal of available data bearing on the present problem; but apparently no systematic effort has been made to utilize this material, the reason probably being that the importance of the relative length of day in affecting plant processes, and, in particular, reproduction, has not been appreciated.  Bailey (*3, 4, 5,*) carried out an extensive series of tests in which daylight illumination was supplemented by the electric arc light applied for different portions of the night.  The addition of the artificial light induced blossoming and seed formation in spinach.  The additional light also favored the growth of lettuce.  Rane (*20*), using the incandescent filament electric light, and Corbett (*7*), employing incandescent gas light, observed that certain flowering plants and some vegetables blossomed somewhat earlier when the normal daylight illumination was supplemented with artificial light.  In most of these tests the artificial light was applied for the entire night, but apparently the results so far as concerns reproduction were essentially the same as when the plants were darkened for a portion of the night.  Tournois (*24, 25*) has reported the results of an interesting experiment with hemp (*Cannabis sativa* L.) and a species of hops (*Humulus japonicus* Sieb. and Zucc.) in which these plants were exposed to sunlight only from 8 a. m. to 2 p. m. daily.  It had been shown by Girou de Buzareingues (*10*) as early as 1831 that when planted in the late winter or very early spring months the hemp plant first develops in the spring a number of abnormal sterile blossoms in the leaf axils and later produces normal flowers at the regular blossoming period.  Following up this fact Tournois concludes from the above-mentioned experiment that the abnormal blossoming period is induced by the short length of day prevailing in the early spring months.

In a few words, previous work on light action clearly indicates that permanent exclusion of light effectually prevents completion of the blossoming and seed-forming processes, while in certain cases lengthening the normal daily period of illumination by the use of artificial light or by propagation in far northern latitudes hastens the approach of the blossoming period, and, in the case of two species, shortening the daily exposure to light induces the formation of precocious blossoms. That the relative length of the day is really a dominating factor in plant reproduction processes, as is demonstrated in the present paper, seems not to have been suspected by previous workers in this field.

## PRELIMINARY OBSERVATIONS

In 1906 there were observed in a strain of Maryland Narrowleaf tobacco (*Nicotiana tabacum*, L.), which is a very old variety, several plants which grew to an extraordinary height and produced an abnormally large number of leaves. As these plants showed no signs of blossoming with the advent of cold weather, some of them were transplanted from the field to the greenhouse and the stalks of others were cut off and the stumps replanted in the greenhouse. These roots soon developed new shoots which blossomed and produced seed, as did also the plants which had been transferred in their entirety. This very interesting giant tobacco, commonly known as Maryland Mammoth, which normally continues to grow till cold weather in the latitude of Washington, D. C., without blossoming, proved to be a very valuable new type for commercial purposes, but the above-mentioned procedure has been the only method by which seed could be obtained. The type bred true from the outset, and no matter how small the seed plant the progeny have always shown the giant type of growth when propagated under favorable summer conditions. It may be remarked at this point that inheritance of gigantism[1] in this tobacco has been studied by one of the present writers (2) and it has been shown that this character acts as a simple Mendelian recessive.

On one occasion it was observed that seedlings of the Mammoth transplanted to 8-inch pots in late winter blossomed in early spring after reaching a height of some 3 feet and developed an excellent crop of seed. From this it was at first concluded that growing the plant under conditions of partial starvation would induce blossoming, but this idea proved to be erroneous. Repeated attempts during the summer months to force blossoming by subjecting the plant to conditions which would permit only limited growth were futile. On the other hand, it was found that

[1] Throughout this paper the term gigantism is used to signify a tendency toward more or less indefinite vegetative activity manifested by plants under certain favorable environmental conditions. Though an inherited characteristic, it may come into expression only under definite conditions of environment; and the present investigation seems to make it clear that the length of the daily light exposure is the controlling factor.

seedlings grown in the greenhouse during the winter months invariably blossomed without regard to the size of the pot containing the seedling or the extent to which the plant was stunted by unfavorable nutrition conditions. The seedlings behaved, therefore, like the summer-grown giant plants which were transferred to the greenhouse late in the fall. Finally, it was observed that the shoots which were constantly developing from the transplanted roots of giant plants transferred to the greenhouse blossomed freely during the winter months, but as early spring advanced blossoming soon ceased and the younger shoots once more developed giant stalks. Obviously, then, the time of year in which the Mammoth tobacco develops determines whether the growth is of the giant character. During the summer months the plants may attain a height of 10 to 15 feet or more and produce many times the normal number of leaves without blossoming, while during the winter months blossoming invariably occurs before the plants attain a height of 5 feet. Naturally it became of interest from both a practical and a scientific standpoint to determine the factor of the environment responsible for the remarkable winter effect in forcing blossoming. It may be added just here that gigantism also has been observed in several distinct varieties of tobacco other than the Maryland—namely, in Sumatra, Cuban, and Connecticut Havana.

Again, in following out an investigation on the relation of the nutrition conditions to the quantity of oil formed in the seeds of such plants as cotton, peanuts, and soybeans, the present writers (9) had occasion to investigate the significance of the observation made by Mooers (19), that successive plantings of certain varieties of soybeans (*Soja max* (L.) Piper) made through the summer months, show a decided tendency to blossom at approximately the same date regardless of the date of planting. In other words, the later the planting the shorter is the period of growth up to the time of blossoming. In the course of the investigation on oil formation it became desirable to study the possible effects of temperature differences on the process. Since it is much simpler and cheaper to maintain temperature differences during the winter by the use of heat than during the summer by means of refrigeration, it was planned to make some tests with soybeans during the winter. It was soon found, however, that the plants began to develop blossoms before they had made anything like a normal growth, and the few blossoms produced were cleistogamous, so that it became necessary to abandon the plan of conducting the tests in question during the winter months. As is the case with the Mammoth tobacco, the time of year in which the plants are grown exerts a very profound influence on growth and reproduction in the soybean.

In seeking a solution of the problem as to why the behavior of these plants is radically different from the normal during the fall and winter months one naturally thinks of light and temperature as possible factors. It was observed, however, that both the Mammoth tobacco and the soybeans still showed the abnormal behavior in the winter even when the temperature in the greenhouse was kept quite as high as prevails out of doors during the summer months. This observation seemed to dispose of temperature as a possible factor of importance in the "winter effect." It is clear that the quantity of solar radiation received by plants is less in winter than in summer, for both the number of hours of sunshine per day and the intensity of the light are reduced during the winter months. The quality of the light also is affected, since the angle of elevation of the sun's path during the winter is less than during the summer and the selective absorptive action of the atmosphere comes into play. It happened that in the investigation on oil formation in seeds a number of experiments had been made with soybeans to determine the effect of light intensity on this process and, incidentally, it was observed that in no case was the date of blossoming materially affected by the intensity of the light. It had been found, also, that partial shading was without decided effect on the blossoming of the Mammoth tobacco. In view of these experiences it hardly seemed likely that the other primary factor controlling the maximum amount of radiation received by the plant— namely, the length of the daily exposure—could be responsible for the effects in question. Nevertheless, the simple expedient of shortening artificially by a few hours the length of the daily exposure to the sun by use of a dark chamber was tried, and some very striking results were obtained, as detailed in the following paragraphs.

## PLAN OF THE EXPERIMENTS

The first experiments with the use of the dark chamber were begun in July, 1918. A small, ventilated, dark chamber with a door which could be tightly closed was placed in the field. The soybeans used in the tests were grown in wooden boxes 10 inches wide, 10 inches deep, and 3 feet long. These containers have been extensively used in growing soybeans and other small plants under controlled conditions, and it has been found that normal plants are easily obtained in this way. The dark chamber and the type of box used for growing soybeans and similar plants are shown in Plate 64, A. Larger plants like tobacco have been grown in large galvanized iron buckets or, in some cases, in ordinary flower pots. When the test plants have attained the desired stage of development the procedure has been to place them in the dark chamber at the selected hour in the afternoon each day. The plants were left in the dark chamber till the hour decided upon in the following morning, when they were again placed in the sunlight. This procedure was followed each day till the test was completed. Appropriate control plants were left in the open throughout the test in each case. By this method the number of hours of exposure to sunlight during the 24-hour period could be reduced as far as desired.

In the preliminary tests of 1918 no special means were provided for moving the boxes and pots containing the plants in and out of the dark chamber.   In the spring of the present year a much larger dark house was constructed, and suitable facilities were installed for easily moving the test plants in or out of the house as often as desired.   The dark house consisted of a rectangular frame structure 30 feet by 18 feet and 6 feet in height to the eaves and 9 feet to the ridgepole.   All crevices by which light could enter were covered, tight-fitting doors were provided, and the interior was painted black.   Means were provided at the bottom and top of the house for free circulation of air without the admission of light.   A series of four steel tracks, each entering through a separate door, was provided;  and on these tracks were mounted a number of trucks carrying the test plants in their containers.   This equipment proved very satisfactory.   A general view of the dark house, the trucks, and the test plants is shown in Plate 64, B.

It has been rather generally assumed that the pronounced changes in plant activities which come on with the approach of fall are due in some way to the lower mean daily temperatures or the wider daily range in temperature caused by cool nights.   It seemed desirable, therefore, to compare the temperatures inside and outside the dark house, and for this purpose thermographs were installed.   It was found that there were only slight differences in temperature.   The temperature inside the dark house tended to run 2° or 3° F. higher than the temperature outside, particularly at night.   Hence, any responses on the part of the plants resembling those appearing in the fall of the year could not be attributed to lower temperatures.   To guard further against possible temperature effects, as soon as the above-mentioned temperature difference was discovered all doors of the dark house were opened as darkness came on each day.

In the various tests the length of the exposure to light was varied from a minimum of 5 hours per day to a maximum of 12 hours, 7 hours and 12 hours being the exposures chiefly used.   For the shortest exposure the plants were placed in the dark house at 3 o'clock p. m. and returned to the light at 10 a. m.; for the 7-hour exposure the plants were darkened at 4 p. m. and returned to the light at 9 a. m.; and for the 12-hour exposure they were in the dark house from 6 p. m. till 6 a. m.   A further modification in exposure consisted in placing the plants in the dark house at 10 a. m. and returning them to the light at 2 p. m.   In most instances the daily treatment began with the germination of the seed or in the earlier stages of growth and continued until maturity, but in some cases the plants were permanently restored to the open as soon as blossoming occurred, and in other cases the artificial shortening of the day was not begun until after blossoming had occurred.   To facilitate discussion it will be convenient to use the expressions "long day" as meaning exposure to light for more than 12 hours and "short day" as referring to an exposure of 12 hours or less.   The term "length of day" as used in this paper refers to the duration of the illumination period for each 24-hour interval.

As a part of the present investigation a series of plantings of soybeans was made in the field at intervals of approximately three days throughout the season, in order that the effects produced by different dates of planting might be compared with those produced by artificially shortening the length of the daily exposure to light.

## BEHAVIOR OF THE PLANTS TESTED

The initial experiment was made in the summer of 1918, and in this instance a box containing the Peking variety of soybeans in blossom and three pots containing Mammoth tobacco plants which had been growing for several weeks were first placed in the dark chamber at 4 p. m. on July 10 and removed therefrom at 9 a. m. the following morning. This treatment was continued each day till the seeds of the beans and tobacco were mature. All subsequent experiments were made during the year 1919. Details of the tests for both years follow.

### SOYBEANS (SOJA MAX (L.) PIPER)

(a) MANDARIN [1] (F. S. P. I. No. 36,653), early maturing:

(1) Exposed to light from 10 a. m. to 3 p. m. Planted May 8, up May 17, and placed in dark house May 20. First blossoms appeared June 12 on test plants and June 15 on controls. Average height of test plants 6 to 7 inches and that of controls 18 to 20 inches. After blossoming, the growth and development of the seed pods was much more rapid in the test plants than in the controls.

(2) Exposed to light from 9 a. m. to 4 p. m. Planted May 8, up May 17, and placed in dark house May 20. First blossoms appeared June 10 on test plants and June 15 on controls. Average height of test plants 9 to 10 inches and that of controls 19 to 20 inches.

(3) Exposed to light from 6 a. m. to 6 p. m. Planted and placed in dark house June 11, up June 16. First blossoms appeared July 7 on test plants and July 14 on controls. Average height of test plants 14 to 15 inches and that of controls 32 to 33 inches. Six weeks after blossoming the seed pods and foliage were still green and the plants stocky, whereas, under the same conditions, the Peking variety, listed below, showed many brown, mature pods, foliage yellowing, and the plants slender.

(b) PEKING [1] (F. S. P. I. No. 32,907), medium maturing:

(1) Exposed to light from 10 a. m. to 3 p. m. Planted May 8, up May 17, and placed in dark house May 20. First blossoms appeared June 12 on test plants and July 21 on controls. Seed pods on test plants

---

[1] Horticultural variety.

were turning brown by July 18, and all were mature before August 10. Average height of test plants 5 to 6 inches and that of controls 42 to 43 inches. Test plants were restored to normal light exposure June 20.

(2) Exposed to light from 9 a. m. to 4 p. m. Planted May 8, up May 17, and placed in dark house May 20. First blossoms appeared June 10 on test plants and July 21 on controls. Average height of test plants 8 inches and that of controls 45 to 48 inches. See Plate 65.

(2a) Exposed to light from 9 a. m. to 4 p. m. Planted May 8, up May 17, placed in dark house June 7. First blossoms June 29. Average height of plants 16 to 17 inches.

(3) Exposed to light from 9 a. m. to 4 p. m. after blossoming. Planted May 7, blossomed July 9, and first placed in dark house July 10. By July 26 there were many full-grown pods on test plants while there were none on controls more than half-grown. By August 29 the leaves had yellowed and were falling, and some pods were fully ripe on test plants while control plants were still green. By September 7 all seeds were fully ripe on test plants, but those on controls did not fully mature till about October 1. See Plate 66.

(4) Exposed to light from 6 a. m. to 6 p. m. Planted and placed in dark house June 11, up June 16. First blossoms July 7 on test plants and August 6 on controls. Average height of test plants 14 to 15 inches and that of controls 39 to 40 inches.

(5) Exposed to light from daylight to 10 a. m. and from 2 p. m. till dark. Planted June 14, up June 19, and placed in dark house June 19. First blossoms July 29 on test plants and August 11 on controls. Average height of test plants 25 to 26 inches and that of controls 41 to 42 inches.

(c) Tokyo,[1] late maturing:

(1) Exposed to light from 10 a. m. to 3 p. m. Planted May 8, up May 17, and placed in dark house May 20. First blossoms appeared June 13 on test plants and July 29 on controls. Average height of test plants 7 to 8 inches and that of controls 49 to 50 inches. Test plants were restored to normal light exposure June 20.

(2) Exposed to light from 9 a. m. to 4 p. m. Planted May 8, up May 17, and placed in dark house May 20. First blossoms appeared June 13 on test plants and July 29 on controls. Average height of test plants 7 to 8 inches and that of controls 49 to 50 inches.

(2a) Exposed to light from 9 a. m. to 4 p. m. Planted May 8, up May 17, and placed in dark house June 7. First blossoms appeared July 4. Average height of plants 23 to 24 inches.

(3) Exposed to light from 6 a. m. to 6 p. m. Planted and placed in dark house June 11, up June 16. First blossoms appeared July 14 on test test plants and August 21 on controls. Average height of test plants 17 to 18 inches and that of controls 42 to 43 inches.

---

[1] Horticultural variety.

(4) Exposed to light from daylight till 10 a. m. and from 2 p. m. to darkness. Planted June 14, placed in dark house June 16, up June 19. First blossoms appeared August 20 on test plants and August 23 on controls. Average height of test plants 24 to 25 inches and that of controls 42 to 43 inches.

(d) BILOXI,[1] very late maturing:

(1) Exposed to light from 10 a. m. to 3 p. m. Planted May 8, up May 17, and placed in dark house May 20. First blossoms appeared June 16 on test plants and September 4 on controls. Average height of test plants 6 to 7 inches and that of controls 57 to 58 inches. See Plate 68, A. Test plants were restored to normal light exposure June 20.

(2) Exposed to light from 9 a. m. to 4 p. m. Planted May 8, up May 17, and placed in dark house May 20. First blossoms appeared June 15 on test plants and September 4 on controls. Average height of test plants 11 inches and that of controls 57 to 58 inches. See Plate 67.

(2a) Exposed to light from 9 a. m. to 4 p. m. Planted June 10, up June 15, and placed in dark house June 24. First blossoms July 22 on test plants and September 15 on controls. Average height of test plants 15 to 16 inches and that of controls 56 to 58 inches.

(3) Exposed to light from 6 a. m. to 6 p. m. Planted and placed in dark house June 11, up June 16. First blossoms appeared July 14 on test plants and September 8 on controls. Average height of test plants 23 to 24 inches and that of controls 54 to 55 inches. See Plate 68, B.

(4) Exposed to light from daylight to 10 a. m. and from 2 p. m. to darkness. Planted June 14, placed in dark house June 16, up June 19. First blossoms appeared September 6 on test plants and September 15 on controls. See Plate 69, A. Average height of test plants 39 to 40 inches and that of controls 47 to 48 inches. In all of the above-described tests with soybeans observations were made on from 20 to 25 individuals.

TOBACCO (NICOTIANA TABACUM AND N. RUSTICA L.)

(a) NICOTIANA TABACUM;[1] MARYLAND MAMMOTH, giant type:

(1) Exposed to light from 10 a. m. to 3 p. m. Observations on 14 test plants and 10 controls. Planted March 6, transplanted to 6-inch pots May 10, and placed in dark house May 14. First blossoms appeared July 8 to August 14 on test plants and in last week of October on controls. Average height of test plants 14 to 16 inches and that of controls 3 to 5 inches.

(2) Exposed to light from 9 a. m. to 4 p. m. Observations on 7 test plants and 10 controls. Planted March 6, transplanted to 6-inch pots May 10, and placed in dark house May 14. First blossoms appeared July 18 to August 1 on test plants and in last week of October on controls.

[1] Horticultural variety.

Average height of test plants 12 to 14 inches and that of controls 5 to 6 inches.

(2a) Exposed to light from 9 a. m. to 4 p. m. Observations on 8 test plants and 8 controls. Planted January 8, transplanted to 8-inch pots May 3, and placed in dark house May 14. First blossoms appeared July 5 to 25 on test plants and October 1 to 25 on controls. See Plate 70.

(2b) Exposed to light from 9 a. m. to 4 p. m. Obseravtions on three test plants and four controls. Planted April 14, transplanted in steam-sterilized soil in 12-quart iron pails and placed in dark house June 10. First blossoms appeared August 1 to 7 on test plants and August 30 to September 8 on controls. Average height of test plants 37 inches and that of controls 39 inches.

(3) Exposed to light from 6 a. m. to 6 p. m. Observations on 6 test plants and 3 controls. Planted April 14 and transplanted to 12-quart iron pails containing steam-sterilized soil and placed in dark house June 11. First blossoms appeared August 26 to September 4 on test plants and September 3 to 20 on controls. Average height of test plants 48 inches and that of controls 49 inches. See Plates 71 and 72, A.

(b) N. tabacum; Stewart 70-Leaf Cuban,[1] giant type:

(1) Exposed to light from 9 a. m. to 4 p. m. Observations on 6 test plants and 5 controls. Planted April 14 and transplanted in steam-sterilized soil in 12-quart iron pails and placed in dark house June 10. First blossoms appeared August 16 to September 2 on test plants and September 24 to October 10 on controls. Average height of test plants 53 to 69 inches and that of controls 73 to 84 inches.

(c) N. tabacum; Connecticut Broafleaf:[1]

(1) Exposed to light from 9 a. m. to 4 p. m. Observations on 11 test plants and 10 controls. Planted April 14 and transplanted to 14-quart iron pails and placed in dark house June 5. First blossoms appeared July 18 to 24 on test plants and July 17 to 22 on controls. Average height of test plants 38 inches and that of controls 34 inches. Average number of nodes on test plants 36 and same number on controls.

(1a) Exposed to light from 9 a. m. to 4 p. m. Observations on 8 test plants and 6 controls. Planted April 5 and transplanted to 14-quart iron pails and placed in dark house May 28. First blossoms appeared July 13 to 20 on test plants and July 7 to 15 on controls. Average height of test plants 37 inches and that of controls 40 inches.

(d) N. rustica:

(1) Exposed to light from 9 a. m. to 4 p. m. Observations on 5 test plants and 3 controls. Planted April 14, transplanted to 14-quart iron pails, and 5 plants placed in dark house on June 2. Test plants blossomed July 5 to 28 and controls July 1 to 12.

---

[1] Horticultural variety.

### ASTER LINARIIFOLIUS L.

A common wild aster found in dry, open situations from Maine to Wisconsin and southward. The normal blossoming period begins about September 1 and extends over a period of two or three months.

(1) Exposed to light from 9 a. m. to 4 p. m. Six individuals taken from the field May 13 and transplanted to boxes of the type used for soybeans, three plants to the box. One box of the plants placed at once in the dark house. The control plants soon resumed vegetative development, throwing out numerous axillary branches on the upper portion of the stems as the normal limit in height was approached, thus following the regular course of development in the field. The test plants, on the other hand, made little additional growth and by June 1 were showing tiny flower heads. First blossoms appeared June 18 on test plants and September 12 on controls. Average height of test plants on June 24, 8 to 10 inches and that of controls 14 to 15 inches. Test plants were permanently returned to normal light on June 20. See Plate 72, B.

(2) Exposed to light from 6 a. m. to 6 p. m. Three individuals transplanted from field to each of two 8-gallon iron cans June 10 and those in one can placed in dark house June 12. Tiny flower heads were showing on the test plants by July 2. First blossoms appeared July 19 on test plants and September 20 on controls. Average height of test plants 8 to 9 inches and that of controls 14 to 15 inches.

(3) Exposed to light from daylight to 10 a. m. and from 2 p. m. to darkness. Three individuals transplanted from the field to each of two 8-gallon iron cans on June 14 and those in one can placed in dark house June 16. Flower heads were showing on both test plants and controls by August 20. First blossoms appeared September 16 on test plants and September 18 on controls. Average height of test plants 11 to 12 inches and that of controls 14 to 15 inches.

### CLIMBING HEMPWEED (MIKANIA SCANDENS, L.)

A climbing composite, ranging from southern Maine to Florida and westward to Ontario, Mississippi, and Texas. The normal blooming period extends from late July to the latter part of September. The aerial summer growth perishes in the fall, and the plants are carried over the winter period by perennial underground shoots.

(1) Exposed to light from 9 a. m. to 4 p. m. A number of roots were transplanted from the field to 6-inch pots and placed in the greenhouse in November, 1918. These roots threw up shoots which made considerable growth during the winter months but did not blossom. On June 3 one plant was transferred to each of six 12-quart iron pails, three of which were placed in the dark house at once. The controls began blossoming in late July and continued to blossom profusely till the latter part of September. Some of the plants which had been left in the green-

house, where the temperature was much higher than out-of-doors, blossomed at the same time. The test plants behaved quite differently, for blossoming was completely inhibited throughout the summer. Moreover, the growth of the controls has been considerably greater than that of the test plants. See Plate 74.

### BEANS (PHASEOLUS VULGARIS L.)

Three lots of seed of a tropical bean—two of which came from Arequipa, Peru, and one from Oruro, Bolivia—were planted together in two boxes measuring 3 feet by 10 inches by 10 inches on June 16, and one box was placed in the dark house June 24. Exposed to light from 9 a. m. to 4 p. m. According to Dr. D. N. Shoemaker this bean when planted in the field at Washington has been found to make a very large growth without blossoming till late in the fall, but when propagated in the greenhouse in the winter months the plant promptly blossoms and sets seed. The test plants blossomed July 21 to 23, and some of the seed pods were mature by August 22, whereas the controls did not blossom till October 11. The average height of the test plants was 4½ to 5 feet and that of the controls 7 to 8 feet. See Plate 73.

### RAGWEED (AMBROSIA ARTEMISIIFOLIA L.)

Exposed to light from 9 a. m. to 4 p. m. Observations based on 6 test plants and 6 controls. Small plants taken from the roadside were transplanted to 6-inch pots on June 3, and a portion of these were immediately placed in the dark house. Staminate heads were showing on the test plants by June 17, and the anthers were shedding pollen freely by July 1. The controls did not begin blossoming till the last week in August, which is the normal period for the appearance of first blossoms on the plant. The average height of the test plants at the time of blossoming was 8 to 9 inches, while that of the controls on the same date was 11 inches and their final height 29 inches. The test plants were returned permanently to normal light exposure on July 1. See Plate 75, A.

### RADISH (RAPHANUS SATIVUS L.)

SCARLET GLOBE:[1]

Exposed to light from 9 a. m. to 4 p. m. Planted May 15, up May 19, and placed in dark house on day of planting. The test plants grew more slowly than the controls for a time and then appeared to grow no further. All but two of the test plants, of which there were a large number, became diseased and finally died without forming seed stalks. The two survivors developed a crown of large leaves, and the roots also reached much larger proportions than those of the controls. Apparently enlargement of the roots had not ceased as late as October 15, when one

---

[1] Horticultural variety.

of them measured nearly 4 inches in diameter while its rosette of leaves measured 30 inches from tip to tip. Flower stems did not develop. The controls grew more rapidly from the outset, and all except three or four to be considered later formed flower stems in June, the first blossom appearing June 21. See Plate 75, B.

### CARROT (DAUCUS CAROTA L.)

OXHEART:[1]
Exposed to light from 9 a. m. to 4 p. m. Planted June 4 and at once placed in dark house. The test plants made a uniform but slow growth, and the roots, which were very small, appeared to be devoid of the yellow pigment, carotin, since they were almost snow-white in color. The controls grew and developed normally, the roots showing the normal yellow color. On August 19 the average height of the test plants was 8 to 9 inches and that of the controls 18 to 20 inches. See Plate 79, B.

### LETTUCE (LACTUCA SATIVA L.)

BLACK SEEDED SUMMER:[1]
Exposed to light from 9 a. m. to 4 p. m. Planted in dark house June 4. Germination was satisfactory, but the seedlings made very little growth, and after a time all died. The controls grew vigorously but under the stimulus of the long day the plants soon sent up flowering shoots and blossomed.

### HIBISCUS MOSCHEUTOS L.

A wild perennial in marshes, ranging from Ontario to Florida and Texas. Normal blooming period July to September. Exposed to light from 9 a. m. to 4 p. m. Planted in November in greenhouse. Seed did not germinate till the following March. Seven plants transferred to 12-quart iron pails on June 6, three of which were placed in dark house June 7. The test plants did not blossom nor did they make any growth during the summer. The controls grew vigorously, and the first blossoms appeared August 22 to September 10. The average height of the test plants was 12 inches and that of the controls 29 inches.

### CABBAGE (BRASSICA OLERACEA CAPITATA L.)

EARLY JERSEY WAKEFIELD:[1]
Exposed to light from 9 a. m. to 4 p. m. Observations based on four test plants and four controls. Transplanted and placed in dark house on June 7. The test plants grew slowly but uninterruptedly throughout the season, although they showed little tendency to form heads. The control plants grew normally and formed large heads which eventually burst open, followed by the formation of new heads of small size.

---

[1] Horticultural variety.

### VIOLETS (VIOLA FIMBRIATULA SM.)

A common wild species ranging from Nova Scotia to Wisconsin and southward and growing in sandy fields and on dry hillsides. The normal blooming period comes in April. Exposed to light from 9 a. m. to 4 p. m. Two lots of six plants were transferred from the field to two boxes measuring 3 feet by 10 inches by 10 inches on June 9, and one of the boxes was placed at once in the dark house. The test plants showed flower buds as early as June 21 and were in blossom early in July, producing purple, petaliferous flowers and also cleistogamous flowers. The control plants produced numerous cleistogamous flowers but none of the purple, petaliferous type.

### EARLY GOLDENROD (SOLIDAGO JUNCEA AIT.)

The earliest species of goldenrod, ranging from New Brunswick to Saskatchewan and south to North Carolina and Missouri. Blossoming normally extends from late June to September. Exposed to light from 9 a. m. to 4 p. m. Two lots of six plants were transplanted to two boxes measuring 3 feet by 10 inches by 10 inches on June 6, and one of the boxes was at once placed in the dark house. The test plants and the controls blossomed at the same time, late in August. The test plants however, were shorter and more compact than the controls. The heights of the test plants averaged 24 inches and those of the controls 38 inches. The test plants advanced toward maturation more rapidly than the controls after the flowering stage had been reached.

### EFFECT OF RESTORING THE TEST PLANTS TO NORMAL LIGHT EXPOSURE AFTER BLOSSOMING HAD OCCURRED

In the experiments with soybeans, aster, and ragweed described above it has been made clear that after blossoming has occurred the effect of shortening the daily exposure to sunlight is to hasten greatly the ripening of the seed. In certain instances, however, as has been recorded under the several experiments, the test plants were restored to the normal light exposure as soon as blossoming had occurred.

Under these conditions seed pods of the soybeans ripened rapidly, the leaves turned yellow, and for a time it appeared that the plants would die as is normal for the soybean. Eventually, however, new branches developed under the influence of the long summer days. The renewed growth was especially well-developed in the Biloxi variety, and the final result was that these plants, still bearing the first crop of ripened seed pods, blossomed for the second time September 4 to 8. This date of blossoming, moreover, is also that for the first blossoming of the control plants which had been planted on the same date as the test plants and had been exposed to the normal daylight period throughout their development.

Like the soybeans, the asters after a time responded to the long-day influence; and by July 20 the plants, though bearing ripened seed, were developing new axillary branches. The new growth finally developed flower heads; and thus the plants blossomed for the second time during the first half of September, which is the time of blossoming of the original

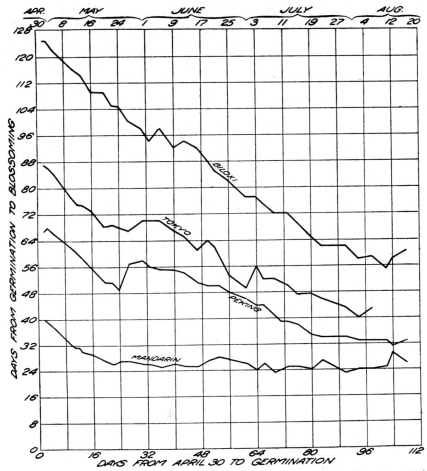

FIG. 1.—Graph showing the shortening of the vegetative period preceding flowering in soybeans which results from progressively later planting during the growing season.

controls exposed to the normal daylight period throughout their development.

The ragweed, likewise, resumed vegetative development after a time, and, in fact, under the influence of the full length of the daylight period the new growth exceeded in size that of the original plants. The plants blossomed the second time during the last week in August, which is also the time of blossoming of the original controls and of ragweed growing in the field. It may be noted, however, that while the original growth produced staminate spikes as well as pistillate flowers in the usual manner, the second growth produced pistillate flowers almost exclusively and the leaves were mostly atypical.

### RELATION OF DATE OF PLANTING TO DATE OF BLOSSOMING IN SOYBEANS

Through the spring and summer of 1919 a series of plantings of soybeans which included the four varieties used in the tests described above were made in the field at regular intervals of three days as nearly as conditions would permit. All plantings of each variety consisted of rows 10 feet in length. The date recorded as that when first blossoms appeared is in each case that when the majority of the individuals in the planting first showed one or two open blossoms. In most instances the

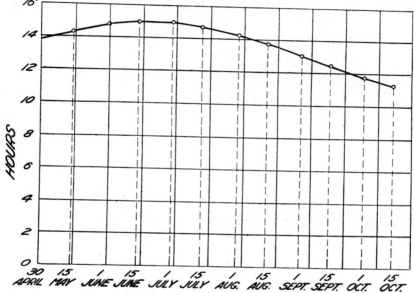

FIG. 2.—Graph showing changes in length of day during the growing season in the latitude of Washington, D. C. Ordinates indicate 2-hour intervals of the day and abscissae indicate 16-day periods of the growing season.

greater number of the individuals in a planting showed their first open blossoms on practically the same date. The dates of planting, germination, and appearance of first blossoms, together with the number of days from germination till blossoming are shown in Table I.

The effect of the date of planting on the length of the period from germination to the blossoming stage for each variety is more easily seen in the curves of figure 1, in the construction of which the number of days from April 30 to dates of germination are used as ordinates and the number of days included in the periods of growth prior to blossoming are used as abscissae. The relative length of the day—that is, the time between sunrise and sunset, expressed in 2-hour periods—also is shown for the same period in figure 2. The relative heights of the plants in the consecutive plantings of the Biloxi variety are shown graphically in figure 3.

160115°—20——2

TABLE I.—*Effect of date of planting on date of blossoming of soybeans grown in field at Arlington, Va., 1919*

| Date of planting. | Date of appearance above ground. | Mandarin. | | Peking. | | Tokyo. | | Biloxi. | |
|---|---|---|---|---|---|---|---|---|---|
| | | Date of first blossoms. | Time from germination to blossoming. | Date of first blossoms. | Time from germination to blossoming. | Date of first blossoms. | Time from germination to blossoming. | Date of first blossoms. | Time from germination to blossoming. |
| | | | *Days.* | | *Days.* | | *Days.* | | *Days.* |
| Apr. 9.. | May 2 | June 11 | 40 | July 8 | 67 | July 28 | 87 | Sept. 4 | 125 |
| 14.. | 3 | 11 | 39 | 10 | 68 | 28 | 86 | 5 | 125 |
| 18.. | 5 | 11 | 37 | 10 | 66 | 28 | 84 | 4 | 122 |
| 22.. | 10 | 11 | 32 | 11 | 62 | 26 | 77 | 4 | 117 |
| 26.. | 11 | 11 | 31 | 11 | 61 | 26 | 76 | 4 | 116 |
| 30.. | 12 | 12 | 31 | 11 | 60 | 26 | 75 | 4 | 115 |
| May 3.. | 13 | 12 | 30 | 11 | 59 | 27 | 75 | 4 | 114 |
| 6.. | 16 | 14 | 29 | 11 | 56 | 28 | 73 | 2 | 109 |
| 9.. | 20 | 16 | 27 | 11 | 51 | 27 | 68 | 6 | 109 |
| 13.. | 22 | 17 | 26 | 12 | 51 | 30 | 69 | 4 | 105 |
| 16.. | 24 | 20 | 27 | 12 | 49 | 31 | 68 | 6 | 105 |
| 20.. | 27 | 23 | 27 | 23 | 57 | Aug. 2 | 67 | 4 | 100 |
| 24.. | 31 | 26 | 26 | 28 | 58 | 9 | 70 | 6 | 98 |
| 27.. | June 2 | 28 | 26 | 28 | 56 | 11 | 70 | 4 | 94 |
| 31.. | 5 | 30 | 25 | 30 | 55 | 14 | 70 | 11 | 98 |
| June 4.. | 9 | July 5 | 26 | Aug. 3 | 55 | 15 | 67 | 11 | 92 |
| 7.. | 12 | 7 | 25 | 5 | 54 | 16 | 65 | 10 | 94 |
| 11.. | 16 | 11 | 25 | 6 | 51 | 16 | 61 | 11 | 92 |
| 14.. | 19 | 16 | 27 | 8 | 50 | 22 | 64 | 15 | 88 |
| 17.. | 22 | 20 | 28 | 11 | 50 | 23 | 62 | 15 | 85 |
| 20.. | 25 | 22 | 27 | 12 | 48 | 17 | 53 | 15 | 82 |
| 23.. | 30 | 26 | 26 | 15 | 46 | 18 | 49 | 15 | 77 |
| 26.. | July 3 | 27 | 24 | 16 | 44 | 26 | 56 | 18 | 77 |
| 30.. | 5 | 31 | 26 | 18 | 44 | 26 | 52 | 18 | 75 |
| July 3.. | 8 | 31 | 23 | 18 | 41 | 29 | 52 | 18 | 72 |
| 7.. | 12 | Aug. 6 | 25 | 20 | 39 | 31 | 50 | 22 | 72 |
| 10.. | 15 | 9 | 25 | 22 | 38 | 31 | 47 | 22 | 69 |
| 14.. | 19 | 12 | 24 | 23 | 35 | Sept. 4 | 47 | 20 | 63 |
| 17.. | 22 | 18 | 27 | 25 | 34 | 6 | 46 | 22 | 62 |
| 25.. | 29 | 21 | 23 | Sept. 6 | 39 | 10 | 43 | 29 | 62 |
| 29.. | Aug. 2 | 26 | 24 | 6 | 35 | 11 | 40 | 29 | 58 |
| Aug. 2.. | 6 | 30 | 24 | 8 | 33 | .......... | .......... | Oct. 4 | 59 |
| 5.. | 10 | Sept. 4 | 25 | 11 | 33 | | | 4 | 55 |
| 8.. | 12 | 10 | 29 | 12 | 31 | | | 9 | 58 |
| 11.. | 16 | 11 | 26 | 18 | 33 | | | 16 | 61 |
| 14.. | 17 | .......... | .......... | 19 | 33 | | | .......... | .......... |
| 20.. | 25 | .......... | .......... | 26 | 31 | | | | |

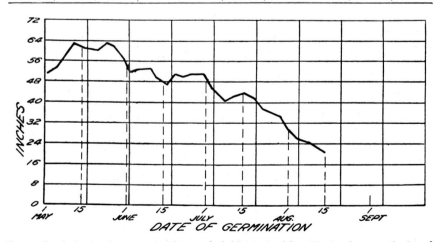

FIG. 3.—Graph showing the progressive decrease in height attained by Biloxi soybeans as the date of planting is delayed beyond late spring

## DISCUSSION OF RESULTS

The results of the experiments which have been described show clearly that both the rate and extent of the growth attained by the plants under study and the time required for reaching and completing the flowering and fruiting stages are profoundly affected by the length of the daily exposure to sunlight. The behavior of some of the plants under the different exposures would seem to indicate that the action on the vegetative phase of development is more or less independent of that on reproduction, but only tentative conclusions can be drawn on these points at the present time. The effects of the different light exposures on these two phases of plant development can best be discussed separately.

### LENGTH OF DAILY LIGHT EXPOSURE IN RELATION TO VEGETATIVE DEVELOPMENT

Under the conditions of the tests it was not possible to secure quantitative data on the various details of vegetative growth and development, but measurements of height and the photographic records will clearly indicate some of the differences resulting from the various light exposures. In general, the extent of growth was proportional to the length of the daily exposure to light; and this held true when the plants received two daily exposures to light, with an intervening period of darkening, as well as when there was only a single daily exposure to the light. Under the shorter exposures the plants were shorter and less stocky, and there were some indications of etiolation or chlorosis. Histological examination of the test plants was not undertaken, but in most species no very striking differences in gross anatomy resulted from the different exposures. Broadly speaking, the extent rather than the character of growth and vegetative development was chiefly affected. Table II is intended to bring out the relationship between size of plant and length of the exposure to the light for soybeans and the aster. This relationship is strikingly brought out for the Biloxi soybean in figure 3, which shows the decreasing heights of progressively later plantings. How length of exposure affects the Mandarin is shown in the foreground of Plate 78, B.

TABLE II.—*Effect of length of daily exposure to light on the height of soybeans and aster*

| Length of daily exposure. | Average heights of plants. | | | | |
|---|---|---|---|---|---|
| | Soybeans. | | | | Aster. |
| | Mandarin. | Peking. | Tokyo. | Biloxi. | |
| 10 a. m. to 3 p. m., 5 hours.... | *Inches.* 6 to 7 | *Inches.* 5 to 6 | *Inches.* 7 to 8 | *Inches.* 6 to 7 | *Inches.* ........... |
| 9 a. m. to 4 p. m., 7 hours..... | 9 to 10 | 8 | 7 to 8 | 11 | 8 to 10 |
| Daylight to 10 a. m. and 2 p. m. till dark, 8½ to 11 hours [a].. | ........... | 25 to 26 | 24 to 25 | 38 to 40 | 11 to 12 |
| 6 a. m. to 6 p. m., 12 hours.... | 14 to 15 | 14 to 15 | 17 to 18 | 23 to 24 | 8 to 10 |
| Full daylight, 12½ to 15 hours. | 18 to 20 | 40 to 44 | 42 to 44 | 54 to 58 | 14 to 15 |

[a] The relatively greater heights in proportion to the number of hours in the total daily exposures under this treatment are due to the fact that in this case the length of the growing period was not materially shortened by forced earliness in blossoming; they are not to be ascribed to an increased rate of growth.

It may be worthy of note that in the tests under controlled conditions the height of the Biloxi plants under a 12-hour light exposure was practically the same as that of the latest field plantings shown in figure 3, while that of the controls was about the same as that of the early field plantings.

Since in many cases the length of the growing period was greatly curtailed by the forcing action of reduced light exposure on reproduction, the amount of growth was necessarily limited thereby in those plants having a determinate type of inflorescence; but, in addition, measurements made when the blossoming stage of the forced plants had been reached show that the rate of growth was greater as the length of the exposure to light increased. The measurements of height recorded under the several tests relate to the final heights attained by the plants. In the species tested no exceptions to the foregoing principle were encountered; but it is possible, of course, that other species will be found to act differently. It has been demonstrated by a number of investigators that when many green plants are transferred from light to darkness the immediate effect is an acceleration in the rate of growth; and, conversely, the first effect of exposure to light is a retarding of growth. These facts, however, bear on necessary relation to the total effect on rate of growth over a considerable period of time produced by differences in the relative length of night and day.

It remains to be pointed out that striking differences in sensitiveness to decreased length of the daily exposure to light were observed in the different species under investigation. Aside from considerable reductions in the rate of growth and slight chlorosis, soybeans, tobacco, aster, and some others showed no ill effects from the reduced length of illumination, while Hibiscus was not able to make any appreciable growth with the illumination period reduced to nine hours, and lettuce was much more seriously affected, all individuals having perished without making any material growth.

LENGTH OF DAILY LIGHT EXPOSURE IN RELATION TO SEXUAL REPRODUCTION

While the rate of growth of the species tested was markedly affected by change in the length of the daily illumination period, the effects on blossoming and fruiting are particularly interesting and important. The experiments with soybeans included four varieties which range from **early to very late in** maturing under normal conditions when grown in the latitude of Washington, D. C. Thus, for plantings in the field extending through the month of May the average number of days from germination to blossoming was approximately 27, 56, 70, and 105, respectively, for the Mandarin, Peking, Tokyo, and Biloxi, the lastnamed showing no open blossoms till early September. Table III brings out several important facts regarding the effects of reduced light exposure **on these** four varieties.

TABLE III.—*Number of days required by soybeans to reach the flowering stage under daily light exposures of different lengths*

| Length of daily exposure. | Mandarin. | | | Peking. | | |
|---|---|---|---|---|---|---|
| | Date of germination. | Date of transfer to dark house. | Time from germination to blossoming. | Date of germination. | Date of transfer to dark house. | Time from germination to blossoming. |
| | | | *Days.* | | | *Days.* |
| 10 a. m. to 3 p. m., 5 hours.............. | May 17 | May 20 | a 23 | May 17 | May 20 | a 23 |
| 9 a. m. to 4 p. m., 7 hours................. | ...do..... | ...do..... | 21 | ...do..... | ...do..... | 21 |
| Do.................. | | | | ...do..... | June 7 | 22 |
| Daylight to 10 a. m. and 2 p. m. till dark, 8½ to 11 hours. | ......... | ......... | ......... | June 19 | 19 | 40 |
| 6 a. m. to 6 p. m., 12 hours............... | June 16 | June 11 | 21 | 16 | 11 | 21 |
| Full daylight, 12½ to 15 hours........... | May 17 | Control | 26 | May 17 | Control | 62 |
| Do........................ | June 16 | ...do..... | 28 | June 16 | ...do..... | 51 |

| Length of daily exposure. | Tokyo. | | | Biloxi. | | |
|---|---|---|---|---|---|---|
| | Date of germination. | Date of transfer to dark house. | Time from germination to blossoming. | Date of germination. | Date of transfer to dark house. | Time from germination to blossoming. |
| | | | *Days.* | | | *Days.* |
| 10 a. m. to 3 p. m., 5 hours.............. | May 17 | May 20 | a 24 | May 17 | May 20 | a 27 |
| 9 a. m. to 4 p. m., 7 hours................. | ...do..... | ...do..... | 24 | ...do..... | ...do..... | 26 |
| Do.................. | ...do..... | June 7 | 27 | June 15 | June 24 | 28 |
| Daylight to 10 a. m. and 2 p. m. till dark, 8½ to 11 hours. | June 19 | 16 | 62 | 19 | 16 | 79 |
| 6 a. m. to 6 p. m., 12 hours............... | 16 | 11 | 28 | 16 | 11 | 28 |
| Full daylight, 12½ to 15 hours........... | May 17 | Control | 73 | May 17 | Control | 110 |
| Do........................ | June 16 | ...do..... | 66 | June 16 | ...do..... | 90 |

[a] In those cases in which the plants were placed in the dark house after they had germinated, only the period elapsing after they had been transferred is taken into account, rather than that beginning with the date of germination.

It is seen that when the daily illumination consists of a single exposure of 12 hours or less, the usual length of the growing period from germination to blossoming is only slightly shortened in the early variety, Mandarin; but the shortening effect is increasingly accentuated as the usual growing period increases, till, in the very late variety, Biloxi, this period is reduced to less than one-fourth that of the control plants grown under full daylight exposure during the summer months. In reality, all varieties become early maturing ones under these conditions, and there is but little difference in the time required by the four varieties to reach the blossoming stage. These tests also show that reducing the length of the illumination period below 12 hours has no further effect in shortening the vegetative period, so that apparently there is a certain minimum period of light exposure, reduction of which is without action in hastening the appearance of the flowering stage. These results seem to indicate further that for each variety a certain minimum period of time (ordinarily one of vegetative activity) must elapse from the inception of the stimulating action resulting from the reduced light exposure before the flowering stage can be attained. The data in Table III suggest that this minimum formative period is approximately 21 days for the Mandarin and Peking varieties, 24 days for the Tokyo, and 26 days for the Biloxi, although under suitable conditions these periods might possibly be somewhat further shortened.

Subjecting the plants to two periods of illumination daily, whereby the total daily exposure averaged 9 or 10 hours, was vastly less effective in inducing early blossoming than a single daily exposure of 12 hours; and, in fact, in the later varieties the effect was of little significance. This is true in spite of the fact that the plants were in darkness during the hours of most intense sunlight—namely, from 10 a. m. to 2 p. m. Obviously it is not merely the total number of hours of sunshine received daily by the plant that may induce such marked shortening of the vegetative period, but the continuity of the exposure also plays an important part. The two plantings of soybeans serving as controls, which first appeared above ground on May 17 and June 16, respectively, did not respond in the same manner to the prevailing seasonal conditions. The vegetative period of the Mandarin was lengthened by two days as a result of the later planting, while the later maturing varieties were affected in the reverse manner. These results are in accord with the fact that the average length of day during the vegetative period was longer for the later planting than for the earlier in the case of the Mandarin, while the reverse is true of the other varieties. The marked action of a decrease in the length of the day, within certain limits, in hastening the arrival of the blossoming stage is equally in evidence throughout the stages of seed formation and maturation. This fact is shown by numerous tests; but experiments (a) (1) and (b) (3) with the Mandarin and Peking varieties, respectively, may be cited specifically.

These tests under controlled conditions clearly show that so far as concerns sexual reproduction the Mandarin soybean is adapted to a relatively long day, since the time required by it to reach the blossoming stage during the long summer days can not be greatly reduced by shortening the length of the daily exposure to light. On the other hand, the Biloxi is distinctively a "short day" variety; and with a daily light exposure of 12 hours or less it blossoms almost as early as the Mandarin, whereas the control plantings show that it refuses to blossom during the long summer days when normally exposed to the light. It is interesting to note that, on the basis of these results, all of the four varieties tested should behave similarly when grown under a 12-hour day such as prevails at the equator. The action of the shortened period of daily light exposure in promoting sexual reproduction offers a satisfactory explanation of the fact that there is a marked progressive shortening of the vegetative period in successive plantings of medium and late maturing varieties of soybeans made during the summer months. In this connection an examination of figure 1, showing graphically this progressive shortening in the vegetative period, is of interest. It should be pointed out here that the progressive decrease in the length of the vegetative period of all varieties apparent in the very early plantings which germinated during the early part of May is probably due to a gradual reduction in the retarding action of relatively low temperatures which prevailed at the time. Again, there is distinct evidence of the retarding influence of lower temperatures on the very latest plantings of the Peking and Biloxi varieties. Eliminating these portions of the curves from consideration, it is evident that the graph for the early variety, Mandarin, is practically horizontal, while there is a marked downward trend in the graphs for the remaining varieties which increases in pitch as we pass toward the later varieties, the drop being quite precipitate in the curve of the very late variety, Biloxi. There is, in short, a marked tendency for the graphs to converge toward a common point as the summer season advances, a fact which is in full accord with the results of the tests under controlled conditions. Another interesting feature of these curves is that for the period around May 25 to June 15 there is a more or less well defined "hump" which is most strongly developed in the curve for the Peking, less prominent in that for the Tokyo, and hardly apparent in the curves for the Biloxi and the Mandarin. A possible explanation of this relative lengthening of the vegetative period of the Peking and Tokyo plantings which germinated during the close of May and early June is to be found in the fact that these plants received the longest possible average light exposure. This would not affect the Mandarin or the Biloxi, since the length of the day is well above the "critical" for the Biloxi and below it for the Mandarin. Apparently field plantings can not be extended through the season in such a way as to bring the plants throughout the vegetative period under a light exposure below the critical in length and at the same time secure throughout the period a sufficiently high temperature (and possibly other favorable factors) to reduce the length of the vegetative period to that which experiments conducted under controlled conditions have established as apparently the physiological minimum requisite for sexual reproduction. There can be no doubt that decreasing temperature, within limits, will retard vital activities of the plant; and the fact should be emphasized that, as a rule, the action of decreasing temperatures as fall approaches must be retarding rather than accelerating in its influence on the attainment of the flowering stage by the plant. It should be pointed out here that the hastening effect of the shorter days on the final maturation of the seed of the soybeans is shown by the fact that in the late plantings there is an evident tendency for the early Mandarin and the later Peking varieties to progress toward maturity at the same rate.

As regards the critical length of day required for furnishing the stimulus which brings into expression the processes of sexual reproduction mentioned above, it should be stated that this has not been determined as yet for any of the plants under study, and it is not possible to state how narrowly defined this maximum length of day capable of inducing sexual reproduction may be. The outstanding fact is that it is quite different for the four varieties of soybeans. In all cases, however, it is in excess of 12 hours.

Coming to tobacco, the contrast in behavior of the Connecticut Broadleaf and the Maryland Mammoth varieties is very striking. Sexual reproduction in the Connecticut Broadleaf is not materially affected by changes in length of day within the seasonal range for the latitude of Washington or southward. On the other hand, the Maryland Mammoth, which is presumably a mutation from a very old variety of Maryland tobacco and appears to be a typical example of gigantism, can not be forced into blossoming during the summer months by any method now known except artificial shortening of the duration of the daily exposure to light, while the character of gigantism is completely suppressed when the plant is grown during the short days of winter. A glance at Table IV shows that shortening the daily light exposure has not materially affected the Connecticut Broadleaf but has been effective in shortening the vegetative period of the Maryland Mammoth. The Cuban type of Mammoth was affected like the Maryland type, but it appears that the former has a somewhat longer vegetative period than the latter under similar conditions. The Maryland type blossoms readily under the influence of a 12-hour light exposure; but there is a suggestion that a time factor is operative here, for the plants seem not to blossom so promptly as when under the 7-hour exposure. It seems probable also that the Cuban Mammoth will blossom under a 12-hour exposure to light. The observation has been made by Lodewijks (17) that a giant type of Sumatra tobacco—grown under the influence of the 12-hour equatorial day—which may reach the extreme height of 24 feet, either does not blossom at all or forms only a few flowers and seeds. Gigantism in tobacco disappears when the plant is brought under the influence of short days such as prevail in the temperate zone during the winter months. *Nicotiana rustica*, so far as tested, behaves like the Connecticut Broadleaf.

*Aster linariifolius*, again, has given clean-cut results under the different light exposures, as is shown in the summarized data of Table IV. Its behavior is strictly comparable with that of the Biloxi soybean and the giant type of tobacco. It is a typical "short-day" flowering perennial. As with the Biloxi soybean, however, this maximum length of day capable of bringing into expression the flowering and seed-formation processes is in excess of 12 hours. Exposure to light twice daily was without effect, for the vegetative period of the test plants, counting from the beginning of the experiment, was 92 days and that of the controls (not shown in Table IV) was 94 days. Here, again, attention is called to the fact that the total daily exposure to light averaged only about 10 hours, and the plants were in darkness during the period of most intense illumination, 10 a. m. to 2 p. m.

TABLE IV.—*Length of the vegetative period of tobacco and aster as affected by the length of the daily exposure to light*

| Length of exposure. | Connecticut Broadleaf. | | Maryland Mammoth. | | Stewart 70-Leaf Cuban. | | Aster linariifolius. | |
|---|---|---|---|---|---|---|---|---|
| | Date of transfer to dark house. | Length of vegetative period. | Date of transfer to dark house. | Length of vegetative period. | Date of transfer to dark house. | Length of vegetative period. | Date of transfer to dark house. | Length of vegetative period. |
| 10 a. m. to 3 p. m., 5 hours. | ......... | *Days.* ......... | May 14 | *Days.* 55 to 61 | ......... | *Days.* ......... | ......... | *Days.* ......... |
| 9 a. m. to 4 p. m., 7 hours. | June 5 | 43 to 49 | ...do.... | 61 to 78 | ......... | ......... | May 13 | 36 |
| Full daylight, 12 to 15 hours. | Controls | 42 to 47 | Controls | 152 to 160 | ......... | ......... | Controls | 122 |
| 9 a. m. to 4 p. m., 7 hours. | ......... | ......... | May 14 | 52 to 72 | ......... | ......... | ......... | ......... |
| Full daylight, 12 to 15 hours. | ......... | ......... | Controls[a] | 140 to 164 | ......... | ......... | ......... | ......... |
| 9 a. m. to 4 p. m., 7 hours. | May 28 | 46 to 53 | June 10 | 52 to 59 | June 10 | 67 to 84 | ......... | ......... |
| Full daylight, 12 to 15 hours. | Controls | 36 to 44 | Controls | 84 to 101 | Controls | 81 to 90 | ......... | ......... |
| Daylight to 10 a. m. and 2 p. m. till dark, 11 to 8½ hours. | ......... | ......... | ......... | ......... | ......... | ......... | June 16 | 92 |
| 6 a. m. to 6 p. m., 12 hours. | ......... | ......... | June 11 | 76 to 85 | ......... | ......... | 12 | 37 |
| Full daylight, 12 to 15 hours. | ......... | ......... | Controls | 84 to 101 | ......... | ......... | Controls | 101 |

*a* These controls and the test plants having a vegetative period of 52 to 72 days were in 8-inch pots.

The composite *Mikania scandens* L. is of interest as presenting a new type of plants so far as concerns behavior under long-day and short-day conditions. Under short-day conditions which were maintained for nearly 12 months this plant lost its power of blossoming. In other words, the plant became sterile. The early varieties of soybeans and the Connecticut Broadleaf tobacco blossom and fruit freely through the range of seasonal changes in the length of the day which obtains for the latitude of Washington, while the late varieties of soybeans, the giant types of tobacco, and the aster are essentially sterile when under the influence of the long summer days; and Mikania, on the other hand, is sterile during all seasons of the year except summer when long days prevail. It is worth noting that the Mikania was unable to develop flowers during the summer months when kept under the influence of a short daily exposure to light, notwithstanding that it had been growing in the greenhouse for several months previously.

The bean from the Tropics, *Phaseolus vulgaris*, included in the tests, brings us a step nearer to complete sterility in the latitude of Washington (approximately 39°), for whether it is able to blossom here will depend on the early or late occurrence of killing frost. Evidently it could not blossom very far northward of Washington. Under the influence of a 7-hour daily illumination this bean blossomed in 28 days, and one month later some of its seed pods were mature; yet under outdoor conditions blossoming did not occur till October 11, 109 days after germination. The fact that this plant does not blossom here till the middle of October indicates that the critical length of day for flowering can not be much in excess of 12 hours; and the physiological minimum for the vegetative period appears to be approximately 28 days, about the same as for the Biloxi soybean. This bean would seem to be admirably adapted to tropical conditions.

The writers are informed by Dr. Shoemaker that in tests made by him at Washington this species in the greenhouse blossomed freely during the winter and developed seed. In the spring some of the plants, having been transferred to pots after the tops had been largely removed, were placed out of doors. New shoots developed, and these grew throughout the summer without blossoming. It is clear that this plant behaves like the Mammoth or giant type of tobacco toward differences in the length of day.

Ragweed is still another example of a short-day plant, for, under a 7-hour exposure, the anthers of the staminate heads were shedding pollen freely within 27 days after the beginning of the test, while under outdoor conditions blossoming did not occur till 7 weeks later. Radish is a good example of the type requiring a long day for attainment of the flowering stage, for, like Mikania, it has not been able to blossom under a 7-hour exposure although the test was continued throughout the summer, while under outdoor conditions blossoms appeared one month after germination. Throughout the test the rosette type of leaf development was maintained under the shortened light exposure, and both leaf and root continued to grow; so here, once more, is apparently a manifestation of gigantism. Under the conditions of the tests, the two biennials, cabbage and carrot, showed no decided response to shortened light exposure so far as concerns flowering; but their behavior under normal conditions indicates that they are to be regarded as typically long-day plants. Hibiscus is a striking example of a long-day plant, for not only is it unable to blossom under a 7-hour light exposure but it is also unable to make any appreciable growth under these conditions. The behavior of Viola is of interest because of the habit of forming both cleistogamous and chasmogamous flowers, the two types appearing at different seasons. It appears that the later developing cleistogamous flowers are to be regarded as forming the more distinctively reproductive organs. Under a 7-hour light exposure, which was not begun till June 7, the plants showed open, purple, petalliferous flowers during the first week in July, although, of course, a previous crop of these blossoms had been produced earlier in the season. The cleistogamous blossoms appeared also on the plants at the usual time, in June. The early goldenrod used in the tests showed no shortening of the vegetative period under a 7-hour exposure.

RELATIONSHIPS BETWEEN ANNUALS, BIENNIALS, AND PERENNIALS

It is well known that there are no hard and fast lines of distinction separating annuals, biennials, and perennials; for plants may change from one of these types to another under influences of environment, although in the past the particular factors of the environment involved have not for the most part been understood. The experiments recorded in this paper make it clear that in any particular region the relative lengths of the days and nights running through the year constitute one of the controlling factors in determining the behavior of plants in this particular. The soybean is commonly regarded as a typical annual in that its entire life cycle is completed in a single season, and coincident with or soon following the maturation of the seed the plant as a whole perishes. As recorded on page 567, however, a suitable change in the length of the daily exposure to light revived the vegetative life of the matured plants. After the first crop of seed had ripened and the foliage had yellowed just as usual immediately before the plant died, new shoots developed on the old stems, vegetative activity was resumed, and, finally, with the approach of the shorter days of autumn, the plants blossomed and fruited a second time. Thus, under controlled conditions the plant simulated the behavior of a flowering perennial except that the two cycles of alternate vegetative and reproductive activity have been crowded into a single season. Ragweed behaved in essentially the same manner. To make the analogy more convincing, attention is directed to the fact that aster, a flowering perennial, under the same treatment gave exactly the same results as the soybeans and ragweed. Thus, the aster readily completed two complete annual cycles within a period of about four months, except that, in the absence of low temperature, the original growth above ground, of course, was not killed. Moreover, in the second period of vegetative activity new shoots were sent up from the roots in addition to the new axillary shoots appearing on the original stems. The first flowering and fruiting of the soybeans, ragweed, and aster were forced by artificially shortening the length of the day. When the plants were restored to the full exposure of the normal summer day, vegetative activity was resumed, and, finally, the natural shortening of the days in August and September resulted in the second flowering and fruiting periods. The factor of the environment which makes the cycle of alternating vegetative and reproductive activities an annual event would thus seem to be the annual periodicity in the length of day. If temperature differences are assumed to be the primary factor, annual periodicity in tropical regions (not including the immediate vicinity of the equator) is not readily explainable.

As has already been pointed out, the Mammoth or giant type of tobacco behaves as a typical flowering annual, like the ordinary tobaccos, when grown under the influence of days not exceeding 12 hours in length. During the winter months the plant blossoms readily and, in fact, becomes practically an ever-blooming type. It is an interesting fact, however, that as the seed capsules mature the seed-bearing stem dies back only to the first node which may have sent up a new branch. This holds true even though the new branch be but a few inches below the seed head. The portion of the stem below the new branch and the root system henceforth function as parts of a new plant. In winter the new branch blossoms and fruits promptly, perishes, and is succeeded by new branches. As spring advances the new branches coming out assume the giant or nonflowering type of growth which continues till fall brings a return of the short days, when blossoms promptly appear. It would seem that the new branch acts as a rejuvenating or a protective agent against the death of the older organs to which it is attached. Obviously the Mammoth tobacco resembles both the annual and the perennial types of plant life. The sharpness with which the new branch controls the extent of the dying-back of the mother stem is shown in Plate 76, A.

In the latitude of Washington the radish is an annual unless planted very late in the season. It has already been shown that under a shortened light exposure, on the other hand, while vegetative development may continue, flowering does not occur. It would appear from this that the radish might not flower in regions where the maximum length of days is relatively short; and, in fact, according to Dr. Walter Van Fleet, of the Bureau of Plant Industry, the radish as a rule does not blossom when grown in the equatorial region. Similarly, the radish blossoms only occasionally as far north as Porto Rico, where the principal growing season is during the winter months (*13*). This behavior of the radish, again, is obviously an approach toward the nonflowering type of perennial. Similarly, Dr. Van Fleet states that a lima bean coming under his observation in the Tropics had continued to grow as a perennial for a number of years, having attained giant proportions, while there was only occasional and sparse fruiting. Conversely, the beet ordinarily is a biennial in the latitude of Washington, but when grown in Alaska where the summer days are very long, it is likely to develop seed and thus complete its life cycle in a single season. The intimate relationship existing between the length of day and the attainment of the reproductive stage is strikingly shown by the behavior of the radish under special conditions. In the box of plants used as controls in the experiment described on page 565 and discussed above, the great majority of the individuals developed normal flowering stalks and seed pods in due season (see Pl. 75, B). A few individuals, however, developed considerably later, because of delayed germination or some other reason; and these delayed plants began the formation of flowering stalks. The length of the day having decreased to the critical length, the growth of the seed stalk was arrested after a height of a few inches was attained; and instead of the normal flower head, a crown of foliage leaves developed, as shown in Plate 69, B, thus indicating the resumption of vegetative activity. What is believed to be another example of the directing action of relative day length is the behavior of certain northern varieties of pepper (Capsicum) when planted in Porto Rico in the spring months (*13*). Under these conditions the peppers imported from the higher latitudes of the United States were able to form only a very few fruits before they began to yellow and shed their foliage, after which the plants soon perished. Also, it is stated that the radish when grown in Porto Rico during the winter months behaves as it does when grown nearer the equator. The above-mentioned experimental results and observations seem to justify the conclusion that the relative length of the day through the year is a factor of the first importance in determining whether many plants behave as annuals, biennials, or perennials, and whether reproduction in such plants is vegetative or sexual or both in any particular region.

The forcing of two flowering periods in a single season under controlled conditions naturally directs attention to another phase of periodicity in plant activity—namely, the appearance of the blossoming period in both spring and fall, or only in one of these seasons in regions outside the Tropics. This question is of special interest with respect to perennials. It is apparent that plants blooming only in the spring or fall or in both seasons are to be regarded as requiring relatively short days for attaining this stage. In annuals, ordinarily a period of vegetative development must necessarily precede flowering, so that the latter stage is likely to be deferred till autumn; but when propagation is by means of bulbs or other reproductive storage organs, blossoming may well occur in the spring. In hardy shrubs and trees a typical condition is that in which the formation of flowers or flower buds is inaugurated in the autumn under the influence of the shortening days, while the flowering process is interrupted before completion through the intervention of cold weather. The result is that actual blossoming usually takes place in the spring; but if the fall or early winter temperatures are abnormally high, the flowering process may be completed before cold weather intervenes. This phenomenon is occasionally observed in the apple. In the spring, temperature would be the chief factor in determining the date of blossoming for this class of plants. It is suggested that the seasonal distribution of flower-bud formation in the lemon which is considered in a recent interesting article by Reed (*22*) may be due to these light and temperature relations. The process is most active during the late fall and again in very early spring, with a winter period of low activity. Throughout the summer period of long days, also, activity is at a minimum.

### LENGTH OF DAY CONTRASTED WITH LIGHT INTENSITY

As early as 1735 Reaumer (*21*) undertook to make accurate comparisons of the total quantities of heat required to bring plants to given stages of maturity. At intervals since that time this idea has been revived, and serious efforts have been made to establish some form of quantitative relationship between plant development and the quantity of heat received from the sun. The work of Linsser (*15, 16*) and of Hoffman (*11, 12*) in this field is worthy of special mention. In this connection, also, Abbe's critical review of investigations having to do with the relations between climates and crops is of interest (*1*). It is believed that the results of the present investigation have an important bearing on the subject. Since the quantity of solar radiation received directly by the plant is the product of the intensity and the length of the exposure, it might be expected that any relationship existing between plant processes and the total quantity of radiation received would be disturbed by changes in either the intensity of the light or the duration of the exposure to its action. It has been shown that the relative length of the day is a factor of the greatest importance in relation to reproductive processes in the plant, and it will be of interest to consider whether the intensity of the solar radiation is also of special significance. At the outset it may be observed that it hardly seems likely that light intensity could exert a controlling influence on reproduction in plants, in view of the extent to which the response of plants to differences in light intensity has been studied by investigators without discovery of any very significant relationships so far as concerns reproduction. In the experiments discussed in preceding paragraphs it was found that where daily exposures of 7 hours and 12 hours, respectively, were equally effective in shortening the vegetative period, a total daily illumination aggregating on an average 9 to 10 hours but consisting of two separate exposures, with a 4-hour period of darkness intervening, was vastly less effective in this respect. This shows at once that the total quantity of radiation received can not be responsible for the shortening of the vegetative period produced by shortening the single daily exposure to light. Furthermore, since in the double daily exposures the intervening period of darkness to which the plants were subjected, 10 a. m. to 2 p. m., was at the time of day when the intensity of the solar radiation reaching the earth's surface is at its maximum, the average intensity of the radiation received by these plants is less than that received by those plants which were exposed continuously from 9 a. m. to 4 p. m.

The Stewart Cuban Mammoth tobacco which requires a day length of 12 hours or less to attain the blossoming stage has been grown commercially to some extent under an artificial shade of coarse cheesecloth estimated to reduce the intensity of the sunlight by approximately one-third. It has been observed that this shade has had no noticable effect on the date of blossoming of the tobacco. Again, the aster used in the present investigation grows in the wild state under a variety of situations, some of which are very shaded, but observation during the past season showed that there was no appreciable difference in dates of flowering under these varied exposures.

Further evidence on this subject is furnished by the following experiments in which soybeans were subjected to different degrees of shading, primarily for determining the effect on oil formation in the seed. Different types of shade were employed, and in some instances shading was combined with regulated differences in the water supply of the soil. In all these experiments the aim has been to use a type of shade which would reduce to a minimum secondary effects, such as modifying the air temperature and the temperature and moisture content of the soil. The object, in short, was to measure, as far as practicable, only the direct action of different light intensities on the plant itself, though, of course, this goal can not be fully attained. With this aim in mind the triangular type of shade, shown in Plate 76, B, was used in a series of tests made in 1916. For this shade the standard cheese-cloth of best grade, extensively used for surgical dressings, was employed (see Pl. 77, E). The opening extending around the shade near the top, with loose overhanging flap, is for the purpose of facilitating ventilation. The arrangement is such that the frame of the shade can be raised from time to time to accommodate the growth of the plants. The width of the frame was 4 inches at the base and 18 inches at the top, and it was 30 inches high. In these as in the later tests the Peking variety of soybean was used. It will be recalled that this variety is quite sensitive to changes in the length of the day.

The simplest and perhaps the most satisfactory type of shade was that employed in 1917 and 1918. A frame of iron pipe, 30 inches high, 40 inches wide, and of the desired length, was used to support the cloth. The shades in all cases extended almost due east and west. The beans in each instance were planted in a row 6 inches to the north of the center line of the shade to allow for the southerly swing of the sun's course through the sky. Comparatively open, loosely woven cloth, of the type used for the commercial culture of cigar-wrapper tobacco in New England and Florida, was used for this shade. Four different weaves of cloth were used—6 by 6, 8 by 10, 12 by 12, and 12 by 20 mesh, these figures indicating the average number of threads to the linear inch. These cloths are shown in natural size in Plate 77, A–D.

In 1918 tests were extended to include differences in water supply in combination with three different degrees of shading (see Pl. 78, A). This was accomplished by planting the beans in wooden boxes 24 feet long, 12 inches wide, and 14 inches deep, each dox being divided by partitions into three 8-foot sections. These boxes were set in the soil so as to extend about 2 inches above the surface and were filled with soil up to 2 inches of the top. Under each degree of shading, three different soil-moisture contents were maintained, designated as wet, medium, and dry. Rainfall was largely excluded by laying boards over the boxes on each side of the plants, the boards having sufficient pitch outward to turn the flow of the water. In addition, control plantings were made in the field, a portion without shade and the remainder covered with the shade cloths; and these received no water except the rainfall. Only those features of the test which relate to shading will be considered here, details of the differences in water supply and their effects pertaining more properly to the next section of the paper. To ascertain whether the simplified form of shade exerted any decided indirect effect through the soil, soil thermographs were installed in the soil at a depth of 3 inches under the 12 by 20 cloth shade, in a position near the plants and in a similar position on the field row receiving no special treatment. No significant differences in the temperature records were obtained.

A matter of special importance, of course, is the degree of shading produced by the different types of shade and different weaves of cloth used. For several reasons only approximations can be had as to the intensity of the light received by the plants under the shades. The positions of different plants and different parts of the same plant with respect to the light necessarily vary, and the shape of the shade involves a constantly changing transmission rate by the shade cloth. The normal daily range in light intensity is magnified by the shade, since the coefficient of transmission of the cloth is greatest at midday and decreases toward sunrise and sunset. In the 1916 type of shade there is a relatively small coefficient of light transmission furnished by the sloping side walls covered with cheesecloth. In the simplified type of shade only the transmission through the top comes into consideration, since there are no side walls. The southward extension of the top is such, however, that only diffuse light reaches the plants from the side, with the exception of their extreme lower portions, which are exposed to the direct sunlight in the early morning and late afternoon. In the open type of shade, diffuse light naturally becomes a larger factor. Observations made by Prof. H. H. Kimball, of the United States Weather Bureau, by means of the pyrheliometer gave transmission coefficients of 0.441, 0.292, 0.452, 0.613, and 0.727, respectively, for cheesecloth 12 by 20, 12 by 12, and 8 by 10, and for 6 by 6 mesh netting when exposed normally to the sun's rays. Formulas also were developed by Prof. Kimball which make it possible to compute the shading effect at any hour of the day and for any date. Since the sun's rays never strike the shade cloth at normal incidence, the maximum intensity of the transmitted light, which is attained at midday, is slightly less than indicated by the above values. The computed shading effect produced by each type of netting at various hours of the day on June 1, July 1, and August 1 is shown in Table V. It is seen that for horizontal exposures the shading effect is almost constant from 10 a. m. to 2 p. m. but increases considerably from 10

a. m. to 8 a. m. and from 2 p. m. to 4 p. m. and increases very rapidly from 8 a. m. to 6 a. m. and from 4 p. m. to 6 p. m. For vertical exposures the reverse relations, of course, obtain.

TABLE V.—*Computed shading effect of netting of various weaves and of cheesecloth at different hours of the day during the summer months, with horizontal exposure of the netting and cheesecloth and also with vertical exposure of the cheesecloth*

[Complete shading represented by unity]

| Kind of material. | June 1. | | | | July 1. | | | | August 1. | | | |
|---|---|---|---|---|---|---|---|---|---|---|---|---|
| | Noon. | 10 a. m. and 2 p. m. | 8 a. m. and 4 p. m. | 6 a. m. and 6 p. m. | Noon. | 10 a. m. and 2 p. m. | 8 a. m. and 4 p. m. | 6 a. m. and 5 p. m. | Noon. | 10 a.m. and 2 p. m. | 8 a.m. and 4 p. m. | 6 a. m. and 6 p. m. |
| 6 by 6 netting.. | 0.30 | 0.30 | 0.37 | 0.66 | 0.29 | 0.31 | 0.36 | 0.61 | 0.30 | 0.31 | 0.37 | 0.69 |
| 8 by 10 netting. | .41 | .43 | .51 | .90 | .41 | .42 | .50 | .83 | .41 | .43 | .51 | .94 |
| 12 by 12 netting. | .56 | .58 | .69 | 1.0 | .55 | .58 | .78 | 1.0 | .56 | .59 | .69 | 1.0 |
| 12 by 20 netting. | .69 | .71 | .83 | 1.0 | .68 | .71 | .81 | 1.0 | .69 | .72 | .84 | 1.0 |
| Cheesecloth (top)......... | .57 | .59 | .70 | 1.0 | .57 | .59 | .69 | 1.0 | .56 | .59 | .71 | 1.0 |
| Cheesecloth (vertical sides). | 1.0 | .78 | .62 | .57 | 1.0 | .81 | .63 | .57 | .96 | .77 | .62 | .57 |

To obtain further information as to the shading effect of the nettings used, a section of the simplified type of shade, without side covering, was set up and covered with the 12 by 12 netting. Under this shade (about 6 inches below the netting) Livingstone standardized black and white spherical atmometer cups were installed, and corresponding control cups were placed in full sunlight in the open air. In general, it was found that satisfactory results could not be secured when the wind was blowing; but when there was no appreciable breeze, readings were obtained which seemed to indicate a coefficient of light transmission reasonably close to that determined by Prof. Kimball. Typical readings obtained on clear, calm days are given in Table VI.

TABLE VI.—*Readings of black and white spherical atmometer cups under 12 by 12 netting and in direct sunlight, and the indicated coefficient of light transmission, 1919*

| Date. | Period of exposure. | Readings. | | | | Difference. | | Indicated coefficient of light transmission for 12 by 12 net. |
|---|---|---|---|---|---|---|---|---|
| | | Under the net. | | In the open. | | Under the net. | In the open. | |
| | | Black cup. | White cup. | Black cup. | White cup. | | | |
| | | *Cc.* | *Cc.* | *Cc.* | *Cc.* | *Cc.* | *Cc.* | |
| Aug. 26 | 10.45 a. m. to 11.45 a. m.. | 4.8 | 3.5 | 6.0 | 3.7 | 1.3 | 2.3 | 0.56 |
| 26 | 10.45 a. m. to 3 p. m...... | 24.3 | 18.0 | 29.6 | 19.5 | 6.3 | 10.1 | .62 |
| 28 | 9 a. m. to 3 p. m......... | 20.7 | 14.4 | 26.9 | 15.8 | 6.3 | 11.1 | .56 |
| 29 | 10.15 a. m. to 3 p. m...... | 18.2 | 12.8 | 22.4 | 13.0 | 5.4 | 9.4 | .57 |

In the 1916 experiments the soybeans were planted June 21, and the shade was placed in position July 5. Detailed observations were made on the growth and development of the shaded plants and of the unshaded controls. There were 93 individuals under the shade and 67 in the

160115°—20——3

control row. The summarized data in Table VII will bring out the comparative behavior of the shaded and unshaded plants.

TABLE VII.—*Effect of shading soybeans with cheesecloth, 1916*

| Treatment. | Average height. | Air-dry weight per stalk, defoliated. | Yield of beans per stalk. | Yield of hulls per stalk. | Percentage of beans in seed pods. | Date of blossoming. |
|---|---|---|---|---|---|---|
| | | *Gr.* | *Gr.* | *Gr.* | | |
| Plants shaded............ | 3 ft. 5 in.... | 5. 4 | 10. 5 | 5. 4 | 66. 1 | Aug. 7 |
| Plants not shaded....... | 2 ft. 3 in.... | 9. 9 | 17. 0 | 9. 0 | 65. 2 | Do. |

The shaded plants show the typical effects of reduced light intensity so often observed—increased elongation of stem, slender growth, enlarged area of leaves, reduced production of dry matter. Besides these effects the yield of seed was considerably reduced. For present purposes the important fact is that although the maximum intensity of the direct light reaching these plants was only about 43 per cent of the normal, the date of blossoming was not affected in the slightest degree. This is a striking contrast with the fact that by reducing the length of the daily light exposure from an average of approximately 14 hours to 12 hours, or about 15 per cent, the length of the period from germination till blossoming was reduced from 51 to 21 days. It was observed, however, that the seeds of the shaded plants were about a week later than those of the control plants in reaching final maturity.

In the 1917 tests the beans were planted June 27 and the shades placed in position a few days after germination for the first series, while in a second series the shades were set up at the time of blossoming. Two grades of netting were used, the 6 by 6 and the 8 by 10 mesh. The general behavior of the plants is shown in Table VIII. Here, again, it is seen that reducing the intensity of the direct sunlight to maxima of about 70 and 59 per cent, respectively, of the normal has shown no effect on the date of flowering.

TABLE VIII.—*Effect of shading soybeans with 6 by 6 and 8 by 10 mesh cotton netting, 1917*

| Treatment. | Number of individuals grown. | Average height. | Air-dry weight per stalk, defoliated. | Yield of beans per stalk. | Yield of hulls per stalk. | Percentage of beans in seed pods. | Date of blossoming. |
|---|---|---|---|---|---|---|---|
| | | *Inches.* | *Gr.* | *Gr.* | *Gr.* | *Gr.* | |
| 6 by 6 netting from germination to maturity.................... | 72 | 28 | 4. 7 | 9. 4 | 5. 1 | 64. 6 | Aug. 17 |
| 6 by 6 netting from blossoming to maturity.................... | 67 | 24 | 5. 8 | 9. 9 | 5. 0 | 66. 8 | Do. |
| 8 by 10 netting from germination to maturity.................. | 75 | 25 | 3. 7 | 7. 0 | 4. 1 | 62. 9 | Do. |
| 8 by 10 netting from blossoming to maturity.................... | 65 | 25 | 5. 4 | 9. 4 | 5 4 | 63. 4 | Do. |
| Not shaded.................... | 305 | 22 | 4. 8 | 8. 5 | 5. 1 | 62. 7 | Do. |

TABLE IX.—*Effect of various degrees of shading in combination with differences in water supply on the growth and development of soybeans, 1918*

| Treatment. | | | Number of plants grown. | Average height. | Weight per stalk defoliated. | Weight of hulls per plant. | Weight of beans per plant. | Percentage of beans in seed pods. |
|---|---|---|---|---|---|---|---|---|
| Shade. | Moisture. | Duration. | | | | | | |
| | | | | *Inches.* | *Gr.* | *Gr.* | *Gr.* | |
| 12 by 12 netting.. | Wet........ | Germination till maturity. | 40 | 31 | 9.8 | 8.4 | 14.5 | 63.5 |
| Do.......... | Medium..... | ....do............. | 42 | 26 | 6.1 | 5.6 | 10.2 | 64.5 |
| Do.......... | Dry........ | ....do........... | 44 | 21 | 3.4 | 3.5 | 6.8 | 65.8 |
| 6 by 6 netting.... | Wet........ | ....do........... | 43 | 31 | 9.6 | 8.5 | 14.3 | 62.8 |
| Do.......... | Medium..... | ....do........... | 47 | 28 | 6.7 | 6.7 | 11.4 | 63.0 |
| Do.......... | Dry........ | ....do........... | 48 | 21 | 3.1 | 3.1 | 6.9 | 69.4 |
| 12 by 12 netting... | Actual rainfall. | ....do........... | 63 | 31 | 9.4 | 8.1 | 15.7 | 66.1 |
| 6 by 6 netting.... | ......do....... | ....do........... | 58 | 31 | 9.6 | 8.6 | 16.0 | 65.6 |
| 12 by 20 netting... | ......do....... | ....do........... | 47 | 34 | 9.3 | 7.5 | 15.0 | 66.9 |
| Not shaded*a*....... | ......do....... | ....do........... | 110 | 26 | 9.9 | 8.5 | 16.4 | 65.7 |
| Not shaded*a*..... | ......do....... | ....do........... | 77 | 31 | 13.5 | 10.5 | 24.8 | 70.2 |
| 12 by 12 netting... | Wet........ | Blossoming till maturity. | 43 | 28 | 7.7 | 6.3 | 14.5 | 69.9 |
| Do.......... | Medium..... | ....do........... | 45 | 30 | 7.8 | 7.1 | 13.1 | 65.0 |
| Do.......... | Dry........ | ....do........... | 45 | 29 | 7.0 | 4.8 | 11.0 | 69.5 |
| 6 by 6 netting.... | Wet........ | ....do........... | 45 | 32 | 8.7 | 6.7 | 13.3 | 66.3 |
| Do.......... | Medium..... | ....do........... | 44 | 31 | 8.3 | 6.5 | 12.7 | 66.3 |
| Do........ | Dry........ | ....do........... | 44 | 25 | 7.3 | 6.5 | 11.6 | 63.9 |
| Not shaded.... | Wet........ | ....do........... | 46 | 31 | 7.8 | 7.2 | 13.1 | 64.9 |
| Do.......... | Medium..... | ....do........... | 48 | 32 | 7.9 | 6.3 | 12.3 | 66.2 |
| Do.......... | Dry........ | ....do........... | 48 | 24 | 5.7 | 4.9 | 8.8 | 64.3 |
| 12 by 12 netting.. | Actual rainfall. | ....do............. | 58 | 30 | 10.1 | 8.2 | 16.8 | 67.3 |
| 6 by 6 netting.... | ......do....... | ....do............. | 62 | 30 | 10.6 | 9.8 | 17.7 | 64.4 |
| 12 by 20 netting... | ......do....... | ....do............. | 49 | 32 | 9.7 | 8.2 | 16.9 | 67.2 |

*a* This planting differed from the control immediately preceding only in that the plants were spaced 5 to 6 inches apart in the row while in all other cases they were spaced 2 to 3 inches apart.

In the 1918 experiments the plantings were made from June 4 to 6. Two different degrees of shading were used in combination with three different rates of water supply in each of two series, one covering the period from germination to maturation and the other extending only from blossoming till maturation. In addition, two corresponding series were run, in each of which three different degrees of shading were employed without variation in the water supply, the plants in this case being grown in open field rows without use of boxes, so that the actual rainfall of the season was received by the soil. As controls, a series was arranged without shade but with the three rates of water supply, which extended only from the blossoming period till maturation, the plantings being in buried boxes as in the other experiments having to do with water supply. An additional control consisted of a planting in the field without any special treatment as to either shade or water supply; and, incidentally, a similar planting was made which differed only in that the plants were spaced 5 to 6 inches apart instead of the standard distance of 2 to 3 inches used in all other cases. The shades for the two periods of shading were placed in position, and the special water treatments were begun on June 12 and 13 and August 9, respectively. The results of the tests are summarized in Table IX. It appears that the effect of the shade on the size, weight, and relative proportions of the plant parts is dependent to a considerable extent on the relative water supply. In general, however, reduction in light intensity during the period from germination till maturity gives results similar to those obtained in the preceding tests; and there is a tendency toward a reduced yield of seed, as previously noted. Reducing the intensity of light during the period between blossoming and final maturity, on the other hand, appears to increase somewhat the yield of seed. Without exception, the plants began blossoming on August 7 under all treatments as to shade and differences in water supply, applied either singly or in combination. In these tests it is estimated that under the heaviest shading the maximum intensity of the direct sunlight reaching the plant was only 32 per cent of the normal, and the average for the day could scarcely have exceeded 25 per cent of the normal.

## RELATION OF OTHER FACTORS OF THE ENVIRONMENT TO REPRODUCTION

Having seen that under the conditions of the experiments described in the previous section differences in light intensity were without effect on the length of the vegetative period which preceeds flowering in soybeans, it is worth while considering whether other factors of the environment, especially water supply and temperature, are of significance. In studying the relation of the water supply to the formation of oil in the seed, a number of tests have been made with soybeans, beginning with 1912; but it will suffice to consider here only the results obtained for the years 1916 and 1918 with the Peking variety. In 1916 plantings were made in a series of four boxes set in the soil and provided with board covers, just as has been described in the preceding section (see p. 583). Each of the boxes was 12 feet long, 12 inches wide, and 12 inches deep. In one of these boxes the soil was maintained in a relatively moist condition from germination to final maturity, and in a second one the soil was kept comparatively dry during this period. In the third box the soil moisture was kept the same as that in the first box till the most active flowering stage was past, after which the moisture content was reduced to that of the second box. In the fourth box the soil was kept relatively dry till the flowering stage was past and thereafter in a relatively moist condition. A control planting receiving the actual rainfall was also made in the field. The beans were planted June 21, and the addition of water to the boxes began July 17. The transition in the moisture relations of the third and fourth boxes was begun August 19. The appearance of the plants in the boxes in the late summer is shown in Plate 79, A. The quantities of water supplied to the boxes each week, together with the rainfall for the period of the tests, are given in Table X. Determinations of the moisture content of the soil in the boxes were made at intervals through the month of August. Experience has shown that in the field the soil used in these tests contains 16 to 18 per cent moisture when in best condition for most crop plants. The results of the moisture determinations in the boxes are shown in Table XI.

TABLE X.—*Quantities of water added to boxes and the rainfall during the period of the tests dealing with effect of differences in soil moisture on the development of soybeans, 1916*

| Week ending— | Box 1, wet from germination to maturity. | Box 2, dry from germination to maturity. | Box 3, wet from germination to blossoming; dry thereafter. | Box 4, dry from germination to maturity; wet thereafter. | Rainfall (on field planting only). |
|---|---|---|---|---|---|
| | Gallons. | Gallons. | Gallons. | Gallons. | Inches. |
| July 24 | 32 | 0 | 32 | 0 | 1.77 |
| 31 | 20 | 0 | 20 | 0 | 2.14 |
| Aug. 7 | 20 | 0 | 20 | 0 | .56 |
| 14 | 18 | 4 | 18 | 4 | .23 |
| 21 | 18 | 4 | 4 | 12 | .38 |
| 28 | 20 | 2 | 8 | 12 | .92 |
| Sept. 4 | 12 | 4 | 6 | 6 | .03 |
| 11 | 37 | 6 | 6 | 26 | .70 |
| Total | 177 | 20 | 114 | 60 | 6.73 |
| Equivalent to | a 23.6 | a 2.7 | a 15.2 | a 8 | ........ |

a Inches.

TABLE XI.—*Moisture content of soil in boxes used for growing soybeans, 1916*

| Date of examination. | Box 1, wet from germination to maturity. | Box 2, dry from germination to maturity. | Box 3, wet from germination to blossoming; dry thereafter. | Box 4, dry from germination to maturity; wet thereafter. |
|---|---|---|---|---|
| | Per cent. | Per cent. | Per cent. | Per cent. |
| July 17 | 13.0 | 13.0 | 13.0 | 13.0 |
| Aug. 1 | 20.6 | 11.0 | 21.2 | 11.0 |
| 8 | 20.6 | 11.0 | 13.8 | 11.0 |
| 15 | 15.5 | 9.5 | 15.5 | 9.5 |
| 22 | 14.7 | 10.0 | 10.8 | 16.8 |

The comparative growth and development of the plants under the different treatments are indicated by the data presented in Table XII. It appears that the control plants in the field were somewhat larger and considerably more productive than the best plants in the boxes, which were those receiving the larger water supply from germination to maturity. These differences were possibly due to the larger volume of soil available to the plants in the field. There are large reductions in the size and productiveness of the plants in the boxes resulting from a deficiency in the water supply. It appears also that a more favorable water supply during the period preceding the flowering stage resulted in greater vegetative development, while a more favorable water supply after the flowering stage gave a larger yield of seed. In spite of the well-defined differences in the size of the plants and their fruitfulness brought about by differences in the water supply, the date of blossoming was not affected at all, the first blossoms in all boxes and in the field appearing August 7. No important differences were observed in the time of final maturation under the different treatments.

TABLE XII.—*Effect of differences in the moisture content of the soil on the growth and development of soybeans, 1916*

| Treatment. | Number of plants grown. | Average height. | Air-dry weight per stalk, defoliated. | Weight of seed hulls per stalk. | Weight of beans per stalk. | Percentage of beans in seed pods. |
|---|---|---|---|---|---|---|
| | | *Inches.* | *Gr.* | *Gr.* | *Gr.* | |
| Soil wet from germination to maturity...... | 64 | 27 | 8. 0 | 7. 3 | 12. 4 | 63. 0 |
| Soil dry from germination to maturity..... | 67 | 19 | 4. 1 | 3. 1 | 8. 1 | 62. 4 |
| Soil wet from germination to blossoming and dry thereafter.................... | 69 | 26 | 7. 4 | 3. 2 | 8. 6 | 62. 2 |
| Soil dry from germination to blossoming and wet thereafter..................... | 57 | 21 | 5. 5 | 6. 6 | 11. 5 | 63. 5 |
| Control, in field under actual rainfall....... | 67 | 27 | 9. 8 | 9. 1 | 17. 0 | 65. 2 |

It may be observed in passing that the wooden boxes placed in the soil as indicated above have been found to be very satisfactory for conducting field tests dealing with the effects of the water supply on plants. By the arrangement of sloping covers on either side of the plants rainfall can be very largely excluded, and losses of soil moisture from causes other than transpiration are reduced to a minimum. Boxes of any convenient size and length may be used, and placing the boxes in the soil insures a close approach to general field conditions.

The general plan as well as a summary of the results of the 1918 tests on the effects of differences in water supply in combination with different degrees of shading have been given in the preceding section on light intensity. It is appropriate to give here further details of the water treatments. The surface area of the soil in each 8-foot section of the boxes was 8 square feet, so that nearly 5 gallons of water would be required to supply the equivalent of a rainfall of 1 inch. The water was applied in measured quantities by means of a garden hose. Although nearly all water lost by the soil was through transpiration, it was found necessary to water heavily each day in periods of hot and dry weather. The quantities of water added weekly and the rainfall during the period of the tests are shown in Table XIII.

Soil moisture determinations were made at intervals during the season from samples taken from the field and from the boxes. The samples were taken to a depth of 12 inches—that is, to the bottom of the soil in the boxes. Composite samples were made up from boxes receiving the same quantities of water. The samples were taken in all cases just before adding water to the soil, so that the average moisture contents would be somewhat higher than these figures. Results of the moisture determinations are shown in Table XIV.

TABLE XIII.—*Quantities of water added weekly to soybeans, and the rainfall during the period of the tests dealing with effect of differences in soil moisture, 1918*

| Week ending— | From germination to maturity. | | | | | | From blossoming to maturity. | | | | | | | | | Rainfall on controls. |
|---|---|---|---|---|---|---|---|---|---|---|---|---|---|---|---|---|
| | 12 by 12 net shade. | | | 6 by 6 net shade. | | | 12 by 12 net shade. | | | 6 by 6 net shade. | | | Not shaded. | | | |
| | Wet. | Medium. | Dry. | Wet. | Medium. | Dry. | Wet. | Medium. | Dry. | Wet. | Medium. | Dry. | Wet. | Medium. | Dry. | |
| | Gall. | Gall. | Gall. | Gall. | Gall. | Gall. | Gall. | Gall. | Gall. | Gall. | Gall. | Gall. | Gall. | Gall. | Gall. | In. |
| June 21 | 0.5 | 0.5 | (a) | 0.5 | 0.5 | (a) | 4.0 | 4.0 | 4.0 | 4.0 | 4.0 | 4.0 | 4.0 | 4.0 | 4.0 | 0.11 |
| 28 | 4.0 | 1.0 | (a) | 4.0 | 1.0 | (a) | 2.0 | 2.0 | 2.0 | 2.0 | 2.0 | 2.0 | 2.0 | 2.0 | 2.0 | .87 |
| July 5 | 9.0 | 7.0 | (a) | 9.0 | 7.0 | (a) | 9.0 | 9.0 | 9.0 | 9.0 | 9.0 | 9.0 | 9.0 | 9.0 | 9.0 | .58 |
| 12 | 3.0 | 2.0 | 1.5 | 3.0 | 2.0 | 1.5 | 4.0 | 4.0 | 4.0 | 4.0 | 4.0 | 4.0 | 4.0 | 4.0 | 4.0 | .57 |
| 19 | 3.0 | 2.0 | 1.0 | 3.0 | 2.0 | 1.0 | 3.0 | 3.0 | 3.0 | 3.0 | 3.0 | 3.0 | 3.0 | 3.0 | 3.0 | 1.41 |
| 26 | 6.0 | 6.0 | 6.5 | 6.0 | 6.0 | 6.5 | 5.0 | 5.0 | 5.0 | 5.0 | 5.0 | 5.0 | 5.0 | 5.0 | 5.0 | .71 |
| Aug. 2 | 8.0 | 8.0 | 3.0 | 8.0 | 8.0 | 3.0 | 7.0 | 7.0 | 7.0 | 7.0 | 7.0 | 7.0 | 7.0 | 7.0 | 7.0 | .69 |
| 9 | 24.0 | 18.0 | 9.0 | 24.0 | 18.0 | 9.0 | 22.0 | 20.0 | 18.0 | 22.0 | 20.0 | 18.0 | 22.0 | 20.0 | 18.0 | .46 |
| 16 | 24.0 | 22.0 | 5.0 | 24.0 | 22.0 | 5.0 | 24.0 | 22.0 | 4.0 | 24.0 | 22.0 | 4.0 | 24.0 | 22.0 | 4.0 | 1.03 |
| 23 | 24.0 | 22.0 | 5.5 | 24.0 | 22.0 | 5.5 | 24.0 | 22.0 | 4.0 | 25.0 | 22.0 | 4.0 | 24.0 | 22.0 | 4.0 | .26 |
| 30 | 30.0 | 20.0 | 8.0 | 30.0 | 20.0 | 8.0 | 30.0 | 20.0 | 5.5 | 30.0 | 20.0 | 5.5 | 30.0 | 20.0 | 5.5 | .44 |
| Sept. 6 | 28.0 | 18.0 | 6.0 | 28.0 | 18.0 | 6.0 | 28.0 | 18.0 | 8.0 | 28.0 | 18.0 | 8.0 | 28.0 | 18.0 | 8.0 | .44 |
| 13 | 8.0 | 6.0 | 2.0 | 8.0 | 6.0 | 2.0 | 8.0 | 6.0 | 6.0 | 8.0 | 6.0 | 6.0 | 8.0 | 6.0 | 6.0 | .51 |
| 20 | 18.5 | 14.0 | 5.5 | 18.5 | 14.0 | 5.5 | 18.5 | 14.0 | 2.0 | 18.5 | 14.0 | 2.0 | 18.5 | 14.0 | 2.0 | 1.77 |
| 27 | 8.0 | 4.0 | 4.0 | 8.0 | 4.0 | 4.0 | 8.0 | 4.0 | 5.5 | 18.5 | 14.0 | 5.5 | 8.0 | 4.0 | 5.5 | .62 |
| Total | 198.0 | 150.5 | 57.0 | 198.0 | 150.5 | 57.0 | 196.5 | 160.0 | 87.0 | 196.5 | 160.0 | 87.0 | 196.5 | 160.0 | 87.0 | 10.47 |
| Equivalent to | b39.6 | b30.1 | b11.4 | b39.6 | b30.1 | b11.4 | b39.3 | b33.3 | b17.4 | b39.3 | b33.3 | b17.4 | b39.3 | b33.3 | b17.4 | ..... |
| For period from Aug. 2 to Sept. 27 | | | | | | | 32.5 | 25.2 | 10.6 | 32.5 | 25.2 | 10.6 | 32.5 | 25.2 | 10.6 | 5.5 |

a None.      b Inches.

TABLE XIV.—*Water content of soil in boxes and in the field during the period of the tests with soybeans, 1918*

| Date of sampling. | From germination to maturity. | | | From blossoming to maturity. | | | | | | In field. |
|---|---|---|---|---|---|---|---|---|---|---|
| | 12 by 12 and 6 by 6 netting. | | | 12 by 12 and 6 by 6 netting. | | | No shade. | | | |
| | Wet. | Medium. | Dry. | Wet. | Medium. | Dry. | Wet. | Medium. | Dry. | |
| | Per ct. | Per ct. | Per ct. | Per ct. | Per ct. | Per ct. | Per ct. | Per ct. | Per ct. | Per ct. |
| June 19 | 19.4 | 18.6 | 17.1 | ...... | ...... | ...... | ...... | ...... | ...... | ...... |
| July 1 | 19.5 | 16.4 | 15.3 | 20.6 | 20.6 | 20.6 | 20.6 | 20.6 | 20.6 | ...... |
| 8 | 20.8 | 16.4 | 12.9 | 19.1 | 19.1 | 19.1 | 19.1 | 19.1 | 19.1 | ...... |
| 15 | 21.9 | 18.0 | 14.6 | 18.8 | 18.8 | 18.8 | 20.2 | 20.2 | 20.2 | ...... |
| 23 | 16.7 | 13.1 | 11.3 | 18.1 | 18.1 | 18.1 | 18.3 | 18.3 | 18.3 | ...... |
| Aug. 2 | 17.1 | 12.5 | 10.4 | 16.6 | 16.6 | 16.6 | 15.5 | 15.5 | 15.5 | ...... |
| 13 | ...... | ...... | 10.3 | ...... | ...... | 10.3 | ...... | ...... | ...... | 10.8 |
| 16 | 19.6 | 18.2 | 12.2 | 19.6 | 18.2 | 12.2 | 19.6 | 18.2 | 12.2 | 11.9 |
| 20 | ...... | ...... | 8.4 | ...... | ...... | 10.0 | ...... | ...... | 11.0 | 11.5 |
| Sept. 5 | 22.1 | 18.7 | 11.5 | 22.1 | 18.7 | 11.5 | 22.1 | 18.7 | 11.5 | 8.3 |

The rainfall was relatively light but extraordinarily uniform in distribution during the season; and, though the field soil was comparatively dry most of the time, the plants did not show wilting at any stage. In the boxes the plants in the "dry" sections were kept in a condition in which wilting frequently occurred in the middle of the day through the season. The boxes were 12 inches in width while the distance between field rows was 3.3 feet. It is interesting to note, therefore, that the plants in the boxes, which had slightly less than a third as much lateral soil area from which to draw their moisture as was available to the plants in the field, required an addition of water equal to three times the rainfall in order to attain the same development as was reached by the field plants. When this condition was attained—that is, in the "wet" boxes— the size of the plants was almost exactly the same as that of the plants in the field.

It is seen at once (Table X) that in the boxes the water supply is the chief limiting factor, for the height of the plants, the size of the stalks, the production of seed, etc., are greatly affected by the quantity of the water supplied. On the other hand, reduction of the water supply, even to the point where almost daily wilting of the plants occurred, did not change the time of flowering by a single day. Changing the water supply after the flowering stage likewise produced decided effects on the further development of the plants, and in the same general direction as noted above, although naturally the changes are not so great as when the differences in water supply are maintained throughout the active period of the plant's life. As regards maturation, it was observed that the plants in the wetter soil of the boxes were perhaps a week later in shedding their leaves and ripening their seed pods than those in the field and in the drier soil of the boxes.

While water supply is the chief factor influencing plant development in the boxes, these tests furnish a clear case of the simultaneous action of two limiting factors, for the different degrees of shading likewise affected the development of the plants. Quantitatively, these two limiting factors are of decidedly unequal significance. Within the limits covered by the tests, the effects of the differences in water supply could be demonstrated in nearly all cases even if the light intensity were an uncontrolled variable. On the other hand, the effects of the differences in light intensity would be completely masked in most instances if the water supply were not rigidly controlled. This experiment illustrates the problems of soil productiveness and crop yields which confront the agronomist and clearly points to the futility of attempting to deal with limiting factors of relatively small significance, such as comparatively narrow distinctions in fertilizer requirements of given soils or crops or in the crop-yielding powers of different strains or varieties of plants.

That temperature is a factor of first importance in influencing and controlling plant activities is well understood, and it needs to be considered here only in its relations to the length of day as a factor playing a dominant rôle in the reproductive processes of the plant. It is well known that in accordance with Vant Hoff's law the speed of chemical reactions is doubled for each increase of 5° to 10° in temperature; and similarly plant activities and processes such as respiration and growth are accelerated by increase in temperature, provided the optimum is not exceeded. Conversely, decrease in temperature may moderately retard the plant's activities, or this effect may increase to more or less complete inhibition. Extremes in either direction, of course, may result in killing the flowering buds or fruits, the vegetative shoots, or the entire plant. It is a matter of common knowledge that low temperature retards the development and the unfolding of flowers. An interesting interrelationship of this action of temperature and that of the seasonal decrease in length of day is seen in the behavior of such trees as the apple, previously referred to. Under the influence of the relatively short days fruit trees of this type might be expected to unfold their flowers regularly in the fall instead of in the spring were it not for the interference of low temperatures. The low temperature of winter would seem to have the effect of changing what would otherwise be among the latest flowering plants of the fall into the early flowering ones of spring.

For the temperate and frigid zones the results of the present investigation have made it clear that in some species, at least, distinctions in the time of flowering and fruiting of different varieties, which may be classed as early maturing, late maturing, nonmaturing, and sterile or nonflowering, are due primarily to responses to different day lengths which come into play as the season advances. Here, again, low temperature becomes a factor of increasing importance as the season advances, and, so far as concerns "short-day" plants, it controls the situation with respect to the conditions of nonripening of fruit and of nonflowering. With increasing latitude this relationship between the opposed action of the length of day and the falling temperature becomes more critical for the later maturing varieties. With decreasing latitude a condition is reached in the subtropics which is much more favorable to late maturing or short-day varieties, for the length of day may fall below the critical maximum for flowering without the inhibiting or destructive action of low temperatures coming into play to prevent successful fruiting. At the equator, annual periodicity of both temperature and length of day cease to play an important rôle in plant processes.

## LENGTH OF DAY AS A FACTOR IN THE NATURAL DISTRIBUTION OF PLANTS

In an intelligent understanding of the natural distribution of plants over a particular area those factors which are favorable or unfavorable to growth and successful reproduction for each species must be given consideration. Heretofore temperature, water, and light intensity relations have been considered the chief external limiting factors governing the distribution or range of plants. In the light of the observations and experimental results presented in this paper it seems probable that an additional factor, the relative length of the days and nights during the growing period, must also be recognized as among those causes underlying the northward or southward distribution of plants. [1] It is evident that the equatorial regions of the earth alone enjoy equal days and nights throughout the entire year. Provided the water relations are favorable, the warm temperatures in these regions favor a continuous growing season for plants. Passing northward from the equatorial regions into higher latitudes, temperatures promoting active vegetative growth and development are restricted to a summer period which, other conditions being equal, becomes progressively shorter as the polar regions are approached. Coincident with these changes from lower to higher latitudes, the summers are characterized by lengthening periods of daylight and the winters by decreasing periods of daylight. We may now consider how these different day and night relations operating during the summer growing period will exercise more or less control upon the northward or southward distribution of certain plants.

It is evident that a plant can not persist in a given region or extend its range in any direction unless it finds conditions not only favorable for vegetative activity but also for some form of successful reproduction. For present purposes only sexual or seed reproduction need be considered. The experiments above described have indicated that for certain plants— for example, ragweed and the aster—the reproductive or flowering phase of development in some way depends upon a stimulus afforded by the shortening of the days and the consequent lengthening of the nights as the summer solstice is passed. It remains to consider more specifically the bearing these facts may have when plants characterized by this type of behavior are subjected to the daylight relations of different latitudes. In the vicinity of Washington, D. C., the ragweeds regularly shed their first pollen about the middle of August. It may be considered that the earliest flowering plants bloom about this date each season because they react to a length of day somewhat less than that of the longest day, which is about 15 hours in this latitude. In other words, as soon as the decreasing length of day falls somewhat below 15 hours, a condition which obtains about July 1, the period of purely vegetative activity is checked, and the flowering phase of development is initiated. Should the seeds of such plants now be carried as far as northern Maine into a latitude of 46° to 47°, these plants would not experience a length of day falling below 15 hours in length, for which it is assumed they are best

---

[1] In this connection the tables showing the time of sunrise and sunset at 10-day intervals through the year for various latitudes in North America, as given in SMITHSONIAN CONTRIBUTIONS TO KNOWLEDGE, v. 21 (1876), p. 114–119, will be found very convenient for reference.

suited, until about August 1. In this latitude, then, provided other conditions did not intrude, flowering would be delayed until about August 1, and the chances of successfully maturing seed before killing frosts intervened would be greatly lessened. If the seed were carried still farther north, the plants might not blossom at all, owing to the fact that even the shortest days of the summer growing period would exceed those to which they were best suited in their normal habitat. Although in such instances these failures naturally would have been explained in the past on the basis of unfavorable temperature relations alone, it is obvious that length of day, primarily, is the limiting factor which has retarded the reproductive period so that unfavorable temperature relations have intervened to prevent the ripening of seed.

Although the Arctic summers are very short, plants have become successfully established under such conditions, largely by the development of specialized perennial types, which find the extremely long days favorable both to vegetative growth and to flower production. Although it has been usually considered that the purely Arctic forms are confined to Arctic conditions because of certain temperature requirements, etc., it is possible that length of day, hitherto overlooked as a factor in plant distribution, may have much to do with their restricted range apart from other factors of the environment.

In tropical regions it is probable that the success of many native plants is more or less closely dependent upon the conditions of equal or nearly equal days and nights which prevail there during the entire year. The varieties of bean coming from Peru and Bolivia appear to be of this type. It is evident that such plants, whose flowering conditions depend more or less closely upon a length of day little if at all exceeding 12 hours, can not attain the flowering stage attended by successful seed production in higher latitudes, at least during the summer season, which would necessarily be characterized by days in excess of 12 hours. It is indicated by the beans in question, however, that some plants of this class may grow and attain successful seed production under day lengths less than 12 hours. This being the case, such plants could at least extend their range beyond the Tropics in so far as the temperature conditions of the winter months in these latitudes were favorable to growth and reproduction.

In any study of the phenological aspects of different species of plants the fact stands out that certain plants bloom at definite seasons of the year. This is quite as marked in subtropical regions as in more northern regions having a definite summer growing season. In this connection it is probable that the relative lengths of the days and nights are of particular significance in many instances. The behavior of the composite *Mikania scandens*, as observed under specially controlled conditions and under winter conditions in the greenhouse, may be more critically considered in relation to its normal blooming season throughout its range.

This plant normally blooms from late July to middle or late September, indicating that blossoming becomes more or less inhibited as the autumnal equinox is passed in late September and the length of the day falls below 12 hours. In the greenhouse at Washington the short days of the winter, ranging around 9 to 10 hours in length, have completely inhibited the flowering phase of development of this plant. The shorter 7-hour daily exposures to light under controlled conditions have produced identical results. Thus it appears that the normal flowering period of *Mikania scandens* even in the warmer portions of its range should not occur much later in the season than the period when the days are not less than 12 hours in length. This seems to be the case in Florida, where the blooming season of Mikania is confined to August and September, as it is in much more northern portions of its range. Plants of this type, attaining their best development under daylight lengths of approximately 12 hours, should also find a more or less congenial environment under truly tropical conditions where the days are never much less than 12 hours in length. It is probable that in the Tropics, however, many plants of this type would not only become perennial in their aerial portions, but would also have a more or less continuous flowering period.

Since it has been shown that the stature of some plants increases in proportion to the length of the day to which the plants are exposed under experimental conditions, this factor should be expected to have some influence upon the stature of such plants in their normal habitat. In general, exceptional stature would be attained in those regions in which a long day period allowed the plants to attain their maximum vegetative expression before the shorter days intervened to initiate the reproductive period. This condition should hold true not only for different latitudes where a plant has an extensive northward and southward range but for different sowings in the same locality at successively later dates during the season. It is a matter of common observation that the rankest growing individuals among such weeds as the ragweed, pigweed (Amaranthus), lamb's quarters (Chenopodium), cocklebur (Xanthium), beggarticks (Bidens), other conditions being equal, are those which germinated earliest in the season, and consequently were afforded the longest favorable period of vegetative activity preceding the final flowering period. It is also a matter of common observation that all these weeds, when germinating very late in the summer and coming at once under the influence of the stimulus of the shortening days, blossom when very small, often at a height of only a few inches.

Many species of plants have an extensive northward and southward distribution. In these instances it may be that such species are capable of reacting successfully to a wide range of different lengths of day, or it is possible that the apparent adjustment to such a wide range of conditions may depend upon slightly different physiological requirements of different types which have been developed as a result of natural selections. It yet remains to be seen whether those individuals of a given species which grow successfully in high latitudes have the same physiological requirements with respect to length of day as those growing quite as successfully near the equator. In any study of the behavior of plants introduced from other regions, with a view to determining certain economic qualities, it is evident that the factor of length of day must be taken into consideration as a matter likely to have great significance.

### LENGTH OF DAY AS A FACTOR IN CROP YIELDS

From the facts which have been developed in this paper it would seem that the seasonal change in length of day is a hitherto unrecognized factor of the environment which must be taken into account when dealing with the problems in crop production. So far as is now known, the length of the day is the most potent factor in determining the relative proportions between the vegetative and the fruiting parts of many crop plants; and, in fact, as already pointed out, fruiting may be completely suppressed by a length of day either too long or too short. In some crop plants the vegetative parts alone are chiefly sought, while in others the fruit or seed only are wanted, and in still others maximum yields of both vegetative and reproductive parts are desired. It is apparent that the merits of different varieties or strains may depend largely on the relative length of day in which they are grown, and, therefore, the date of planting may easily become the decisive factor. These are matters of vital importance to the plant breeder and the agronomist. Obviously, a delay of even two or three weeks in seeding certain crops because of inclement weather conditions or other considerations may bring about misleading results. It is to be remembered, furthermore, that planting too early may be equally inadvisable, for crops requiring relatively short days for blossoming may thus come under the influence of short days in early spring, resulting in "premature" flowering and a restricted amount of growth. An impressive lesson as to the influence of length of day on the size attained by the plant before blossoming is seen in the relative heights of consecutive plantings of the Biloxi soybean, as shown in figure 3. For maximum yields of many crops it is essential that the date of planting be so regulated as to insure exposure of the plant to the proper length of day, due regard being had for the specific light requirements of each crop as well as for the relative values of the vegetative and fruiting portions of the plant.

### RESULTS OBTAINED WITH ARTIFICIAL LIGHT USED TO INCREASE THE LENGTH OF THE DAILY ILLUMINATION PERIOD DURING THE SHORT DAYS OF WINTER

These results are of particular significance, since increasing the duration of the illumination period of the short winter day by the use of electric light of comparatively low intensity has consistently resulted in initiating or inhibiting the reproductive or the vegetative phases of development, depending upon whether the plants employed normally require long or short days for these forms of expression. In these experiments a greenhouse 50 feet long, 20 feet wide, and 12 feet high to the ridge, with side walls of concrete 5 feet high to the eaves, was provided with 34 tungsten filament incandescent lights, each rated at 32 candlepower, evenly distributed beneath the glass roof. As a control, a similar greenhouse without artificial light was used. The long axis of these houses was on a north and south line. The temperature was approximately the same in the two greenhouses, ranging at night around 60° to 65° F. and 75° to 80° during the day. The unlighted greenhouse, however, tended to run two to three degrees higher than the illuminated house. Beginning on November 1 the electric lights were switched on at 4.30 p. m. and turned off at 12.30 a. m., this procedure being followed throughout the course of the experiments. Supplementing the natural length of the winter days with this 8-hour period of artificial illumination has given about 18 hours of continuous daily illumination, approaching in length the summer days of southern Alaska. Under these conditions the following results have been obtained:

A large clump of *Iris florentina* L., with all earth intact, was transplanted October 20, 1919, to each of the two greenhouses. The plants exposed to the long daily period of illumination began growing vigorously at once, soon attaining the normal size for this species, and produced blossoms on December 24 and December 30. The controls remained practically dormant and showed no tendency to blossom as late as February 12, 1920.

Seed of spinach (*Spinacea oleracea* L.), Bloomsdale Curley Savoy,[1] was sowed November 1, 1919, and came up in both houses on November 6. The plants in the control house, 20 to 25 in number, grew very slowly, producing low, compact, leafy growths or rosettes, and gave no evidence of blossoming as late as February 12. The plants in the lighted house elongated very rapidly, soon developing flower stalks, and all blossomed in the period between the dates December 8 and December 23. These have continued to elongate more or less, blossoming and shedding pollen continuously, thus becoming in effect "everblooming" plants.

Seed of cosmos (*Cosmos bipinnata* Cav.) was sowed November 1, 1919, and germinated in both houses November 5. In each greenhouse 40 to 45 plants were grown. The plants in the control house quickly flowered, and all blossomed in the period from December 22 to January 2. The plants in the lighted house grew well but remained in the strictly vegetative stage and were showing no indications of blossoming on February 12. On this date the control plants averaged 30 inches in height and the plants in the lighted house 60 inches.

---

[1] Horticultural variety.

Seed of radish (*Raphanus sativus* L.), Scarlet Globe,[1] was sowep November 1, 1919, coming up in both lighted and unlighted houses November 5. On February 12 the control plants, although more stocky and having larger roots, showed no indications of developing flower stems. In the lighted house, however, the plants had developed smaller roots, and flower buds were plainly in evidence, showing that the plants would soon blossom, as is their normal behavior in response to the long summer days out of doors.

Seedlings of the Maryland Mammoth variety of tobacco were transplanted to 12-quart iron pails on November 10, on which date they were placed in the control and the lighted houses. The control plants, six in number, exhibited the typical behavior of winter-grown Maryland Mammoth plants, all blossoming during the period from December 31 to January 8. The plants in the lighted house, six in number, behaved as typical summer-grown mammoths, becoming very compact, stout and leafy, with no indications of blossoming on February 12. On this date these plants had already produced many more leaves than the control plants.

Bulbs of *Freesia refracta* Klatt were placed in soil in 5-inch pots on July 11, 1919. Four pots of these plants were kept in the large dark house previously described from 4 p. m. to 9 a. m. daily from July 23 till November 15, when they were transferred to the greenhouses, two pots being placed in the control house and two in the lighted house. None of these plants when taken from the dark house on November 15 showed any indications of blossoming. Both lots began blossoming about December 27. In the control house, however, the plants produced many flower stalks and continued to blossom profusely for a long period. The plants in the lighted house, on the contrary, produced but few flower stalks and few blossoms and soon ceased blooming entirely.

Large, robust clumps of wild violets, of the species *Viola papilionacea* Pursh, were transplanted to pots and boxes and placed in the control and lighted houses on October 31, 1919. At the time these plants were removed from the field the abnormally warm autumn weather had forced them into bloom, and many purple, petaliferous blossoms were in evidence. As the winter days continued to shorten naturally in the control house, blossoming was suppressed and no new leaves were produced. These control plants appeared to be almost dormant, except for the production of numerous short, thickened stems which were crowded close to the ground among the old leaves. In the lighted house the production of the purple, petaliferous blossoms also ceased, but vegetative growth was initiated and new leaves appeared in great abundance. Coincident with this marked vegetative activity, the plants continuously produced fertile, cleistogamous flowers in great abundance. This furnishes another example of ever-blooming in response to a favorable length of day. In all respects this behavior of the violets in the lighted house

---

[1] Horticultural variety.

simulates the normal behavior of these plants out of doors under the influence of the long summer days.

Through the kindness of Dr. D. N. Shoemaker, several varieties of Lima beans (*Phaseolus lunatus* L.), F. S. P. I. No. 46153, from Chincha, Peru, and F. S. P. I. No. 46339, from Guayaquil, Eucador, were transplanted from the field to the greenhouse. Several plants of each of these varieties were transplanted into each house on October 18, 1919. Up to that time the plants had not flowered but gave evidence in the field of being extremely late varieties in this latitude. The control plants grew rather slowly but soon became markedly floriferous, setting pods freely. On the other hand, the plants in the artificially illuminated house produced an exceptionally rank growth of vines but did not flower.

Biloxi, Tokyo, Peking, and Mandarin varieties of soybeans were sowed November 1, 1919, and came up November 10 in both houses. In the control house the first blossoms appeared on the Biloxi and Mandarin about December 24, and on the Tokyo and Peking about December 18. In the artificially lighted house the early variety, Mandarin, blossomed on about the same date as in the control house; but under the longer light exposure the plants continued to grow vigorously, and only a very few blossoms appeared, suggesting a tendency toward gigantism. The few blossoms which formed, however, were normal and developed normal pods, while those on the control plants were cleistogamous and sterile. As late as February 12 the other three varieties showed no indications of blossoming. On that date all varieties were much taller in the electrically lighted house than in the control house.

Seed of Beggar-ticks (*Bidens frondosa* L.) were sowed in both houses on November 19, 1919, and came up in each on December 1. When the plants were very small they were transferred from the flats to 5-inch pots. The transplanting took place on December 19, and on January 12 all the control plants in the unlighted house, 8 or 10 in number, were showing tiny flower heads, although these had attained a height of only 1 to 2 inches. These flower heads came into expression as soon as the plants had developed the second pair of foliage leaves above the cotyledons. The plants in the lighted house continued to produce vegetative growth and gave no evidence of producing flower heads as late as February 12. Two of these plants which had attained a height of 8 to 9 inches in response to the lengthened period of illumination in the artificially lighted house were transferred to the control house on December 19. On January 12 both plants had produced flower heads in response to the naturally short winter days prevailing in this house and gave promise of blossoming in a short time. The sister plants remaining in the illuminated house continued to produce vegetative growth, with no evidence of blossoming.

Buckwheat (*Fagopyrum vulgare* Hill) was sowed November 1, 1919, and came up in both houses on November 7. In the control house 28 plants were grown, and 32 plants were grown in the illuminated house. The dates on which the first blossoms appeared on the control plants exposed to the short winter days extended over a range of only about a week—from December 4 to December 10, inclusive. On the other hand, the dates on which first blossoms appeared on the plants exposed to the artificially lengthened day extended over a period of about four weeks—from December 6 to January 2, inclusive. On February 12 the control plants averaged uniformly only 24 inches in height and had practically ceased growing and blooming. The plants in the artificially illuminated house, on the contrary, continued to grow vigorously and to flower freely, having attained an average height of 58 inches, some of the taller being more than 9 feet in height on February 12. These taller plants blossomed much later than the others and produced very few blossoms, thus showing a tendency to become giant forms in response to the artificially produced longer day. The ever-blooming tendency of the plants as a whole, however, was much more marked under the influence of the lengthened illumination period than in the control greenhouse. Again, although the control plants showed very uniform behavior in the range of their earliest blossoming, it is evident that the artificially lengthened period of illumination has in some manner led to a greatly extended range in the time of blossoming. Whether this really represents an unequal response of several more or less distinct, intermingled races to the artificially increased length of day or may be due in part to a more profound physiological variability which has been induced can not be determined until systematic selection and breeding studies have been carried on.

It will be evident that these data dealing with an artificially lengthened illumination period obtained by means of the electric light greatly strengthen the results of the experiments secured during the previous summer by artificially shortening the natural period of illumination through the use of dark houses. The results with the Maryland Mammoth variety of tobacco, the several soybean varieties in question, and the radish are of special significance since they were obtained by methods the direct converse of those used during the summer. Although the intensity of the electric light was undoubtedly far below that of normal sunlight, it was sufficient to initiate or to suppress the reproductive and vegetative activities of these three species as did the long days of the summer time. With respect to the ever-blooming behavior of certain of the plants under study, the results obtained indicate that this behavior is likely to follow when an approximately constant daily illumination period of a duration favorable to both growth and reproduction is maintained for a sufficient length of time. It thus seems possible that the comparatively uniform length of day prevailing in the Tropics accounts for the particular abundance of ever-bloomers in that region.

160115°—20——4

## IS THE RESPONSE TO DIFFERENCES IN THE LENGTH OF DAY A PRINCIPLE OF GENERAL APPLICABILITY IN BIOLOGY?

Experience has abundantly demonstrated the fact that the biologist who attempts to draw sweeping generalizations regarding responses of plants or animals as a whole to conditions of the environment is in serious danger of going astray, even though his observations be based on the behavior of relatively large numbers of species. With this fact clearly in mind, the following suggestions are put forward tentatively but as possibly being of sufficient interest to justify careful consideration on the part of biologists especially concerned in the fields touched upon. It has been clearly brought out in this paper that for a number of plant species the appropriate length of day acts, not merely as an accelerative, but rather as an initiative influence in bringing into expression the plant's potential capacity for sexual reproduction. Perhaps, as an equally satisfactory way of expressing the fact, it may be said that the length of the day exercises a truly determinative influence on plant growth as between the purely vegetative and the (sexually) reproductive forms of development. The response to length of day may be expected to hold for other species, although it would be premature at present to assert that all higher plants will be found to respond to this factor.

One is naturally inclined to inquire whether, also, the length of day is a controlling factor in sexual reproduction among the lower forms of plant life. The observed behavior of some of these lower forms certainly suggests that they come under the influence of the seasonal range in length of day. A single instance will suffice to illustrate the parallelism existing between the vegetative and the reproductive periods of activity, on the one hand, and the periodical change in the length of the day, on the other. Reference is made to the work of Lewis (*14*), in which it is shown that in certain species of red Algae there is a definite seasonal periodicity in the appearance of sexual and asexual forms. In brief, the July growth of these species consists primarily of tetrasporic or asexual individuals, while through August the growth is characterized by a predominance of sexual plants produced from the tetraspores of the July crop of plants. The carpospores of autumn become sporelings which persist through the winter and give rise to the tetrasporic plants of the early summer period. Should it be true that lower plants respond to differences in length of day as do some of the higher species it may be expected that various relationships between annual and perennial forms, differences in sensibility to relatively long and short days, and other facts which have been shown to apply to these higher species would likewise hold true for lower organisms. It is possible, even, that the seasonal activities of some of the parasitic microorganisms are the result of response to changes in day length.

As to animal life nothing definite can be said, but it may be found eventually that the animal organism is capable of responding to the stimulus of certain day lengths. It has occurred to the writers that possibly the migration of birds furnishes an interesting illustration of this response. Direct response to a stimulus of this character would seem to be more nearly in line with modern teachings of biology than are theories which make it necessary to assume the operation of instinct or volition in some form as explaining the phenomena in question.

### CONCLUSION

The results of the experiments which have been presented in this paper seem to make it plain that of the various factors of the environment which affect plant life the length of the day is unique in its action on sexual reproduction. Except under such extreme ranges as would be totally destructive or at least highly injurious to the general well-being of the plant, the result of differences in temperature, water supply, and light intensity, so far as concerns sexual reproduction, appears to be, at most, merely an accelerating or a retarding effect, as the case may be, while the seasonal length of day may induce definite expression, initiating the reproductive processes or inhibiting them, depending on whether this length of day happens to be favorable or unfavorable to the particular species. In broad terms, this action of the length of day may be tentatively formulated in the following principle: Sexual reproduction can be attained by the plant only when it is exposed to a specifically favorable length of day (the requirements in this particular varying widely with the species and variety), and exposure to a length of day unfavorable to reproduction but favorable to growth tends to produce gigantism or indefinite continuation of vegetative development, while exposure to a length of day favorable alike to sexual reproduction and to vegetative development extends the period of sexual reproduction and tends to induce the "ever-bearing" type of fruiting.

The term *photoperiod* is suggested to designate the favorable length of day for each organism, and *photoperiodism* is suggested to designate the response of organism to the relative length of day and night.

### SUMMARY

(1) The relative length of the day is a factor of the first importance in the growth and development of plants, particularly with respect to sexual reproduction.

(2) In a number of species studied it has been found that normally the plant can attain the flowering and fruiting stages only when the length of day falls within certain limits, and, consequently, these stages of development ordinarily are reached only during certain seasons of the year. In this particular, some species and varieties respond to relatively long days, while others respond to short days, and still others are capable of responding to all lengths of the day which prevail in the latitude of Washington where the tests were made.

(3) In the absence of the favorable length of day for bringing into expression the reproductive processes in certain species, vegetative development may continue more or less indefinitely, thus leading to the phenomenon of gigantism. On the other hand, under the influence of a suitable length of day, precocious flowering and fruiting may be induced. Thus, certain varieties or species may act as early- or late maturing, depending simply on the length of day to which they happen to be exposed.

(4) Several species, when exposed to a length of day distinctly favorable to both growth and sexual reproduction, have shown a tendency to assume the "ever-blooming" or "ever-bearing" type of development—that is, the two processes of growth and reproduction have tended to proceed hand in hand for an indefinite period.

(5) The relationships existing between annuals, biennials, and perennials, as such, are dependent in large measure on responses to the prevailing seasonal range in length of day. In many species the annual cycle of events is governed primarily by the seasonal change in length of day, and the retarding or more or less injurious and destructive effects of winter temperatures are largely incidental rather than fundamental. Hence, by artificial regulation of the length of the daily exposure to light it has been found that in certain species the normal yearly cycle of the plant's activities can be greatly shortened in point of time, or, on the other hand, it may be lengthened almost indefinitely. In certain cases, annuals may complete two cycles of alternate vegetative and reproductive activity in a single season under the influence of a suitable length of the daily exposure to light. Similarly, under certain light exposures some annuals behave like nonflowering perennials.

(6) In all species thus far studied the rate of growth is directly proportional to the length of the daily exposure to light.

(7) Although the length of the daily exposure to light may exert a controlling influence on the attainment of the reproductive stage, experiments reported in this paper indicate that light intensity, within the range from full normal sunlight to a third or a fourth of the normal, and even much less, is not a factor of importance. It follows that the total quantity of solar radiation received by the plant daily during the summer season, within the range above indicated, is of little importance directly so far as concerns the attainment of the flowering stage.

(8) In extensive tests with soybeans, variations in the water supply ranging from optimum to a condition of drought sufficient to induce temporary wilting daily and to cause severe stunting of the plants were entirely without effect on the date of flowering, although in some cases drought seemed to hasten somewhat the final maturation of the seed. Similarly, differences in light intensity, in combination with differences in water supply, failed to change the date of flowering in soybeans.

(9) The seasonal range in the length of the day is an important factor in the natural distribution of plants.

(10) The interrelationships between the length of day and the prevailing temperatures of the winter season largely control successful reproduction in many species and their ability to survive in given regions.

(11) The relation between the length of the day and the time of flowering becomes of great importance in crop yields in many instances and in such cases brings to the forefront the necessity for seeding at the proper time.

## LITERATURE CITED

(1) ABBE, Cleveland.
> 1905. A FIRST REPORT ON THE RELATIONS BEWTEEN CLIMATES AND CROPS. U. S. Dept. Agr. Weather Bur. Bul. 36, 386 p. Catalogue of periodicals and authors referred to . . . p. 364–375.

(2) ALLARD, H. A.
> 1919. GIGANTISM IN NICOTIANA TABACUM AND ITS ALTERNATIVE INHERITANCE. *In* Amer. Nat., v. 53, no. 626, p. 218–233. Literature cited, p. 233.

(3) BAILEY, L. H.
> 1891. SOME PRELIMINARY STUDIES OF THE INFLUENCE OF THE ELECTRIC ARC LIGHT UPON GREENHOUSE PLANTS. N. Y. Cornell Agr. Exp. Sta. Bul. 30, p. 83–122, 13 fig., 2 pl.

(4) ——
> 1892. SECOND REPORT UPON ELECTROHORTICULTURE. N. Y. Cornell Agr. Exp. Sta. Bul. 42, p. 133–146, illus.

(5) ——
> 1893. GREENHOUSE NOTES FOR 1892–93. I. THIRD REPORT UPON ELECTRO-HORTICULTURE. N. Y. Cornell Agr. Exp. Sta. Bul. 55, p. 147–157, 2 pl.

(6) BONNIER, Gaston.
> 1895. INFLUENCE DE LA LUMIÉRE ÈLECTRIQUE CONTINUE SUR LA FORME ET LA STRUCTURE DES PLANTES. *In* Rev. Gén. Bot., t. 7, no. 78, p. 241–257; no. 79, p. 289–306; no. 80, p. 332–342; no. 82, p. 409–419. pl. 6–15.

(7) CORBETT, L. C.
> 1899. A STUDY OF THE EFFECT OF INCANDESCENT GAS LIGHT ON PLANT GROWTH. W. Va. Agr. Exp. Sta. Bul. 62, p. 77–110, pl. 1–9.

(8) FLAMMARION, Camille.
> 1898(?). PHYSICAL AND METEOROLOGICAL RESEARCHES, PRINCIPALLY ON SOLAR RAYS, MADE AT THE STATION OF AGRICULTURAL CLIMATOLOGY AT THE OBSERVATORY OF JUVISY. *In* Exp. Sta. Rec., v. 10, no. 2, p. 103–114, 4 fig., 2 col. pl.

(9) GARNER, W. W., ALLARD, H. A., and FOUBERT, C. L.
> 1914. OIL CONTENT OF SEEDS AS AFFECTED BY THE NUTRITION OF THE PLANT. *In* Jour. Agr. Research, v. 3, no. 3, p. 227–249. Literature cited, p. 249.

(10) GIROU DE BUZAREINGUES, Ch.
> 1831. SUITE DES EXPÉRIENCES SUR LA GÉNÉRATION DES PLANTES. *In* Ann. Sci. Nat., t. 24, p. 138–147.

(11) HOFFMAN, H.
> 1882. THERMISCHE VEGETATIONSCONSTANTEN; SONNEN- UND SCHATTENTEM-PERATUREN. *In* Ztschr. Österr. Gesell. Met., Bd., 17, p. 121–131.

(12) ——
> 1886. PHÄNOLOGISCHE STUDIEN. *In* Met. Ztschr., Jahrg. 3 (Ztschr. Österr. Gesell. Met., Bd. 21), p. 113–120, pl. 6.

(13) KINMAN, C. F., and McCLELLAND, T. B.
1916. EXPERIMENTS ON THE SUPPOSED DETERIORATION OF VARIETIES OF VEGE-
TABLES IN PORTO RICO, WITH SUGGESTIONS FOR SEED PRESERVATION.
Porto Rico Agr. Exp. Sta. Bul. 20, 30 p., tab., diagr.

(14) LEWIS, I. F.
1914. THE SEASONAL LIFE CYCLE OF SOME RED ALGAE AT WOODS HOLE. *In*
Plant World, v. 17, no. 2, p. 31–35.

(15) LINSSER, Carl.
1867. DIE PERIODISCHEN ERSCHEINUNGEN DES PFLANZENLEBENS IN IHREM
VERHÄLTNISS ZU DEN WÄRMEERSCHEINUNGEN. *In* Mém. Acad. Imp.
Sci. St. Petersb., s. 7, t. 11, no. 7, 44 p.

(16) ———
1869. UNTERSUCHUNGEN ÜBER DIE PERIODISCHEN LEBENSERSCHEINUNGEN DER
PFLANZEN. ZWEITE ABHANDLUNG. *In* Mém. Acad. Imp. Sci. St.
Petersb., s. 7, t. 13, no. 8, 87 p.

(17) LODEWIJKS, J. A., Jr.
1911. ERBLICHKEITSVERSUCHE MIT TABAK. *In* Ztschr. Induk. Abstam.
u. Vererb., Bd. 5, Heft 2/3, p. 139–172; Heft 4/5, p. 285–323, illus. Lit-
eraturverzeichnis, p. 171–172, 322–323.

(18) MacDOUGAL, Daniel Trembly.
1903. THE INFLUENCE OF LIGHT AND DARKNESS UPON GROWTH AND DEVELOP-
MENT. Mem. N. Y. Bot. Gard., v. 2, 319 p., illus.

(19) MOOERS, Charles A.
1908. THE SOY BEAN. A COMPARISON WITH THE COWPEA. Tenn. Agr. Exp.
Sta. Bul. 82, p. 75–104, illus.

(20) RANE, F. William.
1894. ELECTROHORTICULTURE WITH THE INCANDESCENT LAMP. W. Va. Agr.
Exp. Sta. Bul. 37, 27 p., illus.

(21) REAUMUR, R. A. F. DE
1738. OBSERVATIONS DU THERMOMETRE, FAITES À PARIS PENDANT L'ANNÉE
M.DCCXXXV. COMPARÉES AVEC CELLES QUI ONT ÉTÉ FAITES SOUS LA
LIGNE, À L'ISLE DE FRANCE, À ALGER, ET EN QUELQUES-UNES DE NOS
ISLES DE L'AMERIQUE. *In* Hist. Acad. Roy. Sci. [Paris], Mém. Math.
& Phys., ann. 1735, p. 545–376.

(22) REED, Howard S.
1919. CERTAIN RELATIONSHIPS BETWEEN THE FLOWERS AND FRUITS OF THE
LEMON. *In* Jour. Agr. Research, v. 17, no. 4, p. 153–165, 1 fig.

(23) SCHÜBELER.
1879. THE EFFECTS OF UNINTERRUPTED SUNLIGHT ON PLANTS. (Abstract.)
*In* Nature, v. 21, p. 311–312. 1880. Original article (Studier over
Klimatets Indflydelse paa Plantelivet) in Naturen, Aarg. 3, No. 6,
p. 81–89; No. 8, p. 113–123, illus. 1879.

(24) TOURNOIS, J.
1911. ANOMALIES FLORALES DU HOUBLON JAPONAIS ET CHANVRE DÉTERMINÉES
PAR DES SEMIS HÂTIFS. *In* Compt. Rend. Acad. Sci. [Paris], t. 153,
no. 21, p. 1017–1020.

(25) ———
1912. INFLUENCE DE LA LUMIÈRE SUR LA FLORAISON DU HOUBLON JAPONAIS
ET DU CHANVRE. *In* Compt. Rend. Acad. Sci. [Paris], t. 155, no. 4,
p. 297–300.

(26) WIESNER, J.
1907. DER LICHTGENUSS DER PFLANZEN . . . vii, 322 p., illus. Leipzig.

F. Kidd and C. West. 1930. The gas storage of fruit. II. Optimum temperatures and atmospheres. Journal of Pomology and Horticultural Science 8(1):67–77.

F. Kidd

C. West

Various nineteenth-century research workers reported beneficial effects of high concentrations of carbon dioxide and low oxygen on the preservation of fruits, but it was not until 1918 that the first detailed systematic research on the so-called "gas" storage (later renamed "controlled atmosphere" storage) of fruit was initiated by Franklin Kidd and Cyril West in England. This work culminated in two classic papers on the storage of apples based on research carried out between 1918 and 1927 at the Low Temperature Laboratory for Research in Biochemistry and Biophysics at Cambridge. The second paper, published in 1930 and reproduced here, is representative of the type of research that was done during this period. It gives the results of a definitive experiment carried out in 1926–1927 which sought to bring together all the factors that had been shown in their earlier, 1927, paper (Special Report 30) to be of significance in determining the response of pome fruit to different storage atmospheres and temperatures. It therefore represents an important milestone in the development of controlled atmosphere storage technology.

Franklin Kidd was born at Weston-super-Mare in Somerset in England in 1890 and was educated at Tonbridge School in Kent before going to St. John's College, Cambridge, in 1910. While still at school, Kidd had begun experimenting with seeds and it seemed natural, having obtained a first-class Honours degree in Botany in 1912, that he should then undertake further investigations into the effects of low oxygen and high carbon dioxide concentrations on the dormancy of seeds under the supervision of F. F. Blackman. He quickly impressed with his ability and originality and received recognition for his work in the form of several prizes and university scholarships.

Cyril West was born in 1887 at Forest Hill in southeast London. After education at St. Olave's School, he studied at Imperial College, London, where he graduated in 1910. He then began working on the effects of oxygen and carbon dioxide concentrations in the soil environment on seed germination under the direction of the brother of Kidd's supervisor, V. H. Blackman. As with Kidd, this was a subject that had interested West from his pre-university days, and had it not been for the outbreak of World War I, he might have pursued a career as an academic botanist; instead, he joined the army. He was commissioned in the Royal Artillery, but chance then took a hand. He was seriously injured in a riding accident and in 1915 was invalided out of his regiment and returned to Imperial College to resume his research under V. H. Blackman.

At about this time, the Blackman brothers decided that because of their common interests in seed physiology, it would be useful to bring their two students together to work at Imperial College. Thus began a lifelong partnership between these two very able research workers which was to result in some 46 joint publications between 1917 and 1952.

By 1917, government concern over food shortages during the war led the British Department of Scientific and Industrial Research to establish the Food Investigation Organization under the direction of William B. Hardy. Hardy was one of the first scientific administrators to bring workers from different disciplines into the same laboratory environment and organize them to work on specific problems. Hitherto, grants had normally been given to individual scientists who continued working in their own laboratories. One of the problems that was drawn to Hardy's attention was the excessive wastage which was occurring in the storage of apples, particularly the important English culinary variety, 'Bramley's Seedling'. Hardy asked Blackman if he could find someone to work on this problem, and because of his interest in the effects of modified atmospheres on seed germination, Blackman recommended Kidd. Thus Kidd became one of the first scientists to join the staff of the Food Investigation Board and he quickly persuaded Blackman and Hardy that the investigation justified the additional appointment of his co-worker, Cyril West.

The two men transferred their work to Cambridge in 1918 and the first storage experiments with apples were conducted in the same year using the refrigerated rooms at the medical school. Later the work was continued inside the large John Street cold store in the Port of London. None of these facilities were entirely satisfactory, but by 1922 they were able to set up small-scale experiments in the new Low Temperature Laboratory for Research in Biochemistry and Biophysics (LTRS) in Cambridge. They also conducted some semicommercial scale trials, the first in the 1920–1921 season, using a specially modfied, more-or-less gastight store on John Chivers's farm at Histon just outside Cambridge.

Using their academic studies on seed respiration as a basis, Kidd and West soon demonstrated that the ripening of apples, pears, plums, and various types of soft fruit could be retarded without injury by increasing the carbon dioxide and lowering the oxygen concentrations in the storage atmosphere. Concentrating mainly on apples, they constructed wooden containers to hold about 120 to 150 fruits each. The containers were made gastight by coating all surfaces and joints with petroleum jelly. Control of conditions was first by hand-regulated ventilation, but later they were able to make up their experimental gas mixtures in small gas holders supplied by cylinders of compressed nitrogen, oxygen, and carbon dioxide. By the time of the 1926–1927 experiment three special cabinets had been constructed, each divided into 10 gastight compartments. All gas analyses were carried out by chemical absorption methods and, in these days of modern paramagnetic oxygen and infrared carbon dioxide analyzers, not to mention automatic control of controlled atmosphere stores by computers, it is difficult to imagine the detailed care and attention that had to be devoted simply to the maintenance of the required experimental conditions.

It is apparent from the 1927 report that their first objective was to ascertain the extent to which gas storage could be substituted for storage in air with mechanical refrigeration. At that time refrigerated storage was not generally available on farms, and if the new technique could be shown to be equally effective, it might provide a more cost-effective solution to the problems of extending the marketing season.

They developed reliable sampling and assessment procedures for measuring the commercial storage life (up to 10 percent wastage) and the mean storage life (up to 50 percent wastage) of representative samples of fruit under different conditions. In most trials, wastage was due mainly to rotting, but they noted that the incidence of rotting closely reflected the changes in firmness and background color associated with fruit senescence. At the same time they were careful to measure other disorders, such as "brownheart" (caused by excessive accumulation of carbon dioxide), internal breakdown (caused by storage at low temperatures), and scald (apparently caused by storage in a "stagnant" atmosphere). They recognized that these disorders were more directly associated with the storage conditions and that it would be necessary to set precise limits to the ranges in temperature, carbon dioxide, and oxygen most appropriate for each individual variety. In particular, they demonstrated the increased sensitivity of "gas"-stored apples to low-temperature injury.

By quantifying storage life, they were able to express the differences in the "efficiency" of different storage conditions as a ratio. Thus at mean storage temperatures of about 10°C, gas storage proved to be 1.5 to 1.9 times as effective as air storage, but at warmer ambient temperatures (e.g., 15°C during the autumn of 1921), gas storage provided no advantage. They were able to demonstrate the problem of "self-heating" which occurs when large quantities of fruit are stored together in one chamber. Thus the heat of respiration leads to a very significant increase in the temperature inside the store, and this factor is of greater importance the warmer the weather and the larger the bulk of fruit stored.

These findings convinced Kidd and West of the need for some

refrigeration of gas-stored fruit, particularly during the warmer months of the year. This, in turn, led them to study the relationships between the effective oxygen and carbon dioxide concentrations in solution in apple cells under different storage conditions and the temperature at which the fruit was being stored. They concluded that restricted ventilation was likely to be more effective at intermediate storage temperatures, but at lower storage temperatures, the store atmosphere should contain less oxygen than the equivalent maximum safe concentration of carbon dioxide. The inevitable consequence of this finding was the need to allow for the independent control of the two gases and thus a necessity to remove carbon dioxide without increasing oxygen levels.

Further research was to extend the storage life of other varieties of apples by control of oxygen as well as carbon dioxide and early experiments by Kidd and West led to the development of a system of removing carbon dioxide by absorption with a solution of sodium carbonate and returning the carbon dioxide–free atmosphere to the store, thereby lowering the level of oxygen by respiration. More recently, scrubbing with hydrated lime or adsorption on activated charcoal have been introduced as means for removing carbon dioxide from the atmospheres of fruit stores.

Their first definitive small-scale trial in 1920 on the interactions of oxygen and carbon dioxide demonstrated that at temperatures between 5 and 10°C, the best storage condition for some five different apple varieties was 11 percent $O_2$ with 10 percent carbon dioxide. However, recognizing the important effects of storage temperature, store humidity, variety and season, they embarked on further experiments to produce recommendations that were optimum for specific varieties. The paper reproduced here describes their most comprehensive experiment on 'Bramley's Seedling' apples, in which they compared storage at three temperatures—1, 5, and 10°C—and at 5, 10, and 15 percent $CO_2$ with 0, 5, 10, and 15 percent $O_2$. It provided a model for all subsequent work in this field, in terms of the design and execution of the experiment and the definitive assessment methods that were employed.

As in all pioneer research, some of the concepts subsequently proved to be an oversimplification. For example, they demonstrated that diminished oxygen and high carbon dioxide concentrations reduced the respiration rate of the stored fruit and that this effect was directly proportional to the storage life of the fruit. This led to the tentative suggestion that the availability of respirable substrate might be the factor limiting mean storage life (see Table V of the 1930 paper). It is now known that the higher rate of respiration of air-stored fruit is also associated with other degradative changes, which lead to loss of membrane integrity and function, cell wall softening, and other deteriorative changes well before exhaustion of carbohydrate reserves. Interestingly, the first paper to consider this aspect of senescence was published by F. F. Blackman and P. Parija (1928), working at the Botany School near the LTRS at Cambridge. They noted that the respiration rate of stored apples increased as the fruit aged and attributed this to a lowering of the "organizational resistance" of the tissues. Although the 'Bramley' apples used by Blackman were provided by Kidd and West from their own 1920 storage experiments, this concept of senescence does not appear to have been considered in subsequent fruit storage research at the LTRS, and it was left to others to demonstrate the changes in cell ultrastructure and tissue permeability that accompany ripening in apples.

Kidd and West did, however, note that various volatile substances accumulated in gas storage, and they obtained indirect evidence that their accumulation under restricted ventilation could be harmful to the fruit. Use of oiled wraps alleviated this problem as far as superficial scald was concerned, but their keen observation of the effects of volatiles and the marked differences in potential storage life between individual apple fruits laid the basis for later work on the roles of ethylene in fruit ripening and senescence and of the volatile oxidation products of farnesene in the development of scald.

Once these two papers on the gas storage of apples had been published, growers began to build refrigerated controlled atmosphere stores for apples. The first of these was erected by Spencer Mount of Canterbury in 1929, and by 1932 there were five large commerical "gas" stores in operation in Kent, with a capacity of about 1500 tons. In all these cases the atmosphere was maintained at 8 to 10 percent $CO_2$ by restricted ventilation.

1927 must have been a very busy year for Franklin Kidd. Apart from the comprehensive 'Bramley' experiment which would have continued until midsummer, he spent five months in Australia with shorter visits (three to six weeks each) in Canada, the United States, South Africa, and New Zealand. During this tour he publicized the work of the LTRS and laid the foundations for future collaboration between the group at Cambridge and leading postharvest workers in these five countries.

The success of the work of Kidd and West was recognized in 1929 when the Empire Marketing Board provided the capital for the building of the Ditton Laboratory, close to East Malling Research Station in the county of Kent. As an outstation of the LTRS, it was designed for work on problems encountered in the transport of fruit from the southern hemisphere and also provided improved facilities for other aspects of postharvest research. As the laboratory became established, so staff transferred from Cambridge to Ditton and West was appointed as superintendent there in 1931. He remained in this post for the next 18 years, during which major advances were made in our understanding of the biological and engineering principles essential for practical storage technology.

By 1933, with the Ditton Laboratory now fully operational, Kidd and West were joined by visiting workers from many parts of the world. These scientists later returned home to adapt the ideas and techniques for the commerical storage and transport of fresh fruit produced in their own countries.

The work of Kidd and West was followed with particular interest in North America. F. W. Allen, working at Davis in the early 1930s, conducted preliminary trials on the storage of 'Yellow Newtown' apples. In 1935 he was joined by the late Robert M. Smock, and for about 18 months they broadened their research to include pears and stone fruits. Allen spent some time with Kidd and West in England in 1936–1937, and in the summer of 1938 he was followed by Smock, who spent three months working with West at the Ditton Laboratory. Smock was to prove one of the most distinguished of the many scientific visitors to the laboratory, and during the next 15 years, he was instrumental in introducing controlled atmosphere storage to the fruit-growing regions of New York and New England in the eastern United States. His work provided the example that led to the establishment of controlled atmosphere (CA) technology in other areas of the American continent, and while Kidd and West can be described as the founders of modern CA storage, Bob Smock must surely be credited with the birth of this technology as far as North America is concerned.

The rapid uptake of this research in England and its subsequent export to North America and southern hemisphere Commonwealth countries is often quoted as an example of the successful translation of research into practice. The reasons lie in the care taken by Kidd and West to conduct trials on semicommercial stores in parallel with their laboratory experiments and their detailed appreciation of the engineering requirements of the new technology. Equally important, however, was their readiness to work with growers and shippers and their complementary abilities to communicate with administrators, scientists, and practical operators.

Both men had enquiring minds and showed great enthusiasm for research. Kidd was the more extroverted and a good organizer and leader. Later in his career he became a very capable and imaginative administrator, succeeding Hardy as superintendent of the LTRS in 1933 and becoming director of the Food Investigation Board in 1947. Although West was more retiring than his colleague, he was particularly competent in the laboratory, with a commitment to practical work of the highest standards. The different characters of the two men complemented one another admirably, and whereas Kidd was at his best in generating new ideas, encouraging his staff, and ensuring that the research was adequately funded, West was happy to get on with the day-to-day research, to deal with practical fruit growers and the detailed development of the new storage techniques. Their complementary relationship extended even to their "cold-hardiness." According to West, Kidd was unable to survive in the cold for more than 20 minutes, whereas West claimed that he never minded working long hours in the refrigerated stores at the Port of London. Even in later life West always preferred crisp winter weather to the heat of the summer. In 1963, they were each awarded Kamerlingh Onnes gold medals by the Netherlands Refrigeration Association in recognition of their pioneer work.

It is interesting to note that they rarely met outside the laboratory. Kidd was a keen walker, general naturalist, gardener, and a very enthusiastic beekeeper. He also wrote poetry and painted, particularly abstract works. In contrast, West was profoundly interested in systematic botany and spent all his spare time hunting for rare and interesting plants. He had a phenomenal ability to recognize plants, even at a distance. When asked how he was able to do this, he would generally reply: "The different species look like different people to me." While still a young boy he decided to specialize in the classification of the genus *Hieracium*. He kept this interest in the hawkweeds for over 80 years, both during his professional career and subsequently during his retirement, when he devoted almost all his time to this demanding task. His remarkable contribution to systematic botany was recognized by election as a Fellow *honoris causa* of the Linnaean Society of London. He was also very active in environmental conservation work and became president of the Kent County Trust for Nature Conservation.

Franklin Kidd died in 1974 at the age of 83, but Cyril West, two years his senior, lived on and was still a frequent visitor to the Ditton Laboratory well into his ninety-sixth year. He eventually died in 1986 at the age of 98. Sadly, Bob Smock died later in the same year, bringing to an end an illustrious chapter in the history of horticultural research.

Today, the modern marketing of apples and pears in virtually all fruit-growing regions of the world is highly dependent on controlled atmosphere storage technology. It provides a means of regulating supplies of the main commercial varieties of apples and pears which are harvested over a relatively short period, yet can now be made available to the public for the greater part of the year. In his introduction to the 1927 paper on gas storage of fruit, Sir William Hardy cautioned that "There was nothing in the traditional common-sense of mankind to act as a guide to the fruit industry when attempting to apply the principles of gas storage in commerical practice." He pointed out that because of this, the body of experimental evidence should be all the greater. It is, however, a testimony to the excellence of the work of Kidd and West that their results were found to be directly and immediately applicable for stores containing over 100 tons of fruit and that the recommendations which they issued for the storage of 'Bramley' apples in 1930 are still in general use in the United Kingdom today.

R. O. Sharples
Institute of Horticultural Research
East Malling
Maidstone, Kent ME19 6BJ
England

## REFERENCES

BLACKMAN, F. F., and P. PARIJA. 1928. Analytic studies in plant respiration. I. The respiration of a population of senescent ripening apples. Proc. R. Soc. London Ser. B 103:412–45.

DALRYMPLE, D. G. 1967. The development of controlled atmosphere storage of fruit. U.S. Dep. Agr. Bull. January.

KIDD, F., and C. WEST. 1927. Gas storage of fruits. Special Report 30. Food Investigation Board, DSIR, London, 87 pp.

MONTGOMERY, H. B. S., and A. F. POSNETTE. 1975. Franklin Kidd 1890–1974. Biogr. Mem. Fellows R. Soc. 21:407–430.

SHARPLES, R. O. 1986. Obituary note, Cyril West. J. Hort. Sci. 61(4):558.

SMOCK, R. M. 1938. The possibilities of gas storage in the United States. Refrig. Eng. 36(6):366–368.

SMOCK, R. M. 1960. Controlled atmosphere storage of fruit in the USA. p. 186–189 In: Proceedings of the 10th International Congress on Refrigeration, Copenhagen, 1959, Vol. III.

# THE GAS STORAGE OF FRUIT.

## *II. OPTIMUM TEMPERATURES AND ATMOSPHERES.

By FRANKLIN KIDD AND CYRIL WEST.

*(Low Temperature Research Station, Cambridge.)*

INTRODUCTION.

In the present communication it is proposed to take up the study of the effects of storing apples in atmospheres of regulated composition from the point at which it was left in the first paper in this series*. In that paper it was stated that an artificial atmosphere containing less oxygen and more carbon dioxide than are normally present in air retards the natural ripening processes of certain varieties of apples, in particular the Bramley's Seedling variety. The experiments, however, did not provide precise information as to the optimum temperatures and gas concentrations.

The next step in this investigation, therefore, was a series of storage trials to ascertain the interaction of the three variables (1) temperature, (2) concentration of carbon dioxide, and (3) concentration of oxygen in the storage of Bramley's Seedling apples.

In the previous paper (l.c.) attention was directed to the fact that the effects produced by any given concentrations of oxygen and of carbon dioxide vary with temperature. It seemed also not unlikely that the effects produced at any given temperature and with any given concentration of carbon dioxide varied with the concentration of oxygen used. Similarly with different concentrations of carbon dioxide the effects of any given temperature and oxygen concentration were not the same. The three variables with which we are concerned in the present paper must therefore be regarded as being interdependent as regards their effects upon the ripening processes of the apple fruit.

Since practical interest lies in the magnitude of these effects and since there is so much variation in fruit from year to year and from orchard to orchard, it appeared desirable to include as many combinations as possible of temperature, oxygen concentration and carbon dioxide concentration in a simultaneous series of experiments for which strictly comparable samples of apples could be employed. The equipment required to carry out such a survey is not inconsiderable, and even with the resources of the Low Temperature Research Station available it has only been possible so far to carry it out on a small scale.

* F. Kidd and C. West. Gas Storage of Fruit. Special Report No. 30, Food Investigation Board, London, 1927.

THE STORAGE CONDITIONS.

Twenty-three bushels of Bramley's Seedling apples were gathered on the 4th October, 1926, from an orchard on a chalk soil near Cambridge and were brought into the laboratory on the same day. From this supply 141 comparable trays, each holding twenty-three fruits, were made up in such a way that each tray contained one apple from each of the twenty-three bushels gathered. The following storage conditions were selected :—three temperatures, namely, 1°C. (34°F.), 5°C. (41°F.), 10°C. (50°F.), and at each of these temperatures ten different atmospheres, namely,

*Gas Series* 1-4 :   5% oxygen with 0%, 5%, 10% and 15% carbon dioxide.
*Gas Series* 5-8 :  10% oxygen with 0%, 5%, 10% and 15% carbon dioxide.
*Gas Series* 9 :   15% oxygen with 15% carbon dioxide.
*Gas Series* 10 :  Air.

In all, therefore, the effects of thirty different storage conditions were compared on samples of apples as nearly as possible comparable when put into store. Four trays of fruit were stored under each condition.

The storage conditions were obtained in the following way. Three cabinets, each containing ten gas-tight compartments, were specially designed and constructed. These were placed in constant-temperature rooms at the L.T.R.S. maintained at 1°, 5° and 10°C. respectively. Atmospheres of the desired composition were made up in ten gasometers of approximately 60 cubic feet capacity, the oxygen, nitrogen and carbon dioxide being supplied from pressure cylinders. From each gasometer a pipe was led, branching to each of the three constant-temperature rooms and there connecting to the inlet pipes of the compartments of the cabinets. From the opposite side of each compartment the gas mixtures were led away to the outside of the building. The composition of each storage atmosphere was maintained by constant ventilation of the storage chambers with the artificial mixtures from the gasometers. The rate of ventilation was adjusted to allow approximately a 1% difference in the concentrations of oxygen and carbon dioxide between the actual storage atmosphere and the incoming gas mixture.

The cabinets were constructed of five-ply wood with the exception of the partitions and fronts which were of Kauria pine. In spite of the most careful cabinet-making the compartments were found to be far from gas-tight until all the woodwork had been treated with vaseline which was melted in with a hot iron. This treatment was followed by a coating of odourless mineral oil. The removable fronts were bedded into position with vaseline to ensure gas-tight joints.

From the above it will be seen that the conditions common to all experiments were, in the first place, the avoidance of stagnation by a slow ventilating current of gas uncontaminated with volatile products from the fruit ; secondly, the presence of mineral oil in close proximity to the fruit. Under these circumstances mineral oil wrappers for the prevention of scald, as advocated in our previous Report, for use when "gas" storage is conducted in stagnant atmospheres, were not thought to be necessary and were omitted.

With the appliances to hand it was found impossible to control accurately the humidity of the atmospheres. In order to avoid saturation and to obtain approximately 95% relative humidity in all cases, the incoming ventilating currents of mixed gases were bubbled through tubes containing solutions of calcium chloride. By the end of March the loss of water by evaporation amounted to about 5% of the original fresh weight of the fruit at 10°C. and 5°C. The fruit stored at 1°C. had lost only 4% of its original fresh weight.

The thirty storage compartments were closed and the "gas" conditions established on the 23rd of October, that is, nineteen days after the fruit was gathered.

The concentrations of oxygen and of carbon dioxide in the storage compartments were checked by frequent analyses. Two typical gas records are shown in Figure 1. The oxygen and carbon dioxide conditions were not quite constant neither did they run quite as planned. The average gas conditions for the eight months of storage in the thirty different storage compartments are shown in Table I.

TABLE I.

*Showing average gas concentrations during experiments.*

| Gas series. | Required concentrations | | Actual Concentrations. | | | |
|---|---|---|---|---|---|---|
| | | | Oxygen (in gasometers). | Carbon dioxide (in storage compartments). | | |
| | $O_2$ | $CO_2$ | % | A.(1°C.). % | B. (5°C.). % | C. (10°C.). % |
| 1 | 5 | 0 | 6.4 | 1.0 | 1.4 | 1.6 |
| 2 | 5 | 5 | 6.4 | 5.1 | 5.7 | 5.2 |
| 3 | 5 | 10 | 6.1 | 10.8 | 11.9 | 10.1 |
| 4 | 5 | 15 | 6.3 | 14.9 | 16.9 | 14.2 |
| 5 | 10 | 0 | 11.0 | 1.1 | 1.5 | 2.1 |
| 6 | 10 | 5 | 10.6 | 5.5 | 6.1 | 6.3 |
| 7 | 10 | 10 | 10.6 | 10.3 | 11.6 | 9.9 |
| 8 | 10 | 15 | 10.9 | 15.9 | 16.9 | 14.1 |
| 9 | 15 | 15 | 15.5 | 15.9 | 16.6 | 14.6 |
| 10 | — | — | — | 0.4 | 1.2 | — |

The cabinets were opened and the stored fruit examined for wastage, colour change, taste and general condition at monthly intervals, the dates being 30th of November, 4th of January, 2nd of February, 23rd of March, 22nd of

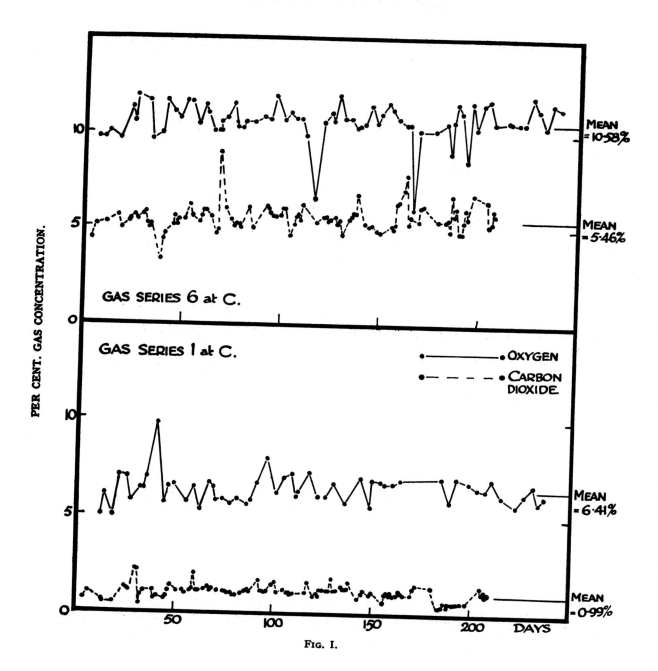

FIG. I.

April, 23rd of May and 22nd of June. The results obtained under the different storage conditions are described in the following sections dealing respectively with effects on commercial storage life, with effects on the colour change from green to yellow, and with effects on low temperature internal breakdown.

EFFECTS ON COMMERCIAL STORAGE LIFE.

Commercial storage life is defined as the interval of time between the storage date of the fruit and the development of 10% wastage in the store. The total percentage wastage to date observed at each examination is plotted against time and the most probable curve drawn through the points. One can then judge with an accuracy depending on the number of observations and the size of the samples stored the commercial storage life of a much larger quantity of fruit.

The wastage at 1°C. was due entirely to the development of low temperature breakdown in the green, unripe fruit. The wastage at 5°C. and 10°C. was due entirely to fungal rotting. No brownheart occurred.

Fungal rotting was first observed in the air-stored controls at 10°C. in January. The first signs of low temperature breakdown at 1°C. occurred in February. The breakdown was then not detectable at the surface of the fruit, but was visible on cutting. Scald developed in certain cases late in the storage life of the fruit, when wastage from internal breakdown or fungal rotting had exceeded 10%.

The results as regards commercial storage life are summarised in Tables II. and III.

TABLE II.

*Duration of commercial storage life (in weeks) in various gas mixture at different temperatures.*

| Gas mixtures (Approximate values ; for actual values, see Table I.) | | | Commercial storage life (in weeks). | | | Life duration at 5° C. / Life duration at 10° C. |
|---|---|---|---|---|---|---|
| Gas series. | Oxygen. % | Carbon dioxide. % | 1°C. | 5°C. | 10°C. | |
| 1 | 5 | 0 | 20 | 26 | 17 | 1.5 |
| 2 | 5 | 5 | 19 | 34 | 21 | 1.6 |
| 3 | 5 | 10 | 18 | 36 | 21 | 1.7 |
| 4 | 5 | 15 | 15 | 37 | 22 | 1.7 |
| 5 | 10 | 0 | 23 | 24 | 16 | 1.5 |
| 6 | 10 | 5 | 23 | 34 | 22 | 1.5 |
| 7 | 10 | 10 | 21 | 39 | 26 | 1.5 |
| 8 | 10 | 15 | 17 | 38 | 21 | 1.8 |
| 9 | 15 | 15 | 15 | 36 | 21 | 1.7 |
| 10 | 21 | 0(air) | 24 | 23 | 14 | 1.6 |

TABLE III.

*Efficiency of various gas mixtures as compared with air at different temperatures.*

| Gas mixtures (Approximate values; for actual values see Table I.) | | | Efficiency of gas mixtures. Air = 1.0. | | |
|---|---|---|---|---|---|
| Gas series. | Oxygen. % | Carbon dioxide. % | 1°C. | 5°C. | 10°C. |
| 1 | 5 | 0 | 0.8 | 1.1 | 1.2 |
| 2 | 5 | 5 | 0.8 | 1.5 | 1.5 |
| 3 | 5 | 10 | 0.7 | 1.6 | 1.5 |
| 4 | 5 | 15 | 0.6 | 1.6 | 1.6 |
| 5 | 10 | 0 | 1.0 | 1.0 | 1.1 |
| 6 | 10 | 5 | 1.0 | 1.5 | 1.6 |
| 7 | 10 | 10 | 0.9 | 1.7 | 1.9 |
| 8 | 10 | 15 | 0.7 | 1.7 | 1.5 |
| 9 | 15 | 15 | 0.6 | 1.6 | 1.5 |
| 10 | 21 | 0 (air) | 1.0 | 1.0 | 1.0 |

In the first of these tables the duration of commercial storage life under each of the thirty different conditions of storage is shown. In the second table the efficiency of each gas mixture as compared with air storage at the same temperature is shown, the figures indicating how much longer or shorter was the commercial storage life under the controlled gas conditions than in air.

A glance at these tables at once reveals the outstanding results of the whole set of trials. They may be summarised as follows :—

1. At 1°C. the controlled artificial atmospheres are in all cases less efficient than air. Low temperature internal breakdown is accelerated by atmospheres containing more carbon dioxide and less oxygen than air. Within the limits investigated, reduced oxygen concentrations accelerate breakdown whether carbon dioxide is present or not. Increased carbon dioxide concentrations accelerate breakdown with any concentration of oxygen.

2. At 5°C., which is above the limit for low temperature breakdown (i.e. 3.5°C. in the case of the Bramley's Seedling varieties), and at 10°C. the controlled artificial atmospheres are in all cases more efficient than air.

The degree of efficiency shown by any given concentration of carbon dioxide depends both upon temperature and upon the oxygen concentration.

The retarding effect upon ripening produced by reduced oxygen concentrations in the storage atmosphere is not so great as that produced by increased concentrations of carbon dioxide.

The interdependence of the three variables, oxygen, carbon dioxide and temperature, in their effects upon storage efficiency, is illustrated in the following diagram (Fig. II.) which is based upon the data presented in Table III. This

DIAGRAM ILLUSTRATING HOW RELATIVE EFFICIENCY OF ATMOSPHERES CONTAINING CO₂
DEPENDS ON TEMPERATURE AND OXYGEN CONCENTRATION.

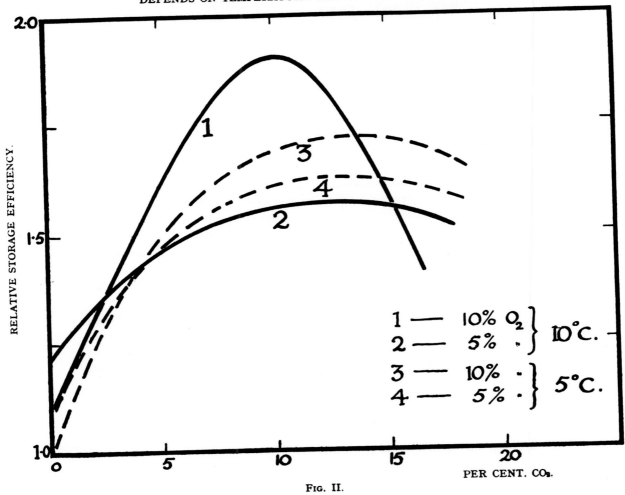

FIG. II.

diagram illustrates, as far as it can be stated at present, how the relationship between storage efficiency and carbon dioxide content of an atmosphere depends upon temperature and oxygen concentration. More exact work with larger samples will be needed to confirm these relations. The diagram indicates also that it is important to ascertain the effects of still higher concentrations of carbon dioxide in the storage atmosphere.

3. From a practical point of view the most efficient combination of conditions is clearly a temperature of 5°C. with an atmosphere containing 10%-15% of carbon dioxide and about 10% of oxygen. The oxygen apparently need not be very exactly regulated as between 5% and 15%. The regulation of the temperature on the other hand is most important.

With this combination of controlled conditions the results obtained as regards commercial storage life are 1.5 times better than the best possible by the use of controlled temperature alone (i.e. cold storage).

It remains to be determined whether one could go closer to the limiting temperature for low temperature breakdown (i.e. 3.5°C.) without danger.

Effects on the Change of Ground Colour from Green to Yellow.

At each examination the colour of the fruit under each storage condition was classified as deep green, green, yellowish green, slightly yellow, green yellow, yellow green, yellow, or deep yellow. This method of observing colour change clearly needs much improvement, and attention has since been devoted to the problem of colour standards and to a systematic study of the whole question of colour change during ripening and in stored fruit.

The observations that were made in the present series of experiments enable one to gain a fairly clear idea of the broad effects of the various conditions investigated upon the yellowing of the Bramley's Seedling apple.

It appears that at 10°C. the effect of 15% carbon dioxide is nearly to double the time taken to reach a given stage of yellowing and that the percentage of oxygen over the range studied does not markedly influence this ratio.

The effect of oxygen concentration upon the rate of yellowing was as pronounced as that of carbon dioxide. With 5% oxygen the fruit at 10°C. took about twice as long to yellow as in air (21% oxygen).

Practical interest attaches to the fact that the ratio between rate of colour change and length of commercial storage life is not the same under different storage conditions. It appears that in general atmospheric control not only extends the commercial storage life of the fruit, but to an even greater extent retards yellowing. This point is made clear in the following table.

TABLE IV.

| Storage condition. | Colour of fruit at end of commercial storage life (i.e. by the time 10% wastage had developed). | |
|---|---|---|
| Air at 10°C. | Yellow | : 14 weeks. |
| Air at 5°C. | Yellow | : 23 weeks. |
| Air at 1°C. | Yellowing | : 24 weeks. |
| 15% CO₂ : 10% O₂ : at 10°C. | Deep green | : 21 weeks. |
| at 5°C. | Deep green | : 38 weeks. |
| at 1°C. | Deep green | : 17 weeks. |
| 10% CO₂ : 10% O₂ : at 10°C. | Greenish yellow | : 26 weeks. |
| at 5°C. | Deep green | : 39 weeks. |
| at 1°C. | Deep green | : 21 weeks. |
| 5% CO₂ : 10% O₂ : at 10°C. | Slight yellowing | : 22 weeks. |
| at 5°C. | Deep green | : 34 weeks. |
| at 1°C. | Deep green | : 23 weeks. |
| 10% CO₂ : 5% O₂ : at 10°C. | Slight yellowing | : 21 weeks. |
| at 5°C. | Deep green | : 36 weeks. |
| at 1°C. | Deep green | : 18 weeks. |
| 5% CO₂ : 5% O₂ : at 10°C. | Slight yellowing | : 21 weeks. |
| at 5°C. | Deep green | : 34 weeks. |
| at 1°C. | Deep green | : 19 weeks. |

BEHAVIOUR OF THE FRUIT AFTER REMOVAL TO AIR.

On June 23rd, that is, after thirty-four weeks in storage, samples of sound fruit were removed to air from gas-series Nos. 2, 3, 4, 7, and 8 at 5°C. These apples were then quite green, firm and perfect in appearance and flavour. At the prevailing summer temperatures they remained in excellent condition for a further six weeks, after which they were all used for culinary purposes.

EFFECTS ON DEVELOPMENT OF LOW TEMPERATURE INTERNAL BREAKDOWN.

It has already been pointed out that at 1°C. all the artificial atmospheres tested were found to be less efficient than air, as shown by the earlier development of internal breakdown. The practical significance of this result lies in the fact that under present conditions of commercial practice carbon dioxide often accumulates in refrigerated fruit stores and in the holds of ships carrying apples.

The results revealed two other facts of interest. In the first place, it appeared that within the limits of the experiments, the presence of carbon dioxide was more injurious to the fruit than a reduction in the concentration of oxygen. The worst results obtained were with a combination of a high concentration of carbon dioxide and a low oxygen concentration.

In the second place, it was found that a reduction in the concentration of oxygen, while accelerating the development of internal breakdown, actually decreased the susceptibility of the fruit to this disease.   The presence of carbon dioxide, on the other hand, appeared to increase it.   For example, in the presence of 5% oxygen and 0% carbon dioxide (gas-series 1) the indication was that about 50% of the apples would never develop breakdown.   In air possibly 20% of the apples would remain immune from this disease.   When carbon dioxide was present every fruit broke down.

From these facts it would appear that the low temperature limit for the cold storage of apples susceptible to low temperature internal breakdown will depend upon the composition of the atmosphere of the store.   Further investigation is needed to answer this question.

## EFFECTS ON THE CHEMICAL CHANGES TAKING PLACE IN THE FRUIT.

A separate experiment was carried out to examine this question.   Samples of Bramley's Seedling apples were used strictly comparable to those employed in the main storage experiment described above.   These were stored at 10°C. in three different artificial atmospheres.   The following were used : (1) 21% $O_2$ and 0% $CO_2$ (i.e. Air) ;  (2) 10% $O_2$ and 9% $CO_2$ ;  and (3) 10% $O_2$ and 0% $CO_2$.   A comparison between (1) and (3) should show the type of effects due to reduced oxygen concentration :  and comparison between (2) and (3) the type of effects due to increased carbon dioxide concentration.   Samples were removed from storage at approximately monthly intervals and the losses in cane sugar, total sugar, acid and total dry matter determined.

The results indicate that the average rate of loss of carbohydrates by respiration was between 1.2 and 1.4 times faster in air (21% $O_2$) than in 10% $O_2$ : and between 1.35 and 1.55 times faster in the absence of carbon dioxide than in the presence of 10% of this gas.

It is interesting to compare these results with the relative durations of commercial storage life observed in these three cases (Table V.).

TABLE V.

| Storage conditions. | Relative average rate of loss of dry material. | Inverse of average rate of loss. | Relative length of storage life. |
|---|---|---|---|
| 21% $O_2$ (Air) | 1.00 | 1.00 | 1.00 |
| 10% $O_2$ | 0.77 | 1.30 | 1.15 |
| 10% $O_2$ and 9% $CO_2$ | 0.53 | 1.90 | 1.85 |
| 10% $O_2$ | 1.00 | 1.00 | 1.00 |
| 10% $O_2$ and 9% $CO_2$ | 0.69 | 1.45 | 1.60 |

From the above table it appears that the effect of reducing the concentration of oxygen is rather more marked on the rate of loss of dry material than it is upon the duration of life.   On the other hand the effect of carbon dioxide is more marked on storage life than it is upon the rate of loss of dry material. This result is attributed to the fact that carbon dioxide exerts a retarding effect not only upon the rate of respiration of the fruit, but also upon the germination and growth of the fungi responsible for rotting.

The chemical results confirmed the interesting fact, noted in our previous paper (l.c.), that the changes occurring are not all proportionately retarded by artificial atmospheres and that in the early stages of storage cane sugar, for example, actually disappears faster in the presence of carbon dioxide than it does in air. The present results also give some indication that during the early stages of storage cane sugar is lost more rapidly in 10% oxygen than in air.

The rate of acid loss does not appear to be markedly affected either by reduction in oxygen concentration or by the presence of 9% carbon dioxide.

*(Received 7/11/29.)*

H. C. Thompson. 1933. Temperature as a factor affecting flowering of plants. Proceedings of the American Society for Horticultural Science 30:440–446.

H. C. Thompson

This seven-page report, and the research required for it, emerged from the depths of the Great Depression. It, along with a summary paper (Thompson 1939), presented before a joint session of the agriculture section of the American Association for the Advancement of Science, at Columbus, Ohio, in 1939, summarized some of the most significant research in the horticultural sciences extending over the decades of the 1920s and 1930s.

At the time of retirement in 1951, Homer Calumbur Thompson had served for over 30 years as head of the department of vegetable crops at Cornell University and had trained more than half of those in the country who had earned advanced degrees in vegetable crops. Advanced degree training under Thompson's direction focused heavily on the effects of temperature, length of day (photoperiod), light intensity, and other environmental variables on flowering and premature seedstalk development of many vegetables, particularly the biennials. Temperature effects were most pronounced. Many graduate students were awarded advanced degrees as a result of their research under his tutelage on the effects of temperature on flowering and the interactions of daylength and crop maturity. Doctoral theses were first published in the memoirs or bulletins of the Cornell Agricultural Experiment Station, and the results then summarized in the *Proceedings of the American Society for Horticultural Science*. Crops included cabbage (Miller 1929), celery (Paltenius 1932), spinach and lettuce (Thompson and Knott 1933), garden beets (Chroboczek 1934), and carrots (Sakr and Thompson 1942).

This intense area of activity arose out of an emergency and the need for solutions to a problem. In the summer of 1924, celery growers in New York State suffered severe financial losses because of premature seeding. Over 60 percent of some fields produced seedstalks and were worthless for market. Thompson's approach in his research and with his students, was initially to solve the problem at hand and not worry too much why something happens, without first making sure that it does happen. It was my privilege to meet Thompson on several occasions and to know him well. Such opportunities were provided during the annual meetings of the American Society for Horticultural Science in the late 1940s and the 1950s. I remember him as a tall, kindly man, and a good listener, always with something positive to offer.

The 1933 paper on factors affecting flowering is a classic of horticultural research. In a few late winter and spring simple experiments, over several years, involving different regimes of temperature control in a greenhouse, Thompson answered all the basic questions as to the causes of premature seedstalk formation and flowering. No statistical estimates of variability were given or appeared necessary at that time, (this was before the science of biometrics in field plot design was in vogue). Numerous studies that followed, using elaborate statistical methods for estimating variability, only confirmed what Thompson had already reported. With celery, low or cool temperatures (40 to 50°F) were the determinant factor. He also noted that some strains of celery produced few if any seedstalks under the most favorable conditions, while with others there was 100 percent flowering. No mention is made in this paper as to the celery cultivars tested. It remained for Emsweller (1934) to demonstrate the great differences that occurred in premature seeding with different lines of celery. 'Cornell 19', introduced at the time Thompson and associates were conducting their studies, was a golden celery of supreme quality but one of the most susceptible to bolting of any cultivar known.

Thompson's classical paper also delineated temperature effects on flowering of cabbage. Cool- or low-temperature results with 'Danish Ballhead', one of the latest maturing of cabbage cultivars, were similar to those for celery. The older and the larger the plants, the more readily they went to seed.

Both young and mature plants of beets (presumably, red garden types) were studied. To induce flowering and seedstalk formation required a longer exposure of beets to cold than either celery or cabbage. There was also a more pronounced effect of photoperiod. Lengthening the photoperiod hastened seeding when plants were grown under cool conditions. Repeated crops of seed were produced in some plants by changing the environment. Also, the effects of cold-temperature exposure, even for extended periods, were nullified if followed by warm temperatures.

Reference in this paper is also made to onions grown from sets that often produce seedstalks. It was observed that the large onion sets (D inch diameter or larger) were prone to bolt, and that most seeders came from sets previously stored at around 40°F.

Finally, Thompson recorded the effects of temperature on two predominately long-day plants: spinach (Magruder 1935) and head lettuce (Thompson and Knott 1933). With 'Virginia Savoy' and 'Old Dominion' spinach, both very high and very low temperatures during the early stages of growth delayed seedstalk development when the plants were grown under long days. Under a short photoperiod, a 30-day low-temperature treatment hastened seedstalk development. With head lettuce (cultivars not given) high temperatures (70 to 80°F) prevented head formation and hastened seeding, even under short days (10 to 12 hours). Temperatures ranging from 60 to 70°F were most satisfactory for head formation. It is now known that great differences in rates of seedstalk formation exist among cultivars of head lettuce as well as spinach, onions, beets, cabbage, and celery.

Thompson's paper is valuable in that not only does it suggest plant growing techniques that will prevent the enormous losses previously encountered by growers from premature seedstalk formation and flowering but also specifies conditions whereby flowering may be accelerated for seed production. The research for this paper was conducted about 30 years before the effects of gibberellin on flowering and seedstalk formation in biennials and long-day plants were discovered (Wittwer and Bukovac 1957). The effects of temperature and gibberellin have since been noted to essentially duplicate each other for all the crops that Thompson and his graduate students examined. Additive effects of the two—temperature and gibberellin—have been noted for hastening the flowering responses in many crops. Cabbage plants 8 feet in height were produced by Thompson with temperature manipulations, quite comparable, indeed, to the 10-foot-high cabbage plants produced three decades later with gibberellin.

Sylvan H. Wittwer
Director Emeritus
Agricultural Experiment Station
Michigan State University
East Lansing, Michigan 48824

## REFERENCES

CHROBOCZEK, E. 1934. Study of some ecological factors influencing seedstalk development in beets. Cornell Univ. Agr. Expt. Sta. Mem. 154.

EMSWELLER, S. L. 1934. Premature seeding in inbred lines of celery. Proc. Am. Soc. Hort. Sci. 31:155–159.

MAGRUDER, R. 1935. The effect of controlled photoperiods on the production of seedstalks in eight varieties of spinach. Proc. Am. Soc. Hort. Sci. 34:502–506.

MILLER, J. C. 1929. A study of some factors affecting seedstalk development in cabbage. Cornell Univ. Agr. Expt. Sta. Bull. 488.

PLATENIUS, H. 1932. Carbohydrate and nitrogen metabolism in the celery plant as related to premature seeding. Cornell Univ. Agr. Expt. Sta. Mem. 140.

SAKR, EL SAYED, and H. C. THOMPSON. 1942. Effects of temperature and photoperiod on seedstalk development in carrots. Proc. Am. Soc. Hort. Sci. 41:343–346.

THOMPSON, H. C. 1939. Temperature in relation to vegetative and reproductive development in plants. Proc. Am. Soc. Hort. Sci. 37:672–679.

THOMPSON, H. C. and J. E. KNOTT. 1933. The effect of temperature and photoperiod on the growth of lettuce. Proc. Am. Soc. Hort. Sci. 30:507–509.

WITTWER, S. H., and M. J. BUKOVAC. 1957. Gibberellins: new chemicals for crop production. Mich. State Univ. Agr. Expt. Sta. Q. Bull. 39(3):1–28.

# Temperature as a Factor Affecting Flowering of Plants

By H. C. Thompson, *Cornell University, Ithaca, N. Y.*

DURING the past several years the writer and some of his co-workers have carried on studies to determine the effects of various ecological factors on seeding of certain vegetable crop plants.

Of the various factors considered, temperature has had the greatest effect on type of growth, vegetative or reproductive, of most of the plants used in the studies. Among the studies on the effect of temperature on seeding made by the Department of Vegetable Crops, Cornell University, are those by Thompson (9) on celery, by Platenius (8) also on celery, by Miller (7) on cabbage, by Knott (5) on spinach, by Chroboczek (2) on beets, by Thompson (10) on onions grown from sets, and by Thompson and Knott (11) on lettuce. A brief summary of the results of these studies is presented in this paper.

## RESULTS OF STUDIES WITH CELERY

Experiments on celery were conducted both in the field and under somewhat controlled conditions in the greenhouse. Plants for the field experiments were started in a medium temperature (60 to 70 degrees F) greenhouse about the middle of February and were kept there until about April 1. At this time some of the plants were placed in a greenhouse where the temperature was held at 40 to 50 degrees F while others were held at medium temperature, as controls, until time for planting in the field. The plants were set in the field about the first of May and were given good culture throughout the season. Treatments were in duplicate and each room contained from 40 to 75 plants. The results for 3 years with two strains of a standard variety are given in Table I.

TABLE I—EFFECT OF EXPOSURE TO RELATIVELY LOW TEMPERATURE (40 to 50 DEGREES F) FOR 10, 20 AND 30 DAYS ON SUBSEQUENT DEVELOPMENT OF THE SEED-STALK IN CELERY

| Preliminary Temperature Treatment | Number of Plants | Per cent Seed-Stalks |
|---|---|---|
| Check 60–70 degrees F | 550 | 0.00 |
| 10 days 40–50 degrees F | 550 | 7.63 |
| 20 days 40–50 degrees F | 550 | 44.36 |
| 30 days 40–50 degrees F | 550 | 74.00 |

The data in Table I show that the plants grown under medium temperature in the greenhouse until they were set in the field produced no seed-stalks. Those given 10 days at 40 to 50 degrees F produced an average of 7.63 per cent of seed-stalks; those with 20 days produced 44.36 per cent; and those with 30 days exposure to this temperature produced an average of 74 per cent of seed-stalks in the 3 years. There was variation in the percentage of seed-stalks produced in different years. For example, 30 days exposure to the

440

relatively low temperature resulted in 100 per cent of seed-stalks one year, 91.33 per cent in another year, and only 45.83 per cent in the third year. The variation is thought to result from difference in conditions prevailing in the field soon after planting. When the conditions are favorable for rapid growth for a few weeks following planting, the percentage of seed-stalks is much higher than when the conditions are not favorable for good growth.

*Greenhouse experiments:*—A large number of greenhouse experiments have been conducted during the past 10 years and all have given essentially the same results with reference to the effects of temperature on bolting. After the plants were given the preliminary treatments as previously described, they were divided into three lots for growth under three ranges of temperature, cool (50 to 60 degrees F), medium (60 to 70 degrees F), and warm (70 to 80 degrees F). Not all of the treatments were included in the warm house. Results of one of these experiments are given in Table II to illustrate the general trend. In this particular experiment the plants were put under the three temperature conditions November 16.

TABLE II—Effect of Exposure to Relatively Low Temperature for 15 and 30 Days on Subsequent Development of Seed-Stalks of Celery Under Three Ranges of Temperature

| Preliminary Treatment | Number of Plants | Per cent of Seed-Stalks on Dates Given | | | |
|---|---|---|---|---|---|
| | | March 20 | April 3 | April 25 | May 8 |
| *In medium-temperature house* | | | | | |
| Check    60–70 degrees F.. | 20 | 0.00 | 0.00 | 0.00 | 0.00 |
| 15 days 50–60 degrees F.. | 20 | 0.00 | 0.00 | 65.00 | 100.00 |
| 30 days 50–60 degrees F.. | 20 | 0.00 | 0.00 | 65.00 | 100.00 |
| 15 days 40–50 degrees F.. | 20 | 0.00 | 85.00 | 100.00 | 100.00 |
| 30 days 40–50 degrees F.. | 20 | 25.00 | 45.00 | 100.00 | 100.00 |
| 30 days 70–80 degrees F.. | 20 | 0.00 | 0.00 | 0.00 | 0.00 |
| *In cool house* | | | | | |
| Check    60–70 degrees F.. | 20 | 0.00 | 0.00 | 100.00 | 100.00 |
| 15 days 50–60 degrees F.. | 20 | 5.00 | 65.00 | 100.00 | 100.00 |
| 30 days 50–60 degrees F.. | 20 | 10.00 | 60.00 | 100.00 | 100.00 |
| 15 days 40–50 degrees F.. | 20 | 88.00 | 100.00 | 100.00 | 100.00 |
| 30 days 40–50 degrees F.. | 20 | 80.00 | 100.00 | 100.00 | 100.00 |
| *In warm house* | | | | | |
| Check    60–70 degrees F.. | 10 | 0.00 | 0.00 | 0.00 | 0.00 |
| 15 days 40–50 degrees F.. | 10 | 0.00 | 0.00 | 0.00 | 0.00 |
| 30 days 40–50 degrees F.. | 10 | 0.00 | 0.00 | 0.00 | 0.00 |

All of the plants that had been given preliminary low-temperature treatment, either 40 to 50 degrees F or 50 to 60 degrees F, went to seed in the medium-temperature house. The plants of the other two lots, that had no temperature treatment below the 60 to 70 degree range, did not produce a single seed-stalk. All of the plants grown, after the preliminary treatment, in the cool house went to seed, but the previous low-temperature treatment hastened seeding. In the warm house, not a single plant went to seed regardless of the previous treatment. The high temperature seems to nullify the effect of the low-temperature treatment given previously. In a large number of experiments conducted by the writer, there has never been an instance of seed-stalk development in high temperature, except where they had started before the plants were subjected to the high temperature. After the stalks have started to develop, they elongate more rapidly in the high temperature (70 to 80 degrees F) than in a cool temperature (50 to 60 degrees F).

Attention should be called to the fact that heredity is involved in premature seeding. Some strains produce few seed-stalks even under the most favorable conditions, and some produce 100 per cent under the same conditions. But regardless of the heredity, no seed-stalks have been produced, in our experiments, unless the plants have been subjected to relatively low temperature. The experiments carried on entirely in the greenhouse show clearly the effects of temperature on type of growth—vegetative or reproductive.

## RESULTS OF STUDIES WITH CABBAGE

Two types of experiments were conducted with cabbage, one with young plants and the other with mature plants. Both types of experiments were carried on largely in the greenhouse, although one experiment on premature seeding was conducted in the field after the preliminary treatments were given.

Three experiments with young plants were conducted, two in the greenhouse and one in the field. The methods followed were similar to those discussed for celery, except that in the greenhouse experiments the cabbage plants were exposed to a lower temperature and for a much longer period than were the celery plants. In general, the results with cabbage plants were similar to those obtained with celery. The larger and older the plants were at the time of the temperature treatment, however, the greater was the percentage of seed-stalks.

*Seed-stalk development in mature cabbage:*—The first experiment with mature cabbage was carried on in the greenhouses during the fall and winter of 1925–26. Plants of a pure line of Danish Ballhead with mature heads were taken from the field on October 8 and 60 were potted immediately; 30 of these were placed in a medium-temperature greenhouse (60 to 70 degrees F) and 30 in a cool greenhouse (50 to 60 degrees F). Fifteen plants in each house were grown under the normal length of day and fifteen under the long day. The day length was increased by use of electric lights from 5 to 10 p. m. each day. The data are presented in Table III.

The plants in the medium-temperature house did not develop flower stalks but continued vegetative growth and each plant produced a second head on the elongated stalk. In the cool house, flower stalks emerged from the old heads and produced flowers in about 80 to 90 days. Increasing the length of day hastened slightly the production of the second head in the medium-temperature house and the emergence of the growing point of the plants in the cool house. Flowering was not hastened materially by increasing the length of day. In an experiment with immature plants, increasing the length of day 5 hours by use of electric lights from November 15 to April 15 appeared to have a slightly depressing effect on seed-stalk development.

TABLE III—Effect of Temperature and Length of Day on Type of Growth in Mature Cabbage

| Temperature and Length of Day | Number of Days for Growing Point to Emerge from Head | Number of Days to Flowering | Number of Days to Produce Second Head |
|---|---|---|---|
| Medium temperature- | | | |
| Normal day........ | 50 | Did not flower | 150 |
| Long day.......... | 49 | Did not flower | 140 |
| Cool temperature | | | |
| Normal day........ | 76 | 154 | ——— |
| Long day.......... | 62 | 152 | ——— |

Five of the plants in the medium-temperature house were kept until the fall of 1926 and these produced three heads each. In moving them to a new greenhouse four of the plants were broken, but the one remaining was kept in the medium-temperature house until the fall of 1927. This plant, during the 2 years, produced six heads and grew to a height of more than 8 feet. In the fall of 1927 it was moved to the cool greenhouse and by April 10, 1928, it had produced flowers.

Two experiments were conducted to determine the effect of a rest period at relatively low temperature on seed-stalk development. In the experiment previously discussed, no seed-stalks developed in the medium-temperature house. Some of the same lot of plants were given a rest period in storage at 40 degrees F for 2 months and these plants produced flowers in the medium-temperature house in 39 days. In the cool greenhouse, the plants produced flowers in 68 days after being given a rest period of 2 months. Rest periods of 15 days and 30 days at 40 degrees F were not very effective.

The marked effect of temperature on type of growth is shown in the results of an experiment on immature plants. By growing them at relatively low temperature, these plants went to seed without forming a head. After the seed had ripened, the seed stems were cut off and the stubs were planted in the garden on June 1. Normal axillary heads were produced and in the fall the plants were placed in cold storage at 40 degrees F for 2 months. They were then planted in the greenhouse where they produced a second crop of seed. The seed-stalk was again cut off after the seed had ripened and the stubs were planted in the garden as before. A second head was produced. Every plant in this lot of ten produced two crops of seed and two crops of heads in 2 years, but the normal order was reversed.

## RESULTS OF STUDIES WITH BEETS

Experiments were conducted with young and mature plants in the greenhouse and field to determine the effects of temperature and of length of photoperiod on seed-stalk development.

In general, the results of studies on premature seeding of beets were similar to those obtained in experiments with celery. However, beet plants required a longer exposure to relatively low temperature for subsequent seed-stalk development under medium temperature. The length of photoperiod is also an important factor in the development of the seed-stalk of the beet plant.

Beet plants grown continuously in a warm greenhouse (70 to 80 degrees F) under the normal length of day, from October to May, did not go to seed at all. Lengthening the day 5 hours by means of artificial light did not cause these plants to go to seed. Plants grown in a medium-temperature (60 to 70 degrees F) house developed a few seed-stalks under a day length of 15 hours. On the other hand, 100 per cent of the beet plants grown in the cool greenhouse (50 to 60 degrees F) produced seed-stalks under an 11-hour day. Even under an 8-hour day the plants grown in the cool house produced seed-stalks although they were barren. Increasing the length of day 5 hours in the medium-temperature house, increased slightly the percentage of seed-stalks. In the cool house, where all plants went to seed under the short day, increasing the length of day hastened seed-stalk formation by several weeks. After the seed-stalk had started to develop, it grew much more rapidly in the medium, or in the warm house, than in the cool house, but if the plants were kept too long at high temperature the seed-stalk became vegetative.

*Effect of exposure to relatively low temperature on subsequent seeding:*—Beet plants were exposed for varying lengths of time to relatively low temperature (40 to 50 degrees F) to determine the effects of this range of temperature on subsequent seeding at higher temperatures. The effects of exposing plants to relatively low temperature varied considerably, depending on the temperature and the photoperiod under which they were grown subsequently. For example, plants grown for 30 days at 40 to 50 degrees F and then grown in a warm house did not produce any seed-stalks under either short- or long-day conditions. The warm temperature seems to nullify the effect produced by the previous low-temperature treatment. The combination of medium temperature and short day also nullified the effects of the low-temperature treatment given previously. Plants grown under medium-temperature and long-day conditions produced a considerable percentage of seed-stalks. The low temperature treatment hastened seeding of plants grown subsequently in the cool house, but all plants in the cool house went to seed.

As mentioned above, 30 days' exposure to relatively low temperature was not sufficient to cause seeding in the warm house, even under a long photoperiod. Lengthening the cold treatment to 60 days resulted in some seeding in the warm house, but even 90 days of exposure to the cold treatment did not bring all of the plants to flowering when they were grown subsequently at a high temperature.

Experimental results with full-grown beet plants were similar to those with seedlings. When full-grown plants were grown in the cool greenhouse (50 to 60 degrees F) they went to seed under the short, medium, or long day, but the lengthening of the day hastened seeding. In the medium and warm greenhouses the plants did not develop seed-stalks under either the normal or long day, but developed a second and a third enlargement above the original one. Some of the plants which did not go to seed in the third year of their growth were transferred to the cool house in the fall of 1931 and by January 12, 1932, seed-stalks began to appear. By March 5, blossoms were in evidence. These plants were kept growing continuously and developed new shoots from the axils of the leaves on the main axis. Flowers were produced on these shoots in July.

Seed-stalks frequently start to develop but, owing to unfavorable conditions such as high temperature, they do not produce flowers but a rosette of leaves at the top of the stalk. If such plants are grown subsequently in a medium or warm house, an enlargement develops at the top of the stalk. This enlargement has the appearance of a normal beet root and may develop to very large size (5½ inches in one instance). If plants with barren stalks are grown subsequently in cool temperature, new seed-stalks develop from the top of the one previously formed. Even plants which have developed normal seed-stalks will produce a second crop of seed when grown under favorable conditions. In fact, the type of growth can be changed several times by changing the environment.

## Results of Studies with Onions

Onions grown from sets frequently develop seed-stalks within a few weeks after planting and such plants seldom, if ever, produce marketable bulbs. Experimental results obtained by the writer indicate that under ordinary field conditions only the large sets, ¾-inch in diameter or larger, produce any considerable percentage of seeders. No experiments have been conducted to determine the effects of growing temperature on seeding but studies have been made on the effects of storage temperature on seed-stalk development in the field. Sets have been stored in cold storage at 30, 32, 40 and 50 degrees F and in common storage at 50 to 60 and 60 to 70 degrees F. There was some seed-stalk development at all temperatures except at 60 to 70 degrees F. The 40-degree storage resulted in the largest percentage and the earliest development of seeders; followed by the 50- and 32-degree temperatures. Sets stored at 30 degrees F produced few stalks and this is considered the best temperature of those used for the storage of sets. In common storage at relatively high temperatures, the sets dry out; in cold storage at 40 to 50 degrees F, the sets sprout.

## Results of Studies with Spinach

Spinach plants of the Virginia Savoy and Old Dominion varieties exposed to 40 to 50, 50 to 60, 60 to 70, and 70 to 80 degrees F for 30 days when grown under a long day (15 hours) go to seed in the following order: 60 to 70, 50 to 60, 70 to 80, and 40 to 50 degrees F.

This order is the same whether the plants are grown subsequently at 50 to 60, 60 to 70, or 70 to 80 degrees F. In other words, both the very high and the very low temperature during the early stage of growth delay seed-stalk development in spinach plants grown under long-day conditions. The seed-stalks appear first in the 70 to 80 degree house, then in the 60 to 70 and last in the 50 to 60 degree F.

When the plants are grown under a short photoperiod, the 30-day low-temperature treatment is effective in hastening seed-stalk development, the order of seed-stalk appearance being 40 to 50, 50 to 60, 60 to 70, and 70 to 80 degrees F. Following the preliminary treatment the order of seed-stalk development under the three ranges of growing temperature is 70 to 80, 60 to 70, and 50 to 60 degrees F. Seed-stalks developed at 70 to 80 and 60 to 70 degrees F when the day length was less than 11 hours and at 50 to 60 degrees F when the photoperiod was 12 hours.

## RESULTS OF STUDIES WITH LETTUCE

Preliminary studies indicate that high temperature (70 to 80 degrees F) prevents head formation and materially hastens seeding. Seed stalks developed a month earlier at 70 to 80 degrees than at 60 to 70 degrees F. At 70 to 80 degrees F lettuce plants went to seed under a short day (10 to 12 hours). A temperature range of 60 to 70 degrees F was most satisfactory for head formation. Heading took place about a month later in the 50 to 60 degree house than in the 60 to 70 degree house. Increasing the length of day increased the size of the heads in the cool greenhouse.

A report on this study is given by Thompson and Knott in another paper at this meeting.

## LITERATURE CITED

1. BOSWELL, VICTOR R. Studies of premature flower formation in wintered-over cabbage. Md. Agr. Exp. Sta. Bul. 313. 1929.
2. CHROBOCZEK, EMIL. Study of some ecological factors influencing seed-stalk development in beets. Cornell Agr. Exp. Sta. Memoir, 154. 1934.
3. CURTIS, O. F., and CHANG, H. T. The relative effectiveness of the temperature of the crown as contrasted to that of the plant upon flowering of celery plants. (Abstract) Amer. Jour. Bot., 17: 1047–1048. 1930.
4. KLEBS, GEORG. Über die Blutenbildung von Sempervivum. Festschrift zumm 70 Geb. V. Ernst Stahl. Jena, 128–151. 1918.
5. KNOTT, J. E. Unpublished data on seeding in spinach. Cornell Agr. Exp. Sta. 1933.
6. KRASAN, F. Studien über die periodishen Lebenserscheinungen der Pflanzen. Verhandl. der k. k. zoog.-bot. Ges. Wein, 20: 265–366. 1870.
7. MILLER, J. C. A study of some factors affecting seed-stalk development in cabbage. Cornell Agr. Exp. Sta. Bul., 488. 1929.
8. PLATENIUS, H. Carbohydrate and nitrogen metabolism in the celery plant as related to premature seeding. Cornell Agr. Exp. Sta. Memoir, 140. 1932.
9. THOMPSON, H. C. Premature seeding of celery. Cornell Agr. Exp. Sta. Bul., 480. 1929.
10. ————. Unpublished data on seeding of onions grown from sets. Cornell Agr. Exp. Sta. 1933.
11. ———— and KNOTT, J. E. The effects of temperature and photoperiod on the growth of lettuce. Proc. Amer. Soc. Hort. Sci. 1933.

Kenneth Post. 1934. Production of early blooms of chrysanthemums by the use of black cloth to reduce the length of day. New York State Experiment Station Bulletin 594.

Kenneth Post

This bulletin was received for publication January 1934, printed in April 1934 (a very quick turnaround) and reprinted in July 1935—obviously a very popular bulletin. It described original studies with chrysanthemum research, which are being refined even today. Perhaps more important, this work started an international chrysanthemum industry which has expanded to thousands of hectares of year-round flowering. In the early 1930s, chrysanthemum flowers were only available during the fall, now known as natural-season flowering. Horticulturists were trying to extend the natural season, both early and late (Laurie and Poesch 1932; Post 1932a,b). They were able to extend the season by pinching the plants to delay flowering, but early flowering was more difficult to achieve. Cultivar selection was the most obvious and practical solution. The work of Garner and Allard (1920) clearly showed that flowering of some species was dependent on the daylength. Kenneth Post at Cornell and Alex Laurie at Ohio State realized that the chrysanthemum was a photoperiodic plant; however, they were struggling with how to transform this basic information into practical cultural procedures for the commercial grower (Post 1931, 1932b, 1950). How could large production areas be covered to keep out light, how much light must be kept out, when should the cover be put on, and for how long—these were just a few of the questions being asked. Laurie and Poesch (1932) reported that sateen (a tightly woven) cloth worked satisfactorily as a cover over plants. Post tested cloth that was not tightly woven, and obtained poor results, and sateen cloth, which worked well, as reported by Laurie. He tested the time of cover application rather than precisely controlling the number of hours of light; he covered the plants at 4 P.M., at 6 P.M., or at dark until 9 in the morning rather than creating a daylength of 12, 13, or 14 hours. His results showed that the 6 P.M. covering was best. It appears that he did not realize the photoperiodic significance of flower bud initiation and flower development, a concept that he and his colleagues later used (Post 1937, 1939, 1948b).

I first read this bulletin 35 years ago when I was a student of Kenneth Post. Rereading it is like viewing an old thriller and knowing who did what and to whom. It is easy to criticize the experiments based on today's knowledge, but the facilities were not up to today's standards and the culture of the plants was primitive.

The difficulties in conducting research in that era are apparent if we compare the plant material available then with today. Post had to produce his own experimental plant material, probably divisions or cuttings of plants from the previous year, kept "dormant" in a cool greenhouse. He started the plants in March and grew them for 4 or more months before the treatments were started. These plants were pinched at least three times. We now know that old plants are not very sensitive to photoperiod, are very reproductive, and would bud in spite of the photoperiod. Today, hundreds of uniform disease- and virus-free cuttings can be ordered from a chrysanthemum propagation specialist to be delivered any day of the year. These stock plants, specially grown to produce cuttings, are exposed to the correct temperature, photoperiod, and are not too old. Post's plant material would make interpretation of the results difficult. The cultivars used in the early 1930s were very different from those used today. They selected cultivars that would flower early, middle, and late in the natural season of flowering. This meant that they were selecting for temperature response as well as photoperiod. Today's cultivars are selected much more critically to respond more accurately to photoperiod and temperature for a particular time of the year. In addition, Post used 8 to 25 plants per treatment. Although he used the expression "statistically significant" (no statistics were shown), I would question whether any significance would be found with such few plants because of the variation between plants.

Despite these experimental deficiencies, this work led to the start of the year-round chrysanthemum industry. Year-round flowering was not yet understood, but the groundwork had been done. It was 15 years later before it became a commercial practice (Post 1946). Horticulturists had yet to learn how to grow stock plants and vegetative cuttings on a year-round basis. They had yet to define the "critical photoperiod" concept for the chrysanthemum, that is, 14.5 hours or less for flower bud initiation, 13.5 hours for flower development, and daylengths longer than 14.5 hours to maintain vegetative growth (Post 1948b). Post's later work indicated that it was the length of the night that was critical and not the day, which led to the use of incandescent lamps to break up the night length (Post 1948b, 1953). It was also shown that very low irradiance was necessary to accomplish the long-day effect (Post 1937). In the 1950s, H. M. Cathey, one of Post's students, showed the relationship between temperature and photoperiod and classified the chrysanthemum cultivars into three response groups (Cathey 1954a,b,c, 1955, 1957). I am sure the original studies in the 1930s were compounded with temperature effects. Post explicitly defined "crown" buds in this publication solely from observations (Chan 1950). His great intuitive understanding of plants can be appreciated by reading this publication.

Post started his career at Michigan State University and came to Cornell in 1930 to work on his Ph.D. degree. It was common in those days to do Ph.D. work as an instructor, which was a way to earn more money, but it took a long time. In Post's case, it was eight years. In addition to his research programs, he realized the value of extension as it served his purpose very well, especially for the time (1930–1940). He knew that information about growing crops could be gleaned from growers and there was a great shortage of knowledge about the cultural requirements of floricultural crops. Growers would be able to utilize this research information through extension to improve their production and he needed the political clout that growers could supply him when he spoke to deans and legislators. Post was a master technician of using people.

At the time of Post's death in 1955, at the age of 51, Cornell and Post were clearly the floricultural leaders of the world. He laid a foundation for expansion of the department facilities, garnered a number of assistantships, and expanded the floriculture faculty.

Nationally, he made major contributions to the field of floriculture by showing the value of scientific research to growers and by developing, with Bill Blauvelt (entomology) and Wat Dimock (plant pathology), a floriculture extension team that was envied throughout the world.

Post visited Europe in the early 1950s and opened up avenues of communication between European and American scientists, and growers. In his address as president of the American Society for Horticultural Science in 1952, he spoke of the knowledge available in Europe.

His personal relationship with people was exceptional. He had the ability to ask people to work hard for a cause and love it, whether talking to growers, students, or colleagues. As a young graduate student, I remember a greenhouse tour when Post spent 10 minutes telling the group what was wrong with the operation, and the grower was all smiles. He was generous with praise. His book *Florists Crop Production and Marketing*, published in 1949, is still a classic.

This work is a gem of applied horticulture. Post adapted the discovery of Garner and Allard and created a practical application that has resulted in a multimillion-dollar industry (Langhans 1964).

Robert W. Langhans
Department of Floriculture and
  Ornamental Horticulture
Cornell University
Ithaca, New York 14853

# REFERENCES

CATHEY, H. M. 1954a. Chrysanthemum temperature study. A. Thermal induction of stock plants of *Chrysanthemum morifolium.* Proc. Am. Soc. Hort. Sci. 64:483–491.

CATHEY, H. M. 1954b. Chrysanthemum temperature study. B. Thermal modification of photoperiod previous to and after flower bud initiation. Proc. Am. Soc. Hort. Sci. 64:492–498.

CATHEY, H. M. 1954c. Chrysanthemum temperature study. C. The effect of night, day and mean temperature upon the flowering of *Chrysanthemum morifolium.* Proc. Am. Soc. Hort. Sci. 64:499–502.

CATHEY, H. M. 1955. Chrysanthemum temperature study. D. Effects of temperature shifts upon the spray formation and flowering time of Chrysanthemum morifolium. Proc. Am. Soc. Hort. Sci. 66:386–391.

CATHEY, H. M. 1957. Chrysanthemum temperature study. F. The effect of temperature upon the critical photoperiod necessary for the initiation and development of flowers of *Chrysanthemum morifolium.* Proc. Am. Soc. Hort. Sci. 69:485–491.

CHAN, A. P. 1950. The development of crown and terminal buds of *Chrysanthemum morifolium.* Proc. Am. Soc. Hort. Sci. 61:555–558.

GARNER, W. W., and H. A. ALLARD. 1920. Effect of length of day and other factors of the environment on growth and reproduction in plants. J. Agr. Res. 18:553–606.

LANGHANS, R. W. 1964. Chrysanthemums: A manual of the culture, disease and insects and economics of chrysanthemums. New York State Extension Service, Ithaca, N.Y.

LAURIE, A., and G. H. POESCH. 1932. Photoperiodism: The value of supplementary illumination and reduction of light on flowering plants in the greenhouse. Ohio Agr. Expt. Sta. Res. Bull. 512.

POESCH, G. H. 1932. Studies of photoperiodism of the chrysanthemum. Proc. Am. Soc. Hort. Sci. 28:389–392.

POST, K. 1931. Reducing the daylength of chrysanthemums for the production of early blooms by the use of black sateen cloth. Proc. Am. Soc. Hort. Sci. 28:382–388.

POST, K. 1932a. Darkening chrysanthemums brings early blooms. Florists Exch. 79(2):11–22.

POST, K. 1932b. Reducing the day length of chrysanthemums for the production of early blooms by the use of black cloth. Proc. Am. Soc. Hort. Sci. 29:545–548.

POST, K. 1937. The determination of the normal date of flower bud formation of short day plants. Proc. Am. Soc. Hort. Sci. 34:618–620.

POST, K. 1939. The relationship of temperature to flower bud formation in chrysanthemums. Proc. Am. Soc. Hort. Sci. 37:1003–1006.

POST, K. 1946. Now you can have chrysanthemums every day of the year. Florists Rev. 98(2530):25–26.

POST, K. 1947. Year-around chrysanthemum production. Proc. Am. Soc. Hort. Sci. 49:417–419.

POST, K. 1948a. Temperature and chrysanthemum flower development. New York State Flower Grow. Bull. 30:3.

POST, K. 1948b. Daylength and flower bud development in chrysanthemum. Proc. Am. Soc. Hort. Sci. 51:590–592.

POST, K. 1950. Controlled photoperiod and spray formation of chrysanthemums. Proc. Am. Soc. Hort. Sci. 55:467–472.

POST, K. 1953. It's a short night you want. New York State Flower Grow. Bull. 99:3–4.

Bulletin 594

April 1934

# Production of Early Blooms
## of Chrysanthemums
## by the Use of Black Cloth to Reduce
## the Length of Day

Kenneth Post

SECTION OF A BENCH OF POMPONS WHICH WERE GIVEN SHORT-DAY TREATMENTS; PLANTS
IN THE BACKGROUND UNTREATED

Published by the
Cornell University Agricultural Experiment Station
Ithaca, New York

Received for publication January 3, 1934

CONTENTS

# PRODUCTION OF EARLY BLOOMS OF CHRYSANTHEMUMS BY THE USE OF BLACK CLOTH TO REDUCE THE LENGTH OF DAY

### KENNETH POST

The effects of the length of day[1] on plant growth and flowering were recorded in the latter part of the seventeenth century. Since that time many experiments have been made, but the work of Garner and Allard (1920, 1923) has formed the basis for practical application.

As a result of their early experiments, these workers classified plants into *long-day*, *short-day*, and *indifferent* types. Long-day plants, under normal conditions, bloom during the summer months or when the days are lengthening. Conversely, short-day plants bloom when the days are shortening. Indifferent types bloom at any time of year regardless of the length of day and night. The length of day necessary for bud formation and flowering of various plants differs with the species and even with the variety of the species.

The chrysanthemum was classified as a short-day plant by Garner and Allard (1920) but this was of little value to the commercial grower until a practical method of reducing the length of day could be devised. In 1930 Laurie reported that black sateen cloth could be used effectively in reducing the length of day to obtain early blooms of chrysanthemum varieties for commercial purposes. Poesch (1931–1932) has investigated certain phases of the applications of photoperiodism to the chrysanthemum.

The purpose of this bulletin is to summarize the results of four years of work with the use of black cloth for reducing the length of day on chrysanthemums for the production of early blooms. The results of the experiments made in 1931 and 1932 have appeared in brief in the *Proceedings of the American Society for Horticultural Science* for those respective years.

## MATERIALS AND METHODS

Both large-flowered and pompon varieties of chrysanthemums were used in this work. Five or more varieties and ten or more plants of each variety were used in each plot in most of the experiments. Unless otherwise noted, the plants were propagated during the month of March and benched from three-inch pots between June 15 and 25. The plants were selected for uniformity at the time of planting and the planting distance of all varieties was 7½ by 9 inches.

The soil used was of the Dunkirk silty clay loam type and had been composted with manure so that the resulting product was about one-fourth organic material. Superphosphate was incorporated into the soil at the rate of five pounds per 100 square feet of soil area at the time the plants were benched. Ammonium sulphate was used at the rate of one pound per 100 square feet of bench area at intervals of two weeks, starting six weeks after the plants were benched and continuing until the buds were showing color.

---

[1] In this bulletin length of day refers to the daily exposure to light.

3

Watering, disbudding, tying, and general culture of the plants were carried out as usually practiced in commercial greenhouses. The temperature was kept as near 50°F. as possible at night. All varieties were pinched when the fifth node was developed and 4 nodes were left above the soil at this time. Large-flowered varieties were grown two stems to a plant. All pinches on pompons, after the first, were made so that 4 nodes were left on the side branches. Usually three or more pinches were given pompon varieties.

Large-flowered varieties were cut when the center petals were well developed. Pompon varieties were cut when the center flower was fully developed and all buds showing color were counted as blooms. The diameter of the flower was measured perpendicular to the stem. The diameter of the terminal bloom of pompon varieties was measured. The stem length of large-flowered forms was measured from the surface of the ground to the base of the flower head. The pompons were cut at the point of union of the branch with the parent stem. The length of stem was measured from the top of the topmost bloom to the point where the cut was made.

Black sateen cloth with 68 by 104 threads to the square inch was used in all experiments unless otherwise noted. The plants were completely enclosed with this material at 6 P.M. and uncovered at 7 A. M., except when length of day or part of the plant treated were the factors to be varied. The cloth was sewed in strips wide enough to cover the top of the bench and to reach from the surface of the soil to a height of three or more feet. The length varied with the plot to be treated. Wires were run lengthwise of the bench on each side at a height of three feet from the soil.

FIGURE 1. PORTION OF A BENCH OF PLANTS WHICH WERE COVERED WITH BLACK CLOTH

These served as supports for the cloth which was pulled over the plants in the afternoon and pushed back in the morning. Small pieces of cloth were used to cover the ends of the plots. These were thumbtacked to small pieces of wood. A section of a bench, darkened as indicated, is shown in figure 1. All treatments were based on Eastern Standard time.

## RESULTS

Preliminary experiments, with the variety Yellow Fellow, were made in 1930. As a result of these experiments several changes in technique were made. Plans for the next three years of work were outlined at this time.

### EFFECTS OF VARIOUS LENGTHS OF DAY

Fifteen plants of each of six varieties were grown in each of four plots in the greenhouse bench during the summer of 1931. Beginning July 15, one plot was given a 9-hour day, a second plot was given an 11-hour day, and a third plot a 13-hour day; the fourth plot was allowed a normal day which was about 15 hours in length at that time of year. In all cases the treatment was applied daily until color was showing in the buds of all varieties. The results with five of the varieties appear in table 1. The remaining variety was Golden Menza which reacted the same as White Menza.

It is evident that all varieties did not react in exactly the same manner, but the order of bloom of any variety was the same. Plants treated from 6 P. M. to 7 A. M. produced blooms earlier than those of any other treatment. With the exception of the variety Sunshine the average number of stems per plant showed a direct relationship to the effectiveness of the treatment. The average number of flowers per stem showed an inverse relationship to the effectiveness of the treatment and the number of stems per plant, but showed a direct relation to the stem length, except in the 9-hour day treatment. The variation of the diameter of the terminal bloom was within the experimental error. The plot indicated as *Tops of plants treated* is discussed under "the part of the plant affected."

Further experiments of a similar nature were made in 1933. Fifteen plants each of the varieties Popcorn, Copper City, Helen Hubbard, and White Menza, and twenty plants of the variety Yellow Fellow were planted in each of four plots. Starting August 1, one plot was treated from 6 P. M. to 7 A. M., a second plot from 5 P. M. to 7 A. M., a third from 5 P. M. to 8 A. M., and a fourth was given no treatment. The results with these varieties appear in table 2.

A fifth plot was treated from 6 P. M. to 7 A. M., and additional light, to make a 16-hour day, was given by means of 75-watt frosted-glass lamps hung at 6-foot intervals, 18 inches above the plants, after they were showing color. The results of this experiment also appear in table 2.

The results show that there is practically no difference in the time of bloom of these varieties whether given a 9-, 10-, or 11-hour day. It will be noticed in comparing these data with table 1, that many factors were different in the two years of work. The seasons may not have been the same. The plants were of different ages when the treatment was started. The 9-hour day in 1931 extended from 7 A. M. to 4 P. M., while in 1933 the 9-hour day extended from 8 A. M. to 5 P. M. In comparing the results

TABLE 1. EFFECTS OF VARIOUS LENGTHS OF DAY

| Variety | Length of day | Treatment | Buds showing | First blooms cut | Period in bloom before check number | Cutting period | Average stems per plant | Average stem length | Average flowers per stem | Average flower diameter |
|---|---|---|---|---|---|---|---|---|---|---|
| | Hours | Time | Date | Date | Days | Days | Number | Inches | Number | Inches |
| Dorothy Turner | 11 | 6 P.M.–7 A.M. | Aug. 7 | Oct. 5 | 49 | 23 | 6.6 | 16.8 | 9.7 | 2.0 |
| | 9 | 4 P.M.–7 A.M. | Aug. 10 | Oct. 14 | 40 | 7 | 4.1 | 18.1 | 7.7 | 2.0 |
| | 13 | 8 P.M.–7 A.M. | Aug. 24 | Nov. 17 | 6 | 1 | 3.0 | 31.7 | 9.9 | 1.9 |
| | | No treatment | Sept. 20 | Nov. 23 | ...... | 5 | 3.9 | 31.6 | 10.9 | 1.9 |
| | 11 | 6 P.M.–7 A.M.* | Aug. 24 | Oct. 23 | 31 | 19 | 3.5 | 24.7 | 12.8 | 2.0 |
| Popcorn | 11 | 6 P.M.–7 A.M. | Aug. 3 | Sept. 11 | 53 | 14 | 11.1 | 15.5 | 7.3 | 1.9 |
| | 9 | 4 P.M.–7 A.M. | Aug. 3 | Sept. 23 | 41 | 1 | 10.4 | 14.8 | 6.8 | 1.9 |
| | 13 | 8 P.M.–7 A.M. | Aug. 24 | Oct. 15 | 19 | 18 | 6.5 | 23.6 | 12.7 | 2.1 |
| | | No treatment | Sept. 15 | Nov. 3 | ..... | 17 | 4.3 | 26.9 | 18.8 | 2.1 |
| | 11 | 6 P.M.–7 A.M.* | Aug. 24 | Oct. 3 | 31 | 3 | 9.9 | 21.3 | 8.9 | 2.1 |
| Sunshine | 11 | 6 P.M.–7 A.M. | Aug. 5 | Sept. 28 | 56 | 9 | 3.7 | 14.4 | 6.6 | 1.9 |
| | 9 | 4 P.M.–7 A.M. | Aug. 10 | Oct. 3 | 51 | 6 | 4.5 | 15.3 | 7.1 | 1.9 |
| | 13 | 8 P.M.–7 A.M. | Sept. 1 | Nov. 6 | 17 | 11 | 7.3 | 24.4 | 10.7 | 2.1 |
| | | No treatment | ...... | Nov. 23 | ..... | 1 | 2.5 | 26.0 | 7.7 | 2.1 |
| | 11 | 6 P.M.–7 A.M.* | Sept. 1 | Nov. 3 | 20 | 10 | 6.0 | 22.4 | 9.0 | 2.0 |
| White Menza | 11 | 6 P.M.–7 A.M. | Aug. 3 | Sept. 21 | 45 | 12 | 8.1 | 19.0 | 5.0 | 2.7 |
| | 9 | 4 P.M.–7 A.M. | Aug. 5 | Sept. 25 | 41 | 3 | 6.0 | 18.9 | 5.6 | 2.8 |
| | 13 | 8 P.M.–7 A.M. | Aug. 24 | Sept. 26 | 40 | 18 | 6.0 | 27.9 | 7.9 | 3.0 |
| | | No treatment | Sept. 15 | Nov. 5 | ..... | 5 | 4.9 | 31.3 | 8.0 | 3.0 |
| | 11 | 6 P.M.–7 A.M.* | Aug. 20 | Oct. 7 | 29 | 14 | 7.4 | 25.4 | 4.8 | 3.0 |
| Yellow Fellow | 11 | 6 P.M.–7 A.M. | Aug. 4 | Sept. 18 | 53 | 14 | 6.7 | 16.6 | 7.3 | 2.0 |
| | 9 | 4 P.M.–7 A.M. | Aug. 11 | Sept. 23 | 48 | 13 | 6.4 | 16.6 | 5.9 | 1.9 |
| | 13 | 8 P.M.–7 A.M. | Aug. 28 | Oct. 30 | 11 | 10 | .. | 24.2 | 7.9 | 2.0 |
| | | No treatment | Sept. 19 | Nov. 10 | ..... | 7 | 4.4 | 29.3 | 11.8 | 2.1 |
| | 11 | 6 P.M.–7 A.M.* | Sept. 1 | Oct. 15 | 26 | 16 | 4.6 | 25.5 | 14.0 | 2.0 |

*Tops of plants treated.

of these experiments, it appears that the earliest blooms were obtained when the cloth was placed over the plants at 5 or 6 P. M. and removed at 7 or 8 A. M. The angle at which the sun's rays strike the cloth apparently determines, to a great extent, the amount of light reaching the plants. When the cloth was placed over the plants as early as four o'clock in the

TABLE 2. EFFECTS OF VARIOUS LENGTHS OF DAY

| Variety | Length of day | Treatment | Buds showing | First blooms cut | Cutting period | Period in bloom before check number | Average stems per plant | Average stem length | Average flowers per stem |
|---|---|---|---|---|---|---|---|---|---|
| | Hours | Time | Date | Date | Days | Days | Number | Inches | Number |
| Copper city | 9 | 5 P. M.–8 A. M. | Aug. 27 | Oct. 12 | 5 | 27 | 7.3 | 22.0 | 8.0 |
| | 10 | 5 P. M.–7 A. M. | Aug. 26 | Oct. 12 | 5 | 27 | 8.5 | 22.0 | 8.1 |
| | 11 | 6 P. M.–7 A. M. | Aug. 26 | Oct. 12 | 5 | 27 | 8.0 | 22.7 | 7.5 |
| | 11 | 6 P. M.–7 A. M.* | Aug. 27 | Oct. 12 | 7 | 27 | 7.7 | 23.0 | 11.9 |
| | | no treatment | Sept. 20 | Nov. 8 | 9 | .. | 5.8 | 28.9 | 6.2 |
| Helen Hubbard | 9 | 5 P. M.–8 A. M. | Aug. 26 | Oct. 9 | 3 | 28 | 6.5 | 24.6 | 7.8 |
| | 10 | 5 P. M.–7 A. M. | Aug. 25 | Oct. 9 | 3 | 28 | 4.4 | 25.4 | 9.6 |
| | 11 | 6 P. M.–7 A. M. | Aug. 25 | Oct. 9 | 3 | 28 | 5.6 | 25.1 | 9.4 |
| | 11 | 6 P. M.–7 A. M.* | Aug. 25 | Oct. 9 | 3 | 28 | 5.2 | 25.2 | 8.5 |
| | | no treatment | Aug. 20 | Nov. 6 | 8 | .. | 5.5 | 28.9 | 7.0 |
| Popcorn | 9 | 5 P. M.–8 A. M. | Aug. 25 | Sept. 25 | 7 | 35 | 7.1 | 20.3 | 8.0 |
| | 10 | 5 P. M.–7 A. M. | Aug. 22 | Sept. 25 | 7 | 35 | 9.3 | 18.9 | 8.5 |
| | 11 | 6 P. M.–7 A. M. | Aug. 19 | Sept. 25 | 7 | 35 | 6.9 | 23.6 | 9.8 |
| | 11 | 6 P. M.–7 A. M.* | Aug. 25 | Sept. 25 | 7 | 35 | 8.4 | 20.3 | 9.1 |
| | | no treatment | Sept. 15 | Oct. 30 | 4 | .. | 7.0 | 27.7 | 13.7 |
| White Menza | 9 | 5 P. M.–8 A. M. | Aug. 27 | Oct. 5 | 7 | 27 | 7.3 | 26.3 | 4.7 |
| | 10 | 5 P. M.–7 A. M. | Aug. 25 | Oct. 5 | 7 | 27 | 7.9 | 25.1 | 4.8 |
| | 11 | 6 P. M.–7 A. M. | Aug. 25 | Oct. 2 | 7 | 30 | 7.0 | 25.3 | 5.9 |
| | 11 | 6 P. M.–7 A. M.* | Aug. 25 | Oct. 5 | 7 | 27 | 7.3 | 26.0 | 4.6 |
| | | no treatment | Sept. 17 | Nov. 1 | 7 | .. | 6.2 | 36.7 | 5.1 |
| Yellow Fellow | 9 | 5 P. M.–8 A. M. | Aug. 25 | Oct. 2 | 7 | 30 | 5.2 | 21.9 | 8.1 |
| | 10 | 5 P. M.–7 A. M. | Aug. 25 | Oct. 2 | 7 | 30 | 5.9 | 22.3 | 9.4 |
| | 11 | 6 P. M.–7 A. M. | Aug. 23 | Oct. 2 | 7 | 30 | 5.7 | 23.4 | 9.5 |
| | 11 | 6 P. M.–7 A. M.* | Aug. 25 | Oct. 2 | 7 | 30 | 5.0 | 23.4 | 9.9 |
| | | no treatment | Sept. 20 | Nov. 1 | 7 | .. | 4.5 | 40.4 | 10.4 |

*Additional light to make a 16-hour day after buds showed color.

afternoon, the rays of light may have struck the cloth at the proper angle so that the action was that of a dense shade rather than that of a shortened day.

With the exception of the variety Helen Hubbard, plants treated from 5 P. M. to 7 A. M. produced the greatest number of stems per plant. The average number of flowers per stem and the average stem length were not consistent in the varieties and for this reason are not considered significant. As a result of these experiments an 11-hour day was selected as a basis for all other work.

Increasing the length of day by the use of electric light after the buds were showing color had no effect on the time of bloom, length of stem, number of flowers, or number of stems per plant. The variety Popcorn faded badly under this treatment.

### PART OF THE PLANT AFFECTED

Working with *Cosmos sulphureus*, Garner and Allard (1925) gave certain parts of plants short-day treatments and obtained blooms on those treated parts, while the sections given normal summer days bloomed in normal season. To determine the effect of treating the terminal part of chrysanthemum plants, an experiment was carried out using the same varieties and numbers of plants, as indicated in table 1. The cloth covered the tops of the plants and extended down the sides so that the terminal growth was covered at the beginning of the experiment. The plants were allowed to grow up under the cloth. The results appear in table 1 and show that the treatment was not so effective as when the cloth covered the entire plant. The data indicate that the plants were not affected by the treatment until they reached such a height as to reduce the reflected light, producing the effect of a short day. The daily period during which the plants were partly covered was from 6 P. M. to 7 A. M.

### TIME OF DAY OF TREATMENT

To determine the effects of darkening at different times of day with black cloth, twenty-five plants each of the varieties Bokhara and Charles Maynard and ten plants of Rodell and Pink Dot were planted in each of five plots. All plots which were treated received approximately eleven hours of daylight from July 15, 1932, until after the buds were showing color. One plot was treated from 6 o'clock at night until 7 o'clock in the morning; a second plot was darkened from 4 o'clock at night until dark; a third plot was darkened from dark at night until 9 o'clock in the morning; and a fourth plot was darkened from 9 o'clock in the morning until 1 o'clock in the afternoon. The results appear in table 3.

TABLE 3. EFFECT OF TREATING AT DIFFERENT TIMES OF DAY

| Variety | Length of day | Darkening | Buds showing | First blooms cut | Cutting period | Period in bloom before check number | Average stems per plant | Average stem length | Average flowers per stem | Average flower diam. |
|---|---|---|---|---|---|---|---|---|---|---|
| | Hours | Time | Date | Date | Days | Days | Number | Inches | Number | Inches |
| Bokhara | 11 | 6 P.M.–7 A.M. | Aug. 1 | Sept. 9 | 13 | 47 | 8.6 | 14.8 | 10.5 | 1.8 |
| | 11 | 4 P.M.–dark | Aug. 3 | Sept. 21 | 18 | 35 | 7.0 | 17.6 | 7.2 | 1.9 |
| | 11 | Dark–9 A.M. | Aug. 18 | Oct. 5 | 21 | 21 | 5.7 | 19.8 | 10.9 | 1.8 |
| | Check | Normal day | Sept. 5 | Oct. 26 | 3 | .. | 4.7 | 27.9 | 19.0 | 1.9 |
| | 11 | 9 A.M.–1 P.M. | Sept. 15 | Oct. 28 | 8 | –2 | 5.1 | 22.9 | 11.3 | 1.9 |
| Chas. Maynard | 11 | 6 P.M.–7 A.M. | Aug. 1 | Sept. 9 | 13 | 47 | 7.5 | 16.4 | 8.2 | 1.7 |
| | 11 | 4 P.M.–dark | Aug. 8 | Oct. 7 | 19 | 19 | 5.2 | 21.0 | 9.6 | 1.9 |
| | 11 | Dark–9 A.M. | Aug. 24 | Oct. 12 | 14 | 14 | 4.8 | 22.1 | 9.4 | 1.8 |
| | Check | Normal day | Sept. 5 | Oct. 26 | 3 | .. | 3.6 | 27.7 | 13.1 | 1.7 |
| | 11 | 9 A.M.–1 P.M. | Sept. 15 | Oct. 24 | 10 | 2 | 2.9 | 24.3 | 8.4 | 1.6 |
| Pink Dot | 11 | 6 P.M.–7 A.M. | Aug. 1 | Sept. 9 | 8 | 47 | 6.9 | 13.5 | 8.8 | 1.6 |
| | 11 | 4 P.M.–dark | Aug. 10 | Sept. 28 | 7 | 28 | 5.5 | 16.7 | 7.2 | 1.8 |
| | 11 | Dark–9 A.M. | Aug. 20 | Oct. 9 | 17 | 17 | 5.1 | 19.2 | 8.4 | 1.6 |
| | Check | Normal day | Sept. 5 | Oct. 26 | 3 | .. | 3.3 | 25.7 | 15.2 | 1.9 |
| | 11 | 9 A.M.–1 P.M. | Sept. 15 | Oct. 28 | 6 | –2 | 2.9 | 20.3 | 9.9 | 1.8 |
| Rodell | 11 | 6 P.M.–7 A.M. | Aug. 1 | Sept. 4 | 21 | 45 | 8.0 | 12.6 | 12.2 | 1.5 |
| | 11 | 4 P.M.–dark | Aug. 10 | Sept. 28 | 10 | 21 | 5.8 | 16.6 | 7.8 | 1.5 |
| | 11 | Dark–9 A.M. | Aug. 18 | Sept. 26 | 11 | 23 | 6.8 | 17.2 | 11.0 | 1.5 |
| | Check | Normal day | Sept. 5 | Oct. 19 | 7 | .. | 6.0 | 21.8 | 16.7 | 1.4 |
| | 11 | 9 A.M.–1 P.M. | Sept. 15 | Oct. 21 | 5 | –2 | 4.6 | 20.3 | 13.0 | 1.4 |

All varieties responded in a similar manner and it is evident that reducing the length of day, by treating from 6 P. M. until 7 A. M., was most effective in producing early blooms.   The relative time of bloom of the various plots is shown in f gure 2.   Treating the plants from 4 P. M. until dark was more effective in reducing the time necessary for blooming than treating from dark until 9 A. M.   Plants which were darkened from 9 A. M. until 1 P. M. bloomed from two to five days later than the normal season.

FIGURE 2.   EFFECTS OF TREATMENTS AT DIFFERENT TIMES OF DAY (VARIETY BOKHARA)
The various periods of treatment were as follows:

1.  4 P.  M.—dark
2.  Dark—9 A.  M.
3.  9 A.  M.—1 P.  M.
4.  Not darkened
5.  6 P.  M.—7 A.  M.

Garner and Allard (1931) made similar observations.   The more effective the treatment the shorter the stems produced and the greater the number of stems per plant.   The stems of plants treated from 9 A. M. to 1 P. M. were weaker than those given no treatment and the foliage was a light green color.   The average diameter of the blooms was not consistent in the varieties and not significantly different.   Espino and Pantaleon (1931) have shown that the morning light was more beneficial to plant growth than afternoon light when the plants were shaded at the respective ends of the day.

### TYPES OF CLOTH AND EFFECTS OF USED CLOTH

To determine the effects of different types of black cloth, sixteen plants each of the varieties Silver Sheen, Gold Lode, October Rose, Mrs. David F. Roy, Gladys Pearson, and Pink Delight, and twelve plants of each of the varieties Mrs. H. E. Kidder and Monument were used in each test.

The plants were grown in a ground bed and darkened from July 25, 1932, until all buds were selected.

The types of cloth used were (a) black sateen, with a count of 68 by 104 threads to the square inch, which had been used the year previously and washed twice; (b) unused black sateen with a count of 68 by 104 threads to the square inch; (c) unused black sateen, with a count of 64 by 88 threads to the square inch; and (d) new black cloth, plain weave, with a count of 68 by 72 threads to the square inch. The results with five varieties appear in table 4.

TABLE 4. TYPES OF BLACK CLOTH

| Variety | Cloth | Thread count per inch | First buds selected | First blooms cut | Cutting period | Period in bloom before check | Average stem length | Average flower diameter |
|---|---|---|---|---|---|---|---|---|
| | *Type* | *Number* | *Date* | *Date* | *Days* | *Days* | *Inches* | *Inches* |
| Gold Lode | Sateen, used | 68 x 104 | Aug. 12 | Sept. 6 | 6 | 39 | 45 | 4.01 |
| | Sateen, new | 68 x 104 | Aug. 12 | Sept. 6 | 3 | 39 | 45 | 4.45 |
| | Sateen, new | 64 x 88 | Aug. 12 | Sept. 6 | 3 | 39 | 44 | 4.42 |
| | Plain | 68 x 72 | Aug. 15 | Sept. 12 | 7 | 33 | 45 | 4.14 |
| | No treatment | ... | Sept. 10 | Oct. 15 | 9 | .. | 66 | 5.45 |
| Mrs. David F. Roy | Sateen, used | 68 x 104 | Aug. 20 | Oct. 7 | 18 | 25 | 36 | 5.32 |
| | Sateen, new | 68 x 104 | Aug. 20 | Oct. 7 | 22 | 25 | 35 | 5.27 |
| | Sateen, new | 64 x 88 | Aug. 17 | Oct. 3 | 26 | 29 | 34 | 5.02 |
| | Plain | 68 x 72 | Aug. 22 | Oct. 12 | 17 | 20 | 34 | 5.01 |
| | No treatment | ... | Sept. 18 | Nov. 1 | 10 | .. | 37 | 5.04 |
| Mrs. H. E. Kidder | Sateen, used | 68 x 104 | Aug. 17 | Sept. 20 | 17 | 34 | 35 | 4.92 |
| | Sateen, new | 68 x 104 | Aug. 12 | Sept. 20 | 13 | 34 | 36 | 4.84 |
| | Sateen, new | 64 x 88 | Aug. 12 | Sept. 20 | 13 | 34 | 36 | 4.92 |
| | Plain | 68 x 72 | Aug. 22 | Sept. 26 | 23 | 28 | 34 | 4.92 |
| | No treatment | ... | Sept. 12 | Oct. 24 | 8 | .. | 42 | 5.18 |
| October Rose | Sateen, used | 68 x 104 | Aug. 17 | Sept. 19 | 14 | 32 | 53 | 5.79 |
| | Sateen, new | 68 x 104 | Aug. 17 | Sept. 19 | 14 | 32 | 55 | 5.61 |
| | Sateen, new | 64 x 88 | Aug. 15 | Sept. 20 | 13 | 31 | 53 | 5.91 |
| | Plain | 68 x 72 | Aug. 17 | Sept. 26 | 11 | 25 | 54 | 5.39 |
| | No treatment | ... | Sept. 12 | Oct. 21 | 8 | .. | 66 | 5.53 |
| Silver Sheen | Sateen, used | 68 x 104 | Aug. 15 | Sept. 6 | 17 | 45 | 46 | 4.78 |
| | Sateen, new | 68 x 104 | Aug. 12 | Sept. 6 | 17 | 45 | 46 | 4.72 |
| | Sateen, new | 64 x 88 | Aug. 12 | Sept. 6 | 20 | 45 | 44 | 5.00 |
| | Plain | 68 x 72 | Aug. 17 | Sept. 20 | 13 | 31 | 46 | 4.93 |
| | No treatment | ... | Sept. 12 | Oct. 21 | 8 | .. | 65 | 4.99 |

The date of bloom was approximately the same with each of the sateen cloths. Plants treated with black cloth plain weave bloomed later than those treated with sateen.

A second experiment was made at the same time using the varieties Gold Lode, October Rose, Snow White, Friendly Rival, Marie De Petris, and Citronella. Black sateen cloth of 68 by 104 threads to the square inch was compared with black costume cloth of 54 by 54 threads to the square inch. The results appear in table 5.

It is evident from these results that black costume cloth was less effective than the black sateen.

### DATE OF PROPAGATION

The date of propagation was considered important for this short-day treatment, and, to determine its effect, fifteen or more plants of each of the varieties Helen Hubbard, White Menza, Golden Menza, Yellow Fellow,

TABLE 5.  Black Costume Cloth Compared with Black Sateen

| Variety | Cloth | Thread count per inch | First buds selected | Blooms cut | Cutting period | Period in bloom before check | Average stem length | Average flower diameter |
|---|---|---|---|---|---|---|---|---|
| | *Type* | *Number* | *Date* | *Date* | *Days* | *Days* | *Inches* | *Inches* |
| Citronella | Sateen | 68 x 104 | Aug. 21 | Sept. 26 | 14 | 24 | 28 | 4.82 |
| | Costume | 54 x 54 | Aug. 21 | Oct. 3 | 11 | 17 | 31 | 5.14 |
| | No treatment | ... | Sept. 12 | Oct. 20 | 11 | .. | 38 | 5.41 |
| Friendly Rival | Sateen | 68 x 104 | Aug. 17 | Sept. 29 | 18 | 43 | 33 | 5.14 |
| | Costume | 54 x 54 | Aug. 25 | Oct. 14 | 23 | 28 | 37 | 5.50 |
| | No treatment | ... | Sept. 21 | Nov. 11 | 7 | .. | 44 | 5.90 |
| Gold Lode | Sateen | 68 x 104 | Aug. 5 | Sept. 2 | 1 | 42 | 21 | 4.89 |
| | Costume | 54 x 54 | Aug. 12 | Sept. 12 | 8 | 30 | 23 | 4.07 |
| | No treatment | ... | Sept. 12 | Oct. 14 | 6 | .. | 38 | 5.45 |
| Marie De Petris | Sateen | 68 x 104 | Aug. 21 | Oct. 11 | 18 | 31 | 35 | 5.53 |
| | Costume | 54 x 54 | Aug. 21 | Oct. 29 | 9 | 13 | 42 | 5.72 |
| | No treatment | ... | Sept. 21 | Nov. 11 | 8 | .. | 46 | 5.45 |
| October Rose | Sateen | 68 x 104 | Aug. 10 | Sept. 9 | 11 | 41 | 31 | 5.26 |
| | Costume | 54 x 54 | Aug. 21 | Oct. 3 | 8 | 17 | 35 | 5.48 |
| | No treatment | ... | Sept. 12 | Oct. 20 | 9 | .. | 51 | 5.80 |
| Snow White | Sateen | 68 x 104 | Aug. 28 | Oct. 3 | 26 | 29 | 39 | 5.21 |
| | Costume | 54 x 54 | Sept. 25 | Oct. 20 | 21 | 12 | 46 | 5.73 |
| | No treatment | ... | Oct. 4 | Nov. 1 | 17 | .. | 49 | 5.79 |

and Popcorn were propagated in 1932 on each of the following days: February 15, March 1, 15, 30, and April 15. The last pinch of all plants was made July 10. The plants propagated February 15 were not treated but all others were darkened in the usual manner, starting July 25. Results of this work appear in table 6.

TABLE 6.  Effect of Date of Propagation

| Variety | Propagation | First blooms cut | Cutting period | Average stems per plant | Average flowers per stem | Average stem length |
|---|---|---|---|---|---|---|
| | *Date* | *Date* | *Days* | *Number* | *Number* | *Inches* |
| Golden Menza......... | Feb. 15* | Nov. 1 | 10 | 6.4 | 5.2 | 33. |
| | Mar. 1 | Sept. 26 | 10 | 4.2 | 5.8 | 24.3 |
| | Mar. 15 | Sept. 26 | 10 | l7.2 | 5.3 | 23.9 |
| | Mar. 30 | Sept. 26 | 10 | 10.1 | 5.3 | 22.2 |
| Helen Hubbard........ | Feb. 15* | Nov. 11 | 6 | 4.7 | 8.7 | 30.9 |
| | Mar. 1 | Oct. 6 | 15 | 6.1 | 8.6 | 25.2 |
| | Mar. 15 | Oct. 6 | 15 | 6.6 | 9.2 | 24.8 |
| | Mar. 30 | Oct. 6 | 15 | 6.3 | 8.8 | 23.8 |
| | Apr. 15 | Oct. 6 | 15 | 6.1 | 8.8 | 23.0 |
| Popcorn.............. | Feb. 15* | Nov. 1 | 10 | 6.3 | 8.7 | 27.9 |
| | Mar. 1 | Sept. 20 | 6 | 6.9 | 7.4 | 16.9 |
| | Mar. 15 | Sept. 20 | 16 | 10.1 | 7.1 | 19.4 |
| | Mar. 30 | Sept. 20 | 9 | 13.1 | 7.3 | 18.4 |
| White Menza.......... | Feb. 15* | Nov. 8 | 3 | 7.0 | 5.6 | 37.5 |
| | Mar. 1 | Sept. 26 | 10 | 7.9 | 4.6 | 21.6 |
| | Mar. 15 | Sept. 26 | 10 | 8.9 | 5.3 | 24.2 |
| | Mar. 30 | Sept. 26 | 10 | 9.2 | 6.0 | 22.6 |
| | Apr. 15 | Sept. 26 | 10 | 6.0 | 5.9 | 22.4 |
| Yellow Fellow........ | Feb. 15* | Nov. 5 | 6 | 4.5 | 10.3 | 30.7 |
| | Mar. 1 | Sept. 23 | 28 | 5.9 | 5.5 | 20.5 |
| | Mar. 15 | Sept. 23 | 17 | 5.5 | 6.5 | 19.6 |
| | Mar. 30 | Sept. 29 | 22 | 7.8 | 5.7 | 19.2 |
| | Apr. 15 | Sept. 29 | 12 | 3.5 | 5.7 | 18.1 |

*Not darkened.

The time of bloom was not affected by the time of propagation when the plants were all given the same short-day treatment. The number of branches was increased on all plants by the treatment. The number of branches cut was usually greater when the plants were propagated March 15 or 30 than when propagated at any other time. The stem length and number of blooms per stem were not appreciably affected by the time of propagation, although the longer stems and greater number of flowers per stem usually occurred on the early-propagated plants.

DATE OF THE LAST PINCH

To determine the effect of the date the last pinch was made on the stem length, time of bloom, number of stems, and number of blooms per stem, fifteen plants each of the varieties Sunshine, Izola, Norine, Mrs. William Buckingham, and Bronze Buckingham were planted in each of four plots. Three plots were darkened each day starting July 15, 1932, and continuing until buds were showing color on all varieties. The plants were given the last pinch in the various plots on June 25, July 5, and July 15. The data for this experiment appear in table 7.

TABLE 7. EFFECT OF DATE OF LAST PINCH

| Variety | Last pinch | First blooms cut | Cutting period | Average stems per plant | Average flowers per stem | Average stem length |
|---|---|---|---|---|---|---|
| | *Date* | *Date* | *Days* | *Number* | *Number* | *Inches* |
| Bronze Buckingham........... | June 25 | Sept. 20 | 3 | 2.7 | 7.3 | 12.4 |
| | July 5 | Sept. 20 | 28 | 4.8 | 4.1 | 10.4 |
| | July 15 | Sept. 19 | 7 | 5.2 | 4.1 | 9.8 |
| | July 15* | Nov. 1 | 7 | 4.3 | 4.6 | 19.6 |
| Izola....................... | June 25 | Sept. 20 | 9 | 8.0 | 9.7 | 17.2 |
| | July 5 | Sept. 20 | 13 | 10.5 | 7.7 | 15.8 |
| | July 15 | Sept. 19 | 14 | 11.4 | 7.1 | 13.5 |
| | July 15* | Nov. 7 | 14 | 5.2 | 13.8 | 29.3 |
| Mrs. William Buckingham...... | June 25 | Sept. 20 | 3 | 6.7 | 4.5 | 12.7 |
| | July 5 | Sept. 20 | 6 | 8.7 | 3.7 | 10.6 |
| | July 15 | Sept. 19 | 7 | 8.7 | 3.6 | 10.0 |
| | July 15* | Nov. 1 | 7 | 5.5 | 3.3 | 20.4 |
| Norine...................... | June 25 | Sept. 23 | 13 | 9.5 | 13.7 | 18.4 |
| | July 5 | Sept. 26 | 19 | 12.9 | 7.9 | 14.4 |
| | July 15 | Sept. 26 | 14 | 11.4 | 7.7 | 13.1 |
| | July 15* | Nov. 15 | 10 | 6.6 | 11.8 | 28.6 |
| Sunshine.................... | June 25 | Sept. 29 | 19 | 6.2 | 6.6 | 18.4 |
| | July 5 | Sept. 26 | 19 | 8.8 | 5.3 | 15.4 |
| | July 15 | Oct. 3 | 15 | 9.4 | 4.8 | 13.5 |
| | July 15* | Nov. 15 | 6 | 5.2 | 6.5 | 26.2 |

*Not darkened.

The work was repeated, using different varieties the following season, and the results of this experiment appear in table 8.

From these results it appears that the time of bloom was not affected by the date the last pinch was made. The later the final pinch was made, the greater the number of stems produced per plant, but the shorter the stems. The fewer the number of stems per plant, the greater the number of flowers per stem. It is evident from these data that each variety reacted differently even though all were pinched on the same date. The more rapidly growing varieties produced longer stems after the short-day treat-

TABLE 8. EFFECT OF THE DATE THE LAST PINCH WAS MADE

| Variety | Last pinch | First blooms cut | Cutting period | Average stems per plant | Average flowers per stem | Average stem length |
|---|---|---|---|---|---|---|
| | Date | Date | Days | Number | Number | Inches |
| Bokhara..................... | June 25 | Sept. 23 | 5 | 5.9 | 21.4 | 14.7 |
| | July 5 | Sept. 23 | 5 | 9.0 | 16.3 | 10.5 |
| | July 15 | Sept. 23 | 5 | 7.0 | 15.9 | 11.2 |
| | July 15* | Oct. 23 | 8 | 7.6 | 22.5 | 10.6 |
| Chas. Maynard............... | June 25 | Sept. 23 | 5 | 3.1 | 23.7 | 11.6 |
| | July 5 | Sept. 23 | 9 | 7.4 | 16.9 | 6.4 |
| | July 15 | Sept. 25 | 3 | 7.2 | 16.5 | 6.0 |
| | July 15* | Oct. 23 | 3 | 4.3 | 25.4 | 8.1 |
| Pink Dot.................... | June 25 | Sept. 23 | 9 | 10.8 | 20.3 | 8.2 |
| | July 5 | Sept. 25 | 10 | 10.7 | 17.9 | 8.0 |
| | July 15 | Sept. 28 | 7 | 6.4 | 16.7 | 7.6 |
| | July 15* | Oct. 23 | 8 | 10.1 | 21.6 | 6.3 |
| Rodell..................... | June 25 | Sept. 23 | 9 | 8.7 | 18.6 | 11.8 |
| | July 5 | Sept. 23 | 5 | 12.2 | 15.3 | 8.0 |
| | July 15 | Sept. 23 | 5 | 12.2 | 15.4 | 7.7 |
| | July 15* | Oct. 19 | 2 | 9.7 | 19.0 | 8.9 |
| Uvalda..................... | June 25 | Sept. 18 | 5 | 7.0 | 20.5 | 8.1 |
| | July 5 | Sept. 20 | 3 | 7.8 | 16.8 | 6.0 |
| | July 15 | Sept. 20 | 3 | 7.6 | 16.8 | 6.0 |
| | July 15* | Oct. 16 | 3 | 6.8 | 19.2 | 5.3 |

*Not darkened.

ment was started than the slower growing types. From this it appears that in order to obtain stems of sufficient length for commercial purposes, it is necessary to pinch the slower-growing varieties sooner than those which grow more rapidly.

### EFFECTS OF LENGTH OF SHORT-DAY TREATMENT ON DISBUDDED VARIETIES

In order to determine the length of time necessary to continue the short-day treatment, sixteen plants each of the varieties Gold Lode, Quaker Maid, October Rose, Silver Sheen, Rose Chocard, Chas. Rager, and Friendly Rival were planted in a ground bed in each of five plots (1931). Starting July 15, one plot was given 10 successive short days, a second plot 20 short days, a third 30 short days, a fourth continuous short days until the blooms were cut, and a fifth was not treated. The results appear in table 9.

From this table it is evident that plants given ten successive short days formed buds which eventually developed into flowers. Earlier varieties responded with a shorter length of treatment than did later types. The date of bloom, length of stem, and size of flower were not very different, if the plants were treated each successive day until all the buds were selected or for a longer period of time. When the treatment was discontinued before all buds were selected, the blooming period of the variety was lengthened and the date of bloom delayed. The greater the interval of time between the termination of the treatment and the selecting of the buds, the later the variety bloomed. Late varieties, such as Chas. Rager, which were treated for a short time, formed buds with all the characteristics of "terminals." These buds did not mature normally but eventually assumed all the characteristics of "crown" buds. The blooms resulting

TABLE 9. EFFECT OF THE LENGTH OF SHORT-DAY TREATMENT ON LARGE-FLOWERED VARIETIES

| Variety | Treatment from July 15 until— | Period of treatment | First buds selected | First color showed | First flowers cut | Period in bloom before check | Average stem length | Average flower diameter |
|---|---|---|---|---|---|---|---|---|
| | | *Days* | *Date* | *Date* | *Date* | *Days* | *Inches* | *Inches* |
| Chas. Rager | Bloom | 60 | Aug. 10 | Aug. 28 | Sept. 14 | 57 | 33.9 | 4.3 |
| | Aug. 14 | 30 | Aug. 10 | Aug. 26 | Sept. 18 | 53 | 33.9 | 4.2 |
| | Aug. 4 | 20 | Aug. 16 | Sept. 19 | Oct. 5 | 36 | 36.5 | 5.1 |
| | July 25 | 10 | Aug. 16 | Oct. 5 | Nov. 10 | 0 | 55.2 | 4.7 |
| | No treatment | ....... | Sept. 21 | Oct. 24 | Nov. 10 | ....... | 65.0 | 5.3 |
| Friendly Rival | Bloom | 67 | Aug. 10 | Aug. 28 | Sept. 21 | 47 | 21.1 | 4.6 |
| | Aug. 14 | 30 | Aug. 10 | Aug. 28 | Sept. 23 | 45 | 23.2 | 4.7 |
| | Aug. 4 | 20 | Aug. 10 | Sept. 9 | Sept. 26 | 42 | 27.4 | 4.6 |
| | July 25 | 10 | Aug. 16 | Sept. 19 | Oct. 15 | 23 | 34.1 | 4.8 |
| | No treatment | ....... | Sept. 20 | Oct. 24 | Nov. 7 | ....... | 59.5 | 5.4 |
| Gold Lode | Bloom | 42 | Aug. 5 | Aug. 18 | Aug. 26 | 48 | 23.8 | 3.9 |
| | Aug. 14 | 30 | Aug. 5 | Aug. 20 | Aug. 26 | 48 | 26.0 | 4.4 |
| | Aug. 4 | 20 | Aug. 5 | Aug. 20 | Aug. 26 | 48 | 26.0 | 4.4 |
| | July 25 | 10 | Aug. 5 | Aug. 24 | Sept. 8 | 35 | 28.4 | 4.7 |
| | No treatment | ....... | Sept. 11 | Sept. 29 | Oct. 13 | ....... | 58.0 | 4.7 |
| October Rose | Bloom | 54 | Aug. 5 | Aug. 20 | Sept. 8 | 45 | 35.7 | 5.0 |
| | Aug. 14 | 30 | Aug. 5 | Aug. 21 | Sept. 8 | 45 | 37.5 | 5.3 |
| | Aug. 4 | 20 | Aug. 5 | Aug. 23 | Sept. 14 | 39 | 38.5 | 5.0 |
| | July 25 | 10 | Aug. 16 | Sept. 14 | Oct. 23 | 0 | 46.3 | 4.8 |
| | No treatment | ....... | Sept. 14 | Oct. 5 | Oct. 23 | ....... | 66.6 | 4.8 |
| Quaker Maid | Bloom | 44 | Aug. 5 | Aug. 17 | Aug. 28 | 38 | 37.3 | 4.4 |
| | Aug. 14 | 30 | Aug. 5 | Aug. 20 | Aug. 28 | 38 | 37.2 | 4.5 |
| | Aug. 4 | 20 | Aug. 5 | Aug. 23 | Sept. 4 | 31 | 38.6 | 4.7 |
| | July 25 | 10 | Aug. 10 | Aug. 26 | Sept. 14 | 21 | 41.9 | 5.0 |
| | No treatment | ....... | Sept. 14 | Oct. 5 | Oct. 5 | ....... | 69.4 | 4.4 |
| Rose Chocard | Bloom | 60 | Aug. 10 | Aug. 26 | Sept. 14 | 42 | 20.6 | 3.6 |
| | Aug. 14 | 30 | Aug. 10 | Aug. 26 | Sept. 14 | 42 | 21.1 | 3.9 |
| | Aug. 4 | 20 | Aug. 10 | Aug. 30 | Sept. 26 | 30 | 22.9 | 3.4 |
| | July 25 | 10 | Aug. 16 | Sept. 14 | Oct. 13 | 13 | 26.3 | 2.9 |
| | No treatment | ....... | Sept. 14 | Oct. 10 | Oct. 26 | ....... | 38.5 | 4.1 |
| Silver Sheen | Bloom | 54 | Aug. 5 | Aug. 21 | Sept. 8 | 48 | 28.9 | 4.5 |
| | Aug. 14 | 30 | Aug. 5 | Aug. 24 | Sept. 8 | 48 | 31.1 | 4.8 |
| | Aug. 4 | 20 | Aug. 5 | Aug. 26 | Sept. 18 | 38 | 32.5 | 5.0 |
| | July 25 | 10 | Aug. 16 | Sept. 15 | Oct. 23 | 3 | 46.2 | 5.2 |
| | No treatment | ....... | Sept. 14 | Oct. 5 | Oct. 26 | ....... | 61.1 | 4.7 |

from these buds were larger and appeared to have more florets than did those which bloomed in normal season or those which bloomed early as a result of continued treatment. On many plants the florets developed irregularly and those which opened first were faded before the majority had developed. (Figures 3 and 4.)

### EFFECTS OF LENGTH OF SHORT-DAY TREATMENT ON POMPON VARIETIES

It was considered advisable to determine the effects of various lengths of short-day treatment on pompon varieties. Fifteen plants of each of the varieties Bokhara, Chas. Maynard, Pink Dot, Rodell, and Uvalda were planted in each of 5 plots. The last pinch was made July 5 and 4 plots were given short days starting July 15, 1932 and continuing: (1) until

FIGURE 3.  BUDS OF THE VARIETY OCTOBER ROSE, WHICH WAS GIVEN 10 SUCCESSIVE SHORT DAYS FOLLOWED BY NORMAL DAYS

color was showing in the buds of all varieties, (2) 15 successive days, (3) 10 successive days, and (4) 5 successive days.  The fifth plot was not treated.  All plots were given normal day length after the period of treatment.  The results with these varieties are given in table 10.

Plants which were treated for fifteen or fewer days formed buds somewhat later than those treated every day until color was showing in the buds. The date of bloom for most of the plants given fifteen or fewer short days was little or no earlier than that of the plants which were given no treatment.  Plants which were given this short period of treatment, and started to bloom earlier than the check plants, bloomed over a much longer period and the last flowers were cut at approximately the same time in both cases. Five short days was sufficient to cause the terminal flower buds to form. Most of these remained undeveloped until the normal flowering season, but vegetative growth continued below them.  This growth produced flower buds and blooms in normal season.  The undeveloped buds were present when the blooms were cut.

Plants darkened for ten successive days formed flower buds to about half way down the stem; the remainder grew vegetatively and bloomed in normal season.  The early buds were undeveloped at the time the flowers from the normal buds were fully open.  Plants treated for fifteen days formed flower buds nearly all the way down the stem.  Most of these remained dormant until the days were of the proper length for normal blooming at which time some of them opened.  Plants given short days until color was showing in the buds produced normal blooms.  In the majority of the plants the number of stems per plant was increased by the treatments.  The greatest number developed when the plants were darkened until color was showing in the buds.  The number of blooms per stem was decreased by the treatment.  The greatest decrease occurred when the plants were treated for fifteen days.  This may be accounted

FIGURE 4. BLOOMS OF THE VARIETY OCTOBER ROSE, WHICH WAS GIVEN 10 SUCCESSIVE SHORT DAYS FOLLOWED BY NORMAL DAYS

TABLE 10. EFFECT OF THE LENGTH OF THE SHORT-DAY TREATMENT ON POMPON VARIETIES

| Variety | Treatment | First buds | Color in first buds | First blooms cut | Period of cutting | Period in bloom before check | Average stems per plant | Average flowers per stem | Average stem length |
|---|---|---|---|---|---|---|---|---|---|
| | *Length* | *Date* | *Date* | *Date* | *Days* | *Days* | *Number* | *Number* | *Inches* |
| Bokhara | Until colored | Aug. 3 | Aug. 25 | Sept. 12 | 7 | 39 | 7.2 | 7.6 | 15.8 |
| | 15 nights | Aug. 5 | Sept. 12 | Oct. 28 | 8 | −7 | 7.4 | 5.2 | 16.8 |
| | 10 nights | Aug. 10 | Oct. 10 | Oct. 24 | 7 | −3 | 6.9 | 15.5 | 25.1 |
| | 5 nights | Aug. 20 | Oct. 10 | Oct. 24 | 8 | −3 | 5.9 | 10.5 | 20.0 |
| | No treatment | Sept. 7 | Oct. 10 | Oct. 21 | 8 | ..... | 6.8 | 17.2 | 27.7 |
| Chas. Maynard | Until colored | Aug. 3 | Aug. 26 | Sept. 14 | 2 | 40 | 3.5 | 7.9 | 18.1 |
| | 15 nights | Aug. 5 | Oct. 10 | Oct. 24 | 7 | 0 | 7.2 | 5.9 | 20.5 |
| | 10 nights | Aug. 10 | Oct. 10 | Oct. 24 | 7 | 0 | 4.7 | 8.7 | 26.9 |
| | 5 nights | Aug. 20 | Oct. 10 | Oct. 24 | 7 | 0 | 5.0 | 9.4 | 22.7 |
| | No treatment | Sept. 7 | Oct. 10 | Oct. 24 | 7 | ..... | 5.0 | 7.6 | 28.1 |
| Pink Dot | Until colored | Aug. 3 | Aug. 23 | Sept. 14 | 6 | 44 | 12.2 | 6.3 | 15.1 |
| | 15 nights | Aug. 5 | Oct. 10 | Oct. 24 | 10 | 4 | 9.0 | 5.0 | 18.2 |
| | 10 nights | Aug. 10 | Oct. 10 | Oct. 28 | 4 | 0 | 6.6 | 10.4 | 22.2 |
| | 5 nights | Sept. 12 | Oct. 10 | Oct. 21 | 11 | 7 | 8.8 | 5.6 | 21.4 |
| | No treatment | Sept. 7 | Oct. 10 | Oct. 28 | 4 | ..... | 5.6 | 10.0 | 25.9 |
| Rodell | Until colored | Aug. 3 | Aug. 23 | Sept. 6 | 13 | 42 | 11.3 | 8.5 | 16.3 |
| | 15 nights | Aug. 5 | Sept. 6 | Sept. 23 | 30 | 25 | 11.2 | 4.5 | 18.1 |
| | 10 nights | Aug. 10 | Oct. 10 | Oct. 21 | 3 | −3 | 7.6 | 11.8 | 25.1 |
| | 5 nights | Sept. 7 | Oct. 10 | Oct. 18 | 6 | 0 | 9.6 | 7.9 | 20.7 |
| | No treatment | Sept. 7 | Oct. 10 | Oct. 18 | 6 | ..... | 8.1 | 11.8 | 24.9 |
| Uvalda | Until colored | Aug. 3 | Aug. 23 | Sept. 5 | 14 | 43 | 9.1 | 6.8 | 17.2 |
| | 15 nights | Aug. 5 | Sept. 12 | Oct. 6 | 18 | 12 | 7.0 | 4.2 | 18.2 |
| | 10 nights | Aug. 10 | Oct. 8 | Oct. 18 | 6 | 0 | 6.2 | 9.5 | 26.1 |
| | 5 nights | Sept. 7 | Oct. 8 | Oct 18 | 6 | 0 | 8.0 | 4.2 | 19.9 |
| | No treatment | Sept. 7 | Oct. 8 | Oct. 18 | 6 | ..... | 6.2 | 7.2 | 26.4 |

for by the fact that a number of buds which formed as a result of the treatment failed to open. When the treatment was continued for five successive days, a great decrease in number of blooms per stem also occurred. An increase in the number of blooms per stem occurred when the plants were treated for ten successive nights, as compared with other treated plots. This was probably because many of the buds which formed early, as a result of the treatment, developed at the same time as those produced in normal season. The type of growth which resulted from these short-time treatments is shown in figure 5. The buds which formed under the short periods of treatment developed into what appeared to be crown buds. The foliage below the bud was not normal for the varieties; the buds became extremely large and often malformed. The resulting flowers were larger than normal and usually imperfect. On some plants the development was similar to that of one bud within another. Flowers produced on plants treated for 10 and 15 days were not salable in most instances, because of the presence of the undeveloped buds.

1             2             3             4             5

FIGURE 5.  EFFECTS OF A FEW SHORT DAYS (VARIETY PINK DOT)

The treatments were as follows:
   1.  Short-day treatment until bloom        3.  No treatment
   2.  15 short days                          4.  10 short days
                          5.  5 short days

Another experiment was made to determine the effect of darkening until the buds were well developed but not showing color. The varieties Sunshine, Izola, Norine, and Mrs. William Buckingham were used. The last pinch was given July 15, and the treatment was started the same date. One group of plants were treated until color showed (September 7) while another plot was treated until the buds were well developed but not showing color (August 7).

The results (table 11) show that the plants from both treatments bore buds at the same time, but when the treatment was continued only until August 7 blooms were cut at about the same time as those from the check. Plants which were darkened until color showed bloomed more than 40 days before the checks. Many distorted blooms and undeveloped buds were produced on plants treated until buds were well developed but not showing color. Varieties which started to bloom before the normal season, as a result of treating until August 7, flowered over an extremely long period. The number of stems per plant was increased by the treatment while the average number of blooms per stem and stem length were decreased.

The data show that the time necessary for flower production was lengthened and the quality of bloom was inferior when the period of darkening the plants was discontinued before color showed in the buds.

### TABLE 11. Effect of Darkening Until Buds Were Formed

| Variety | Treatment discontinued | First buds | First cutting | Period cutting | Period in bloom before check | Average stems per plant | Average flowers per stem | Average stem length |
|---|---|---|---|---|---|---|---|---|
| | *Time* | *Date* | *Date* | *Days* | *Days* | *Number* | *Number* | *Inches* |
| Izola.......... | Sept. 7 | Aug. 5 | Sept. 19 | 14 | 49 | 11.4 | 7.1 | 13.5 |
| | Aug. 7 | Aug. 5 | Sept. 26 | 42 | 42 | 9.4 | 3.2 | 16.0 |
| | No treatment | Sept. 12 | Nov. 7 | 14 | ....... | 5.2 | 13.8 | 29.3 |
| Mrs. Wm. Buckingham. | Sept. 7 | Aug. 5 | Sept. 19 | 7 | 43 | 8.7 | 3.6 | 10.0 |
| | Aug. 7 | Aug. 5 | Sept. 20 | 40 | 42 | 8.9 | 2.2 | 11.3 |
| | No treatment | Sept. 16 | Nov. 1 | 7 | ....... | 5.5 | 3.3 | 20.4 |
| Norine........ | Sept. 7 | Aug. 3 | Sept. 26 | 14 | 50 | 11.4 | 7.7 | 13.1 |
| | Aug. 7 | Aug. 3 | Nov. 11 | 11 | 4 | 8.2 | 8.5 | 21.1 |
| | No treatment | Sept. 16 | Nov. 15 | 10 | ....... | 6.6 | 11.8 | 28.6 |
| Sunshine...... | Sept. 7 | Aug. 5 | Oct. 3 | 15 | 43 | 8.8 | 4.8 | 13.5 |
| | Aug. 7 | Aug. 5 | Nov. 11 | 24 | 4 | 8.8 | 4.4 | 17.4 |
| | No treatment | Sept. 16 | Nov. 15 | 6 | ....... | 5.2 | 6.5 | 26.2 |

FIGURE 6. EFFECTS OF SHORT-DAY TREATMENTS CONTINUED UNTIL BUDS WERE FORMED BUT NOT SHOWING COLOR (VARIETY NORINE)

The two plants at the left were treated; those at the right were not treated

## A PERIOD OF SHORT DAYS ALTERNATED WITH NORMAL LENGTH OF DAY

Garner and Allard (1931) report that short-day plants show a reaction to alternate 10- and 15-hour days, similar to their reaction to constant 10-hour days. This part of the investigation was carried on to determine the relative effect of (1) one, two, three, four, and five, eleven-hour days, followed by one normal day; (2) one, two, three, four, and five, eleven-hour days followed by two normal days; (3) one, two, three, four, and five, eleven-hour days followed by three normal days; (4) one, two, three, four, and five, eleven-hour days followed by four normal days; and two and five eleven-hour days followed by five normal days. Nine plants of the variety Bokhara were used in each test. The plants were grown in the field under "aster" cloth and darkened as indicated with black sateen. The treatments were started July 15, 1932.

TABLE 12. EFFECT OF SHORT DAYS ALTERNATED WITH NORMAL DAYS

| Sequence grouping | Successive short days | Successive normal days | First buds | First cutting |
|---|---|---|---|---|
| | *Number* | *Number* | *Date* | *Date* |
| 2 or more short days followed by 1 or less normal days.... | 2 | 1 | Aug. 6 | Sept. 6 |
| | 3 | 1 | Aug. 6 | Sept. 6 |
| | 4 | 1 | Aug. 6 | Sept. 6 |
| | 5 | 1 | Aug. 6 | Sept. 6 |
| | All | 0 | Aug. 3 | Sept. 6 |
| Number of short days greater than normal days and normal days greater than 1...................... | 5 | 2 | Aug. 6 | Sept. 12 |
| | 5 | 3 | Aug. 6 | Sept. 15 |
| | 5 | 4 | Aug. 6 | Sept. 15 |
| | 4 | 2 | Aug. 6 | Sept. 15 |
| | 4 | 3 | Aug. 6 | Sept. 24 |
| Number of short days equal to the number of normal days | 1 | 1 | Aug. 6 | Sept. 20 |
| | 3 | 3 | Aug. 6 | Sept. 20 |
| | 4 | 4 | Aug. 6 | Sept. 20 |
| | 5 | 5 | Aug. 6 | Sept. 20 |
| Number of short days less than the number of long days.. | 2 | 4 | Aug. 6 | Oct. 20 |
| | 2 | 5 | Aug. 12 | Oct. 20 |
| | 1 | 2 | Aug. 19 | Oct. 20 |
| | 1 | 3 | Aug. 19 | Oct. 20 |
| | 1 | 4 | Aug. 19 | Oct. 27 |
| | No treatment | ........ | Sept. 1 | Oct. 29 |

The data (table 12) show that after the plants had been subjected to two or more eleven-hour days, they could be exposed to one full day without affecting the time of bloom. Exposure of the plants to more than one but less than five normal days, after they had been exposed to five short days, retarded blooming, when compared with the continuous treatment. The time of blooming was further delayed when a given number of short days were followed by an equal number of normal days. When the number of normal days exceeded the number of short days the time of bloom was delayed to a still greater extent and approached the normal time of bloom. The date of bud formation under the various treatments did not vary so greatly as the date of bloom.

These experiments were partly repeated under the same conditions, excepting seasonal differences, in 1933. In addition to these treatments some plants were given a definite number of short days followed by a given number of normal days, for a period of time; then the treatments were varied for the remainder of the period until color was showing. In table 13 the last columns of short and normal refer to the sequence of treatment during the remainder of the period; the second treatment consisted of five short days followed by one normal day, continued in this sequence until color was showing in the buds. The ninth treatment consisted of one short and three normal days, continued until four short days had been given, then the plants were given continuous short days until color was showing.

TABLE 13.  ALTERNATIONS OF SHORT AND LONG DAYS (VARIETY BOKHARA)

| Original sequence | | Times original sequence was repeated | Final sequence followed until color showed | | Date first buds appeared | Date of first cutting |
|---|---|---|---|---|---|---|
| Short | Normal | | Short | normal | | |
| *Days* | *Days* | *Number* | *Days* | *Days* | | |
| All | 0 | Until bloom | ......... | ......... | Aug. 1 | Sept. 5 |
| 5 | 1 | Until bloom | ......... | ......... | Aug. 4 | Sept. 11 |
| 4 | 1 | Until bloom | ......... | ......... | Aug. 4 | Sept. 18 |
| 3 | 1 | Until bloom | ......... | ......... | Aug. 7 | Sept. 20 |
| 2 | 1 | Until bloom | ......... | ......... | Aug. 7 | Sept. 25 |
| 10 | 10 | Until bloom | ......... | ......... | Aug. 4 | Sept. 25 |
| 7 | 7 | Until bloom | ......... | ......... | Aug. 4 | Sept. 25 |
| 5 | 2 | Until bloom | ......... | ......... | Aug. 4 | Sept. 25 |
| 1 | 3 | 4 | All | 0 | Aug. 14 | Sept. 25 |
| 15 | 4 | 1 | 1 | 4 | Aug. 1 | Oct. 5 |
| 5 | 3 | ............ | ......... | ......... | Aug. 7 | Oct. 5 |
| 5 | 4 | ............ | ......... | ......... | Aug. 7 | Oct. 5 |
| 5 | 5 | ............ | ......... | ......... | Aug. 7 | Oct. 5 |
| 5 | 1 | 1 | 1 | 1 | Aug. 7 | Oct. 5 |
| 2 | 1 | 4 | 2 | 3 | Aug. 11 | Oct. 5 |
| 1 | 3 | 4 | 2 | 1 | Aug. 14 | Oct. 5 |
| 10 | 4 | 1 | 1 | 4 | Aug. 4 | Oct. 5 |
| 2 | 1 | 3 | 2 | 3 | Aug. 11 | Oct. 12 |
| 1 | 1 | Until bloom | ......... | ......... | Aug. 14 | Oct. 12 |
| 1 | 3 | 4 | ......... | 1 | Aug. 14 | Oct. 12 |
| 1 | 5 | 3 | 3 | 2 | Aug. 14 | Oct. 12 |
| 5 | 2 | 2 | 1 | 2 | Aug. 7 | Oct. 19 |
| 5 | 4 | 1 | 2 | 4 | Aug. 11 | Oct. 19 |
| 5 | 3 | 1 | 1 | 3 | Aug. 11 | Oct. 25 |
| 5 | 4 | 3 | 1 | 4 | Aug. 14 | Oct. 25 |
| 2 | 1 | 3 | 1 | 4 | Aug. 14 | Oct. 25 |
| No treatment | ......... | ............ | ......... | ......... | Sept. 5 | Oct. 25 |

The dates of bloom obtained in 1933 do not agree entirely with those of 1932. The two seasons were exactly opposite, from the growing standpoint, 1932 was dry and warm during the latter part of the growing season and 1933 was moist and cool. This probably accounts for the difference in dates of bloom of the same treatments in the two seasons. However, the sequence of bloom resulting from the treatments is similar.

A comparison of these data show that normal day length given after a period of short days tends to retard the time of bloom when compared with a continuation of the short-day treatment. The date of bud formation did not vary so much as did the date of bloom. It is evident that changing the sequence of treatments affects the time of bloom depending upon the later sequence; plants given 2 short days followed by one normal day were budded August 7 and bloomed September 25, while plants given alternations of two short days and one normal day for three successive treatments, then a treatment of two short days and three normal days until color was showing, were budded August 11 and bloomed October 12.

When more long than short days occurred in the sequence, the time of bloom was later, as compared with more short than long days. All the buds which formed under any treatment eventually produced blooms, which indicates that after buds were formed an occasional short day caused them to continue development; the rate of development was dependent upon the frequency of occurrence of the short days.

## OUTDOOR CULTURE

During the seasons of 1931 and 1932, large-flowered varieties were grown under aster cloth in the field and given short-day treatments. The response to the treatment was similar to that of the same varieties grown in the greenhouse, but almost no salable blooms were obtained either year with any variety. The moisture condensed in the blooms and subsequent rotting occurred.

Pompon varieties grown under like conditions in 1931, 1932, and 1933 responded similarly to the same varieties grown in the greenhouse and the blooms were equal or superior to greenhouse-grown blooms. The date of bloom varied to a greater extent from year to year under outdoor conditions, owing, probably, to the difference in climatic conditions.

## A SUCCESSION OF BLOOM OF THE SAME VARIETY

During the seasons of 1931, 1932, and 1933, several varieties of pompon and large-flowered varieties were grown and the short-day treatment started at different times during the year to determine the effects upon the time of bloom and the resulting flower size and stem length. The results during the three years were approximately the same with the several varieties, and the results with the two types of plants are shown in tables 14 and 15.

These data show that the tendency with both large-flowered and pompon varieties was for the buds to form, show color, and bloom on any one variety in about the same succession as that in which the treatments were started. It is evident, however, that the later varieties responded to the treatments more slowly earlier in the season. In table 14 the earliest variety used was Gold Lode, and the time intervals between the dates of bloom were exactly the same as the intervals between the time of starting the treatments. Marie De Petris, on the other hand, was one of the latest varieties used and the time interval between blooming dates was three and seven days when the treatments were started at ten-day intervals. All late varieties did not respond in this manner, however, and no variety responded exactly the same each year. The length of stem and size of bloom were usually greater from plants which bloomed later in the season.

The average number of flowers per stem and the average stem length were usually increased on pompon varieties in direct relationship to the length of the growing period of the plant (table 15). The average number of stems per plant showed an inverse relationship to the length of the growing period.

TABLE 14. EFFECTS OF TIME OF STARTING THE TREATMENT (LARGE-FLOWERED TYPES)

| Variety | Beginning of treatment | Buds selected | First blooms cut | Cutting period | Period in bloom before check | Average stem length | Average flower diameter |
|---|---|---|---|---|---|---|---|
| | *Date* | *Date* | *Date* | *Days* | *Days* | *Inches* | *Inches* |
| Chas. Rager....... | July 20 | Aug. 19 | Sept. 20 | 21 | 52 | 25 | 4.83 |
| | July 30 | Aug. 28 | Oct. 3 | 8 | 39 | 34 | 5.32 |
| | Aug. 10 | Sept. 5 | Oct. 14 | 6 | 28 | 37 | 5.25 |
| | No treatment | Sept. 27 | Nov. 11 | 7 | ........ | 51 | 5.23 |
| Citronella......... | July 20 | Aug. 21 | Sept. 26 | 14 | 24 | 28 | 4.82 |
| | Aug. 10 | Sept. 5 | Oct. 11 | 13 | 9 | 32 | 4.95 |
| | No treatment | Sept. 12 | Oct. 20 | 11 | ........ | 38 | 5.41 |
| Detroit News...... | July 20 | Aug. 10 | Sept. 12 | 25 | 47 | 32 | 4.68 |
| | July 30 | Aug. 23 | Sept. 29 | 8 | 30 | 43 | 5.00 |
| | Aug. 10 | Aug. 28 | Oct. 11 | 13 | 18 | 44 | 5.63 |
| | No treatment | Sept. 12 | Oct. 29 | 9 | ........ | 46 | 5.03 |
| Friendly Rival..... | July 20 | Aug. 17 | Sept. 29 | 18 | 43 | 33 | 5.14 |
| | July 30 | Aug. 28 | Oct. 11 | 27 | 31 | 39 | 5.50 |
| | Aug. 10 | Aug. 28 | Oct. 11 | 13 | 31 | 44 | 6.10 |
| | No treatment | Sept. 21 | Nov. 11 | 7 | ........ | 44 | 5.90 |
| Golden State...... | July 20 | Aug. 21 | Oct. 3 | 7 | 39 | 38 | 4.54 |
| | July 30 | Aug. 28 | Oct. 11 | 9 | 31 | 43 | 5.39 |
| | Aug. 10 | Sept. 9 | Oct. 24 | 5 | 18 | 46 | 4.98 |
| | No treatment | Oct. 4 | Nov. 11 | 5 | ........ | 59 | 4.69 |
| Gold Lode........ | July 20 | Aug. 5 | Sept. 2 | 1 | 42 | 21 | 4.89 |
| | July 30 | Aug. 23 | Sept. 12 | 8 | 32 | 30 | 4.50 |
| | Aug. 10 | Aug. 28 | Sept. 23 | 14 | 21 | 34 | 4.69 |
| | No treatment | Sept. 12 | Oct. 14 | 6 | ........ | 38 | 5.45 |
| Marie De Petris.... | July 20 | Aug. 21 | Oct. 11 | 18 | 31 | 35 | 5.53 |
| | July 30 | Aug. 28 | Oct. 14 | 17 | 28 | 44 | 6.08 |
| | Aug. 10 | Sept. 6 | Oct. 21 | 17 | 21 | 42 | 5.72 |
| | No treatment | Sept. 21 | Nov. 11 | 8 | ........ | 46 | 5.45 |
| October Rose...... | July 20 | Aug. 10 | Sept. 9 | 11 | 41 | 31 | 5.26 |
| | July 30 | Aug. 23 | Sept. 26 | 1 | 24 | 37 | 5.31 |
| | Aug. 10 | Aug. 28 | Oct. 11 | 9 | 9 | 44 | 5.65 |
| | No treatment | Sept. 12 | Oct. 20 | 9 | ........ | 51 | 5.08 |
| Smith Early White | July 20 | Aug. 10 | Sept. 9 | 1 | 41 | 31 | 4.70 |
| | Aug. 10 | Aug. 28 | Sept. 29 | 4 | 21 | 37 | 5.22 |
| | No treatment | Sept. 12 | Oct. 20 | 4 | ........ | 44 | 5.67 |
| Snow White....... | July 20 | Aug. 28 | Oct. 3 | 26 | 29 | 39 | 5.25 |
| | July 30 | Aug. 28 | Oct. 7 | 25 | 25 | 37 | 5.79 |
| | Aug. 10 | Sept. 5 | Oct. 11 | 28 | 21 | 41 | 5.77 |
| | No treatment | Oct. 4 | Nov. 1 | 17 | ........ | 49 | 5.79 |

TABLE 15.   EFFECTS OF TIME OF STARTING THE TREATMENTS (POMPON VARIETIES)

| Variety | Treatment started | Buds showing | First flowers cut | Cutting period | Period in bloom before check | Average stems per plant | Average flowers per stem | Average stem length |
|---|---|---|---|---|---|---|---|---|
| | *Date* | *Date* | *Date* | *Days* | *Days* | *Number* | *Number* | *Inches* |
| Bronzetto...... | July 10 | Aug. 6 | Sept. 20 | 8 | 52 | 9.4 | 10.1 | 17 |
| | July 20 | Aug. 15 | Sept. 24 | 11 | 48 | 6.9 | 9.3 | 17 |
| | Aug. 10 | Aug. 25 | Oct. 5 | 7 | 37 | 7.8 | 12.9 | 19 |
| Copper City.... | July 10 | July 28 | Sept. 23 | 5 | 39 | 9.8 | 10.1 | 15 |
| | July 20 | Aug. 15 | Sept. 28 | 4 | 34 | 8.1 | 8.3 | 19 |
| | Aug. 10 | Aug. 27 | Oct. 9 | 8 | 22 | 7.6 | 10.1 | 20 |
| Dorothy Turner | July 10 | Aug. 12 | Oct. 2 | 7 | 52 | 5.6 | 9.3 | 17 |
| | July 20 | Aug. 15 | Oct. 5 | 7 | 49 | 6.0 | 10.8 | 18 |
| | Aug. 10 | Aug. 29 | Oct. 19 | 4 | 35 | 5.4 | 12.1 | 21 |
| Gold Coin...... | July 10 | Aug. 6 | Oct. 5 | 11 | 46 | 8.3 | 7.8 | 16 |
| | July 20 | Aug. 15 | Oct. 8 | 9 | 43 | 7.0 | 9.4 | 18 |
| | Aug. 10 | Aug. 30 | Oct. 19 | 7 | 32 | 7.8 | 8.9 | 20 |
| Helen Hubbard | July 10 | Aug. 6 | Sept. 23 | 5 | 44 | 6.3 | 7.9 | 19 |
| | July 20 | Aug. 15 | Sept. 24 | 8 | 43 | 4.7 | 9.5 | 22 |
| | Aug. 10 | Aug. 27 | Oct. 9 | 7 | 28 | 4.0 | 10.5 | 24 |
| Norma......... | July 10 | July 28 | Sept. 28 | 7 | 41 | 7.9 | 6.20 | 21 |
| | July 20 | Aug. 12 | Sept. 28 | 11 | 41 | 7.5 | 7.29 | 23 |
| | Aug. 10 | Aug. 25 | Oct. 12 | 5 | 27 | 6.5 | 7.95 | 25 |
| Popcorn........ | July 10 | July 28 | Sept. 5 | 3 | 57 | 8.4 | 7.3 | 13 |
| | July 20 | Aug. 11 | Sept. 13 | 1 | 49 | 8.1 | 8.8 | 16 |
| | Aug. 10 | Aug. 23 | Sept. 28 | 4 | 34 | 8.1 | 10.1 | 19 |
| Red Wilcox..... | July 10 | Aug. 5 | Oct. 2 | 7 | 49 | 6.5 | 8.3 | 16 |
| | July 20 | Aug. 12 | Oct. 2 | 7 | 49 | 6.9 | 8.0 | 18 |
| | Aug. 10 | Aug. 25 | Oct. 16 | 3 | 35 | 6.3 | 8.9 | 21 |
| Roman Bronze.. | July 10 | July 28 | Sept. 2 | 6 | ....... | 5.7 | 7.4 | 15 |
| | July 20 | Aug. 11 | Sept. 8 | 3 | ....... | 4.9 | 9.1 | 20 |
| | Aug. 10 | Aug. 25 | Sept. 23 | 5 | ....... | 5.4 | 9.6 | 20 |
| White Menza... | July 10 | Aug. 6 | Sept. 18 | 2 | 44 | 7.2 | 3.8 | 18 |
| | July 20 | Aug. 12 | Sept. 23 | 5 | 39 | 5.9 | 5.6 | 21 |
| | Aug. 10 | Aug. 29 | Oct. 2 | 10 | 30 | 5.1 | 6.5 | 25 |
| Yellow Fellow... | July 10 | July 28 | Sept. 14 | 10 | 48 | 6.9 | 7.2 | 16 |
| | July 20 | Aug. 11 | Sept. 18 | 6 | 44 | 7.0 | 8.2 | 20 |
| | Aug. 10 | Aug. 25 | Oct. 2 | 7 | 30 | 5.6 | 11.7 | 22 |

A summary of the dates of bloom of the varieties used during the three years is given in table 16.   If one variety was found to bloom on different dates when treated the same period of time for more than one year, the dates of bloom are given for both seasons.   Slight variation from these dates of bloom may be expected if the same varieties are grown in different sections of the country, because of local and seasonal variations in climate.

It is evident from this table that the later in the season the treatment is started the more rapid the response.   The variation in the time the first flower of one variety was cut from year to year was usually not greater than ten days when the treatment was started at the same time during different seasons.

GENERAL DISCUSSION

The amount of light which passed through the black cloth used in these experiments was probably governed to a great extent by the angle at which the light rays struck the cloth.   The maximum amount of light passed through the cloth when the angle of incidence was 90 degrees. This may be of some consequence in determining the effectiveness of a given treatment at either end of the day, as contrasted with a treatment of similar length part of which was given at both ends of the day.

When the variety Bokhara (table 3) was given 11-hour days but the dark period occurred from 4 P. M. until dark or from dark until 9 A. M., the plants were in bloom 35 and 21 days, respectively, before the checks; an advance of 47 days was obtained if part of the treatment was in the afternoon and part in the morning (6 P. M.–7 A. M.). If this is compared with the results of treatments to permit various day lengths (table 1, variety Dorothy Turner) a 13-hour day reduced by six days the time necessary for blooming; an 11-hour day reduced this by 49 days.

A second example of this appears in a comparison of tables 1 and 2. When the 9-hour day extended from 7 A. M. to 4 P. M. it was found to be less effective than an 11-hour day, but if the 9-hour day extended from 8 A. M. to 5 P. M., the plants bloomed at the same date as did those given an 11-hour day. The treatment from 4 until 5 o'clock apparently retarded the development of the flowers.

Cloth which did not reduce the light intensity as much as the black sateen did, had an effect on the plants similar to that of a dense shade, and the time of bloom was not affected to so great an extent. Probably the angle of incidence of the light rays on the cloth was of some importance in determining the relative effectiveness of the more loosely woven materials. The plants treated with these less opaque materials responded similarly to plants given a somewhat longer day than the treatment would provide. Plants which were shaded continued to grow vegetatively with

TABLE 16. RELATION OF DATE OF BLOOM TO DATE WHEN TREATMENT WAS STARTED

| Pompon varieties | | | | | | | | | |
|---|---|---|---|---|---|---|---|---|---|
| | Date treatment was started | | | | | | | | |
| | June 20 | July 1 | July 10 | July 15 | July 20 | July 25 | Aug. 1 | Aug. 5 | Check |
| Bokhara* | | | | Sept. 9, 12 | | | Sept. 23 | | Oct. 21, 26 |
| Bronzetto | | | Sept. 20 | | | | | | Nov. 11, 15 |
| Buckingham | Aug. 8 | | | Sept. 24 | | | Oct. 4 | | Nov. 1 |
| Captain Cook* | | | Sept. 20 | | | | | | Oct. 23 |
| Chas. Maynard | | | | Sept. 9, 14 | | | Sept. 23 | | Oct. 23, 24 |
| Copper City | | | Sept. 23 | Sept. 18 | Sept. 28 | | | | Nov. 8 |
| Dorothy Turner* | | | Oct. 2 | Oct. 5 | Oct. 5 | | | Oct. 19 | Nov. 23 |
| Gold Coin* | | | Oct. 5 | | Oct. 8 | | Oct. 19 | | Nov. 20 |
| Helen Hubbard* | | | Sept. 23 | | Sept. 24 | Oct. 6 | Oct. 8 | Oct. 9 | Nov. 6, 11 |
| Izola | Aug. 27 | | | Sept. 7, 19 | | | Oct. 2 | | Nov. 1, 7 |
| Menza* | | | Sept. 18 | Sept. 21 | Sept. 23 | Sept. 26 | Oct. 25 | Oct. 2 | Nov. 1, 8 |
| Norine | | | | Sept. 23 | | | | | Nov. 15 |
| Norma | | | Sept. 28 | | Sept. 28 | | Oct. 9 | Oct. 12 | Nov. 8 |
| Pink Dot* | Aug. 22 | Sept. 5 | | Sept. 9, 14 | | | Sept. 23 | | Nov. 1 |
| Popcorn* | | | Sept. 5 | Sept. 11 | Sept. 23 | | Sept. 25 | Sept. 28 | Nov. 1 |
| Red Wilcox* | | | Oct. 2 | | Oct. 2 | | | Oct. 16 | Nov. 20 |
| Redell | Aug. 8 | | | Sept. 4, 6 | | | Sept. 23 | | Oct. 18, 19 |
| Roman Bronze | | Sept. 2 | | | Sept. 8 | | Sept. 23 | | |
| Sunshine* | | | Sept. 28, Oct. 3 | | Sept. 8 | | Oct. 12 | | Nov. 23 |
| Uvalda* | Aug. 6 | | Sept. 5 | | | | | | Oct. 6, 8 |
| White Wings | | Aug. 23 | Aug. 28 | | Sept. 4 | | Sept. 20 | Sept. 18 | Oct. 2 |
| Yellow Fellow* | Aug. 22 | | Sept. 14 | Sept. 11 | Sept. 18 | Sept. 23 | Oct. 2 | Oct. 2 | Nov. 1, 10 |

(Continued on page 26)

### TABLE 16. (Concluded)

| | June 20 | July 10 | July 15 | July 20 | July 25 | July 30 | Aug. 5 | Aug. 10 | Aug. 15 | Check |
|---|---|---|---|---|---|---|---|---|---|---|
| **Standard varieties** | | | | | | | | | | |
| **Date treatment was started** | | | | | | | | | | |
| Ambassador* | | Sept. 5 | | Sept. 13 | | Sept. 20 | | Sept. 28 | | Oct. 23 |
| Betsy Ross* | | | | Sept. 18 | | | | Oct. 2 | | Oct. 21 |
| Celestra | | | Sept. 5 | | | | | Sept. 24 | | Oct. 12 |
| Chas. Rager | | | Sept. 14, 18 | Sept. 20 | Sept. 23 | Oct. 3 | Oct. 29 | Oct. 14 | | Nov. 7, 10, 11 |
| Chrysolara* | | | | Sept. 14 | | | | Oct. 2 | | Oct. 23 |
| Citronella* | | | | Sept. 20, 26 | Sept. 26 | | | Oct. 5, 11 | | Oct. 16, 20 |
| C. W. Johnson* | | | | Sept. 24 | | | | Oct. 8 | | Oct. 21 |
| Detroit News* | | Sept. 11 | | Sept. 12, 24 | Sept. 24, 29 | | | Oct. 5, 11 | | Oct. 29, Nov. 1 |
| Friendly Rival | | | Sept. 21 | Sept. 29 | Sept. 29 | Oct. 11 | | Oct. 11 | | Nov. 7, 11 |
| Gladys Pearson* | | Oct. 2 | | Oct. 5 | Oct. 12 | Oct. 8 | | Oct. 16 | | Nov 11, 14 |
| Golden Bronze* | | | | Sept. 14 | | | | Oct. 2 | | Oct. 30 |
| Golden Glory* | | | | Sept. 14 | | | | Sept. 24 | | Oct. 23 |
| Golden State* | | | | Oct. 3 | Oct. 11 | | | Oct. 24 | | Nov. 11 |
| Gold Lode | Aug. 5 | | Aug. 26 | Sept. 2 | Sept. 2, 6 | Sept. 12 | | | Sept. 23 | Oct. 13, 15 |
| Hilde Bergen | | Oct. 9 | | Oct. 9 | Oct. 16 | | | Oct. 23 | | Nov. 17 |
| Honey Dew* | | | | Sept. 24 | | | | Oct. 5 | | Nov. 8 |
| Indianala | | | | Sept. 8 | | | | Sept. 28 | | Oct. 12 |
| Justrite | | | | Sept. 11 | | | | Oct. 2 | | Oct. 16 |
| J. W. Prince* | | | | Sept. 18 | | | | Oct. 2 | | Oct. 30 |
| Marie De Petris* | | Oct. 12 | | Oct. 9, 11 | Oct. 9, 11 | Oct. 13 | | Oct. 16, 21 | | Nov. 1, 11 |
| Mrs. David F. Roy | | | Sept. 21 | Sept. 26 | Oct. 5 | | | | Oct. 15 | Nov. 1, 3 |
| Mrs. H. E. Kidder* | | Sept. 11 | | Sept. 14 | Sept. 20 | Sept. 24 | | Oct. 2 | | Nov. 23, 24 |
| Monument* | | | | | Oct. 3 | | | | | Nov. 11 |
| Muskoka | | Sept. 18 | | Sept. 20 | Sept. 26 | Sept. 28 | | Oct. 5 | | Nov. 1 |
| October Rose* | Aug. 22 | Sept. 5 | Aug. 28 | Sept. 9, 14 | Sept. 9, 19 | Sept. 23, 26 | | Oct. 2, 11 | | Oct. 21 |
| Pink Delight* | | | | Sept. 20 | | | | Oct. 2 | | Oct. 25, Nov. 7 |
| Quaker Maid | | | Sept. 1 | Sept. 8 | | Sept. 21 | | Oct. 2 | | Oct. 13 |
| Rose Glory | | | | Sept. 5 | | | | Sept. 28 | | Oct. 8 |
| Silver Sheen | Aug. 22 | Sept. 5 | Sept. 8 | Sept. 6, 13 | Sept. 18 | Sept. 24 | Sept. 23 | Sept. 28 | Oct. 5 | Oct. 21, 26 |
| Smith's Early White | Aug. 12 | | | Sept. 9 | | | | Sept. 29 | | Oct. 20 |
| Snow White | | | | | | Oct. 7 | Oct. 5 | Oct. 5, 11 | Oct. 15 | Nov. 1, 7 |
| Sun Glow | | | | Sept. 14 | | | | Oct. 2 | | Oct. 19 |
| Thanksgiving Pink* | | | | | Sept. 26 | Oct. 5 | | | Oct. 26 | Oct. 23 |

*Best varieties.

the same reaction to the length of day as when under normal conditions. The photosynthetic rate was probably reduced because light was a limiting factor. The vegetative growth was generally lighter green in color and the nodes farther apart than on plants under normal light intensity. Such a reaction as this was not noticed in any of the treatments with black sateen cloth unless this treatment was given during the middle of the day. Since the plants responded the same as when under short-day conditions, it appears that the treatment should be designated as *reduced length of day* or *darkening* rather than as shading of the plants.

The evidence presented shows that covering the plants with cloth at 5 or 6 o'clock in the afternoon and removing it at 7 or 8 o'clock in the morning resulted in earliest blooms of any time or length of day of treatment. This time of application and removal of the cloth can be regulated at the grower's convenience. The complete covering of the plants for this period of time had no apparent detrimental effects on their growth. The blooms appeared to be normal in every respect except that the color of pink and bronze varieties was generally less intense when they bloomed in advance of their normal season.

Shortening the growing period by forcing the plants to bloom earlier than normal was probably the factor responsible for shorter stems. It is evident that the time of propagation and the time the last pinch was made had no effect on the time of bloom of plants of similar size; length of day was the factor of major importance in determining this. The difference in rate of development of the bud at various times during the growing period may be due to variation in light intensity, temperature, size of plant, or leaf area. Length of day varying from ten to sixteen hours after buds were showing color had no apparent effect on the time of bloom.

The number of branches which matured on pompon varieties was dependent upon the time of year they were forced to bloom. The earlier-treated plants produced more branches. This was so nearly every time a variety was treated to produce blooms at different times during the growing season. This condition has been observed by Garner and Allard (1923) and Tincker (1925). The exact cause for it has not been definitely established but in the present experiments it appeared that more of the weaker branches developed when the plants were forced to bloom early. Varieties which bloomed in normal season produced many weak stems which did not produce flowers and in many cases they were almost completely crowded out by the stronger growing stems. When the period of competition for light and food was shortened by forcing the plants to bloom there appeared to be fewer of these weak stems on the plants at maturity.

Flower buds which developed during a given period of short days resembled the typical terminal bud of the chrysanthemum which is normally surrounded by other flower buds. The number of axillary buds which develop into flower buds was governed to a great extent by the number of successive short days. It was evident that terminal-flower-bud differentiation took place before the time of differentiation of the buds on the lower part of the stem. The succession of differentiation of the flower buds was from the top to the base of the stem.

Since a crown bud of a chrysanthemum plant is surrounded by vegetative buds while a terminal bud is surrounded by flower buds, it appears that the production of a crown bud is dependent upon the number of short days which occur in succession. *It is entirely possible that there is no difference between the two at the time of their formation, but the crown bud is nothing more than a terminal bud differentiated under short-day conditions and developed to some extent under long day conditions.* This idea is substantiated by the experiment with several varieties given reduced length of day for a short period and followed by normal length of day. When the length of day was reduced until the buds were showing color, they matured with all the characteristics of terminal buds. When the short-day treatment was for five or ten successive days, the later varieties produced flowers with all the characteristics of a crown. Treatments which produced normal terminal buds and were followed by ten or fewer normal days resulted in flowers which resembled those from terminal buds.

Observations of chrysanthemum varieties blooming in normal season also substantiate this view. Earlier varieties are frequently flowered from crown buds rather than from terminals. These occur as single flower buds at the terminal growing points of the plants. They are surrounded by vegetative buds and appear exactly like buds and vegetative growth produced by treating for five days, as is evident in figure 6. These crown buds may be the result of one or a few short days produced naturally because of cloudy mornings or afternoons during the growing period. If the continuation of the short days is for a sufficient period all vegetative buds change to flower buds and a terminal bud results. During some seasons crown buds are produced on many varieties while in other seasons crown buds do not appear. Some varieties are inclined to produce crown buds early in the growing season and during the longest days of the year. There is no doubt that considerable difference exists in the sensitivity of varieties to length of day; the more sensitive the variety the earlier the blooms and the greater the number of crown buds produced.

## SUMMARY

1. Chrysanthemum plants of both large-flowered and pompon types were made to bloom as much as seventy days before their normal season, by reducing the length of day to eleven hours. Early varieties were more sensitive to any treatment than were late varieties.

2. Of the materials used, black sateen was found to be the most effective for reducing the length of day.

3. Darkening the plants at 5 or 6 P. M. and exposing them to light at 7 or 8 A. M. was the most effective of any of the periods of treatment tried.

4. Plants forced to bloom in advance of the normal season produced shorter stems than did those given normal length of day until bloom.

5. The number of branches which developed on pompon varieties was increased by the short-day treatment. Usually the number showed an inverse relationship to the age of the plants.

6. The size of bloom of the large-flowered forms was reduced or increased by the treatment, depending upon the variety.

7. The size of the blooms of pompon varieties was not altered by the treatment.

8. The number of blooms per stem which developed on pompon varieties showed a direct relationship to the stem length and the length of the growing period of the plants.

9. Pink and bronze varieties were lighter in color when forced to bloom in advance of their normal season.

10. A continuation of an eleven-hour day or increasing the day length to sixteen hours by the use of 75 watt lamps did not change the time of bloom of the varieties if the treatment was given after the buds were showing color.

11. Increasing the length of day to sixteen hours after the buds were showing color produced lighter-colored flowers in the pink variety than did the normal length of day.

12. Five successive short days, followed by normal length of day, caused chrysanthemum plants to produce flower buds which did not open until normal season.

13. Discontinuation of short-day treatments on chrysanthemum plants, before the buds were selected on disbudded varieties and before color was showing in pompons, caused the buds to remain undeveloped until the length of day was proper for the flowering of that variety.

14. A succession of a few short days followed by normal summer days produced distorted buds and imperfect blooms.

15. The treatment was effective only when the cloth completely enclosed the terminal buds of the plants.

16. The time of propagation had no effect on the time of bloom of plants of similar size when the treatment was started. The stem length and number of blooms per stem were affected to a greater extent.

17. The time the last pinch was made on pompon varieties had no effect on the time of bloom of the varieties used, even though this was made on the date the treatment was started. The length of stem and number of stems per plant were governed to some extent by the date of the last pinch.

18. In comparison with a continuous short day, alternate short- and long-day treatments in various sequence caused buds to form, but the date of flowering was retarded by the presence of normal days in the sequence. The retardation showed a direct relationship to the number of normal days in the sequence.

19. A sequence of bloom of any one variety was obtained by starting the treatments at various times during the growing period. The interval between the blooming dates was approximately the same as the interval between the dates of starting the treatments.

20. Seasonal differences and size of plant altered the time of bloom of varieties given the same treatment from year to year.

21. A table has been compiled which shows the dates of bloom of the varieties used when the treatment was started at various times during the season.

22. The treatment should be designated as *darkening*, or reduced length of day.

23. Crown and terminal buds were differentiated under the same conditions but crown buds started to develop under long-day conditions while terminal buds developed under short-day conditions.

## REFERENCES CITED

ALLARD, H. A.    Daylight a factor in flowering.    U. S. Agr. Dept. Year-book **1926** : 306–309.  1927.

ESPINO, R. B., AND PANTALEON, F.    Influence of light upon growth and development of plants with special reference to the comparative effects of the morning and of the afternoon light.    Philippine agriculturist **19** : 563–579.  1931.

Garner, W. W., and Allard, H. A.   Effects of length of day and other factors of the environment on growth and reproduction in plants. Journ. agr. res. **18** : 553–606. 1920.

——————. Flowering and fruiting of plants as controlled by the length of day. U. S. Agr. Dept. Yearbook **1920** : 377–400. 1921.

——————. Further studies in photoperiodism, the response of the plant to relative length of day and night. Journ. agr. res. **23** : 871–920. 1923.

——————. Localization of the response in plants to relative length of day and night. Journ. agr. res. **31** : 555–566. 1925.

——————. Effects of short alternating periods of light and darkness on plant growth. Science **66** : 40–42. 1927.

——————. Effect of abnormally long and short alternations of light and darkness on growth and development of plants. Journ. agr. res. **42** : 629–651. 1931.

Laurie, Alex.   Photoperiodism—Practical application to greenhouse culture. Amer. Soc. Hort. Sci. Proc. **27** : 319–322. 1930.

——————, and Poesch, G. H.   Photoperiodism: the value of supplementing illumination and reduction of light on flowering plants in the greenhouse. Ohio Agr. Exp. Sta. Bul. **512** : 1–42. 1932.

Poesch, G. H.   Studies of photoperiodism of the chrysanthemum. Amer. Soc. Hort. Sci. Proc. **28** : 389–392. 1931.

——————. Further studies of photoperiodism of the chrysanthemum. Amer. Soc. Hort. Sci. Proc. **29** : 540–543. 1932.

Post, Kenneth.   Reducing the day length of chrysanthemums for the production of early blooms by the use of black sateen cloth. Amer. Soc. Hort. Sci. Proc. **28** : 382–388. 1931.

——————. Further results with black cloth for the production of early blooms of the chrysanthemum. Amer. Soc. Hort. Sci. Proc. **29** : 545–548. 1932.

Tincker, M. A. H.   Effect of length of day upon the growth and reproduction of some economic plants. Ann. bot. **39** : 721–754. 1925.

——————. On the effect of length of daily period of illumination upon the growth of plants. Royal Hort. Soc. Journ. **54** : 354–378. 1929.

Arthur J. Heinicke and Norman F. Childers. 1937. The daily rate of photosynthesis during the growing season of 1935, of a young apple tree of bearing age. Cornell University Experiment Station Memoir 201.

Arthur J. Heinicke

Norman F. Childers

The 1930s can be described as one of the most important periods in horticulture and plant science in general. Technological advancements in chemistry and biology were such that earlier theories could be tested on a field scale. Plant breeding flourished, plant protection took large steps ahead, and plant nutrition for the first time could base its recommendations on scientifically obtained analytical data. Of course, plant physiology was not stagnant. At the beginning of the century, the British plant physiologist F. F. Blackman recognized that in light-saturated leaves factors other than available light could limit photosynthetic rate. Hoover, Johnston, and Brackett of the Smithsonian Institution, perceived the importance of $CO_2$ concentrations, and Manning observed the interaction between temperature and photosynthesis. All of this work was performed with individual leaves or with small potted plants. There was no information on how photosynthesis occurred in nature. This was the area where Heinicke needed and wanted information.

Arthur Heinicke was a recognized plant physiologist as well as a good pomologist. He received his early training under William H. Chandler, a great pomologist, and from the early physiologists at the

University of Missouri. These associations shaped his interest in physiological (light, water, heat, sprays) and cultural factors that affected the performance of fruit trees. He had good funding support and did things in a big and technical way.

Arthur Heinicke came from St. Louis, Missouri. It was here during a Christmas vacation that Norman Childers visited him seeking an assistantship in his department at Cornell. Heinicke awarded him the assistantship on the spot and Childers was elated. He could not possibly have known at that time how much work he was facing later. As it turned out, Childers was the person who carried the burden of the photosynthesis research Heinicke had in mind. He was the one who would have the responsibility of taking $CO_2$ measurements every 4 hours of every day during an entire growing season, literally running back and forth between the orchard and the laboratory. Practically living and sleeping on laboratory desks for the entire summer of 1935.

There were other circumstances which made the time ripe for a relatively bold series of investigations to probe something others did not dare to try. At the time of Heinicke's arrival at Cornell, there was no useful method to determine atmospheric $CO_2$ exchange in the measurements of photosynthesis by plants. The best gasometric, volumetric, electrometric, and gravimetric methods all left much to be desired, either being inconvenient or inaccurate for photosynthetic measurements. Since the atmosphere contains only about 310 ppm of $CO_2$, available gasometric and gravimetric methods were inadequate for low-level detection, and volumetric and electrometric apparatus needed more efficient absorbers to permit the use of larger air samples. The first order of business was to develop better absorbers. Heinicke, with another graduate student, M. B. Hoffman, developed an apparatus that could absorb sufficient amounts of $CO_2$ under natural conditions. This work was published in 1933. Heinicke then had the farm crew build a glasshouse out of windowpanes, nailed and glued together, around an 8-year-old fruiting apple tree, installed "sirocco fans," gas-proofed the structure, and put Childers to work day and night for a full growing season.

The result was a remarkable series of measurements. They determined the level of irradiation, leaf area, transpiration, and of course, carbon assimilation for the entire season. Although some of the values they obtained are low (see Proctor et al. 1976), they are still the only available data today for an entire fruit tree. They demonstrated the effect of light on assimilation of $CO_2$ and the daily variation in photosynthesis. They went even further, accounting for the utilization of carbohydrates by measuring the dry weight partitioning among the various tree organs. All of this research was performed about 40 years before it became fashionable to do it with fruit trees (see a series of papers, HortScience 13:640–652, 1978) or crops in general.

The work was not without its rewards. Heinicke was invited to give numerous seminars on the results. The research was the subject of a Sigma Xi lecture tour, dissemination of the results influenced other scientists. It is hard to say how much this work inspired a similar series of investigations in Utah with alfalfa and wheat which became the forerunner of photosynthetic measurements in agronomic field plots. The American Smelting and Refining Company was brought under suit for fumes emanating from a mining operation which were suspected to be the cause of losses to farmer's crops and pastures. This needed to be investigated. In 1933, Moyer D. Thomas, the physiologist for the American Smelting and Refining Company, also published on an absorption apparatus which was slightly different from that used by Heinicke and Hoffman. This was followed by photosynthetic studies carried out in field cages which looked much like those designed by Heinicke and Childers. The work was done in 1936, a year later than Heinicke and Childers. Moyer D. Thomas and George R. Hill published their findings in 1937, the year that Heinicke and Childers's work appeared.

The photosynthetic work of Heinicke did not stop with this classical work. He evaluated the effect of pesticide sprays on photosynthesis. Lime sulfur sprays, which were commonly used at this time, greatly decreased photosynthesis, and his work eventually led to the development of better pesticides that did not affect photosynthesis and obviously increased yield. Childers completed his Ph.D. degree and took a job as assistant professor at Ohio State University. He maintained his interest in photosynthesis for several years. He built environmental growth chambers, determined the deleterious effect of another commonly used spray, Bordeaux mixture, and evaluated the harmful effects of insects, especially leafhoppers, on photosynthesis of apple leaves, and the importance of water supply in maintaining a high photosynthetic rate. As World War II approached, this classical era of tree fruit photosynthesis slowly came to an end.

Miklos Faust
Fruit Laboratory
U.S. Department of Agriculture
Beltsville, Maryland 20705

## REFERENCES

BRODY, H. W., and N. F. CHILDERS. 1938. The effect of dilute liquid lime–sulphur sprays on the photosynthesis of apple leaves. Proc. Am. Soc. Hort. Sci. 36:205–209.

CHILDERS, N. F., and H. W. BRODY. 1939. An environmental-control chamber for study of photosynthesis, respiration and transpiration of horticultural plants. Proc. Am. Soc. Hort. Sci. 37:384.

CHILDERS, N. F., and F. F. COWART. 1936. The photosynthesis, transpiration and stomata of apple leaves as affected by certain nutrient deficiencies. Proc. Am. Soc. Hort. Sci. 33:160–163.

CHILDERS, N. F., and D. G. WHITE. 1942. Influence of submersion of the roots on transpiration, apparent photosynthesis and respiration of young apple trees. Plant Physiol. 17(4):608–618.

HEINICKE, A. J. 1932. Assimilation of carbon dioxide by apple leaves as affected by ringing the stem. Proc. Am. Soc. Hort. Sci. 29:225–229.

HEINICKE, A. J. 1933. A special air chamber for studying photosynthesis under natural conditions. Science 77:516–517.

HEINICKE, A. J. 1934. Photosynthesis in apple leaves during late fall and its significance in annual bearing. Proc. Am. Soc. Hort. Sci. 32:77–80.

HEINICKE, A. J. 1937. How lime sulphur spray affects the photosynthesis of an entire ten-year-old apple tree. Proc. Am. Soc. Hort. Sci. 35:256–259.

HEINICKE, A. J. 1938. Influence of sulphur dust on the rate of photosynthesis of an entire apple tree. Proc. Am. Soc. Hort. Sci. 36:202.

HEINICKE, A. J., and N. F. CHILDERS. 1937. Influence of respiration on the daily rate of photosynthesis of entire apple trees. Proc. Am. Soc. Hort. Sci. 34:142–144.

HEINICKE, A. J., and M. B. HOFFMAN. 1933a. The rate of photosynthesis of apple leaves under natural conditions, I. Cornell Univ. Agr. Expt. Sta. Bull. 577:1–32.

HEINICKE, A. J. and M. B. HOFFMAN. 1933b. An apparatus for determining the absorption of carbon dioxide by leaves under natural conditions. Science 77:55–58.

MARSHALL, G., N. F. CHILDERS, and H. W. BRODY. 1942. The effects of leafhopper feeding injury on apparent photosynthesis and transpiration of apple leaves. J. Agr. Res. 65(6):265–281.

PROCTER, J. T. A., R. L. WATSON, and J. J. LANDSBERG. 1976. The carbon budget of a young apple tree. J. Am. Soc. Hort. Sci. 101:579–582.

SCHNEIDER, G., and N. F. CHILDERS. 1941. Influence of soil moisture on photosynthesis, respiration and transpiration of apple leaves. Plant Physiol. 16:3.

SOUTHWICK, F. W., and N. F. CHILDERS. 1941. Influence of Bordeaux mixture and its component parts on transpiration and apparent photosynthesis of apple leaves. Plant Physiol. 16:4.

THOMAS, MOYER D., and R. HILL. 1937. The continuous measurement of photosynthesis, respiration and transpiration of alfalfa and wheat growing under field condition. Plant Physiol. 12:285–307.

# THE DAILY RATE OF PHOTOSYNTHESIS, DURING THE GROWING SEASON OF 1935, OF A YOUNG APPLE TREE OF BEARING AGE

## A. J. Heinicke and N. F. Childers

A thorough knowledge of the photosynthetic activity of the leaves of fruit plants is of fundamental importance to the pomologist. The fruit-grower realizes that many of his practices, such as pruning, cultivation, fertilization, and spraying, may influence the activity of the leaves as well as the extent of the leaf surface, and thus affect the amount of food that the plant can manufacture for itself.

The amount of carbohydrates that the entire leaf surface of a plant can synthesize from day to day, or at different periods during the season, is one of the most important influences in determining such vital matters in fruit production as the amount of shoot and root growth, the initiation of flower buds for the next year's crop, the development of size, color, and quality of the fruit, the development of proper maturity and resistance of the tissue of the plant to cold during the winter, and the provision of an abundance of storage materials required for early growth activity in the following spring.

In spite of the obvious importance of the subject, knowledge regarding the normal rate of photosynthesis and the factors that influence the efficiency of the foliage of entire plants under natural conditions is still very limited. Much of the commonly accepted information on photosynthetic activity is based on studies carried on to a considerable extent with small aquatic plants or with detached leaves of larger land species. Because of the difficulty of manipulating large samples, the amount of tissue involved in such studies usually has been conveniently small and not necessarily representative of the foliage of the plant as a whole. Generally the determinations have lasted for relatively short periods of time—varying from a few minutes to a few hours, or at most a day or so. Extended bibliographies on this subject are given by Willstätter and Stoll (1918), McLean (1920), Stiles (1925), Spoehr (1926), Miller (1931), Gassner and Goeze (1932), and Stocker (1935).

If investigators are to have a more adequate basis for evaluating the knowledge of photosynthesis gained in this way, especially with respect to its application to natural or cultural conditions, they need to know more about the daily course of carbon-dioxide assimilation, throughout the season, of a large population of leaves attached to the plant under study and functioning more or less normally as a part of the organism as a whole. To be sure, the individual leaves on a large plant will have widely differing exposure to light. They will carry on activity under varying meteorological and other environmental conditions. Furthermore, they will be subject to the influences of the many complex and continuously changing internal as well as external factors which may affect not only carbon-dioxide assimilation but other metabolic processes as well. The measurement of the rate of photosynthesis of the foliage as a whole, however, will give a picture of the average behavior of all the leaves on the plant. If the determinations are continued day in and day out

3

throughout the growing season for several years in succession, and especially if each day is divided into a number of shorter periods, there is a fair probability of frequent replications of definite sets of conditions.

This paper presents a report of the daily rate of photosynthetic activity of the entire leaf surface of a young apple tree of bearing age growing under approximately natural conditions during the season of 1935, from May 14 to November 17 inclusive. The entire plant, with more than 10,000 leaves having an area of more than 330,000 square centimeters, is regarded as a unit. The determinations were carried on continuously for 188 days in succession. The influence of light and other meteorological conditions on the activity of the leaf surface as a whole is given special consideration in the report.

## METHODS USED IN THE STUDY

The method of determining the apparent photosynthesis used for individual leaves by Heinicke and Hoffman (1933) was modified to suit the requirements of a much larger leaf area. The principle of the method, which has been used by many workers since Kreusler (1885) first suggested it, is based on the fact that green leaves exposed to light have the ability to remove $CO_2$ from the air. The leaf surface is inclosed in an assimilation chamber, which is continuously supplied with adequate quantities of fresh air. The amount of $CO_2$ in the incoming and in the outgoing air is accurately determined, and the difference is regarded as apparent photosynthesis if there has been a decrease during contact of the air with the leaves, or as apparent respiration if there has been an increase.

Among the important features of the method are, first, an assimilation chamber which should be practically air-tight except for the inlet and outlet points, and which should provide as nearly as possible the same conditions with respect to light, temperature, humidity, and carbon-dioxide supply as prevail in the open. In the second place, the sampling and analysis of the atmosphere must be carried out with great accuracy, since the amount of $CO_2$ originally present in normal air is very small and since the differences to be determined ordinarily should be less than 20 per cent of the total amount (Heinicke and Hoffman, 1933). The amount of air moving through the chamber must therefore be accurately known, and the atmosphere within the chamber must be thoroughly mixed so as to prevent dead-air spaces and to make it possible to obtain a representative sample for analysis.

### THE ASSIMILATION CHAMBER

A diagram of the assimilation chamber used in these determinations is presented in figure 1. The dimensions of the chamber, or cage, were 7 by 7 by 11.5 feet, with a volume of 563.5 cubic feet. In constructing the cage, a substantial, thoroughly anchored, level platform was first built under the tree and about 1 foot above the surface of the ground. The floor of the platform was covered with linoleum to prevent entrance of air except at the opening provided for the air intake around the trunk of the tree.

The sides and the top of the cage were made of glazed window-sash held together and supported by wooden strips. A thin layer of a plastic-rubber adhesive compound was spread between the adjoining edges of the windows

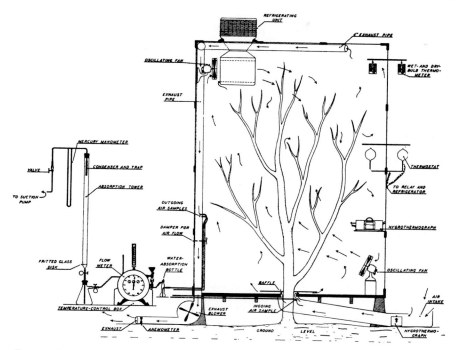

FIGURE 1. DIAGRAM OF ASSIMILATION CHAMBER, SHOWING ACCESSORIES AND ONE OF THE CO₂-ABSORPTION UNITS

and between the sash and the wooden supports. This compound sealed all joints, and allowed for expansion and contraction without cracking, thus effectively preventing air leaks. The windows in each of the four sides and the top were assembled to form five units. These units were subsequently put into position on the platform, and were held in place by being fastened to each other along the edges and drawn together tightly by long wood screws. The top was supported by the upper edges of the sides and by two rigid cross members. Plastic sealing compound was used along all edges of the units. The structure required no further support, and it withstood a severe hail-and-wind storm a few days after it was completed. One of the windows on the north side could be removed to allow access to the interior of the cage. This window was carefully fitted so that practically no air could pass through its unsealed edges.

## AIR SUPPLY

The incoming air was supplied to the tree through an opening around the trunk in the platform (figure 1). Air was led to this opening through a duct made of sheet metal which extended several feet beyond the south side of the platform and which opened on the top. This arrangement gave an air supply derived from a few feet above the surface of the soil. A baffle was used around the trunk of the tree to help divert the air to all sides of the cage rather than have it move directly upward from the opening. Two oscillating fans on opposite sides of the chamber and at different levels helped to keep the air thoroughly mixed while it was gradually moving from the intake to the outlet opening.

The air was pulled through the chamber by means of an electrically operated exhaust blower connected to a 4-inch galvanized iron pipe. The blower was located just below the platform. The opening for exhausted air was on the north side of the cage. The exhaust pipe extended through the platform to the top of the chamber, and was provided with two laterals with openings at opposite corners of the cage about 1 foot below the top. The exhaust pump had a capacity of about 200 cubic feet per minute, so that it was possible to renew the air within the chamber once every three minutes, on the average. The pump was operated at a constant speed, but the amount of air which passed through the cage could be controlled by a damper in the exhaust pipe.

The rate of air-flow was checked every few hours by means of a carefully standardized anemometer. This instrument was attached to a pipe on the blower side of the exhaust pump in such a manner that all the air had to pass through the recording device. The velocity of the air moving through the anemometer for several one-minute intervals was thus recorded. The unobstructed area of the cylinder of the anemometer having previously been determined, the volume of air which passed out of the cage during a given interval was easily calculated.

The standard rate of air-flow varied within about 2 per cent of 11,000 cubic feet per hour. Slight changes in the rate of air movement were observed, due to extremes of temperature, humidity, wind, and barometric pressure. A slower rate, ranging from one-third to two-thirds of the standard, was used before the leaf surface was fully developed and after about half of the leaves had been shed. If the rate were too rapid, the differences between incoming and outgoing air would be very small and hence more difficult to determine accurately; if too slow, the leaves would be functioning with a much lower $CO_2$ supply than is normally present in fresh air. The rate chosen was rapid enough so that the $CO_2$ removed from the amount available in fresh air seldom exceeded 13 per cent. A slower rate than was used would be desirable for nighttime and on very cloudy days.

### TEMPERATURE CONTROL

A special thermostat was devised for controlling the temperature within the cage so that it would fluctuate in approximately the same manner as that on the outside.[1] The thermostat consisted of two glass bulbs, each having a capacity of about 500 cc, which were connected by a small glass U-tube about 30 cm wide and 10 cm high (figure 1). A platinum wire was sealed in one of the arms of the U-tube, and another platinum wire was inserted through the glass into the base of the U-tube. The connecting tube was filled with mercury to within a millimeter of the wire in the upright tube. Before being sealed, the bulbs were exhausted and filled with nitrogen gas to prevent oxidation of the mercury. The thermostat was placed on the north wall of the chamber in such a way that one bulb was on the out-

---

[1]The authors are indebted to J. G. Waugh, formerly student assistant in pomology at Cornell University, for the construction of the thermostat.

side and the other on the inside of the cage. Whenever the air within the cage became warmer than the outside air, the gas in the inside bulb would tend to expand and force the mercury toward the cooler outside bulb. As soon as the mercury came in contact with the platinum wire, an electric circuit was closed, actuating a relay which, in turn, threw the switch that supplied current to operate the refrigeration unit.

The refrigeration unit, which was placed near the north side of the top of the cage, was taken from an ordinary household refrigerator. Before this unit was installed, during the first week in June, the temperature inside the cage was from 3 to 5 degrees (centigrade) higher than that on the outside whenever the sun shone brightly. It was found possible to lower the temperature a few degrees by wetting the outside and the top of the cage. Moisture would condense on the inside of the glass, however, and would form lenses in which the sun's rays were concentrated enough to burn spots on some of the leaves. After the refrigeration unit was installed, there was seldom a difference of more than 1 or 2 degrees between the inside and the outside air. It should be kept in mind that the air within the chamber was renewed once every three minutes, on the average, so that the temperature could be kept reasonably low in spite of the large amount of glass surface.

The temperatures on the wet-bulb and dry-bulb thermometers exposed to moving air on the inside and on the outside of the cage were recorded at intervals of about one hour. The values for relative and absolute humidity on the inside and on the outside of the cage were subsequently determined from standard psychrometric tables (Marvin, 1915). The difference in weight of a cubic meter of aqueous vapor on the inside and one on the outside of the cage was ascribed to transpiration. The rate of transpiration could also be checked directly by absorbing the moisture in an air sample from the chamber in a water-absorption bottle (figure 1). Hygrothermographs placed on the inside and on the outside of the cage gave continuous records of temperature and humidity.

### AIR SAMPLING

Since it was obviously impractical to analyze all the air that passed through the assimilation chamber, the method used was to withdraw an aliquot continuously and at uniform rates throughout each determination. Thus, if there was any variation in $CO_2$ content of the atmosphere at different times during the interval, the method of sampling would nevertheless give a true average of the prevailing conditions. As is shown by the diagram (figure 1), the samples of fresh air were obtained from the end of the air duct leading to the chamber. The samples of outgoing air were withdrawn from the exhaust pipe just above the damper device. Duplicate samples were withdrawn for analysis in each case.

The samples were conveyed from the assimilation chamber through flexible aluminum tubes, having a bore of 5 mm, to the metering and absorbing instruments housed in small huts several meters distant.

Wet test meters were used to determine the amount of air withdrawn in each sample. The rate of air withdrawn was about 40 to 45 liters per hour, or a total of about 200 to 225 liters for a five-hour daylight interval. A slower rate of withdrawal was used for the night intervals so that the solution would remain efficient throughout the longer period. The meters were carefully calibrated and adjusted to a United States Bureau of Standards liter. They were frequently checked to ascertain whether they gave strictly comparable readings. The temperature and the barometric pressure were taken at frequent intervals so that the meter readings could be adjusted to standard conditions if desired. During the cooler periods of the season, a special thermostat and heating unit was used to keep the temperature of the meters approximately at 20° C. The path of the air through the drying tube, the meters, and the absorption towers, is shown in figure 1.

### CARBON-DIOXIDE ABSORPTION

The $CO_2$-absorption apparatus was essentially the same as that described by Heinicke and Hoffman (1933). A few minor modifications may be noted. A special condenser consisting of a glass tube filled with glass wool was used at the top of the absorption tower to prevent the loss of minute particles of solution which occasionally passed over with the air stream. The rate of air movement through the towers was slower than in the earlier work, and this tended to increase the efficiency of absorption. A small quantity of normal butyl alcohol was used in the absorption solution to decrease the surface tension and to make for better absorption (Thomas, 1933). Frequent tests showed that practically no $CO_2$ was left in the air after it had passed through the absorption tower. During the cooler part of the year, heating units were required to prevent the contraction of the fritted glass plates and the freezing of water used in washing the towers.

### STANDARD ROUTINE PROCEDURE

For the sake of convenience, each day was divided into three standard intervals as follows: (a) 7:30 p.m. to 9:30 a.m.; (b) 9:30 a.m. to 2:30 p.m.; (c) 2:30 p.m. to 7:30 p.m. On a number of occasions the first interval was subdivided into several four- to five-hour periods.

The procedure for determining the carbon dioxide in the duplicate samples of incoming and of outgoing air was essentially the same as that described by Heinicke and Hoffman (1933). By means of a refill pipette connected to a $CO_2$-free system, exactly 100 cc of an approximately 1/15 N KOH solution was withdrawn from the stock bottle and placed in each of five 500-cc specially equipped suction flasks, one of which served as a blank. These flasks were taken to the field and inserted in the assembly as shown in figure 1. After the meter readings were recorded, a suction pump for taking air samples from the assimilation chamber was started. The air-flow through each tower was adjusted to the desired rate by means of a stop watch.

Just before the run was terminated, such data as wet- and dry-bulb-thermometer readings, maximum and minimum temperatures during the run, barometric pressure, velocity of the wind, condition of the sky, and rate of air flow through the chamber, were recorded. A few minutes before the scheduled termination of the run, the sampling tubes leading to the chamber were closed and the suction pump was shut off. The readings for each meter were taken quickly, and the condensers and absorption towers were rinsed thoroughly with distilled water. As soon as the rinsing was completed, the pinchcocks were closed and the flasks containing the solution were disconnected and removed from the towers and from the tubes leading to the chamber. These were immediately replaced by another set of flasks containing fresh KOH solution. Before the next run was started, the meter readings were rechecked. The time required for the change of solution flasks was about six minutes.

The flasks which were removed from the assembly were taken to the laboratory, where the dilute KOH solution was carefully transferred to 500-cc volumetric flasks. To each flask, 10 cubic centimeters of a 10-percent $BaCl_2$ solution and sufficient distilled water to bring the volume up to 500 cc were added. The flasks were then stoppered tightly and shaken vigorously. In about half an hour, 200 cc of the clear supernatant liquid was drawn off with a pipette and titrated against the standardized hydrochloric acid (approximately 1/10 N), phenolphthalein being used as an indicator.

The amount of $CO_2$ absorbed by each of the four units during the run was determined by the following formula:

Mg $CO_2$ per sample = cc acid used in titrating 2/5 aliquot of blank − cc acid used in titrating 2/5 aliquot of sample × 10/4 × $CO_2$ equiv. of 1 cc of standardized acid

The following illustration indicates what other calculations were involved in each determination. The values obtained by the above formula for two samples of incoming air were found to be 119.885 mg and 117.831 mg of $CO_2$, respectively, and the readings of the flow meters indicated that 8.088 and 7.946 cubic feet of fresh air, respectively, bubbled through each tower. The amount of $CO_2$ contained in a cubic foot of air would then be $\frac{119.885}{8.088}$, or 14.823 mg, and $\frac{117.831}{7.946}$, or 14.829 mg, respectively, with an average of 14.826 mg. By the same procedure the average $CO_2$ content per cubic foot of outgoing air was found to be 13.096 mg. The difference between these two values, namely, 1.730 mg, is the average amount of $CO_2$ which the tree removed from each cubic foot of air that entered and left the chamber during the run. In the tables included in this report, the figures for $CO_2$ content are given on a cubic-meter basis at standard temperature and barometric pressure.

By calculation from anemometer readings, as previously indicated, the rate of air-flow through the chamber during the five-hour period was found to be 11,078 cubic feet per hour. The grams of $CO_2$ absorbed by the tree per hour would therefore be $\frac{11,078 \times 1.730}{1000}$, or 19.165 gm. Since the determination lasted five hours, this figure would be multiplied by 5, which gives 95.825 gm, the amount of $CO_2$ assimilated by the apple tree during the entire interval.

### LIGHT MEASUREMENT

The sky-radiation measurements were made with an Eppley pyrheliometer located on the top of the Plant Science Building at Cornell University, about a mile distant from the orchard. The measurements were automatically recorded every three seconds, and gave curves similar to those shown in figure 2. The radiation during any interval, in gram-calories per square centimeter of horizontal surface, could be determined by measuring the corresponding area with a planimeter and multiplying by a suitable factor (Kimball, 1931). In order to give some idea of the approximate values represented by the curves on the graph, the value of gram-calories per hour was divided by 60 to convert it to a minute basis. Ac-

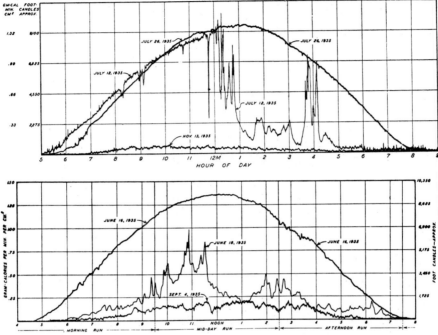

FIGURE 2.   REPRODUCTION OF THE AUTOGRAPHIC PYRHELIOMETER RECORDS, SHOWING FLUCTUATIONS IN SOLAR RADIATION ON SEVERAL DAYS AND AT DIFFERENT TIMES OF THE DAYS GIVEN

The total amount of radiation for each period during which photosynthesis was determined is obtained from these records

tually, the curves represent about twenty separate readings for each minute. These values are converted to approximate values for foot-candles by dividing the gram-calories per hour by 60 and multiplying by 6900 (Kimball, 1925).

## DESCRIPTION OF THE TREE USED IN PHOTOSYNTHESIS DETERMINATIONS

A photograph of the apple tree which was used, made soon after the tree was inclosed in the assimilation chamber, is shown in figure 3. The tree was a healthy, vigorous specimen of the variety McIntosh, planted in the spring of 1928. It was growing in an experimental planting of the Cornell University orchard at Ithaca. The trees in this planting were set 10 by 10

FIGURE 3.  VIEW OF ASSIMILATION CHAMBER INCLOSING AN ENTIRE TREE
The photograph was made soon after the tree was inclosed and before all the accessories were installed

feet apart, and were growing in sod which received an application of 5 pounds of ammonium sulfate applied uniformly over the entire area allotted to each tree.

The profile of the soil in which this tree grew is described as follows: Layer $A_1$: depth, 0–8 inches; medium brown silt loam, loose, crumb structure. Layer $A_2$: depth, 8–11 inches; yellow-brown silt loam, loose, friable. Layer B: depth, 11–33 inches; compact silty clay, olive brown mottled gray and rusty, cloddy structure; in one of the samples, the lower 3 or 4 inches of this layer was found to be less compact and somewhat siltier than the upper part; some manganese concretions occur in the upper part of the layer. Layer $C_1$: depth, 33–40 inches; calcareous silt to silty clay, olive brown mottled gray and rusty, somewhat laminated, with white carbonate accumulations apparent in some of the laminations; less compact than the B layer. Layer $C_2$: depth, 40–48 inches; calcareous silt, olive brown with no mottling, laminated, no white carbonate accumulations.

The moisture equivalents in the first and the second foot below the surface were 30.4 and 25.2 per cent, respectively. The wilting coefficients were 16.5 and 13.7 per cent, respectively. The percentages of moisture in the soil at the end of October were 22.4 and 18.1 per cent, respectively, which was about as low as the moisture content was at any time during the season. While no rain could fall directly on the soil immediately under the platform, the drainage from the roof was diverted in such a way that the area protected by the platform was kept moist.

A detailed description of the tree and its growth response during 1935 is given in table 1. In general, the tree made a more vigorous vegetative growth than did neighboring trees outside the cage. This may have been owing in part to a somewhat higher relative humidity caused by confining the moisture lost in transpiration within the cage for an average of about three minutes, and by a somewhat higher average temperature. Even though ultra-violet rays are not concerned directly in photosynthesis, the fact that the ordinary window-glass used in the cage keeps out most of the ultra-violet rays of the sunlight may have had something to do with the characteristics of the growth. That the ultra-violet rays were kept out of the cage was demonstrated by the almost complete absence of red color on the mature apples which had been growing in well-exposed parts of the tree.

The tree was sprayed with lime-sulfur once before it was inclosed in the cage. There was relatively little difficulty with insects, and the few minor infestations that occurred were easily controlled (table 1). No fungous diseases were noticed on the tree.

Information as to the amount and character of the foliage shed during different periods is given in table 2. The data indicate that the tree retained more than 95 per cent of all its foliage until about the middle of October. Within twenty-four hours after the temperature dropped to −5° C. on October 18, about one-third of the leaf area of the tree was lost.

Practically all of the leaves shed up to the middle of October had been formed in the early part of the season, and it is estimated that they were fully expanded by early June. When these leaves were first formed they were well exposed to light, but, since they grew in a basal part of the spurs or shoots, they naturally received less light after new leaves appeared in the terminal part of the new growth. The table shows clearly that the leaves which fell first had a lower dry weight per unit area than had those which adhered to the trees longer. They were less than half as heavy, per given area, as those that remained on the tree until November.

TABLE 1. Statistics and Phenological Notes Pertaining to McIntosh Tree Used in Photosynthesis Determination

| Date (1935) | Notation |
|---|---|
| May 13 | Circumference of trunk 24.5 cm; tree 3.1 meters high and 2 meters spread. 233 twigs made 10 cm or more growth during 1934; 394 spurs made less than 10 cm growth; 331 flower clusters with an average leaf area of 20.1 cm²; 324 leaf clusters with an average leaf area of 5 cm²; total leaf area when determinations began, 8273 cm²; largest leaves 3 cm long; most of the leaves smaller. |
| May 14 | Flowers beginning to separate from cluster. |
| May 16 | Petals fully expanded on several central flowers of cluster. |
| May 19 | 190 flower clusters fully expanded, mostly on spurs; 141 flower clusters from lateral buds or from terminal buds on long twigs still closed; 296 vigorous leaf clusters from terminal buds; 312 smaller leaf clusters arising from lateral buds; about 100 flower clusters hand-pollinated with pollen of Delicious variety. |
| May 20 | First petals beginning to fall. |
| May 22 | Last flowers in top part of tree open. |
| May 23 | 2 teaspoons nicotine vaporized 6–7 p.m. Dense smoke enveloping tree during night. Ventilated one-half hour before beginning of next run. |
| May 24 | No live aphids or other insects remaining after treatment; no apparent injury to tree. |
| May 25 | About 75 per cent of flowers with petals dropped; (first bloom on trees outside of cage); very rapid enlargement of leaf area during past few days. |
| May 28 | All petals fallen; largest leaf 62 mm long; most of the leaves small; vigorous terminal shoots 10 cm long, with 7–9 leaves unfolded. |
| June 2 | Non-pollinated flowers turning yellow and beginning to abscise; those pollinated about 1 cm in diameter; refrigeration unit installed to control temperature; terminal buds beginning to form on weakest spurs. |
| June 5 | Vigorous terminal growth 15 cm long; most shoots about 10 cm; terminal buds formed on less vigorous spurs; leaf area estimated at about 100,000 cm². |
| July 15 | Many shoots still growing; (trees outside of cage showing terminal buds on all growths); leaves larger than on trees in open; foliage of healthy color, free from pests. |
| August 25 | European red mite (*Paratetranychus pilosus* Canestrini & Fanzago) established; tree treated with spray applied with 3-gallon knapsack sprayer between 7:30 and 8:30 p.m., using "Lethane Junior" 1–400, with 40-per-cent pine-tar soap 1–200 as a spreader. |
| September 15 | A few basal leaves on spurs turned yellow and falling; vigorous terminals still forming new leaves. |
| October 1 | 25 full-sized fruits harvested; flavor normal but color poor. |
| October 5 | Basal leaves on spurs and shoots beginning to lose dark green color. |
| October 19 | Heavy leaf-fall following temperature of –5° C. on morning of October 18. |
| November 15 | Tree with more than 80 per cent of its leaf surface gone; remaining leaves the ones formed on long shoots; these leaves large, and appearing healthy even though they were frozen stiff several times. |
| November 17 | Experiment concluded; remaining leaves, constituting about 10 per cent of total leaf area, removed from tree even though still green and adhering tightly. |
| November 18 | Circumference of trunk 27.2 cm; total length of 803 shoots grown in 1935, 25,175 cm; 6 shoots more than 100 cm long; 32 shoots 75–100 cm long; 78 shoots 50–75 cm long; 361 shoots 25–50 cm long; 241 shoots 10–25 cm long; 85 shoots less than 10 cm long. |

There was also a marked difference in shape and size of leaves formed at different times. The younger leaves in the terminal part of the shoots were larger, and relatively much wider, than the older leaves. For example, the average area of leaves that fell on October 18 was 16.4 cm², while those that remained on the tree until November 18 averaged 28.6 cm². The largest leaves in both cases were about 10 cm long, but the width of those

TABLE 2.   CHARACTER AND AMOUNT OF FOLIAGE SHED DURING DIFFERENT PERIODS,
AND TOTAL LEAF AREA OF TREE USED IN PHOTOSYNTHESIS DETERMINATIONS

| Period of leaf fall | Dry weight of foliage (grams) | Equivalent of 1 gram dry weight in square centimeters of leaf area | Amount of foliage shed | |
|---|---|---|---|---|
| | | | Area (cm²) | Per cent of total |
| Up to Oct. 6.... | 18.5 | 206.5 | 3,820.3 | 1.1 |
| Oct. 7–13...... | 34.4 | 203.2 | 6,990.1 | 2.1 |
| Oct. 14–20*.... | 618.5* | 201.3 | 124,504.0 | 37.6 |
| Oct. 21–27..... | 290.0 | 174.2 | 50,518.0 | 15.3 |
| Oct. 28–Nov. 4. | 366.0 | 142.6 | 52,191.6 | 15.8 |
| Nov. 5–11..... | 373.2 | 82.8 | 30,901.0 | 9.3 |
| Nov. 12–17.... | 911.3† | 68.2 | 62,150.7 | 18.8 |
| Total........ | 2,611.9 | | 331,075.7 | 100.0 |

*More than 80 per cent of this loss occurred on October 18 following the sudden drop in temperature to −5° C.

†About half of this foliage was still green and adhered tightly when the experiment was concluded.

that fell on the former date was about 5 cm as compared with 7.5 cm for those that fell on the latter date. The young leaves retained their chlorophyll until they were frozen. As has been shown previously, such green leaves are likely to exhibit a higher rate of activity, even at a given light intensity, as compared with older leaves (Heinicke, 1935). These young leaves also had the best exposure to light, and probably had a better opportunity than had the older leaves to obtain water and soil nutrients whenever there was severe competition.

## RESULTS OF THE STUDY

The results of each day's determinations are given in detail for each month, from May 14 to November 17, in tables 12 to 18 in the appendix. The tables contain the following data for each interval during which determinations were made: the average hourly rate of apparent photosynthesis or apparent respiration, and the total assimilation during the interval; the carbon-dioxide content in a cubic meter of air at standard temperature and pressure, and the percentage of $CO_2$ removed by the leaves; the solar radiation per interval in gram-calories per cm²; the minimum, maximum, and mean temperatures; the relative humidity inside the chamber at the beginning, the middle, and the end of the run; the grams of water transpired per interval; the ratio of transpiration to assimilation; and the atmospheric pressure. There are given also, in these tables, the total carbon dioxide assimilated during each day, the total solar radiation during the day, the total daily transpiration, and the daily assimilation–transpiration ratio.

### DAILY RATES OF PHOTOSYNTHESIS AND TRANSPIRATION

The rate of apparent photosynthesis for the tree varies from less than nothing (apparent respiration) to an assimilation of almost 250 grams of $CO_2$ for the 24-hour interval beginning at 7:30 p.m. The light during the day varies from as low as 22 to as high as 726 gram-calories per cm² per day. The average $CO_2$ content per cubic meter of air during periods of five hours or longer ranged from 525 to 865 mg at standard conditions of temperature and pressure. Usually, less than 12 per cent of the available $CO_2$ was removed by the tree during any interval, although in one case as much as 22 per cent was absorbed.   The temperature surrounding the leaves of the tree on different days ranged from about 5° below zero (centigrade) to as high as 38° above.   Transpiration varied from practically none to a maximum of more than 37,000 grams a day for the entire tree.

The data for the total apparent photosynthesis, the total light, the minimum, maximum, and mean temperatures, and the total transpiration, for each day throughout the growing season, are shown in figure 4.   It is

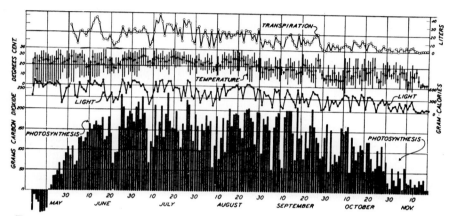

FIGURE 4.   DAILY FLUCTUATIONS IN APPARENT PHOTOSYNTHESIS, LIGHT, TEMPERATURE, AND TRANSPIRATION

The extremes of photosynthesis generally coincide with corresponding extremes in light during the greater part of the season

apparent that there are marked fluctuations in photosynthetic activity from day to day, with a minimum of activity at the beginning and at the end of the season.   During the first week or so after the determinations were begun, the tree was not able to assimilate as much $CO_2$ as was given off in respiration.   At this time of the year the tree was producing new shoot growth, expanding more foliage, maturing flower parts, and carrying on other metabolic activities associated with rapid growth.   Much of these activities obviously occurred at the expense of stored food.   The energy release involved exceeded the total photosynthesis which could be carried on by the relatively small amount of foliage present in the early weeks.

Soon after bloom, however, and with a rapidly increasing leaf surface, the tree began to assimilate more $CO_2$ than it had produced during respiration.   Even though the leaf surface was not yet fully developed by June 10, the tree was able to carry on a relatively high rate of activity at that time.   It may be noted that approximately the maximum daily rate was obtained as early as June 25, even though new leaves continued to expand until about September 1.   Photosynthesis continued to fluctuate between high and low values throughout the summer months until the middle of September.

After the beginning of October there was a noticeable decline, and this became very marked during the first part of November.   A minimum temperature of several degrees below zero was recorded on October 7 but the leaf tissue was not frozen stiff until the night of October 17–18. The older leaves at the base of the twigs and spurs began to lose their dark green color in the latter part of September, and some of them began to fall in early October.   As is shown in table 2, a large part of the foliage had fallen before the experiment was concluded.

The trend of apparent photosynthesis at the beginning and toward the end of the season is obviously determined in large part by the amount of active leaf surface available.   As is indicated in table 1, new leaves were being formed on this tree on terminal shoots until the middle of September, but the new foliage caused more shade and necessitated more competition for water and nutrients for the older leaves in the interior of the tree. Thus the effective leaf area was probably not increased greatly after about July 1.   These shaded leaves at the base of the shoots and spurs were the first leaves to lose their green color and drop, beginning in early October. Such leaves probably lost most of their ability to carry on active photosynthesis long before they abscised (Heinicke, 1935).

A general relationship between light and photosynthesis is clearly indicated in figure 4.   On days when there was little photosynthetic activity there was little light, and most of the high daily rates of assimilation coincide with the high points on the light curve.   The relationships in intermediate rates are not always so clear-cut.   There is also a general trend of decreasing photosynthesis as the season advances, which corresponds roughly to the gradually decreasing daily light intensity.

The daily transpiration curve bears a fairly close relationship to the daily light curve.   It is considerably higher during the early part of the season than during the latter part.   In general, the low points in transpiration correspond to the low points in photosynthesis, but the transpiration may sometimes be high without a corresponding high rate of photosynthesis.

The relationship between daily variations in temperature and in photosynthesis is not very apparent from the graph.   That air temperature does have, however, at least an indirect influence in determining the average daily rate of photosynthesis, is shown in another part of this paper.

### PHOTOSYNTHESIS IN DIFFERENT PERIODS DURING THE SEASON

The amounts of apparent photosynthesis of the entire tree are summarized by months and by selected periods in table 3, and these data are shown graphically in figures 5 and 6.   The peak in average daily activity of the tree was reached during the month of July, with an average of 168.8 grams of $CO_2$ assimilated each day in excess of respiration.   It may be noted that during the month of September the eight-year-old apple tree accumulated more carbohydrates than it did during the month of June. The rate of activity in October was only about 55 per cent as much as that in July.   No doubt this was owing in large part to a heavy loss of leaf surface after the middle of the month, but in addition there were very few clear days in late fall.   The activity of the whole tree during November, when most of the leaf surface was off and when there was very poor light, was, of course, very low.   It may be noted, however, that on the few days with fairly good light—for example, November 6 and 9—the small leaf surface that remained gave a good account of itself.

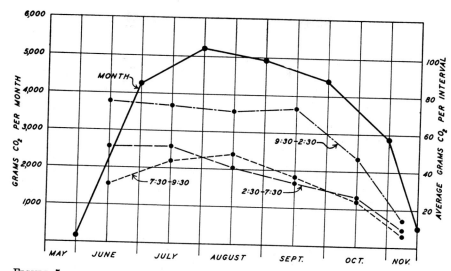

FIGURE 5.   TOTAL AMOUNT OF APPARENT PHOTOSYNTHESIS FOR EACH MONTH OF THE
GROWING SEASON, AND AVERAGE RATE PER INTERVAL

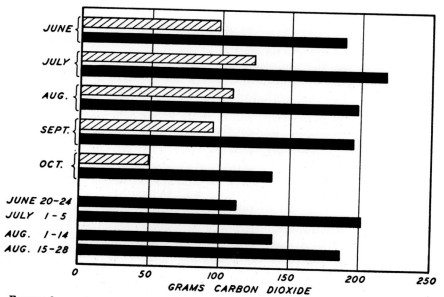

FIGURE 6.   AVERAGE RATE OF APPARENT PHOTOSYNTHESIS PER DAY DURING DIFFERENT
PERIODS

In the upper part of the graph the cross-hatched bars represent the ten poorest days, and the solid bars
the ten best days, for the respective months

TABLE 3.  APPARENT PHOTOSYNTHESIS OF ENTIRE TREE BY MONTHS
AND DURING SELECTED DAYS

| Period selected | CO₂ assimilated in excess of respiration | |
| --- | --- | --- |
| | Total grams | Average grams per day |
| May 14–31.................................... | 155.1* | ..... |
| June 1–30..................................... | 4,278.8 | 142.6 |
| July 1–31..................................... | 5,234.0 | 168.8 |
| August 1–31.................................. | 4,952.0 | 159.7 |
| September 1–30............................... | 4,411.2 | 147.0 |
| October 1–31................................. | 2,906.5 | 93.8 |
| November 1–17............................... | 498.0 | ..... |
| Period A: 188 days, May 14 to November 17......... | 22,436.0 | 119.3 |
| Period B: 141 days, June 3 to October 21............. | 20,940.0 | 148.5 |
| Days better than period-B average: | | |
| 16 in June................................. | 2,511.1 | 179.4 |
| 24 in July................................. | 4,308.1 | 187.3 |
| 22 in August.............................. | 4,023.9 | 182.9 |
| 18 in September........................... | 2,908.1 | 181.8 |
| 4 in October.............................. | 469.1 | 156.4 |
| Best and poorest 10 days, respectively, in: | | |
| June................................... | 1,881.6;  973.0 | 188.2;  97.3 |
| July................................... | 2,152.2; 1,226.2 | 215.2; 122.6 |
| August................................. | 1,973.1; 1,077.6 | 197.3; 107.8 |
| September.............................. | 1,947.9;  940.5 | 194.8;  94.1 |
| October................................ | 1,381.6;  492.3 | 138.2;  49.2 |
| Period B................................ | 2,207.0;  609.0 | 220.7;  60.9 |
| June 20–24 and July 1–5, respectively............... | 561.5; 1,005.0 | 112.3; 201.0 |
| August 1–14 and August 15–28, respectively.......... | 1,962.8; 2,612.4 | 140.2; 186.6 |

*Excess of total apparent photosynthesis over total apparent respiration.   517.2 − 362.1 = 155.1.

The total assimilation of $CO_2$ during the 188 days from May 14 to November 17 was 22,436 grams, with an average of 119.3 grams per day. During the time when the tree had more than half of its foliage, from June 3 to October 21, the total apparent photosynthesis was 20,940 grams, with an average of 148.5 grams per day.   For the days in June on which the activity was better than this average, the rate exceeded the normal by about 21 per cent; in July, by 26 per cent; in August, by 23 per cent; in September, by 23 per cent; but only by about 5 per cent in October. The data in table 3 show also that the best 10 days of each month account for 40 to 48 per cent of the total assimilation for that month.   There were 5 consecutive days, from July 1 to July 5, when the tree absorbed 79 per cent more $CO_2$ than it did during the 5-day period from June 20 to June 24. In the last half of August the tree accumulated about 33 per cent more carbohydrates than it did in the first part of the month.

If the cumulative totals are plotted at 5-day intervals, a fairly straight curve is obtained in spite of these violent fluctuations, as is shown in figure 7.   The graph shows clearly that the tree did not begin to add to its dry weight until several weeks after the buds opened; but from about June 3 to October 10, when there was a good leaf surface, the rate of accumulation during 5-day intervals was fairly constant.

The total apparent photosynthesis per unit of average leaf area for that part of the season when the tree had about its full leaf surface amounted to 677.6 mg per 100 cm² of leaf surface.   This is about 42.5 mg per day, or about 4.5 mg per hour of daylight.   This rate is considerably less than the average based on determinations of single leaves exposed to good

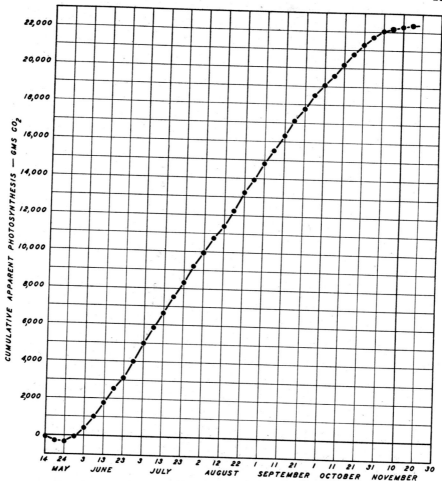

FIGURE 7.   APPARENT PHOTOSYNTHESIS FOR THE ENTIRE SEASON
The dots represent amounts of $CO_2$ accumulated at 5-day intervals

light (Heinicke and Hoffman, 1933).   For the best 10 days during the season, the average was 66.5 mg per day per 100 cm² of leaf area.   That these values are not too low for the tree as a whole, however, is indicated by table 11 (page 30).   Obviously, much of the leaf surface, probably more than 90 per cent of the total, is functioning at a very low rate on many days during the season.   The details of utilization of the carbohydrates produced are considered more fully in subsequent paragraphs.

### CONSIDERATION OF THE DATA FOR DIFFERENT INTERVALS OF EACH DAY

While the daily fluctuations in apparent photosynthesis and transpiration are evidently related to corresponding fluctuations in light, the influences of the various meteorological factors on leaf activity are shown more clearly if the data for the several intervals of each day are considered. There are very few times, during the season, when the external conditions prevailing throughout the day are nearly alike. The chances of having many sets of comparable conditions are much greater when shorter intervals are involved.

The average apparent photosynthesis and transpiration, together with the average meteorological conditions prevailing, in different months from June 17 to November 17 during three periods each day, are given in table 4. The first interval, from 7:30 p.m. to 9:30 a.m., includes eight hours or more of darkness, during which, ordinarily, no photosynthesis, but only respiration, occurred. A greater number of hours of daylight would be included in this interval in June and July than in subsequent months. The second period, from 9:30 a.m. to 2:30 p.m., covers the best part of the day so far as average light is concerned, and contains five daylight hours throughout the season. In the third period, from 2:30 p.m. to 7:30 p.m., as in the first, the number of daylight hours is gradually reduced in each succeeding month. These differences in exposure to daylight should be kept in mind when a given interval is compared for the different months.

TABLE 4. Average Apparent Photosynthesis, Average Transpiration, and Average Meteorological Conditions Prevailing in Different Months during Three Periods Each Day

| Month | Photosynthesis (grams of $CO_2$) | Light (gram-calories per cm²) | Mean temperature (degrees centigrade) | $CO_2$ supply (mg of $CO_2$ per m³ of air) | Transpiration (grams $H_2O$) | Transpiration-assimilation ratio |
|---|---|---|---|---|---|---|
| 7:30 p.m. to 9:30 a.m. | | | | | | |
| June.............. | 32.0 | 91 | 17.8 | 630 | 2,591 | 81 |
| July.............. | 43.1 | 102 | 21.1 | 678 | 3,124 | 72 |
| August........... | 48.4 | 97 | 18.2 | 645 | 3,070 | 63 |
| September........ | 38.3 | 63 | 12.7 | 598 | 1,181 | 31 |
| October.......... | 22.9 | 57 | 9.2 | 587 | 722 | 32 |
| November........ | 5.7 | 18 | 6.7 | 574 | .... | .. |
| Average........ | 31.7 | 71 | 14.3 | 619 | 2,138 | 56 |
| 9:30 a.m. to 2:30 p.m. | | | | | | |
| June.............. | 76.1 | 232 | 25.8 | 561 | 7,828 | 103 |
| July.............. | 74.2 | 299 | 28.2 | 560 | 12,915 | 174 |
| August........... | 70.5 | 254 | 25.7 | 577 | 9,624 | 137 |
| September........ | 73.3 | 174 | 19.3 | 566 | 6,298 | 86 |
| October.......... | 46.4 | 163 | 16.5 | 574 | 3,941 | 85 |
| November........ | 14.8 | 74 | 11.0 | 564 | .... | ... |
| Average........ | 59.2 | 199 | 21.1 | 567 | 8,121 | 117 |
| 2:30 p.m. to 7:30 p.m. | | | | | | |
| June.............. | 51.2 | 100 | 23.5 | 557 | 4,699 | 92 |
| July.............. | 51.5 | 123 | 27.7 | 578 | 8,027 | 156 |
| August........... | 40.8 | 98 | 24.6 | 570 | 5,251 | 129 |
| September........ | 35.5 | 48 | 18.6 | 567 | 3,678 | 104 |
| October.......... | 24.7 | 25 | 14.3 | 578 | 2,410 | 98 |
| November........ | 9.0 | 12 | 9.1 | 564 | .... | ... |
| Average........ | 35.5 | 68 | 19.6 | 569 | 4,813 | 116 |

Considering the season as a whole, the tree assimilated about 47 per cent of its daily average between 9:30 a.m. and 2:30 p.m., the five hours of greatest light intensity. In July and August, this midday interval accounts for about 44 per cent of the total for the month; while in September, October, and November, 50 per cent or slightly more of the day's apparent photosynthesis takes place during this time. The fluctuation of the leaf activity of the tree during the midday interval is shown in figure 8.

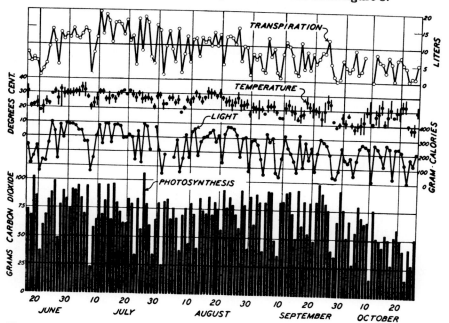

FIGURE 8. FLUCTUATIONS IN APPARENT PHOTOSYNTHESIS, LIGHT, TEMPERATURE, AND TRANSPIRATION DURING THE INTERVAL FROM 9:30 A.M. TO 2:30 P.M. EACH DAY FROM JUNE 17 TO OCTOBER 26

As can be readily seen from figure 5 (page 17), the apparent photosynthesis during the midday interval (9:30 a.m. to 2:30 p.m.) remains at approximately the same high average level from June through September, after which there is a rapid decline. This decline, as previously noted, is due in part to the reduced efficiency of the leaves because of loss of chlorophyll, and in part to the heavy drop of the leaves in the latter part of October.

The average assimilation of $CO_2$ during the night–morning period (7:30 p.m. to 9:30 a.m.) reaches a peak in August, while the average during the afternoon period (2:30 p.m. to 7:30 p.m.) is highest in the early summer months and declines steadily thereafter. The activity of the leaf surface of this tree was less from 7:30 p.m. to 9:30 a.m. than it was from 2:30 p.m. to 7:30 p.m. in all months except August and September. It must be remembered, however, that in the first period the tree had to make up the deficit caused by respiration during the night, before there would be an accumulation for apparent photosynthesis. A study of the detailed data for intervals from 7:30 p.m. to midnight or to sunrise, indicates that some assimilation undoubtedly occurred after 7:30 p.m. There is also evidence suggesting that on clear nights during the period of full moon the apparent respiration was less than on totally dark nights.

Since, in the method used, the air-flow was usually as rapid during the night as during the day, the difference in $CO_2$ of the ingoing and of the outgoing air during the periods of darkness was very small. The determinations for respiration, therefore, are not so dependable as where larger differences exist. This should be kept in mind in evaluating the respiration determinations carried out during the early part of the season (table 12), when new leaf tissues were still being developed and when much of the shoot growth was still in a succulent condition. At the relatively low average mean temperature of 14° C. the respiration during the night amounted to 0.89 gram per hour for the entire tree top, and at the relatively high mean temperature of 18.9° C. the rate was 1.3 grams per hour. The highest mean temperature prevailing in the respiration determinations was only 20°. During July and August, with a higher night temperature, the respiration would no doubt average more; and in September, October, and November, it would be less because of the low temperature and the retarded growth activity.

To put the first interval (7:30 p.m. to 9:30 a.m.), involving a period of darkness, on a more nearly comparable basis with the afternoon interval, probably about 20 per cent should be added to the average value of the first interval in June, July, and August. Even with this credit, however, the activity during the early morning hours would still not greatly exceed that during the afternoon runs, except possibly in the month of August. The data seem to indicate that the tree as a whole was carrying on photosynthesis about as actively, on the average, in the afternoon as in the morning.

### Average light intensity and average photosynthesis during different intervals

It is clear, from the data in table 4, that the low and the high average rates of assimilation during the different intervals are associated, respectively, with low and with high average light intensity. The highest average rates, however, do not always coincide with the highest average light intensity. For example, in the midday period in July there was greater light intensity, on the average, than in the same period in September, but there was practically no difference in average photosynthesis. During the months of September, October, and November, the tree was slightly more active, relatively, at the lower light intensities in the afternoon as compared with the morning. The relationship between light and photosynthesis is discussed more fully in another section of this memoir.

### Average temperature, light, and photosynthesis during different intervals

For any one of the three intervals, there is a striking relationship between light intensity and mean temperature. The higher the average light intensity during any interval, the higher is the average mean temperature. On comparing one interval with another, however, it is seen that a given light intensity in the afternoon is accompanied by a higher average mean temperature than prevailed at the same light intensity during the morning or during the midday period.

Referring again to table 4, it is apparent from a study of the data for any one interval in June, July, August, or September that the temperature as well as the light plays a part in determining the rate of apparent photosynthesis. For example, the average photosynthetic activity during the midday period in July was not so high as it was in June with a somewhat lower light intensity, but the average mean temperature in July was several degrees higher. On the other hand, the average assimilation during the midday period in September, when the mean temperature was relatively low, was practically the same as in July even though the light intensity was much lower. The higher temperatures would normally favor higher respiration rates and would tend to reduce the amount of apparent photosynthesis. With cooler weather there would be less utilization of carbohydrates by respiration during the period of determination, and hence the rate of assimilation would appear to be higher. There is also, of course, the possibility that the high temperature itself may interfere with the mechanism of photosynthesis, and thus directly reduce the rate of assimilation.

### Average transpiration and photosynthesis during different intervals

A further consideration of the data in table 4 indicates that other factors in addition to light and temperature are evidently operating to establish the rate of apparent photosynthesis. For example, the average apparent photosynthesis during the midday period in June was higher than in August in spite of the fact that the light intensity averaged somewhat greater in the latter month, and even though the mean temperatures were about the same in the two cases. The average transpiration rate in August, however, was about 23 per cent greater than in June, and about one-third more water was lost per gram of $CO_2$ assimilated. At a given temperature there is usually a greater loss of water with higher light intensity. The greater the transpiration–assimilation ratio, other conditions being approximately the same, the lower is the rate of apparent photosynthesis.

The greatest amount of water was lost by the tree in July, during each of the different intervals. More than half of the water transpired by the tree was lost during the midday interval (9:30 a.m. to 2:30 p.m.)—the period when the average light and temperature were at the maximum. In general, more than twice as much water was lost by the tree during the afternoon as during the night and morning. The temperature seemed to be the most important factor in accounting for this difference in transpiration between these two intervals, although the higher relative humidity in the morning no doubt also reduced water loss. This tree was especially economical in the water utilization in the night–morning intervals of September and October.

The total transpiration during July amounted to 746 liters, or approximately 200 gallons. This is equivalent to about 3.5 inches of rain on an area 10 by 10 feet in size, the space allotted to each tree in the orchard in which the tree under experiment grew.

Even though the stomates showed a tendency to close in the afternoon, they did not bring about a reduction in the loss of water from the leaves below that recorded in the morning. It is possible, however, that the closing of the stomates did in fact tend to retard the rate of transpiration, but the effect of this influence was overcome by the higher temperature and lower humidity.

Since the tree manufactures, on the average, about as much food during the morning hours as it does in the afternoon hours, it is obvious that less water is consumed for each gram of $CO_2$ assimilated in the morning as compared with the afternoon. During the midday interval in July, the water lost per gram of $CO_2$ assimilated was greater than it was during the same period in any other month. In September and October the transpiration–assimilation ratio in the afternoon period was greater than that in the midday period.

### Carbon dioxide and photosynthesis during different intervals

As is shown by table 4, the average $CO_2$ content of the air is much greater during the night–morning interval than in the remaining part of the twenty-four hours. The values given for the period from 7:30 p.m. to 9:30 a.m. would be even higher if only the hours of darkness were considered. Even so, in some cases the average $CO_2$ content of a cubic meter of air sampled during a 14-hour period reached the high point of 865 mg per cubic meter (table 14, page 39). It may be of interest to note in this connection that an unusually high $CO_2$ content persisted for several days preceding the heaviest rain (July 7 and 8, 1935) that has ever been recorded in the vicinity of Ithaca.

As a rule, the $CO_2$ content during the night was highest on the warm, cloudy nights, and lowest on the cool, clear, and windy nights. On the average, the $CO_2$ content during the night reached a peak in July, after which it gradually dropped until November. During the midday and afternoon periods there was relatively little difference in the $CO_2$ content for different months. However, there seems to have been a peak in October corresponding with the reduction in the average rate of photosynthesis and a rapid decay of vegetation. The difference in $CO_2$ content as between day and night was much less marked at the end of the season than during the summer months.

The influence of the average $CO_2$ supply on apparent photosynthesis during the different intervals is not easy to establish, since most of the higher concentrations occurred when the light was likely to be limiting. In a few individual cases—as, for example, during the midday interval on July 26, when the light was good and there happened at the same time to be a high $CO_2$ content in the air—the amount of apparent photosynthesis was high also.

### AVERAGE RATE OF PHOTOSYNTHESIS AT DIFFERENT LEVELS OF LIGHT INTENSITY

In order to show more clearly the general relationship between light intensity and photosynthesis, the average rates of assimilation obtained at the different levels of solar radiation are grouped together in table 5. The data are presented graphically in figure 9. It is evident that, on the whole, the amount of apparent photosynthesis increases with an increase in light intensity. It may be noted that, with the same light intensity prevailing, the photosynthetic activity in the afternoon is as good as, or, on the average, slightly better than, the activity in the middle of the day. Of course the light from 2:30 p.m. to 7:30 p.m. does not exceed 200 gram-calories per interval. The interval from 7:30 p.m. to 9:30 a.m. was not included in the table and the figure, since the respiration values during the night periods were not determined separately in all cases.

TABLE 5.  APPARENT PHOTOSYNTHESIS OF THE ENTIRE TREE AT DIFFERENT LEVELS
OF LIGHT INTENSITY, JUNE TO OCTOBER 1935

| Light intensity (gram-calories per cm² per interval) | Assimilation (grams of CO₂) | |
|---|---|---|
| | 9:30 a.m. to 2:30 p.m. | 2:30 p.m. to 7:30 p.m. |
| 10–49. | | |
| 50–99. | 18.1 ±1.83 | 16.7 ±3.66 |
| 100–149. | 37.0 ±3.19 | 37.1 ±1.44 |
| 150–199. | 41.2 ±3.86 | 52.2 ±1.54 |
| 200–249. | 56.6 ±2.90 | 63.2 ±1.77 |
| 250–299. | 70.9 ±2.16 | ......... |
| 300–349. | 78.0 ±1.71 | ......... |
| 350+. | 82.8 ±1.44 | ......... |
| | 85.2 ±1.61 | ......... |

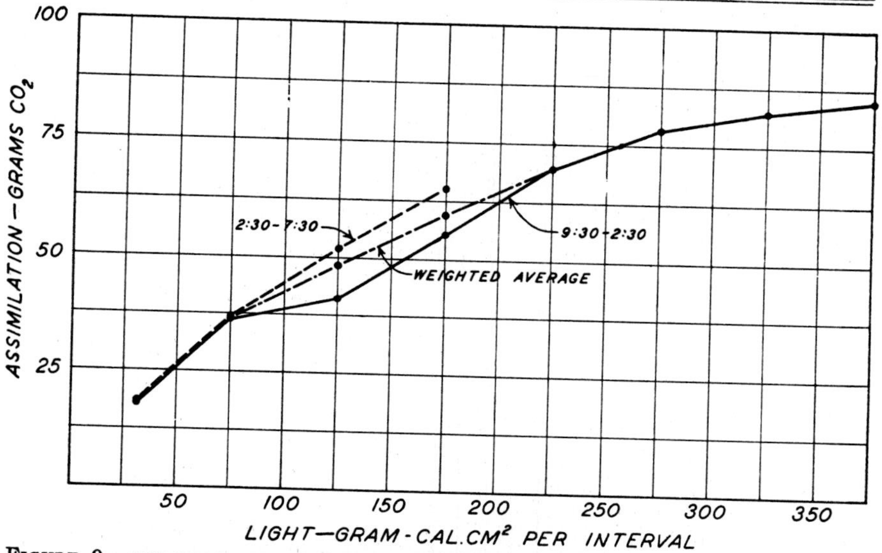

FIGURE 9.  RELATIONSHIP BETWEEN AVERAGE ASSIMILATION OF CO₂ AND AVERAGE
LIGHT INTENSITY FOR THE MIDDAY AND AFTERNOON INTERVALS

The distribution of all the individual values of apparent photosynthesis with respect to light intensity is shown in table 6.  The data for the table were obtained by comparing the results of each interval with the monthly average for that interval.  For example, the data on June 20 for the interval from 9:30 a.m. to 2:30 p.m. were compared with the average for that interval for the month of June.  This method tended to minimize somewhat the fluctuations due to seasonal variations in light and photosynthetic activity.

On the basis of these data, it appears that for about 3 out of 4 days on which the rate of apparent photosynthesis is 25 per cent or more above the average for that period, the light intensity will also be 25 per cent or more above the average; in 85 per cent of the cases, the light will be at least 10 per cent better than the average.  On the other hand, with the rate of photosynthesis 25 per cent or more below the average, the light will also be correspondingly low in about 70 per cent of the cases.  Low rates may occur, however, about 5 times in 100 even though the light is 25 per cent

TABLE 6. DISTRIBUTION OF LIGHT INTENSITY PREVAILING IN INDIVIDUAL CASES AT DIFFERENT LEVELS OF PHOTOSYNTHESIS*

| Relative level of photosynthesis | Number of cases | Per cent of cases with relative light intensities | | | | |
|---|---|---|---|---|---|---|
| | | Light above average | | Light within 10 per cent of average | Light below average | |
| | | 25 + per cent | 10–25 per cent | | 10–25 per cent | 25 + per cent |
| 25 + per cent above average...... | 126 | 72.2 | 13.5 | 1.6 | 5.6 | 7.1 |
| 10–25 per cent above average.... | 93 | 51.7 | 29.0 | 6.4 | 5.4 | 7.5 |
| Within 10 per cent of average.... | 40 | 35.0 | 20.0 | 7.5 | 10.0 | 27.5 |
| 10–25 per cent below average.... | 63 | 22.2 | 6.3 | 6.3 | 20.7 | 44.4 |
| 25 + per cent below average...... | 134 | 5.2 | 9.7 | 5.2 | 9.0 | 70.9 |

*The averages are calculated for each interval in each month, and every case is related to its appropriate average.

above the average, and high rates may occur about 7 times in 100 with the light 25 per cent below the average.

CONDITIONS PREVAILING AT DIFFERENT LEVELS OF PHOTOSYNTHESIS

The conditions in the midday interval are grouped in table 7 according to different levels of apparent photosynthesis. The average and also the range of conditions prevailing at these levels are given in the table.

TABLE 7. SOME CONDITIONS PREVAILING AT DIFFERENT LEVELS OF APPARENT PHOTOSYNTHESIS DURING THE MIDDAY INTERVAL, JUNE TO OCTOBER 1935

| Group | Photosynthesis (grams of $CO_2$ per interval) | | Light (gram-calories per cm²) | | Mean temperature (degrees C.) | | $CO_2$ supply (milligrams per m³ of air) | | Transpiration–assimilation ratio | |
|---|---|---|---|---|---|---|---|---|---|---|
| | Average | Range | Average | Range | Average | Range | Average | Range | Average | Range |
| 1 | 15.7 | 15.1–16.2 | 20 | 10–23 | 12.7 | 11.0–14.0 | 580 | 568–603 | 33 | 15–52 |
| 2 | 24.5 | 23.8–25.2 | 54 | 48–60 | 17.1 | 8.0–22.8 | 570 | 560–580 | 41 | 2–60 |
| 3 | 36.3 | 30.6–39.7 | 65 | 34–143 | 18.0 | 9.0–27.2 | 561 | 540–582 | 64 | 6–133 |
| 4 | 45.1 | 41.0–49.2 | 106 | 42–173 | 21.2 | 13.0–28.1 | 567 | 537–591 | 91 | 47–171 |
| 5 | 57.2 | 51.5–59.6 | 134 | 73–170 | 22.4 | 14.4–27.5 | 571 | 548–664 | 88 | 42–142 |
| 6 | 66.0 | 60.5–69.8 | 227 | 85–357 | 25.3 | 11.0–32.0 | 562 | 538–608 | 149 | 19–274 |
| 7 | 75.3 | 70.2–78.9 | 295 | 148–380 | 27.8 | 19.4–32.5 | 570 | 540–607 | 150 | 43–221 |
| 8 | 83.6 | 80.2–89.2 | 287 | 162–396 | 24.9 | 11.0–33.6 | 566 | 543–606 | 130 | 41–223 |
| 9 | 95.1 | 90.5–105.6 | 307 | 160–388 | 24.6 | 15.0–31.4 | 573 | 547–652 | 122 | 60–202 |

The data are presented graphically in figure 10. It is evident that the curves for temperature, light, and transpiration–assimilation ratio, are more or less parallel to the assimilation curve up to the level of 70–80 grams. At the higher levels of apparent photosynthesis the light curve flattens, and there is a drop in both the temperature and the transpiration–assimilation curve. The highest average rates of apparent photosynthesis are associated with high light intensities and with moderate mean temperatures and relatively low transpiration–assimilation ratios. At any given light intensity a relatively low transpiration rate seems to be associated with increased photosynthetic activity. There seems to be little relation, on the average, between the $CO_2$ content of the air during the midday interval and the rate of apparent photosynthesis.

In table 8 are given the average conditions prevailing at relatively high, medium, and low rates of apparent photosynthesis at each of several different levels of light intensity between 9:30 a.m. and 2:30 p.m. This

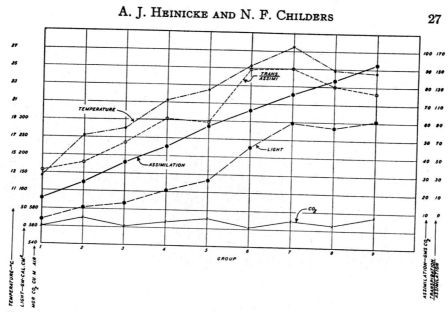

FIGURE 10.   AVERAGE CONDITIONS AFFECTING DIFFERENT LEVELS OF APPARENT
PHOTOSYNTHESIS
Each group represents about twenty determinations

grouping of the data provides opportunity for one to see how temperature,
transpiration, and the $CO_2$ supply might, on the average, influence the
apparent photosynthesis when the light is approximately constant.   It is
evident from these figures that the highest light group has the highest
average photosynthesis, although the differences between group 1 and
group 2 are only slight.

In each group, lot "a" has a higher rate of apparent photosynthesis

TABLE 8.   CONDITIONS ACCOMPANYING RELATIVELY HIGH, MEDIUM, AND LOW RATES
OF APPARENT PHOTOSYNTHESIS AT GIVEN LEVELS OF LIGHT INTENSITY BETWEEN
9:30 A.M. AND 2:30 P.M.

| Light (gram-calories per cm²) | Group | Photosynthesis (grams of CO₂) | Light (gram-calories per cm²) | CO₂ supply (milligrams per m³ of air) | Temperature (degrees C.) | Transpiration-assimilation ratio |
|---|---|---|---|---|---|---|
| 350–400 | 1a | 95.4 | 381 | 585 | 28.5 | 166 |
|  | 1b | 81.8 | 377 | 568 | 26.8 | 178 |
|  | 1c | 72.8 | 367 | 593 | 29.8 | 190 |
| 300–349 | 2a | 93.9 | 324 | 569 | 25.1 | 119 |
|  | 2b | 82.8 | 325 | 572 | 26.9 | 145 |
|  | 2c | 71.6 | 316 | 566 | 28.9 | 180 |
| 250–299 | 3a | 89.9 | 275 | 561 | 23.5 | 109 |
|  | 3b | 78.9 | 270 | 565 | 25.5 | 136 |
|  | 3c | 64.0 | 273 | 570 | 23.6 | 186 |
| 200–249 | 4a | 89.2 | 225 | 579 | 20.3 | 89 |
|  | 4b | 76.9 | 234 | 563 | 24.9 | 115 |
|  | 4c | 54.2 | 232 | 576 | 22.1 | 115 |
| 150–199 | 5a | 84.3 | 171 | 556 | 19.7 | 75 |
|  | 5b | 67.9 | 165 | 556 | 19.8 | 80 |
|  | 5c | 51.7 | 159 | 564 | 20.9 | 99 |

than has lot "c"; but the temperature is slightly lower, and there is also a much smaller amount of water lost per gram of $CO_2$ assimilated, in lot "a" than in lot "c". On comparing lots "a" in groups 1, 2, 3, and 4, in which the values for average photosynthesis are approximately the same, a marked decrease in the temperature at the lower light levels is found, and an even more marked reduction in the transpiration–assimilation ratio. This last relationship holds also for lots "b" in groups 1, 2, 3, and 4.

In general, then, it seems that the rate of apparent photosynthesis at a given light intensity may be expected to be highest when the temperature is lowest, and, conversely, it will be relatively low at a given light intensity if the temperature is high. Furthermore, where approximately the same average amount of apparent photosynthesis is found at two different light levels with the same average mean temperature—as, for example, 1b and 2b—there is less water lost at the lower light level.

The relatively high $CO_2$ concentration is found mainly at the higher light levels, probably because the accompanying temperature is favorable for higher $CO_2$ production owing to increases in respiration and decomposition processes. There is no marked relationship, on the average, between the different rates of apparent photosynthesis and the $CO_2$ content of the air at any given level of light intensity. It may be, however, that the relatively high $CO_2$ content of the air in lot 1c had some influence in counteracting the very high temperature and the very high transpiration–assimilation ratio.

In table 9 the cases falling within a given level of photosynthesis are separated into two groups—those having relatively high and those having relatively low light intensities. This arrangement of the data again indicates that high temperature and high transpiration–assimilation ratio tend to retard the rate of apparent photosynthesis. While the level of photosynthesis remains practically the same within each numbered group, the "a" lots in each case have better light than have the "b" lots. In all

TABLE 9. TEMPERATURE AND OTHER CONDITIONS ACCOMPANYING A GIVEN LEVEL OF APPARENT PHOTOSYNTHESIS AT RELATIVELY HIGH AND AT RELATIVELY LOW LIGHT INTENSITIES BETWEEN 9:30 A.M. AND 2:30 P.M.

| Group* | Photosynthesis (grams of $CO_2$) | | Light (gram-calories per cm²) | | Mean temperature (degrees C.) | | $CO_2$ supply (milligrams per m³ of air) | | Transpiration–assimilation ratio | |
|---|---|---|---|---|---|---|---|---|---|---|
| | Average | Range | Average | Range | Average | Range | Average | Range | Average | Range |
| 1a | 96.1 | 91.3–105.6 | 359 | 324–388 | 27.4 | 17.2–31.4 | 577 | 552–652 | 149 | 110–202 |
| 1b | 94.1 | 90.5–103.2 | 260 | 160–304 | 22.0 | 15.0–29.4 | 568 | 547–593 | 97 | 60–143 |
| 2a | 83.9 | 80.2–89.2 | 353 | 312–396 | 27.9 | 22.2–33.6 | 573 | 551–606 | 161 | 117–223 |
| 2b | 83.4 | 80.6–89.0 | 233 | 162–280 | 22.3 | 11.0–28.6 | 559 | 543–580 | 102 | 41–174 |
| 3a | 75.0 | 70.2–79.4 | 349 | 308–380 | 29.1 | 25.0–32.5 | 574 | 540–607 | 181 | 132–221 |
| 3b | 75.7 | 73.1–78.9 | 234 | 148–304 | 25.8 | 19.4–30.5 | 564 | 551–590 | 102 | 43–160 |
| 4a | 66.4 | 62.4–69.8 | 294 | 248–357 | 29.0 | 25.6–32.0 | 569 | 538–608 | 200 | 134–274 |
| 4b | 65.5 | 60.5–69.3 | 140 | 85–212 | 20.1 | 11.0–25.6 | 552 | 545–577 | 77 | 19–109 |
| 5a | 87.3 | 71.8–97.9 | 378 | 368–388 | 28.7 | 25.6–31.1 | 584 | 551–592 | 185 | 130–223 |
| 5b | 89.8 | 82.0–94.7 | 285 | 241–331 | 21.4 | 15.0–25.8 | 568 | 558–578 | 108 | 67–149 |
| 6a | 42.0 | 23.8–58.3 | 115 | 60–152 | 23.6 | 20.6–27.2 | 602 | 560–664 | 119 | 60–142 |
| 6b | 49.9 | 30.6–73.1 | 76 | 34–148 | 15.7 | 9.0–20.6 | 568 | 538–590 | 49 | 6–111 |

*Groups 1 to 4 inclusive—All days between June 17 and October 10. Lots 5a and 5b—10 days with best light in July and September, respectively. Lots 6a and 6b—10 days with less light than 150 gram-calories per cm² in July and September, respectively.

groups, however, the mean temperatures and the transpiration–assimilation ratios are considerably higher in the first lot than in the second. Group 4 is especially interesting, since the marked difference in light is accompanied by a great difference in temperature and in transpiration-assimilation ratio. Lots 1a, 2a, and 3a have about the same average light intensity, but the level of photosynthesis decreases as the temperature and, especially, the transpiration–assimilation ratio increase. Lot 4a, with a relatively high average light intensity, shows a lower level of photosynthesis than do lots 2b, 3b, and 5b, all of which have less light, on the average, but also have much lower temperatures and lower transpiration–assimilation ratios.

Even though the light during the 10 brightest days in July averages considerably greater than the average light intensity for the 10 best days in September, there is practically no difference in the amount of photosynthesis, as is shown by the data in group 5. But the average temperature and the transpiration–assimilation ratio are much lower in September than in July. On cloudy days in September the tree has accumulated more $CO_2$, with less light, than on cloudy days in July, but at a lower temperature and with less transpiration (group 6).

In no group has the higher $CO_2$ supply in the "a" lots offset the effect of the higher temperature and the higher relative transpiration. While the influence of variations in $CO_2$ concentration which occur under natural conditions is not sufficiently pronounced to show in the average values, this factor nevertheless plays an obvious part in establishing the values of individual cases. This is indicated by the data in table 10. When the level of photosynthesis is 25 per cent above the average, the $CO_2$

TABLE 10.   Distribution of $CO_2$ Concentration in Individual Cases at Different Levels of Photosynthesis*

| Relative level of photosynthesis | Number of cases | Per cent of cases with relative $CO_2$ concentration | | | | |
| --- | --- | --- | --- | --- | --- | --- |
| | | $CO_2$ above average | | $CO_2$ within 5 per cent of average | $CO_2$ below average | |
| | | 10 per cent | 5–10 per cent | | 5–10 per cent | 10 per cent |
| 25 + per cent above average....... | 126 | 18.3 | 25.4 | 29.4 | 22.2 | 4.8 |
| 10–25 per cent above average...... | 93 | 7.5 | 31.2 | 24.7 | 30.1 | 6.4 |
| Within 10 per cent of average...... | 40 | 10.0 | 22.5 | 20.0 | 37.5 | 10.0 |
| 10–25 per cent below average....... | 63 | 14.3 | 19.0 | 15.9 | 34.9 | 15.9 |
| 25 + per cent below average....... | 134 | 8.2 | 14.9 | 21.6 | 36.6 | 18.7 |

*The averages are calculated for each interval in each month, and every case is related to its appropriate average.

concentration is likely to be better than the average in about 45 per cent of the cases, and below the average in about 27 per cent. At the other extreme, if the level of apparent photosynthesis is 25 per cent below the average, the $CO_2$ concentration is likely to be below the average in about 55 per cent of the cases, and above the average in about 23 per cent of the cases. When these data are considered in connection with those presented in table 6 (page 26), it is clear that variations in light have a much greater influence than have variations in $CO_2$ in determining the rate of apparent photosynthesis under natural conditions.

While the results in general clearly indicate a relationship between apparent photosynthesis and the factors of light, transpiration, tempera-

than has lot "c"; but the temperature is slightly lower, and there is also a much smaller amount of water lost per gram of $CO_2$ assimilated, in lot "a" than in lot "c". On comparing lots "a" in groups 1, 2, 3, and 4, in which the values for average photosynthesis are approximately the same, a marked decrease in the temperature at the lower light levels is found, and an even more marked reduction in the transpiration–assimilation ratio. This last relationship holds also for lots "b" in groups 1, 2, 3, and 4.

In general, then, it seems that the rate of apparent photosynthesis at a given light intensity may be expected to be highest when the temperature is lowest, and, conversely, it will be relatively low at a given light intensity if the temperature is high. Furthermore, where approximately the same average amount of apparent photosynthesis is found at two different light levels with the same average mean temperature—as, for example, 1b and 2b—there is less water lost at the lower light level.

The relatively high $CO_2$ concentration is found mainly at the higher light levels, probably because the accompanying temperature is favorable for higher $CO_2$ production owing to increases in respiration and decomposition processes. There is no marked relationship, on the average, between the different rates of apparent photosynthesis and the $CO_2$ content of the air at any given level of light intensity. It may be, however, that the relatively high $CO_2$ content of the air in lot 1c had some influence in counteracting the very high temperature and the very high transpiration–assimilation ratio.

In table 9 the cases falling within a given level of photosynthesis are separated into two groups—those having relatively high and those having relatively low light intensities. This arrangement of the data again indicates that high temperature and high transpiration–assimilation ratio tend to retard the rate of apparent photosynthesis. While the level of photosynthesis remains practically the same within each numbered group, the "a" lots in each case have better light than have the "b" lots. In all

TABLE 9.   TEMPERATURE AND OTHER CONDITIONS ACCOMPANYING A GIVEN LEVEL OF APPARENT PHOTOSYNTHESIS AT RELATIVELY HIGH AND AT RELATIVELY LOW LIGHT INTENSITIES BETWEEN 9:30 A.M. AND 2:30 P.M.

| Group* | Photosynthesis (grams of $CO_2$) | | Light (gram-calories per cm²) | | Mean temperature (degrees C.) | | $CO_2$ supply (milligrams per m³ of air) | | Transpiration-assimilation ratio | |
|---|---|---|---|---|---|---|---|---|---|---|
| | Average | Range | Average | Range | Average | Range | Average | Range | Average | Range |
| 1a | 96.1 | 91.3–105.6 | 359 | 324–388 | 27.4 | 17.2–31.4 | 577 | 552–652 | 149 | 110–202 |
| 1b | 94.1 | 90.5–103.2 | 260 | 160–304 | 22.0 | 15.0–29.4 | 568 | 547–593 | 97 | 60–143 |
| 2a | 83.9 | 80.2–89.2 | 353 | 312–396 | 27.9 | 22.2–33.6 | 573 | 551–606 | 161 | 117–223 |
| 2b | 83.4 | 80.6–89.0 | 233 | 162–280 | 22.3 | 11.0–28.6 | 559 | 543–580 | 102 | 41–174 |
| 3a | 75.0 | 70.2–79.4 | 349 | 308–380 | 29.1 | 25.0–32.5 | 574 | 540–607 | 181 | 132–221 |
| 3b | 75.7 | 73.1–78.9 | 234 | 148–304 | 25.8 | 19.4–30.5 | 564 | 551–590 | 102 | 43–160 |
| 4a | 66.4 | 62.4–69.8 | 294 | 248–357 | 29.0 | 25.6–32.0 | 569 | 538–608 | 200 | 134–274 |
| 4b | 65.5 | 60.5–69.3 | 140 | 85–212 | 20.1 | 11.0–25.6 | 552 | 545–577 | 77 | 19–109 |
| 5a | 87.3 | 71.8–97.9 | 378 | 368–388 | 28.7 | 25.6–31.1 | 584 | 551–592 | 185 | 130–223 |
| 5b | 89.8 | 82.0–94.7 | 285 | 241–331 | 21.4 | 15.0–25.8 | 568 | 558–578 | 108 | 67–149 |
| 6a | 42.0 | 23.8–58.3 | 115 | 60–152 | 23.6 | 20.6–27.2 | 602 | 560–664 | 119 | 60–142 |
| 6b | 49.9 | 30.6–73.1 | 76 | 34–148 | 15.7 | 9.0–20.6 | 568 | 538–590 | 49 | 6–111 |

*Groups 1 to 4 inclusive—All days between June 17 and October 10. Lots 5a and 5b—10 days with best light in July and September, respectively. Lots 6a and 6b—10 days with less light than 150 gram-calories per cm² in July and September, respectively.

groups, however, the mean temperatures and the transpiration–assimilation ratios are considerably higher in the first lot than in the second. Group 4 is especially interesting, since the marked difference in light is accompanied by a great difference in temperature and in transpiration-assimilation ratio.    Lots 1a, 2a, and 3a have about the same average light intensity, but the level of photosynthesis decreases as the temperature and, especially, the transpiration–assimilation ratio increase.    Lot 4a, with a relatively high average light intensity, shows a lower level of photosynthesis than do lots 2b, 3b, and 5b, all of which have less light, on the average, but also have much lower temperatures and lower transpiration–assimilation ratios.

Even though the light during the 10 brightest days in July averages considerably greater than the average light intensity for the 10 best days in September, there is practically no difference in the amount of photosynthesis, as is shown by the data in group 5.    But the average temperature and the transpiration–assimilation ratio are much lower in September than in July.    On cloudy days in September the tree has accumulated more $CO_2$, with less light, than on cloudy days in July, but at a lower temperature and with less transpiration (group 6).

In no group has the higher $CO_2$ supply in the "a" lots offset the effect of the higher temperature and the higher relative transpiration.    While the influence of variations in $CO_2$ concentration which occur under natural conditions is not sufficiently pronounced to show in the average values, this factor nevertheless plays an obvious part in establishing the values of individual cases.    This is indicated by the data in table 10.    When the level of photosynthesis is 25 per cent above the average, the $CO_2$

TABLE 10.    Distribution of $CO_2$ Concentration in Individual Cases at Different Levels of Photosynthesis*

| Relative level of photosynthesis | Number of cases | Per cent of cases with relative $CO_2$ concentration | | | | |
| | | $CO_2$ above average | | $CO_2$ within 5 per cent of average | $CO_2$ below average | |
| | | 10 per cent | 5–10 per cent | | 5–10 per cent | 10 per cent |
|---|---|---|---|---|---|---|
| 25 + per cent above average....... | 126 | 18.3 | 25.4 | 29.4 | 22.2 | 4.8 |
| 10–25 per cent above average...... | 93 | 7.5 | 31.2 | 24.7 | 30.1 | 6.4 |
| Within 10 per cent of average..... | 40 | 10.0 | 22.5 | 20.0 | 37.5 | 10.0 |
| 10–25 per cent below average...... | 63 | 14.3 | 19.0 | 15.9 | 34.9 | 15.9 |
| 25 + per cent below average....... | 134 | 8.2 | 14.9 | 21.6 | 36.6 | 18.7 |

*The averages are calculated for each interval in each month, and every case is related to its appropriate average.

concentration is likely to be better than the average in about 45 per cent of the cases, and below the average in about 27 per cent.    At the other extreme, if the level of apparent photosynthesis is 25 per cent below the average, the $CO_2$ concentration is likely to be below the average in about 55 per cent of the cases, and above the average in about 23 per cent of the cases.    When these data are considered in connection with those presented in table 6 (page 26), it is clear that variations in light have a much greater influence than have variations in $CO_2$ in determining the rate of apparent photosynthesis under natural conditions.

While the results in general clearly indicate a relationship between apparent photosynthesis and the factors of light, transpiration, temperature, and, to some extent, $CO_2$, there are special cases recorded in tables 12 to 18 which seem not to conform to the average behavior.   Many of these cases might be explained more satisfactorily if more were known about the internal conditions of the plant tissue, as well as the external factors, which prevailed during the determinations.

### THE UTILIZATION OF CARBOHYDRATES

In order to determine whether the methods used in measuring the total photosynthetic activity of the entire tree were reasonably accurate, an attempt was made to estimate the dry weight of all the tissues formed by the tree during 1935.   The data are presented in table 11.

TABLE 11.   ACCOUNTING OF UTILIZATION OF $CO_2$ ASSIMILATED IN 1935 BY AN EIGHT-YEAR-OLD MCINTOSH APPLE TREE

| Item no. | Description | Dry weight (grams) | |
|---|---|---|---|
| 1 | Flowers | 80 | |
| 2 | Fruits | 656 | |
| 3 | Entire leaf surface | 2,612 | |
| 4 | 1935 shoots | 1,540 | |
| 5 | 1935 tissue on older wood | 3,933 | |
| 6 | New root growth | 2,737 | |
| 7 | Total of items 1 to 6 | 11,558 | |
| 8 | Deducting 5 per cent of item 7 for nitrogen and ash | 578 | |
| 9 | Balance | 10,980 | |
| 10 | Respiration of roots | 3,011 | |
| 11 | Total of above items less nitrogen and ash | 13,991 | |
| 12 | Excess ( = "increased reserve") | 592 | |
| 13 | Total carbohydrates accounted for | | 14,583 |
| 14 | Total $CO_2$ assimilation | 22,436 | |
| 15 | Carbohydrates (65 per cent of $CO_2$) | | 14,583 |

The flowers and partially developed fruits were collected and dried as soon as they turned yellow and showed signs of separating from the tree. The dry weight of mature fruits was determined by estimating an aliquot sample of the fresh tissue.   All the leaves that fell during weekly intervals beginning in October (table 2, page 14) were oven-dried.   After the determination of apparent photosynthesis was concluded for the season on November 17, the tree was cut down.   All 1935 shoot growth was removed from the tree and dried in a ventilated oven at 100° C.

The older part of the tree was cut into sections varying in length from 30 to 100 cm and corresponding to a year's growth or to a part of the larger limbs free from lateral branches.   The length of each section and the diameter of the xylem at the end of 1934 and at the end of 1935 were recorded.   From these data the volume of the woody cylinder in each section at the end of the two years was easily calculated.   The volume of wood formed during 1935 was then determined by the difference.   The total volume of all sections obtained in this manner amounted to 6242.8 cc.

The average weight of 1 cc of dry xylem formed by the tree in 1935 on branches older than one year, was obtained by dividing the difference in dry weight by the difference in volume of one- and two-year-old twigs or of two- and three-year-old twigs which made approximately the same length growth and which grew on the same branch.   The bark was removed from the samples before drying.   The average figure thus obtained from a number of cases was 0.63. The figure for the dry weight of the 1935 xylem of twigs and branches older than one year, namely, 3933 grams, was obtained by multiplying 6242.8 by 0.63.

The dry weight of new root growth was estimated as one-half of the total dry weight of the 1935 shoots plus the 1935 tissue on older parts of the top of the tree. This proportion of roots to top was chosen because this is approximately the average figure obtained from a study of the top-root ratio of 900 young McIntosh trees growing in a near-by field (Heinicke, 1921).

The value for respiration of the roots was based on a determination of $CO_2$ production by the root system of several three-year-old apple trees which had been growing in the greenhouse in 3-gallon containers. After being carefully washed for removal of all dirt, the entire root system from each of several trees was placed in a large moist respiration chamber. A stream of fresh air was drawn slowly through the chamber for a three-day period at a temperature of about 20° C. All the air which passed through the inclosure during a five-hour interval was analyzed for $CO_2$ by the method primarily used for photosynthesis for single leaves (Heinicke, 1933). The average $CO_2$ content in a continuous stream of fresh air was determined simultaneously. The increase in $CO_2$ content of the air as it passed through the chamber, was ascribed to respiration of the roots. After the determinations were ended, the small rootlets and the new tissue on older roots were removed and oven-dried. In calculating the results, it was assumed that the older parts of the root system which are completely surrounded by young tissues contribute relatively little to the total respiration. The average value for respiration of the root tissue was found to be 0.25 milligram of $CO_2$ per hour, or 6 milligrams per day, per gram of dry weight. If it is assumed that this rate will hold for the entire root system of the tree grown in the orchard, the total amount of respiration for the 188 days during which assimilation was determined would amount to 1.1 grams per gram of dry matter. If it is assumed further that the weight of the actively respiring root tissue was equivalent to the weight of the new dry matter of the roots, then the value for total respiration of the roots is $2737 \times 1.1$, or 3011 grams of $CO_2$.

There is no separate entry in table 11 for the respiration of the top, since this item is provided for in the determinations for apparent photosynthesis. It is estimated that the average respiration of the top amounted to about 12 grams during the night and about 24 grams during the day, or 36 grams for a 24-hour interval.

The total of the items listed in table 11 does not equal the amount of carbohydrates produced, if it is assumed that about 65 per cent of the $CO_2$ assimilated was converted into carbohydrates. In order to balance the account, the difference, amounting to about 4 per cent of the total, is itemized as "increased reserve". This material would be stored in the 1934 tissues and in older tissues of the tree, in excess of the amount found in the tissues in the spring of 1935. If the value for respiration of the roots or that of the conversion factor for $CO_2$ is changed, corresponding changes must be made in the reserve item. On the whole, the figures indicate that the methods used in determining the apparent photosynthesis were reasonably accurate.

## SUMMARY

The continuous record of photosynthesis of a young apple tree throughout the season of 1935 indicates that there are marked fluctuations in the rate of food manufactured from day to day. The factor that most frequently limited photosynthesis of the full leaf surface of the tree, under conditions prevailing in the Cornell University orchard at Ithaca, New York, was light. In general, the greater the total amount of sunlight during a short or a long interval, the more food did the tree accumulate. While individual leaves with the best exposure to light might function at full capacity with as little as one-third to one-fourth of full sunlight (Heinicke and Hoffman, 1933), a large part of the leaf surface of a tree is normally in the shade for the greater part of the day.[2] Such shaded foliage evidently cannot function at its best except during the midday hours on the very bright days.

The foliage that develops within a month after the buds have opened is capable of relatively high rates of apparent photosynthesis early in the season. The older basal leaves, however, soon have to compete for light, water, and nutrients with the younger leaves that develop on the terminal parts of the shoots. These younger leaves undoubtedly account for most of the food manufactured by the tree during the greater part of the season.

Leaves which are well exposed to light, when abundantly supplied with water and soil nutrients, are evidently capable of carrying on photosynthesis over a wide range of temperatures. Such leaves continue to assimilate $CO_2$ throughout the warmest days of the season, and also on days when the temperature is just above freezing even though they may have been frozen stiff previously. They continue to function as long as they are alive.

The most pronounced influence of temperature on apparent photosynthesis is probably its effect on the rate of respiration and transpiration. The rate of respiration is evidently influenced more by temperature than is the rate of photosynthesis. The varying amounts of carbohydrates required for respiratory processes at the different temperatures undoubtedly affect the amount of apparent photosynthesis. The higher temperatures which frequently accompany the highest light intensities favor higher rates of respiration, and thus tend to counteract the influences of good light. When the intensity of the sun is below the average for any interval during a given month, the temperature also is likely to be below average, and a smaller amount of the $CO_2$ assimilated would be required for respiration.

In general, the factors that influence photosynthesis influence transpiration also; but if the transpiration is higher than normal for a given temperature and light intensity, apparent photosynthesis is likely to be lower than might otherwise be expected.

The closing of the stomates, which normally occurs in the afternoon, evidently does not reduce the water loss below that found in the morning, whenever temperature and light are favorable. Certainly such stomatal behavior did not greatly reduce the rate of apparent photosynthesis of the entire tree in this study. It should be pointed out, however, that in no case were the leaves on this tree actually wilted.

[2] On July 26, 1936, at noon, the light intensity recorded in foot-candles as determined by a Weston illuminometer was as follows: light target toward sun 9200, horizontal 8100; shaded leaves 1 foot in from south side of tree 350, shaded leaves north side 320, leaves in interior of tree 8 feet above surface of ground 80, 4 feet above surface 40.

While variations in $CO_2$ concentration have some influence in determining the rate of apparent photosynthesis, other factors are more important under natural conditions.

In this paper the emphasis is placed on the daily rate of apparent photosynthesis and the external factors that influence the processes. Only a brief account is given of the final disposition of the products of the season's assimilation. The picture will not be complete until more is known about the daily utilization by the plant of the material manufactured every twenty-four hours during the season.

It seems obvious that the daily fluctuations of apparent photosynthesis must have an effect on the chemical balance of the tissues of a tree with respect to carbohydrate–nitrogen–water–mineral-nutrient ratios. If there are several cloudy days or several clear days in succession at critical periods in the seasonal development of the tree—as, for example, during the period when flower buds are initiated—the amount of carbohydrates available will vary considerably and it is probable that the behavior of the tree will be profoundly influenced. The response of the fruit plant to any soil treatments that may be given, may be determined very definitely by the number of successive days favorable and the number unfavorable for photosynthesis. Likewise, the rate of photosynthesis is an important factor in determining the susceptibility of a tree to the attacks of certain pests such as aphids or blight, which do more or less damage depending on the internal chemical balances of the tissues of the tree.

Just how rapidly growth and other metabolic processes adjust themselves to the internal chemical environment of a plant is not definitely known. In most cases the adjustment is probably so gradual that the responses are determined to a greater extent by the total amount of carbohydrates manufactured during intervals of a week or more, than by the daily fluctuations in apparent photosynthesis. However, without a specific knowledge of the daily fluctuations in apparent photosynthesis, one would be at a disadvantage in attempting to evaluate definitely the effects of all the environmental factors or treatments that influence the metabolism of a plant.

## REFERENCES

Arnold, August. Der Verlauf der Assimilation von *Helodea canadensis* unter konstanten Aussenbedingungen. Planta **13**:529–574. 1931.

Bauer, Philipp. Geben abgeschnittene Blätter physiologisch richtige Assimilationswerte? Planta **24**:446–453. 1935.

Blackman, F. F. Optima and limiting factors. Ann. bot. **19**:281–295. 1905.

Blackman, F. Frost, and Matthaei, Gabrielle L. C. A quantitative study of carbon-dioxide assimilation and leaf-temperature in natural illumination. Roy. Soc. London. Proc., ser. B, **76**:402–460. 1905.

Boysen-Jensen, P., and Müller, D. Die maximale Ausbeute und der tägliche Verlauf der Kohlensäureassimilation. Jahrb. wiss. Bot. **70**:493–502. 1929.

Gassner, G., and Goeze, G. Über den Einfluss der Kaliernährung auf die Assimilationsgrösse von Weizenblättern. Deut. Bot. Gesell. Ber. **50 A**:412–482. 1932.

HEINICKE, A. J.  Some relations between circumference and weight, and between root and top growth of young apple trees.  Amer. Soc. Hort. Sci.  Proc. **18**:222–227.  1921.

———————— A special air-chamber for studying photosynthesis under natural conditions.  Science **77**:516–517.  1933.

———————— Photosynthesis in apple leaves during late fall and its significance in annual bearing.  Amer. Soc. Hort. Sci.  Proc. **32** (1934): 77–80.  1935.

HEINICKE, A. J., AND HOFFMAN, M. B.  The rate of photosynthesis of apple leaves under natural conditions.  Part I.  Cornell Univ. Agr. Exp. Sta.  Bul. 577:1–32.  1933.

KIMBALL, HERBERT H.  Records of total solar radiation intensity and their relation to daylight intensity.  U. S. Agr. Dept., Weather Bur. Monthly weather rev. **52** (1924):473–479.  1925.

———————— Pyrheliometers and pyrheliometric measurements.  U. S. Agr. Dept., Weather Bur.  Circ. Q.  1931.

KOSTYTSCHEW, S., BAZYRINA, K., AND TSCHESNOKOV, W.  Untersuchungen über die Photosynthese der Laubblätter unter natürlichen Verhältnissen.  Planta **5**:696–724.  1928.

KREUSLER, U.  Ueber eine Methode zur Beobachtung der Assimilation und Athmung der Pflanzen und über einige diese Vorgänge beeinflussende Momente.  Landw. Jahrb. **14**:913–965.  1885.

McLEAN, F. T.  Field studies of the carbon-dioxide absorption of coconut leaves.  Ann. bot. **34**:367–389.  1920.

MARVIN, C. F.  Psychrometric tables, p. 1–87.  U. S. Agr. Dept., Weather Bur.  1915.

MILLER, EDWIN C.  The formation of carbohydrates by the green plant. *In* Plant physiology, p. 404–499.  1931.

SCHANDERL, HUGO.  Untersuchungen über die Photosynthese einiger Rebsorten, speziell des Rieslings, unter natürlichen Verhältnissen.  Wiss. Arch. Landw., Abt. A, **3**:529–560.  1930.

SCHODER, ANNEMARIE.  Über die Beziehungen des Tagesganges der Kohlensäureassimilation von Freilandpflanzen zu den Aussenfaktoren. Jahrb. wiss. Bot. **76**:441–484.  1932.

SPOEHR, H. A.  Photosynthesis, p. 1–393.  1926.

STILES, WALTER.  Photosynthesis, p. 1–268.  1925.

STOCKER, OTTO.  Assimilation und Atmung westjavanischer Tropenbäume.  Planta **24**:402–445.  1935.

TAGEEVA, SOPHIE.  Zur Frage des Zusammenhangs zwischen Assimilation und Ertragsfähigkeit.  Planta **17**:758–793.  1932.

THOMAS, MOYER D.  Precise automatic apparatus for continuous determination of carbon dioxide in air.  Indus. and engin. chem., Analyt. ed., **5**:193–198.  1933.

WILLSTÄTTER, RICHARD MARTIN, AND STOLL, ARTHUR.  Untersuchungen über die Assimilation der Kohlensäure, p. 1–448.  1918.

A. C. Bobb and M. A. Blake. 1938. Annual bearing in the Wealthy apple was induced by blossom thinning. Proceedings of the American Society for Horticultural Science 36:321–327.

A. C. Bobb

M. A. Blake

The cause and control of biennial bearing in fruit trees has been written about and investigated for a century or more. The literature is voluminous and to some extent contradictory. But one observation has been consistent, and is woven throughout the early writings—based on both conjecture and experimentation—that overcropping is a competitive and exhaustive process. This debilitates the tree and results in small fruit size and an alternating bearing habit. In the introduction to a U.S. Department of Agriculture (USDA) technical bulletin reporting on research during the decade of the 1930s on cause and control of biennial bearing, J. R. Magness wrote as follows:

> The first and most important [problem] is the tendency of many apple varieties to bear heavy crops one year and little or no fruit the year following. This alternate or biennial bearing tendency is recorded in the earliest horticultural writings, and the condition can be found today to a greater or less degree in practically every orchard.

The research reported in that bulletin (Harley et al. 1942) provided some important information on biennial bearing through detailed studies of leaf/fruit ratios and time of hand fruit thinning in relation to the normal "June drop," but did not elucidate the hoped-for control concept.

A. C. McCormick, working in the Pacific Northwest around 1930, recognized the serious biennial problem with 'Yellow Newtown' and 'Ortley' apples, which he described as follows:

> Enormous crops of comparatively small sized fruit are produced during the on-year but little or no fruit during the off-year. After producing these heavy crops the trees show evidence of near exhaustion. Many attempts have been made to modify this habit through some particular system of pruning or thinning and by the use of fertilizers. Such attempted corrections have proved of little or no avail (McCormick 1934).

McCormick took an ingenious approach to the problem. He decided to make a detailed study of any and all annual-bearing trees in the hope of finding a clue to the factors involved in annual and biennial bearing. For this purpose he selected the 'Anjou' pear because he noted that the trees blossomed heavily each year and set just the right amount of fruit to produce a full crop consistently year after year without loss of vitality. No thinning was necessary. The studies showed that 2 to 3 percent of the 'Anjou' blossoms matured fruit while another 6 to 8 percent of blossoms apparently set but dropped during the early stage of development within 12 to 18 days following bloom. So his experiment on 'Yellow Newtown' and 'Ortley' apples was to reproduce artificially the natural fruiting habit of the annual 'Anjou' pear. This necessitated hand removal of flowers on the on-year apple trees. Six weeks later, at the regular fruit thinning time, the flower thinned and the control trees were reduced to the same comparative number of fruits per tree—approximately 3 percent of the original blossoms. It is interesting to note that McCormick chose to call these blossom removal treatments as "pre-thinning" to emphasize the break with the traditional hand-fruit-thinning concept. The result of this pre-thinning in 1930, in addition to increasing fruit size and yield that year, was the formation of ample flower buds for the following year. In 1931, the check trees returned to the characteristic off-year condition, whereas the pre-thinned apple trees produced abundant bloom.

So the stage was set for A. C. Bobb and M. A. Blake to run their classic hand blossom thinning or pre-thinning concept through a time course of several years *on the same trees* to get the convincing, practical proof of converting the distinctly biennial 'Wealthy' apple into an annual bearing entity. Arthur Bobb had the right background and the right partner for the job, which was to involve long hours of tedious work. Bobb was born in Pennsylvania but spent his early life on a general farm in southern New Jersey where the family had apple and peach orchards and raised poultry and hogs. His first encounter with Rutgers University was attending the poultry short course in 1933—with the depression in full swing. Jobs were scarce or nonexistent when he completed the course, so he enrolled in the four-year program at Rutgers and came under the influence of M. A. Blake in horticulture. Although Blake is best known for his long and successful peach breeding program in New Jersey, he had been concerned, along with many other pomologists, about the lack of response of biennial apple cultivars such as 'Wealthy' to conventional hand thinning. During his early college days, Bobb was working with Blake in the university orchards on this thinning problem. On one occasion they recorded the end product of a conventional hand-thinning job on a mature biennial 'Wealthy' tree as follows: ". . . it was necessary to remove 4,575 fruits which filled 8 bushel baskets and weighed 350 pounds. This illustrates the tremendous waste of carbohydrate which may occur in the period following blossoming where the set of fruit is excessive and allowed to develop to the usual time of thinning." To add insult to injury, a tree so thinned will still be largely alternating in bearing habit, being unlikely to differentiate many flower buds during the period of fruit competition and depletion of carbohydrate between bloom and the traditional hand-thinning period.

So, taking the cue from McCormick, Bobb and Blake teamed up in the spring of 1935 and set about converting two 19-year-old 'Wealthy' apple trees to annual-bearing status with a four-year master plan. The task seemed enormous—a heavily blossoming 'Wealthy' apple tree in the "on" year had, according to their estimates 20,000 spurs that blossomed. In round numbers, the tree developed 100,000 flowers, excluding any terminal and axillary bloom on the shoots. An average 20-bushel crop requires only about 2000 flowers. Bobb did most of the hand flower removal himself, with Blake supervising the week-long procedure. New Jersey sunshine and favorable temperatures prevailed through the pollination and fruit-setting period, and each of the well-spaced flowers chosen to remain produced a fruit—to the extent that the "June drop" was largely bypassed and the ultimate fruit density was approximately 10 to 12 inches apart, similar to McCormick's 'Anjou' pear example. The intervening spurs, relieved of their debilitating fruit load, could in due time form flower primordia for the following year. A full crop of good-sized fruit was harvested from the flower-thinned trees. The following year (1936) the control trees were in their "off" cycle and failed to bloom, whereas the flower-thinned trees, with adequate repeat bloom needed some flower thinning again. With some minor adjustments this thinning sequence was continued through 1938. The yields of a blossom-thinned tree and the control tree on which the fruit was thinned at the conventional time after the June drop for the four-year period of the experiment (1935 through 1938) are shown graphically in Figure 1 of their paper. During the four-year period the blossom-thinned tree produced 76.2 bushels; the control tree, with two "off" years, produced only 41.3 bushels.

Subsequently, M. B. Hoffman (1947) accomplished this same feat in a commerical 'Wealthy' apple orchard in New York with chemical flower thinning using a dinitro compound in the four-year period 1943 through 1946. The yields for chemically thinned and control trees during this period are almost identical with the four-year 'Wealthy' yield data from Bobb and Blake for their hand-blossom-thinned trees (Edgerton 1973).

A few years after the Bobb and Blake paper was published, a report appeared in the *ASHS Proceedings* (1941) by C. L. Burkholder and M. McCown of Purdue University on the general subject of apple fruit set and thinning. Their trials may have seemed inconsequential at the time from the standpoint of early fruit thinning and biennial bearing. After all, they were apparently expecting to show that napththaleneacetic acid (NAA) would *increase* the set of apples when applied as a spray during the bloom or early postbloom period. NAA had just been shown to delay fruit abscission and control harvest drop of apples (Gardner et al. 1939). So might it not be reasonable to expect that applications on 'Delicious' flowers and young fruits could enhance set? Instead, they found a reduced fruit set on the NAA-treated units. Perhaps this was disappointing and trivial to the authors at the time, and it was several years before the full potential of NAA as a chemical thinner was demonstrated. Then its usefulness as a means of fruit thinning in the early postbloom period became recognized and it has been widely used along with other chemicals to correct the biennial bearing habit of apples and assure annual production.

The controversy as to the precise *cause* of biennial bearing continues. Basically, the problem is why the presence of developing reproductive units on a perennial plant such as the apple interferes with the simultaneous differentiation of flower buds for a crop the following year. Probably involved are nutritional and competitive factors; environmental factors are probably involved to the extent that carbohydrate supply is affected; hormonal conditions?—there is increasing evidence that they are involved, perhaps through the inhibitory effect of developing seeds in the pollinated and fertilized flowers (Chan and Cain 1967). But the *control* of biennial bearing was demonstrated by Bobb and Blake on a few 'Wealthy' apple trees in New Jersey in the late 1930s: establish the ultimate fruit distribution or spacing early in the bloom or postbloom period, then sit back and watch the apple boxes fill up at harvest year after year.

Louis J. Edgerton
Department of Pomology
Cornell University
Ithaca, New York 14853

## REFERENCES

BURKHOLDER, C. L., and M. McCOWN. 1941. Effect of scoring and of $\alpha$-napthal acetic acid and amide spray upon fruit set and of the spray upon pre-harvest fruit drop. Proc. Am. Soc. Hort. Sci. 38:117–120.

CHAN, B. C., and J. C. CAIN. 1967. The effect of seed formation on subsequent flowering in apple. Proc. Am. Soc. Hort. Sci. 91:63–68.

EDGERTON, L. J. 1973. Chemical thinning of flowers and fruits. p. 435–474. In: T. T. Kozlowski (ed.). Shedding of plant parts. Academic Press, New York.

GARDNER, F. E., P. C. MARTH, and L. P. BATJER. 1939. Spraying with plant growth substances to prevent apple fruit dropping. Science 90:208–209.

HARLEY, C. P., J. R. MAGNESS, M. P. MEASURE, L. A. FLETCHER, and E. S. DEGMAN. 1942. Investigations on the cause and control of biennial bearing of apple trees. U.S. Dep. Agr. Tech. Bull. 792.

HOFFMAN, M. B. 1947. Further experience with the chemical thinning of 'Wealthy' apples during bloom and its influence on annual production and fruit size. Proc. Am. Soc. Hort. Sci. 49:21–25.

McCORMICK, A. C. 1934. Control of biennial bearing in apples. Proc. Am. Soc. Hort. Sci. 30:326–329.

Reprinted from Proceedings of the
AMERICAN SOCIETY FOR HORTICULTURAL SCIENCE
Vol. 36, 1938

# Annual Bearing in the Wealthy Apple was Induced by Blossom Thinning

By A. C. BOBB and M. A. BLAKE, *Agricultural Experiment Station, New Brunswick, N. J.*

ANNUAL fruit production of commercial apple orchards is of marked economic importance to the fruit grower. Young bearing apple trees of many varieties often produce some fruit annually. However, there is a general tendency towards biennial bearing after the trees produce their first very heavy crop. This habit is quite pronounced with such varieties as Baldwin and Wealthy in the East. It is a well established fact that cultural operations such as fertilization, soil management and pruning have not proved to be consistent remedies for such a condition. In other words, climatic factors over a considerable area not infrequently tend to exert a dominating influence in determining the growth status of apple trees.

Excessive blooming and fruit production tend to result in biennial and irregular bearing. Formerly it was quite generally believed that this effect was due in a large measure to the development of a heavy crop of fruit. Well cared for trees in commercial orchards in the on-year, commonly set such large numbers of fruits that thinning is necessary not only to secure fruits of good commercial size and color, but also to prevent severe breakage to the trees.

McCormick (1) states that Yellow Newtown and Ortley apples from the blossoming to the thinning period in 1930 and 1932, carried a load of green apples 300 per cent greater than necessary to produce a full crop.

In thinning a 19-year-old Wealthy tree at New Brunswick having a height and spread of approximately 25 feet in 1935, it was necessary to remove 4,575 fruits which filled 8 bushel baskets and weighed 350 pounds. This illustrates the tremendous waste of carbohydrates which may occur in the period immediately following blooming where the set of fruit is excessive and allowed to develop to the usual time of thinning.

## EXCESSIVE BLOOMING INHIBITS SPRING GROWTH

The fact is now becoming established that even an excessive number of flowers markedly inhibits the leaf development and the total wood growth of the tree in early spring. Theis (2) found that defloration of the McIntosh apple increased the leaf area 41 to 54 per cent by May 25th, 1932. Chandler (3) states that, "the mere formation of many blooms greatly inhibits spring growth". The fact has also been observed many times that when frost destroys the bloom upon a tree in a normal on-year this same tree tends to bloom again the following year.

These facts suggest the question, to what extent does a vigorous bearing tree develop more flowers than are required for annual production? It is obvious that this will vary with the variety, the age and size of the tree and other factors. It raises the further question as to whether all of the flowers developed by a tree are equally dependable

321

and valuable for production or whether some selection is desirable. Varieties including Baldwin and Wealthy in the East, quite commonly develop spur, terminal and axillary flower buds. Wealthy sometimes produces its first fruits from terminal flower buds. When the trees attain full bearing condition and tend to become biennial, axillary and terminal buds add to the excess formed or set on spurs. It has further been observed that fruit production on the terminals of Baldwin and Wealthy tends to inhibit fruit bud formation on spurs. Speaking of Baldwin in New York, Chandler (4) states, "Sometimes the bloom and the crop will be mainly on the terminal of twigs; much the larger number of spurs being without bloom, yet there will be little or no bloom in the succeeding year". Similar observations have been made upon Baldwin at New Brunswick. Observations of Wealthy show that terminal fruit bearing tends to inhibit fruit bud formation for some distance back of the terminal upon the older wood.

## Studies at New Brunswick

The data later reported in this paper was obtained from 19-year-old Wealthy trees in good vigor, having a height and spread of approximately 25 feet and planted at a density of 50 trees per acre. The annual growth at the tips of vigorous branches was commonly 8 to 10 inches long and relatively thick in diameter; nevertheless, the trees were biennial in bearing habit. In the on-year with a commercial 6 to 8 inch thinning they commonly produced 20 to 25 bushels of apples mostly 2¾ inches and above in diameter. Wealthy trees of this character at New Brunswick differentiate fruit buds profusely upon spurs, terminals and axillaries in the off-year. One such tree in 1935 had a total of approximately 20,000 spurs that bloomed. The number of individual flowers per spur cluster averaged in excess of five. In round numbers, therefore, the tree developed 100,000 flowers, exclusive of terminals and axillaries. A crop of 25 bushels of apples per tree would require a maximum of approximately 2,500 specimens of 2½ inch apples or 2,000 apples of 2¾ inch size. These figures give some measure of the excess of bloom in an on-year over what is actually required for a crop.

## Classification of Fruit Buds on Spurs

Several of the Wealthy trees at New Brunswick possessed an excessive number of dormant fruit buds on spurs in the early spring of 1935. In order to promote annual bearing, plans were formulated for reducing the number of bud clusters prior to full bloom. This brought up the question of the relative quality of the individual spur buds. Previous observations in New Jersey (5) of the Delicious apple had shown that the individual spur buds upon a tree may vary greatly in vigor and ability to set and develop fruit. It was found that they could be grouped in four classes as follows: Class I and Class II comprise large buds which bloom and set fruit well under favorable conditions. Class III is comprised of buds of a slightly smaller diameter than Class II. A percentage of such buds tend to bloom and set fruit but the large majority drop and never mature. Class IV is comprised of buds smaller than those of Class III. A percentage of these may bloom but they will seldom set any fruit that is retained.

Observations upon Wealthy indicated that a somewhat similar relationship existed between individual spur buds of this variety except that a somewhat larger percentage of Class III spurs may set, retain and mature fruit. Since Class III and IV spur buds may bloom but produce little, if any, fruit they largely serve to waste carbohydrates and promote biennial bearing. It, therefore, became apparent in planning the studies to promote annual bearing that the number and distribution of the Class I and II spurs should be made the basis of the studies. Throughout this experiment all well developed dormant spur buds with a minimum diameter of .19 to .21 inches were considered Class II and buds exceeding .22 inches in diameter were regarded as Class I. Examination by sectioning of a considerable number of buds of these two sizes demonstrated that the great majority were differentiated as flower buds. Buds that were not differentiated could usually be detected by their distinct form.

## DEGREE OF BLOSSOM THINNING

Two blossom thinning treatments in 1935 were based upon distribution of Class I and II buds. Treatment 1, blossom clusters were thinned to a distance apart of 6 to 8 inches. Treatment 2, blossom clusters thinned to 10 to 12 inches apart. All Class III and IV spur flower clusters and all axillary buds were removed. Any terminal buds which were not removed in the dormant season pruning were blossom thinned. Treatment 3 received no blossom thinning but did receive the normal winter pruning given all of the trees which included the removal of terminals and some axillaries. The work of blossom removal was completed as near as possible at the early pink bud stage which was also found to be the most efficient time for their removal. The cluster of buds upon a spur could be quickly pinched off with the thumb and fore-finger of either hand without causing any injury to the newly forming leaves at the base of the developing spur.

## EARLY RESULTS IN 1935

The effect of blossom thinning in promoting increased leaf development was apparently almost immediate. A second outstanding effect was the increase in the number of fruits set per spur and in the increased rate of growth compared to those upon the tree not blossom thinned. All trees received fruit thinning to one apple per spur. The check tree was thinned so that no two fruits were closer than 6 to 8 inches apart, which amounted to 10 to 12 leaves per fruit. On July 11th, the fruits upon the check tree averaged 1¾ inches in diameter, while those upon the tree blossom thinned to 10 to 12 inches averaged 2¼ inches in diameter.

## YIELDS 1935

The two blossom thinned trees and the check tree each brought to maturity slightly more than 2,300 apples. A higher percentage, however, of the fruits upon the tree blossom thinned 10 to 12 inches were 3 inches and over in size and this resulted in an increase in yield. Blossom thinning, so that Class I and II spurs were permitted to fruit

6 to 8 inches apart, did not prove to be much different in effect from commercial thinning to 6 to 8 inches. The classified yields are given in Table I.

TABLE I—RESULTS IN YIELDS, 1935 TO 1938 INCLUSIVE

| Year and Tree Number | Size Classified in Inches | | | Total Fruits | Total Yield (Bu) |
|---|---|---|---|---|---|
| | 2½ and Under | 2½ to 3 | 3 and Over | | |
| 1935  1...... | 181 | 1,347 | 792 | 2,320 | 17.75 |
|       2...... | 84 | 704 | 1,529 | 2,317 | 22.12 |
|       3...... | 218 | 1,729 | 545 | 2,492 | 19.50 |
| 1936  1...... | 83 | 340 | 196 | 609 | 4.87 |
|       2...... | 153 | 1,216 | 1,341 | 2,600 | 26.12 |
|       3...... | — | — | — | none | none |
| 1937  2...... | 31 | 691 | 1,572 | 2,294 | 23.75 |
|       3...... | 23 | 1,331 | 830 | 2,184 | 20.50 |
| 1938  2...... | 34 | 194 | 1,214 | 1,442 | 17.00 |
|       3...... | 17 | 66 | 102 | 185 | 1.25 |

Tree 1, blossom thinned to 6 inches.
Tree 2, blossom thinned to 10 to 12 inches.
Tree 3, check, no blossom thinning.
1935 and 1937 normal "on-years".

## TREATMENT AND RESULTS SECOND YEAR, 1936

During the winter of 1935–36 all trees received a light general pruning wherever there was any crowding, crossing or breakage of branches. The tips of annual growth of all trees were cut back to remove any terminal buds. Since this was the normal off-year for these trees the check tree produced no blossoms. Tree 1 produced only a few blossom clusters and most of these were on one large limb. Tree 2 produced 1,654 spur blossom clusters with an added number of 418 axillary buds which were removed. On June 27th, 1,424 apples were thinned off tree 2, leaving not more than two apples to a cluster.

The yields obtained for 1936 appear in Table I. It is quite apparent that blossom thinning so as to leave Class I and II spurs 6 inches apart did not prove to be sufficient reduction in bloom to insure annual bearing. This treatment was therefore dropped from the experiment after 1936. Blossom thinning so that Class I and II spurs were spaced 10 to 12 inches apart resulted in production of 26.12 bushels of large apples in what would normally be the off-year.

## STATUS SPRING 1937

Although tree 2 produced a good crop of apples in 1936, it also developed a total of 15,670 fruit buds on spurs. This was considered too large a number for the continuance of annual bearing so 11,281 spurs received the blossom removal treatment and 4,389 Class I and II spurs were allowed to complete blossoming. The check tree developed a heavy set of bloom which was not reduced except in the form of terminal buds removed in dormant pruning.

## EARLY RESULTS SPRING 1937

This was the on-year for the check, tree 3. In order to compare its relative leaf surface with the blossom thinned tree and another in the off-year it was necessary to make observations on a third Wealthy tree namely No. 4. The length and width of leaves upon a number of criterion branches of each tree was made the basis for estimating the relative amount of leaf surface of the three trees. These estimates were made in the middle of May. Considering the leaf development upon non-blooming tree 4, as 100 per cent, the regular check, tree 3, had only one-half as much leaf area or 50 per cent, while the blossom thinned tree, number 2, had a leaf area of 75 per cent and enough fruit buds differentiated for a crop in 1938.

## NUMBER AND DIAMETER OF FRUITS PER SPUR

In order to secure definite data upon the relative set and size of fruits per spur in May 1937, a number of criterion spurs were selected from the blossom thinned tree, number 2, and the on-year tree, number 3, and counts and measurements made. Results are given in Table II.

It is quite clear from the data in Table II that the spurs upon the blossom thinned tree set and retained an average of more fruits per spur than the tree not blossom thinned, and yet the average size of the

TABLE II—AVERAGE DIAMETERS OF GREEN FRUITS AND FRUIT SET (1937)

|  | Largest Fruit Per Spur | | Average All Fruits Per Spur | | Number Fruits Set Per Spur | |
|---|---|---|---|---|---|---|
|  | Tree 2 (Inches) | Check (Inches) | Tree 2 (Inches) | Check (Inches) | Tree 2 | Check |
| May 27.... | 0.75 | 0.62 | 0.66 | 0.63 | 5 | 3 |
| June 8..... | 1.31 | 1.06 | 1.25 | 1.03 | 3.5 | 1 |

fruits in the clusters was greater. These differences were so marked that they were readily noted by the eye. These results agree with the observations made in 1935, the first year of the experiment.

## YIELDS FOR 1937

The results in Table I show that the check tree produced 20.5 bushels in its on-year, 1937. The bulk of its crop was of a satisfactory size. The reason for this was due to a rather heavy natural thinning. It may be noted that the total number of fruits produced by this tree was 2,184 or a little less than produced by tree 2 which was blossom thinned. The check tree, however, only developed a few scattered Class I and II spur buds for 1938, while tree 2 produced a greater yield of larger apples in 1937 and enough Class I and II fruit buds for a good crop in 1938.

## 1938 STATUS

In 1938, the check tree was in the off-year. In contrast the blossom thinned tree had enough fruit buds for a crop. There were 2,888 spurs on tree 2 which developed flower buds in the early spring of 1938. These were reduced to 1,743. This reduction was more severe than necessary. It resulted in a yield of only 17.0 bushels which is below the set goal. The apples were exceptionally large in size. It may be noted in Table I, that, of the 1,442 apples produced in 1938, 1,214 or 84.0 per cent were 3 inches or over in diameter, of these, 415 apples or 28.0 per cent of the total number were 3.5 inches or over in diameter. At this date, December, 1938, tree 2 has enough Class I and II spurs for a crop in 1939.

## SUMMARY

Blossom thinning of a 19-year-old Wealthy tree was begun in 1935. Class I and II spur clusters were thinned at the pink bud stage so that they were distributed 10 to 12 inches apart over the tree. All Class III and IV clusters and all axillary and terminal clusters were removed. The effect was to promote immediately increased leaf development upon all spurs, increase the number of fruit set per spur, and the rate of fruit enlargement. By July 15, 1935, a considerable number of the Class I, II and III spurs deflorated at the pink bud stage, had formed and differentiated flower buds for 1936.

The yield of the check and blossom thinned trees previous to and after treatment are shown in Fig. 1. It may be noted that both trees were biennial in bearing at the time the blossom thinning treatment was started in the spring of 1935.

The maximum number of spurs permitted to bloom annually was approximately 4,500. The maximum number of fruits permitted to develop was approximately 2,600 and the crop goal was 20 to 25 bushels. The majority of the fruits produced on the blossom thinned tree during this 4 year period exceeded 3 inches in diameter. In fact, the 1938 yield records show that 84.0 per cent of the total crop was 3 inches or over in size. This fact together with other indicators supports the conclusion that the blossom thinning in certain seasons was more severe than was necessary to maintain annual bearing, and that somewhat larger yields might have been permitted and annual bearing maintained.

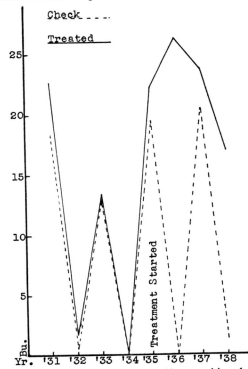

FIG. 1. Yields of check and blossom thinned Wealthy, before and after treatment.

The success of the blossom thinning treatment at New Brunswick is considered to be due to several factors. These include, the growth status of the trees, the removal of terminal flower buds, the removal of Class III and IV flower clusters, the proper number and distribution of the Class I and II clusters allowed to fruit, and some later thinning of the fruits.

### LITERATURE CITED

1. McCORMICK, A. C.   Control of biennial bearing in apples.   *Proc. Amer. Soc. Hort. Sci.* 30: 326–329.   1933.
2. THEIS, W. H.   Effect of defloration on spur leaf area in the McIntosh apple. *Proc. Amer. Soc. Hort. Sci.* 30: 309–312.   1933.
3. CHANDLER, W. H.   Fruit Growing.   p. 223.   Houghton Mifflin Co., New York.   1925.
4. ———— *Ibid.*   p. 228.
5. BLAKE, M. A., and DAVIDSON, O. W.   The New Jersey standard for judging the growth status of Delicious apple.   *N. J. Agr. Exp. Sta. Bul.* 559. 1934.

 Paul C. Marth and John W. Mitchell. 1944. 2,4-Dichlorophenoxyacetic acid as a differential herbicide. Botanical Gazette 106:224–232.

Paul C. Marth

John W. Mitchell

Weed control has been a major horticultural and agricultural problem since the first crop was planted. The discovery in 1896 that a spray of copper sulfate could control wild mustard (*Sinapsis arvensis* L.) without damaging oats served as the beginning point for the practical use of selective herbicides (Shaw 1911).

During the 1930s extensive research was conducted to characterize and develop the potential for a wide range of plant growth regulators. Over 5000 research papers about this subject were published between 1934 and 1937 alone (Tukey 1947). Among the centers of plant growth regulator research in the United States were the botany department at the University of Chicago under the direction of E. J. Kraus; the Boyce Thompson Institute in Yonkers, New York, led by the team of P. W. Zimmerman and A. E. Hitchcock; the University of California System, including A. S. Crafts (at Davis), J. Bonner, F. W. Went, J. van Overbeek (at Pasadena), and F. Skoog and K. V. Thimann (later at Harvard) at Berkeley; and the U.S. Department of Agriculture Bureau of Plant Industry located at Beltsville, Maryland, which included Paul C. Marth and John W. Mitchell.

Paul Charles Marth, Sr. was born in 1909 in Easton, Maryland, where he grew up on a farm. He attended the University of Maryland receiving a B.S. degree in 1930, an M.S. degree in 1933, and a Ph.D. degree in 1942, all in horticulture and plant physiology. Marth then joined the U.S. Department of Agriculture Horticulture Station at Beltsville, Maryland, where he cooperated on a wide range of horticultural research with F. E. Gardner. One of their early projects evaluated indoleacetic acid (IAA), indolebutyric acid (IBA), indolepropionic acid (IPA), and naphthaleneacetic acid (NAA) as sprays to induce pathenocarpy on American holly, apple, peach, and strawberry (Gardner and Marth 1937, 1939).

Marth's career entered a very productive period prior to and during World War II. In 1937, Neil Stuart and Marth reported an easy way to propagate American holly from stem cuttings using IBA, a breakthrough that helped to expand the use of plant growth regulators as a propagation tool using techniques that are still standard industry practices. In the late 1930s he was part of a team that developed plant growth regulator uses for reducing preharvest drop of apples (Gardner et al. 1939a,b), controlling bud sprouting of roses and other ornamentals during winter storage (Marth 1942), for propagation of rubber substitutes, including *Taraxacum* and *Solidago* species (Marth and Hamner 1943; Hamner and Marth 1943), and for increasing the number and size of blackberry fruit (Marth and Meader 1944).

John William Mitchell was born in 1905 in Hornell, New York, but grew up in Parma, Idaho. He graduated from the University of Idaho, where he received a B.S. degree in botany in 1928 and an M.S. degree in plant physiology in 1929. He then moved to the botany department, chaired by E. J. Kraus, at the University of Chicago, where he worked as a research assistant with Charles L. Shull. Mitchell received his Ph.D. degree in plant physiology in 1932 and remained until 1938 at the University of Chicago, where he designed and built a controlled-environment facility for growing plants which contained one of the first effective uses of carbon arc lights. In cooperation with William H. Martin, he researched the influence of IBA on the growth of etiolated bean plants (Mitchell and Martin 1937).

In 1938, Mitchell moved to Beltsville and continued intensive investigations of auxin effects on growth and metabolism of bean (Mitchell and Hamner 1938; Mitchell and Stuart 1939; Kraus and Mitchell 1939; Mitchell et al. 1940). From 1940 to 1943, Mitchell was assigned to the Emergency Rubber Project, where he worked on guayule and became well acquainted with Paul C. Marth.

In 1944 the team of Marth and Mitchell were well poised to develop a plant growth regulator for use as a selective herbicide. The personable Marth was the consummate idea man, being an outstanding observer and successful problem solver in many horticultural areas. He could visualize a practical use for a scientific concept before, during, and after the research was conducted. He employed the best methodology and freely shared results. Mitchell was the trained plant physiologist, whose keen analytical powers allowed him to survey and analyze a difficult problem, reduce it to its least complex form, and initiate and complete precise experiments designed to yield meaningful results.

The discovery and the progression of the development of plant growth regulators as herbicides during the World War II period is an exciting story. A series of articles published during the early 1940s demonstrated that plant growth regulators could work as herbicides if purposefully applied to weeds in toxic doses. Kraus and Mitchell began research on various regulators as herbicides in 1941. Among the compounds they tested, NAA had the most potential as a herbicide, but for practical applications a more potent compound was needed (Peterson 1967). Kraus and Mitchell became interested in the potential of substituted phenoxy and benzoic acids as herbicides after reading Zimmerman and Hitchcock's (1942) description of their biological activity on plants. Zimmerman sent them samples of these compounds [one of which was 2,4-dichlorophenoxyacetic acid (2,4-D)], which were tested at the University of Chicago and at Beltsville. In April 1944, Marth joined Mitchell on a series of experiments testing the sensitivity of various common weeds to 2,4-D, research that led to their 1944 paper on the differential selectivity of 2,4-D.

The first public announcement of the herbicidal effects of 2,4-D was made by Mitchell and Hamner (1944). Shortly thereafter, Hamner and Tukey (1944a,b) reported that 2,4-D killed field bindweed within 10 days of application. Marth and Mitchell published the results of their 2,4-D selectivity work in December 1944 in the *Botanical Gazette*. This paper explored the potential specificity of 2,4-D as a herbicide in grass crops and was the catalyst for further investigations. It opened the door to selective broadleaf weed control in agriculture and turf management (Mitchell et al. 1944) and 2,4-D is still the primary herbicide used to control broadleaf weeds without fear of injury in turf and maize. Although the herbicide has been widely used since the 1940s, those species Mitchell and Marth described as being controlled by 2,4-D are still susceptible. The results described in this paper were impressive. 2,4-D killed broadleaf plants at a concentration of 0.025 to 0.01 percent, over a thousandfold more effective than existing inorganic compounds such as sodium chlorate.

This classic paper involved the classic herbicide. Marth and Mitchell described, in a clear and concise manner, proper methods to employ when conducting weed science research. No one at that time seriously thought that weeds could be controlled selectively and easily by spraying them with a chemical. The hoe and plow were the primary tools for weed control. The development of 2,4-D as a selective herbicide opened the chemical age of weed control in agriculture and began the discipline of weed science.

Stephen C. Weller
Department of Horticulture
Purdue University
West Lafayette, Indiana 47907

J. Ray Frank
U.S. Department of Agriculture, ARS
Fort Detrick
Fredrick, Maryland 21701

## REFERENCES

GARDNER, F. E., and P. C. MARTH. 1937. Parthenocarpic fruits induced by spraying with growth promoting compounds. Bot. Gaz. 99:184–195.

GARDNER, F. E., and P. C. MARTH. 1939. Effectiveness of several growth substances on parthenocarpy in holly. Bot. Gaz. 101:226–229.

GARDNER, F. E., P. C. MARTH, and L. P. BATJER. 1939a. Spraying with plant growth substances to prevent apple fruit dropping. Science 90:208–209.

GARDNER, F. E., P. C. MARTH, and L. P. BATJER. 1939b. Spraying with plant growth substances for control of the preharvest drop of apples. Proc. Am. Soc. Hort. Sci. 37:415–428.

HAMNER, C. L., and P. C. MARTH. 1943. Effects of growth-regulating substances on propagation of goldenrod. Bot. Gaz. 105:182–192.

HAMNER, C. O., and H. B. TUKEY. 1944a. The herbicidal action of 2,4-dichlorophenoxyacetic and 2,4,5-trichlorophenoxyacetic acid on bindweed. Science 100:154–155.

HAMNER, C. L., and H. B. TUKEY. 1944b. Selective herbicidal action of midsummer and fall applications of 2,4-dichlorophenoxyacetic acid. Bot. Gaz. 106:232–245.

KRAUS, E. J., and J. W. MITCHELL. 1939. Histological responses of bean plants to alpha napthalene acetamide. Bot. Gaz. 101:688–699.

MARTH, P. C. 1942. Effects of growth regulating substances on shoot development of roses during common storage. Bot. Gaz. 104:26–49.

MARTH, P. C., and C. L. HAMNER. 1943. Vegetative propagation of *Taraxacum kok-saghyz* with the aid of growth substances. Bot. Gaz. 105:35–48.

MARTH, P. C., and E. M. MEADER. 1944. Influence of growth-regulating chemicals on blackberry fruit development. Proc. Am. Soc. Hort. Sci. 45:293–299.

MITCHELL, J. W., and C. L. HAMNER. 1938. Stimulating effect of beta β indoleacetic acid on synthesis of solid matter by bean plants. Bot. Gaz. 99:569–583.

MITCHELL, J. W., and C. L. HAMNER. 1944. Polyethylene glycols as carriers for growth regulating substances. Bot. Gaz. 105:474–483.

MITCHELL, J. W., and W. E. MARTIN. 1937. Effect of indoleacetic acid on growth and chemical composition of etiolated bean plants. Bot. Gaz. 99:171–183.

MITCHELL, J. W., and N. W. STUART. 1939. Growth and metabolism of bean cuttings subsequent to rooting with indoleacetic acid. Bot. Gaz. 100:627–650.

MITCHELL, J. W., E. J. KRAUS, and M. R. WHITEHEAD. 1940. Starch hydrolysis in bean leaves following spraying with alpha napthalene acetic emulsion. Bot. Gaz. 102:97–104.

MITCHELL, J. W., F. F. DAVIS, and P. C. MARTH. 1944. Turf weed control with plant growth regulators. U.S. Golf Assoc. Publ., Golfdom 18 (Oct.):34–36.

PETERSON, G. E. 1967. The discovery and development of 2,4-D. Q. Agr. Hist. Soc. 41:243–254.

SHAW, T. 1911. Weeds and how to eradicate them, 3rd ed. Webb, St. Paul, Minn.

STUART, N. W., and P. C. MARTH. 1937. Composition and rooting of American holly cuttings as affected by treatment with indolebutyric acid. Proc. Am. Soc. Hort. Sci. 34:839–844.

TUKEY, H. B. (ed.). 1947. Plant regulators in agriculture. Wiley, New York.

ZIMMERMAN, P. W., and A. E. HITCHCOCK. 1942. Substituted phenoxy and benzoic acid growth substances and the relation of structure to physiological activity. Contrib. Boyce Thompson Inst. 12:321–343.

# 2,4-DICHLOROPHENOXYACETIC ACID AS A
## DIFFERENTIAL HERBICIDE

PAUL C. MARTH[1] AND JOHN W. MITCHELL[2]

## Introduction

In studying plant responses to growth-regulating substances, it has since 1936 been observed repeatedly that some of these substances are toxic to plant tissues when applied in relatively large amounts. Thus, localized application of lanolin paste containing 2% or more of indoleacetic acid or of naphthoxyacetic acid often results in the death of tissues near the treated regions; and in connection with vegetative propagation of plants, overtreatment with these compounds has often resulted in greatly reduced growth or in death of the cuttings. For some years, responses to growth regulators, which were associated with decrease in plant growth, were considered disadvantageous in crop improvement, and they received relatively little attention. However, KRAUS (3) in 1941 proposed that the growth-inhibiting properties of these compounds be utilized in connection with weed control.

Recently, experimental evidence was obtained indicating that 2,4-dichlorophenoxyacetic acid, a potent substance whose activity was discovered by ZIMMERMAN (6), possesses herbicidal properties when applied in a solution of Carbowax (5) or when dispersed as a liquefied-gas aerosol (1, 2).

In April, 1944, experiments were undertaken at Beltsville, Maryland, to determine the sensitivity of various common weeds to treatment with 2,4-dichlorophenoxyacetic acid. These experiments have since been extended in order to determine the possibility of using the acid as a differential herbicide, and the results are reported here.

## Methods

Two methods of application were used, aqueous sprays and the liquefied-gas aerosol. The aqueous spray solutions were prepared by dissolving the required amount of the acid in Carbowax 1500 (polyethylene glycol), then adding this solution, with stirring, to the required amount of tap water. The Carbowax served as a carrier and spreading agent (5) and was added in amounts sufficient to give a 0.5% solution. Sprays were applied to plants or small plots with a precision sprayer of 1-quart capacity and carrying an air pressure of 60–100 pounds, or by means of a knapsack 3-gallon sprayer when somewhat larger plots were treated.

In dispersing the acid as an aerosol, a metal cylinder equipped with outlet and valve was evacuated. One hundred fifty grams of dimethyl ether and 25 gm. of engine oil (SAE 40) containing 3.15 gm. of 2,4-dichlorophenoxyacetic acid was drawn into the vessel. The mixture was then dispersed as an aerosol by releasing the valve, and treatments were applied by directing the mist toward the plants for several seconds. In some instances the plants were first covered with a cloth tent and the aerosol released within the enclosure, where it was allowed to remain for a period of several seconds.

PRELIMINARY EXPERIMENTS.—In preliminary experiments under field condi-

[1] Associate Physiologist, [2] Physiologist; Bureau of Plant Industry, Soils, and Agricultural Engineering, Agricultural Reasearch Administration, United States Department of Agriculture, Beltsville, Maryland.

tions, weeds on plots of ground 2 × 2 feet were treated with the aerosol and the behavior of the plants later compared with that of others of like species on an adjacent untreated plot. In other exploratory experiments, weeds of uniform size and in the seedling stage were transplanted from the field to 4-inch pots and grown under greenhouse conditions until well established. The species used included *Plantago lanceolata, P. major, Datura stramonium,* and *Capsella bursa-pastoris.* Aerosol treatments were applied to these plants in varying amounts and their subsequent behavior noted.

EXPERIMENTS WITH DANDELION.—On September 2, an area of Kentucky bluegrass lawn 32.5 feet long and 13 feet wide was selected in which an even, dense stand of dandelion (*Taraxacum officinale*) was growing. Six adjacent plots 6.5 feet square were laid out, two replications for each of three treatments. The following treatments were applied: (*a*) untreated control, (*b*) 500 p.p.m. of 2,4-dichlorophenoxyacetic acid applied as a spray at the rate of 4 gallons per 1000 square feet, and (*c*) 1000 p.p.m. of the acid applied in the same manner and rate. The spray treatments were repeated 26 days after the initial treatment. On September 21, Milorganite fertilizer was applied at the rate of 20 pounds per 1000 square feet of lawn surface. During the experiment the lawn was cut at regular intervals and given usual lawn care.

EXPERIMENTS WITH PLANTAIN.—On August 28, an area of Kentucky bluegrass lawn heavily infested with narrow-leaf plantain (*Plantago lanceolata*) was selected for uniformity of both grass and plantain. Three parallel tiers of plots 6 feet square were laid out in this area, with twenty plots in each tier—a total of sixty plots. A 1-foot border was left around each plot. The three tiers of plots

were separated into four equal blocks or replicates, each being five plots long and three plots wide. Aqueous sprays containing 2,4-dichlorophenoxyacetic acid in concentrations of 0, 125, 250, 500, and 1000 p.p.m. were applied. Five plots in each block were selected at random, and on August 28 four of these received a single spray treatment at each respective concentration level. Five other plots in each block received two applications of the respective solutions, the first on August 28 and the second on September 7. The five remaining plots in each block were reserved for treatment during the spring season.

All spray applications were made by means of a 3-gallon knapsack sprayer at the rate of approximately 4 gallons per 1000 square feet. The grass was cut when needed during the experiments and the lawn given usual care.

## Results

PRELIMINARY EXPERIMENTS.—Plants of *Datura, Plantago,* and *Capsella* transplanted from outdoors and grown under greenhouse conditions during April were found to be very sensitive to application of 2,4-dichlorophenoxyacetic acid. Seedlings of *Datura* were killed by aerosol treatments of 5 seconds' duration. Exposure for 1 second to the aerosol resulted in the death of plants of *Plantago* (fig. 1) and *Capsella.* These treatments were applied after the plants had developed visible flower buds but before any flowers had opened. Within 48 hours following treatment the plants showed severe leaf-curl, and—in the case of *Datura*—stem bending. There was no evidence of burning of the leaves, but growth of the plants was inhibited, the leaves became light green in color, and within 3 weeks the sprayed plants were dead. European bindweed (*Convolvulus arvensis*) grown from seed during early spring under greenhouse conditions was also killed by a 5 seconds' exposure to the aerosol.

In preliminary field experiments, treatment applied on August 11 to ma-

FIG. 1.—Eradication of narrow-leaf plantain (*Plantago anceolata*) with 2,4-dichlorophenoxy-acetic acid applied by aerosol method: *1*, untreated; *2*, exposed to the aerosol for 1 second; *3*, for 5 seconds; and *4*, for 20 seconds. Photographed approximately 1 month after treatment.

ture dandelion plants resulted in complete killing by August 29. Likewise, applications of aqueous sprays containing 1000 p.p.m. of the acid resulted in almost complete eradication of several hundred dandelions from a lawn area 35 feet long and 10 feet wide. Two plots containing forty-five to fifty plants each of ragweed were treated on August 12 by means of the aerosol method, while two similar adjacent plots received a thorough coverage with an aqueous spray containing 800 p.p.m. concentration of the acid. Periodic examination of these plots indicated that application of 2,4-dichlorophenoxyacetic acid, both as an aerosol and in water, had resulted in complete inhibition of vegetative growth. Maturation of flowers was inhibited as the result of both methods of treatment, and the shedding of pollen was prevented. On August 29, all the plants in the aerosol plots were dead.

Ninety per cent of the plants that received the aqueous-spray treatment were dead at this time, and the remainder were severely distorted. After death of the ragweed, volunteer crabgrass present at the time of treatment rapidly covered bare areas in all plots.

EXPERIMENT WITH DANDELION.—A more detailed experiment was undertaken, as previously described, to determine the effectiveness of 2,4-dichloro-phenoxyacetic acid when used to eradicate dandelion plants in a Kentucky bluegrass lawn. The spray treatments listed in table 1 were applied to the treated plots on September 2 and the original number of dandelions per plot determined 3 days later. On the third day following treatment, the plants sprayed with the acid were severely twisted and had turned from dark green to a pale yellowish green color.

TABLE 1

ERADICATION OF DANDELION IN KENTUCKY BLUEGRASS SOD BY AQUEOUS SOLUTIONS OF 2,4-DICHLOROPHENOXYACETIC ACID AT 500 AND 1000 P.P.M. CONCENTRATIONS. FIRST SPRAY APPLICATION ON SEPTEMBER 2; SECOND ON SEPTEMBER 28

| SPRAY CONCENTRA-TION (P.P.M.) | NO. OF DANDELION PLANTS PER PLOT (AV. OF 3 PLOTS) | | | PERCENTAGE ORIGINAL PLANTS KILLED | |
|---|---|---|---|---|---|
| | Sept. 7 | Sept. 26 | Oct. 19 | Sept. 26 | Oct. 19 |
| Unsprayed control. | 118.0 | 213.5 | 215.0 | 0 | 0 |
| 500...... | 94.6 | 22.0 | 3.0 | 75.6 | 95.8 |
| 1000...... | 117.0 | 4.6 | 1.5 | 96.0 | 98.3 |

During an interval of 24 days following treatment, the 500 p.p.m. spray reduced the original number of dandelion plants by 75.6%, while 1000 p.p.m. reduced the dandelion population by 96.0% (table 1; fig. 2). The number of dandelion plants in control plots more than doubled during the same interval, owing to unrestricted germination of the seeds. Later examination showed that many of the dandelions killed by the spray treatments had completely disintegrated as the result of attacks by soil organisms. Apparently, application of

killing as obtained previously by a single application of a spray containing 1000 p.p.m. of the acid (table 1). On October 19, the newly planted grass had germinated and was growing vigorously. The bare areas had been almost completely covered with grass in the reseeded portions of the treated plots.

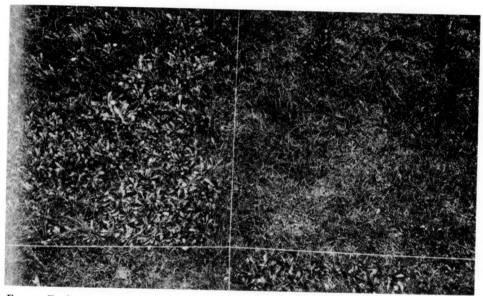

FIG. 2.—Eradication of dandelion from Kentucky bluegrass sod by spraying at rate of 4 gallons per 1000 square feet of sod with aqueous solutions of 1000 p.p.m. concentration of 2,4-dichlorophenoxyacetic acid and 0.5% Carbowax. Sprayed plot at right; untreated control plot at left. Spray applied September 2 and plot photographed 24 days later.

the acid did not greatly inhibit the subsequent development of microörganisms in the tissues after the plants died.

Eradication of the dandelions left numerous bare areas, but the grass continued to grow following treatment, and regular weekly mowings were necessary to maintain it at a height of 1½ inches. The plots were resprayed and one-half of each plot was reseeded on September 28. Dandelion counts were again made on October 19. Retreatment with the 500 p.p.m. concentration resulted in approximately the same percentage of

EXPERIMENT WITH NARROW-LEAF PLANTAIN.—On August 28, experiments were undertaken, as previously described, to determine the effectiveness of 2,4-dichlorophenoxyacetic acid when used to eradicate narrow-leaf plantain growing in an area of well-established lawn. Heavily infested plots sprayed twice with a solution containing 500 p.p.m. of the acid, or that received either one or two applications of 1000 p.p.m., were practically free of plantain 3 weeks after treatment (94.0–99.0% eradicated). There was no appreciable change in the

Fig. 3.—Eradication of narrow-leaf plantain from Kentucky bluegrass sod. Above: aqueous solution of 1000 p.p.m. concentration of 2,4-dichlorophenoxyacetic acid and 0.5% Carbowax. Below: untreated control plot. Spray applied August 28 and plots photographed 23 days later.

population of plantain in untreated plots during the 52 days immediately following the first treatment, at which time final readings were made (table 2; figs. 3, 4). Lower concentrations (125 and 250 p.p.m.) were less effective, and some of the injured plants appeared to recover in these treatments. It is noteworthy that a single treatment at the 500 p.p.m.

TABLE 2

EFFECT OF ONE AND OF TWO AQUEOUS SPRAY APPLICATIONS OF VARIOUS CONCENTRATIONS OF 2,4-DICHLOROPHENOXYACETIC ACID IN ERADICATING NARROW-LEAF PLANTAIN IN PLOTS 6×6 FEET OF KENTUCKY BLUEGRASS SOD

| SPRAY CONCENTRATION (P.P.M.) | NO. OF APPLICATIONS* | NO. OF PLANTAIN PLANTS PER PLOT (AV. OF 4 PLOTS) | | | PERCENTAGE ORIGINAL PLANTS KILLED | |
|---|---|---|---|---|---|---|
| | | Original count on Aug. 28 | Count on Sept. 29 | Count on Oct. 19 | Sept. 29 | Oct. 19 |
| Unsprayed control.. | 0 | 62.4 | 66.1 | 69.7 | 0 | 0 |
| 125...... | 1 | 68.2 | 69.0 | 64.0 | 0 | 0.6 |
| | 2 | 58.7 | 42.7 | 41.0 | 22.3 | 20.9 |
| 250...... | 1 | 52.5 | 47.2 | 39.5 | 10.0 | 24.7 |
| | 2 | 54.2 | 18.5 | 26.5 | 65.9 | 51.1 |
| 500...... | 1 | 53.7 | 14.5 | 17.2 | 73.0 | 67.6 |
| | 2 | 63.0 | 3.5 | 0.7 | 94.0 | 98.8 |
| 1000...... | 1 | 45.5 | 2.0 | 3.0 | 95.6 | 93.4 |
| | 2 | 50.2 | 0.5 | 0.2 | 99.0 | 99.5 |

* Single spray treatment on August 28; duplicate sprayings on August 28 and September 7.

concentration killed only 67.6% of the plantain, while the double application at this concentration was as effective as the 1000 p.p.m. spray applied either once or twice. These data indicate that, although thorough coverage may be of considerable importance, it was also necessary to deposit the acid on the weed foliage in sufficient amounts per unit surface area to bring about adequate selective killing. Apparently, repeated sprayings with relatively low concentrations would give about the same effects as fewer applications of sprays containing relatively high concentrations, provided approximately equivalent amounts of the acid are deposited in each instance. Further experiments are necessary to determine which type of spray treatment is more desirable. At present, grass in all the plots is growing vigorously. Grass that received a spray containing 1000 p.p.m. is a somewhat darker, richer green than that in other plots which received the more dilute sprays. Intensification of green col-

FIG. 4.—Effect of 2,4-dichlorophenoxyacetic acid (1000 p.p.m. spray) on mature field-grown plants of plantain. Left: two applications, August 29 and September 7. Center: single application, August 29. Right: untreated control.

oration has also been noted with other experimental plants, such as bean and tomato, when these were sprayed with solutions containing growth-regulating substances. Whether in this instance it may or may not be desirable has not yet been determined.

A number of common weeds other than those already listed occurred in the treated plots in many of the experiments. The following were very sensitive and readily killed by aqueous sprays at 500–1000 p.p.m. concentrations: woodsorrel (*Oxalis* sp.), chickweed (*Stellaria media*), pigweed (*Amarantus retroflexus*), knotweed (*Polygonum avicular*), and morning glory (*Convolvulus sepium*).

Other weeds—such as oxeye daisy (*Chrysanthemum leucanthemum* var. *pinnatifidum*), yarrow (*Achillea millefolium*), broad-leaf plantain (*Plantago major*), sheep sorrel (*Rumex acetosella*), dock (*Rumex obtusifolius*), as well as two

United States Golf Association. These concern the eradication of Dutch white clover from bluegrass and bentgrass sods. In these experiments clover was found to be very sensitive to aqueous sprays containing 2,4-dichlorophenoxy-

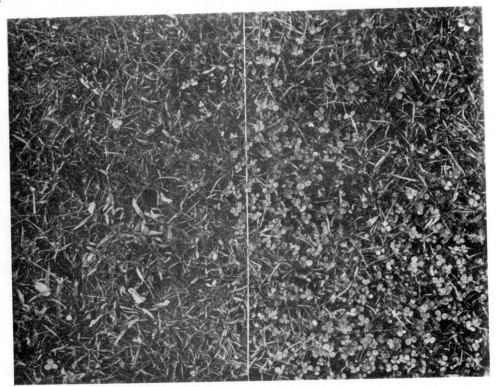

FIG. 5.—Eradication of Dutch white clover from Kentucky bluegrass sod. Left: 1000 p.p.m. concentration of 2,4-dichlorophenoxyacetic acid and 0.5% Carbowax applied as spray in water. Right: untreated control. Bluegrass and crabgrass in treated plot apparently uninjured.

species of *Rubus*—a blackberry and a trailing dewberry—appeared to be more resistant to the sprays and tended to recover from single applications at 1000 p.p.m. concentration. Preliminary results, however, indicate that these weeds may be killed by repeated sprays at this concentration.

A series of similar experiments reported elsewhere (4) has been in progress in co-operation with the Greens Section,

acetic acid (fig. 5). In the same series of experiments, shiny pennywort (*Hydrocotyle rotundifolia*), which had been crowding out the bluegrass in a lawn for a period of several years, was found to be even more readily killed than clover. A single spray with the 500 p.p.m. concentration of the acid at 5 gallons per 1000 square feet resulted in its complete eradication from bluegrass sod, without apparent injury to the grass.

## Discussion

Chemicals such as sodium chlorate, sodium thiocyanate, and various arsenical preparations have been used extensively for weed eradication during the last decade. These compounds are extremely toxic to plants when applied in relatively large amounts (1–10% solutions), and their killing action is generally characterized by a localized burning effect which may later extend through the plant for some distance from the point of contact. In contrast, 2,4-dichlorophenoxyacetic acid is effective in concentrations of approximately 0.025–0.1%. It does not cause rapid superficial burns, but it brings about morphological and physical responses at the point of application or in parts of the plant at some distance from this region. These growth responses usually occur soon after treatment (within 24 hours), and apparently before the concentration of the acid or of its derivatives reaches a lethal level in the plant, and they are characterized by distorted stem and leaf growth, by inhibition of bud growth (particularly the terminal ones), and sometimes by the formation of galls or roots in the main stem. These responses are followed by death of the plant and disintegration of its tissues. Succulent plants that are susceptible to treatment with the acid often exhibit distorted growth for a period of 2–3 weeks following treatment, and they eventually die and are then readily digested by soil microörganisms.

There are marked differences in sensitivity between closely related plants when treated with 2,4-dichlorophenoxyacetic acid. For instance, *Plantago lanceolata* is extremely sensitive to treatment, while *P. major* is much less so. Bluegrass and crabgrass were relatively insensitive, while certain closely cut bentgrasses were relatively sensitive. Little is known concerning the biological effects of growth-regulating chemicals when applied to soil in varying amounts. Accordingly, caution should be exercised in treating lawn or pasture areas until further information has been obtained as to the relative sensitivity of various kinds of plants and the effects of the compounds when present in the soil.

## Summary

1. 2,4-Dichlorophenoxyacetic acid was effective as a differential herbicide when applied as an aqueous spray in concentrations of from 250 to 1000 p.p.m. or more.

2. Plants killed by these sprays were dandelion, narrow-leaf plantain, Dutch white clover, chickweed, pigweed, woodsorrel, knotweed, broad-leaf dock, bindweed, and shiny pennywort. Other plants, such as broad-leaf plantain, sheep sorrel, daisy, yarrow, and various species of *Rubus*, were relatively insensitive to the acid.

3. Two applications of either 500 or 1000 p.p.m. concentration of the acid in aqueous solution were made to well-established Kentucky bluegrass sod without apparent injury to the grass. In addition, Kentucky bluegrass seed planted under a light top dressing of soil, which was then sprayed with the acid concentrations, germinated and readily became established in these lawn areas.

4. It was possible to obtain 95% control of dandelion and narrow-leaf plantain by a single spray application of a solution containing 1000 p.p.m. of 2,4-dichlorophenoxyacetic acid or with two

applications at 500 p.p.m. concentration.

5. Caution should be exercised in the use of sprays containing the acid until more information is obtained concerning the effects of its presence in the soil.

Appreciation is expressed for assistance and helpful suggestions from L. W. KEPHART.

BUREAU OF PLANT INDUSTRY, SOILS
AND AGRICULTURAL ENGINEERING
BELTSVILLE, MARYLAND

## LITERATURE CITED

1. GOODHUE, L. D., Insecticidal aerosol production. Spray solutions in liquefied gases. Ind. and Eng. Chem. 34:1456–1459. 1942.

2. HAMNER, C. L., and TUKEY, H. B., The herbicidal action of 2,4-dichlorophenoxyacetic acid and 2,4,5-trichlorophenoxyacetic acid on bindweed. Science 154–155. 1944.

3. KRAUS, E. J., Correspondence, 1941.

4. MITCHELL, J. W., DAVIS, F. F., and MARTH, P.

C., Clover and weed control in turf with plant growth regulators, Turfdom: October, 1944.

5. MITCHELL, J. W., and HAMNER, C. L., Polyethylene glycols as carriers for growth-regulating substances. BOT. GAZ. 105:474–483. 1944.

6. ZIMMERMAN, P. W., and HITCHCOCK, A. E., Substituted phenoxy and benzoic acid growth substances and the relation of structure to physiological activity. Contr. Boyce Thompson Inst. 12: 321–344. 1942.

J. P. Nitsch. 1950. Growth and morphogenesis of the strawberry as related to auxin. American Journal of Botany 37:211–215.

J. P. Nitsch

Jean Nitsch was born in Mulhouse, France, in 1921. His father was an architect, and perhaps this gave Jean his sense of style, for his experiments were elegant. His love for learning was evident early in life. He left his native Alsace during the Nazi occupation to study botany in Grenoble, Switzerland but for fear of reprisals did not tell his family that he was leaving. World War II had ended by the time he completed his *licence*, and he continued his studies at the Institut National Agronomique in Paris, where he graduated at the top of his class. After a brief period of work at the Institut National de la Recherche (INRA) at Versailles, he accepted an offer of a fellowship to tour botanical laboratories in the United States. For this he had first to learn English, although subsequently he became so fluent that some mistook him for an American. A three-month stay with F. G. Gustafson at the University of Michigan introduced him to an area of work in which he was to make many contributions—the induction of parthenocarpy with growth regulators. From Michigan he traveled to California to work with Frits Went at the California Institute of Technology. Here his talents soon became evident, and he was encouraged to remain at Pasadena. A four-year leave of absence from INRA permitted him to complete his Ph.D. degree with Went.

At Cal Tech he began the work on fruit growth for which he was to become renowned. One of his first breakthroughs was the successful culture of tomato ovaries on agar; adding auxin ($\beta$-naphthoxyacetic acid) to the medium induced rooting of the pedicel and promoted growth and normal ripening, although the resultant fruits were but 15 to 20 mm in diameter. A second was his discovery that removal of the achenes stopped the development of strawberry receptacles, and that auxin application renewed it; it is this paper that has been selected as a classic contribution.

On completing his doctorate, Nitsch spent two years as a postdoctoral fellow with R. H. Wetmore and Kenneth Thimann at Harvard. Here he developed paper chromatographic techniques for separating plant hormones. He then returned to Versailles and married Colette Kirchner, who was to be his companion in research for the next 18 years. Another postdoctoral fellowship permitted the Nitsches to return to Harvard, where they perfected the oat mesocotyl test, which was to become widely used for auxin assay. In 1955, Nitsch accepted an appointment as assistant professor in floriculture and ornamental horticulture at Cornell, where Jean and Colette continued their research on endogenous growth substances in relation to dormancy and photoperiodic induction of flowering and measured the responses of woody plants to photoperiod. In September 1959, Jean returned to France as assistant director of the Phytotron at the CNRS laboratory at Gif-sur-Yvette near Paris. This greenhouse and growth chamber complex, patterned after the phytotron at Pasadena, was the pride and joy of Pierre Chouard, professor of botany at the Sorbonne, and Nitsch's job was to make it work! Here Jean and Colette continued their studies of environmental effects upon vegetative and reproductive growth and of the hormonal changes thought to be responsible. In time, the emphasis changed to tissue culture, including studies of flowering *in vitro,* and in 1967 Nitsch was appointed the director of a separate Laboratory for Pluricellular Physiology. Much of the effort here was directed toward the production of haploid plants via anther culture.

Nitsch continued making major contributions to plant physiology and horticulture until his tragic death in a scuba-diving accident in July 1971. Collette, who shared his desire to unravel the mysteries of plant growth, his high standards for research, and his drive to accomplish the goals he had set for himself, is well-known for her current research with organogenesis *in vitro.*

At the time Nitsch began his work, the correlation between fruit size and seed number was well established (Muller-Thurgau 1898; Heinicke 1917). Gustafson (1936, 1938) and others had demonstrated that synthetic auxins were capable of inducing fruit development in several species, including strawberry (Gardner and Marth 1937; Wong 1941), in the absence of pollination, suggesting that auxin was responsible for the effects of seeds. "The problem," in Nitsch's words, "was to find out whether or not the achenes (containing seeds) have an influence upon the growth of the fleshy receptacle, and whether such an effect is correlated with differences in the auxin content of the achenes and receptacle.''

Nitsch proceeded to demonstrate that receptacle growth was indeed dependent on the presence of fertilized achenes; fruits of various shapes and sizes could be produced at will by removing the achenes from certain areas of the receptacle, and total receptacle weight increased logarithmically with the number of developed achenes. He then did a crucial experiment to determine if synthetic auxin would substitute for the effects of the achenes in stimulating fruit development. The rest is history! $\beta$-Naphthoxyacetic acid proved to be very effective for this purpose, and the importance of auxin in fruit development was further confirmed.

Nitsch subsequently used the *Avena* curvature test to measure the auxin content of ether extracts of the receptacle and the achenes from 3 days after pollination to maturity. The achenes proved to be rich sources of auxin, whereas receptacle extracts had no activity. Later work with physical methods of analysis has shown that the receptacle does indeed contain IAA, but that the concentrations present are considerably lower than those in achenes except for a brief period of time (Archbold and Dennis 1985).

Why is this paper a classic? Primarily because of its elegance and simplicity. Remove the achenes and receptacle growth ceases; replace them with a lanolin paste containing auxin and growth resumes. Achenes contain auxin; therefore, auxins are responsible for the effects of the achenes. This, of course, is probably an oversimplification, but the elegance remains. As in all scientific work, the tools had to be available before the discovery could be made. Had synthetic auxins not been available, only half the story would be known. Nitsch recognized the experimental value of a species in which the ''seeds'' (achenes) are external to the ''fruit'' (receptacle). Thus seed removal does not require major surgery, with consequent injury to the developing organ.

Jean Nitsch was a dreamer who believed in the ability of plant hormones to control many (all?) developmental processes. The skeptics among us envied him his enthusiasm—and his ability to do the impossible!

Frank G. Dennis, Jr.
Department of Horticulture
Michigan State University
East Lansing, Michigan 48824

## REFERENCES

ARCHBOLD, D. D., and F. G. DENNIS, JR. 1985. Quantification of free abscisic acid and free and conjugated indoleacetic acid in strawberry (*Fragaria* × *ananassa* Duch.) achene and receptacle tissue during fruit development. J. Am. Soc. Hort. Sci. 109:330–335.

GARDNER, F. E., and P. C. MARTH. 1937. Parthenogenic fruits induced by spraying with growth promoting compounds. Bot. Gaz. 99:184–195.

GUSTAFSON, F. G. 1936. Inducement of fruit development by growth-promoting chemicals. Proc. Natl. Acad. Sci. (USA) 22:628–636.

GUSTAFSON, F. G. 1938. Induced parthenocarpy. Bot. Gaz. 99:840–844.

HEINICKE, A. J. 1917. Factors affecting the abscission of flowers and partially developed fruits of the apple (*Pyrus malus*). Cornell Univ. Agr. Expt. Sta. Bull. 393.

MULLER-THURGAU, H. 1898. Abhangigkeit der Ausbildung der Traubenbeeren und einiger anderer Früchte von der Entwicklung der Samen. Landwirtsch. Jahrb. Schweiz 12:135–205.

WONG, C. Y. 1941. Chemically induced parthenocarpy in certain horticultural plants, with special reference to the watermelon. Bot. Gaz. 103:64–86.

# GROWTH AND MORPHOGENESIS OF THE STRAWBERRY AS RELATED TO AUXIN [1]

## J. P. Nitsch

UPON POLLINATION, a fruit may be looked at as consisting of two growing entities, the ovules and the ovary. Yet little is known of the interaction between them. Many fruits do not develop if seeds have not been formed, but, in some instances, growth of fruits without seeds has been obtained by the application of growth substances. This fact has led to the belief that the factor supplied by the seeds to this ovary is auxin. Several workers have provided experiments suggesting such a possibility. For example, Gustafson (1939) extracted different parts of tomatoes and found that the

[1] Received for publication August 4, 1949.

It is a real pleasure to thank Professor F. G. Gustafson, of the University of Michigan, and Professor F. W. Went, of the California Institute of Technology, for the facilities and suggestions they provided the writer. Acknowledgment is due to Dr. S. Wildman for his great help and criticism during preparation of the manuscript.

seeds and the region surrounding them contain the highest concentrations of auxin. Attempts were made by Dollfus (1936) to obtain with auxin the same growth effect as the seeds exert on the ovary. He cut open fruits of several species including *Symphoricarpus* and *Rosa,* removed the seeds and replaced them with an indoleacetic acid lanolin paste. A certain amount of growth resulted in the ovaries supplied with indoleacetic acid. However the necessary wounding prior to such a treatment causes extensive injury to the fruits, so that they very often become deformed and show extensive necrosis. The strawberry has been found to be a much better material for investigating the rôle of the ovules on fruit development. Since the seeds are distributed on the outside, they can be removed easily without such a drastic injury to the remaining tissue. While the strawberry is not a true fruit —the botanical fruits being the small achenes dis-

Fig. 1.—A. Growth of the strawberry receptacle induced by one single fertilized achene. (Magnified three times). —B. Growth induced by three fertilized achenes. (Magnified three times). Note the difference in size between the fertilized and the non-fertilized achenes.—C. (*Right*), growth induced by three rows of achenes left in a vertical position, the others having been removed. (*Left*), control.—D. (*Left*), strawberry having had all its achenes removed but two rows in a horizontal position. (*Right*), control.—E. Strawberries of the same age: (*left*), control (strawberry No. 1 of fig. 5), (*middle*), strawberry No. 3 of fig. 5 which had all its achenes removed and replaced with lanolin alone, (*right*), strawberry No. 2 which had all its achenes removed at the same time, but replaced by a lanolin paste containing 100 p.p.m. of beta-naphthoxyacetic acid.

posed all around it—nevertheless the receptacle accumulates sugars and vitamins and ripens like a true fleshy fruit, such as a cherry or a tomato.

Each achene contains a single ovule and can therefore be treated as a single unit. The problem was then to find out whether or not the achenes have an influence upon the growth of the fleshy receptacle, and whether such an effect is correlated with differences in the auxin content of the achenes and the receptacle.

MATERIAL AND METHODS.—*Plant material.*—The Marshall variety of strawberry was used in this investigation. The plants were grown in a greenhouse, in sand, and watered daily with a mineral nutrient solution. Each flower was pollinated by hand with a camel-hair brush. Diameter measurements of the strawberries were made every third day with a Vernier caliper. Two perpendicular diameters were measured to the nearest 0.1 mm., and the average of the two readings recorded.

*Auxin determinations.*—After separation of the achenes from the receptacles, both were frozen, ground in a mortar and dried in the frozen state by lyophil. When dry, the tissues were analyzed for free auxin, according to the method of Wildman and Muir (1949). Twenty-five to 200 mg. of dry material were extracted with redistilled wet ether for 6 hr. at 0°C. After decanting, the ether extract was analyzed for auxin by the *Avena* test (Went and Thimann, 1937). The amount of auxin extracted was calculated in gammas of indoleacetic acid by comparison with standard rows of oat seedlings run each time with 25 and 50 γ of indoleacetic acid per liter.

EXPERIMENTAL RESULTS. — *The effect of the achenes on the growth of the receptacle.*—Achenes can be removed from a strawberry with the tip of a knife without apparent injury to the receptacle. However, when all of the achenes are removed from strawberries 4 days after pollination, the growth of the receptacle completely stops. This result also occurs when the achenes are removed 7, 12, 19, and 21 days after pollination. In the last instance (21 days), the "berries" turned red at the usual time (26–30 days after pollination), although no further growth in diameter occurred. It seems likely, therefore, that the achenes control the growth of the receptacle at any time of the development.

In nature no growth of strawberries occurs unless the ovules contained in the achenes are fertilized. However, fertilization of one ovule is sufficient to cause some growth in the area of the receptacle immediately surrounding that achene, as shown in fig. 1A. Similarly, fertilization of three ovules causes three small areas of growth (fig. 1B). As a matter of fact, the weight of the fleshy part of a strawberry is roughly proportional to the number of fertilized ovules. This may be seen in fig. 2, where the number of developed achenes is plotted against the weight of the receptacles which had grown for 18 days after pollination.

Further evidence in favor of the view that fertilized achenes directly control the growth of the receptacle lies in the following experiments. Achenes were removed in such a way as to leave only two or three narrow rows, forming a ring around the whole "berry." When a vertical ring of

achenes was left, only a long flat strawberry developed (fig. 1C and 3). If the achene ring was left in a horizontal position, a short thick strawberry resulted (fig. 1D and 4). Depending on the position of the remaining achenes, almost any shape of "berry" can be obtained. Control strawberries on the same plant which did not have the achenes removed developed normally.

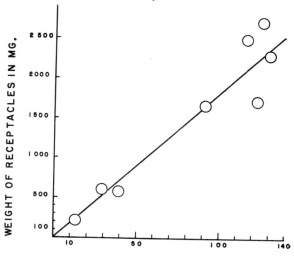

Fig. 2. Proportionality between number of developed achenes and weight of receptacles of strawberries of the same age.

*The effect of synthetic growth substances on the growth of the receptacle.*—The question arises as to what the fertilized achene contributes to the receptacle to cause it to grow. It has been suggested (Nitsch, 1949) that some substance of a hormone nature diffuses from the achene to the receptacle and stimulates its growth. Hunter (1941) and Swarbrick (1943) have been able to induce the growth of unpollinated strawberries by spraying the flowers with synthetic growth substances. Since unfertilized achenes do not induce any growth around them, and yet the receptacle grew in the cited experiments as a result of the hormone treatment, it seems probable that the function of the achenes in causing growth of the receptacle is to supply it with auxin.

This view is supported by the following experiment. Strawberries were pollinated and allowed to develop for 9 days. At this time, all the achenes were removed from some of the "berries" which were then coated with a lanolin paste containing 100 p.p.m. of beta-naphthoxyacetic acid. Others were coated with lanolin alone. Strawberries with intact achenes developed and ripened normally. Strawberries in which the achenes were removed and replaced by lanolin alone failed to grow or to ripen. However, strawberries in which the achenes were, replaced by lanolin containing the growth substance developed and ripened like the controls.

TABLE 1. *Free auxin content of the Marshall strawberry and its variation with the stage of development.*

| Age in days after pollination | Auxin per 100 mg. dry weight (in gammas × 10⁻³ of IAA) | | Auxin per strawberry (in gammas × 10⁻³ of IAA) |
|---|---|---|---|
| 3 days | achenes | 35 | 3.6 |
| | receptacles | 0 | .... |
| 6 days | achenes | 204 | 43.5 |
| | receptacles | 0 | .... |
| 12 days | achenes | 320 | 127.0 |
| | receptacles | 0 | .... |
| 17 days | achenes | 113 | 65.3 |
| | receptacles | 0 | .... |
| 20 days | achenes | 80 | 58.3 |
| | receptacles | 0 | .... |
| 30 days | achenes | 50 | 45.0 |
| | receptacles | 0 | .... |

The results are shown photographically in fig. 1E and graphically in fig. 5, where the growth in diameter of the "berries" is plotted against days after pollination. Beta-indolebutyric (0.3 per cent in lanolin) worked as well as naphthoxyacetic acid. Indoleacetic acid was not used because of its rapid inactivation.

Fig. 3–4.—Fig. 3 (*above*). The three rows of achenes which have been left on the strawberry (*left*) induce the receptacle to grow into a long flat fruit (*right*).—Fig. 4 (*below*). If three rows of achenes are left in a horizontal position (*left*), a short thick strawberry results (*right*).

These results show that growth substances can replace the achenes in regulating the growth of the receptacle. It was therefore of interest to investigate the auxin content of achenes and receptacles.

*Auxin content of achenes and receptacles.*— Achenes were separated from the receptacles 3, 6, 12, 17, 20, and 30 days after pollination, and analyzed for free auxin. As shown by the data of table 1, no free auxin could be extracted from the receptacles at any stage of development. Indoleacetic acid added to the receptacle tissue before ether extraction was subsequently quantitatively recovered, indicating the absence of ether extractable inhibitors of the *Avena* test.

In contrast to the lack of free auxin in the receptacles, the achenes were rich sources of auxin. Analysis revealed small amounts of free auxin shortly after pollination. Auxin concentration increased rapidly until approximately 12 days after pollination, then decreased again toward a constant level. It appears, therefore, that the achenes produce the auxin which is essential for the growth of the receptacle.

DISCUSSION.—The auxin experiments with strawberries are in general agreement with other work on the auxin content of developing ovules, which have always been found to contain a relatively high amount of growth substances. Actually, developing ovules constitute such a good source of auxin that Haagen-Smit *et al.* (1946) used immature corn kernels as a material for extraction of a large quantity of natural plant hormone.

The free auxin level in ovules does not remain constant. This was already known to be true in monocotyledons. Avery *et al.* (1942) followed the auxin production in the corn kernel and found that the amount of free auxin reaches a peak around the milk stage. These observations have been confirmed by Stehsel (1949). Hatcher (1945) found a similar picture in rye. Both the monocotyledons and the herbaceous dicotyledons so far investigated seem to show the same type of fluctuations in free auxin concentrations during ovule development.

The experiments related in the present paper lend support to previous observations on fruit

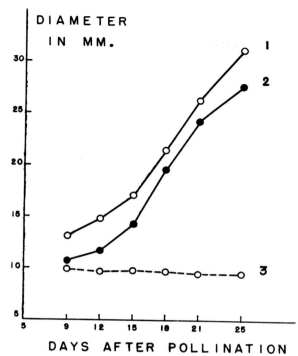

Fig. 5. Growth curves of the three strawberries shown in fig. 1E: No. 1: control, No. 2: strawberry where all achenes were removed on the ninth day and replaced by a lanolin paste containing 100 p.p.m. of beta-naphthoxyacetic acid. No. 3: strawberry where all achenes were removed and replaced by lanolin alone.

growth. Sinnott (1942) has shown that in cucurbits the volume of the inner region increases at a higher rate than the volume of the outer region of the fruit. On the contrary, Havis (1943) found that in strawberries the outer part grows more rapidly than the inner. This apparent contradiction might be explained by the fact that seeds producing auxin are located inside of the fruit in the case of cucurbits, but outside in the case of strawberries. It is possible that a gradient in auxin concentration is established, such that the nearest tissues to the seeds receive the highest amounts of auxin, the most remote regions the lowest. Thus, the regions which are the best supplied with auxin are expected to grow the more rapidly.

### SUMMARY

The action of the achenes upon the growth of the strawberry has been studied. Total removal of the achenes completely stops further growth of the fleshy part. Partial removal of the achenes results in fruits of abnormal shape because only the parts of the receptacle adjacent to the remaining achenes continue to grow. Only fertilized achenes are active. The weight of the fleshy part of a strawberry is a function of the number of developed achenes. The fertilized achenes can be replaced by synthetic growth substances in their action upon the growth of the receptacle. Beta-naphthoxyacetic acid and beta-indolebutyric acid in lanolin paste have induced strawberries of normal shape and size in the absence of achenes. The achenes have been found to contain relatively large amounts of free auxin, in contrast to the receptacles which did not yield any free auxin. The concentration of auxin in the achenes varies greatly with the stage of development. Free auxin fluctuations in the developing strawberry ovules are similar to those encountered in corn and rye kernels.

KERCKHOFF BIOLOGICAL LABORATORIES,
  CALIFORNIA INSTITUTE OF TECHNOLOGY,
    PASADENA 4, CALIFORNIA

## LITERATURE CITED

AVERY, G. S., JR., J. BERGER, AND BARBARA SHALUCHA. 1942. Auxin content of maize kernels during ontogeny, from plants of varying heterotic vigor. Amer. Jour. Bot. 29:765–772.

DOLLFUS, H. 1936. Wuchsstoffstudien. Planta 25:1–21.

GUSTAFSON, F. G. 1939. Auxin distribution in fruits and its significance in fruit development. Amer. Jour. Bot. 26:189–194.

HAAGEN-SMIT, A. J., W. B. DANDLIKER, S. H. WITTWER, AND A. E. MURNEEK. 1946. Isolation of 3-indoleacetic acid from immature corn kernels. Amer. Jour. Bot. 33:118–120.

HATCHER, E. S. J. 1945. Studies in the vernalisation of cereals. IX. Auxin production during development and ripening of the anther and carpel of spring and winter rye. Ann. Bot. N.S. 9:235–266.

HAVIS, A. L. 1943. A developmental analysis of the strawberry fruit. Amer. Jour. Bot. 30:311–314.

HUNTER, A. W. S. 1941. The experimental induction of parthenocarpic strawberries. Canadian Jour. Res. C 19:413–419.

NITSCH, J. 1949. Influence des akènes sur la croissance du réceptacle du fraisier. Compt. Rend. Acad. Sci., Sci., Paris 228:120–122.

SINNOTT, E. W. 1942. An analysis of the comparative rates of cell division in various parts of the developing cucurbit ovary. Amer. Jour. Bot. 29:317–323.

STEHSEL, M. 1949. Thesis. University of California, Berkeley, Calif.

SWARBRICK, T. 1943. Progress report on the use of naphthoxyacetic acid to increase the fruit set of the strawberry variety Tardive de Leopold. Ann. Rept. Long Ashton Res. Sta.

WENT, F. W., AND K. V. THIMANN. 1937. Phytohormones. Macmillan Co. New York.

WILDMAN, S. G., AND R. M. MUIR. 1949. Observations on the mechanism of auxin formation in plant tissues. Plant Physiol. 24:84–92.

H. A. Borthwick, S. B. Hendricks, M. W. Parker, E. H. Toole, and Vivian K. Toole. 1952. A reversible photoreaction controlling seed germination. Proceedings of the National Academy of Science 38(8):662–666.

H. A. Borthwick

S. B. Hendricks

M. W. Parker

E. H. Toole

Vivian K. Toole

Eben Henry Toole joined the U.S. Department of Agriculture (USDA) Seed Testing Laboratory, later known as the Bureau of Seed Investigations (BSI), after completing graduate work at the University of Wisconsin in 1920. Vivian Kearns, later Mrs. V. K. Toole, came from the University of North Carolina at Greensboro to join the bureau in 1929. Although the seed laboratory was originally concerned with the problems and standards of seed testing, it was soon apparent that development of seed testing procedures depended on research into the physiology of seed germination. About 1933 some nonrecurring funds were allocated to BSI, and Toole, now assistant chief, was able to invite L. H. Flint to work on the way light affects germination. A publication (Flint 1934) reporting work done during that period showed that lettuce seeds which failed to germinate in darkness would germinate 100 percent after a single exposure to red light. If the red light, measured in footcandles, was accompanied by an equal illuminance of blue light, no seeds germinated. The blue light had counteracted the promotive effect of the red radiation. Moreover, seeds of a lettuce cultivar that normally germinated in darkness, and therefore were not light requiring, could be made light sensitive by irradiating imbibed seeds with blue light, then drying them in darkness. Upon reimbibition the seeds would germinate only after an exposure to light, responding exactly as the light-requiring cultivars. This, Flint said, showed that the reaction was reversible.

BSI did not have adequate equipment for pursuing more detailed studies into these responses to light, so Flint, with Eben Toole's concurrence, sought the cooperation of the Smithsonian Institution, specifically the Divison of Radiation and Organisms that Charles Greeley Abbot had established in 1929 shortly after becoming secretary. This division, chiefly through the expertise of L. B. Clark, had constructed excellent facilities for measuring and controlling radiation, and studies were being conducted on a wide range of plant responses to light (Hoover et al. 1933; Meier 1934), including what were later to be called action spectra (Johnston 1934; Hoover 1937). The limitations of the seed germination work up to that time were discussed with E. D. McAlister and he agreed that a detailed spectral sensitivity response was needed if the light-absorbing system was to be identified. McAlister constructed a prismatic spectrograph for this purpose that produced a spectrum about 30 cm long. The resulting action spectrum for germination provided a detailed view of the promotive action in the red and the inhibitory response to blue radiation (Flint and McAlister 1935; Flint 1936). It also showed a very pronounced inhibition of germination between 700 and 800 nm.

A later report (Flint and McAlister 1937) made little mention of the far-red inhibition but noted that "the radiation most effective in promoting germination is that most abundantly absorbed by chlorophyll in the same region." They also pointed out that the principal critical wavelengths in the blue that inhibited germination "were identical with those reported by Johnston (1934) for the phototropic response of oat coleoptiles." Thus, they said, the blue absorption was not due to chlorophyll alone but probably involved carotenes. No comments were made about the inhibition in the far-red region of the spectrum. Later, after Flint moved to Louisiana State University, he stated: "In accordance with long established principle the material responding to 7600Å radiation might be expected to contain a pigment absorbing this radiation. Up to the present time no satisfactory data indicating such an absorption have been obtained" (Flint and Moreland 1939).

In 1938 after the testing and research sections of BSI were separated, Eben and Vivian Toole transferred to the division of fruit and vegetable crops and moved to Beltsville in 1939 or 1940. During the war years they were engaged primarily in seed storage and vegetable seed production research but retained their interest in light effects on germination. About 1945 the Seed Laboratory was fortuitously moved into the same building as H. A. Borthwick's Photoperiod Project, later (1957) becoming the Pioneering Research Laboratory for Plant Physiology. The Photoperiod Project resulted from a USDA decision about 1936 to set up a small group to investigate further the fundamental discoveries of Garner and Allard and do research in depth into the nature of photoperiodism and dormancy. Both Garner and Allard were still at Beltsville at this time, and in fact would not retire until nearly 10 years later, but neither was invited to join the Photoperiod Project. Instead, Harry A. Borthwick, a Stanford Phi Beta Kappa working on seed development at the University of California at Davis, and Marion W. Parker, a plant physiologist from the University of Maryland, were invited to conduct these photoperiod studies. Borthwick's work at Davis concerned harvest injury to bean seeds, development of vegetable seeds from fertilization to maturity, and their subsequent germination. His outstanding study of the development of carrot seeds was a factor in his selection to the photoperiod project, since one of the objectives was to examine the morphological aspects of the photoperiodically induced change from the vegetative to the reproductive state. Parker's initial role seems to have been the study of nutrition, particularly nitrogen, in relation to photoperiodism. It was soon clear that knowledge of the nature of the photoreceptor was essential to understanding photoperiodism, so about 1943 Borthwick and Parker established a cooperative program with Sterling B. Hendricks, a graduate of Cal Tech who had joined the USDA Fixed Nitrogen Laboratory in 1928. Hendricks was a physical chemist known for his work on X-ray diffraction analysis, especially of rock phosphate, and the demonstration of the crystalline nature of soil colloids. His active interest in plants developed through studies into the uptake of phosphorus by plants and he was intrigued by the photoperiodism phenomenon, particularly the fact that flowering could be prevented by a low-energy irradiation of brief duration near the middle of the dark period.

The initial objectives of the joint program were to construct a large prismatic spectrograph adjacent to the Seed Laboratory and conduct studies on action spectra of the photoperiodic response. Toole avidly followed the development of this spectrograph and the research that was conducted with it. The size, 183 cm long and 8 cm high, and spectral purity, scattered light less than $10^{-3}$ of the incident light, of the spectrum produced by this instrument exceeded anything used in plant research up to that time and in Toole's view was especially suitable for exposing seeds to very narrow spectral regions.

One day as Borthwick passed by the Seed Lab, Eben Toole asked if he would have time to discuss cooperative research between the two laboratories. At a meeting held that same day, Eben and Vivian Toole reviewed with Borthwick the general status of knowledge on how light affected seed germination. They pointed out two areas of special interest: the suggestion by Flint that one region of the spectrum was reversing the action of another region, and the far-red inhibition of germination shown by Flint and McAlister, which did not fit the characteristics of chlorophyll, carotenes, or any other known pigment. Flint and McAlister had in fact shown clearly that blue, and especially far red, reversed the potential germination induced by a previous exposure to red radiation, but failed to comment on it. Moreover, these far-red effects were not confined to seed germination but had been reported to inhibit growth of dark-grown lettuce hypocotyls (Flint and Moreland 1939) and to stimulate movement of *Mimosa* leaflets (Burkholder and Pratt 1936). After discussing the problem with Parker and Hendricks, Borthwick agreed to establish a cooperative research program. This cooperative program formed the nucleus of what came to be widely known as the Beltsville Group.

The first effort produced by this cooperation between people from three different laboratories was the now classical paper "A Reversible Photoreaction Controlling Seed Germination." This paper established beyond any doubt that the germination of light-sensitive seeds promoted by red radiation could be countermanded by a subsequent exposure to a different part of the spectrum, the far red, at a wavelength only 100 nm longer. Most important, the system could be reversed many times by alternate exposures to red and far-red light with the plant response being determined by the quality of the radiation received last. A subsequent publication (Borthwick et al. 1954) reporting the work in more detail revealed that both the red and far-red reactions were temperature independent. This temperature independence was additional evidence that the red and far-red responses were brought about by two forms of the same photoreceptor rather than two separate ones, and made it extremely unlikely that another reactant was involved in the conversion of one pigment form to the other.

The quantitative action spectrum, in contrast to the qualitative one Flint and McAlister obtained using saturating light levels, made possible a comparison with action spectra for other photoregulated plant responses. The similarity of action spectra for flowering of long-day plants (Borthwick et al. 1948) and short-day plants (Parker et al. 1946), and for leaf-size regulation in dark-grown peas (Parker et al. 1949) had led Borthwick, Parker, and Hendricks to the conclusion that a single-pigment system controlled these plant responses and probably most responses of a regulatory nature. Conclusive proof of a ubiquitous pigment was now available by demonstrating the reversibility of each of the many plant responses to the regulatory action of light. Biological control by this reversible photoreaction was soon extended to other kinds of light-requiring seeds, photoperiodic control of flowering, coloration of fruits, de-etiolation of dark-grown seedlings, and etiolation of light-grown ones. (Toole et al. 1955, 1956; Downs and Butler 1960; Borthwick and Hendricks 1961; Toole 1961; Hendricks and Borthwick 1963; Borthwick 1965).

The presence of the photoreversible pigment, now known as phytochrome, in dark-grown seedlings where chlorophyll would not interfere raised the possibility of using spectrophotometric methods to observe the pigment in living tissue by means of its reversibility. If this could be done, an assay for the isolation of phytochrome was at hand. Gerald Birth in Karl Norris's Instrumentation Research Laboratory had been developing a dual-monochromator spectrophotometer for studying light-transmittance properties of agricultural commodities. A preliminary test with this instrument in still another cooperative effort, this time with Karl Norris and Warren Butler, showed the pigment in living tissue clearly evident from the optical density changes (Butler et al. 1959). Thus by 1959 phytochrome had been measured in several dark-grown plant tissues and had been extracted and partially purified.

As soon as phytochrome was extracted, laboratories worldwide began to expend a great deal of research effort on the phytochrome molecule. Today, highly purified phytochrome is routinely obtained in a number of laboratories investigating the structure of the chromoprotein complex. As noted by Piringer in his introduction to Garner and Allard's paper, it seems that the amino acid sequence of the protein is, or shortly will be, known and a three-dimensional structure of phytochrome will soon be established (Briggs and Rice 1972; Mitrakos and Shropshire 1972; Smith 1975; Pratt 1979; Rudiger 1980; Smith 1982; Shropshire and Mohr 1983; Smith and Holmes 1984).

Although much less research effort has been directed toward phytochrome action at the plant level than that expended on the phytochome molecule, many horticulturists make use of the knowledge of photomorphogenesis that originated from the discovery of the photo-reversible reaction (Vince-Prue and Canham 1983). That discovery resulted in a renewed interest in seed dormancy and germination, and in better control of photoperiodism. Horticulturists today routinely consider the light factor when planting seeds of many bedding plants, floricultural crops, and even woody ornamentals. Where winter light levels seem to be limiting production of protected crops, electric lighting systems are used to supplement natural light. Electric light is also used in growing rooms for seedling establishment and commercial, controlled environment, continuous flow, plant growth systems for lettuce and other salad crops. The photomorphogenic effects of the light sources selected for these lighting tasks is predictable from our knowledge of the influence of the percentage of phytochrome converted to the far-red absorbing form by the incident spectral energy distribution—in modern parlance the phytochrome photoequilibrium. We would also expect to predict the most effective light source for photoperiod control while reducing or eliminating the undesirable elongation that often accompanies the use of long photoperiods. Unfortunately, this is not the case, and more research by those depending on photoperiod control is required.

Marion Parker left research for administration but the others continued to work together for many years. Despite the interesting molecular approaches opened up by the purification of phytochrome, most of their subsequent research continued to be concerned with phytochrome physiology because they felt this route would be the most productive in developing an understanding of how phytochrome functioned. Eben Toole retired in 1959 but continued as part of the Beltsville Group until his death. H. A. Borthwick retired in 1969 and S. B. Hendricks in 1970 and also continued to work with each other and with Mrs. Toole until their demise in 1974 and 1981. Vivian Toole[1] retired in 1976 but continued as a collaborator with the Seed Research Lab until 1986, when she returned to Greensboro, North Carolina.

Robert J. Downs
Southeastern Plant Environment Laboratories
North Carolina State University
Raleigh, North Carolina 27695

[1] The cooperation of Vivian K. Toole in supplying much of the early history of the seed laboratory is gratefully acknowledged.

# REFERENCES

BORTHWICK, H. A. 1965. Light effects with particular reference to seed germination. Proc. Int. Seed Test. Assoc. 30:15–27.

BORTHWICK, H. A., and S. B. HENDRICKS. 1961. Effects of radiation on growth and development. Encycl. Pflanzenphysiol. 16:299–330.

BORTHWICK, H. A., S. B. HENDRICKS, and M. W. PARKER. 1948. Action spectrum for the photoperiodic control of floral initiation of a long-day plant, Wintex barley. Bot. Gaz. 110:103–118.

BORTHWICK. H. A., S. B. HENDRICKS, E. H. TOOLE, and V. K. TOOLE. 1954. Action of light on lettuce seed germination. Bot. Gaz. 115:205–225.

BRIGGS, W. R., and H. V. RICE. 1972. Phytochrome: Chemical and physical properties and mechanism of action. Annu. Rev. Plant Physiol. 23:293–334.

BURKHOLDER, P. R., and R. PRATT. 1936. Leaflet movement of Mimosa pudica in relationship to the intensity and wavelength of the incident radiation. Am. J. Bot. 23:46–52, 212–220.

BUTLER, W. L., K. H. NORRIS, H. W. SIEGELMAN, and S. B. HENDRICKS. 1959. Detection, assay and preliminary purification of the pigment controlling photoreversible development in plants. Proc. Natl. Acad. Sci. (USA) 45:1703–1708.

DOWNS, R. J., and W. L. BUTLER. 1960. The effect of light. Chapter 6. In: P. Chouard and J. P. Nitsch (eds.). Life and man. René Kister, Paris.

FLINT, L. H. 1934. Light in relation to dormancy and germination in lettuce seed. Science 80:38–40.

FLINT, L. H. 1936. The action of radiation of specific wave lengths in relation to the germination of light-sensitive lettuce seed. Proc. Int. Seed Test. Assoc. 8:1–4.

FLINT, L. H., and E. D. McALISTER. 1935. Wave lengths of radiation in the visible spectrum inhibiting the germination of light-sensitive lettuce seed. Smithson. Misc. Collect. 94(5):1–11.

FLINT, L. H., and E. D. McALISTER. 1937. Wave lengths of radiation in the visible spectrum promoting the germination of light-sensitive lettuce seed. Smithson. Misc. Collect. 96(2):1–8.

FLINT, L. H., and C. F. MORELAND. 1939. Response of lettuce seedlings to 7600Å radiation. Am. J. Bot. 26:231–233.

HENDRICKS, S. B., and H. A. BORTHWICK. 1963. Control of plant growth by light. p. 233–261. In L. T. Evans (ed.). Environmental control of plant growth. Academic Press, New York.

HOOVER, W. H. 1937. The dependence of carbon dioxide assimilation in a higher plant on wave length of radiation. Smithon. Misc. Collect. 95(21):1–13.

HOOVER, W. H., E. S. JOHNSTON, and F. S. BRACKETT. 1933. Carbon dioxide assimilation in a higher plant. Smithson. Misc. Collect. 87(16):1–19.

JOHNSTON, E. S. 1934. Phototropic sensitivity in relation to wavelength. Smithson. Misc. Collect. 92(11):1–17.

MEIER, FLORENCE. 1934. Effects of intensities and wave lengths of light on unicellular green algae. Smithson. Misc. Collect. 92(6):1–27.

MITRAKOS, K., and W. SHROPSHIRE (eds.). 1972. Phytochrome. Academic Press, New York.

PARKER, M. W., S. B. HENDRICKS, H. A. BORTHWICK, and N. J. SCULLY. 1946. Action spectrum for the photoperiodic control of floral initiation of short-day plants. Bot. Gaz. 108:1–26.

PARKER, M. W., S. B. HENDRICKS, H. A. BORTHWICK, and F. W. WENT. 1949. Spectral sensitivities for leaf and stem growth of etiolated pea seedlings and their similarity to action spectra for photoperiodism. Am. J. Bot. 36:194–204.

PRATT, L. H. 1979. Phytochrome: Function and properties. Photochem. Photobiol. Rev. 4:59–124.

RUDIGER, W. 1980. Phytochrome, a light receptor of plant photomorphogenesis. Struct. Bonding 40:101–140.

SHROPSHIRE, W., and H. MOHR (eds.). 1983. Photomorphogenesis. In: A. Pirson and M. H. Zimmerman (eds.). Encyclopedia of plant physiology, New series, Vols 16A, 16B. Springer-Verlag, Berlin.

SMITH, H. 1975. Phytochrome and photomorphogenesis. McGraw-Hill, Maidenhead, Berkshire, England.

SMITH, H. 1982. Light quality, photoreception and plant strategy. Annu. Rev. Plant Physiol. 33:481–518.

SMITH, H., and M. G. HOLMES. 1984. Techniques in photomorphogenesis. Academic Press, London.

TOOLE, E. H. 1961. The effects of light and other variables on the control of seed germination. Proc. Int. Seed Test. Assoc. 26:659–673.

TOOLE, E. H., S. B. HENDRICKS, H. A. BORTHWICK, and V. K. TOOLE. 1956. Physiology of seed germination. Annu. Rev. Plant Physiol. 7:299–324.

TOOLE, E. H., V. K. TOOLE, H. A. BORTHWICK, and S. B. HENDRICKS. 1955. Photocontrol of *Lepidium* seed germination. Plant Physiol. 30:15–21.

VINCE-PRUE, D., and A. E. CANHAM. 1983. Horticultural significance of photomorphogenesis. p. 518–544. In: W. Shropshire and H. Mohr (eds.). Photomorphogenesis. Vol. 16B of A. Pirson and M. H. Zimmerman (eds.). Encyclopedia of plant physiology. Springer-Verlag, Berlin.

Reprinted from the Proceedings of the NATIONAL ACADEMY OF SCIENCES
Vol. 38, No. 8, pp. 662–666.   August, 1952

# A REVERSIBLE PHOTOREACTION CONTROLLING SEED GERMINATION

H. A. BORTHWICK, S. B. HENDRICKS, M. W. PARKER, E. H. TOOLE AND VIVIAN K. TOOLE

BUREAU OF PLANT INDUSTRY, SOILS AND AGRICULTURAL ENGINEERING, BELTSVILLE, MARYLAND

Communicated June 7, 1952

Effect of light upon germination of some seed is one manifestation of a very general, possibly universal, phenomenon controlling living processes. Some of the details of this effect on lettuce seed (*Lactuca sativa* L.) are given here.   Their bearing on the nature of a *reversible* photoreaction involved is developed.

Germination of one variety of lettuce seed was found by Flint and McAlister[1] and Flint[2] to be promoted by radiation in the region of 5250 to 7000 Å.   The greatest response for promotion resulting from a given irradiance was in the region of 6600 Å.   Germination was inhibited by radiation in the region of 7000 to 8200 Å. with the maximum inhibition between 7100 and 7500 Å.   We have verified the observations of Flint and McAlister for wave-lengths greater than 5200 Å. and have quantitatively determined the action spectrum (Fig. 1).

The action spectrum is expressed in terms of the incident energy required at various wave-lengths to promote or to inhibit germination to half its maximum value.   The results were obtained with a two-prism (glass) spectrograph having a dispersion of about 35 Å./cm. at 6000 Å.[3]   The slit width was about 90 Å. at this wave-length and the incident energy was 0.100 $\times$ $10^{-3}$ joules/cm.$^2$ sec. Lettuce seeds for these experiments were placed on wetted blotters in dishes that were immediately covered and maintained in darkness at 20°C. for various times before removal for irradiation. Those to be inhibited in germination were first exposed to radiation between 5800 and 6600 Å., as obtained with filters and a 4500° white fluorescent source. The exposure was such as to lead to germination of more than 98% of the viable seeds in controls. Following this treatment the seeds were irradiated in various wave-length regions to test for inhibition of germination. Those to be promoted were taken directly from darkness and irradiated. After irradiation, all seeds were held for two days at 20° in darkness and the ones then germinated were counted.

Two pigments apparently are involved with other reactants in the action of light on germination. Evidence will be given that these materials are directly related by the reactions.

$$(1) \quad \underset{\substack{\text{6500 Max.} \\ \text{Dormant}}}{\text{Pigment + RX}} \underset{\longleftarrow}{\overset{\text{Red}}{\longrightarrow}} \underset{\substack{\text{7300 Max.} \\ \text{Germinating}}}{\text{Pigment X + R}}$$

$$(2) \quad \text{Pigment X + R} \underset{\substack{\text{Infra Red} \\ \text{Darkness}}}{\overset{\longrightarrow}{\longleftarrow}} \text{Pigment + RX}$$

$$(3) \quad \text{with (1)} \quad \text{Sub. + RX} \longrightarrow \text{R + Sub. X}$$

$$(4) \quad \text{with (2)} \quad \text{Sub. X + R} \longrightarrow \text{RX + Sub.}$$

Two equations are given since no entirely trustworthy method has been found to associate Pigment and Pigment X with their respective absorption maxima. The nature of the pigment and the possibility of other reactants will not be examined since we do not wish to obscure the immediate deduction of a reversible photoreaction with any speculation.

The sample of Grand Rapids lettuce seed used was equally sensitive to promotion by red radiation for 4 to 16 hr. after imbibition. Fifty per cent germination required 2.11 $\times$ $10^{-3}$ joules/cm.$^2$ of incident energy between 6530 and 6635 Å. and 10.1 $\times$ $10^{-3}$ joules/cm.$^2$ gave 90 per cent germination with a linear variation of the probit with log energy. Seeds that had been exposed to red radiation were equally sensitive to suppression of germination by radiation at wave lengths greater than 7000 Å. over this period of 4 to 16 hr. After 16 hr. of imbibition, the radiant energy required for promotion to a given per cent germination increased and after 31 hr. had changed by about threefold. During this period, the requirement for inhibition decreased by about the same amount. This was the first indication of a relation between the photoreceptive pigments. It probably arises, however, from a thermal conversion of RX or R toward the condition on the right side of the equations.

The dependence of the germination response solely upon the last of a sequence of exposures to red and infra-red radiation is striking evidence of the immediate formation of the one pigment from the other. Seven lots of 200 lettuce seed were exposed to red radiation (5800 to 6800 Å.) for one minute after 16 hr. of imbibition of water in darkness. One lot was removed to the dark and the remaining ones were exposed to infra-red radiation of wave-lengths longer than 7000 Å. for four minutes. Again one lot was removed to the dark and the rest were exposed to red radiation for one minute. The alternations of red and infra-red irradiations were continued until all lots were treated. After this the treated and control lots of seed were held in darkness at 20°C. for two days. The germinated seeds were then counted (table 1).

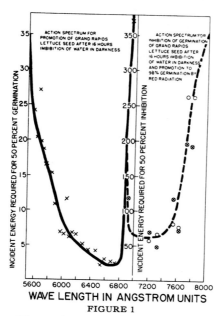

FIGURE 1

The action spectrum for promotion (solid line, ordinate on the left) and inhibition (dotted line, ordinate on the right) for germination of Grand Rapids lettuce seed. The several symbols give results from different experiments. Ordinates are in joules $\times 10^{-3}$/cm$^2$.

Energy greatly in excess of that required for full germination does not appreciably influence the sensitivity to inhibition by infra-red radiation. Thus a thirty-two fold increase in irradiance between 5800 and 6800 Å. over that required for 98 per cent germination of seed soaked for 16 hr. raised by less than twofold the energy required in the infra-red for suppressing the germination to a particular value. The result is in harmony with displacement of the equilibrium in a photoreaction and indicates that the pigment does not serve merely to transfer energy. Rather a certain energy is absorbed to activate the change of the pigment from one form to another.

The increased energy required for a given per cent germination, as the time of imbibition prior to irradiation is prolonged to periods greater than 16 hr., indicates that RX in equation (1) or R in equation (2) is limiting. In other words after long periods of imbibition, energy absorbed by the pigment is dissipated in some way, such as fluorescence, before reaction (1) or (2) takes place. That the pigment or pigment X is not limiting is shown by the simultaneous decrease in the energy required for inhibition. In fact this change in response is one evidence that a reactant other than the pigment is involved in the photoreaction.

After displacement by radiation does the reaction return toward an equilibrium in darkness? Experiments to answer this question must avoid confounding the sequence of germination following the displacement of the reaction to the right with a possible back reaction having about the same time constant. Seed of Grand Rapids lettuce was held fully imbibed at 20°C. for 16 hr. and then irradiated with energy adequate for full promotion of germination. This treatment was followed 7 hr. later by irradiation in the infra-red adequate to inhibit germination fully when given immediately after promotion. About 50 per cent of the seeds germinated. This indicates that in 7 hr. the germination process in half the promoted seeds had proceeded beyond determination by light.

An answer to the question was afforded by germination of a white-seeded lettuce variety that was thought to be more sensitive to temperature than Grand Rapids. Seeds of this variety germinate in light or in darkness at 20°C. but are dormant at 30°. After having been held for several days in

TABLE 1

LETTUCE SEED GERMINATING IN POPULATIONS OF 200 AFTER EXPOSURE TO RED AND INFRA-RED (IR) RADIATION IN SEQUENCE

| IRRADIATION | GERMINATION, % |
|---|---|
| None (dark control) | 8.5 |
| Red | 98 |
| Red + IR | 54 |
| Red + IR + Red | 100 |
| Red + IR + Red + IR | 43 |
| Red + IR + Red + IR + Red | 99 |
| Red + IR + Red + IR + Red + IR | 54 |
| Red + IR + Red + IR + Red + IR + Red | 98 |

darkness at 30° these seeds remain dormant if returned to 20° in darkness. However, if they are first irradiated in the region 5800 to 6800 Å. before returning to 20° in darkness, full germination is promoted. It follows that the reaction for the pigment was displaced from right to left in darkness at 30°. Existence of such a dark reaction indicates that the radiant energy is chiefly required for pigment activation rather than for the free energy change in the reaction.

The reaction can be poised at any point for a time adequate for germination; that is, a photochemical steady state can be established. The per cent of germination depends upon the integrated absorption of the two pigments. Thus in some wave band between 6900 and 7100 Å., 50 per cent germination is effected by continuous irradiation.

We have not thoroughly studied the action spectrum in the region below 5200 Å. An irradiance of $90 \times 10^{-3}$ joules/cm.$^2$ in the region of 4600 Å. resulted in about 60 per cent germination of Grand Rapids lettuce seed that had been held imbibed for 16 hr. in darkness. Less than $0.75 \times 10^{-3}$

joules/cm.[2] gave 50 per cent germination. This indicates that blue radiation can be very effective for promotion of germination. Failure of promotion to full germination as well as absence of an inhibitory action of blue radiation on seed that had been irradiated in the red adequate for full germination requires more detailed study for understanding.

The immediate general extension of these observations is to the varied response of seed germination to radiation.[4, 5] Germination of some seed is promoted by continuous light in the visible,[6] while that of others such as *Phacelia tanacetifolia* Benth. and *Amaranthus caudatus* L. is inhibited.[7] This is a result of the integrated effect of the photoreaction and suggests that the amount of the pigments and reactants R and RX vary among seeds. Many seeds that apparently are insensitive to control of germination by light nevertheless have the reaction for control. Thus the white-seeded lettuce variety previously mentioned germinates under radiation from an incandescent-filament lamp or in darkness but fails when exposed to radiation having wave-lengths in the region of 7000 to 8000 Å.

Dormancy in a seed is a suspended animation for the small plant that is present. Dormant imbibed seeds are known to have been viable for at least 70 years.[8] In these cases the time is so great as to raise serious question about maintenance of organization for future germination by slow but continued expenditure of energy in respiration. Imbibed seeds that have been held for long periods in darkness germinated after they were finally examined in light. If, following examination, the seeds had been irradiated in the region from 7000 to 8000 Å, the dormancy would likely have been re-established. This, for seeds, is the control of the living process that was mentioned in the first paragraph.

[1] Flint, L. H., and McAlister, E. D., *Smithsonian Miscellaneous Pub.*, **94**, No. 5 (1935).

[2] Flint, L. H., *Compt. rend. assoc. intern., essais semences Copenhagen*, **8**, 1 (1936).

[3] Parker, M. W., Hendricks, S. B., Borthwick, H. A., and Scully, N. J., *Bot. Gaz.*, **108**, 1 (1946).

[4] Crocker, W., *Biological Effects of Radiation*, Duggar, New York, Vol. **2**, (1936), p. 791.

[5] Evenari, M., *Radiation Biology*, **3**, (in course of publication)

[6] Meischke, D., *Jahr. Wiss. Bot.*, **83**, 359 (1936).

[7] Resühr, B., *Planta*, **30**, 471 (1939).

[8] Darlington, H. T., *Am. J. Bot.*, **38**, 379 (1951).

Martin J. Bukovac and Sylvan H. Wittwer. 1958. Reproductive responses of lettuce (*Lactuca sativa*, variety Great Lakes) to gibberellin as influenced by seed vernalization, photoperiod, and temperature. Proceedings of the American Society for Horticultural Science 71:407–411.

Martin J. Bukovac

Sylvan H. Wittwer

This paper is an important bridge between early studies of the physiology of bolting and flowering as governed by environmental factors, and subsequent studies of the regulation of flowering by the gibberellins, a class of plant hormones which were discovered in Japan and which, when this paper was published, was generating a great deal of excitement in Europe and the United States. The compound was perceived as commercially so promising that a major pharmaceutical company advertised gibberellins for sale over two pages of the *New York Times*. In an atypical burst of enthusiasm, the same newspaper published an editorial entitled "Ten-Foot Cabbages," and numerous articles appeared in popular magazines such as the *Reader's Digest* and *Time*.

It was an exciting time because the gibberellins were the first hormones known consistently to promote stem elongation of plants. Moreover, the number of reports of other developmental responses to gibberellins was increasing dramatically. Notable among these were the effects of gibberellins in promoting flowering (Lang 1956; Wittwer and Bukovac 1957), shortening of the rest period (Rappaport 1956), and increased fruit enlargement in seedless grapes (Weaver 1956, 1958). The research by Weaver led to the most important commercial application of the hormone in use today. Virtually all 'Thompson Seedless' and other seedless cultivars grown in California and in many other parts of the world are sprayed with the hormone to increase berry size. Wittwer and Bukovac had published a significant paper entitled "Gibberellins: New Chemicals for Crop Production" in 1957, and a more complete compilation of "The Effects of Gibberellin on Economic Crops" in 1958. This was the opening of the floodgate, as it were, and it heralded the publication of thousands of papers on the subject.

The paper under review is exceptional in that it addressed an old and particularly difficult horticultural problem, positioning it rapidly for a solution by use of a hormonal treatment. The basic observations were that when 'Great Lakes', a firm-headed "iceberg-type" lettuce, was treated with gibberellin (a mixture of gibberellins $A_1$ and $A_3$), the plants initiated a seedstalk (i.e., bolted) both at low (10 to 13° C) and higher (18 to 21° C) temperatures, and at short (8-hour) and long (16-hour) photoperiods. This observation led to the alleviation of a problem inherent in such cultivars, namely, mechanical impedance of seedstalk emergence which limits commercial seed production. Up to that time, the only practical method of dealing with the problem was slashing or mechanically shredding the top of the head to facilitate release of the seedstalk. The treatment frequently leads to mechanical injury of the stem apex resulting in a reduction in seed yield. This problem had been addressed earlier by Thompson and Knott (1933), Knott et al. (1937), Rappaport et al. (1956), and Rappaport and Wittwer (1956), all of whom were interested in the role of environmental factors in flowering of firm-headed cultivars. Much of this work was based on the hypothesis that a better understanding of the physiology of flowering would lead to a practical solution to the problem of facilitating bolting in head lettuce.

Head lettuce is a somewhat unusual annual. Whereas all plants will ultimately flower, regardless of the environmental conditions under which they are grown, low-temperature treatment (vernalization) of the seed followed by plant growth at high temperatures promotes bolting and flowering in the absence of any heading. Thus it appeared that for seed production a seed vernalization treatment might be used just prior to planting, to avoid the heading stage. However, 'Great Lakes' head lettuce was shown to have a surprising response to temperature following seed vernalization. Low temperatures and short days following seed vernalization actually devernalized the plants (Rappaport and Wittwer 1956). Thus even were the vernalization treatment to prove effective under controlled greenhouse conditions, the plants would subsequently be subject to the vagaries of the environment. In fact, in field experiments I conducted at Michigan State with vernalized lettuce seed planted in the winter, none of the resulting plants exhibited the vernalization effect; they all produced heads prior to bolting. Although these results provided me with a Ph.D. thesis (1956), they did not yield a satisfactory solution to the problem, although our understanding of the environmental requirements for flowering of head lettuce was somewhat enhanced (Lang 1965).

Bukovac and Wittwer's definitive paper was an important forerunner to a paper by Harrington (1960), who provided excellent evidence from well-replicated field experiments to show the efficacy of gibberellin $A_3$ in promoting bolting without heading in 'Great Lakes' head lettuce. The sprayed plants produced five times as much seed as the untreated control plants, and twice as much as those that were slashed with a knife to release the seedstalks. This reduction in yield probably resulted from physical damage to the seedstalk.

While gibberellin treatment of young head lettuce plants to promote bolting is a useful commercial application today, Harrington (1960) reported an important drawback. Since the plants do not head, it is impossible for seed producers to rogue off-types in a standard population of plants grown for seed. The method is useful only for foundation plant materials which are very uniform in growth habit.

Although the paper by Bukovac and Wittwer was not addressed directly to the physiological role of gibberellins in the flowering process, it was nevertheless a pivotal contribution, since it combined relevant physiological and horticultural information. Lang (1956) had previously shown that gibberellin promotes bolting and flowering in *Hyoscyamus niger*, a biennial plant that requires a low-temperature period for induction. Similar effects were demonstrated by Wittwer and Bukovac (1957) and Wittwer et al. (1957) with a number of annual and biennial plants.

An important outcome of the paper by Wittwer and Bukovac (1957) was that it helped focus research on the role of the gibberellins in the regulation of bolting and flowering as separate events. This separation was intimated by Lang's finding that gibberellin promotes bolting first and then flowering in *Hyoscyamus niger*. It was left to Sachs et al. (1959), using inhibitors of gibberellin biosynthesis, to show that the hormone promotes cell division and elongation in the subapical meristem, resulting in promotion of stem elongation. Gibberellin, they found, did not affect cell division appreciably in the apical meristem. This, coupled with earlier information (Wittwer and Bukovac, 1957) that gibberellin can only supplement low temperature and favorable photoperiods for flowering in most biennials, helped to anull the hypothesis that gibberellins are flowering hormones.

I feel particularly privileged to write this Introduction because I worked with Sylvan Wittwer on my doctoral degree at Michigan State and was contemporary with John Bukovac. Wittwer is best known for his outstanding contributions to our understanding of plant growth regulation relative to virtually all aspects of plant development. His studies of regulation of dormancy, flowering, fruit set, and plant productivity and quality are found in numerous publications. He was also among the first in the United States to undertake studies of the then very mysterious field of radioisotopes in agriculture, especially the nature of ion transport. He is a remarkable horticulturist in that not only does he understand a great deal about the physiology of plants, but he knows how to grow them remarkably well. His numerous, acclaimed publications reflect the diversity of his interests and the ardor with which he approaches his investigations.

Sylvan's exhuberance about his profession is legend. He has a conviction about everything he undertakes, which leaves ordinary mortals breathless. As director of the Michigan Agricultural Experiment Station, he encouraged the development of modern technologies and the acquisition of forefront faculty. He also became a vigorous advocate of agriculture, especially horticulture, both in Michigan and elsewhere. He has championed his cause within the university and to the state legislature, and by participating on important national committees which helped establish national priorities for agricultural research.

His interest in international agriculture led him to visit more than 100 countries. He has argued vigorously the cause of international research and teaching as part of the solution to the chronic problem of world hunger. Inevitably, when I have retraced (some of) his steps on foreign visits, I found a large residue of goodwill and affection for Wittwer, because of his concern for people as well as his dedication to the improvement of food production and quality in those countries. Sylvan Wittwer is indeed the international ambassador-at-large for agriculture, and his many contributions will be felt for years to come.

John Bukovac's brilliance, talents, and interest in science were glaringly evident as a graduate student at Michigan State. Bukovac was among the first to undertake critical research utilizing radioactive isotopes to study ion transport in plants. When he joined the faculty at Michigan State in 1957 he began a lifelong interest in the complex interaction between plant regulators important in horticulture and their transport through the leaf cuticle. This research has provided a basic understanding of their effects on productivity and quality of harvested crops, notably cherries. His publications form the major body of modern horticultural literature on the subject, and his work is a cornerstone of current practices of mechanical harvesting.

In addition, Bukovac made other very significant findings, not the least of which was the subject of this introduction. His publications epitomize the valuable balance between disciplinary analysis and practical application so valuable in horticultural research. His performance has been recognized by his peers at Michigan State, by his fellow horticulturists, and by the scientific community at large. Bukovac served as president of the American Society for Horticultural Science in 1974–1975 and in 1983 was elected to the National Academy of Sciences.

Lawrence Rappaport
Department of Vegetable Crops
University of California, Davis
Davis, California 95616

## REFERENCES

HARRINGTON, J. F. 1960. The use of gibberellic acid to induce bolting and increase seed yield of tight-heading lettuce. Proc. Am. Soc. Hort. Sci. 75:476–479.

KNOTT, J. E., O. W. TERRY, and E. M. ANDERSEN. 1937. Vernalization in lettuce. Proc. Am. Soc. Hort. Sci. 35:644–646.

LANG, A. 1956. Induction of flower formation in biennial Hyoscyamus by treatment with gibberellins. Naturwissenshaften 43:234–285.

LANG, A. 1965. Physiology of flower initiation. p. 1380–1532. In: A. Lang (ed.). Encyclopedia of plant physiology, Part I, Vol. 15. Springer-Verlag, Berlin.

RAPPAPORT, L. 1956. Growth regulating metabolites. Calif. Agr. 10:4, 11.

RAPPAPORT, L., and S. H. WITTWER. 1956. Flowering in head lettuce as influenced by seed vernalization, temperature and photoperiod. Proc. Am. Soc. Hort. Sci. 67:429–437.

RAPPAPORT, L., S. H. WITTWER, and H. B. TUKEY. 1956. Seed vernalization and flowering in lettuce (*Lactuca sativa*). Nature 178:51.

SACHS, R. M., C. F. BRETZ, and A. LANG. 1959. Shoot histogenesis: The early effects of gibberellin upon stem elongation in two rosette plants. Am. J. Bot. 46:376–384.

THOMPSON, H. C., and J. E. KNOTT. 1933. The effect of temperature and photoperiod on the growth of head lettuce. Proc. Am. Soc. Hort. Sci. 30:507–509.

WEAVER, J. 1956. Gibberellins on grapes. Blue Anchor 34:10–11.

WEAVER, R. S. 1958. Effect of gibberellic acid on fruit set and berry enlargement in seedless grapes (*Vitis vinifera*). Nature 181:851–852.

WITTWER, S. H., and M. J. BUKOVAC. 1957. Gibberellins: new chemicals for crop production. Mich. State Univ. Agr. Expt. Sta. Bull. 39:661–672.

WITTWER, S. H., and M. J. BUKOVAC. 1958. The effects of gibberellin on economic crops. Econ. Bot. 12:213–255.

Reprinted from Proceedings of the
AMERICAN SOCIETY FOR HORTICULTURAL SCIENCE
Vol. 71, 1958

# Reproductive Responses of Lettuce (*Lactuca sativa*, variety Great Lakes) to Gibberellin as Influenced by Seed Vernalization, Photoperiod and Temperature[1]

By M. J. BUKOVAC and S. H. WITTWER, *Michigan State University, East Lansing, Michigan*

## INTRODUCTION

PRIOR to 1956 no detailed reports appeared relative to the effects of gibberellin on flowering of plants (5). It has now been established that one of the consistent effects of gibberellin has been the acceleration of flowering in long day annuals grown under short photoperiods (2, 7, 8). In lettuce, a long day annual, flowering is influenced by temperature as well as by photoperiod (4, 6) and may be decidedly altered by seed vernalization (3). A recent report by Harrington *et al.,* (1) for endive *(Cichorium endivia),* a plant botanically and culturally related to lettuce, suggests that under long days and at high temperatures, earlier flowering may be induced with gibberellin in plants grown from non-vernalized seed, and flowering may be greatly accelerated with gibberellin when the seed is vernalized. An analysis of the effects of gibberellin on flowering of lettuce plants grown from vernalized and non-vernalized seed, and maintained at inductive and non-inductive temperatures and photoperiods, was the objective of the present study.

## MATERIALS AND METHODS

All plants were grown from seed of the head lettuce variety Great Lakes 6238 (Stock No. 20215 of the Ferry Morse Seed Co., Detroit, Michigan). The vernalization procedure consisted of placing the seed on moist filter paper in Petri dishes at about 20 degrees C for 24 to 36 hours or until the radicals emerged. The partially germinated seed was then transferred to a temperature of 4 degrees C in a water-saturated atmosphere for 16 days. Non-vernalized seed was also germinated on moist filter paper and when planted was at a comparable stage of growth to that which was vernalized.

Seed of both vernalized and non-vernalized lots were sown in moist vermiculite on October 12, 1956. The seedlings were subsequently transplanted into standard 4-inch clay pots of loam soil, and later to 6-inch pots, and grown for three weeks at a night temperature of 10 to 15 degrees C and at the prevailing photoperiod. High, but not excessive, levels of fertility were maintained by periodic additions of di-ammonium and mono-potassium phosphates in the irrigation water. Beginning November 1, when the first true leaves had appeared, equal numbers (24) of vernalized and non-vernalized

[1]Received for publication August 9, 1957. Journal article No. 2096 from the Michigan Agricultural Experiment Station. The gibberellin used in these studies consisted of gibberellic acid $[\alpha] \, _\mathrm{D}^{25} + 62.7$ and was procured through the courtesy of Dr. Curt Leben, Lilly Research Laboratories, Indianapolis, Indiana.

407

plants were subjected thereafter to each of the following greenhouse environments:

1.—A night temperature of 10 to 13 degrees C and a 9 hour photoperiod.

2.—A night temperature of 10 to 13 degrees C and an 18 hour photoperiod.

3.—A night temperature of 18 to 21 degrees C and a 9 hour photoperiod.

4.—A night temperature of 18 to 21 degrees C and an 18 hour photoperiod.

(daytime temperatures averaged 3 to 5 degrees C higher).

Aqueous stock solutions of gibberellin were prepared as needed and 20 micrograms of gibberellin was delivered onto the growing tip or apex of one half (12) of the vernalized and non-vernalized plants in each of the above four environments on December 17, 1956. This treatment was repeated January 21, 1957. At the time of the initial treatment (December 17, 1956) the plants had 8 to 10 fully developed leaves; no heads had formed, and no stem elongation was evident.

The 12 plants which constituted each treatment were completely randomized within each of the four environments listed above so that direct comparisons within each environment were possible for gibberellin-treated and non-treated plants. For each experimental plant, heights of stems (seedstalks) were recorded weekly and daily records taken of the appearance of visible flower primordia.

On April 30, 1957 the experiment was terminated. Night temperatures of 10 to 13 degrees C within the greenhouse were no longer possible because of high outdoor temperatures, and all lettuce plants capable of flowering under the designated environments and gibberellin treatments had achieved their final seedstalk heights.

## RESULTS AND DISCUSSION

The effects of gibberellin on the reproductive responses of Great Lakes head lettuce, as influenced by temperature, photoperiod and seed vernalization are recorded in Table 1 and illustrated in Figure 1. One of the most striking effects was that, with the exception of a single plant grown from non-vernalized seed, and at a temperature of 10–13 degrees C and a 9 hour photoperiod, all gibberellin-treated plants produced seedstalks (Figures 1–C and 1–D) and eventually flowered (Table 1) irrespective of growing conditions. The mean increases in percentages of plants which flowered following treatment with gibberellin were highly significant. In this respect the effects of gibberellin were far more pronounced than were those of seed vernalization.

Gibberellin also hastened flowering when expressed as days from seeding to first flower primordia (Table 1). These values were significant at both temperatures and photoperiods and were further accentuated by seed vernalization. The only instance where seed vernalization failed to hasten flowering was when lettuce was grown at a night temperature of 10–13 degrees C and at a 9 hour photo-

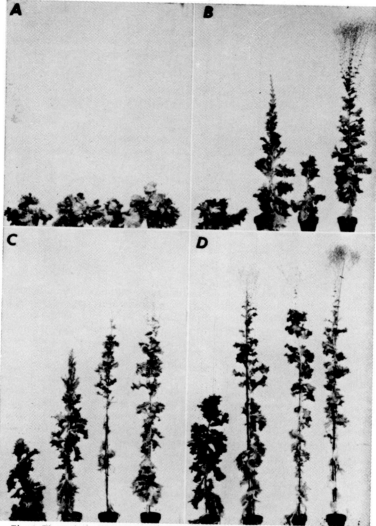

*Fig. 1.* Flowering and seedstalk development in Great Lakes lettuce plants as influenced by gibberellin, seed vernalization, temperature and photoperiod. A—Controls, not vernalized and no gibberellin; B—vernalized and no gibberellin; C—not vernalized plus gibberellin; D—seed vernalized plus gibberellin.

Plants within each (A, B, C, D) group from left to right were grown as follows: 10–13 degrees C and 9 hour photoperiod; 10–13 degrees C and 18 hour photoperiod; 18–21 degrees C and 9 hour photoperiod; 18–21 degrees C and 18 hour photoperiod.

*Table 1.*—Reproductive responses of *Lactuca sativa* (variety, Great Lakes) to gibberellin, seed vernalization, photoperiod and temperature.

| Treatment | Per cent of plants which flowered | | Days from seeding to first flower primordia | | Final seedstalk heights (centimeters) | |
|---|---|---|---|---|---|---|
| | Gibberellin | No Gibberellin | Gibberellin | No Gibberellin | Gibberellin | No Gibberellin |
| **Plants grown at 10–13°C** | | | | | | |
| *9 hour photoperiod* | | | | | | |
| Vernalized | 100 | 0 | 148 | No flowering | 125 | No flowering |
| Not vernalized | 92 | 0 | 144 | No flowering | 51 | No flowering |
| *18 hour photoperiod* | | | | | | |
| Vernalized | 100 | 83 | 106 | 123 | 219 | 99 |
| Not vernalized | 100 | 33 | 126 | 154 | 191 | 114 |
| Means | 98** | 29 | 131** | 139ᵃ | 147** | 107ᵃ |
| **Plants grown at 18–21°C** | | | | | | |
| *9 hour photoperiod* | | | | | | |
| Vernalized | 100 | 57 | 112 | 151 | 216 | 126 |
| Not vernalized | 100 | 17 | 122 | 154 | 176 | 110 |
| *18 hour photoperiod* | | | | | | |
| Vernalized | 100 | 89 | 95 | 105 | 253 | 166 |
| Not vernalized | 100 | 55 | 104 | 145 | 218 | 57 |
| Means | 100** | 55 | 108** | 139 | 216** | 115 |

ᵃMeans only of plants which flowered.
**Means of gibberellin-treated plants significantly greater than means of non-treated plants at 1% level.

period. This confirms the report of Rappaport and Wittwer (3) that the effects of lettuce seed vernalization were not manifest when the plants were subsequently grown under a short (9 hour) photoperiod and at a low (10–13 degrees C) temperature. Thus it appears that the flower promoting effects of gibberellin are more general than are those of seed vernalization when lettuce is grown under non-inductive environments.

Final heights in all gibberellin-treated plants were greatly increased. Plant heights increased (left to right, Figures 1–C and 1–D) as temperatures and photoperiods became more favorable for flowering. As with other criteria of reproductive development, the effects of gibberellin on seedstalk heights were more striking than were those of any other factor affecting flowering. At temperatures and daylengths most conducive to flowering, seedstalk heights were almost doubled. As has been observed repeatedly in other experiments, gibberellin-treated plants formed no heads (7, 8) and flower formation was preceded by extensive stem elongation, while the two processes, flower formation and stem elongation, occurred simultaneously in non-giberellin-treated plants which flowered.

These results indicate that gibberellin, in addition to promoting flowering of lettuce grown under non-inductive environments, also hastens reproductive development under the most favorable conditions of seed vernalization, temperature, and photoperiod. Of all factors affecting the various flowering responses of lettuce herein studied, the effects of gibberellin were most consistent and uniform. The firm head which often forms a mechanical barrier to the emergence of the seedstalk was completely eliminated, and the percentage of plants which flowered was significantly increased by treatment with gibberellin. Seed harvested from gibberellin-treated plants germinated normally and the seedlings did not differ from those grown from seed harvested from plants not treated. Further evaluation of the seed and progeny is in progress. It appears that areas heretofore not climatically adapted might in the future be utilized for the production of lettuce seed crops.

## Summary

Gibberellin promoted flowering in head lettuce grown under short (9 hour) and long (18 hour) photoperiods, and at either low (10–13 degrees C) or high (18–21 degrees C) temperatures. The effects of gibberellin were additive to those of long days, high temperatures and seed vernalization. Of the criteria used for measuring flowering responses which included per cent of plants which flowered, days from seeding to first flower primordia and seedstalk heights, the effects of gibberellin were more pronounced and consistent than those of temperature, photoperiod, or seed vernalization. Gibberellin-treated plants formed no heads and produced elongated stems prior to the appearance of visible flower primordia. Possible use of gibberellin is suggested for the commercial production of lettuce seed crops.

### Literature Cited

1. HARRINGTON, J. F., L. RAPPAPORT, and K. J. HOOD. 1957. Influence of gibberellins on stem elongation and flowering of endive. *Science* 125:601–602.
2. LANG, A. Gibberellin and flower formation. 1956. *Die Naturwiss.* 43:544.
3. RAPPAPORT, L., and S. H. WITTWER. 1956. Flowering in head lettuce as influenced by seed vernalization, temperature and photoperiod. *Proc. Amer. Soc. Hort. Sci.* 67:429–437.
4. ———— and ————. 1956. Night temperature and photoperiod effects on flowering of leaf lettuce. *Proc. Amer. Soc. Hort. Sci.* 68:279–282.
5. STOWE, B. B., and T. YAMAKI. 1957. The history and physiological action of the gibberellins. *Ann. Rev. Plant Physiol.* 8:181–216.
6. THOMPSON, H. C., and J. E. KNOTT. 1933. The effect of temperature and photoperiod on the growth of lettuce. *Proc. Amer. Soc. Hort. Sci.* 30:507–509.
7. WITTWER, S. H., M. J. BUKOVAC, H. M. SELL, and L. E. WELLER. 1957. Some effects of gibberellin on flowering and fruit setting. *Plant. Physiol.* 32:39–41.
8. WITTWER, S. H., and M. J. BUKOVAC. 1957. Gibberellin and higher plants: III. Induction of flowering in long-day annuals grown under short days. *Michigan Agr. Exp. Sta. Quart. Bul.* 39:661–672.

S. P. Burg and E. A. Burg. 1967. Molecular requirements for the biological activity of ethylene. Plant Physiology 42:144–152.

S. P. Burg

E. A. Burg

Dimitry Nikolayevich Neljubov was a graduate student in St. Petersburg, Russia, at the turn of the century when he first noticed that illuminating gas affected the growth of pea seedlings. His subsequent experiments demonstrated that ethylene and acetylene were the active principles in illuminating gas and that these and other gases effected the growth and development of many plants (Neljubov 1901). He used the "triple response" in etiolated legume seedlings: reduced stem growth, thickened stems, and horizontal growth as a bioassay to study ethylene-like activity of a number of compounds. Similar types of bioassays which exploit the characteristic responses of a number of plants and tissues to ethylene remained the predominant analytical method for over half a century.

For the next 30 years, Neljubov's results were confirmed and expanded on by other plant scientists. Studies on the basic physiological effects of ethylene were eclipsed when experiments by Sievers and True (1912), Denny (1924), and others showed that ethylene possessed the economically important ability to ripen fruit. Although experiments using bioassays had indicated that ethylene was produced by plants and effected their growth and development as early as 1910, it was not until 1934 that Gane used chemical methods to prove that ethylene was actually produced by ripening fruit (Gane 1934). After this discovery, research became focused on practical means of using ethylene to hasten ripening, to remove excess acidity, to increase sugar content, to improve aroma, and to hasten the removal of tannins and other objectionable substances from fruit and vegetables.

In the middle 1930s outstanding papers by W. Crocker, F. E. Denny, A. E. Hitchock, F. Wilcoxon, and P. W. Zimmerman, researchers at the Boyce Thompson Institute, elaborated most of the physiological effects of ethylene that are currently known (Crocker et al. 1932, 1935). Using bioassays, they showed that other gases had ethylene-like activity, that auxins produced responses similar to ethylene, that ethylene was produced by vegetative and fruit tissue, and that ethylene affected such diverse processes as flowering, rooting, dormancy, and responses to gravity. Many of these discoveries were neglected by other researchers until the early 1960s. For example, Hall and Morgan (1964) showed that naturally occurring auxin stimulated ethylene production and that ethylene increased auxin catabolism. Other auxin–ethylene interrelationships are described by Burg and Burg (1966).

This accumulating body of information on the biological activity of ethylene was sufficient to convince a number of plant physiologists that ethylene should be considered a plant hormone that affected fruit ripening and vegetative growth (Crocker et al. 1935, Pratt and Goeschl 1969). However, because ethylene is a gas and dissimilar to other compounds termed plant hormones, debate continued for over two decades over a question of semantics: was the word "hormone" (as derived from animal physiology) appropriate for ethylene? In retrospect a great deal of effort could have been saved by adopting the term "plant growth substance" or "plant growth regulator."

Other researchers conceded that ethylene was produced by plants and had pronounced effects on them, but doubted that ethylene was a natural regulator of plant growth and development. Biale et al. (1954) thought that ethylene was a by-product of ripening since ripening of some fruit started hours to days before ethylene production increased. When more sensitive instruments became available it was possible to show that increases in ethylene production were coincident with increases in respiration and that these increases often occurred before other ripening associated changes in the fruit.

This debate has lost much of its impact today because we now know that many ripening parameters can be separated from the respiratory and ethylene climacteric. Pears are an example in which softening and the loss of chlorophyll can be separated from the respiratory and ethylene climacterics. In many fruits, ripening is regulated by the relationship between the internal concentration of ethylene and the tissue's sensitivity to ethylene. As the fruit matures, the tissue's sensitivity to ethylene decreases, while the internal concentration of ethylene may increase. Exogenous ethylene treatments hasten the increase in tissue sensitivity while increasing internal levels. When the tissue's sensitivity becomes responsive to the internal concentration, autocatalytic ethylene production and ripening are initiated. Although the role of ethylene in fruit ripening is fairly well resolved, its role in other aspects of plant growth and development (e.g., responses to stress, senescence) still stimulates vigorous debates among researchers.

For many years, studies of ethylene physiology, synthesis, and action did not engender as much interest as did similar studies on other plant hormones. Papers published by the Boyce Thompson researchers showing that auxin and ethylene were interrelated displeased some prominent researchers of the day who endeavored to relegate ethylene's role to a by-product of auxin action. Apart from a few fruit and postharvest physiologists such as H. K. Pratt (see Pratt and Goeschl 1969), there was little interest in basic research on ethylene physiology, synthesis, or action until the early 1960s.

Research in ethylene physiology was long hampered by chemical and bioassay methods of analysis that were insensitive, time consuming, and incapable of detecting the parts per million and parts per billion concentrations of ethylene that regulate plant growth and development. The number of research papers on ethylene physiology increased dramatically after analytical instruments became available that incorporated refinements based on the gas chromatographic techniques described in Burg and Stolwijk's (1959) paper. Ethylene is now the easiest plant hormone to study.

When the Burgs published their classical paper, "Molecular Requirements for the Biological Activity of Ethylene," in 1967, Stanley P. Burg was an assistant professor at the University of Miami School of Medicine, and Ellen A. Burg (now Ellen M. Brown) was his technician and wife. Stanley met Ellen, a student at Radcliffe, during his pursuit of a Ph.D. degree at Harvard University under K. V. Thimann; his dissertation was entitled "Biogenesis of Ethylene." Stanley has subsequently held various posts at the University of Miami and the University of Florida, and has been a leader in applying basic research in the areas of postharvest physiology and low-pressure storage to the commercial storage of agricultural commodities. Ellen has since left the field of ethylene physiology and is now a professional psychotherapist.

Their 1967 paper is characterized by a thorough and systematic approach typical of all their publications. Gases were highly purified by gas chromatography before use, and rigorous strictures were applied to determine biological activity. The first part of the paper covered the determination of the biological activity of various compounds in the pea growth test. It was known from published studies that several gases could elicit an ethylene-like physiological response. However, the likelihood of impurities in the gases and vapors used in those experiments left some question as to whether the compound or contaminants (e.g., traces of ethylene) caused the physiological response. Over a period of a few years, the Burgs used recently developed gas chromatographic techniques to purify various gaseous compounds and then tested these pure compounds for biological activity in the etiolated pea stem straight growth assay. Some gases were also tested for their ability to ripen bananas. Many of the gases tested had about the same efficacy in inhibiting the elongation of pea stems, in promoting banana ripening, and in causing epinasty in tomato plants.

The second section dealt with the requirements for ethylene action. Five requirements were deduced for ethylene-like biological activity: (1) only unsaturated aliphatic compounds are active, (2) activity is inversely related to molecular size, (3) substitution causing electron delocalization reduced activity, (4) the unsaturated position must be adjacent to a terminal carbon, and (5) the terminal carbon must not be positively charged. This codification of properties for ethylene-like activity has led to the formulation of various models for ethylene action. One model, which still awaits definitive testing, was proposed in the third section of their paper.

The third section of the paper presented evidence that biological activity required metal binding. The empirical requirements for biological activity outlined in the second part shared a marked resemblance to the rules for silver binding to olefins. The stability constants for various olefin–silver complexes were determined and a close correlation was found between biological activity and the stability constants. Additionally, the relationship between the activity of carbon monoxide as an analogue of ethylene and as an inhibitor of metal containing enzymes further suggested that ethylene had to bind to a metallic receptor. After a decade, this model still remains one of the currently accepted models for ethylene action.

The fourth section dealt with the competitive inhibition of ethylene action by carbon dioxide. It was known that carbon dioxide counteracted some ethylene effects. In studying the requirements for ethylene action, the Burgs noticed that carbon dioxide and allene were analogous in structure but that carbon dioxide was inactive while allene was active in both the pea growth and fruit-ripening assay. Since carbon dioxide possessed all the molecular requirements for ethylene action, except a terminal carbon and being negatively charged on both ends, it could possibly bind to the receptor without inducing a physiological response (i.e., it could be a competitive inhibitor of ethylene action). This hypothesis was confirmed using double reciprocal plots of the growth of pea stem sections exposed to various concentrations of ethylene and carbon dioxide. Many empirical observations concerning the effect of carbon dioxide on ethylene action were explained by this discovery.

The fifth and final section of the paper showed the oxygen requirement for ethylene action. Double reciprocal plots of the effect of various concentrations of oxygen and ethylene on respiration and pea stem elongation demonstrated that ethylene action was reduced at oxygen levels that had no effect by themselves.

Data presented in the Burg's 1967 paper clarified previously observed effects of ethylene with other gases, and provided a firm, theoretical basis for studying ethylene action. The physiological relationships reported in this paper between ethylene action and the concentration of oxygen and carbon dioxide provided a physiological basis for the empirical observations of the beneficial effect of reduced oxygen and elevated carbon dioxide during the controlled atmosphere storage of horticultural crops. Most research on controlled and modified atmosphere storage continues to be empirical in nature, and does not further our understanding of the physiological processes involved. After 20 years, this paper remains one of the most cited research papers in the field of ethylene physiology.

The most recent milestone in ethylene physiology was the elucidation of ethylene biosynthesis. Starting with the discovery that methionine was a precursor of ethylene (Lieberman et al. 1966), research by Abeles, Burg, Kende, Lieberman, Lurssen, and Yang culminated in the discovery of 1-aminocyclopropane-1-carboxylic acid (ACC) as the direct precursor of ethylene in the laboratory of S. F. Yang (Adams and Yang 1979) and K. Lurssen (Lurssen et al. 1979). The development of a simple assay for determining ACC in slightly purified aqueous plant extracts greatly assisted studies of its control and involvement in ethylene production, and was used to show that ACC levels parallel ethylene production during the climacteric associated with fruit ripening.

Ethylene action now remains the major problem to be solved in ethylene physiology (Sisler and Goren 1981). Most scientists agree that ethylene must be bound to a specific receptor before it can exert its biological effect. However, there is disagreement over whether the bound ethylene must also be metabolized. This remains one of the most exciting areas of research in ethylene physiology.

It took over 30 years from the elucidation of the physiological effects of ethylene by Boyce Thompson researchers to the discovery by Lieberman et al. (Lieberman et al. 1966) that methionine gave rise to ethylene, and over an additional 10 years for ACC to be identified as the immediate precursor of ethylene. About half this length of time has passed since the Burgs's paper was published. Let us hope that an additional 20 years is not needed to resolve the vital problem of ethylene action. Since ethylene is produced by many biological organisms (e.g., decay microorganisms) and chemical processes (e.g., internal combustion engines) and is difficult to economically eliminate from the storage environment, the effects of ethylene on many horticultural commodities could be more successfully modulated by effecting ethylene action rather than ethylene synthesis. Reduced sensitivity to ethylene action could prolong the storage life of commodities in which ethylene promotes either ripening (e.g., orange and tomato) or physiological disorders or senescence (e.g., lettuce and carnations).

Continued interest in ethylene physiology has been partially the result of many outstanding reviews (e.g., Burg and Burg 1973, Lieberman 1979, Pratt and Goeschl 1969, Yang and Hoffman 1984). Fred Abeles' 1973 Book *Ethylene in Plant Biology* has had an enormous impact on this field and has been cited far more frequently than any other work on ethylene. Research on ethylene physiology continues to be a topic of major horticultural interest and intense scientific study.

Mikal E. Saltveit, Jr.
Department of Vegetable Crops
Mann Laboratory
University of California, Davis
Davis, California 95616

## REFERENCES

ABELES, F. B. 1973. Ethylene in plant biology. Academic Press, New York.

ADAMS, D. O., and S. F. YANG. 1979. Ethylene biosynthesis: Identification of 1-aminocyclopropane-1-carboxylic acid as an intermediate in the conversion of methionine to ethylene. Proc. Natl. Acad. Sci. (USA) 76:170–174.

BIALE, J. B., R. E. YOUNG, and A. J. OLMSTEAD. 1954. Fruit respiration and ethylene production. Plant Physiol. 29:168–174.

BURG, S. P., and E. A. BURG. 1966. The interaction between auxin and ethylene and its role in plant growth. Proc. Natl. Acad. Sci. (USA) 55:262–269.

BURG, S. P., and E. A. BURG. 1973. Ethylene in plant growth and development. Proc. Natl. Acad. Sci. (USA) 70:591–597.

BURG, S. P., and J. A. J. STOLWIJK. 1959. A highly sensitive kathrometer and its application to the measurement of ethylene and other gases of biological importance. J. Biochem. Microbiol. Technol. Eng. 1:245–259.

CROCKER, W., A. E. HITCHCOCK, and P. W. ZIMMERMAN. 1935. Similarities in the effects of ethylene and the plant auxins. Contrib. Boyce Thompson Inst. 7:231–248.

CROCKER, W., P. W. ZIMMERMAN, and A. E. HITCHCOCK. 1932. Ethylene-induced epinasty of leaves and the relation of gravity to it. Contrib. Boyce Thompson Inst. 4:177–218.

DENNY, F. E. 1924. Hastening the coloration of lemons. J. Agr. Res. 27:757–769.

GANE, R. 1934. Production of ethylene by some ripening fruits. Nature 134:1008.

HALL, W. C., and P. W. MORGAN. 1964. Auxin–ethylene interrelationships. p. 727–745. In: J. P. Nitsch (ed.). Régulateurs naturels de la croissance végétale. C.N.R.S., Paris.

LIEBERMAN, M. 1979. Biosynthesis and action of ethylene. Annu. Rev. Plant Physiol. 30:533–591.

LIEBERMAN, M., A. T. KUNISHI, L. W. MAPSON, and WARDALE. 1966. Stimulation of ethylene production in apple tissue slices by methionine. Plant Physiol. 41:376–382.

LURSSEN, K., K. NAUMAN, and R. SCHRODER. 1979. 1-Amino-cyclopropane-1-carboxylic acid—an intermediate of the ethylene biosynthesis in higher plants. Z. Pflanzenphysiol. 92:285–294.

NELJUBOV, D. 1901. Uber die horizontale Nutation der Stengel von *Pisum sativum* und einiger anderen. Pflanzen. Beih. Bot. Zentralbl. 10:128–138.

PRATT, H. K., and J. D. GOESCHL. 1969. Physiological roles of ethylene in plants. Annu. Rev. Plant Physiol. 20:541–584.

SIEVERS, A. F., AND R. H. TRUE. 1912. A preliminary study of the forced curing of lemons as practiced in California. U.S. Dep. Agr. Bur. Plant Ind. Bull. 232.

SISLER, E. C., and R. GOREN. 1981. Ethylene binding—the basis for hormone action in plants? What's New in Plant Physiol. 12:37–40.

YANG, S. F., and N. E. HOFFMAN. 1984. Ethylene biosynthesis and its regulation in higher plants. Annu. Rev. Plant Physiol. 35:155–189.

# Molecular Requirements for the Biological Activity of Ethylene[1]

## Stanley P. Burg and Ellen A. Burg

University of Miami School of Medicine, Miami, Florida 33136

Received October 17, 1965.

*Summary.* The molecular requirements for ethylene action were investigated using the pea straight growth test. Biological activity requires an unsaturated bond adjacent to a terminal carbon atom, is inversely related to molecular size, and is decreased by substitutions which lower the electron density in the unsaturated position. Evidence is presented that ethylene binds to a metal containing receptor site. $CO_2$ is a competitive inhibitor of ethylene action, and prevents high concentrations of auxin (which stimulate ethylene formation) from retarding the elongation of etiolated pea stem sections. It is suggested that $CO_2$ delays fruit ripening by displacing the ripening hormone, ethylene, from its receptor site. Binding of ethylene to the receptor site is also impeded when the $O_2$ concentration is lowered, and this may explain why fruit ripening is delayed at low $O_2$ tensions.

Several gases in addition to ethylene elicit the triple response in etiolated seedlings, and the concentration of each required to produce a just discernible effect has been determined (9, 25). Included are propylene, butylene (isomer unknown), acetylene, and CO. Since these vapors also substitute for ethylene in causing epinasty (9, 10), fruit ripening (11, 27), and other effects (8, 9, 29), it would appear that the molecular requirements for biological action are similar in each case. We have attempted to delineate these requirements using as an assay for ethylene action the effect which the gas has on the growth and tropistic behavior of etiolated pea stem sections.

## Materials and Methods

Peas (*Pisum sativum*, var. Alaska) were soaked in water for 5 hours and germinated in moist vermiculite. The seedlings developed in darkness at 23° receiving an occasional exposure to dim red illumination, and were used on the seventh day. Sections 10 mm long were cut from the third internode just below the leaf hook and incubated in a medium containing 1 $\mu$M indole-3-acetic acid (IAA), 2 % sucrose (w/v), 0.05 M potassium phosphate buffer (pH 6.8), and 5 $\mu$M $CoCl_2$ in glass distilled water. Ten ml of media and 10 sections were placed in a 125 ml Erlenmeyer flask, this was sealed with a vaccine cap, the gas phase ad-

justed as described below, and the tissue slowly shaken at 23° in the dark. After 3 hours the sections were visually scored in red light for curvature and 15 hours later the gas phase was analyzed by gas chromatography before the tissue was removed, blotted dry, weighed and measured. Details of the procedure have been published elsewhere (4).

*Gas Chromatography and Preparation of Gas Mixtures.* All gases were chromatographed on a 3 foot alumina column to determine whether ethylene or other low molecular weight contaminants were present. The instrument, a Perkin-Elmer flame ionization detector-Cary Model 31 electrometer combination, could detect a few ppb ethylene in a 5 ml air sample. If a gas was contaminated several 5 ml samples of it were purified by passage through the alumina column, or a shorter length of column packed with alumina or silica gel in the case of compounds with retention times much greater than that of ethylene. A stream splitter inserted between the column and detector diverted about 95 % of the desired chromatographic band which was collected by displacement of water. This sample was recovered admixed with $N_2$ carrier gas in a ratio of about 1:4 and an aliquot of the mixture was analyzed to assure its purity and determine the exact proportions of $N_2$ and sample. Then a measured amount of the mixture was added to the air phase above the tissue by injecting it through the vaccine cap. When high concentrations were required about half of the air in the flasks was removed before a measured amount of the mixture was added, followed by a known volume of $O_2$, and enough $N_2$ to restore atmospheric pressure. In this way mixtures containing 20 % $O_2$, up to 15 % sample gas and the balance $N_2$ were prepared.

[1] This investigation was supported by research grants EF-00214 and EF-00782 from the United States Public Health Service, Division of Environmental Engineering and Food Protection, and was carried out while S. P. Burg was the recipient of Career Research Development Award 1-K3-GM-6871 from the USPHS.

144

$CO_2$ was chromatographed on a 1 foot silica gel column, and analyzed with a thermal conductivity detector. To adjust the $CO_2$ level in the gas phase above the tissue 20 ml of air was removed from the sealed container, an appropriate amount of $CO_2$ added, and air readmitted until atmospheric pressure was restored. The $CO_2$ concentration was checked at the start and conclusion of the incubation, and the pH of the media at the end by injecting pH indicator dye into the flask. The initial $O_2$ content was adjusted by flushing the sealed flasks with $N_2$ for several minutes to instate anaerobic conditions, after which a measured amount of $N_2$ was removed and the same volume of $O_2$ added. The $O_2$ concentration was determined at the start and finish of the incubation by means of gas chromatography, using a 6 foot 5A molecular sieve column and thermal conductivity detector.

As a measure of respiration (except when $CO_2$ was added) the $CO_2$ content of the flasks was determined by gas chromatography at the end of the incubation period. If an effect on respiration was indicated or if it was particularly important to determine whether any had occurred, $O_2$ consumption and $CO_2$ production were measured using a Warburg respirometer.

*Determination of Henry's Law Constants.* The Henry's law constants for many of the gases studied in these experiments are not available, so a simple procedure was devised to determine them. A 125 ml flask (A) and a 1000 ml flask (B) were sealed with vaccine caps and their volumes calibrated. Into (A) was injected 10 ml of distilled water and a 0.3 ml gas sample, the flask was shaken for an hour, and the concentration of gas in the air phase determined by gas chromatography. Flask (A) was then inverted and 5 ml of water removed with a hypodermic syringe. This was injected into (B), the flask shaken for an hour, and the concentration of gas in the air phase determined. Because the vapors tested are relatively insoluble in water, and since the volume of the air phase in (B) is very large by comparison to that of the liquid phase, at equilibrium essentially all of the gas is found in the air phase. Therefore the amount of gas dissolved in 5 ml of the liquid phase of (A) is equal to the total gas content of the air phase in (B), and by comparing this value to that for the concentration of gas in the air phase of (A) the Henry's law constant was calculated. Reliability of the procedure is evidenced by the fact that experimentally determined values for ethylene ($0.80 \times 10^7$) and acetylene ($0.0985 \times 10^7$) at 23° are within a few percent of published figures (20).

*Stability Constants of Metal Complexes.* The methods of Muhs and Weiss (28) and Gil-Av and Herling (18) were used. In the first a stationary phase consisting of a measured weight of 0, 0.85 and 1.59 M $AgNO_3$ dissolved in ethylene glycol was coated on 60/80 mesh Gas Chrom R at a concentration of 40 % (w:w) and packed into a 6 foot column. The temperature, flow rate, pressure at the column inlet and outlet, density and amount of liquid phase were measured, and from these values and the retention time of a compound on each column a value for the stability constant of the silver-olefin complex was calculated (28). In the second method retention time on an ethylene glycol column containing 1.59 M $AgNO_3$ was compared to that on a column containing ethylene glycol and 1.59 M $NaNO_3$ (18).

*Testing Gases for Their Ability to Ripen Fruits.* Gros Michel bananas were kindly supplied by a local shipper (Banana Supply Co. of Miami). A single hand of green fruit was divided into 5 lots of 4 fingers and each lot placed in a 10 liter desiccator at 15°. One desiccator served as a control, into another was injected 1 ml of ethylene, and into the remaining chambers quantities of the gas to be tested in the ratio 1:10:100 respectively. After 18 hours the fruits were removed, and thereafter they were kept at 15° until ripening occurred. No attempt was made to purify the gases by chromatography because of the large volumes required. However, the ethylene content of each was determined and results are only included for gases whose activity cannot be due to contamination with ethylene.

*Use of Michaelis-Menten Kinetics and Lineweaver-Burk Plots.* These familiar techniques and their associated symbols, interpretation, and nomenclature have been applied to other physiological problems (1, 15, 26), and therefore we have adopted them without modification. It is assumed, although not proven, that the specific terminology employed (16) has the same meaning in the present physiological context as it does in the study of enzyme kinetics.

## Results
## Discussion and Conclusions

The morphogenetic changes which occur when ethylene is present during the pea straight growth test have been described in detail (4), and are only briefly reviewed here. Within the first few hours of the assay, before the gas begins to retard elongation, control sections curve by about 30° whereas segments exposed to ethylene remain straight. Subsequently ethylene treated sections begin to swell, and ultimately they elongate 50 % less than controls. Both elongation and curvature display the same dependence upon ethylene concentration, half-maximum inhibition resulting when ethylene is present at a concentration of about 0.1 ppm. A double reciprocal plot of the effect (V = % change in rate of elongation) vs ethylene concentration (A = ppm ethylene) yields a straight line (fig 1, lower curve) with an intersect on the ordinate at $1/V_m$ where $V_m$ is the maximum percent inhibition caused by ethylene and $1/V_m$ equals 0.02. Any substance which substitutes for ethylene in the pea straight growth test should produce the same maximum effect as ethylene, so that a double reciprocal plot describing its activity at various concentrations must intersect the ordinate at 0.02. The inhibition of elongation must not be accompanied by a significant change in fresh weight however, for ethylene causes pea tissue to swell but does not inhibit its increase in volume (4). Upon applying an optimal or supra-optimal concentration of the vapor, the ratio between percentage increase in weight vs percentage increase in length should change from the control value, 1.3, to a maximum of 2.6, just as it does when the ethylene concentration exceeds 1 ppm (4). In addition an active compound must prevent the nastic curvature from developing during the first few hours of the assay. Growth inhibitors do not prevent curvature development unless they are present at sufficient concentration to retard elongation by more than 70 %, and substances such as benzimidazole and colchicine which retard elongation without decreasing the fresh weight of pea tissue do not prevent curvature development. That concentration of an active compound which half-inhibits curvature should also produce a half-maximum retardation of elongation (4). Finally an active compound should not decrease either $O_2$ consumption or $CO_2$ evolution when it causes a maximum effect on elongation and curvature, for ethylene has no effect on the respiration of pea tissue (table I).

The compounds listed in table II demonstrate

Table I. *Effect of Ethylene on the Respiration of Etiolated Pea Stem Sections*

Measurements were made in a Warburg manometer using the dual flask method for $CO_2$. Ten 5 mm stem sections were incubated in 3 ml of media containing 2 % sucrose (w/v), 5 $\mu M$ CoCl, 0.05 M potassium phosphate buffer (pH 6.8), and in some cases 1 $\mu M$ IAA. Half the samples were gassed with air containing 0.1 % ethylene for 2 minutes before the flasks were sealed. Rates are calculated on the basis of the starting weights, and each value in the table is an average derived from 7 duplicate flasks. At the end of the incubation the flasks were opened and the sections weighed and measured. It was found that ethylene had caused its characteristic inhibition of elongation.

| | | No IAA | | 1 $\mu M$ IAA | |
| | Hours | Control | Ethylene | Control | Ethylene |
|---|---|---|---|---|---|
| $Q_{O_2}$ ($\mu L/g/hr$) | 0–8 | 279 | 268 | 569 | 555 |
| $Q_{O_2}$ ($\mu L/g/hr$) | 18–20 | ... | ... | 281 | 278 |
| Respiratory quotient | 0–8 | ... | ... | 0.95 | 0.95 |

Table II. *Biological Activity of Ethylene and Other Unsaturated Compounds as Determined by the Pea Straight Growth Test*

| Compound* | K'$_A$ relative to ethylene** | ppm in gas phase for half-maximum activity |
|---|---|---|
| Ethylene | 1 | 0.1 |
| Propylene | 130 | 10 |
| Vinyl chloride | 2370 | 140 |
| Carbon monoxide | 2900 | 270 |
| Vinyl fluoride | 7100 | 430 |
| Acetylene | 12,500 | 280 |
| Allene | 14,000 | 2900 |
| Methyl acetylene | 45,000 | 800 |
| 1-Butene | 140,000 | 27,000 |
| Vinyl bromide | 220 000 | 1600 |
| Ethyl acetylene | 765,000 | 11 000 |
| Vinyl methyl ether | 1 175,000 | 10 000 |
| Butadiene | 1,200,000 (?) | 500,000 (?)*** |
| 1, 1 Difluoro ethylene | 2,060,000 | 350 000 |
| Vinyl ethyl ether | 5,440,000 | 30 000 |

\* Ethane, trans-2-butene, cis-2-butene, isobutene, and nitrous oxide were inactive at the highest concentration which could be tested (300,000 ppm) and acetonitrile at 0.1 M. The following were inactive at all concentrations below the toxic level which is indicated in parenthesis: dichloromethane, cis-dichloroethylene, trans-dichloroethylene, trichloroethylene, tetrachloroethylene, $H_2S$, ethylene oxide, allyl chloride (all approximately 10,000 ppm) ; $CO_2$ (100,000 ppm) ; HCN (0.03 mM) ; acrylonitrile (0.17 mM) ; $H_2O_2$ (0.01 M) ; $KN_3$ (4 mM) ; allyl alcohol (1 mM) ; and HCHO (0.1 mM).

\*\* The Michaelis-Menten constant (K'$_A$) for ethylene at 24° has a value of 0.62 m$\mu M$ in the presence of an infinite concentration of $O_2$. The term K'$_A$ is used to indicate that ethylene is an activator (A) rather than a substrate (S).

\*\*\* This activity may be due to contamination of the butadiene with 0.2 ppm ethylene. So much butadiene was required to produce an effect that it was not possible to purify the requisite amount of gas by chromatography.

an ethylene-like action judged by all the criteria listed above. There is fairly good agreement between results obtained with pea stem sections (table II) and those reported for seedlings (9, 25). Several of these compounds were also tested in the banana ripening assay and it was found that 10 times the half-maximal concentration shown in table II was just as effective as any amount of ethylene, whereas the half-maximal concentration itself gave variable results and one-tenth that level never hastened ripening. Included in these tests were ethylene, propylene, CO, acetylene, methyl acetylene, ethyl acetylene, and allene. Therefore, it is clear that each compound has about the same efficacy in promoting fruit ripening, inhibiting the elongation of pea tissue, and causing epinasty (9). This suggests that triggering a single primary mechanism with ethylene gives rise to a wide variety of responses, just as light and the phytochrome system control a multiplicity of events.

*Requirements for Ethylene Action.* Based on the data in table II we propose the following requirements for ethylene action:

A) Only unsaturated aliphatic compounds are active (also see 9, 10, 25). The double bond (ethylene) confers far more activity than the triple bond (acetylene) whereas the single bond (ethane) is ineffective.

B) Activity is inversely related to molecular size. In the olefinic series the concentrations of ethylene, propylene, and 1-butene causing half-maximal responses are in the ratio 1 : 130 : 140,000. In contrast, the loss in activity associated with alkyl substitution is far less marked in the acetylenic series and the concentrations required are in the ratio 1 : 3.7 : 61 respectively for acetylene, methyl acetylene, and ethyl acetylene. These differences can be explained by assuming that biological activity is inversely related to molecular size because the unsaturated end of the molecule has to attach to a position of limited access. In the alkene series an alkyl substituent is positioned obliquely with respect to the carbon associated with the double bond and hence it would sterically hinder approach to such a site, whereas in the acetylenic series an alkyl substituent is held in line with the triple bond where it would not cause steric hindrance.

C) Substitutions which lower the bond order of the unsaturated position by causing electron delocalization reduce biological activity. On the basis of size the order of biological activity should be vinyl fluoride > vinyl chloride $\cong$ propylene > vinyl bromide. Instead it is observed that propylene is considerably more active than any of the vinyl halides, and vinyl chloride slightly more active than vinyl fluoride. The resonance form $(-):CH_2-CH = X(+)$ contributes to the structure of the vinyl halides, producing electron delocalization which might influence the biological activity of these compounds. The inductive effect is in the order F > Cl > Br so the fluoride

derivative has the lowest bond order and highest dipole even though it has the smallest size. Electron delocalization may also alter the activity of alkyl substituted alkenes and alkynes since both undergo hyperconjugation; alkenes to form $(-):CH_2CH = R(+)$ and alkynes to $(-):CH = C = R(+)$. However, the hyperconjugated form does not contribute very strongly to the structure of the alkyl substituted olefines so their dipole is low, between 0.3 and 0.4 debye units, in contrast to values in excess of 1.4 debye units for the vinyl halides. Assuming that the biological activity of these unsaturated compounds is inversely related to their dipole (a measure of electron delocalization) the order of activity should be approximately opposite to that based on a consideration of molecular size alone. We suggest, therefore, that the observed activity is a compromise between these opposing tendencies and that both steric interference with the approach to the double bond and a lowering of the electron density in the double bond reduce biological activity. The low activity of butadiene relative to 1-butene could then be attributed to extensive electron delocalization throughout the conjugated system of butadiene, and the closely similar activities of methyl acetylene and allene might reflect not only nearly identical sizes, but also the fact that the hyperconjugated form of methyl acetylene has a structure which superficially resembles that of allene.

D) The unsaturated position must be adjacent to a terminal carbon atom. This is indicated by the activity of 1-butene as contrasted with the behavior of cis and trans 2-butene. Presumably in spite of the favorable position of its double bond isobutene is inactive because of extensive steric hindrance. Since acetonitrile is completely inactive, whereas methyl acetylene is highly active, it would appear that nitrogen will not substitute for the carbon adjacent to the double bond. Acetonitrile undergoes extensive hyperconjugation however, and this also might explain its inactivity.

E) The terminal carbon must not be positively charged. This requirement is suggested to account for the activity of CO and inactivity of formaldehyde. The resonance form $(-):C = O(+)$ contributes to the structure of CO, overcoming the inherent polarity of the C-O bond to yield a molecule in which there is a slight negative charge on the carbon. The polarity of the carbonyl group in formaldehyde causes its carbon to be strongly positive in charge.

*Evidence that Biological Activity Requires Metal Binding.* The ability to form complexes with metals is a property of unsaturated aliphatic compounds which is lacking in other aliphatic and all but a few aromatic molecules. The bonding in the complex is influenced by the availability of electrons in the filled $\pi$-orbitals of the unsaturated compound, and also by the ease of overlap of these orbitals with those of the metal as determined by

PLANT PHYSIOLOGY

Table III. *Comparison of Biological Activity of Olefines with the Stability of Various Silver-Olefin Complexes at 23°*

| Compound | Stability constant $(K_1)$* | | Relative concentration needed for half-maximal biological response |
|---|---|---|---|
| | Method 1 | Method 2 | |
| Ethylene | 27.0 | 34.0 | 1 |
| Propylene | 14.4 | 17.0 | 130 |
| 1-Butene | 13.4 | . . . | 140,000 |
| Vinyl methyl ether | 9.5 | 9.2 | 1,175,000 |
| Butadiene | 6.9 | 7.8 | 1,200,000 (?)** |
| Vinyl ethyl ether | 7.2 | . . . | 5,400,000 |
| Cis-2-butene | 6.4 | . . . | inactive |
| Trans-2-butene | 1.9 | 3.2 | inactive |
| Allene | 1.4 | 1.5 | 14,000 |
| Isobutene | 0 | 0 | inactive |
| Methane | 0 | 0 | inactive |

\* The stability constant $K_1 = (U.AgNO_3)_L / (U)_L (AgNO_3)_L$ where $(U)_L$ is the concentration of the compound in the liquid phase, $(AgNO_3)_L$ the concentration of $AgNO_3$ in the liquid phase, and $(U.AgNO_3)_L$ the concentration of the metal complex in the liquid phase. Method 1 is that of Muhs and Weiss (28), and values reported by them at a temperature of 40° are only slightly lower than those in the table. Method 2 is that of Gil-Av and Herling (18). The constants for acetylenic compounds and vinyl halides could not be determined as the alkynes did not pass through the column, and the halogenated olefines apparently were influenced by a strong salting out effect which was not corrected for by the methods employed.
\*\* This activity may be due to a slight contamination with ethylene (see footnote to table II).

steric factors (28). Similarly, biological activity requires unsaturation, is affected by steric factors, and influenced by electron delocalization, and this suggests that ethylene and its analogues might bond to a metal containing receptor site. The empirical requirements for biological activity outlined in the preceding section bear a marked resemblance to the rules for silver binding to olefines (28) and accordingly there is a close correlation between biological activity and the stability constants of various olefin-silver complexes (table III). Even in the case of allene, ostensibly an exception (table III), the data on metal binding explains why the biopotency is only one-hundreth that of propylene, whereas just the opposite result might be predicted from rules 2 and 3 above. Another reason for postulating a metal containing receptor site is the fact that CO produces ethylene-like symptoms at concentrations as low as a few hundred ppm, whereas due to competition with $O_2$ it does not inhibit the respiration of pea and other plant tissues unless it is applied at a much higher concentration. Perhaps by coincidence the $K_A$ for CO binding to the ethylene receptor (table II) and the $K_I$ for CO binding to plant cytochrome oxidase (22) are closely similar. As CO and ethylene must attach to the same receptor to produce the same effect, and since CO characteristically inhibits only enzymes containing metal, it can be concluded that ethylene must bond to a metallic receptor when it produces a biological effect. It will be shown subsequently that the attachment of these gases to the receptor is enhanced by $O_2$, a further indication that the receptor contains metal, for proteins which bind $O_2$ characteristically have this property and CO inhibits particularly those enzymes which

react directly with $O_2$. Finally it should be noted that none of the inactive substances (10, 25, and table II, footnote) possess the ability to complex metal except cyanide, and it was not really possible to test this compound in the pea straight growth test since it prevented almost all growth and respiration at a relatively low concentration.

The nature of the metal involved is not known, but there is circumstantial evidence that it may be zinc. We find that month old zinc deficient tomato plants hardly respond to ethylene even overnight, whereas plants deficient in copper, iron, phosphorous or nitrogen show strong epinastic symptoms within a few hours.

*Competitive Inhibition of Ethylene Action by $CO_2$.* To account for the observation that $CO_2$ delays and ethylene hastens fruit ripening, Kidd and West (24) proposed that $CO_2$ interferes with the action of ethylene. At the molecular level such an interaction is not difficult to envision because $CO_2$ is a close structural analogue of allene, a compound which substitutes for ethylene in both the pea growth and fruit ripening assays:

$O = C = O \quad H_2C = C = CH_2 \quad H_2C = CH_2$
carbon dioxide     allene     ethylene

Since $CO_2$ possesses the essential structural features needed for ethylene action, except that it lacks the terminal carbon atom and is negatively charged on both ends, might it not act as a competitive inhibitor of ethylene action? To test this possibility the growth of pea stem sections was studied in the presence of differing concentrations of $CO_2$ and ethylene. With ethylene omitted, $CO_2$ did not affect growth except that concentrations greater than 10 % caused a progressive inhibition which was not accounted for by a pH change in

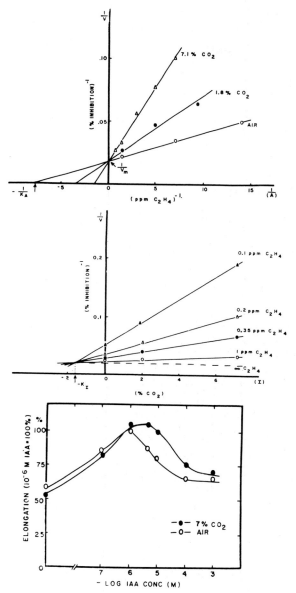

the buffered medium. However, concentrations less than 1.8 % competitively inhibit the action of ethylene (figs 1, 2). The $K_I$ for $CO_2$ in this process is 0.49 mM (1.55 % in the gas phase) based on the data in figure 2. Similar results have been obtained with pea root sections (7).

Many physiological effects can be explained in terms of competition between $CO_2$ and ethylene. For example, ethylene causes flower fading and $CO_2$ reverses this response and also delays normal fading (31, 32); ethylene accelerates and $CO_2$ delays abscission (21); ethylene induced epinasty is not expressed at high concentrations of $CO_2$ (13); and fruit ripening is hastened by ethylene and delayed by $CO_2$ (23, 34). It has recently been reported that the elongation of etiolated pea and sunflower stem sections is inhibited at supra-optimal IAA concentrations not because of any direct action of auxin but because of auxin-induced ethylene production (4). If this is true $CO_2$ should be able to reinstate growth at high concentrations of auxin, and the data in figure 3 shows this to be the case. $CO_2$ is highly effective in reversing the inhibition of elongation caused by 5 μM IAA, and becomes progressively less effective at higher auxin levels because it cannot counteract the action of more than a ppm ethylene (fig 1), the ethylene content of the tissue at 50 μM IAA (4). Similarly, $CO_2$ reinstates auxin inhibited growth in pea roots (7), and it has been demonstrated that auxin inhibits growth in this tissue by inducing ethylene formation (7). These numerous examples suggest that a response to $CO_2$ may be used as a fairly specific diagnostic test for the participation of ethylene in a physiological process.

It is not likely that $CO_2$ frequently functions as a natural growth regulator because the concentration required to cause such an effect is probably higher than that present in most tissues. However, amounts of $CO_2$ in excess of a few percent typically accumulate in the intercellular spaces of preclimacteric fruits, and the concentration may approach 10 % during ripening and the postclimacteric phase. It is probably this endogenous $CO_2$ which raises the threshold for ethylene action in fruits to a slightly higher level than that in vegetative tissue (2). According to this view preclimacteric fruits, with their ethylene content of 0.04 to 0.2 ppm (3), have an internal atmosphere in which a just stimulatory quantity of ethylene is carefully held in check through competitive inhibition by endogenous $CO_2$.

*Oxygen Requirement for Ethylene Action.* To explain why fruit ripening is delayed when the $O_2$ concentration is reduced, Kidd and West (24) proposed that $O_2$ depletion might interfere with the production and action of ethylene. This suggestion is supported by the observation that fruits evolve less ethylene (6) and have a reduced sensitivity to applied ethylene (5, 11, 23) at low $O_2$ partial pressures, but it has also been argued that

FIG. 1 (upper). A Lineweaver-Burk plot (1/V vs 1/A) relating the percent inhibition of growth (V) to the concentration of ethylene (A) at various levels of $CO_2$ (I). $V_m$ is the maximum effect occurring when the ethylene concentration is infinite (1/A = 0).

FIG. B (middle). A plot of 1/V vs (I) at various concentrations of (A). The Michaelis-Menten constant for $CO_2$, $K_I$, has a value of 0.49 mM (1.55 % in the gas phase). The data in figure 1 shows $1/V_m$ to be independent of (I) so that the behavior of tissue at an infinite concentration of ethylene (1/A = 0 can be predicted by the lower dashed line.

FIG. 3 (lower). Reinstatement of auxin inhibited elongation in etiolated pea stem sections by $CO_2$. The fresh weight of the tissue was not increased by $CO_2$ at any auxin concentration in accord with the fact that neither auxin nor ethylene inhibit the water uptake of etiolated pea stem sections (4).

$O_2$ depletion preserves fruits by reducing their rate of respiration (34). To distinguish between these possibilities pea sections were incubated in the presence of differing ethylene and $O_2$ concentrations and their growth, respiration and ability to respond to ethylene determined. $O_2$ consumption (14), $CO_2$ production, and elongation (19) are unaffected at 5 % $O_2$ and half-inhibited at 3 % $O_2$, whereas the efficacy of ethylene is markedly reduced even at 5 % $O_2$ (fig 4). Thus it is possible to lower this tissue's sensitivity to ethylene

without affecting its respiration rate. The same mechanism probably accounts for the retardation of fruit ripening at low $O_2$ tensions, for 5 or 10 % $O_2$ has little effect on the rate of respiration and ethylene formation in apples (6, 23) and bananas (17, 34) but significantly delays the onset of the climacteric in both cases. Pea stem sections were also treated with various concentrations of fluoride, azide and ethylene oxide in order to determine whether a change in growth and respiration alters the tissue's ability to respond to ethylene. Even

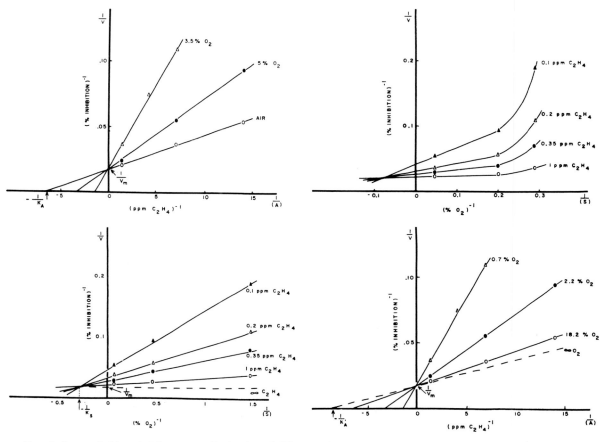

FIG. 4 (upper left). A Lineweaver-Burk plot (1/V vs 1/A) relating the percent inhibition of growth (V) to the concentration of ethylene (A) at various levels of $O_2$ (S). $V_m$ is the maximum effect occurring when the concentration of ethylene is infinite (1/A = O). The data has not been corrected for an $O_2$ deficit in the tissue.

FIG. 5 (upper right). A plot of 1/V vs 1/s at various concentrations of (A). The data has not been corrected for an $O_2$ deficit in the tissue.

FIG. 6 (lower left). A plot of 1/V vs 1/S at various concentrations of (A). The $O_2$ values have been corrected for a 2.8 % deficit in the tissue. The data in figures 4 and 7 show $1/V_m$ to be independent of (S) so that the behavior of the tissue at an infinite concentration of ethylene (1/A = O) can be predicted by the lower dashed line. $K_S$, the Michaelis-Menten constant for $O_2$, has a value of 40 $\mu M$ (2.8 % in the gas phase), which is comparable to values reported for many other oxidases (33) but considerably higher than that for cytochrome oxidase (33).

FIG. 7 (lower right). A plot of 1/V vs 1/S at various concentrations of (A). The $O_2$ values have been corrected for a 2.8 % $O_2$ deficit in the tissue. The inhibition of elongation (V) caused by each concentration of ethylene in the presence of an infinite amount of $O_2$ (1/S = O) is given by the ordinate intersect of each curve in figure 6, and has been used to construct the lower dashed line. $K'_A$, the Michaelis-Menten constant for ethylene at an infinite concentration of $O_2$, equals 0.62 m$\mu$M.

at inhibitor concentrations which retarded growth and respiration by more than 50 % the sensitivity to ethylene remained unchanged and the double reciprocal plot was indistinguishable from the control curve (fig 4, AIR).

The kinetic interpretation of the interaction between $O_2$ and ethylene is complicated by the fact that a plot of 1/V vs 1/S (where S = % $O_2$) at different ethylene concentrations does not yield intersecting straight lines, but rather intersecting curves which are skewed upward at low $O_2$ partial pressures (fig 5). Although many factors can lead to such an effect, in this case it is most likely due to an $O_2$ deficit in the center of the tissue. The factors which give rise to this deficit are well known (33) and it is only necessary here to note that on theoretical grounds the diffusion coefficient for $CO_2$ ($Dco_2$) should be 20 times $Do_2$ (30) so that while a correction may be needed in the case of the $O_2$ data, none is required for the $CO_2$ data in figures 1 and 2. The magnitude of the $O_2$ deficit can be estimated from the fact that the terminal oxidase in pea stem sections is cytochrome oxidase (14, 19), which is half-saturated at 0.2 % $O_2$ (33), whereas the $O_2$ consumption of the tissue is half-inhibited at 3 % $O_2$ (14). When a similar discrepancy has arisen with other tissues it has been possible ultimately to obtain half-inhibition of respiration at 0.2 % $O_2$ by eliminating the causes of the $O_2$ deficit (33). Thus it is clear that under the conditions of the pea straight growth test the stem sections have an internal $O_2$ content approximately 2.8 % below that of the ambient gas mixture. When the $O_2$ concentrations in figure 5 are corrected by subtracting 2.8 % from each, the resultant lines are now straight (fig 6).

A steady state kinetic model exactly accounting for the data in figures 6 and 7 is that termed coupling activation of the second type by Friedenwald and Maengwyn-Davies (16). The equations governing this model are derived by expansion of the Michaelis-Menten equation in a manner analogous to that used in the case of a dissociable inhibitor. Criteria for the model are that the substrate (S = oxygen) must bind to the receptor (metal containing site) before the dissociable activator (A = ethylene) can attach. When this is the case a plot of 1/V vs 1/A (fig 7) yields straight lines intersecting the ordinate at $1/V_m$, and a curve having a positive slope when S = ∞. A plot of 1/V vs 1/S (fig 6) yields straight lines intersecting beyond the ordinate at a point having a value of $-1/K_s$ on the abscissa, and a curve which is parallel to the abscissa when A = ∞. Various other steady state models describing interactions between a dissociable activator, a substrate, and receptor have been developed (16) but none are consistent with the data in figures 6 and 7.

It is not necessary to envision a simultaneous attachment of $O_2$ and ethylene to the metal of the receptor as previously proposed (3). The same

kinetics should result if the metal of the receptor can be reversibly oxidized and reduced, provided that oxidation occurs rapidly after $O_2$ has attached and ethylene only binds to and activates the oxidized form. In fact it is even conceivable that $O_2$ does not attach directly to this receptor, for it might bring about oxidation of the receptor indirectly through a coupled O/R system. In any event we envision the ethylene receptor to be a metal containing compound which is oxidized directly or indirectly through the intervention of molecular $O_2$, and while in the oxidized form activated by ethylene to produce some fundamental change in the metabolism of plant tissue.

## Literature Cited

1. BIDLER, L. M. 1954. A theory of taste stimulation. J. Gen. Physiol. 38: 133–39.
2. BURG, S. P. AND E. A. BURG. 1962. Role of ethylene in fruit ripening. Plant Physiol. 37: 179–89.
3. BURG, S. P. AND E. A. BURG. 1965. Ethylene action and the ripening of fruits. Science 148: 1190–96.
4. BURG, S. P. AND E. A. BURG. 1966. The interaction between auxin and ethylene and its role in plant growth. Proc. Natl. Acad. Sci. 55: 262–69.
5. BURG, S. P. AND E. A. BURG. 1966. Fruit storage at subatmospheric pressures. Science 153: 314–15.
6. BURG, S. P. AND K. V. THIMANN. 1959. The physiology of ethylene formation in apples. Proc. Natl. Acad. Sci. 45: 335–44.
7. CHADWICK, A. V. AND S. P. BURG. 1966. Role of ethylene in root geotropism Plant Physiol. 41: liii.
8. COOPER, W. C. AND P. C. REECE. 1942. Induced flowering of pineapples under Florida conditions. Proc. Florida State Hort. Soc. 54: 132–38.
9. CROCKER, W., A. E. HITCHCOCK, AND P. W. ZIMMERMAN. 1935. Similarities in the effects of ethylene and the plant auxins. Contrib. Boyce Thompson Inst. 7: 231–48.
10. CROCKER, W., P. W. ZIMMERMAN, AND A. E. HITCHCOCK. 1932. Ethylene-induced epinasty of leaves and the relation of gravity to it. Contrib. Boyce Thompson Inst. 4: 177–218.
11. DENNY, F. E. 1924. Hastening the coloration of lemons. J. Agr. Res. 27: 757–69.
12. DENNY, F. E. 1935. Leaf-epinasty tests with chemical vapors. Contrib. Boyce Thompson Inst. 7: 191–95.
13. DENNY, F. E. 1935. Testing plant tissue for emanations causing leaf epinasty. Contrib. Boyce Thompson Inst. 7: 341–47.
14. EICHENBERGER, E. AND K. V. THIMANN. 1957. Terminal oxidases and growth in plant tissues. IV. On the terminal oxidases of etiolated pea internodes. Arch. Biochem. Biophys. 67: 466–78.
15. FOSTER, R. G., D. H. MCRAE, AND J. BONNER. 1952. Auxin induced growth inhibition, a natural consequence of two point attachment. Proc. Natl. Acad. Sci. 38: 1014-22.

16. FRIEDENWALD, J. S. AND G. D. MAENGWYN-DAVIES. 1954. In a Symposium on the Mechanism of Enzyme Action. W. D. McElroy and B. Glass, eds. Johns Hopkins University Press, Baltimore. p 154.

17. GANE, R. 1936. A study of the respiration of bananas. New Phytologist 35: 383–402.

18. GIL-AV, E. AND J. HERLING. 1962. Determination of the stability constants of complexes by gas chromatography. J. Phys. Chem. 66: 1208–09.

19. HACKETT, D. P. AND H. A. SCHNEIDERMAN. 1953. Terminal oxidases and growth in plant tissues. I. The terminal oxidase mediating growth of *Avena* coleoptile and *Pisum* stem sections. Arch. Biochem. Biophys. 47: 190–204.

20. HANDBOOK OF CHEMISTRY AND PHYSICS. 1962. The Chemical Rubber Company, Cleveland. p 1709.

21. HOLM, R. E. AND F. B. ABELES. 1966. Enhancement of protein and RNA synthesis by ethylene. Plant Physiol. 41: liii–liv.

22. IKUMA, H., F. J. SCHINDLER, AND W. D. BONNER. 1964. Kinetic analysis of oxidases in tightly coupled plant mitochondria. Plant Physiol. 39: lx.

23. KIDD, F. AND C. WEST. 1934. The influence of the composition of the atmosphere upon the incidence of the climacteric in apples. Gt. Brit. Dept. Sci. Ind. Res. Rept. Food Invest. Bd. 1933: 51–57.

24. KIDD, F. AND C. WEST. 1945. Respiratory activity and duration of life of apples gathered at different stages of development and subsequently maintained at a constant temperature. Plant Physiol. 20: 467–504.

25. KNIGHT, L. I., C. ROSE, AND W. CROCKER. 1910. Effect of various gases and vapors upon etiolated seedlings of the sweet pea. Science 31: 635–36.

26. LENHOFF, H. 1965. Some physical chemical aspects of the macro- and micro environment surrounding Hydra during activation of their feeding behavior. Amer. Zool. 5: 515–24.

27. LOESECKE, H. W. VON. 1950. Bananas. Interscience Publishing, Inc., New York. p 53.

28. MUHS, M. A. AND F. T. WEISS. 1962. Determination of equilibrium constant of silver-olefin complexes using gas chromatography. J. Am. Chem. Soc. 84: 4698–705.

29. RICHARDS, H. M. AND D. T. MACDOUGAL. 1904. The influence of carbon monoxide and other gases upon plants. Bull. Torrey Botan. Club 3: 57–66.

30. SCHMIDT, C. F. 1956. In: Medical Physiology, P. Bard, ed. C. V. Mosby Co., St. Louis. p 314.

31. SMITH, W. H. AND J. C. PARKER. 1966. Prevention of ethylene injury to carnations by low concentrations of carbon dioxide. Nature 211: 100–01.

32. SMITH, W. H., J. C. PARKER, AND W. W. FREEMAN. 1966. Exposure of cut flowers to ethylene in the presence and absence of carbon dioxide. Nature 211: 99–100.

33. YOCUM, C. S. AND D. P. HACKETT. 1957. Participation of cytochromes in the respiration of the *Aroid spadix*. Plant Physiol. 32: 186–91.

34. YOUNG, R. E., J. ROMANI, AND J. B. BIALE. 1962. Carbon dioxide effects on fruit respiration. II. Responses of avocados, bananas and lemons. Plant Physiol. 37: 416–22.

Bock G. Chan and John C. Cain. 1967. The effect of seed formation on subsequent flowering in apple. Proceedings of the American Society for Horticultural Science 91:63–68.

Bock G. Chan

John C. Cain

Bock G. Chan was born in Kwangtung, China, in 1935 and emigrated in 1950 to the United States to join his father in Los Angeles. While an undergraduate at the University of California, Davis, he was advised by Julian C. Crane, whose research work on the application of plant hormones on apricots and figs kindled young Bock's interest in plant morphogenesis. Following graduation in 1958, Bock moved east to Cornell University to undertake graduate studies in the department of pomology, which at that time contained an unusually high density of plant growth regulator researchers, including J. C. Cain, F. G. Dennis, L. J. Edgerton, C. G. Forshey, M. B. Hoffman, and L. E. Powell. I was a graduate student at that time. Chan received his Master's degree in 1960, spent 1961–1963 in the U.S. Army, and returned to work at Geneva as an experimentalist with John Cain.

John Carlton Cain grew up in Blakely, Georgia, graduated from the University of Florida in 1935, and worked at the Florida Agricultural Experiment Station until 1940. After service in the armed forces he took graduate studies at Cornell University, received a Ph.D. degree in 1946, and immediately joined the staff of Cornell stationed in the department of pomology at the New York State Agricultural Experiment Station at Geneva. Cain made outstanding contributions to plant nutrition, mechanical pruning and harvesting, and tree spacing and worked closely with many graduate students, including Bock Chan.

The Chan and Cain work had its beginnings in 1959 when Karl Brase introduced Bock Chan to the 'Spencer Seedless' apple tree in the Geneva apple collection—a cultivar he had used in connection with his M.Sc. thesis work (Brase 1937). 'Spencer Seedless' bears annual crops of pathenocarpic fruits but if hand pollinated, produces seeded fruits. The Chan and Cain connection came to an end in 1976 when Bock started Ph.D. studies with L. L. Creasy at Ithaca. Chan is now a supervisory plant physiologist with the U.S. Department of Agriculture in Berkeley, California, and Cain spends much of his retirement happily fishing in Florida.

Biennial bearing in apple trees has been observed since early times but is no longer the serious problem to the commercial apple grower that it once was. The increased planting of annual-bearing cultivars, and more important, the development and use of suitable chemical thinning programs, now regulate flower production. Although we know how to regulate cropping, the precise cause of biennial bearing still eludes us.

A concentrated effort was made in the 1930s to understand and regulate biennial bearing since it was a condition ". . . found today to a greater or less degree in practically every orchard . . ." (Harley et al. 1942). Inspired by McCormick's flower thinning work (McCormick 1934), Bobb and Blake (1938) blossom thinned, by hand, the notorious biennial-bearing cultivar 'Wealthy' over several years and induced annual bearing. The next step was to demonstrate that flower thinning could be done with chemicals rather than by hand. Hoffman (1947) successfully thinned 'Wealthy' trees with a dinitro compound, and Burkholder and McCown (1941) inadvertently did the initial experiments with naphthaleneacetic acid that led to its widespread use as an early postbloom fruit thinner. These pioneering experiments provided the foundation for today's excessive use of exogenously applied chemicals for regulating biennial bearing.

In 1953, Nitsch demonstrated that fruit setting and growth in strawberry were controlled by the seed (achene) and presented evidence that auxin originating in the seed was a key factor. Luckwill (1948) demonstrated the presence of auxins in apple seeds, but it remained for Chan and Cain (1967) to demonstrate the inhibiting effect of seeds on flower bud initiation. They showed that more than 90 percent of the spurs that bore seedless fruit flowered the next year, whereas relatively few spurs bearing seeded fruits did so.

Chan and Cain's work (1967) stimulated work on hormonal control of biennial bearing by endogenous growth regulators emanating from the seeds. Luckwill (1948) had found auxins in apple seeds, but their role in flower bud formation and annual bearing is still far from clear (for a recent review, see Buban and Faust 1982). Luckwill's pioneering research became somewhat easier to pursue in the 1960s with the development of suitable techniques for the purification of plant extracts and the chemical identification of endogenous hormones. Different classes of hormones were identified in seeds and their quantitative and qualitative changes studied during fruit growth and flower initiation (Hoad 1979). At the time of Chan and Cain's work, gibberellins had been implicated in cropping in apple. Guttridge (1962) had inhibited flower formation with spray application of gibberellins, and Dennis and Nitsch (1966) had identified the gibberellins $GA_4$ and $GA_7$ in apple seeds. Luckwill et al.

(1969) confirmed the presence of gibberellins in apple seeds and showed that their concentration builds up rapidly from 5 to 6 weeks after full bloom until about 9 weeks, after which the concentrations fall.

Much attention has been given to the implication of gibberellins in the chemical control of morphogenetic processes in apple buds, but it is likely that other hormones, such as cytokinins, auxins, and abscisic acid, are involved. Chan and Cain simply and elegantly demonstrated that seed development in apple inhibits flower bud formation. Twenty years later we know much about the role of endogenous hormones, but much remains to be learned about how they control flowering.

John T. A. Proctor
Department of Horticultural Science
Ontario Agricultural College
University of Guelph
Guelph, Ontario, Canada N1G 2W1

## REFERENCES

Bobb, A. C., and M. A. Blake. 1938. Annual bearing in Wealthy apple was induced by blossom thinning. Proc. Am. Soc. Hort. Sci. 36:321–327.

Brase, K. D. 1937. The vascular anatomy of the flowers of *Malus domestica* Borkh. f. *apetala* Van Eseltine. M.S. thesis, Cornell University, Ithaca, N.Y.

Buban, T., and M. Faust. 1982. Flower bud induction in apple trees: Internal control and differentiation. Hort. Rev. 4:174–203.

Burkholder, C. L., and M. McCown. 1941. Effect of scoring and of $\alpha$-naphthyl acetic acid and amide spray upon fruit set and of the spray upon pre-harvest fruit drop. Proc. Am. Soc. Hort. Sci. 38:117–120.

Dennis, F. G., Jr., and J. P. Nitsch. 1966. Identification of gibberellins $A_4$ and $A_7$ in immature apple seeds. Nature 211:781–782.

Guttridge, C. G. 1962. Inhibition of fruit bud formation in apple with gibberellic acid. Nature 196:1008.

Harley, C. P., J. R. Magness, M. P. Masure, L. A. Fletcher, and E. S. Degman. 1942. Investigations on the cause and control of biennial bearing of apple trees. U.S. Dep. Agr. Tech. Bull. 792.

Hoad, G. V. 1979. Growth regulators, endogenous hormones and flower initiaton in apple. Annu. Rep. Long Ashton Res. Sta. 1979, p. 199–206.

Hoffman, M. B. 1947. Further experience with the chemical thinning of Wealthy apples during bloom and its influence on annual production and fruit size. Proc. Am. Soc. Hort. Sci. 49:21–25.

Luckwill, L. C. 1948. The hormone content of the seed in relation to endosperm development and fruit drop in the apple. J. Hort. Sci. 24:32–44.

Luckwill, L. C., P. Weaver, and J. MacMillan. 1969. Gibberellins and other growth hormones in apple seeds. J. Hort. Sci. 44:413–424.

McCormick, A. C. 1934. Control of biennial bearing in apples. Proc. Am. Soc. Hort. Sci. 30:326–329.

Nitsch, J. P. 1953. The physiology of fruit growth. Annu. Rev. Plant Physiol. 4:199–236.

# The Effect of Seed Formation on Subsequent Flowering in Apple[1]

By Bock G. Chan and John C. Cain,[2] *New York State Agricultural Experiment Station, Cornell, University, Geneva, New York*

*Abstract.* Inhibition of flower bud formation by seed development was demonstrated in apple, cv. 'Spencer Seedless' and 'Ohio 3'. Removal of seeded fruits at intervals after bloom indicated that 65% of the inhibition occurred within the first 3 weeks after pollination. Seedless fruits did not have any effect on subsequent flowering. Therefore, seed formation appeared to be the controlling factor in flower initiation and not nutritional competition by developing fruits. It is suggested that correlative inhibition acts through a hormonal process.

## INTRODUCTION

THE problem of alternate bearing in fruit trees, particularly apples, has been thoroughly reviewed (4, 16, 18). Since Kraus and Kraybill (10) proposed the C-N hypothesis for fruitfulness in tomatoes, the concept of nutritional competition between developing fruits and the formation of flower buds has received the most attention in explaining the inhibition of flower bud formation by abundant fruiting in apples. However, the results of experiments based on this concept have been inconclusive. After the discovery of indoleacetic acid, the hormone concept of flowering was again suggested. However, no definitive evidence for the influence of hormones on alternate bearing has been reported. The cause of alternate bearing is still obscure.

The degree of alternate bearing varies with variety and size of crop. The apetalous varieties 'Spencer Seedless' and 'Ohio 3', bear annual crops of parthenocarpic fruits, but following hand pollination they produce seeded fruits (2). Therefore, they provide unique material to study the effect of fruiting and seed formation on subsequent flowering.

The objective of this work was to determine the effect of seed development on the inhibition of flower bud formation.

## MATERIALS AND METHODS

A 29-year-old 'Spencer Seedless' tree at the New York State Agricultural Experiment Station was used in 1963–66. This tree has produced large annual crops of seedless fruit. Several well distributed branches with many spurs were selected. On each branch, groups of 3 spurs, 2–5 cm apart, were selected in May of 1963, tagged, and treated as follows: 1) the flowers of one spur were not pollinated, and parthenocarpic fruits developed since insects do not visit the apetalous variety, 2) the flowers of another spur

[1]Received for publication December 30, 1966. Approved by the Director of the New York State Agricultural Experiment Station for publication as Journal Paper No. 1538, December 19, 1966.

[2]It is a pleasure to acknowledge the invaluable help of Charlotte Pratt in the preparation of this manuscript.

63

were hand-pollinated with 'McIntosh' pollen and seeded fruits developed and, 3) all flowers were removed from a third spur. All fruits were harvested at maturity, and seeds were counted. In May of 1964, flowering of these spurs was noted.

In 1964–65 and 1965–66 all of the experimental flowers were either pollinated with 'McIntosh' pollen or not treated. Some of the fruits resulting from pollination were removed at random on one or more dates during the season, and these spurs were tagged. The remaining fruits were weighed at maturity, and seeds were counted. Flowers were counted in the following spring. A tree of 'Ohio 3', another apetalous variety, was treated similarly in 1965–66.

In greenhouse experiments of 1963–66, bench grafts of 'Spencer Seedless', started in 1961 in sub-irrigated sand cultures (3) were used. Pollination was done with 'McIntosh' pollen the first week of April, and fruits were harvested in early October and weighed. Seeds were counted. The axillary shoots were measured, and spurs were tagged for flower counts the following April.

## RESULTS

Bearing of seeded fruits greatly reduced the percentage of spurs flowering the following year (Table 1). In 1965–66, the inhibition of flowering in 'Spencer Seedless' was not as great as in the two previous seasons. Spurs with seedless fruits formed flowers as frequently as spurs from which flowers had been removed.

Fig. 1 shows the effect of number of seeds per spur of 'Spencer Seedless' on the percentage of spurs forming flowers. The large numbers of seeds per spur were due to the presence of several fruits on some spurs as well as two whorls of carpels in the fruits (2). There were no differences in inhibition of flowering by different numbers of seeds in 1963–64 and 1964–65, in which very high percentages of inhibition of flowering occurred. However, in 1965–66 there was some increase in percentage of inhibition of flowering with increasing number of seeds. However, the greatest influence was the presence or absence of seeds.

*Table 1.* The effect of seedless and seeded fruits of facultatively parthenocarpic varieties of apples on flowering of the same spurs in the following year.

| | Number of spurs | | | Percentage of spurs flowering | | | |
|---|---|---|---|---|---|---|---|
| | 1963–64 | 1964–65 | 1965–66 | 1963–64 | 1964–65 | 1965–66 | Mean |
| *Spencer Seedless* | | | | | | | |
| Seedless fruit... | 50 | 50 | 20 | 96.0 | 100.0 | 90.0 | 95.3 |
| Seeded fruit... | 52 | 165 | 49 | 5.8 | 6.7 | 26.7 | 13.1 |
| No fruit....... | 82 | — | — | 97.6 | — | — | 97.6 |
| *Ohio 3* | | | | | | | |
| Seeded fruit... | — | — | 86 | — | — | 31.4 | 31.4 |
| No fruit....... | — | — | 187 | — | — | 98.0 | 98.0 |

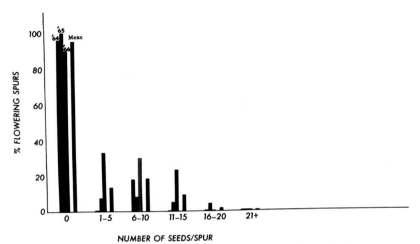

*Fig. 1.* The relation of the number of seeds in the fruit produced per spur of 'Spencer Seedless' to the percentage of flowering spurs in the following year.

Fig. 2 shows the effect of time removal of seeded fruits on the inhibition of flowers in the spurs. The differences in the percentages of flowering spurs at maturity, 150 days after pollination, were greater between seasons than between varieties; however, 30 days after pollination, the percentages were all near the 50% point and most of the inhibition of flowering had occurred by that time.

Table 2 shows that the quantity of fruits produced per spur did not markedly influence the flowering response. Although the seeded fruits were heavier than parthenocarpic fruits, some of the spurs with several parthenocarpic fruits bore as much or more fruit than spurs with seeded fruits. The former produced flowers, but the latter were vegetative the following year.

The results of the greenhouse experiment on 'Spencer Seedless' did not show as clear an inhibition of flowering by seed formation as observed in the field experiments. For example, in 1965–66, 38% of the spurs which had borne seeded fruits were vegetative compared to 15% of the spurs which had borne seedless fruits. Of the spurs which had borne seeded fruits, 67% were vegetative if the axillary shoot was less than 5 cm., but flowering occurred in the axillary shoot longer than 5 cm. Of 1966–67 experiment, only the spurs with axillary growing points less than 2 cm were observed; 65% of the spurs which had borne seeded fruits were vegetative as compared to 28% of the spurs which had borne seedless fruits.

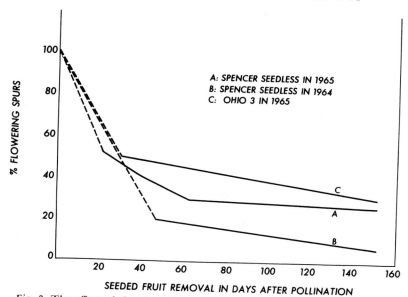

Fig. 2. The effect of the time of removal of seeded fruits on the percentage of flowering spurs in the following year.

## DISCUSSION

The data clearly show a correlative inhibition of flowering by seed formation in 'Spencer Seedless'. 'Ohio 3' did not produce seedless fruits in 1965 although it has produced seedless fruits. The spurs of 'Ohio 3' which had borne seeded fruits were mostly vegetative while those spurs without any fruits were mostly flowering. Nutritional competition by fruit development alone does not appear to be a controlling factor. Parthenocarpic fruits did not inhibit flower formation on the same spurs; removal of seeded fruits 3 weeks after pollination did not overcome the inhibitory effect of seed formation. The fruits removed were still small, and

Table 2. The effect of quantity of fruits of 'Spencer Seedless' on flowering of the same spur in the following year.

| | Year | No. of spurs | Fruits/ spur | Gm/ fruit | Gm of fruit/ spur |
|---|---|---|---|---|---|
| Seeded fruit on spurs vegetative following year...... | 1964 | 152 | 1.38 | 98.2 | 135.1 |
| | 1965 | 34 | 1.50 | 84.5 | 111.7 |
| Seeded fruit on spurs flowering following year...... | 1964 | 11 | 1.27 | 119.6 | 152.2 |
| | 1965 | 12 | 1.33 | 87.7 | 116.8 |
| Seedless fruit on spurs vegetative following year..... | 1964 | 0 | 0.0 | 0.0 | 0.0 |
| | 1965 | 2 | 2.0 | 57.0 | 114.0 |
| Seedless fruit on spurs flowering following year..... | 1964 | 50 | 1.28 | 65.2 | 83.5 |
| | 1965 | 18 | 1.06 | 70.5 | 74.4 |

seeds had only begun to develop. At this stage of development 'McIntosh' fruits normally had less than 2% of the volume of mature fruits[3], and the young seed consisted mostly of nucellus and a very small free-nuclear endosperm (13). In addition the quantity of fruits produced per spur did not influence subsequent flowering. This suggests that seed development rather than nutritional competition may be a major factor in alternate bearing.

In the greenhouse experiments, the axillary buds of the young spurs were often forced out and grew into axillary shoots in the same season, contrary to the field experiments in which the axillary bud remained in bud form. The longer the axillary shoots grew the more of them flowered. Limited observations in the field showed that the axillary shoots of spurs which had borne seeded fruits also flowered. This could be an effect of space between the differentiating meristem of the axillary shoot and the developing seeds in the fruit, or of the time when the growing meristem was formed in relation to the developing seeds in the fruits. The inhibitory effect of seed formation on flowering also appeared to be localized, as was shown in the 1963 experiment in Table 1. It was found only in the spurs which had borne the seeded fruits and not in the adjacent spurs at a distance of 2 to 5 cm, which had produced either seedless fruit or no fruit.

The hormonal concept for alternate bearing has also been suggested (4, 9). Immature apple seeds have been found to contain auxin (13) and gibberellins (5). It should be noted that spray application of gibberellins have inhibited flower formation (8). However, the greatest amount of gibberellins was found when the endosperm was of maximum size, about 60 days after bloom. Most of the inhibition of flowering by seed formation as shown in Fig. 2 occurred in the first month after pollination. The inhibition of flowering may be caused either by the developing seeds using the florigen from the leaves or liberating an inhibitor when flower initiation occurs the first month after bloom (1, 6). Auxin is known to move out from the achene into the receptacle of strawberry (14). A flowering inhibitor was found to be translocated through the stolon from mother strawberry plant to daughter plant (17). An extract of Xanthium and sunflower leaves induced flowering in Xanthium (12). The concept of the flowering process being regulated by hormones is well supported by evidence obtained from experiments on photoperiodism and vernalization, but the complex mechanism of the physiology of flowering is not fully understood (11, 15).

---

[3]Personal communication with O. F. Curtis, Jr.

### LITERATURE CITED

1. BRADFORD, F. C. 1915. Fruit bud development in apples. *Ore. Agr. Exp. Sta. Bul.* 129.
2. BRASE, K. D. 1937. The vascular anatomy of the flowers of *Malus domestica* Borkh. f. *apetala* Van Eseltine. (Thesis) Cornell University, Ithaca, New York.

3. CAIN, J. C. 1963. Automatic sub-irrigation equipment for sand cultures. *Proc. Amer. Soc. Hort. Sci.* 82:631–636.

4. DAVIS, L. D. 1957. Flowering and alternate bearing. *Proc. Amer. Soc. Hort. Sci.* 70:545–556.

5. DENNIS, F. G., Jr., and J. P. NITSCH 1966. Identification of gibberellins A₄ and A₇ in immature apple seeds. *Nature* 211:781–782.

6. GOFF, E. S. 1899. The origin and early development of the flower in the cherry, plum, apple and pear. *Wisc. Agr. Exp. Sta. Ann. Rep.* 16:289–303.

7. GUTTRIDGE, C. G. 1959. Evidence for a flower inhibitor and vegetative growth promoter in strawberry. *Ann. Bot.* 23:351–360.

8. ————. 1962. Inhibition of fruit bud formation in apple with gibberellic acid. *Nature* 196:1008.

9. HARLEY, C. P., L. A. FLETCHER, E. S. DEGMEN, J. R. MAGNESS and M. P. MASURE. 1942. Investigations on the cause and control of biennial bearing of apple trees. *U.S. Dept. Agr. Tech. Bul.* 792.

10. KRAUS, E. J. and H. R. KRAYBILL. 1918. Vegetation and reproduction with special reference to the tomato. *Ore. Agr. Exp. Sta. Bul.* 149.

11. LANG, A. 1965. Physiology of flower initiation. In Encyclopedia of Plant Physiol. Ed. W. Ruhland. Springer-Verlag, Berlin. Vol. XV, Part 1: 1380–1536.

12. LINCOLN, R. G., D. L. MAYFIELD, R. O. HUTCHINS, A. CUNNINGHAM, K. C. HAMNER and B. H. CARPENTER. 1962. The floral initiation of Xanthium in response to application of an extract from a day-neutral plant. *Nature* 195:918.

13. LUCKWILL, L. C. 1948. The hormone content of the seed in relation to endosperm development and fruit drop in the apple. *J. Hort. Sci.* 24:32–44.

14. NITSCH, J. P. 1950. Growth and morphogenesis of the strawberry as related to auxin. *Amer. J. Bot.* 37:211–215.

15. ————. 1965. Physiology of flower and fruit development. In Encyclopedia of Plant Physiol. Ed. W. Ruhland. Springer-Verlag, Berlin. Vol. XV, part 1:1537–1647.

16. SINGH, L. B. 1948. Studies in biennial bearing II. A review of the literature. *J. Hort. Sci.* 24:45–65.

17. THOMPSON, P. A., and C. G. GUTTRIDGE. 1959. Effect of gibberellic acid on the initiation of flowers and runners in the strawberry. *Nature* 184: B.A. 72–73.

18. WIGGANS, C. C. 1918. Some factors favoring or opposing fruitfulness in apple. *Mo. Agr. Exp. Sta. Res. Bul.* 32.

S. D. Goldberg and M. Shmueli. 1970. Drip irrigation: A method used under arid and desert conditions of high water and soil salinity. Transactions of the American Society of Agricultural Engineers 13(1):38–41.

S. D. Goldberg

M. Shmueli

Irrigation has played a central role in agriculture since the dawn of recorded history. Evidence of irrigated agriculture has been traced back as far as 8000 years. The development of irrigated agriculture allowed agrarian societies of stable communities to become established in areas of limited rainfall, such as the Middle East, the "cradle of Western civilization." Today, over 220 million hectares of agricultural land are irrigated worldwide, attesting to the continued importance of this technology to human welfare (Celestre 1973).

Irrigation methods changed very little until the twentieth century. For most of that time, irrigation techniques were limited to gravity-flow systems applying large volumes of water to replenish soil reserves. Even today, flood and furrow irrigation are still the most widely used methods of agricultural water application in the world.

With high-volume irrigation methods, a significant percentage of applied water is normally lost through evaporation, runoff, and/or leaching. This problem of water loss is particularly serious in arid and semiarid regions, where suitable water supplies are limited and water costs are often high. Recognition of the seriousness of this problem led to attempts beginning over a century ago to reduce irrigation water losses with new, lower-volume techniques. The first recorded attempt to reduce irrigation losses with new methodology was undertaken in Germany in the 1860s (Halevy et al. 1973). As early as 1874, a patent was issued in the United States for a device very similar to the modern trickle irrigation emitter (Goldberg 1981). In the 1920s, various trials were carried out using porous or perforated pipe (Bucks and Davis 1986). In the 1930s, fruit growers in the Goulborn valley of Australia chiseled small holes in galvanized pipe to make small, so-called "trickle"-irrigation systems (Black 1976). By the time of World War II, numerous attempts had been made to develop both surface and underground low-volume irrigation methods, with little success. Serious problems of clogging, nonuniform water distribution, and high material costs stymied progress.

The discovery of relatively inexpensive, durable plastics during and after World War II ushered in a new era in irrigation technology (Shoji 1977). By the 1950s, low-volume trickle-irrigation systems based on narrow-bore "spaghetti tubes" had been developed for greenhouse use (Waterfield 1973).

The first successful modern trickle-irrigation emitter was designed and developed by Symcha Blass in Israel, beginning around 1959 (Anonymous 1971; Gustafson 1973). This new low-volume irrigation system was first evaluated under field conditions of brackish water and sandy soils in the early 1960s in the Negev Desert by Y. Zohar of the Israel Ministry of Agriculture Extension Service. His experimental irrigation systems were placed underground to reduce water losses to the minimum. By 1965 Zohar concluded that low-volume irrigation had considerable promise but that subirrigation (underground installation) produced significant problems of clogging as well as plugging by root growth.

Thus the stage was set for what proved to be the decisive period in the evolution of modern trickle irrigation in the field. In the early 1960s, S. D. Goldberg was a senior lecturer in irrigation at the faculty of agriculture of Hebrew University in Rehovot. Goldberg, born in Jerusalem in 1915, had been trained in hydraulic and irrigation engineering in England (B.Sc. degree, 1936, University of London) and California (M.Sc. degree, 1939, University of California). He began his research with Hebrew University in 1950. Menachem Shmueli, born in Amsterdam, Holland, in 1937, began graduate studies under Goldberg in 1962 after completion of high school and military service in Israel. Shmueli finished his M.Sc. degree in 1964.

About the time Shmueli completed his graduate training, the Hebrew University became involved in development work aimed at encouraging settlement in the Arava desert, a harsh and uninviting environment in southern Israel. High temperatures, low relative humidity, saline soils, and lack of good-quality water made this area seem most unpromising for agriculture. In 1964 an experiment station was established under the auspices of the Hebrew University at Yotvata, near Eilat. Shmueli moved there with his family in the summer of 1964 to become the first station director. The mandate of this station was to explore ways to encourage greenery to grow in the bleak desert environment. It was to be here that Goldberg, Shmueli, and their coworkers were to give modern trickle irrigation its first, and perhaps harshest, field evaluation.

The studies of Zohar in the Negev had shown serious problems with subsurface trickle-irrigation installations. It was therefore decided that further trials should be carried out with the Blass irrigation system, but with the emitters installed on the soil surface rather than underground. In addition, climatic records suggested that temperature conditions of the winter season in the Arava (the off-season for Europe) might favor the production of vegetable crops. Goldberg and Shmueli initiated their trickle-irrigation trials in the winter of 1964–1965. During the first several years of research, they carried out numerous tests with crops, including tomatoes, muskmelons, cucumbers, and grapes. Trials were conducted with saline water (electrical conductivity of 3000 microhms per cm, at 25° C; chloride concentration of about 600 mg per liter and sometimes higher; and sulphate concentration of about 700 mg per liter) applied to the very sandy soils of the area. In their many trials, they examined emitter spacing, irrigation frequency, water quality, irrigation efficiency, fertilizer application through trickle-irrigation system, effects of irrigation-borne salts on crop performance, and salt distribution in the soil profile, and compared trickle irrigation with sprinkler and furrow irrigation (Shmueli et al. 1967).

Their striking success in growing vegetable crops with saline water under very limiting soil conditions led to rapid expansion of commercial trickle-irrigation systems in Israel. By 1970, several station reports and publications on drop (later drip) irrigation, largely in Hebrew, had sparked considerable interest among irrigation scientists in other arid regions (Goldberg and Shmueli 1969, Halevy et al. 1973, Shmueli et al. 1967). By this time a few early reports of results from other locations had appeared as well, but were not widely known (Black 1969, Celestre 1973, Woudt 1968).

In early 1970, Goldberg and Shmueli summarized their array of trials in the Arava desert in their article in the *Transaction of the American Society of Agricultural Engineers*. This publication is noteworthy for several reasons. It represented the first comprehensive, widely disseminated, journal-type article in English on the application of modern trickle-irrigation methodology under field conditions. The paper covered in some detail system design, operation, yield responses, salinity effects, and the advantages of trickle irrigation under arid and desert environmental conditions. For the first time, details of this new technology were readily available to the scientific community.

Within a short time after the appearance of this publication, there was a veritable explosion of interest and research on the application of trickle irrigation to a wide variety of both vegetable and tree-fruit crops. Interest spread rapidly from arid zones to more humid agricultural regions of the world. Within four years after this publication appeared, there had already been two international congresses devoted exclusively to the subject of trickle irrigation. Today, trickle irrigation is an accepted and rapidly expanding irrigation method used worldwide. By 1982, over 400,000 hectares of agricultural land in the world were being irrigated with trickle systems (Bucks and Davis 1986).

Both Goldberg and Shmueli continued to be active in trickle-irrigation research, with numerous publications to their credit (Elfving 1982). Goldberg became a full professor in the Department of Irrigation in 1976 and continues to be active in teaching and international consulting activities. Shmueli left the public service of Israel in 1977 to work in private enterprise. He died in 1982.

As with most technological innovation, many people contributed to the development of trickle irrigation. The work of Goldberg and Shmueli is noteworthy because they took what was in fact an old idea in irrigation and developed it, along with new materials, equipment, and approaches, into a successful and practical field methodology. In the 19 years since their 1970 publication, there have been many improvements in trickle irrigation theory, equipment, and methods. Its application has spread across the globe to crops and into areas not imagined by those who first succeeded in farming the inhospitable Arava desert. As agricultural water resources become increasingly strained by expanding global population, trickle irrigation and its offshoots are destined to play an increasingly important role in agriculture.

D. C. Elfving
Horticultural Research Institute of Ontario
Simcoe, Ontario, Canada

# REFERENCES

ANONYMOUS. 1971. How trickle came about. p. 29. In: B. Larkman (ed.). Trickle irrigation. Imperial Chemical Industries Australia, Ltd., Melbourne, Australia.

BLACK, J. D. F. 1969. Daily flow irrigation for fruit trees and row crops. Victoria Dep. Agri. Leaflet H191.

BLACK, J. D. F. 1976. Trickle irrigation—a review. Hort. Abstr. 46:1–7, 69–74.

BUCKS, D. A., and S. DAVIS. 1986. Historical development. p. 1–26. In: F. S. Nakayama and D. A. Bucks (eds.). Trickle irrigation for crop production—design, operation and management. Elsevier, New York.

CELESTRE, P. 1973. Report on drip irrigation and similar methods. p. 121–146. In: Trickel irrigation. European Commission on Agriculture, Working Party on Water Resources and Irrigation, Bucharest, Romania, 1972. Water Resources and Development Service, Land and Water Development Division, Food and Agriculture Organization of the United Nations. FAO Irrigation and Drainage Paper 14. FAO, United Nations, Rome.

ELFVING, D. C. 1982. Crop response to trickle irrigation. p. 1–48. In: J. Janick (ed.). Horticultural reviews, Vol. 4. AVI, Westport, Conn.

GOLDBERG, S. D. 1981. Israel and her wilderness. News Isr. 28(4):8–15.

GOLDBERG, S. D., and M. SHMUELI. 1969. Trickle irrigation, a method for increased agricultural production under conditions of saline water and adverse soils. In: Proceedings of the Conference on Arid Lands in a Changing World, Tucson, Ariz.

GUSTAFSON, C. D. 1973. The history of drip irrigation. Irrig. Age 7(11):4–6.

HALEVY, I., M. BOAZ, Y. ZOHAR, M. SHANI, and H. DAN. 1973. Trickle irrigation. p. 75–119. In: Trickle irrigation. European Commission on Agriculture, Working Party on Water Resources and Irrigation, Bucharest, Romania, 1972. Water Resources and Development Service, Land and Water Development Division, Food and Agriculture Organization of the United Nations FAO Irrigation and Drainage Paper 14. FAO, United Nations, Rome.

SHMUELI, M., D. GOLDBERG, and Y. ZOHAR. 1967. Field trials with drop irrigation in the southern Arava of Israel (1965–1967) (in Hebrew). Volcani Inst. Agr. Res. Div. Sci. Publ. Pam. 122.

SHOJI, K. 1977. Drip irrigation. Sci. Am. 237(5):62–68.

WATERFIELD, A. E. 1973. Trickle irrigation in the United Kingdom. p. 147–153. In: Trickle irrigation. European Commission on Agriculture, Working Party on Water Resources and Irrigation, Bucharest, Romania, 1972. Water Resources and Development Service, Land and Water Development Division, Food and Agriculture Organization of the United Nations. FAO Irrigation and Drainage Paper 14. FAO, United Nations, Rome.

WOUDT, B. D. VAN'T. 1968. Trickle irrigation—a promising second tool for a breakthrough in food production in tropical, subtropical and desert areas. p. 88–96. In: International Commission on Irrigation and Drainage Bulletin, July 1968.

# Drip Irrigation—A Method Used Under Arid and Desert Conditions of High Water and Soil Salinity

## D. Goldberg and M. Shmueli
MEMBER ASAE

IN recent years a new method of water application — drip irrigation — has aroused considerable interest in Israel. The method is a lateral spread of water on the irrigated surface by conducting the water under pressure to a relatively closely-spaced grid of outlets, and discharging the water through these outlets at virtually zero pressure.

This method has many advantages, especially in arid agricultural regions characterized by poor saline soil, saline irrigation water, and high evapotranspiration rates. The advantages are: (a) marked increases in crop yields, often by providing double or more than that obtained with sprinkler or furrow irrigation; (b) crop growth which could not be obtained under normal irrigation conditions due to salinity damage; (c) shortening the growing season and producing earlier crops.

The paper includes a brief technical description of the system, a summary of results from several experiments conducted in Israel's desert region, and a discussion of certain soil aspects related to the drip irrigation method.

## General Technical Description of the System's Components

The drip irrigation system consists of:

1. A "head" connected to the main water supply to the plot which includes filters, valves, couplings, water meter, pressure gage and connections for a fertilizer apparatus (Fig. 1).

2. Conducting pipes of the proper diameter according to distance and discharge.

3. Distribution tubes ("branches") of small diameter (generally ½-⅔ in.), connected in parallel to the conducting pipes.

4. Nozzles which can be built in many forms. They have a small perforation and an arrangement to reduce the pressure so that the water leaves the system in the form of drops. The discharge rate of a nozzle usually ranges between 0.4-2.2 gal per hr.

5. A fertilizer apparatus, connected to the "head," through which ⅓-¼ of

The paper was prepared expressly for publication in the TRANSACTIONS of the ASAE.

The authors—D. GOLDBERG and M. SHMUELI—are, respectively, head of irrigation department, Faculty of Agriculture, Hebrew University, Rehovot, Israel; head of soil and water laboratory, Irrigation Department, Yotvata, Arava Desert, Israel.

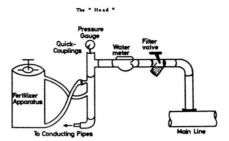

FIG. 1 The "head" connected to the main water supply includes filters, valves, couplings, water meter, pressure gage and fertilizer apparatus.

the total water flow passes. This water carries with it the dissolved fertilizer. (See 1-4).

The different components of the system, other than the "head" and the fertilizer apparatus, are generally of plastic construction. The "branches" and the nozzles are laid on the soil surface or are buried to a depth not exceeding 2-4 in.

The system described is especially suited for row crops, where each distribution tube with nozzles irrigates a single row or a larger number of closely-spaced rows.

## Details of the System and Its Operation

Under the conditions in which most of the experiments were conducted, the optimum length of a ½ in. diameter distribution pipe was 100-140 ft. The distance between the nozzles depends on the nozzle discharge and the soil type. Under conditions of sandy to loamy-sand soil, the optimum results were obtained with a spacing of 20 in. between the nozzles along the branch. Nozzle discharge rate was 0.4 gal per hr.

Experiments in a vineyard showed

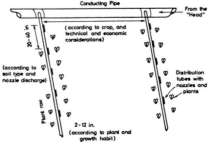

FIG. 2 Conducting and distribution network.

that it is possible to place one nozzle beside each vine, that is, every 60 in. In one experimental plot, the nozzle discharge was 2.2 gal per hr while in another plot it was 7.7 gal per hr.

The distances between the branches depends on the crop and on agrotechnical aspects (cultivations, lighting, harvesting, etc.) as well as economic aspects (cost of the branches and nozzles compared to cost of the conducting pipe). The customary distances generally range from 4-6 ft.

In annual crops, the system is assembled and put in position after the field is prepared for seeding. On occasion, for technical reasons, the equipment is laid down only after the seeding is completed. Operating the system is simple and involves the opening and closing of a valve for each plot of any desired size. At the end of the growing season, the equipment is gathered up, leaving the field free to be prepared for the succeeding crop.

## Irrigation and Fertilization

At least two factors determine the achievements attained with drip irrigation in an arid region: frequent (even daily) irrigations, and frequent applications of nitrogen fertilizer together with the irrigation water.

With most crops irrigations are given daily, and in all cases the interval between drip irrigations is no greater than 3 days. The fertilizer is given with the irrigation water at every irrigation during the main season of vegetative growth.

The central technical factor in proper operation of the system is the matter of blocking or plugging of the nozzle. The structure of the nozzle and the manner of its operation demands that no extraneous matter be allowed to enter with the water and the fertilizer. This is achieved by using a suitable filter at the "head" of the system and after the fertilizer apparatus. The nozzle's structure and its placing (close to or on the ground surface) guarantee that no roots enter and block the perforations. The system is based on the principle of drip irrigation on the soil surface.

## Location of the Crop

The proper distance of the crop from the nozzles is 2 to 12 in. in sandy to

**FIG. 3 Actual installation of head to water supply.**

loamy-sand soils. At a greater distance the rate of plant development is slow and the yield is reduced. In every case, the crop should not be planted beyond the wetted strip which can be seen on the soil surface. This wetted area may be as much as 20 in. from the nozzle in the coarser-textured soils.

## Use of the Method in Producing Protected Crops

The method of drip irrigation is especially suited to protected crops in plastic tunnels or similar structures. It permits irrigation during periods and days when opening the tunnels might otherwise result in injury to the crop.

### SOME RESULTS OF FIELD EXPERIMENTS

#### Water, Soil, and Climate at the Site of the Experiments

All the experiments were conducted in the Arava desert of Israel. The agriculture of the region is based on off-season crop production. The main growing season is between August and April. Average climatological data for this period is presented in Table 1. The annual rainfall ranges from 0.8-1.6 in., and the amount of effective rainfall is essentially zero.

The irrigation water contains sulphate salts, and is classed as $C_4S_2$ according to the classification of the U.S.

Salinity Laboratory at Riverside. The electrical conductivity is 3000 micromhos per cm (at 25 C), the chloride concentration is about 600 mg per $l$ and sometimes higher, and the sulphate concentration is about 700 mg per $l$.

The texture of the experimental soil is sandy to loamy-sand. The field capacity is 12-15 percent by volume. The pH of a saturated soil paste is 7.7-8.2. The $CaCO_3$ content is 10-18 percent. The cation exchange capacity ranges from 2.5-5.5 meq per 100 g soil. The hygroscopic water content is 0.5-3 percent (on the basis of oven-dry soil). The salt content of a saturated soil paste under normal conditions of irrigation varies between 5-25 millimhos per cm.

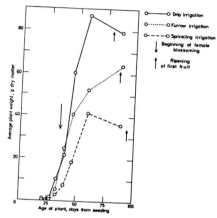

**FIG. 5 Growth curves of melon plants in Experiment 1.**

### Experiment 1

The comparative effect of three irrigation methods — sprinkling, furrows, and drip — were tested on musk-melons. The growing season was from September to the middle of December.

The growth curves of melon plants from all three irrigation methods are shown in Fig. 5. There is a marked difference between drip irrigation and the other two methods: rate of growth and the total vegetative growth were higher. Furthermore, with drip irrigation fruit ripened about 2 weeks earlier than it did under sprinkling, and one week earlier than in the case of furrow irrigation. The yields obtained in the experiment are shown in Table 2. Fig. 6 shows fruit bearing of plants in a drip-irrigated plot.

### Experiment 2

A comparison was made between sprinkler irrigation (the conventional method in the region) and drip irrigation on a crop of cucumbers. The growing season was from November to February.

With sprinkling no yield was obtained as the foliage was severely burned due to salinity, and the plants

**TABLE 1. AVERAGE CLIMATOLOGICAL DATA FOR THE EXPERIMENTAL AREA (SOUTHERN ARAVA DESERT)**

| Month | Average air temp., deg F | | Average daily relative humidity, percent | Class A pan evaporation, in. per month |
|---|---|---|---|---|
| | Min. | Max. | | |
| August | 73.0 | 102.7 | 30 | 18.90 |
| September | 67.5 | 98.1 | 53 | 15.45 |
| October | 63.3 | 91.9 | 57 | 11.94 |
| November | 56.5 | 82.9 | 68 | 8.34 |
| December | 45.5 | 74.5 | 62 | 6.33 |
| January | 44.8 | 68.0 | 61 | 5.98 |
| February | 43.5 | 71.0 | 51 | 6.93 |
| March | 49.1 | 77.0 | 43 | 10.22 |
| April | 57.9 | 87.3 | 38 | 13.10 |

**TABLE 2. MELON YIELDS IN EXPERIMENT 1**

| Irrigation method | Yield, tons/acre | | Yield, kg per 1 in. water | |
|---|---|---|---|---|
| | Total | Export grade | Total | Export grade |
| Sprinkling | 9.52 | 5.18 | 95 | 55 |
| Furrows | 9.68 | 6.68 | 98 | 68 |
| Drip | 17.20 | 13.96 | 173 | 140 |

produced no fruit whatsoever. On the other hand, the drip irrigated plants gave 15.76 tons per acre of cucumbers.

### Experiment 3*

A comparison was made between irrigation waters of various qualities applied by sprinkling and by the drip irrigation method. The test crop was tomatoes, and the growing season extended from October to May. The results are summarized in Table 3.

The electrical conductivity of the "good" water was 400 micromhos per cm, and the chloride concentration was 60 mg per $l$.

There was a greater difference between the two irrigation methods when saline water rather than good water was used. Actually, the yield obtained by drip irrigation with saline water was almost the same as that obtained by the same method using good water.

### Experiment 4

The effect of different frequencies of water application by the drip method was tested on various crops. The results are presented in Fig. 7.

For all three crops there was a reduction in yield as the intervals between irrigations became greater. The highest yields, for both nozzle discharge rates (0.4 and 2.2 gals per hr), were obtained with daily water applications.

**FIG. 4 Conducting pipe is positioned between two crop rows.**

**FIG. 6 Fruit respond to drip irrigation.**

* The results of this experiment were placed at our disposal by Y. Zohar of the Irrigation Extension Service, Ministry of Agriculture.

TABLE 3. TOMATO YIELDS (TONS PER ACRE) IN EXPERIMENT 3

| Irrigation method | Saline water (3000 micromhos per cm) | Non-saline water (400 micromhos per cm) |
|---|---|---|
| Sprinkling | 15.72 | 20.80 |
| Drip | 25.96 | 26.68 |

## Experiment 5

The salinity profile was examined in a vineyard irrigated for 2 years by the drip irrigation method. Vine development was very good, and values of growth rate were five times and more greater than those in a vineyard irrigated by flooding (see Fig. 8). The experiment was conducted in a young vineyard before it began bearing.

The salinity profiles are illustrated in Fig. 9. The profile is typical of the conditions existing under drip irrigation and can be divided into three main zones: an upper zone where the salinity increases as the distance from the nozzle and the soil surface decreases; a wide intermediate zone where salinity values are low; and a lower zone where the salinity level increases with depth and with the distance from the nozzle.

The plant roots were concentrated in the intermediate, more leached zone.

In all the experiments conducted, no negative effect was found resulting from the salt concentration near the soil surface which came in contact with the lower part of the plant stems but not with the root system itself.

### SEVERAL SOIL ASPECTS OF DRIP IRRIGATION

In drip irrigation only part of the soil surface is wetted, but at highly frequent intervals. The agricultural crop is located in this wetted zone. The rate of horizontal movement of the water in the soil and the soil moisture tension at different distances from the nozzle are functions of soil type and nozzle discharge rate. The final width of the wetted zone, parallel to the row of nozzles, is a function of the relation between the size of water application

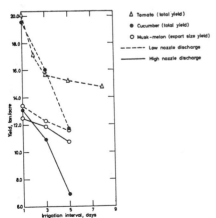

FIG. 7 Effect of drip irrigation interval on yield of various crops.

and evapotranspiration, as well as of the soil type. Experiments to arrive at quantitative values of the water regime which will express these functions are presently being conducted. The necessary quantitative data is being collected on the conditions of salinity, fertilizer and aeration — factors which depend primarily on the water regime.

The advantages of the drip irrigation system are: (a) the establishment of "desirable growing conditions" from the point of view of soil growth factors. (b) the continuity of these "desirable conditions."

When drip irrigation is applied at a certain rate of discharge, it is possible to detect a defined zone in the vicinity of the nozzle in which the growing conditions are optimal. This optimum depends mainly on the water concentrations which are greater than "field capacity" accepted as the basis for other irrigation methods, but less than the soil's water content at saturation.

The conventional methods of irrigation by sprinkling or furrows are typified, from the point of view of the plant's root zone, by certain characteristics: a rapid change from saturation during irrigation to field capacity, and a gradual change from field capacity to a minimum water content reached prior to the succeeding irrigation. This latter change is accompanied by wide fluctuations in the soil's moisture tension. At the same time similar fluctuations occur in the condition of other growth-affecting soil factors which depend on the moisture regime.

With drip irrigation, a high moisture content is reached which is greater than field capacity but close to it. It is possible to maintain an almost uniform moisture content as well as growth-affecting soil factors related to the water. With drip irrigation the soil no longer serves as a "reservoir" for water since the depleted water is returned to the root zone almost continuously. And the soil type is no longer a determining factor insofar as establishing the interval between the irrigations.

Since the fluctuations of the concentration of growth-affecting plant factors is wider when irrigating by the

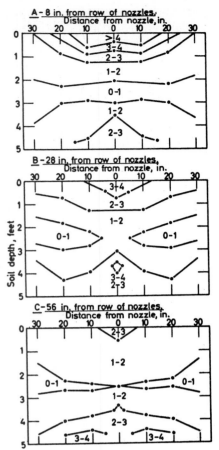

FIG. 9 Soil salinity profiles mmho per cm at 25C in a 1:1 Soil: Water extract, parallel to the nozzle rows in a young vineyard. Sampling 1½ years after installation of drip irrigation system. Nozzle discharge is 15.4 gal per hr.

conventional methods, a greater response is to be expected in the crop being drip irrigated. From the point of view of soil factors and irrigation, such a situation exists especially under conditions of sandy soil, saline water, or low soil fertility levels. From a climatological point of view, the situation is prevalent when evapotranspiration is high. It seems, therefore, that the method of drip irrigation is particularly suited to agriculture in arid and desert conditions.

### DISCUSSION

1. **Poor soil (coarse-textured)** As the drip irrigation method is based on re-

FIG. 8 Drip irrigation in young vineyard showed five times vine growth rate over flooding.

plenishing the water deficit daily (or at a very frequent interval), the soil has ceased to be a factor in holding the water between irrigations. Thus, a soil having a poor structure fulfills the function well, if not better, than a soil with good structure. The fact is that the soil acts merely as a support for the plant roots and as a structure to constantly receive the irrigation water.

2. **Highly saline water (exclusive of water containing salt in amounts which are physiologically toxic)** Due to the low soil moisture tension (between that at field capacity and saturation) resulting from almost continuous irrigation, the tension created by the osmotic effect brought about by the salt concentration in the water is insufficient to reduce drastically the plant's ability to absorb water. On the contrary, a situation is created where under conditions of high salt concentration in the water and low moisture tension in the soil a plant has better growing conditions than those existing close to the date of an irrigation for a plant irrigated by conventional methods with water having a low salt content. In the latter situation, the effective salt concentration of the soil in the vicinity of the plant rootlets is extremely high by the time the irrigation date approaches.

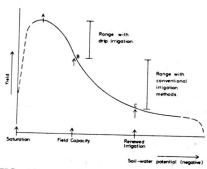

**FIG. 10 Schematic description of the relationship between crop yield and moisture tension of soil.**

During the interval between irrigations, not only do the plant and climate reduce the soil moisture content to a point which demands the application of water, but due to this process the salts in the remaining water become more and more concentrated until the next irrigation is given. This situation does not exist with drip irrigation because there is a continuous supply of water with the same salinity level.

3. **Drip irrigation is a significant improvement over furrow irrigation** (a) It is furrow irrigation with no water flowing in the furrows. The significance of this is that there is no need for accurate levelling — a complicated and expensive practice. Furthermore, there is no erosion due to flowing water. All the well-known difficulties connected with water flowing in a furrow are eliminated. (b) Water distribution is uniform and accurate. The problem of uneven water distribution along the furrow is a common one, especially in the coarser textured soils. The drip irrigation method guarantees that each nozzle discharges exactly the same amount of water. Thus, there is a saving of water and a certainty in its uniform application. (c) In drip irrigation there is no surplus water at the end of the furrow, which in furrow irrigation continues to increase as the soil's infiltration rate decreases.

Fig. 10 illustrates, diagramatically, our achievements with the drip irrigation method which are actually apparent from the description of the method, the experimental results and the previous discussion.

The moisture content with the drip irrigation method is constantly between A and B, and so the yield is expressed by a high value, as opposed to the soil moisture level with conventional methods between A and C (actually between B and C from the point of view of time) with a yield expressed by a lower value.

# part III

# Breeding and Genetics

 William Hooker. 1818. Account of a new pear (with a figure) called Williams' Bon Chretien; in a letter to Joseph Sabine, Esq. Secretary. Transactions of the Horticultural Society of London 2:250–251.

The selection, identification, and description of promising genotypes of useful plants has always been of intense interest to horticulturists, for the cultivar is the keystone of horticultural progress. In various parts of the world, entire industries have been based on unique cultivars (e.g., 'Smooth Cayenne' pineapple in Hawaii, the 'Gros Michel' banana in Central America). This sorting out of useful fruit cultivars was especially prominent in the nineteenth century, when a great literature accumulated in journals and special publications. This tradition continued in the early part of the twentieth century and is still strong in Italy and Hungary, where monographs of fruit cultivars are issued regularly. Before the advent of color photography the depiction of specimens also became a specialized art form.

One of the significant activities of the Horticultural Society of London, formed in 1804, and forerunner of the Royal Horticultural Society, was to establish a collection of drawings of fruits and flowers. A drawing committee was organized and William Hooker (1779–1832) was hired to execute the commissions. Hooker had studied under Francis Bauer, a master of botanical drawing, and was a horticulturist in his own right. He contributed at least five articles to the *Transactions* of the society, including one on the cultivation of pineapple.

The 158 paintings completed for the society between 1815 and 1821 are masterpieces of horticultural illustrations. Incredibly, when the society came upon hard times in the 1850s, they were sold at auction by Sotheby and Wilkinson, along with the entire library (the entire sale brought in £1112 1s 6d, including £49 10s for 10 folders of drawings). Fortunately, the Hooker paintings were restored to the library in 1927.

The communication contained in the second volume of the society transactions by Hooker describing a new pear, 'Williams' Bon Chretien', includes a hand-colored plate, but only a portion of the original drawing was reproduced. Hooker's note engendered a response from Mr. Stair of Aldermaster in Berkshire, correcting an error concerning the age of the original tree. He states (Transactions of the Horticultural Society III:357, 1818) "the tree was a very small plant in the year 1770 at which time *Mr. Wheeler* disposed of the garden in which it grew to *Mr. Stair* who considered that it sprang from the seed of a tree in an adjoining nursery ground, belonging to *Mr. Pendar*, which produced fruit similar in shape, though not as large as the young plant. The garden in which the original seedling exists, is now the property of *William Congreve*, Esq. of Aldermastom."

The cultivar described has in the two centuries of its existence become the most famous pear in the world, where it is known under about 150 synonyms. It was imported into the United States in 1797 or 1799 by James Carter for Thomas Brewer of Roxbury, Massachusetts. His lands and trees were subsequently acquired by Enoch Bartlett of Dorchester, Massachusetts, and unaware of the true name, he released it under his own. It has been known as 'Bartlett' ever since in the United States, and in 1985 accounted for about 80 percent of total U.S. pear production. This cultivar is also the leading pear of Europe and is still widely grown in England.

Jules Janick
Department of Horticulture
Purdue University
West Lafayette, Indiana 47907

## REFERENCES

ROYAL HORTICULTURAL SOCIETY. 1954. Exhibition of manuscripts, books, drawings, portraits, medals and congratulatory addresses on the occasion of the Society's 150th Anniversary celebration.

SMITH, MURIEL W. G. 1976. Catalog of British pears. Ministry of Agriculture. Fishenes & Ford, Agricultural Development and Advisory Service. p. 50–51.

THOMAS, G. WYNNE. 1967. The Williams' pear. J. R. Hort. Soc. 92:28–31.

LXV. *Account of a New* Pear, (*with a Figure*) *called* Williams' Bon Chretien; *in a Letter to* Joseph Sabine, *Esq. Secretary.* By William Hooker, *Esq. F.H.S.*

### Read December 3, 1816.

Dear Sir,

I beg leave to lay before the Horticultural Society, an account which I have obtained, at your request, of a variety of *Pear*, specimens of which were communicated to the Society in *August* last, by Mr. Richard Williams of Turnham Green, and much approved.

This *Pear*, which has been called by Mr. Aiton, (in his Epitome of the Hortus Kewensis,) *Williams' Bon Chretien*, appears to have sprung from seed, in the garden of Mr. Wheeler, a schoolmaster at Aldermaston in Berkshire, about twenty years ago, and was suffered to remain, in order to prove the value of its fruit. Subsequently grafts have been extensively dispersed, and many trees are now in Mr. Williams' nursery and other gardens around London. I have added the following description.

The trees of this variety are of vigorous growth, and fertile habit; their branches remarkably erect and straight, until bent by the weight of fruit, *Leaves* broad, deep green, very sharply serrated. *Fruit* of an irregular pyramidal and somewhat truncated form; large, being from 3 to 4½ inches in length, and 2 to 3 inches in width at the widest part near the head. The *Eye* is inserted on the summit, and never sunk in a hollow cavity, as in the other varieties called *Bon Chretiens*. The *Stalk* is very gross or fleshy, about ¾ of an inch in length. The *Colour* of the fruit is pale green, spotted over with a mixture of darker green and russet brown, becoming yellowish, and faintly tinged with red on the side next the sun when fully ripe. The *Flesh* is whitish, very tender and delicate, abounding with juice, which is sweet, and agreeably perfumed. Ripens in *August* when trained to a west wall, but on standard trees it is from three weeks to a month later.

This *Pear* I would recommend to the notice of the Horticultural Society as superior to any of its season with which I am acquainted. It immediately succeeds the *Jargonelle*, and is earlier than, as well as much superior to, the *Doyenné* or *White Beurrée*, and resembles in flavour the *Summer Musked Bon Chretien*. Its merits over the latter variety are, that on standard trees, as well as when trained, it seldom fails to produce fruit in abundance.

The drawing which accompanies this, was taken from specimens which ripened on a west wall, and may be considered an average size; but I have seen fruit of this variety weighing from ten to twelve ounces.

I remain, Dear Sir,

most respectfully and sincerely yours,

William Hooker.

*No. 5, York Buildings, New Road,*
*30th Nov. 1816.*

Gregor Mendel. 1865. Experiments on plant hybrids. (Versuche uber Pflanzen-Hybriden). Verhandlungen des naturforschenden den Vereines in Brunn 4:3–47.

Gregor Mendel

Mendel's paper is a victory for human intellect, a beacon cutting through the fog of bewilderment and muddled thinking about heredity. The story of its origin, neglect, and so-called rediscovery has become a legend in biology: an obscure monk working alone in his garden discovers a great biological phenomenon, but the report is ignored for a third of a century only to be resurrected simultaneously by three scientists working independently. Recognition, alas, comes too late, for the gentle amateur died 16 years earlier.

In contrast to most romantic revisions of history where the sugar coating melts under the heat of scrutiny, the drama of Mendel intensifies rather than subsides. Indeed, the stature of Mendel as a creative thinker with profound scientific foresight grows with the passage of time. He is revealed as a nineteenth-century Leonardo da Vinci with broad scientific interests and shares with Albert Einstein the distinction of having failed an accreditation examination, not once but twice, and that for certification as a high school teacher. His paper, instead of being revealed fully by the discoverers, is subtly altered in interpretation and a generation of scientific endeavors is necessary to understand it completely. It yields insights still.

The prevailing view of heredity during the middle of the nineteenth century assumed two gross misconceptions: the acceptance of a blending of hereditary factors and the heritability of acquired characters. Evidence for a particulate basis for inheritance, such as the reappearance of ancestral traits, was common knowledge but considered exceptional. In fact, all the "discoveries" attributed to Mendel, such as the equivalence of reciprocal crosses, dominance, uniformity of hybrids, and segregation in the generation following hybridization, are gleanable from the pre-Mendelian literature.

Incredibly, Charles Darwin's explanation of evolution by natural selection became a well-establishd theory in the years following publication of *Origin of Species* in 1859 despite any factual evidence to explain either the nature or the transmission of hereditary variation. Darwin, aware that blending inheritance led to the disappearance of variation, relied on the inheritance of acquired characters to generate the variability essential to his theory. His clear but inaccurate formulation of a model of inheritance (pangenesis) which involved particles (gemmules) passing from somatic cells to reproductive cells exposed the lack of any factual basis. The concept was merely a restatement of views dating from Hippocrates in 400 B.C. and endlessly reformulated.

Mendel's paper describes a series of experiments involving inheritance of traits in peas and beans. Preliminary results with peas allowed Mendel to formulate and then to test laws of inheritance that involved segregation and recombination of particulate elements. The proof of the particulate nature of the elements was made possible by the nature of the plants and traits studied. The traits chosen were contrasting (e.g., yellow versus green cotyledon, tall versus dwarf plant) and constant (i.e., true breeding) after normal self-pollination in the original lines. In seven of the eight characters chosen for study, the hybrid trait resembled one of the parents; in one character, bloom date, the hybrid trait was intermediate. Mendel called traits that pass into hybrid association "entirely or almost entirely unchanged" as *dominating* and the latent trait as *recessive* because the trait reappeared in subsequent crosses. The phenomenon of dominance was clearly not an essential part of the particulate nature of the genetic elements but was important to classify the progeny of hybrids. The disappearance of recessive traits in hybrids and their reappearance, unchanged, in subsequent generations was striking proof that the elements responsible for the traits (now called genes) were unaffected or contaminated in their transmission through generations.

When the hybrids of plants differing by a single trait were self-pollinated, three-fourths of the progeny displayed the dominating trait and one-fourth the recessive. In the next self-generation, progeny of plants displaying the recessive trait remained constant (nonsegregating), but those with the dominant trait produced one of two patterns of inheritance. One-third were constant, as in the original parent with the dominating trait, and two-thirds were segregating as in the hybrid. Thus the 3:1 ratio in the first segregating generation (now called the $F_2$ generation) was broken down into a ratio of 1 (true breeding dominant) : 2 (segregating dominant as in the hybrid) :1 (true breeding recessive). The explanation proposed was that the parental plants had paired elements (e.g., *AA* or *aa*, respectively) and the hybrids of such a cross were of the constitution *Aa*. Further, elements were distributed individually to germinal cells (gametes), which recombined at random in the formation of new individuals. This assumption was predictive for crosses of hybrids to each parent and in continual self-pollinations of the hybrid over several generations. Further, in crosses of plants differing by two or three characters, the segregation of each character was independent and the proportion of progeny expected was predictable with algebraic certainty. The complications brought about by linkage were, fortunately, not encountered.

Mendel's paper extended his analysis to another genus, but reduced fertility in crosses between *Phaseolus nanus* and *P. multiflorus* prevented a complete analysis. Only 49 seeds were obtained of the hybrid which upon selfing produced 44 plants, of which 31 flowered. He reported a 3:1 segregation for stem length and pod shape, but a "remarkable color change in the blossom." Of 31 plants, 30 had colored flowers with gradations from crimson to pale violet, and only 1 was white flowered. The experiment was continued for two more self-generations, indicating that hybrids follow the same law as that developed for peas. The deviation from expected 3:1 ratio in flower color was explained by a brilliant suggestion that this trait may be composed of independent colors each behaving as a single trait—in short, multiple factors. Thus Mendel suggested that discontinuous variation is also explained by discrete inheritance.

In the discussion of his results, Mendel expanded his data to account for "species transformations" via hybridization and backcrossing, a technique that was to have enormous utilization in plant breeding. He developed the concept of progeny testing. There are other nuggets of information, such as heterotic effect in crosses involving extremes in plant length, interactions between genetic and environmental factors, and insights into evolutionary theory.

Gregor Mendel's contribution has been trivialized as a "discovery of genetic ratios," the "beanbag" view of heredity. His enduring fame, however, rests on his ability to formulate a completely new approach to the problem of heredity, to carry out a complicated series of experiments successfully, and to conceptualize a genetic theory that was to create a new biology. The standard approach to unravel the mysteries of heredity was to analyze the complex of characters usually from wide crosses, a method that had failed for 2000 years and was to continue to fail even when applied by the combined talents of Francis Galton and Karl Pearson.

Mendel succeeded because of his approach. His goal was grand, being no less than to obtain a "generally predictable law" of heredity. His previous crosses with ornamentals had indicated predictable patterns, and his assumption was that laws of heredity must be universal. He had reviewed the literature and noted that "of the numerous experiments, no one had been carried out to an extent or in a manner that would make it possible to determine the number of different forms in which hybrid progeny appear, permit classification of these forms in each generation with certainty, and ascertain their numerical relationships."

Mendel clearly knew what he wanted to do and his method was precisely appropriate. His achievement was no serendipitous discovery. He was the consummate experimentalist, fully aware that the "value and validity of any experiment is determined by the suitability of the means as well as by the way they are applied." He defined the system. Plant material must possess constant differing traits which do not disturb fertility in further generations and must remain free from pollen contamination (outcrossing). All the progeny, without expectation, must be observed.

The experimental organism, the garden pea, was a perfect choice. Mendel procured 34 cultivars from seedsmen, tested their uniformity over two years, and selected 22 for hybridization experiments. Results from preliminary crosses indicated that common traits were transmitted unchanged to progeny but that contrasting traits may form a new hybrid trait that changes in subsequent generations. A series of experiments followed traits carefully selected for discontinuity to permit definite and sharp classification rather than "more-or-less" distinctions. This was a key decision.

Mendel restricted his attention to individual traits for each cross, avoiding the "noise" of extraneous characters. His analysis was quantitative and he displayed a mathematical sense in the analysis of data and the design of experiments. Mendel had a clear feeling for probability and was not put off by large deviations in small samples.

Mendel was a meticulous researcher. The sheer mass of his data is impressive and his experiments build from the simple to the complex. His clarity of thought is mirrored by a felicity of expression. His prose is straightforward, free of unnecessary arguments and obfuscation, faults that make many of the papers of his contemporaries almost incomprehensible.

The story of his paper has some interesting twists and turns that are important because they bear on the relation of progress in science and scientific publication. Mendel presented his paper in two oral sessions of the Brunn Society for the Study of Natural Sciences (February 8 and March 8, 1865) and it was published in the *Proceedings of the Brunn Society for the Study of Natural History* in 1865, which appeared in 1866. The paper made no impact on the scientific establishment. In 1966 it was discovered that the two lectures received an enthusiastic but anonymous review in a daily newspaper in Brunn, the only positive feedback that Mendel was to receive. Mendel's published paper was distributed to about 120 libraries throughout the world through the exchange list of the Brunn Society and was available in England and the United States. The paper was listed in the Royal Society (England) *Catalogue of Scientific Papers* for 1866 and referred to without comment in a paper on beans by H. Hoffmann in 1869. The only substantial reference was in the 590-page treatise on plant hybrids by W. O. Focke in 1881; Mendel's name is mentioned 17 times, but it is clear that Focke did not understand Mendel. In the critical passage he writes: "Mendel's numerous crossings gave results which were quite similar to those of [Thomas Andrew] Knight but Mendel believed he found constant numerical relationships between the types of the crosses." R. A. Fisher's comment on Focke's treatise in his famous 1936 paper ("Has Mendel's Work Been Rediscovered?") is priceless: "The fatigued tone of the opening remarks would scarcely arouse the curiosity of any reader, and in all he has to say, Focke's vagueness and caution have eliminated every point of scientific interest."

Focke's mention of Mendel did have repercussions. It was probably the basis for Mendel being listed as a plant hybridizer in the 9th edition of the *Encyclopaedia Britannica* (1881–1885) in an article by G. J. Romanes. Mendel's paper was also cited from Focke, but unread, by Liberty Hyde Bailey in an 1892 paper, "Cross Breeding and Hybridization," which according to one account by Hugo De Vries, may be the source of De Vries's introduction to Mendel. Focke's reference to Mendel was also picked up by Carl Correns and Erich von Tschermak when they began their research with peas.

In a book published in 1885 by Nageli and Peter entitled *Die Hieracien Mitteleuropas*, Mendel's paper is cited out of context, lumped with his other inheritance paper on *Hieracien*. The relation between Mendel and Nageli (analogous in many ways to Mozart and Salieri) is a shameful episode for academic science. Carl Wilhelm Nageli (1817–1891), a distinguished botany professor at the University of Munich, corresponded with Mendel and received Mendel's reprints, a reformulated explanation, packets of seed of peas with notes by Mendel, but he could not or would not understand the paper. His eternal punishment is that he may only be remembered for this fact.

There are other curious references. The Russian botanist I. F. Shmalhausen (1849–1894) appears to have read and appreciated Mendel's results. Mendel is cited as a footnote to a literature review of his 1874 Master's thesis but incorporated in the thesis *after* printing! The thesis was translated into German and appeared in *Botanishe Zeitung*, but the chapter with the historical review was omitted.

Finally, Hugo Iltis, Mendel's biographer, admits to having read Mendel's paper in 1899 when a high school student. "Amazed and puzzled" by the mixture of botany and mathematics, he brought the paper to an unnamed professor of natural history, who also proved uncomprehending.

Mendel received 40 separates of his paper, of which three have been traced. One went to Nageli at Munich, one to Anton Kerner von Marilaun at Innsbruck (found after his death, uncut), and one turned up in the hands of M. J. Beijerinck, who sent it to Hugo De Vries, undoubtedly the true source of De Vries' introduction to Mendel.

The so-called rediscovery of Mendelism by Hugo De Vries (1848–1935), Carl Correns (1864–1933), and Erich von Tschermak (1871–1962) was a consequence of independent investigation, although the almost simultaneous publications were related by events. The first published indication that Mendel's paper was understood occurred when Correns, ironically, a student of Nageli, obliquely cited both Mendel and Darwin in a paper on xenia published January 25, 1900. In March, De Vries completed three papers on his research, submitting them to three journals! Incredibly, the first paper to appear in print, a note in the *Comptes Rendus* of the Paris Academy of Science, did not mention Mendel but used the terms "dominance" and "recessive." Receipt of the reprint by Correns (April 21) triggered an immediate rejoinder in the form of a paper (April 22!) giving due credit to Mendel. Although the other two papers by De Vries did in fact mention Mendel, there is evidence that these references were second thoughts and were made in proof. Tschermak's first paper, which appeared in June 1900, refers to Mendel but showed a weaker grasp of its essentials than De Vries or Correns; both the De Vries and Correns papers were cited in a postscript added in proof.

None of the rediscovers' papers was in the class of Mendel's paper in terms of either analysis or style. De Vries was vague on the role of dominance and Corren was convinced that the law of segregation cannot be applied universally. Tschermak's paper reported the 3 : 1 ratio in the first segregation generation but did not interpret the backcross of the hybrid to the recessive parent as a 1 : 1 ratio, casting doubt of his complete understanding at that time.

Why was Mendel's paper, despite its clarity and incisiveness, ignored for 35 years? The best explanation is that Mendel was ahead of his time and it took that long for the scientific community to catch up. Remarkably, Mendel's paper was precytological and the cytological discoveries that were to provide a physical basis for heredity were published between 1882 and 1903. Nineteenth-century biology was not ready for Mendel. Part of the reason is that science then, as now, is conservative. New ideas are absorbed with difficulty and old ones discarded only reluctantly. One paper is not enough. The human qualities that made Mendel admirable as a person—modesty and reticence—worked against his receiving personal acclaim and fame during his lifetime.

The origins of the modern science of genetics are to be found in a small monastery garden cultivated by one who would worship at the altar of horticulture and science. Johann Mendel (the name "Gregor" was taken at his ordination), the child of peasants, was born in 1822 in Heinzendorf, a small village in a corner of Moravia. The boy, enamored of learning, was attracted to the monastery out of financial considerations. According to his autobiographical essay submitted for entrance to St. Thomas, the Augustine monastery of Brunn (now Brno, Czechoslovakia), his choice of vocation was influenced by his desire to be freed from the "perpetual anxiety about a choice of livelihood." He was accepted as a novice in 1843 and received a classical theological education at the local seminary. After ordination in 1847, he spent a year as a parish priest but proved to be emotionally unsuited, becoming physically ill in the presence of sickness and pain. He was offered a post to teach mathematics and Greek in the local school, a position for which although technically unqualified, he performed with distinction. Despite his lack of university training, he took the qualifying examination for teachers but failed—the question that tripped him up was the classification of mammals—the specialty of the examiner.

His monastery sent Mendel to the University of Vienna to spend three years studying science. In 1854 he was appointed supply (substitute?) teacher in Brunn Modern School, teaching physics and natural history to the lower school, a position he retained for 14 years. Incredibly, he again failed his accreditation exam, probably because of a dispute with his botany examiner, an event speculated to be connected with the initiation of his intensive experimental activities with peas the same year. His talents, however, were not lost on his fellow monks, who elected him prelate (for life) in 1868. The move proved to be a personal tragedy, for it caused him to withdraw from scientific work, and a bitter tax dispute between his monastery and the Austrian government embittered the last decade of his life. He died in 1884.

It is clear that St. Thomas was not a cloistered retreat with silent, tonsured monks in sandals and hairshirts but a vibrant community of scholars and artists. Mendel was neither ascetic nor reclusive and his writings are devoid of any religiosity, whatsoever. Politically, he was antiauthoritarian—a nineteenth-century liberal. He loved good food (he grew quite corpulent) and fine cigars (20 a day). As prelate of a wealthy monastery, he lived the busy life of an administrator, housed in elegant style and traveling widely, serving on committees and boards. He managed farms, became chairman of the Moravian mortgage bank, and even founded a volunteer fire department. He served as officer of the Brunn Society for the Study of Natural Sciences, but after 1870 he switched allegiance to the Royal Agriculture Society, serving for two years as acting chairman. One of his duties was that of examiner for fruit and vegetable growers. Mendel's passions were science and agriculture. In addition to being a superb horticulturist, he was an accomplished meteorologist. He wrote a definitive description of a tornado that tore through the monastery. Ironically, this published report was also ignored; a 1917 treatise by A. Wegener which describes 258 tornadoes in Europe contains no reference to Mendel. Mendel kept records of such diverse phenomena as groundwater, sunspots, and ozone levels. Tragically, most of his scientific notes and correspondence were burned at his death.

Although his scientific career was unrecognized in his lifetime—due in part to his self-effacing style—his fame soared after the rediscovery and the legend of Mendel contributed much to the feverish genetic activity from 1900 to 1925, probably making up for lost time. His subsequent fame as the Father of Genetics has obscured the fact that Mendel was primarily a horticultural scientist. His paper, the most famous horticultural paper written, may be the most famous single paper in biology.

Jules Janick
Department of Horticulture
Purdue University
West Lafayette, Indiana 17907

## REFERENCES

ILTIS, HUGO. 1932 (reprinted 1966). Life of Mendel. George Allen & Unwin, London. (Translated by Eden and Cedar Paul from Gregor Johann Mendel, Leben, Werk und Wirkung. 1924. Julius Springer, Berlin.)

OLBY, ROBERT C. 1966. Origins of Mendelism. Schocken Books, New York.

RIDLEY, MARK. 1984. The horticultural abbot of Brunn. New Sci. 5(Jan.):24–27.

SOSNA, MILAN (ed.). 1966. G. Mendel memorial symposium 1865–1965. Proceedings of a symposium held in Brno, Aug. 4–7, 1965. Academia, Publishing House of the Czechoslovak Academy of Sciences, Prague, Czechoslovakia.

STERN, CURT, and EVA R. SHERWOOD (eds.). 1966. The origin of genetics. A Mendel source book. W.H. Freeman, San Francisco.

STURTEVANT, A. H. 1965. A history of genetics. Harper & Row, New York.

ZIRKEL, CONWAY. 1935. The beginnings of plant hybridization. University of Pennsylvania Press, Philadelphia.

# Part One 1866-1873

## Experiments on Plant Hybrids

Versuche über Pflanzen-Hybriden

GREGOR MENDEL

*Read at the Meetings of 8 February and 8 March, 1865*

*Translated by Eva R. Sherwood*[1]

### INTRODUCTORY REMARKS

Artificial fertilization undertaken on ornamental plants to obtain new color variants initiated the experiments to be discussed here. The striking regularity with which the same hybrid forms always reappeared whenever fertilization between like species took place suggested further experiments whose task it was to follow the development of hybrids in their progeny.

Numerous careful observers, such as Kölreuter, Gärtner, Herbert, Lecoq, Wichura, and others, have devoted a part of their lives to this problem with tireless persistence. Gärtner, especially, in his work "Die Bastarderzeugung im Pflanzenreich" (Hybrid Production in the Plant Kingdom) has recorded very estimable observations, and Wichura has very recently published the results of his thorough investigations of willow hybrids. That no generally applicable law of the formation and development of hybrids has yet been successfully formulated can hardly astonish anyone who is acquainted with the extent of the task and who can appreciate the difficulties with which experiments of this kind have to contend. A final decision can be reached only when the results of detailed experiments from the most diverse plant families are available. Whoever surveys the work in this field will come to the conviction that among the numerous experiments not one has been carried out to an extent or in a manner that would make it possible to determine the number of different forms in which hybrid progeny appear, permit classification of these forms in each generation with certainty, and ascertain their numerical interrelationships. It requires a good deal of courage indeed to undertake such a far-reaching task; however, this seems to be the one correct way of finally reaching the solution to a question whose significance for the evolutionary history of organic forms must not be underestimated.

This paper discusses the attempt at such a detailed experiment. It was expedient to limit the experiment to a fairly small group of plants, and after a period of eight years it is now essentially concluded. Whether the plan by which the

[The original paper was published in Verhandlungen des naturforschenden Vereines in Brünn 4 (1865), Abhandlungen, pp. 3–47, which appeared in 1866.]

[1] [An attempt has been made to preserve Mendel's style; frequently however, it was necessary to add a word or change the order of an English sentence to permit the reader insight into the clarity with which the German version was written. The abbreviations used correspond to those in the printed publication; they are not always identical with the ones in the handwritten manuscript. A historian in genetics would, of course, want to consult the original paper.

Throughout this translation "differierend" has been translated by "differing," "Merkmal" by "trait," "Character" by "character," or "characteristic," "Entwicklungsreihe" by "series," "je zwei" by "one pair," "dominierend" by "dominating," and "Glied" by "term" when used in an algebraic series, by "member" when the word applies to a series of genotypes.

While this terminology needs no further comment, some explanation of the use of "angular" for "kantig" is necessary. In 1901 the Journal of the Royal Horticultural Society of London published a trans-individual experiments were set up and carried out was adequate to the assigned task should be decided by a benevolent judgment.

## SELECTION OF EXPERIMENTAL PLANTS

The value and validity of any experiment are determined by the suitability of the means used as well as by the way they are applied. In the present case as well, it can not be unimportant which plant species were chosen for the experiments and how these were carried out.

Selection of the plant group for experiments of this kind must be made with the greatest possible care if one does not want to jeopardize all possibility of success from the very outset.

The experimental plants must necessarily

1. Possess constant differing traits.

2. Their hybrids must be protected from the influence of all foreign pollen during the flowering period or easily lend themselves to such protection.

3. There should be no marked disturbances in the fertility of the hybrids and their offspring in successive generations.

Contamination with foreign pollen that might take place during the experiment without being recognized would lead to quite erroneous conclusions. Occasional forms with reduced fertility or complete sterility, which occur among the offspring of many hybrids, would render the experiments

lation of Mendel's publication, in which the word "wrinkled" describes the shape of seeds whenever "kantig" is the only adjective. English scientific literature has generally followed this precedent and referred to round vs. wrinkled peas. But Mendel describes the seeds of *P. quadratum* as both angular and wrinkled, using two different German words. Classifying the seeds of hybrid plants he uses only the word for "angular" to contrast with that for "round." A search through English botanical texts confirms the accuracy of this description: the seeds are irregularly shaped, asymmetrically compressed, with smooth surfaces meeting at an angle. E.R.S.]

very difficult or defeat them entirely. To discover the relationships of hybrid forms to each other and to their parental types it seems necessary to observe *without exception all* members of the series of offspring in each generation.

From the start, special attention was given to the *Leguminosae* because of their particular floral structure. Experiments with several members of this family led to the conclusion that the genus *Pisum* had the qualifications demanded to a sufficient degree. Some quite distinct forms of this genus possess constant traits that are easily and reliably distinguishable, and yield perfectly fertile hybrid offspring from reciprocal crosses. Furthermore, interference by foreign pollen cannot easily occur, since the fertilizing organs are closely surrounded by the keel, and the anthers burst within the bud; thus the stigma is covered with pollen even before the flower opens. This fact is of particular importance. The ease with which this plant can be cultivated in open ground and in pots, as well as its relatively short growth period, are further advantages worth mentioning. Artificial fertilization is somewhat cumbersome, but it nearly always succeeds. For this purpose the not yet fully developed bud is opened, the keel is removed, and each stamen is carefully extracted with forceps, after which the stigma can be dusted at once with foreign pollen.

From several seed dealers a total of 34 more or less distinct varieties of peas were procured and subjected to two years of testing. In one variety a few markedly deviating forms were noticed among a fairly large number of like plants. These, however, did not vary in the following year and were exactly like another variety obtained from the same seed dealer; no doubt the seeds had been accidentally mixed. All other varieties yielded quite similar and constant offspring; at least during the two test years no essential change could be noticed. Twenty-two of these varieties were selected for fertilization and planted annually throughout the entire experimental period. They remained stable without exception.

Their systematic classification is difficult and uncertain. If one wanted to use the strictest definition of species, by which only those individuals that display identical traits under identical conditions belong to a species, then no two could be counted as one and the same species. In the opinion of experts, however, the majority belong to the species *Pisum sativum*; while the remaining ones were regarded and described either as sub-species of *P. sativum*, or as separate species, such as *P. quadratum*, *P. saccharatum*, and *P. umbellatum*. In any event, the rank assigned to them in a classification system is completely immaterial to the experiments in question. Just as it is impossible to draw a sharp line between species and varieties, it has been equally impossible so far to establish a fundamental difference between the hybrids of species and those of varieties.

## ARRANGEMENT AND SEQUENCE OF EXPERIMENTS

When two plants, constantly different in one or several traits, are crossed, the traits they have in common are transmitted unchanged to the hybrids and their progeny, as numerous experiments have proven; a pair of differing traits, on the other hand, are united in the hybrid to form a new trait, which usually is subject to changes in the hybrid's progeny. It was the purpose of the experiment to observe these changes for each pair of differing traits, and to deduce the law according to which they appear in successive generations. Thus the study breaks up into just as many separate experiments as there are constantly differing traits in the experimental plants.

The various forms of peas selected for crosses showed differences in length and color of stem; in size and shape of leaves; in position, color, and size of flowers; in length of flower stalks; in color, shape, and size of pods; in shape and size of seeds; and in coloration of seed coats and albumen. However, some of the traits listed do not permit a definite and sharp separation, since the difference rests on a "more

or less" which is often difficult to define. Such traits were not usable for individual experiments; these had to be limited to characteristics which stand out clearly and decisively in the plants. The result should ultimately show whether in hybrid unions the traits all observe concordant behavior, and whether one can also make a decision about those traits which have minor significance in a classification.

The traits selected for experiments relate:

1. *To the difference in the shape of the ripe seeds.* These are either round or nearly round, with depressions, if any occur on the surface, always very shallow; or they are irregularly angular and deeply wrinkled (*P. quadratum*).

2. *To the difference in coloration of seed albumen* (endosperm).[2] The albumen of ripe seeds is either pale yellow, bright yellow and orange, or has a more or less intense green color. This color difference is easily recognizable in the seeds because their coats are transparent.

3. *To the difference in coloration of the seed coat.* This is either white, in which case it is always associated with white flower color; or it is grey, grey-brown, leather-brown with or without violet spotting, in which case the color of the standard is violet, that of the wings is purple, and the stem bears reddish markings at the leaf axils. The grey seed coats turn black-brown in boiling water.

4. *To the difference in shape of the ripe pod.* This is either smoothly arched and never constricted anywhere, or deeply constricted between the seeds and more or less wrinkled (*P. saccharatum*).

5. *To the difference in color of the unripe pod.* It is either colored light to dark green or vivid yellow, which is also the coloration of stalks, leaf-veins, and calyx.[3]

---

[2] ["Mendel uses the terms 'albumen,' and 'endosperm,' somewhat loosely to denote the cotyledons, containing food-material, within the seed." W. Bateson.]

[3] One variety has a beautiful brownish-red pod color which tends to

6. To *the difference in position of flowers*. They are either axillary, that is, distributed along the main stem, or they are terminal, bunched at the end of the stem and arranged in what is almost a short cyme; if the latter, the upper part of the stem is more or less enlarged in cross section (*P. umbellatum*).

7. To *the difference in stem length*. The length of the stem varies greatly in individual varieties; it is, however, a constant trait for each, since in healthy plants grown in the same soil it is subject to only insignificant variations. In experiments with this trait, the long stem of 6 to 7' was always crossed with the short one of ¾' to 1½' to make clear-cut distinction possible.[4]

Each of the two differing traits listed above as pairs were united by fertilization.
For the

| 1st experiment | 60 fertilizations on | 15 plants were undertaken. |
| --- | --- | --- |
| 2nd " | 58 " " | 10 " " " |
| 3rd " | 35 " " | 10 " " " |
| 4th " | 40 " " | 10 " " " |
| 5th " | 23 " " | 5 " " " |
| 6th " | 34 " " | 10 " " " |
| 7th " | 37 " " | 10 " " " |

From a fairly large number of plants of the same kind only the most vigorous were chosen for fertilization. Weak plants always give uncertain results, because many of the offspring either fail to flower entirely or form only few and inferior seeds even in the first generation of hybrids, and still more do so in the following one.

a violet and blue around the time of ripening. The experiment with this trait was started only in the past year.

[4] [Presumably the symbol ' stands for Viennese foot, equal to 0.316 meter, or 12.44 inches. E.R.S.]

Furthermore, in all experiments reciprocal crosses were made in such a manner that that one of the two varieties serving as seed plant in one group of fertilizations was used as pollen plant in the other group.

The plants were grown in garden beds—except for a few in pots—and were maintained in their natural upright position by means of sticks, twigs, and taut strings. For each experiment a number of the potted plants were placed in a greenhouse during the flowering period; they were to serve as controls for the main experiment in the garden against possible disturbance by insects. Among the insects that visit pea plants the beetle *Bruchus pisi* might become dangerous to the experiment should it appear in fairly large numbers. It is well known that the female of this species lays her eggs in the flower and thereby opens the keel; on the tarsi of one specimen caught in a flower some pollen cells could clearly be seen under the hand lens. Mention must also be made here of another circumstance that might possibly lead to the admixture of foreign pollen. For in some rare cases it happens that certain parts of an otherwise quite normally developed flower are stunted, leading to partial exposure of the fertilization organs. Thus, defective development of the keel, which left pistil and anthers partly uncovered, was observed. It also sometimes happens that the pollen does not fully mature. In that event the pistil gradually lengthens during the flowering period until the stigma protrudes from the tip of the keel. This curious phenomenon has also been observed in hybrids of *Phaseolus* and *Lathyrus*.

The risk of adulteration by foreign pollen is, however, a very slight one in *Pisum*, and can have no influence whatsoever on the overall result. Among more than 10,000 carefully examined plants there were only a very few in which admixture had doubtlessly occurred. Since such interference was never noticed in the greenhouse, it may be assumed that this interference in the greenhouse... *Bruchus pisi*, and perhaps also the cited abnormalities in floral structure, are to blame.

## THE FORM OF THE HYBRIDS

Experiments on ornamental plants undertaken in previous years had proven that, as a rule, hybrids do not represent the form exactly intermediate between the parental strains. Although the intermediate form of some of the more striking traits, such as those relating to shape and size of leaves, pubescence of individual parts, and so forth, is indeed nearly always seen, in other cases one of the two parental traits is so preponderant that it is difficult, or quite impossible, to detect the other in the hybrid.

The same is true for *Pisum* hybrids. Each of the seven hybrid traits either resembles so closely one of the two parental traits that the other escapes detection, or is so similar to it that no certain distinction can be made. This is of great importance to the definition and classification of the forms in which the offspring of hybrids appear. In the following discussion those traits that pass into hybrid association entirely or almost entirely unchanged, thus themselves representing the traits of the hybrid, are termed *dominating,* and those that become latent in the association, *recessive.* The word "recessive" was chosen because the traits so designated recede or disappear entirely in the hybrids, but reappear unchanged in their progeny, as will be demonstrated later.

All experiments proved further that it is entirely immaterial whether the dominating trait belongs to the seed or pollen plant; the form of the hybrid is identical in both cases. This interesting phenomenon was also emphasized by Gärtner, with the remark that even the most practiced expert is unable to determine from a hybrid which of the two species crossed was the seed plant and which the pollen plant.

Of the differing traits utilized in the experiments the following are dominating:

1. The round or nearly round seed shape with or without shallow depressions.

2. The yellow coloration of seed albumen [cotyledons].
3. The grey, grey-brown, or leather-brown color of the seed coat, associated with violet-red blossoms and reddish spots in the leaf axils.
4. The smoothly arched pod shape.
5. The green coloration of the unripe pod, associated with the same color of stem, leaf veins, and calyx.
6. The distribution of flowers along the stem.
7. The length of the longer stem.

With respect to this last trait it must be noted that the stem of the hybrid is usually longer than the longer of the two parental stems, a fact which is possibly due only to the great luxuriance that develops in all plant parts when stems of very different lengths are crossed. Thus, for instance, in repeated experiments, hybrid combinations of stems 1' and 6' long yielded, without exception, stems varying in length from 6' to 7½'. *Hybrid seed coats* are often more spotted; the spots sometimes coalesce into rather small bluish-purple patches. Spotting frequently appears even when it is absent as a parental trait.

The hybrid forms of *seed shape* and *albumen* develop immediately after artificial fertilization merely through the influence of the foreign pollen. Therefore they can be observed in the first year of experimentation, while the remaining traits do not appear in the plants raised from fertilized seeds until the following year.

### THE FIRST GENERATION FROM HYBRIDS

In this generation, *along with the dominating* traits, the *recessive* ones also reappear, their individuality fully revealed, and they do so in the decisively expressed average proportion of 3:1, so that among each four plants of this generation three receive the dominating and one the recessive characteristic. This is true, without exception, of all

| Plant | Experiment 1 | | Experiment 2 | |
| --- | --- | --- | --- | --- |
| | Shape of Seeds | | Coloration of Albumen | |
| | Round | Angular | Yellow | Green |
| 1 | 45 | 12 | 25 | 11 |
| 2 | 27 | 8 | 32 | 7 |
| 3 | 24 | 7 | 14 | 5 |
| 4 | 19 | 10 | 70 | 27 |
| 5 | 32 | 11 | 24 | 13 |
| 6 | 26 | 6 | 20 | 6 |
| 7 | 88 | 24 | 32 | 13 |
| 8 | 22 | 10 | 44 | 9 |
| 9 | 28 | 6 | 50 | 14 |
| 10 | 25 | 7 | 44 | 18 |

traits included in the experiments. The angular, wrinkled seed shape, the green coloration of the albumen, the white color of seed coat and flower, the constrictions on the pods, the yellow color of the immature pod, stalk, calyx, and leaf veins, the almost umbellate inflorescence, and the dwarfed stem all reappear in the numerical proportion given, without any essential deviation. *Transitional forms were not observed in any experiment.*

Since the hybrids resulting from reciprocal crosses were of identical appearance and showed no noteworthy deviation in their subsequent development, the results from both crosses may be totaled in each experiment. The numerical proportions obtained for each pair of differing traits are as follows:

*Experiment 1.* Seed shape. From 253 hybrids 7324 seeds were obtained in the second experimental year. Of them, 5474 were round or roundish and 1850 angular wrinkled. This gives the ratio 2.96:1.

*Experiment 2.* Albumen coloration. 258 plants yielded 8023 seeds, 6022 yellow and 2001 green; their ratio, therefore, is 3.01:1.

In these two experiments each pod usually yielded both kinds of seed. In well-developed pods that contained, on the average, six to nine seeds, all seeds were fairly often round (Experiment 1) or all yellow (Experiment 2); on the other hand, no more than 5 angular or 5 green ones were ever observed in one pod. It seems to make no difference whether the pods develop earlier or later in the hybrid, or whether they grow on the main stem, or on an axillary one. In the pods first formed by a small number of plants only a few seeds developed, and these possessed only one of the two traits; in the pods developing later, however, the proportion remained normal. The distribution of traits also varies in individual plants, just as in individual pods. The first ten members of both series of experiments may serve as an illustration:

Extremes observed in the distribution of the two seed traits in a *single* plant were, in Experiment 1, one instance of 43 round and only 2 angular, another of 14 round and 15 angular seeds. In Experiment 2 there was found an instance of 32 yellow and only 1 green, but also one of 20 yellow and 19 green seeds.

These two experiments are important for the determination of mean ratios, which make it possible to obtain very meaningful averages from a fairly small number of experimental plants. However, in counting the seeds, especially in Experiment 2, some attention is necessary, since in individual seeds of some plants the green coloration of the albumen is less developed, and can be easily overlooked at first. The cause of partial disappearance of the green coloration has no connection with the hybrid character of the plants, since it occurs also in the parental plant; furthermore, this peculiarity is restricted to the individual and not inherited by the offspring. In luxuriant plants this phenomenon was noted quite frequently. Seeds damaged during their development by insects often vary in color and shape; with a little practice in sorting, however, mistakes are easy to avoid. It is almost superfluous to mention that the pods

Those forms that receive the recessive character in the first generation do not vary further in the second with respect to this trait; they remain *constant* in their progeny.

The situation is different for those possessing the dominating trait in the first generation. Of these, *two parts* yield offspring that carry the dominating and the recessive trait in the proportion of 3:1, thus showing exactly the same behavior as the hybrid forms; only *one part* remains constant for the dominating trait.

The individual experiments yielded results as follows:

*Experiment 1.* Among 565 plants raised from round seeds of the first generation, 193 yielded only round seeds, and therefore remained constant in this trait; 372, however, produced both round and angular seeds in the proportion of 3:1. Therefore, the number of hybrids compared to that of the constant breeding forms is as 1.93:1.

*Experiment 2.* Of 519 plants raised from seeds whose albumen had yellow coloration in the first generation, 166 yielded exclusively yellow, while 353 yielded yellow and green seeds in the proportion 3:1. Therefore, a partition into hybrid and constant forms resulted in the proportion 2.13:1.

For each of the subsequent experiments, 100 plants that possessed the dominating trait in the first generation were selected, and in order to test this trait's significance 10 seeds from each plant were sown.

*Experiment 3.* The offspring of 36 plants yielded exclusively grey-brown seed coats; from 64 plants came some grey-brown and some white coats.

*Experiment 4.* The offspring of 29 plants had only smoothly arched pods; of 71, on the other hand, some had smoothly arched and some constricted ones.

*Experiment 5.* The offspring of 40 plants had only green pods; those of 60 plants had some green, some yellow.

*Experiment 6.* The offspring of 33 plants had only axillary flowers; of another 67, on the other hand, some had axillary, some terminal flowers.

14  *Gregor Mendel*

must remain on the plants until they are completely ripe and dry, for only then are the shape and color of the seeds fully developed.

*Experiment 3.* Color of seed coat. Among 929 plants 705 bore violet-red flowers and grey-brown seed coats; 224 had white flowers and white seed coats; this yields the proportion 3.15:1.

*Experiment 4.* Shape of pods. Of 1181 plants 882 had smoothly arched pods, 299 constricted ones. Hence the ratio 2.95:1.

*Experiment 5.* Coloration of unripe pods. The experimental plants numbered 580, of which 428 had green and 152 yellow pods. Consequently, the former stand to the latter in the proportion 2.82:1.

*Experiment 6.* Position of flowers. Among 858 cases, 651 had axillary flowers and 207 terminal ones. Consequently, the ratio is 3.14:1.

*Experiment 7.* Length of stem. Of 1064 plants, 787 had the long stems, 277 the short stems. Hence a relative proportion of 2.84:1. In this experiment the dwarfed plants were tenderly lifted and transferred to beds of their own. This precaution was necessary because they would have become stunted growing amidst their tall brothers and sisters. Even in the earliest stages they can be distinguished by their compact growth and thick dark-green leaves.

When the results of all experiments are summarized, the average ratio between the number of forms with the dominating trait and those with the recessive one is 2.98:1, or 3:1.

The dominating trait can have *double significance* here—namely that of the parental characteristic or that of the hybrid trait. In which of the two meanings it appears in each individual case only the following generation can decide. As parental trait it would pass unchanged to all of the offspring; as hybrid trait, on the other hand, it would exhibit the same behavior as it did in the first generation.

*Experiments on Plant Hybrids* 13

*Experiment 7.* The offspring of 28 plants received the long stem, those of 72 plants partly the long, partly the short one.

In each of these experiments a certain number of plants with the dominating trait become constant. In evaluating the proportion in which forms with the constantly persisting trait segregate, the first two experiments are of special importance, since in these a fairly large number of plants could be compared. The ratios 1.93:1 and 2.13:1 taken together give almost exactly the average ratio of 2:1. Experiment 6 gave an entirely concordant result; in other experiments the proportion varies more or less, as was to be expected with the small number of 100 experimental plants. Experiment 5, which showed the greatest deviation, was repeated, and, instead of 60:40, the ratio of 65:35 was obtained. Accordingly, *the average ratio of 2:1 seems ensured.* It is thus proven that of those forms which possess the dominating trait in the first generation, two parts carry the hybrid trait, but the one part with the dominating trait remains constant.

The ratio of 3:1 in which the distribution of the dominating and recessive traits takes place in the first generation therefore resolves itself into the ratio of 2:1:1 in all experiments if one differentiates between the meaning of the dominating trait as a hybrid trait and as a parental character. Since the members of the first generation originate directly from the seeds of the hybrids, *it now becomes apparent that of the seeds formed by the hybrids with one pair of differing traits, one half again develop the hybrid form while the other half yield plants that remain constant and receive the dominating and the recessive character in equal shares.*

THE SUBSEQUENT GENERATIONS FROM HYBRIDS

The proportions in which the descendants of hybrids develop and split up in the first and second generations are probably valid for all further progeny. Experiments 1 and 2 have by now been carried through six generations, 3 and

7 through five, and 4, 5 and 6 through four without any deviation becoming apparent, although from the third generation on a small number of plants were used. In each generation the offspring of the hybrids split up into hybrid and constant forms according to the ratios 2:1:1.

If A denotes one of the two constant traits, for example, the dominating one, $a$ the recessive, and $Aa$ the hybrid form in which both are united, then the expression

$$A + 2Aa + a$$

gives the series for the progeny of plants hybrid in a pair of differing traits.

The observation made by Gärtner, Kölreuter, and others, that hybrids have a tendency to revert to the parental forms, is also confirmed by the experiments discussed. It can be shown that the numbers of hybrids derived from one fertilization decrease significantly from generation to generation as compared to the number of newly constant forms and their progeny, yet they can never disappear entirely. If one assumes, on the average, equal fertility for all plants in all generations, and if one considers, furthermore, that half of the seeds that each hybrid produces yield hybrids again while in the other half the two traits become constant in equal proportions, then the numerical relationships for the progeny in each generation follow from the tabulation below, where A and $a$ again denote the two parental traits and $Aa$ the hybrid form. For brevity's sake one may assume that in each generation each plant supplied only four seeds.

| Generation | A | $Aa$ | $a$ | Expressed in terms of ratios $A : Aa : a$ |
|---|---|---|---|---|
| 1 | 1 | 2 | 1 | 1 : 2 : 1 |
| 2 | 6 | 4 | 6 | 3 : 2 : 3 |
| 3 | 28 | 8 | 28 | 7 : 2 : 7 |
| 4 | 120 | 16 | 120 | 15 : 2 : 15 |
| 5 | 496 | 32 | 496 | 31 : 2 : 31 |
| $n$ | | | | $2^{n}-1 : 2 : 2^{n}-1$ |

In the tenth generation, for example, $2^n - 1 = 1023$. Therefore, of each 2048 plants arising in this generation, there are 1023 with the constant dominating trait, 1023 with the recessive one, and only 2 hybrids.

## THE OFFSPRING OF HYBRIDS IN WHICH SEVERAL DIFFERING TRAITS ARE ASSOCIATED

In the experiments discussed above, plants were used which differed in only one essential trait. The next task consisted in investigating whether the law of development thus found would also apply to a pair of differing traits when several different characteristics are united in the hybrid through fertilization.

The experiments demonstrated throughout that in such a case the hybrids always resemble more closely that one of the two parental plants which possesses the greater number of dominating traits. If, for instance, the seed plant has a short stem, terminal white flowers, and smoothly arched pods, and the pollen plant has a long stem, violet-red lateral flowers, and constricted pods, then the hybrid reminds one of the seed plant only in pod shape, and the remaining traits resemble those of the pollen plant. Should one of the two parental types possess only dominating traits, then the hybrid is hardly or not at all distinguishable from it.

Two experiments were carried out with a larger number of plants. In the first experiment the parental plants differed in seed shape and coloration of albumen; in the second in seed shape, coloration of albumen, and color of seed coat. Experiments with seed traits lead most easily and assuredly to success.

To simplify a survey of the data, the differing traits of the seed plant will be indicated in these experiments by A, B, C, those of the pollen plant by a, b, c, and the hybrid forms of these traits by Aa, Bb and Cc.

*First Experiment:*

AB seed plant,        ab pollen plant,
A shape round,        a shape angular,
B albumen yellow,     b albumen green.

The fertilized seeds were round and yellow, resembling those of the seed plant. The plants raised from them yielded seeds of four kinds, frequently lying together in one pod. From 15 plants a total of 556 seeds were obtained, and of these there were:

315 round and yellow,
101 angular and yellow,
108 round and green,
32 angular and green.

All were planted in the following year. Eleven of the round yellow seeds did not germinate and three plants from such seeds did not attain fruition. Of the remaining plants:

| | |
|---|---|
| 38 had round yellow seeds | AB |
| 65 had round yellow and green seeds | ABb |
| 60 had round yellow and angular yellow seeds | AaB |
| 138 had round yellow and green and angular yellow and green seeds | AaBb |

Of plants grown from angular yellow seeds, 96 bore fruit; of these

| | |
|---|---|
| 28 had only angular yellow seeds | aB |
| 68 had only angular yellow and green seeds | aBb |

Of plants grown from 108 round green seeds, 102 bore fruit; of these

35 had only round green seeds  . . Ab
67 had round and angular green seeds . Aab

The angular green seeds yielded 30 plants with identical seeds throughout; they remained constant . . ab

Thus the offspring of hybrids appear in nine different forms, some of them in very unequal numbers. When these are summarized and arranged in order, one obtains:

| 38 | plants with the designation | AB |
|----|------------------------------|------|
| 35 | "    "    "                 | Ab |
| 28 | "    "    "                 | aB |
| 30 | "    "    "                 | ab |
| 65 | "    "    "                 | ABb |
| 68 | "    "    "                 | aBb |
| 60 | "    "    "                 | AaB |
| 67 | "    "    "                 | Aab |
| 138 | "    "    "                | AaBb |

All forms can be classified into three essentially different groups. The first comprises those with designations AB, Ab, aB, ab; they possess only constant traits and do not change any more in following generations. Each one of these forms is represented 33 times on the average. The second group contains the forms ABb, aBb, AaB, Aab; these are constant for one trait, hybrid for the other, and in the next generation vary only with respect to the hybrid trait. Each of them appears 65 times on the average. The form AaBb occurs 138 times, is hybrid for both traits, and behaves exactly like the hybrid from which it is descended.

Comparing the numbers in which the forms of these groups occur, one cannot fail to recognize the average proportions of 1:2:4. The numbers 33, 65, 138 give quite satisfactory approximations to the numerical proportions 33, 66, 132.

Accordingly, the series consists of nine terms. Four of them occur once each and are constant for both traits; the forms AB, ab resemble the parental types, the other two represent the other possible constant combinations between the associated traits A, a, B, b. Four terms occur twice each and are constant for one trait, hybrid for the other. One term appears four times and is hybrid for both traits. When, therefore, two kinds of differing traits are combined in hybrids, the progeny develop according to the expression:

$$AB + Ab + aB + ab + 2ABb + 2aBb + 2AaB + 2Aab + 4AaBb.$$

Indisputably this series is a combination series in which the two series for the traits A and a, B and b are combined term by term. All the terms of the series are obtained through a combination of the expressions:

$$A + 2Aa + a$$
$$B + 2Bb + b$$

*Second Experiment:*

| ABC seed plant, | abc pollen plant. |
|------------------|-------------------|
| A shape round, | a shape angular. |
| B albumen yellow, | b albumen green. |
| C seed coat grey-brown, | c seed coat white. |

This experiment was conducted in a manner quite similar to that used in the preceding one. Of all experiments it required the most time and effort. From 24 hybrids a total of 687 seeds was obtained, all of which were spotted, colored grey-brown or grey-green, and round or angular. Of the plants grown from them, 639 bore fruit in the following year and, as further investigations showed, they comprised:

supply all terms of the series. The constant associations encountered in it correspond to all possible combinations of the traits A, B, C, a, b, c; two of them, ABC and abc, resemble the two parental plants.

In addition, several more experiments were carried out with a smaller number of experimental plants in which the remaining traits were combined by twos or threes in hybrid fashion; all gave approximately equal results. Therefore there can be no doubt that for all traits included in the experiment this statement is valid: *The progeny of hybrids in which several essentially different traits are united represent the terms of a combination series in which the series for each pair of differing traits are combined.* This also shows at the same time that *the behavior of each pair of differing traits in a hybrid association is independent of all other differences in the two parental plants.*

If $n$ designates the number of characteristic differences in the two parental plants, then $3^n$ is the number of terms in the combination series, $4^n$ the number of individuals that belong to the series, and $2^n$ the number of combinations that remain constant. For instance, when the parental types differ in four traits the series contains $3^4 = 81$ terms, $4^4 = 256$ individuals, and $2^4 = 16$ constant forms; stated differently, among each 256 offspring of hybrids there are 81 different combinations, 16 of which are constant.

All constant associations possible in *Pisum* through combination of the above-mentioned seven characteristic traits were actually obtained through repeated crossing. Their number is given by $2^7 = 128$. At the same time this furnishes factual proof that constant traits occurring in different forms of a plant kindred can, by means of repeated artificial fertilization, enter into all the associations possible within the rules of combination.

Experiments on the flowering time of hybrids are not yet finished. However, it can already be reported that it is almost exactly intermediate between that of the seed plant and that of the pollen plant, and that development of the

| 8 plants | ABC | 22 plants | ABCc | 45 plants | ABbCc |
|---|---|---|---|---|---|
| 14 " | ABc | 17 " | AbCc | 36 " | aBbCc |
| 9 " | AbC | 25 " | aBCc | 38 " | AaBCc |
| 11 " | Abc | 20 " | abCc | 40 " | AabCc |
| 8 " | aBC | 15 " | ABbC | 49 " | AaBbC |
| 10 " | aBc | 18 " | ABbc | 48 " | AaBbc |
| 10 " | abC | 19 " | aBbC | | |
| 7 " | abc | 24 " | aBbc | | |
| | | 14 " | AaBC | 78 " | AaBbCc |
| | | 18 " | AaBc | | |
| | | 20 " | AabC | | |
| | | 16 " | Aabc | | |

The series comprises 27 members. Of these 8 are constant for all traits, and each occurs 10 times on the average; 12 are constant for two traits, hybrid for the third; each appears 19 times on the average; 6 are constant for one trait, hybrid for the other two; each of these turns up 43 times on the average; one form occurs 78 times and is hybrid for all traits. The ratios 10:19:43:78 approach the ratios 10:20:40:80, or 1:2:4:8, so closely that the latter doubtlessly represent the correct values.

The development of hybrids whose parents differ in three traits thus takes place in accord with the expression:

ABC + ABc + AbC + Abc + aBC + aBc + abC + abc + 2ABCc + 2AbCc + 2aBCc + 2abCc + 2ABbC + 2ABbc + 2aBbC + 2aBbc + 2AaBC + 2AaBc + 2AabC + 2Aabc + 4ABbCc + 4aBbCc + 4AaBCc + 4AabCc + 4AaBbC + 4AaBbc + 8AaBbCc.

Here, too, is a combination series in which the series for traits A and a, B and b, C and c are combined with each other. The expressions:

A + 2Aa + a
B + 2Bb + b
C + 2Cc + c

hybrids probably proceeds in the same manner with respect to this trait as it does for the remaining traits. The forms chosen for experiments of this nature must differ by at least 20 days in the mean date of blooming; it is also necessary that all the seeds be planted at an equal depth in order to achieve simultaneous germination, and, furthermore, that throughout the flowering period any relatively large temperature fluctuations with consequent acceleration or delay in blooming be taken into account. It is obvious that this experiment has various difficulties to overcome and demands great attention.

When we try to summarize briefly the results obtained, we find that those differing traits that permit easy and certain differentiation in the experimental plants *show completely concordant behavior in hybrid association.* Half of the progeny of plants hybrid in one pair of differing traits are hybrid again, while the other half become constant, with the characteristics of the seed and pollen plants in equal proportion. When, through fertilization, several differing traits become united in a hybrid, its progeny represent the terms of a combination series, in which the series for any one pair of differing traits are combined.

The complete agreement shown by all characteristics tested probably permits and justifies the assumption that the same behavior can be attributed also to the traits which show less distinctly in the plants, and could therefore not be included in the individual experiments. An experiment on flower stems of different lengths gave on the whole a rather satisfactory result, although distinction and classification of the forms could not be accomplished with the certainty that is indispensable for correct experiments.

THE REPRODUCTIVE CELLS OF HYBRIDS

The results of the previously cited investigations suggested further experiments whose outcome would throw light on the composition of seed and pollen cells in hybrids. An im-

portant clue is the fact that in *Pisum* constant forms appear among the progeny of hybrids and that they do so in all combinations of the associated traits. In our experience we find everywhere confirmation that constant progeny can be formed only when germinal cells and fertilizing pollen are alike, both endowed with the potential for creating identical individuals, as in normal fertilization of pure strains. Therefore we must consider it inevitable that in a hybrid plant also identical factors are acting together in the production of constant forms. Since the different constant forms are produced in a *single* plant, even in just a *single* flower, it seems logical to conclude that in the ovaries of hybrids as many kinds of germinal cells (germinal vesicles), and in the anthers as many kinds of pollen cells are formed as there are possibilities for *constant* combination forms and that these germinal and pollen cells correspond in their internal make-up to the individual forms.

Indeed, it can be shown theoretically that this assumption would be entirely adequate to explain the development of hybrids in separate generations if one could assume at the same time that the different kinds of germinal and pollen cells of a hybrid are produced on the average in equal numbers.

In order to test this hypothesis experimentally, the following experiments were chosen. Two forms which differed constantly in seed shape and albumen coloration were combined by fertilization.

If the differing traits are again designated by A, B, a, b, then:

AB seed plant,       ab pollen plant,
A shape round,       a shape angular,
B albumen yellow,    b albumen green.

The artificially fertilized seeds were sown, together with several seeds of the two parental plants, and the most vigorous specimens were chosen for reciprocal crosses. Fertilized were:

1. The hybrid with pollen from *AB*.
2. The hybrid   "   "   "   *ab*.
3. *AB*   "   "   "   the hybrid.
4. *ab*   "   "   "   the hybrid.

For each of these four experiments all flowers on three plants were fertilized. If the above assumption were correct, then germinal and pollen cells of the forms *AB*, *Ab*, *aB*, *ab* should develop in the hybrids, and combined would be:

1. Germinal cells *AB*, *Ab*, *aB*, *ab* with pollen cells *AB*.
2.   "   "   *AB*, *Ab*, *aB*, *ab*   "   "   "   *ab*.
3.   "   "   *AB*   "   "   "   *AB*, *Ab*, *aB*, *ab*.
4.   "   "   *ab*   "   "   "   *AB*, *Ab*, *aB*, *ab*.

From each of these experiments, therefore, only the following forms could result:

1. *AB*, *ABb*, *AaB*, *AaBb*.
2. *AaBb*, *Aab*, *aBb*, *ab*.
3. *AB*, *ABb*, *AaB*, *AaBb*.
4. *AaBb*, *Aab*, *aBb*, *ab*.

Furthermore, if the individual forms of the hybrid's germinal and pollen cells are, on the average, formed in equal numbers, then in each experiment the four combinations listed must stand in equal ratio to each other. Complete agreement of the numerical values was not to be expected, however, since in every fertilization, even in a normal one, some germinal cells fail to develop or die later, and even some well-developed seeds do not succeed in germinating after being planted. Moreover, the assumption requires only that there be a tendency to approach equality in the number of different kinds of germinal and pollen cells produced, yet this number does not have to be attained by each hybrid with mathematical exactness.

The primary objective of the *first* and *second* experiments

was to test the composition of hybrid germinal cells; that of the *third* and *fourth* experiments was to determine the composition of pollen cells. As shown by the above compilation, the first and third experiments, like the second and fourth, had to give identical combinations. The results should have been partly observable in the second year in the shape and coloration of the artificially fertilized seeds. In the first and third experiments the dominating traits of shape and color, A and B, occur in every combination, constant in some, in hybrid association with the recessive characters *a* and *b* in others, and must, for that reason, imprint their characteristics on all seeds. Therefore, if the premise were correct, all seeds should appear round and yellow. In the second and fourth experiments, on the other hand, one association is hybrid in shape and color, and therefore the seeds are round and yellow; another is hybrid in shape and constant in the recessive trait of color, therefore the seeds are round and green. The third association is constant in the recessive trait of shape and hybrid in color, therefore the seeds are angular and yellow. The fourth is constant in both recessive traits; therefore the seeds are angular and green. Hence, one could expect four different kinds of seeds from these two experiments, namely: round yellow, round green, angular yellow, angular green. The yield was in complete agreement with these expectations. There were obtained in the

First experiment, 98 exclusively round yellow seeds;
Third   "   94   "   "   "   "   ;
Second experiment, 31 round yellow, 26 round green, 27 angular yellow, 26 angular green seeds;
Fourth experiment, 24 round yellow, 25 round green, 22 angular yellow, 27 angular green seeds.

A favorable result could hardly be doubted any longer, but the next generation would have to provide the final decision. From the seeds sown, in the first experiment 90 plants, and in the third 87 plants, bore fruit in the following year; these yielded in

*Experiments*

1. 3.

| 20 | 25 | round yellow seeds | . | . | . | . | AB. |
| 23 | 19 | round yellow and green seeds | . | . | ABb. |
| 25 | 22 | round and angular yellow seeds | . | AaB. |
| 22 | 21 | round and angular, yellow and green seeds | . | AaBb. |

In the second and fourth experiments the round yellow seeds produced plants with round and angular, yellow and green seeds . . . . . . . . . . . . . . . . . . . . . . . *AaBb.*

From the round green seeds plants were obtained with round and angular green seeds . . . . . . . . . . . *Aab.*

The angular yellow seeds produced plants with angular yellow and green seeds . . . . . . . . . . . . . . . . *aBb.*

From the angular green seeds plants were raised which again yielded only angular green seeds . . . . . . . . . . . . *ab.*

Though in these two experiments also some seeds did not germinate, no change could be effected in the figures found in the preceding year, since each kind of seed produced plants that were, with respect to their seeds, alike among themselves and different from the others. Thus in the

*Second experiment*

*Fourth experiment*

| 31 | 24 plants yielded seeds of form | AaBb. |
| 26 | 25 "  "  "  " | Aab. |
| 27 | 22 "  "  "  " | aBb. |
| 26 | 27 "  "  "  " | ab. |

In all experiments, therefore, all forms postulated by the preceding hypothesis appeared, and did so in nearly equal numbers.

In a further test the traits of *flower color* and *stem length* were included in the experiments and selection was made in such a way that in the third year of experimentation every trait had to appear in *half* of all plants if the above assump-

tion were correct. A, B, a, b serve again as designation of the different traits.

A flowers purplish-red,    a flowers white.
B stem long,    b stem short.

Form Ab was fertilized by ab, producing hybrid Aab. In addition, aB was also fertilized with ab, yielding hybrid aBb. For further fertilization in the second year the hybrid Aab was used as seed plant, the hybrid aBb as pollen plant.

Seed plant Aab,    Pollen plant aBb.
Possible germinal cells Ab, ab,    Pollen cells aB, ab.

From fertilization involving the possible germinal and pollen cells, four combinations had to result, namely:

$$AaBb + aBb + Aab + ab.$$

From this it becomes apparent that, according to the above assumption, of all plants in the third year of experimentation

| Half should have violet-red flowers (Aa) | . | terms 1, 3 |
| "  "  " white flowers (a) | . | "  2, 4 |
| "  "  " a long stem (Bb) | . | "  1, 2 |
| "  "  " a short stem (b) | . | "  3, 4 |

Out of 45 fertilizations of the second year, 187 seeds were obtained, from which 166 plants reached the flowering stage in the third year. Among them the individual terms appeared in the following numbers:

| Term | Color of flower | Stem | |
| --- | --- | --- | --- |
| 1 | violet-red | long | 47 times |
| 2 | white | long | 40 " |
| 3 | violet-red | short | 38 " |
| 4 | white | short | 41 " |

Therefore,

| | | | | |
|---|---|---|---|---|
| violet-red flower color (Aa) occurred in 85 plants | | | | |
| white | " | (a) | " | 81 | " |
| long stem | " | (Bb) | " | 87 | " |
| short | " | (b) | " | 79 | " |

In this experiment, too, the proposed hypothesis finds adequate confirmation.

Experiments on a small scale were also made on the traits of *pod shape, pod color,* and *flower position,* and the results obtained were in full agreement: all combinations possible through union of the different traits appeared when expected and in nearly equal numbers.

Thus experimentation also justifies the assumption *that pea hybrids form germinal and pollen cells that in their composition correspond in equal numbers to all the constant forms resulting from the combination of traits united through fertilization.*

The difference of forms among the progeny of hybrids, as well as the ratios in which they are observed, find an adequate explanation in the principle just deduced. The simplest case is given by the series for *one pair of differing traits.* It is known that this series is described by the expression: $A + 2Aa + a$, in which A and a signify the forms with constant differing traits, and Aa the form hybrid for both. The series contains four individuals in three different terms. In their production, pollen and germinal cells of form A and a participate, on the average, equally in fertilization; therefore each form manifests itself twice, since four individuals are produced. Participating in fertilization are thus:

Pollen cells A + A + a + a
Germinal cells A + A + a + a

It is entirely a matter of chance which of the two kinds of pollen combines with each single germinal cell. How-

ever, according to the laws of probability, in an average of many cases it will always happen that every pollen form A and a will unite equally often with every germinal-cell form A and a; therefore, in fertilization, one of the two pollen cells A will meet a germinal cell A, the other a germinal cell a, and equally, one pollen cell a will become associated with a germinal cell A, the other with a.

$$
\begin{array}{lcccc}
\text{Pollen cells} & A & A & a & a \\
 & \downarrow & \times & \times & \downarrow \\
\text{Germinal cells} & A & A & a & a
\end{array}
$$

The result of fertilization can be visualized by writing the designations for associated germinal and pollen cells in the form of fractions, pollen cells above the line, germinal cells below. In the case under discussion one obtains:

$$\frac{A}{A} + \frac{A}{a} + \frac{a}{A} + \frac{a}{a}$$

In the first and fourth terms germinal and pollen cells are alike; therefore the products of their association must be constant, namely A and a; in the second and third, however, a union of the two differing parental traits takes place again, therefore the forms arising from such fertilizations are absolutely identical with the hybrid from which they derive. *Thus, repeated hybridization takes place.* The striking phenomenon, that hybrids are able to produce, in addition to the two parental types, progeny that resemble themselves is thus explained: $\frac{A}{a}$ and $\frac{a}{A}$ both give the same association, Aa, since, as mentioned earlier, it makes no difference to the consequence of fertilization which of the two traits belongs to the pollen and which to the germinal cell. Therefore

$$\frac{A}{A} + \frac{A}{a} + \frac{a}{A} + \frac{a}{a} = A + 2Aa + a.$$

This represents the *average* course of self-fertilization of hybrids when two differing traits are associated in them. In individual flowers and individual plants, however, the ratio in which the members of the series are formed may be subject to not insignificant deviations. Aside from the fact that the numbers in which both kinds of germinal cells occur in the ovary can be considered equal only on the average, it remains purely a matter of chance which of the two kinds of pollen fertilizes each individual germinal cell. Therefore, isolated values must necessarily be subject to fluctuations, and even extreme cases are possible, as mentioned earlier in experiments on seed shape and albumen coloration. The true ratios can be given only by the mean calculated from the sum of as many separate values as possible; the larger their number the more likely it is that mere chance effects will be eliminated.

The series for hybrids in which *two kinds of differing traits* are associated contains 16 individuals representing 9 different forms, namely: $AB + Ab + aB + ab + 2ABb + 2aBb + 2AaB + 2Aab + 4AaBb$. Among the different traits of the parental plants $A$, $a$ and $B$, $b$, 4 constant combinations are possible; therefore the hybrid produces the 4 corresponding forms of germinal and pollen cells, $AB$, $Ab$, $aB$, $ab$, and each of these will fertilize or be fertilized 4 times on the average, since the series contains 16 individuals. Participating in fertilization are thus

$$
\begin{aligned}
\text{Pollen cells:}\quad & AB + AB + AB + AB + \\
& Ab + Ab + Ab + Ab + \\
& aB + aB + aB + aB + \\
& ab + ab + ab + ab.
\end{aligned}
$$

$$
\begin{aligned}
\text{Germinal cells:}\quad & AB + AB + AB + AB + \\
& Ab + Ab + Ab + Ab + \\
& aB + aB + aB + aB + \\
& ab + ab + ab + ab.
\end{aligned}
$$

In fertilization every pollen cell unites, on the average, equally often with each form of germinal cell; thus each of the 4 pollen cells $AB$ once with each of the germinal cell

forms $AB$, $Ab$, $aB$, $ab$. In precisely the same manner the union of the remaining pollen cells of types $Ab$, $aB$, $ab$ with all the other germinal cells takes place. Thus one obtains:

$$
\frac{AB}{AB} + \frac{AB}{Ab} + \frac{AB}{aB} + \frac{AB}{ab} + \frac{Ab}{AB} + \frac{Ab}{Ab} + \frac{Ab}{aB} + \frac{Ab}{ab} +
$$
$$
\frac{aB}{AB} + \frac{aB}{Ab} + \frac{aB}{aB} + \frac{aB}{ab} + \frac{ab}{AB} + \frac{ab}{Ab} + \frac{ab}{aB} + \frac{ab}{ab},
$$

or

$$
AB + ABb + AaB + AaBb + ABb + Ab +
$$
$$
AaBb + Aab + AaB + AaBb + aB + aBb +
$$
$$
+ Aab + aBb + ab
$$
$$
= AB + Ab + aB + ab + 2ABb + 2aBb +
$$
$$
2AaB + 2Aab + 4AaBb.
$$

The series of hybrids in which *three kinds of differing traits* are combined can be explained in quite similar fashion. The hybrid produces 8 different forms of germinal and pollen cells: $ABC$, $ABc$, $AbC$, $Abc$, $aBC$, $aBc$, $abC$, $abc$, and again each pollen form unites once, on the average, with each germinal-cell form.

The law of combination of differing traits according to which hybrid development proceeds thus finds its basis and explanation in the proven statement that hybrids produce germinal and pollen cells that correspond in equal numbers to all the constant forms resulting from the combination of traits united through fertilization.

EXPERIMENTS ON HYBRIDS OF OTHER PLANT SPECIES

The object of further experiments will be to determine whether the law of development discovered for *Pisum* is also valid for hybrids of other plants. Several experiments were started quite recently for this purpose. I have completed two fairly small experiments with species of *Phaseolus*, which might be mentioned here.

An experiment with *Phaseolus vulgaris* and *Phaseolus nanus* L. gave fully concordant results. *Ph. nanus*, in addition to a dwarf-like stem, had green smoothly arched pods; *Ph. vulgaris*, on the other hand, had a stem 10–12' long and yellow pods, constricted at maturity. The numerical relationships in which different forms occurred in individual generations were the same as in *Pisum*. The formation of constant associations also proceeded according to the law of simple combination of traits, exactly as in *Pisum*. Obtained were:

| Constant combination | Stem | Color of unripe pod | Shape of ripe pod |
|---|---|---|---|
| 1 | long | green | arched |
| 2 | " | " | constricted |
| 3 | " | yellow | arched |
| 4 | " | " | constricted |
| 5 | short | green | arched |
| 6 | " | " | constricted |
| 7 | " | yellow | arched |
| 8 | " | " | constricted |

The green pod color, arched pod shape, and tall stem were dominating traits, as in *Pisum*.

Another experiment with two very different *Phaseolus* species was only partly successful. Serving as *seed plant* was *Ph. nanus* L., a very constant species with white blossoms in short racemes and small white seeds in straight, arched, smooth pods; as *pollen plant Ph. multiflorus W.* with tall winding stem, crimson blossoms in very long racemes, rough sickle-like crooked pods and large seeds with black flecks and splashes on a peachblossom-red background.

The hybrid bore the greatest resemblance to the pollen plant, but the blossoms seemed less intensely colored. Its fertility was very limited; from 17 plants that developed a total of many hundreds of blossoms, only 49 seeds were

harvested. These were of medium size and bore a design similar to that of *Ph. multiflorus*; the background color also did not differ basically. In the following year they produced 44 plants of which only 31 reached the flowering stage. The traits of *Ph. nanus*, which all became latent in the hybrid, reappeared in various combinations; their proportion to the dominating traits, however, fluctuated greatly because of the small number of experimental plants; but for some traits, like stem and pod shape, it was, as in *Pisum*, almost exactly 1:3.

Limited as the results of this experiment may be for the determination of ratios in which the various forms occurred, yet it provides a case of a *remarkable color change* in the blossoms and seeds of hybrids. It is known that in *Pisum* the traits of blossom and seed color appear unchanged in the first and in later generations, and that the offspring of hybrids carry exclusively one or the other of the two parental traits. The situation is different in the present experiment. True, the white flower and seed color of *Ph. nanus* appeared immediately in the first generation on one fairly fertile plant, but the remaining 30 plants developed flower colors that represented several gradations from crimson to pale violet. The coloration of the seed pod was no less varied than that of the flower. No plant could be considered fully fertile: some set no fruit at all; in others fruit was produced only by the last blossoms and did not have time to ripen. Well-formed seeds were harvested from only 15 plants. The greatest tendency toward infertility appeared in predominantly red-flowering forms; out of 16 such plants only 4 yielded ripe seeds. Three of these had a seed pattern similar to that of *Ph. multiflorus*, but a more or less pale background color, the fourth plant yielded only one seed, of plain brown coloration. Forms with preponderantly violet flower color had dark-brown, black-brown, and totally black seeds.

The experiment was continued for two more generations under equally unfavorable conditions, since even among the progeny of fairly fertile plants there were again some that were poorly fertile or completely sterile. No flower and seed colors other than those mentioned appeared. Forms receiving one or more of the recessive traits in the first generation remained constant in those traits without exception. Also, among the plants with violet blossoms and brown or black seeds, a few showed no further change in flower and seed color in the next generation, but the majority yielded, in addition to identical offspring, some with white flowers and similarly colored seed coats. Red-flowering plants remained so poorly fertile that nothing can be said with certainty about their further development.

Despite the many obstacles with which the observations had to contend, this experiment still establishes that development of hybrids follows the same law as in *Pisum* with respect to those traits concerned with the shape of the plant. Concerning the color traits, however, it seems difficult to find sufficient agreement. Besides the fact that a union of white and crimson coloration produces a whole range of colors from purple to pale violet and white, it is also striking that out of 31 flowering plants only one received the recessive trait of white coloration, while in *Pisum* this is true of every fourth plant on the average.

But these puzzling phenomena, too, could probably be explained by the law valid for *Pisum* if one might assume that in *Ph. multiflorus* the color of flowers and seeds is composed of two or more totally independent colors that behave individually exactly like any other constant trait in the plant. Were blossom color A composed of independent traits $A_1 + A_2 + \ldots$, which produce the overall impression of crimson coloration, then, through fertilization with the differing trait of white color $a$, hybrid associations $A_1a + A_2a + \ldots$ would have to be formed; and the situation with the corresponding coloration of the seed coat would be similar. According to the above assumption, each of these hybrid color combinations would be independent, and, therefore, would develop entirely independently from the rest. Then

it is easily seen that from the combination of the individual series a complete color range should result. If, for instance, $A = A_1 + A_2$, then the series that correspond to hybrids $A_1a$ and $A_2a$ are

$$A_1 + 2A_1a + a,$$
$$A_2 + 2A_2a + a.$$

The terms of these series can enter into 9 different combinations, each of which represents the designation for another color:

| | | | | | | | |
|---|---|---|---|---|---|---|---|
| 1 | $A_1$ | $A_2$ | 2 | $A_1a$ | $A_2$ | 1 | $A_2$ | $a,$ |
| 2 | $A_1$ | $A_2a$ | 4 | $A_1a$ | $A_2a$ | 2 | $A_2a$ | $a,$ |
| 1 | $A_1$ | $a$ | 2 | $A_1a$ | $a$ | 1 | $a$ | $a$ |

The numbers preceding the individual combinations indicate how many plants of corresponding coloration belong to the series. Since their sum is 16, all colors are distributed over each 16 plants on the average, but, as the series itself shows, in unequal proportions.

If color development really occurred in this manner, then the above-mentioned case of white blossom and seed-coat color appearing only once among 31 plants of the first generation would have an explanation. This coloration occurs only once in the series and, therefore, could be expressed only in every 16 plants, on the average; for three color traits once only even among 64 plants.

One must not forget, however, that the explanation attempted here rests on a mere supposition, with nothing more to commend it than the very incomplete results of the experiment just discussed. It would be a worthwhile task, though, to follow color development in hybrids further by similar experiments, because it is probable that through this approach we can learn to understand the extraordinary diversity in the *coloration of our ornamental flowers*.

Up to now, hardly more is known with certainty than that the flower color in most ornamental plants is an extremely variable trait. The opinion has often been expressed that, through cultivation, species stability is greatly upset or entirely shattered, and there is a strong inclination to describe the development of cultivated forms as devoid of rules and subject to chance; usually the coloration of ornamental plants is pointed out as a model of instability. However, it is not clear why mere transplantation into garden soil should have such thorough and persistent revolution in the plant organism as its consequence. No one would seriously want to maintain that plant development in the wild and in garden beds was governed by different laws. Here as well as there changes in the type must appear when living conditions are changed and when a species has the ability to adapt itself to the new environment. Granted willingly that cultivation favors the formation of new varieties and that by the hand of man many an alteration has been preserved which would have perished in nature, but nothing justifies the assumption that the tendency to form varieties is so extraordinarily increased that species soon lose all stability and their progeny diverge into an infinite number of extremely variable forms. If the change in living conditions were the sole cause of variability one could expect that those cultivated plants that have been grown through centuries under almost identical conditions should have regained stability. This is known not to be the case, for it is precisely among them that not only the most different but also the most variable forms are found. Only Leguminosae, such as *Pisum*, *Phaseolus*, and *Lens*, whose organs of fertilization are protected by the keel, represent notable exceptions. During more than 1000 years of cultivation under the most diversified conditions, numerous varieties have arisen, yet these maintain stability under constant living conditions, just as do species growing wild.

It remains more than probable that a factor that so far has received little attention is involved in the variability

of cultivated plants. Various experiences force us to accept the opinion that our cultivated plants, with few exceptions, are *members of different hybrid series* whose development along regular lines is altered and retarded by frequent intraspecific crosses. It should not be overlooked that cultivated plants are usually raised in fairly large numbers in close proximity to each other, a condition most favorable for reciprocal fertilization among the varieties present and between the species themselves. The likelihood that this opinion is correct is supported by the fact that among the large array of variable forms one finds always some single ones that remain constant for one or the other trait if all extraneous influence is carefully excluded. These forms develop exactly like certain members of the composite hybrid series. Even with respect to the most sensitive of all traits, that of color, it cannot escape careful observation that a tendency to variability exists in the individual forms to a very different degree. Among plants originating from a *single* spontaneous fertilization there are frequently some whose progeny diverge widely in the type and disposition of colors, while others produce forms that deviate little, and, if the number of plants is fairly large, some are encountered that transmit the color of their flowers unchanged to their progeny. Cultivated *Dianthus* species are an instructive example of this. A white-flowering specimen of *Dianthus caryophyllus*, itself derived from a white-flowered variety, was isolated in a greenhouse during the flowering period; its numerous seeds grew into plants with flowers of exactly the same shade of white. A similar result was obtained from a red, slightly violet glistening sport and from a white one with red stripes. On the other hand, many others, protected in the same manner, produced more or less differently colored and patterned progeny.

Anyone surveying the shades of color that appear in ornamental plants as a result of like fertilization cannot easily escape the conviction that here, too, development proceeds according to a certain law which possibly finds its expression through the *combination of several independent color traits*.

## CONCLUDING REMARKS

A comparison of the observations made on *Pisum* with the experimental results obtained by Kölreuter and Gärtner, the two authorities in this field, cannot fail to be of interest. Both concur in the opinion that, in external appearance, hybrids either maintain a form intermediate between the parental strains or they approach the type of one or the other, sometimes being barely distinguishable from them. Various forms that diverge from the normal type usually arise from the seeds of hybrids that were fertilized by their own pollen. As a rule the majority of individuals produced from such a fertilization maintain the form of the hybrid, a few become more like the seed plant, and an occasional individual very nearly matches the pollen plant. This, however, is not valid for all hybrids without exception. Among the offspring of certain individuals some are more like one original stock plant, some more like the other, or they all tend more to one side than the other; but those from a few remain *exactly like the hybrid* and propagate unchanged. The hybrids of varieties behave like species hybrids, but possess a still greater inconstancy and a more pronounced tendency to revert to the original forms.

With respect to the *features* of hybrids and their regular *development*, consistency with the observations made on *Pisum* is unmistakable. This is not so in the exceptional cases mentioned. Gärtner himself admits that precise determination of whether a form bears a greater resemblance to one or the other of the two parental types often presents great difficulties, since much depends on the subjective viewpoint of the observer. And there is yet another circumstance that could contribute to making the results variable and uncertain in spite of the most careful observation and discrimination. For the most part plants which are considered to be good species and that differ in a rather large number of traits were used in the experiments. When one is dealing in a general way with degrees of similarity, then account must be taken not only of the traits that stand out sharply,

but also of those that are often difficult to put into words, yet, as everyone familiar with plants knows, are sufficiently pronounced to give such forms the appearance of a stranger. If it is assumed that development of hybrids follows the law valid for *Pisum*, then the series obtained in each separate experiment must comprise very many forms, because the number of terms is known to increase with the number of differing traits as a power of three. Thus with a relatively small number of experimental plants the result could be only approximately correct and occasionally could deviate not inconsiderably. If, for instance, the two original stocks differed in 7 traits, and if 100 to 200 plants were raised from the seeds of their hybrids for an evaluation of the offsprings' degree of relationship, we can easily understand how uncertain such judgment must be, since the series for 7 differing traits contains 16,384 individuals appearing in 2187 different forms. Sometimes one relationship, sometimes another, would assert itself more strongly, depending on whether the observer found, by chance, a larger number of this or of that form.

Furthermore, when the differing traits include *dominating* ones that are passed on to the hybrid totally or almost totally unchanged, then the one of the two parental types having the larger number of dominating traits must always be the more prominent among the members of the series. In the experiment with three differing traits in *Pisum* described earlier, all of the dominating characters belonged to the seed plant. Although the members of the series tend equally toward both original parents in their internal makeup, the appearance of the seed plant was so preponderant in this experiment that 54 plants out of every 64 in the first generation looked exactly like it, or differed from it in only one trait. One sees how risky it can sometimes be to draw conclusions about the internal kinship of hybrids from their external similarity.

Gärtner mentions that in cases where development was regular the two parental types themselves were not repre-

able to develop into an independent organism through incorporation of matter and the formation of new cells. This development proceeds in accord with a constant law based on the material composition and arrangement of the elements that attained a viable union in the cell. When the reproductive cells are of the same kind and like the primordial cell of the mother, development of the new individual is governed by the same law that is valid for the mother plant. When a germinal cell is successfully combined with a *dissimilar* pollen cell we have to assume that some compromise takes place between those elements of both cells that cause their differences. The resulting mediating cell becomes the basis of the hybrid organism whose development must necessarily proceed in accord with a law different from that for each of the two parental types. If the compromise be considered complete, in the sense that the hybrid embryo is made up of cells of like kind in which the differences are *entirely and permanently mediated*, then a further consequence would be that the hybrid would remain as constant in its progeny as any other stable plant variety. The reproductive cells formed in its ovary and anthers are all the same and like the mediating cell from which they derive.

One could perhaps assume that in those hybrids whose offspring are *variable* a compromise takes place between the differing elements of the germinal and the pollen cell great enough to permit the formation of a cell that becomes the basis for the hybrid, but that this balance between the antagonistic elements is only temporary and does

---

spring of hybrids? If the influence of the germinal cell on the pollen cell were only external, if it merely played the role of a foster mother, then the outcome of each artificial fertilization would have to be that the resulting hybrid resembled the pollen plant exclusively or very closely. Experiments have in no way confirmed this up to now. A thorough proof for complete union of the content of both cells presumably lies in the universally confirmed experience that it is immaterial to the form of the hybrid which of the parental types was the seed or pollen plant.

sented among the offspring of the hybrids, only occasional individuals closely approximating them. Indeed, it cannot be otherwise in very extensive series. For 7 differing traits, for instance, each parental form occurs only once in more than 16,000 offspring of the hybrid. Therefore there is not much likelihood of finding them among a small number of experimental plants, yet, with a reasonable degree of probability, one may count on the appearance of a few forms that approximate those in the series.

We encounter an *essential difference* in those hybrids that remain constant in their progeny and propagate like pure strains. According to Gärtner these include the *highly fertile* hybrids *Aquilegia atropurpurea-canadensis, Lavatera pseudolbia-thuringiaca, Geum urbano-rivale,* and some *Dianthus* hybrids; according to Wichura it includes the hybrids of willow species. This feature is of particular importance to the evolutionary history of plants, because constant hybrids attain the status of *new species.* The correctness of these observations is vouched for by eminent observers and cannot be doubted. Gärtner had the opportunity of following the *Dianthus Armeria-deltoides* to its tenth generation, since that plant propagated itself regularly in the garden.

It was proven experimentally that in *Pisum* hybrids form *different kinds* of germinal and pollen cells and that this is the reason for the variability of their offspring. For other hybrids whose offspring behave similarly, we may assume the same cause; on the other hand, it seems permissible to assume that the germ cells of those that remain constant are identical, and also like the primordial cell of the hybrid. According to the opinion of famous physiologists, propagation in phanerogams is initiated by the union of one germinal and one pollen cell into one single cell,[5] which is

[5] It is presumably beyond doubt that in *Pisum* a complete union of elements from both fertilizing cells has to take place for the formation of a new embryo. How else could one explain that both parental types recur in equal numbers and with all their characteristics in the off-

not extend beyond the lifetime of the hybrid plant. Since no changes in its characteristics can be noticed throughout the entire vegetative period, we must further conclude that the differing elements succeed in escaping from the enforced association only at the stage at which the reproductive cells develop. In the formation of these cells all elements present participate in completely free and uniform fashion, and only those that differ separate from each other. In this manner the production of as many kinds of germinal and pollen cells would be possible as there are combinations of potentially formative elements.

This attempt to relate the important difference in the development of hybrids to a *permanent or temporary association* of differing cell elements can, of course, be of value only as a hypothesis which, for lack of well-substantiated data, still leaves some latitude. Some justification for the opinions expressed lies in the proof cited here that in *Pisum* the behavior of a pair of differing traits in hybrid union is independent of any other differences between the two parental plants and that, furthermore, the hybrid produces as many kinds of germinal and pollen cells as there are possible constant combination forms. The distinguishing traits of two plants can, after all, be caused only by differences in the composition and grouping of the elements existing in dynamic interaction in their primordial cells.

Yet even the validity of the laws proposed for *Pisum* needs confirmation, and a repetition of at least the more important experiments is therefore desirable: for instance, the one on the composition of hybrid fertilizing cells. An individual observer can easily overlook a distinguishing point that seems unimportant in the beginning but can grow to such proportions that it may not be neglected in the final analysis. Whether variable hybrids of other plant species show complete agreement in behavior also remains to be decided experimentally; one might assume, however, that no basic difference could exist in important matters since *unity* in the plan of development of organic life is beyond doubt.

Finally, the experiments performed by Kölreuter, Gärtner, and others on *transformation of one species into another by artificial fertilization* deserve special mention. Particular importance was attached to these experiments; Gärtner counts them as among "the most difficult in hybrid production."

When species *A* was to be transformed into *B*, the two were combined by fertilization and the resulting hybrids once more fertilized with pollen from *B*; from among their various descendants those closest to species *B* were then chosen and repeatedly fertilized by pollen from *B*, and so on, until finally a form that was like *B* and remained constant in its progeny was obtained. Thus species *A* was transformed into the other species, *B*. Gärtner himself has carried out 30 experiments of this kind with plants from genera *Aquilegia, Dianthus, Geum, Lavatera, Lychnis, Malva, Nicotiana,* and *Oenothera.* The length of time needed for transformation was not the same with all species. Although three successive fertilizations were sufficient for some, with others fertilizations had to be repeated five to six times; even with the same species fluctuations were observed in different experiments. Gärtner ascribes these differences to the circumstance that "the characteristic force toward change and transformation of the maternal type—that a species exerts in reproduction is very different in different plants, and consequently the length of time required for one species to become transformed into another, and the number of generations it takes, must also be different; transformation is accomplished after more generations in some species, after fewer in others." The same observer notes further "that in the process of transformation much depends on which type and which individual was chosen for further transformation."

If one may assume that the development of forms proceeded in these experiments in a manner similar to that in *Pisum*, then the entire process of transformation would have a rather simple explanation. The hybrid produces as many kinds of germinal cells as there are constant combinations made possible by the traits associated within the hybrid, and one of these is always just like the fertilizing pollen cells. Thus there is the possibility that in such experiments a constant form identical to the pollen parent will result from the second fertilization. Whether one is actually obtained depends on the number of plants in each experiment as well as on the number of differing traits that were united by the fertilization. Let us assume, for example, that the plants chosen for the experiment differ in three traits and that species ABC is to be transformed into species abc by repeated fertilization with pollen from the latter. The hybrid resulting from the first fertilization forms 8 different kinds of germinal cells, namely:

ABC, ABc, AbC, aBC, Abc, aBc, abC, abc.

In the second year of the experiment these are again combined with pollen cells abc and one obtains the series:

AaBbCc + AaBbc + AabCc + AaBbc + Aabc
+ aBbc + abCc + abc.

Since the form abc occurs once in the series of 8 terms there is little likelihood that it would be missing among the experimental plants, even if only a fairly small number were raised, and transformation would thus be complete after two fertilizations. If, by chance, no transformation was obtained, fertilization would have to be repeated on one of the closest related combinations, Aabc, aBbc, abCc. It becomes obvious that *the smaller the number of experimental plants and the larger the number of differing traits in the two* parental species the longer an experiment of this kind will last, and that furthermore, a delay of one or even two generations could easily occur with these same species, which is what Gärtner has observed. The transformation of widely divergent species cannot be completed before the fifth or

sixth experimental year because the number of different germinal cells formed in the hybrid increases with the number of differing traits as a power of two.

Gärtner found by repeated experiments that the *reciprocal* period of transformation varies for some species, so that quite frequently species A can be transformed into species B a generation earlier than species B into species A. From this he deduces that Kölreuter's opinion that "the two natures in hybrids are in perfect equilibrium" is not entirely tenable. It seems, however, that Gärtner has overlooked an important point, to which he himself draws attention elsewhere, namely, that it "depends on which individual is chosen for further transformation." Experiments set up for this purpose with two *Pisum* species indicate that in the selection of individuals best suited for the purpose of further fertilization it could make a great difference which of the two species is to be transformed into the other. The two experimental plants differed in five traits; those of species A were all dominating, those of species B were all recessive. To effect mutual transformation A was fertilized with pollen from B and B with pollen from A, and the same procedure was repeated on both hybrids in the following year. In the first experiment, $\frac{B}{A}$, 87 plants *in the 32 possible forms* were available in the third experimental year from which to choose individuals for further fertilization; in the second experiment, $\frac{A}{B}$, the external appearance of all 73 plants obtained completely *coincided with that of the pollen plant*, although their internal constitution must have been just as varied as the forms from the other experiment. Intentional selection was therefore possible only in the first experiment; in the second one a few plants had to be chosen purely at random. Only a few flowers of the latter were fertilized with pollen from A, the rest were allowed to self-fertilize.

Among each five plants used for fertilization in the two experiments, the next year's culture showed the following agreement with the pollen plant:

| *First Experiment* | *Second Experiment* | In all traits |
|---|---|---|
| 2 plants | — | |
| 3 " | — | " 4 " |
| | 2 plants | " 3 " |
| | 2 " | " 2 " |
| | 1 plant | " 1 trait |

In the first experiment transformation was thus completed; in the second, which was not continued, two more fertilizations would probably have been necessary.

Though it is infrequently true that the dominating traits belong exclusively to one or the other parental plant, it will always make a difference *which* of the two possesses them in larger number. When the pollen plant has the majority of dominating traits, the choice of forms for further fertilization will afford a lesser degree of certainty than in the opposite case. A delay in the length of time needed for transformation will be the consequence if the experiment be considered complete only when a form is obtained that not only resembles the pollen plant in appearance but, like it, remains constant in its progeny.

The success of transformation experiments led Gärtner to disagree with those scientists who contest the stability of plant species and assume continuous evolution of plant forms. In the complete transformation of one species into another he finds unequivocal proof that a species has fixed limits beyond which it cannot change. Although this opinion cannot be adjudged unconditionally valid, considerable confirmation of the earlier expressed conjecture on the variability of cultivated plants is to be found in the experiments performed by Gärtner.

Among the experimental species were cultivated forms

such as *Aquilegia atropurpurea* and *canadensis, Dianthus Caryophyllus, chinensis* and *japonicus, Nicotiana rustica* and *paniculata,* and these, too, lost none of their stability after 4 to 5 repetitions of hybrid association.

G. D. Karpechenko. 1928. Polyploid hybrids of *Raphanus sativus* L. × *Brassica oleracea* L. Zeitschrift für Induktive Abstammungs- und Vererbungslehre 48:1–85.

G. D. Karpechenko

G. D. Karpechenko did not conceive of the idea that hybridization of diverse species followed by polyploidy could be a means of speciation, nor did he devise and conduct the first experiment showing that it could be done. Otto Winge formulated the hypothesis, but he was apparently unaware that spontaneous creation of *Primula kewensis* from a sterile interspecific hybrid between *P. verticillata* and *P. floribunda* had been observed by L. Digby. She noted the association of the doubled chromosome number of the hybrid and its fertility, but probably failed to grasp its evolutionary significance.

Winge's hypothesis was confirmed experimentally when R. E. Clausen and T. H. Goodspeed crossed *Nicotiana tabacum* and *N. glutinosa* and obtained an amphidiploid, which was fertile, and which they named *N. digluta.*

Of the several other confirmations obtained in quick succession, Karpechenko's was, in the words of G. L. Stebbins, "the most spectacular," as it involved the crossing of two species in different genera which were morphologically quite distinct. Both parents, *Raphanus sativus* and *Brassica oleracea,* are major vegetable crops and therefore important horticulturally.

Karpechenko himself differentiates his own work from that of his contemporaries on the basis of his observation that chromosome doubling had occurred following hybridization. This was not so noted in the other works. Either the doubling had occurred somatically as with *P. kewensis* or the process was unclear, as in the *Nicotiana* experiment. Fertile hybrids in *Rosa* and from an *Aegilops* × *Triticum* cross were also evidently amphidiploids, but the doubling process itself was not observed.

The making of *Raphanus* × *Brassica* crosses had been a fairly common activity by the time Karpechenko performed his experiments. The first recorded cross was by A. Sageret in 1826, who got flowers and a few fruits and only two plants. F. Grawatt described an enormous hybrid with many flowers but no fruits or seeds. K. Moldenhawer obtained one $F_1$ plant which produced four $F_2$ plants; one was sterile, the other three produced fruits and seeds.

The first of two steps leading to the hypothesis of speciation by polyploidy was the realization of the existence of polyploids, for example, the recognition by A. M. Lutz in 1907 and by R. R. Gates in 1909 that one of De Vries's mutated forms, *Oenothera lamarckiana* (*gigas*) was a tetraploid. Still more revealing was the observation that in a number of genera, an arithmetic series of chromosome numbers existed in the species array. For example, in *Rosa*, $2n = 14, 28, 42, 56$; in *Triticum*, $2n = 14, 28, 42$; and in *Chrysanthemum*, $2n = 18, 36, 54, 72, 90$.

Winge perceived that mere repeated chromosome doubling would lead to a geometric progression of chromosome numbers (e.g., from 6 to 12 to 24 to 48, etc.). He proposed instead that hybridization followed by doubling was necessary to achieve an arithmetic progression. For example, a hybrid between a 12-chromosome species and a 24-chromosome species, doubled, gives a 36-chromosome species, and the progression is arithmetic.

Experimental evidence, such as the case of *Primula kewensis*, the *Nicotiana glutinosa* × *N. tabacum* hybrid, and *Raphanobrassica*, demonstrated the validity of Winge's hypothesis. A thorough exposition of the evidence and the ramifications and consequences of the hypothesis are given by Stebbins in his book and provide the basis for elevation of the hypothesis to a full-fledged theory.

Georgi D. Karpechenko (1899–1943) graduated from the Moscow Agricultural Academy. He began his cytological investigation of the Cruciferae there in 1922. He worked at the All Union Institute of Plant Raising, Detskoje Selo, Leningrad, where he headed the genetics laboratory from 1925 to 1941. From 1932 to 1941, he also was head of the subdepartment of plant genetics at Leningrad State University. He was a close collaborator of N. I. Vavilov and suffered the same fate at the hands of Soviet authorities: arrest in 1941 and death in prison in 1942 or 1943.

Karpechenko's work may be summarized in five phases. First was his description of the cytology of the $F_1$ plants. It must be noted that all crosses, both by Karpechenko and others cited in his work, were made with radish (*Raphanus sativus* L.) as the female parent. The male parents were forms of *Brassica oleracea* L., common and Savoy cabbage, Brussels sprouts, and kohlrabi). The hybrids had the diploid number of chromosomes ($2n = 18$ for both species). However, the reduction division was abnormal and the plants were sterile. In a second flowering, however, a few seeds were formed and cytological analysis of plants grown from those seeds showed that they were polyploid.

Karpechenko showed that fertility resulted when, at meiosis, the chromosomes failed to separate in anaphase I. A single group of 36 chromosomes was formed. These split normally in the second division, leading to formation of a dyad instead of a tetrad. Each cell and each subsequent gamete had 18 chromosomes instead of 9. This variation was observed in the division of pollen mother cells only, but Karpechenko concluded that the same phenomenon must have occurred in the formation of female gametes.

In the hybrid progeny, that is, those plants produced from the few seeds on the original hybrids, 93 percent of the plants had 36 chromosomes in the somatic cells. They were tetraploids. The remainder had either some deviation from 36 chromosomes, 27 chromosomes, or deviations from 42 or 54 chromosomes.

Morphologically, the tetraploids were, in general, larger than the diploids. In his words: ". . . we have to deal with a positive influence of the multiplication of the chromosome set on the dimensions of the organs." Some polymorphism was exhibited and is ascribed to the original polymorphism of the parents, both of which are cross-pollinated species, and therefore, genetically variable. The most interesting and informative aspect of hybrid morphology is the appearance of the fruits, as displayed elegantly in Fig. 17 of the paper. The radish has a reduced two-valved lower section which is sterile or nearly so, and a large, spongy upper section, which is filled with seeds. The *Brassica* fruit has an elongated two-valved lower section with many seeds and a reduced, nearly sterile upper section. The diploid hybrid has a fruit with equal-sized upper and lower sections. Total length of the hybrid fruit is slightly less than that of the radish and about one-third that of the *Brassica*. The tetraploid fruit is similar to the diploid in its equal division into two sections, but it is about two-thirds of the length of the cabbage fruit and considerably stouter than either the cabbage fruit or the diploid.

Meiosis in the tetraploids was entirely normal, demonstrating what may be called coexistence of the radish and *Brassica* genomes, each taking part in the event of meiosis, independently of the other and quite normally.

Finally, Karpechenko concluded that hybrid fertility was possible only when the $F_1$ gametes had a minimum of nine radish and nine *Brassica* chromosomes. He emphasized, however, that not all sex cells of this quality were fertile and suggested that genetic abnormalities may also play a part.

Karpechenko suggested that the new tetraploid form would reproduce itself and remain constant, with deviant forms likely to be relatively unfertile and to be selected against. He also suggested that it was taxonomically a new species, or even a new genus, although he did not himself use the name *Raphanobrassica* until later.

In citing the previous work in *Primula, Rosa, Nicotiana,* and *Triticum–Aegilops,* Karpechenko stressed that the hybrid origin of the polyploid forms was uncertain and emphasized the importance of his own work in providing evidence for the validity of Winge's hypothesis. The key difference is that in the *Raphanus* × *Brassica* cross, the chromosome doubling was shown to have taken place on a demonstrated sterile diploid $F_1$ plant, with the production of a second set of flowers, some of which bore seeds. The change could not, therefore, have been an independent mutational event.

Why is this a classic horticultural paper? First, Karpechenko recognized the importance of demonstrating the occurrence of chromosome doubling, and did so. Second, it was a spectacular accomplishment involving *two* important horticultural crops from different genera. Third, the work was elegant, particularly with the detailed and precise demonstration of the relationship between genome distribution and capsule dimensions.

Edward J. Ryder
Agricultural Research Service
U.S. Department of Agriculture
1636 East Alisal Street
Salinas, California 93905

## REFERENCES

BLACKBURN, K. B., and J. W. H. HARRISON. 1924. Genetical and cytological studies in hybrid roses. 1. The origin of a fertile hexaploid form in the *Pimpinellifoliae–Villosae* crosses. Br. J. Expt. Biol. 1:557–569.

CLAUSEN, R. E., and T. H. GOODSPEED. 1925. Interspecific hybridization in *Nicotiana*. II. A tetraploid glutinosa–tabacum hybrid, an experimental verification of Winge's hypothesis. Genetics 10: 279–284.

DIGBY, L. 1912. The cytology of *Primula kewensis* and of other related *Primula* hybrids. Ann. Bot. 26:357–388.

GATES, R. R. 1909. The stature and chromosomes of *Oenothera gigas* De Vries. Arch. Zellforsch. 3:525–552.

GRAVATT, F. 1914. A radish–cabbage hybrid. J. Hered. 5:269–272.

LUTZ, A. M. 1907. A preliminary note on the chromosomes of *Oenothera lamarckiana* and one of its mutants, *O. gigas*. Science (n.s.) 26:151–152.

MEDVEDEV, Z. A. 1978. Soviet science. W.W. Norton, New York.

MOLDENHAWER, K. 1924. On the generic cross *Raphanus* × *Brassica* (in German). Mem. Inst. Genet. Ecole. Sup. Agr. Varsovie 2:191–196.

PROKHOROV, A. M. (ed.). 1976. Great Soviet encyclopedia, 3rd ed., Vol. 11, p. 459. MacMillan, New York. (English trans.)

SAGERET, A. 1826. Considerations on the production of hybrids, variants and varieties in general and on those of the *Cucurbit* family in particular (in French). Ann. Sci. Nat. 8:294–314.

STEBBINS, G. L. 1950. Variation and evolution in plants. Columbia University Press, New York.

TSCHERMAK, E., and H. BLEIER. 1920. On fertile *Aegilops*–wheat hybrids (Examples of the creation of new species through hybridization) (in German). Bot. Gaz. 44:110–132.

WINGE, O. 1917. The chromosomes. Their numbers and general importance. C. R. Trav. Lab. Carlsberg 13:131–275.

# Polyploid Hybrids of *Raphanus sativus* L. × *Brassica oleracea* L.

## (On the problem of experimental species formation)

### By G. D. KARPECHENKO

Institute of Applied Botany, Section of Genetics, Detskoe Selo, U. S. S. R.

(With 12 Tables, 40 Textfigures and 3 Plates)

(Eingegangen 25. Juli 1927)

## Contents

# Introduction

Closely allied species are often known to possess numbers of chromosomes which are related to each other as common multiples. In the genus of *Chrysanthemum*, for instance, there occur species with 18, 36, 54, 72 and 90 chromosomes (2 n, TAHARA, 1921), in that of *Rosa*—species with 14, 21, 28, 35, 42 and 54 chromosomes (TÄCKHOLM, 1922), in *Rumex* of the group *Eulapathum*—species with 20, 40, 60, 80, 100 and 200 chromosomes (KIHARA and ONO, 1926) etc.

There exists an opinion that the occurence of such relations is due to hybridisation, that polyploid multichromosome species originate from intercrosses between species with a small number of chromosomes.

According to WINGE (1917, 1924), for instance, it may be assumed that in the hybrid from the crossing of two species, whose chromosomes show no affinity whatever, a longitudinal splitting of all them takes place, resulting in the usual paired homology of the chromosomes in the complex, and the hybrid having acquired the capability for normal reduction division and constancy, becomes an independent species. Thus, for instance, from the cross of two nine chromosome species there may arise an 18-chromosome species, from the cross of the latter with a new nine chromosome species may result a 27 chromosome one etc.

The present investigation stands in direct relation to this idea. It might be considered as an experimental establishing of the theory of hybrid origin of the polyploid species.

# Part I

## 1. Material and technics

The object of our investigation are the hybrids from crosses of *Raphanus sativus* L. × *Brassica oleracea* L. This cross aroused our interest as early as 1922, when by cytological investigation of the cultivated *Cruciferae*, was established that both *Raphanus sativus* and *Brassica oleracea* possessed the same diploid number of chromosomes—18 (KARPECHENKO, 1922). At that same time the common kitchen garden radish (*Raphanus sativus* L., *prol. niger* PERS.) was being crossed with several cabbages—common cabbage, savoy, Brussels' sprouts and kohlrabi (*Brassica oleracea* L.: *var. capitata* L., *sabauda* L., *gemmifera* DC., *gangyloides* L.), while in 1923 from 202 pollinated flowers were obtained and cultivated 123 hybrids. The latter were studied rather in detail the first year of their vegetation and the results have been published (KARPECHENKO, 1924).

The hybrids possessed the diploid number of chromosomes—18, their reduction division was quite abnormal and they proved sterile. We did not lay great hope on further work with them, nevertheless their stumps were dug out, kept in a cellar and in spring of the following year (1924) they were replanted.

All at once profuse flowering began and towards autumn on some of them seed shedding commenced. We immediately proceeded to a new fixation of the buds and rootlets for cytological analysis and carefully gathered all the seeds. In 1925 the latter were sown out and a cytological analysis of the plants grown from them, showed that the hybrids of the second generation were polyploids i. e. that they possessed a multiple number of chromosomes in relation to $F_1$. This is the circumstance that induced us to undertake a manifold investigation of the radish × cabbage hybrids of the first as well as of following generations. The material proved extensive enough as in the second generation we have over 400 plants (Fig. 1).

Of course, neither *Raphanus* nor *Brassica* are plants capable of selfpollinating, and with a short vegetative period, with a considerable number of readily analized characters and a small number of individual chromosomes, i. e. are plants quite perfect for genetic investigations. Our species possessed rather adverse qualities: they are biennials, crosspollinating plants, their chromosomes are not individual and most of their characters are diffuse. Nevertheless they proved to have some good features. Their number of chromosomes is not large—n = 9, the chromosomes themselves are short and in meiosis almost round, which is convenient for reckoning. To produce crossings, presents no difficulty whatever. Segregation in the hybrids is easily analized by the colour of the flowers and the structure of the siliquas. The biennial mode of life transmitted to the hybrids, presents the convenience that the hybrids, having been morphologically and cytologically investigated during the first year, can be most suitably used for crossing in their second year of vegetation. The ability of all our plants to propagate by way of cuttings, proved very valuable for obtaining interesting forms in sufficient quantity.

The seeds of the hybrids as well as of the parental forms were usually sown into eight cm. pots in a green-house where the plants

1 *

Fig. 1. General view of the plot with radish x cabbage hybrids of the second generation. Detskoe Selo, 1926

remained until the 5-6th leaf had appeared, after which they were removed for several weeks into warm beds and hence transplanted into the ground in the open air. When sown in late March or early April towards the end of summer, many of the hybrids were already beginning to bloom. When it was desirable to conserve the plants for a second year, we cut short all the stalks and kept the stumps in a cellar, covering them with earth. Being planted into the ground in spring, the stumps immediately covered with green leaves, they began to form stalks and soon blossomed. In this way we were able to conserve the plants in some cases for a third year and every year they covered with blossoms. It also proved possible to prolong the life of some hybrids and to propagate it by way of growing the cuttings. There was no difficulty in obtaining the latter; development, blossoming and fruit bearing proceeded with them quite normally.

In every kind of crossings we used only parchment insulators covering the shoot in the place where it was tied up with cotton wool. The flowers from which was taken the pollen for pollination were also covered with insulators before they began blossoming. For castration were left 5-6 big buds in the bunch. Pollination was usually performed on the second-third day after castration.

Under the conditions of our Northern latitude the siliquas obtained from the cross often enough were not able to ripen properly before the cold set in; in such cases the branches with the siliquas were cut off and were placed for some time in a green-house in pots with KNOP's dilution where they attained full maturity.

The fixation of rootlets for cytological investigation was usually made when the plants were in pots. When the plant is removed from the pot together with the earth, it is easy to find a great quantity of rootlets good for fixation. Sometimes we used for fixation adventitious roots from the cuttings or stumps and sometimes we fixed the roots of seeds having germinated on filtering paper.

The fixation of the buds was carried on during the whole summer usually several times for the same plant. Whether division was occuring, had previously been determined by BELLING's (BELLING, 1926) aceto-carmine. In order to induce a more rapid penetration of the fixing liquid into the buds, incisions were made at the tips and bases of the latter.

We generally used a fixative provided by the laboratory of Prof. S. G. NAVASHIN (10 parts of 1% chromic acid [$H_2CrO_4$] + 4 parts

40 % formaline + 1 part glacial acetic acid) and for the buds sometimes ZENKER's fixative.

The microtome sections were made 7—11 mm. thick, in accordance with the size of the cells of the plant under examination.

For staining the preparations iron haematoxylin and gentian violet were used.

Anatomical examination was made on fresh materials or such treated with 70° alcohol. For staining haematoxylin after DELAFIELD was used.

## 2. Terminology

Concerning our hybrids and their gametes, we shall further on employ the terms *diploid*, *tetraploid* etc. If according to WINKLER (1920, pp. 165—166) the haploid set of chromosomes is termed *genom*, all our hybrids will be *heterogenomous*, as they always contain different genoms of *Raphanus* and of *Brassica*; but since the radish and cabbage genoms are indentical as to the number of their chromosomes, our *digenomous* hybrids with respect to number exclusively, will be also *diploid*, those having three genoms—*triploid* etc. This is what we mean when speaking of their polyploidy. Concerning the hybrids and gametes which, for instance, are in their number of chromosomes close to tetraploid, but do not exactly correspond, we use the terms *hypotetraploid* or *hypertetraploid* according to the number of chromosomes included in the hybrid or gamete, it being a more or less considerable one than that of the tetraploid (compare WINKLER, 1916, p. 422).

## 3. Cytological investigation of F₁

We shall permit ourselves to be but brief in giving the cytological data on $F_1$, since everything that remained to be said in this respect in addition to our work of 1924, has been already published recently in a special paper (KARPECHENKO, 1927).

### a) Number of chromosomes in the soma and usual course of the meiosis

The somatic number of chromosomes in the hybrids of $F_1$ is 18. We have examined the roots not only in the young plants but have as well counted the number of chromosomes in adventitious roots obtained from cuttings of the stalks of adult plants, and have reckoned the number of chromosomes in the cones of growth of the branches and in the somatic tissues of the buds.

We have come to the conviction that $F_1$ is a diploid during all its life-time and in all its organs until the formation of sexual cells.

The usual course of the meiosis of hybrids $F_1$ is the following: 9 radish and 9 cabbage chromosomes do not conjugate with one another, and in the diakinesis and metaphase of the 1st division are counted 18 chromosomes. The threads of the spindle are scarcely discernible. Sometimes the spindle seems not to form at all. The univalent chromosomes usually move towards the poles without splitting, the daughter nuclei often receiving a different number of chromosomes. In the second division the univalens split, and the distribution of the splitted chromosomes proceeds again irregularly. Instead of tetrads there often arise groups with other number of cells up to 7 inclusively. The number of chromosomes in these cells is of great variety, but in most of cases it is from 6 to 12.

## b) Formation of diploid gametes

From the just mentioned usual course of meiosis in our hybrids, deviations are met with, on which we shall now stop our attention.

It has been sometimes observed, that in the anaphases of the 1st division, the 18 univalent chromosomes do not distribute towards the poles in order to form two nuclei, but remain lying together and are again inclosed in one nuclear membrane. In such cells with "mono-nuclear telophases of the 1st division" or, in other words, in cells where the first division has dropped away, is subsequently formed one large spindle, in which all univalents split longitudinally, so that the number of chromosomes in the anaphase is 36. There occurs a dyad as consequence of such a division and if the distribution of the chromosomes has proceeded evenly, each cell of the dyad receives 18 chromosomes. These cells develop directly into pollen grains and consequently thus in our hybrids arise the gametes with a diploid number of chromosomes.

The formation of diploid gametes is frequently disturbed by the fact that in the above described peculiar telophases of the first division, not all chromosomes are included into the nucleus, but one or two of them remain in the plasma to lie apart. Then these chromosomes usually form in the second homotypical division a separate small spindle what not unfrequently leads to the formation of a tetrade consisting of cells with different number of chromosomes. In those cases also when in the first division one nucleus consisting of 18 chromosomes is formed, an irregular distribution of the latter in the homotypical division may result in some other complex of cells instead of a dyad, or else several chromosomes may simply not get into the forming nuclei and eliminate. In consequence of all such irregularities, gametes arise, showing not a diploid, but a considerably smaller number of chromosomes, which approaches them to the general lot of gametes formed in our hybrids in result of the usual course of division.

Besides diploid and hypodiploid sexual cells, hyperdiploid gametes may also form in our hybrids. These arise in consequence of an unequal distribution of the chromosomes between both poles in the above described 36 chromosome anaphases, and because of the fact, that certain univalents are apt to split twice. We have had the opportunity of observing anaphases of the 1st division with some splitting chromosomes and of reckoning sometimes more than 18 chromosomes in the "one-nuclear telophases" of the first division; and finally in the anaphases of the second division the number of chromosomes now and then exceeded 36, which can of course be explained only by the fact, that some univalents which had split in the first division, split in the second one also.

### c) Formation of tetraploid gametes

In our hybrids during the formation of sexual cells, phenomena are observed suggesting the idea that in them gametes may be formed with a tetraploid number of chromosomes or a number approaching it.

In the cells of the archesporium, at the last prereduction division, sometimes occur a division of the nucleus without a division of the cell. The cell thus becomes binuclear with each of the nuclei containing 18 chromosomes. In the stage of diakinesis the chromosomes in both nuclei remain nonconjugated and form in the first division a common spindle consisting of 36 chromosomes. If the chromosomes further on will distribute between the 2 poles, each daughter nucleus will receive about 18 chromosomes, and in the second division a tetrad with similar nuclei will be formed. But if as we have observed all 36 chromosomes are enclosed in one nuclear membrane, and subsequently one homotypical spindle is formed, there must arise a dyad with 36 chromosomes in each cell. But in reality matters are complicated by a very unequal distribution of the chromosomes, by the elimination of some of them out of the sphere of division, by double splitting of the univalents etc.; more frequently hypertetraploid and hypotetraploid gametes are formed, of the latter ones a larger number, owing to the same causes which also lead to the formation of many hypodiploid gametes.

### d) Conclusion

Summing up all that was said on the formation of the sexual cells in our hybrids, we come to the conclusion that a lot of them possesses a number of chromosomes more or less approaching the haploid one, then forms an unconsiderable number of diploid, hypo- and hyperdiploid gametes and finally, already only as an exception, there arise tetraploid, hypo- and hypertetraploid gametes. Deviations in the process of formation of the three mentioned groups of sexual cells may be so considerable, that the appearance of gametes with intermediate chromosome complexes may be conceived.

---

All these data were obtained in result of a study on the division of the pollen mother cells in our hybrids. In spite of the great number of preparations examined by us, we were not able however, to elucidate the processes of division in the mother cells of the embryo sac. Single observations in this respect, conducted already in 1923 (KARPECHENKO, 1924) afford no definite data. But judging by the results of a cytological investigation of our $F_2$ hybrids, the male and female gametes in $F_1$ must be equal in their chromosome complex, and it is probable that the processes of their formation are also in most cases identical.

### 4. Fertility of $F_1$ hybrids

The majority of pollen grains in the hybrids of $F_1$, degenerate in the one or the other stage of development, and thus full maturity is reached only by very few of them (see fig. 36, b, p. 58).

As has been already mentioned, the hybrids proved entirely sterile in their first year of vegetation, although many of them flowered. The second year some of them produced seeds, but only in small quantity. Out of 90 individuals planted in the second year, having flowered continually and abundantly, only 19 showed a partial fertility, producing the total amount of 821 seeds. The greatest number of seeds collected from one plant was 189, but some individuals produced not more than 2—4 seeds (see second column in tables I and II, p. 11). Thus the degree of fertility appeared to be different with different hybrids, but on the whole it was low, and no sharp line of demarcation could be traced between such poorly fertile and sterile hybrids.

Just as well no difference was observed between the fertility of branches within the limits of one and the same plant; usually seeds were set on most different branches in the number of 1—2 per siliqua.

These data on the fertility of the hybrids speak in favour of the fact that, as might be concluded already from the pollen investigation, only a very limited number of gametes in our plants can take part in reproduction of progeny.

## 5. Results of investigation on the somatic number of chromosomes in 302 hybrids $F_2$

Let us now pass on to the results of the investigation on somatic chromosome number in hybrids of second generation.

The determination of this number of chromosomes in the $F_2$ plants was performed chiefly in the cells of the cone of growth of the rootlets, the chromosomes being counted in different rootlets and on several plates. Usually no special difficulties were observed in reckoning the number of chromosomes. As in this work were taking part several persons, we often restorted to mutual control to avoid subjectivity in counting. In spite of great number of plates examined by us, we were not able to establish the exact number of chromosomes in certain plants, and in such cases we were obliged to be content with approximate data.

The somatic numbers of the $F_2$ hybrids are given in two tables, as the hybrids had originated on two different plots — in Petrovsko-Rasumovskoje and in Gribovo (see table I and II). On both plots the hybrids grew side by side with seed cabbage, but on the plot in Petrovsko-Rasumovskoje there also grew specimens of flowering radish, while on that in Gribovo there was no radish, and therefore the hybrids were isolated from it.

When the data of the tables are examined together, it must be stated that all hybrids of the second generation show chromosome complexes corresponding either exactly, or more or less closely, to triploid, tetraploid and hexaploid numbers. No other complexes exist (see fig. 2, p. 12 and fig. 17, p. 27).

Here we have evidently to deal only with the progeny of gametes with 9, 18 and 36 chromosomes, or of such approaching the above mentioned ones in number of chromosomes.

Table I. Results of cytological investigations of hybrids of the second generation, obtained under conditions of isolation of $F_1$ from *Raphanus sativus* (Gribovo)

| No No F$_1$ hybrids having produced seeds | Number of collected seeds | Number of F$_2$ hybrids grown | Number of cyto- logically investi- gated F$_2$ hybrids | What number of chromosomes and in how many hybrids F$_2$ found | | | | | | | |
|---|---|---|---|---|---|---|---|---|---|---|---|
| | | | | 27 chrom. | 36 chrom. | 36—37 chrom. | 36—38 chrom. | 38 chrom. | 41 chrom. | 51 chrom. | 51—53 chrom. |
| 4-3 | 56 | 24 | 10 | — | 9 | — | — | 1 | — | — | — |
| 7-3 | 58 | 30 | 23 | — | 22 | — | 1 | — | — | — | — |
| 7-4 | 189 | 143 | 109 | 1 | 97 | 3 | 2 | 2 | 1 | 1 | 2 |
| 12-8 | 172 | 71 | 47 | — | 47 | — | — | — | — | — | — |
| 13-7 | 124 | 86 | 37 | — | 35 | 1 | 1 | — | — | — | — |
| 13-10 | 13 | 7 | 3 | — | 3 | — | — | — | — | — | — |
| 6 plants | 612 | 361 | 229 | 1 | 213 | 4 | 4 | 3 | 1 | 1 | 2 |

Table II. Results of cytological investigation of hybrids of the second generation obtained from $F_1$ not isolated from *Raphanus sativus* (Petrovsko-Rasumovskoje)

| No No of F$_1$ hybrids having produced seeds | Number of collected seeds | Number of F$_2$ hybrids grown | Number of cyto- logically investi- gated F$_2$ hybrids | What number of chromosomes and in how many hybrids F$_2$ found | | | | | | | |
|---|---|---|---|---|---|---|---|---|---|---|---|
| | | | | 27 chrom. | 27—28 chrom. | 29 chrom. | 36 chrom. | 36—38 chrom. | 40—42 chrom. | 45 chrom. | 51—53 chrom. |
| 2-3 | 2 | — | — | — | — | — | — | — | — | — | — |
| 3-2 | 4 | 1 | 1 | 1 | — | — | — | — | — | — | — |
| 12-1 | 15 | 6 | 3 | — | — | — | 2 | 1 | — | — | — |
| 12-6 | 2 | 2 | 1 | — | — | — | 1 | — | — | — | — |
| 14-4 | 4 | 2 | — | — | — | — | — | — | — | — | — |
| 14-5 | 37 | 14 | 12 | 10 | — | — | 2 | — | — | — | — |
| 15-1 | 1 | — | — | — | — | — | — | — | — | — | — |
| 15-3 | 10 | 6 | 5 | 5 | — | — | — | — | — | — | — |
| 17-1 | 72 | 39 | 34 | 29 | 1 | 1 | — | — | 1 | 1 | 1 |
| 17-5 | — | — | — | — | — | — | — | — | — | — | — |
| 21-1 | 33 | 11 | 9 | 9 | — | — | — | — | — | — | — |
| 21-2 | 26 | 8 | 6 | 6 | — | — | — | — | — | — | — |
| 25-4 | 2 | 2 | 2 | 1 | — | — | — | — | — | — | 1 |
| 13 plants | 209 | 91 | 73 | 61 | 1 | 1 | 5 | 1 | 1 | 1 | 2 |

The fusion of a 9 chromosome gamete with one possessing 18, gives rise to a zygote with 27 chromosomes; the fusion of two gametes with 18 chromosomes, produce a zygote with 36; that of a gamete possessing 9 chromosomes with one having 36, gives rise to a zygote with 45 chromosomes, while the fusion of a gamete possessing 18 chromosomes with one showing 36, produces a zygote with 54 chromosomes.

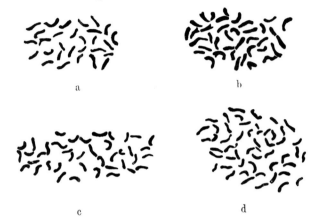

Fig. 2. Somatic plates in hybrids $F_2$. a. Hypertriploid hybrid — 29 chr.  b. Hyper-tetraploid hybrid — 38 chr.  c. Hypopentaploid hybrid — 41 chr.  d. Hypohexaploid hybrid (7-85) — 52 chr.  $\times$ 1725

When several chromosomes in the above mentioned gametes were lost or, on the contrary, a surplus of chromosomes was observed, the chromosome complex of the zygotes changed respectively. Such hybrids as, for instance, hypopentaploids with 41 chromosomes, could certainly have originated from the fusion of a haploid gamete with a hypotetraploid one, in the given case with 32 chromosomes, as well as from the fusion of two hyperdiploid gametes, in the given case, for instance, with 21 and 20 chromosomes. It might be still supposed that a hypopentaploid hybrid is the result of the fusion of a tetraploid gamete with a hypohaploid one, which however is very little probable—hypohaploid gametes are evidently lethal, as we have not one hypotriploid zygote.

In examining the tables separately, we observe that the seeds obtained from Petrovsko-Rasumovskoje, where the $F_1$ hybrids were cultivated in the vicinity of *Raphanus*, have given rise almost exclusively to triploid and hypertriploid plants, and that the single exactly penta-ploid plant is of the same origin, while among the numerous hybrids of

Gribovo, where no radish was grown in the neighbourhood, only one triploid plant has been observed. As will be seen further on, this plant is peculiar in its morphology and cannot be included in the general examination.

This difference makes it very probable, that all our triploids and the pentaploid are the result of back-crosses between the $F_1$ hybrids and *Raphanus*, i. e. that the gametes with 9 chromosomes, having taken part in the formation of the zygotes have always originated in the radish and not in the $F_1$ plants themselves. As will be shown further on, this supposition is confirmed as by the morphological pecularities of the hybrids, so also by their reduction division.

Thus it must be accepted that of the sexual cells of $F_1$ only cells showing 18 and 36 chromosomes, as well as such approaching them, have taken part in the reproduction of the second generation. As may be seen from the tables, the overwhelming majority of gametes had 18 chromosomes, while gametes with other complexes were an exception.

Not one sexual cell with a number of chromosomes less than 18 was realized in the offspring, though such was the majority of gametes formed in $F_1$. They are, evidently, all lethal and produce the lot of the degenerating pollen grains mentioned above; it may be however that some of these gametes are vital, but are not able to compete in fertilization with the diploid and tetraploid sexual cells.

This question we shall discuss yet in future. The study of the reduction division in the $F_2$ hybrids and their progeny, enables us to judge in a more detailed way of the cytological conditions, on which depends the vitality of the gametes in our plants.

## 6. Morphological description of the hybrids $F_1$ and $F_2$

### a) General characteristics

In describing the $F_1$ hybrids in 1924, we pointed out their extreme polymorphousness. In the first generation we had plants of vigorous development as well as dwarf ones, plants with leaves suggesting radish, as well as with such, suggesting cabbage, etc. Only in the structure of the fruit and in the shape of the flower parts, the hybrids were more or less homogeneous.

In the second year of cultivation the collection of $F_1$ hybrids was already less polymorphous, as almost all plants distinguished by dwarfy growth and a great similarity with *Raphanus* in their vegetative characters, did not stand the winter and were excluded from the experiment. The 19 individuals from which seeds were harvested, belonged to the usual hybrid type, appearing as normally developed plants which in the majority of their characters were intermediate between *Raphanus* and *Brassica*. They were partly hybrids of radish with common white cabbage, partly—with brussels' sprouts; one of them was a hybrid between radish and kohlrabi.

The second hybrid generation in an early stage of development already could be easily divided into three groups. The first group consisted of the plants of Gribovo, which in colour, pubescence and dissection of the leaf blades were perfectly similiar to $F_1$ (plate I, fig. 1); the second group included almost all plants of Petrovsko-Rasumovskoje, which in regard to the above mentioned characters occupied an intermediate position between $F_1$ and *Raphanus* (plate I, fig. 2). A cytological analysis revealed afterwards that the plants of the first group were tetraploids and hexaploids, those of the second group—triploids and pentaploids. In the third group, only one hybrid could be singled in regard to its morphology (plate I, fig. 3). As has been established afterwards, this hybrid was the single triploid plant from Gribovo mentioned above. We suppose now that this plant had arisen from an accidental cross of $F_1$ with *Raphanus raphanistrum* L., as characters of the latter were distinctly marked in its progeny. *R. raphanistrum* could certainly grow somewhere in the vicinity of the hybrids; this species as well as *R. sativus* has a diploid number of chromosomes = 18.

Two hypohexaploids obtained from the hybrid 7-4 (see table I) and both plants from the hybrid 25-4 (see table II), as well as some others, perished in an early stage of development, while the remainder of the plants were vegetating fairly well. The greater part of triploids, as well as several tetraploids flowered in the first year, but the majority of the plants remained up to the end of the vegetation period in the stage of "leaf-rosettes" or formed only short stem shoots.

In the second generation, as well as in $F_1$, no traces of the formation of a head or of axil hearts could be detected, characters that might have been transmitted to them by the paternal forms—common cabbage and brussels' sprouts.

In many hybrids of the second generation, especially in the triploids, a strong thickening of the head root was observed, though the lateral rootlets also always developed. On the whole the root system of $F_2$, as well as that of $F_1$, must be regarded as intermediate between

*Raphanus* and *Brassica*. On the roots of the tetraploids especially, not seldom outgrowths and leafy shoots could be observed (see plate I, fig. 4). This phenomenon which up to date cannot be considered as finally

Fig. 3.   Tetraploid hybrid at the end of flowering

elucidated, has been pointed out by us in describing F₁ (KARPECHENKO, 1924); it has been observed by a series of authors (LUND and KJÄRSKOU, 1885; WILSON, 1911; KAJANUS, 1913, 1917) on the hybrids of *Brassica Napus* L. $\times$ *Brassica Rapa* L., and has been met with lately in our experiments with the hybrids of *Brassica Napus* L. $\times$

Fig. 4. Triploid hybrid at the end of flowering

*Brassica juncea* CZERN. This phenomenon is evidentliy of ordinary occurrence in interspecific hybridization among *Cruciferae*.

Towards autumn in the second year of vegetation, the difference in the chromosome complexes of the F₂ hybrids showed itself morphologically completely. The main mass of hybrids-the tetraploids, was exhibited by plants reproducing the common big type of hybrids of first generation. These were on the whole, large, strongly branched bushes, with intermediate leaves, and blooming profusely with white flowers. In distinction to $F_1$, many of the tetraploids were very fruitful (fig. 3). Such were also the hypertetraploid plants. The triploids and hypertriploids

Fig. 5.　Hypopentaploid hybrid at the end of flowering

were in general of a comparatively shorter growth, their leaves were more pubescent and of a lighter hue (fig. 4). They also differed from the other hybrids in their stems, which were hollow nearly as in radish.

The pentaploid and hypopentaploids were somewhat intermediate between triploids and tetraploids. One of the hypopentaploids, with 40—42 chromosomes, that has originated in Petrovsko-Rasumovskoje was characterized by a certain depression of growth (fig. 5). Two

Induktive Abstammungs- und Vererbungslehre. XLVIII　　　　2

hypohexaploids which had survived the winter, remained tetraploids by their foliage, but also showed depressed growth (fig. 6 and fig. 7). One of them especially, distinguished itself by its disharmonious development and depression (fig. 7). This was a small monstrous plant, with only few leaves and very peculiar glabrous shoots

Fig. 7. Hypohexaploid hybrid 7-85 at the end flowering

Fig. 6. Hypohexaploid hybrid 7-225 at the end of flowering

arising from the leaf axils (fig. 8). The first year of its life this plant did not blossom; the second it formed only few buds having soon dropped off[1]).

Among the tetraploids, as well as triploids originating from different plants $F_1$, some differences in size, pubescence, shape of leaves, colour, etc., are naturally manifested. Also, plants obtained from seeds selected from one and the same $F_1$ hybrid, differ

___

[1]) All hybrids in figures 3—7 are taken at a time and on one scale. The etiquettes on the photos seem of different sizes because of their being disposed at different distances from the plants and consequently from the 'apparatus. The photographed tetraploid plant reached up to $1^1/_2$ mt. approximately in height.

from one another in such characters. Whether the latter plants are a result of one and the same cross, i. e. whether they have the same parental forms, is impossible to tell, as the $F_1$ hybrids were pollinated uncontrolled and the paternal plants are unknown to us.

The polymorphousness of triploids and tetraploids, mentioned by us, though concerning the same characters by which $F_1$ was polymorphous, is considerably less expressed than in the latter one.

The great diversity of $F_1$ as to the size of the plants, as well as that we do not know exactly from which paternal plant $F_1$ the one or the other $F_2$ hybrid has originated, obstruct the elucidation of the question, as to the influence of the number of chromosomes on the dimensions of the vegetative organs of our hybrids. A close study of our plants however, leaves the definite impression that tetraploids are larger than diploids and triploids, i. e., that here we have to deal with a positive influence of the multiplication of the chromosome set on the dimensions of the organs. As to the further increase of the number of chromosomes in our hybrids, its influence might be already depressive, as it has been just shown on the hypopentaploid and hypohexaploids.

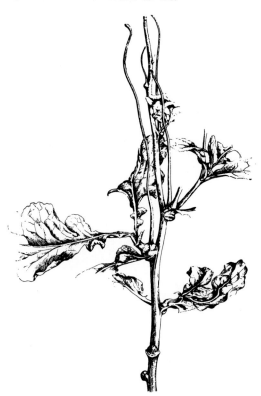

Fig. 8. Branch of hypohexaploid hybrid 7-85 with glabrous shoots. Drawn from nature. $\times \frac{1}{2}$

### b) Flowers

Let us now fix our attention on to the flowers of the hybrids. By the shape of their petals the $F_1$ hybrids are intermediate between the

2*

parental species (fig. 9). The tetraploids, pentaploid, and all plants like them, appear to be also intermediate. In triploid hybrids the shape of the petals is more close to the radish type, while in hypohexaploid it is more close to the cabbage type.

As to dimensions of petals, and the whole flower in general, the hybrids are polymorphous. For instance, in different specimens in F₁,

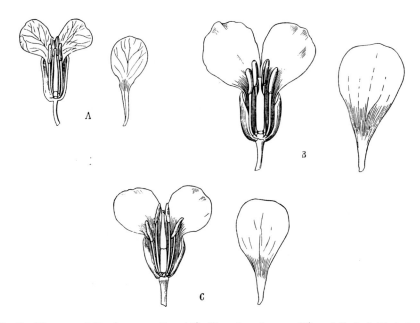

Fig. 9. Flowers of *Raphanus sativus* (A), *Brassica oleracea* (B) and F₁ hybrids between them (C). Drawn from nature. × 2

the length of the petals varied from 1 to 2 cm.; in hybrids of second generation exhibiting the same number of chromosomes, great differences in the size of the flowers were likewise observed.

Selecting by sight among the triploids and tetraploids a small-flowered specimen, a large flowered, and an intermediate one, and considering the size of their flowers for each of the above mentioned types, we measured their petals in length and width of 100 flowers of each, thus determining the arithmetical average of these characters for each specimen. Measurement was carried out on flowers of the same age (as soon as their anthers were beginning to dehisce) from branches of equal nutrition. All four petals were measured, and from the obtained data the average was calculated; of this we took advantage, arranging the flowers into classes for the purpose of determining the arithmetical average for all 100 flowers.

The results of these measurements together with the respective data for pentaploid and hypohexaploid, are given on tables III and IV. These data make evident that, for instance, a large-flowered triploid in size of flowers exceeds the small-flowered tetraploid; the large flowered tetraploid exceeds in size of its flowers the pentaploid and the hypohexaploid etc.

That some of the triploids exceed the tetraploids in size of flowers, may be seen without any measurement, by simply comparing by sight the flowers of different specimens of the ones and of the other hybrids (see plate II, fig. 8 and 9).

This circumstance does not mean, however, that the number of chromosomes in our hybrids does not affect the size of the flowers. If we take average, most typical specimens of a diploid, a triploid and a tetraploid and we compare their flowers, we see that with an increase of the number of chromosomes, the dimensions of the flower increase also. If we arrange the flowers of a typical diploid, triploid and tetraploid in one series adjoining to them the flowers of our single pentaploid and hypohexaploid plants, we obtain a series of flowers more or less regularly increasing in size; a certain deviation, flowers not sufficiently large, will be observed only in the pentaploid (see plate II, fig. 7).

In 1923 taking as basis a series of measurements, we approximately determined the average length and width of the petals of *Raphanus sativus, Brassica oleracea* and of our $F_1$ hybrids. For cabbage the length of the petals was—17·7 mm., the width being 8·8 mm.; for radish the length was 12·0 mm., the width—5·0 mm.; for the $F_1$ hybrids—15·0 mm. and 7·3 mm. respectively.

If we take into consideration that the number of small-flowered and large-flowered plants in the triploids as well as in the tetraploids, is more or less the same, the average length and width of the petals for all triploids and tetraploids may be reckoned, but this only approximately of course, from the data referring to these plants, which will be found in tables III and IV. For triploids we obtain average length of petals 16·3 mm., width—7·2 mm.; for tetraploids length—18·4 mm., width 8·5 mm.

Table III.  Length of petals in the hybrids of the second generation,
in mm.

| Plants in which the flowers were measured | n | M ± m |
|---|---|---|
| Triploid 14-16 . . . . . . . . | 100 | 14·54 ± 0·002 |
| Triploid 21-11 . . . . . . . . | 100 | 16·70 ± 0·012 |
| Triploid 17-31 . . . . . . . . | 100 | 17·76 ± 0·014 |
| Tetraploid 7-131 . . . . . . . | 100 | 15·90 ± 0·012 |
| Tetraploid 7-93 . . . . . . . | 100 | 18·01 ± 0·014 |
| Tetraploid 7-229 . . . . . . . | 100 | 21·36 ± 0·016 |
| Pentaploid 7-28 . . . . . . | 100 | 17·12 ± 0·018 |
| Hypohexaploid 7-255 . . . . . | 100 | 19·40 ± 0·015 |

Table IV.  Width of petals in the hybrids of the second generation,
in mm.

| Plants in which the flowers were measured | n | M ± m |
|---|---|---|
| Triploid 14-16 . . . . . . . . | 100 | 6·26 ± 0·008 |
| Triploid 21-11 . . . . . . . . | 100 | 7·36 ± 0·007 |
| Triploid 17-31 . . . . . . . . | 100 | 7·94 ± 0·008 |
| Tetraploid 7-131 . . . . . . . | 100 | 6·96 ± 0·008 |
| Tetraploid 7-93 . . . . . . . | 100 | 7·81 ± 0·01 |
| Tetraploid 7-229 . . . . . . . | 100 | 10·62 ± 0·019 |
| Pentaploid 7-28 . . . . . . . | 100 | 8·37 ± 0·010 |
| Hypohexaploid 7-255 . . . . . | 100 | 9·48 ± 0·01 |

On the basis of all these averages, we have traced our curves expressing the dependence of the length and width of the petals on the number of chromosomes of their respective hybrids (fig. 10). These curves make evident that in triploids, in comparison to diploids, the length of the petals increases, while the width remains approximately as before, so that on the whole the flower becomes larger but its petals are narrower than in $F_1$, i. e., their shape approaches more the radish type; in tetraploids the length and width of petals is more considerable than in $F_1$ and triploids.

Thus, the influence of the number of chromosomes on the size of the flowers in our hybrids is doubtless, although it is obscured by a strong segregation of the hybrids with regard to this character.

The colour of the flowers is yellow in cabbage and white or purple in radish. In the $F_1$ hybrids, the flowers were white, in some plants purple and only one plant out of 123 exhibited pale cream coloured flowers. Thus, in regard to the colour of the flowers, $F_1$ cannot be

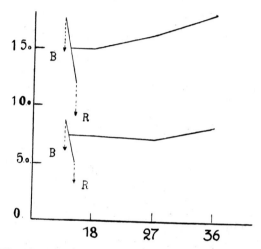

Fig. 10. Curves showing the increase of length and width in the petals of the hybrids with increase of their number of chromosomes. On a horizontal are given the numbers of chromosomes, of a vertical — length and width of the petals in mm. The dots B and R indicate the length and width of petals in *Brassica* and *Raphanus*

Fig. 11. Flower of tetraploid hybrid with 8 stamens. Drawn from nature. × $^3/_2$

considered as homogeneous; on the whole, it is evident that the colour of radish is a dominant character. In all hybrids of the second generation the colour of the flowers remained the same as in $F_1$—white, sometimes purple (more frequently in triploids); splitting off of plants with yellow flowers was not observed.

As to the nervation of the petals, the degree of development of the nerves and the pigmentation of the latter, hybrids of second generation as well as those of $F_1$, showed a considerable diversity.

Some deviations in the structure of the flower in $F_2$ plants are of interest. In some tetraploids, flowers with a number of

Fig. 12. Flowers of triploid hybrid with considerable
number of petals

*a*

Fig. 13. Flower of tetraploid with considerable number
of stamens and with bud cluster instead of pistil. **a.** Side
view. **b.** View from above. Drawn from nature. $\times$ 1

stamens exceeding six were
repeated. Fig. 11, for instance, represents a flower
with eight stamens. This
increase in the number
of stamens not seldom
disturbs entirely their usual
disposition of the circles.
In an early stage of blooming all primary flowers
in the racemes of one triploid had more than four
petals, while the number
of the sepals, the stamens
and pistil were normal
(fig. 12).

In some tetraploids from
time to time the for-
mation of very large flowers
with a great number of
stamens and a cluster of
buds instead of a pistil,
were observed. Fig. 13
represents such a flower
in its natural size in side
view, as well as seen from
above, while fig. 14 shows
its parts separately; the
flower had an asymetric
corolla of four petals and
only two sepals, some of
the stamens appearing as
if grown together from
two.

The buds in the middle
of such flowers developed
later on into independent
flowers, one of the buds
giving rise to a new flower
branch; thus after bloom,

a group of pistils with a flowering shoot in the centre was formed (fig. 15).

Sometimes among our hybrids appeared flowers with a great number of sepals, petals, stamens and pistils, these not unfrequently grown together. These flowers, one of which is represented in fig. 16, are evidently

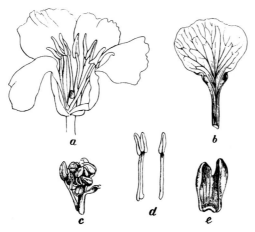

Fig. 14. Parts of flower represented in fig. 13. a. Arrangement of stamens in the flower. b. Petal. c. Cluster of buds cut from the middle of the flower. d. Stamens. e. Petal. Drawn from nature. × 1

Fig. 15. The growth of a flower similar to that represented in fig. 13. Drawn by nature. × 1/8

Fig. 16. Flower of tetraploid formed by the together growth of several normal flowers. Drawn from nature. × 1

the result of the growing together into one of several ordinary flowers.

All above mentioned formations are still of rare occurrence among our hybrids, the structure of the flowers usually being perfectly normal.

### c) Fruits

Now let us pass on towards the most interesting and characteristic feature of our hybrids—the fruits.

On fig. 17 are given drawings from nature of the latter; in the first row are represented fruits of *Raphanus*, of $F_1$, and *Brassica*; in the second—those of triploid, tetraploid, pentaploid and hypohexaploid hybrids. Beneath these drawings of fruits are given somatic plates of chromosomes of the respective plants, made after preparations with the aid of the apparatus Abbé, as well as the formula of the plates, meaning the number of radish and cabbage chromosomes exhibited by the plant; the radish chromosomes are marked with the letter $R$—from *Raphanus*, those of cabbage—with $B$, from *Brassica*.

In radish the siliqua is spindle-like, inflated, does not dehisce in valves, and consists of two parts: of a lower one very short, inversely cone-like, mostly sterile, less frequently one-seeded, and an upper one— much thickened by a spongy mesocarp, contracting towards the end, and more or less regularly stretched over between the seeds. In cabbage the siliqua is linear, tubercular, dehisces in two valves and gradually ending in a sterile or one-seeded non dehiscent apex. Botanists having studied the *Cruciferae* particularly (POMEL, 1883; HAYEK, 1911; SCHULZ, 1919 and others) suggest that in radish the small limb at the foundation of the siliqua is a reduced two-valved part of the fruit ("Valvarglied"), while the siliqua of radish in itself represents a strongly developed many-seeded apex ("Stylarglied"). In cabbage on the contrary, the two-valved part of the fruit is strongly developed, while the apex is small.

In the $F_1$-hybrids a very unconsiderable number of siliquas was formed as soon after fertilization, a great part of ovaries dropped off. The siliquas usually attained not more than half their normal size containing sometimes 1—2 seeds; in most cases, however, only dried ovules could be detected in them. By their structure these siliquas were of great interest: both apex and two-valved part were equally developed, so that the siliqua appeared to consist of two parts of similar length. The upper part showed a non dehiscent whole, its walls being strongly

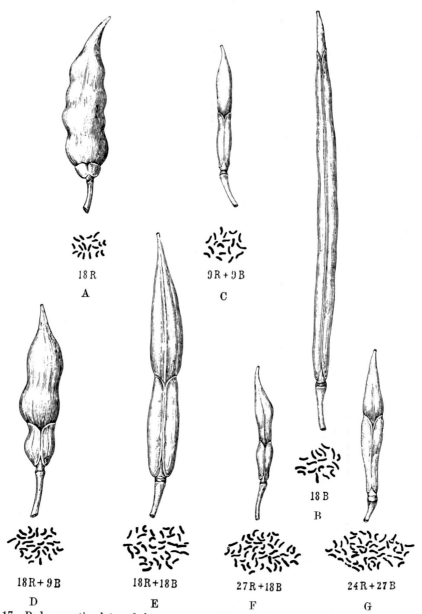

Fig. 17. Pods, somatic plates of chromosomes and formula for *Raphanus* (A), *Brassica* (B), diploid (C), triploid (D), tetraploid (E), pentaploid (F) and hypohexaploid (G) hybrids. The drawings of the pods are done from nature and are of natural size. The drawings of the chromosomes are made with aid of the apparatus Abbé, magn. 1200. Drawn by Miss M. P. Lobanowa

thickened by a spongy mesocarp; the lower part dehisced in two valves with not thickened walls. Evidently such structure of the siliqua was a result of the fact that the $F_1$ plants contained in their diploid complex 9 radish and 9 cabbage chromosomes; by this the radish chromosomes secured the development of their apex, while the cabbage chromosomes—that of the two-valved part of the siliqua.

In triploids the non dehiscent upper part of the siliqua exceeds twice or more the lower two-valved one. As has been already mentioned, it is very probable that the triploids are a result of back crosses of $F_1$ and *Raphanus*. In this cross $F_1$ supplies a gamete containing 9 radish and 9 cabbage chromosomes, i. e. the formula $9 R + 9 B$, while *Raphanus* participates with 9 radish chromosomes—$9 R$; thus the zygotes receive $18 R$ and $9 B$. Such a correlation of the radish and cabbage chromosomes, evidently, bestows the strong development of the non-dehiscent part of the siliqua in the triploids.

In tetraploids originating from the fusion of two equal gametes of the formula $9 R + 9 B$, the zygote has $18 R$ and $18 B$, so that the relation between the radish and the cabbage chromosomes remains the same as in $F_1$, i. e., $1 : 1$. Through this, therefore, is easily understood, that the structure of the siliqua in tetraploids is perfectly identical to that in diploids. The only difference consists in the size of the siliqua which in most tetraploids is quite normal, as they are fertile.

A pentaploid, as triploids originates most probably from a back cross of $F_1$ with *Raphanus*, but in this case we have from the side of $F_1$ a tetraploid gamete of the formula $18 R + 18 B$, which, combined with the $9 R$ of *Raphanus*, produces a pentaploid zygote of the formula $27 R + 18 B$. The correlation of the length of the non dehiscent part of the fruit with the two-valved one in our pentaploid, corresponds to the indicated formula. The siliquas of this plant are small, its fertility being a medium one.

The hexaploid plants in our crosses may arise, as we suggested, from the fusion of tetraploid gametes $18 R + 18 B$, and diploid ones $9 R + 9 B$; in this case the zygotes must possess $27 R + 27 B$ and the siliqua of the hexaploids must agree in structure with the siliquas of the tetraploids and the diploids.

Our plant 7-255 exhibits not 54 but 51 chromosomes, and its two-valved part of the siliqua is somewhat longer than the upper one. This plant might have possibly arisen from the fusion of a diploid gamete, $9 R + 9 B$ and a hypotetraploid gamete of the formula $15 R + 18 B$;

so that the somatic set of its chromosomes is 24 $R$ and 27 $B$, in consequence of which the two-valved part of the siliqua is longer than the apex.

The above stated dependence of the correlation of the radish and cabbage chromosomes on one hand, and of the dehiscent and the non dehiscent part of the siliqua on the other, is to be observed in all our hybrids. Few exceptions to this rule, as will be shown further on, appear to be only apparent, and have their explanation.

Among the different varieties of *Raphanus sativus* L. some difference is to be seen in the degree of reduction of the two-valved limb at the foundation of the siliqua, and in *Brassica oleracea* L. in the degree of development of the apex. Likewise, among our hybrids of one and the same chromosome complex, a certain difference in the correlation of the two-valved part and the non-dehiscent one of the siliqua is to be observed.

In order to ascertain the degree of these differences, we have undertaken a series of measurements of the fruits greatly deviating and typical in structure of triploids and tetraploids, as well as of pentaploid, hypohexaploid and of two tetraploid hybrids quite peculiar as to their fruits, i. e. hybrids 7-13 and 7-150.

We have determined in all these plants the percent ratio of the two-valved part of their pistil to the whole length of the latter. To produce measurements on pistils of the flowers just blooming, appeared more profitable than on the siliquas, as the latter doubtlessly might have more altered by occasional modification effects, while the structure of the future fruit was already fully manifested in the pistils.

The results of the above mentioned determinations are given in table V, p. 31. We see that in triploids the relation of the lower two-valved part of the pistil to its whole length in different specimens varies from 22·5 % to 28·1 %; in a pentaploid this percentage is 31·7; in tetraploids fluctuations are observed from 39·6 % to 47·6 %; in a hypohexaploid the length of the two-valved part of the pistil in regard to its whole length constitutes 58·5 %. Concerning pistils of hybrids 7-13 and 7-150 we shall speak elsewhere, for the present let us leave them unobserved.

If we take a triploid with its highest percentage and reckon for it and for a pentaploid the difference $D + d$, we shall see that $3 d$ is much less than $D$, i. e., that our extreme triploid in structure of its pistil is still far behind the pentaploid; the same is true, as may be seen from the tables, in reckoning of $D \pm d$ in regard to a pentaploid and a tetraploid with lowest percentage, and to a hypohexaploid and a tetraploid with the highest percentage.

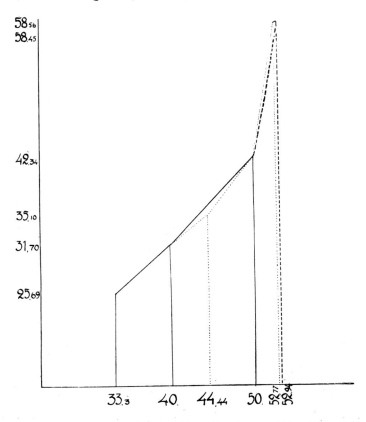

Fig. 18. Correlation between the length of two-valved part of the pistil and the number of cabbage chromosomes in the hybrids. On a vertical are disposed numbers exhibiting the length of the two-valved part of the pistil in percent of its whole length; on a horizontal — numbers of cabbage chromosomes in percent of the general number of chromosomes

Fig. 18 shows a curve being traced on the basis of the data of table V. In horizontal direction we find the ratio, taken in percentage of the cabbage chromosomes to the total number of such in different hybrids; in vertical direction we see the ratio likewise in percentage of the two-valved "cabbage-like" part of the pistil to the whole length

Table V.  Length of the two-valved part of the pistil in different hybrids, expressed in the percentage of the whole length of the pistil

| Plants in which the pistils were measured | n | M + m | C + c | D + d |
|---|---|---|---|---|
| Triploid 17-27 | 50 | 22·50 + 0·348 | 10·84 + 1·084 | |
| Triploid 14-16 | 100 | 23·60 + 0·300 | 12·74 + 0·901 | |
| Triploid 17-8 | 100 | 24·45 + 0·346 | 14·17 + 1·000 | |
| Triploid 17-35 | 50 | 25·90 + 0·367 | 9·26 + 0·996 | |
| Triploid 17-31 | 100 | 27·45 + 0·280 | 10·20 + 0·721 | |
| Triploid 21-11 | 100 | 27·85 + 0·224 | 8·08 + 0·572 | |
| Triploid 17-34 | 50 | 28·10 + 0·388 | 9·75 + 0·975 | 3·60 + 0·500 |
| Pentaploid 17-28 | 100 | 31·70 + 0·307 | 9·808 + 0·624 | 7·90 + 0·539 |
| Tetraploid 7-131 | 100 | 39·60 + 0·445 | 11·23 + 0·798 | |
| Tetraploid 13-83 | 100 | 39·60 + 0·306 | 7·74 + 0·507 | |
| Tetraploid 7-188 | 100 | 40·10 + 0·255 | 6·37 + 0·451 | |
| Tetraploid 12-3 | 100 | 41·00 + 0·335 | 8·17 + 0·577 | |
| Tetraploid 7-93 | 100 | 41·75 + 0·357 | 8·55 + 0·605 | |
| Tetraploid 7-127 | 100 | 46·75 + 0·443 | 9·47 + 0·670 | |
| Tetraploid 7-131 | 100 | 47·60 + 0·475 | 9·98 + 0·708 | |
| Hypohexaploid 7-225 | 100 | 58·50 + 0·525 | 8·96 + 0·634 | 10·90 + 0·700 |
| Hybrid 7-13 | 100 | 35·10 + 0·320 | 9·11 + 0·645 | |
| Hybrid 7-150 | 100 | 58·45 + 0·604 | 10·34 + 0·731 | |

of the latter in the same plants. With that for triploids and tetraploids the arithmetical average of all 7 plants in which the pistils had
been measured, was calculated.

In discussing about the curve for the moment we are not going
to touch the dotted line, as we think more proper to discuss about it
later on. Let us now examine but the chief line concerning the structure
of the siliqua in a triploid, pentaploid, tetraploid and a hypohexaploid.
In a triploid the percentage of cabbage chromosomes is 33·3; in a pentaploid it is 40 %; in a tetraploid — 50, and in a hypohexaploid, as we
suppose, 52·9 % (24 radish and 27 cabbage chromosomes).

We see that in these plants with an increase of the percentage of
cabbage chromosomes, the length of the two-valved part of the pistil
increases also, which takes place in due proportion to the ratio of
chromosomes up to the tetraploid inclusively; in a hypohexaploid the
two-valved part of the pistil is greatly enlarged, though the increase in
the percentage of cabbage chromosomes, compared to the tetraploid is
inconsiderable. If our formula $24\,R + 27\,B$ applied to this plant
approaches more or less reality, one might think, that the considerable
increase of the two-valved part of its pistil, is a result of the absence
of but few radish chromosomes, and that therefore the regular dependence of the structure of the pistil on the ratio of the radish to the
cabbage chromosomes, is here already disturbed. To be judged by its
siliqua, the hypohexaploid is such as to exhibit very great excess of cabbage
chromosomes in regard to radish chromosomes, which in reality doubtlessly
is not the case. Such a nonconformity we shall have the occasion to
state further on in one more hybrid.

In triploids and tetraploids we have determined the relation of the
two-valved part to their whole length for the siliqua as well. Three
plants, extreme and typical in regard to the above mentioned character,
were chosen from each group for measurements. The obtained data are
given in table VI (p. 33). We see that the coefficient of variability is
considerably higher in the siliquas than in the pistils. The reckoning
$D + d$ still shows that no overlapping between triploids and tetraploids
on siliquas exists.

In plate II, fig. 5 and 6 photographs of siliquas of a tetraploid and
of a triploid are given, illustrating the variability of the pod within the
limits of a plant.

One might suggest, that a greater or slighter development of
the two-valved part of the siliqua or of its apex, depends on where

Table VI. Length of the two-valved part of the siliqua in different hybrids, expressed in percentage of the whole length of the siliqua

| Plants in which the pods were measured | n | M ± m | C ± c | D ± d |
|---|---|---|---|---|
| Triploid 14-16 . . . . . | 100 | 16·60 ± 0·487 | 29·355 ± 2·076 | |
| Triploid 17-18 . . . . . | 100 | 22·50 ± 0·600 | 26·666 ± 1·885 | |
| Triploid 17-31 . . . . . | 100 | 29·05 ± 0·865 | ⎫ 29·755 ± 2·104 | 13,85 ± 1,072 |
| Tetraploid 7-93 . . . . | 100 | 44·25 ± 1·007 | ⎭ 22·745 ± 1·608 | |
| Tetraploid 7-131 . . . . | 100 | 46·49 ± 0·743 | 15·985 ± 1·130 | |
| Tetraploid 7-229 . . . . | 100 | 47·60 ± 0·658 | 9·726 ± 0·687 | |

and how many seeds have been set, but as have indicated our reckonings, the coefficient of correlation appears here very inconsiderable. We brought out our coefficient for two tetraploid plants; from each plant hundred siliquas were taken and in each was being calculated the percentage of the length of the two-valved part to the total length of the siliqua, as well as the relation of the percentage of the number of seeds set in the two-valved part to that set in the siliqua as a whole. The reckoning of correlation coefficients for these two characters were $r = + 0·23$ for the first plant, $r = + 0·27$ for the second. Thus is evident that the number of seeds is apt to affect the total size of the fruit in our hybrids, as well may its shape be altered through the development of seeds in the apex or in the two-valved part, but as to its structure, i. e. relative length of the two-valved part and the apex, the influence of the seeds is but very inconsiderable.

There still remain a few words to say concerning the fruits of our other hypo- and hyperploid plants, as well as of hybrids 7-13 and 7-150, but we think it proper to expose these data in connection with the description of the reduction division in the plants, therefore we confine ourselves to the remark that in all above mentioned plants, the structure of the siliquas is to be determined by the correlation of the radish and cabbage chromosomes.

### c) Conclusion

Summing up the above stated, it may be said that the morphological peculiarities of our hybrids are closely connected to their chro-

mosome sets. This being quite definitely established on a great number
of plants (as has been mentioned, 302 individuals were investigated
cytologically), in the remaining hybrids not subjected to cytological
analysis, we were able to determine the number of chromosomes on the
basis of their morphological characteristics.

Altogether were morphologically described, partly only in the
rosette stage, 417 plants of the $F_2$. Out of them 322 plants, ought to
be regarded as tetraploids and hypertetraploids (to separate them morphologically is impossible), 82 plants as triploids and hypertriploids[1]). The
remaining ten plants morphologically do not take part in these groups;
they were all investigated as to the number of chromosomes and proved
to be those hypopentaploids, pentaploid and hypohexaploids mentioned
above, as well as those "particular" tetraploids which shall be discussed
further on. It is quite possible, of course, that if the fruits of all the
cytologically not investigated hybrids were known to us, we should be
able to single out a few more such deviating plants.

Tetraploids and triploids are less polymorphous as to characters
in which $F_1$ appears polymorphous; as to characters in which $F_1$ is
homogeneous enough in the way of siliqua, colour and shape of petals, root
system,—the tetraploids and triploids are also homogenous. Consequently,
we must come to the conclusion that in crosses of $F_1$ plants among
themselves, as well as in back crosses of $F_1$ with *Raphanus* the sexual
cells in plants of the first generation bear no segregation as to the
enumerated characters. For instance, in crossing species with yellow
flowers and species with white ones in *Raphanus* (FOCKE, 1881;
TROUARD RIOLLE, 1916)[2]) as in *Brassica* (TIMOFEYEFF)[3]), in $F_1$ the
white colour of flowers is dominant, and in $F_2$ even with a small
number of plants, some occur with yellow flowers. In our hybrids in
$F_1$, the white colour is also dominant, but in the second generation
among a great number of tetraploids, not one with yellow is to be
found—the hybrids appear in this case nonsegregating.

[1]) All these triploid plants, with the exception of one, have originated in
Petrovsko-Rasumovskoje (comp. pp. 12—13).

[2]) FOCKE and TROUARD RIOLLE crossed the species *Raphanus sativus* (white
flowers) × *Raphanus raphanistrum* (yellow flowers).

[3]) N. N. TIMOFEYEFF crossed *Brassica alboglabra* BAILEY (white flowers) × *Brassica
oleracea capitata* (yellow flowers). The work has not been published. I quote from
written data kindly placed at my disposal by the author.

## 7. Anatomical investigation of the hybrids $F_1$ and $F_2$

An anatomical investigation of the hybrids and their parental forms, undertaken by us with the aim to elucidate the influence of an increased number of chromosomes on the anatomical peculiarities of our plants, was carried out by us chiefly on the flowers, as organs in which indispensable conditions for the comparability of the results obtained might be easily realized. From parental forms and hybrids we chose flowers of the same age when their anthers were just beginning to dehisce, and for investigation we tore off from petals and sepals a bit of epidermis between the central nerve and margin in the middle. Next we investigated cross-sections of the pedicels, made at a distance of a 2 mm. from the base of the flower. In selecting the flowers for the investigation, we took care to choose branches of equal nutrition and light conditions. Fig. 19, showing the epidermis of the petals, fig. 20—the epidermis of the sepals and fig. 21—the cross sections of the pedicels, enable us to form a clear conception of anatomical relations in our hybrids.

We see that with an increase of number of chromosomes, the size of the cells increases in all investigated tissues as well[1]).

We measured the stomata on the epidermis of the sepals; in table VII are shown the results obtained in measuring the width of the stomata, in table VIII—the length. Fig. 22 reproduces the curves based on the data supplied by tables VII and VIII; on a horizontal are ranged the chromosome numbers in a diploid, triploid, tetraploid, pentaploid and hypohexaploid. The points marked with $R$ and $B$ determine the length and width of the petals for *Raphanus* and *Brassica*; the upper curve refers to the length of the stomata, the lower one to the width.

Table VII.   Width of stomata in mm., while magnified 1350 times

|  | Plants in which the stomata were measured | n | $M \pm m$ |
|---|---|---|---|
| 1 | Cabbage . . . . . . . | 100 | $5\cdot29 \pm 0\cdot049$ |
| 2 | Radish . . . . . . . . | 100 | $5\cdot47 \pm 0\cdot054$ |
| 3 | Diploid . . . . . . . . | 100 | $5\cdot38 \pm 0\cdot051$ |
| 4 | Triploid . . . . . . . . | 103 | $6\cdot70 \pm 0\cdot043$ |
| 5 | Tetraploid . . . . . . . | 100 | $7\cdot19 \pm 0\cdot055$ |
| 6 | Pentaploid . . . . . . . | 100 | $7\cdot78 \pm 0\cdot049$ |
| 7 | Hypohexaploid . . . . . | 100 | $7\cdot85 \pm 0\cdot056$ |

[1]) By the way from the epidermis of the petal may be seen, that in hybrids the sinuous cells of *Raphanus* are dominant.

3*

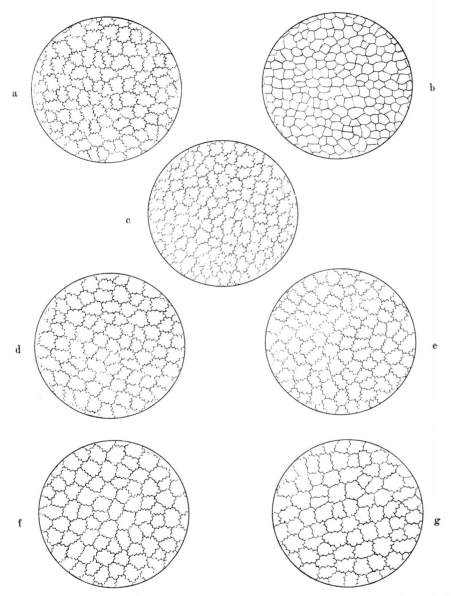

Fig. 19. Epidermis of the petals in *Raphanus sativus* (a), *Brassica oleracea* (b) and of their hybrids: diploid (c), triploid (d), tetraploid (e), pentaploid (f) and hypohexaploid (g).
× 575

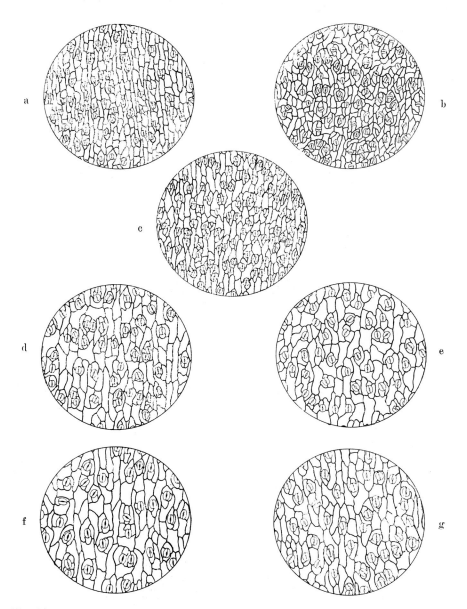

Fig. 20.  Epidermis of the sepals in *Raphanus sativus* (a), *Brassica oleracea* (b) and their hybrids: diploid (c), triploid (d), tetraploid (e), pentaploid (f) and hypohexaploid (g).
× 320

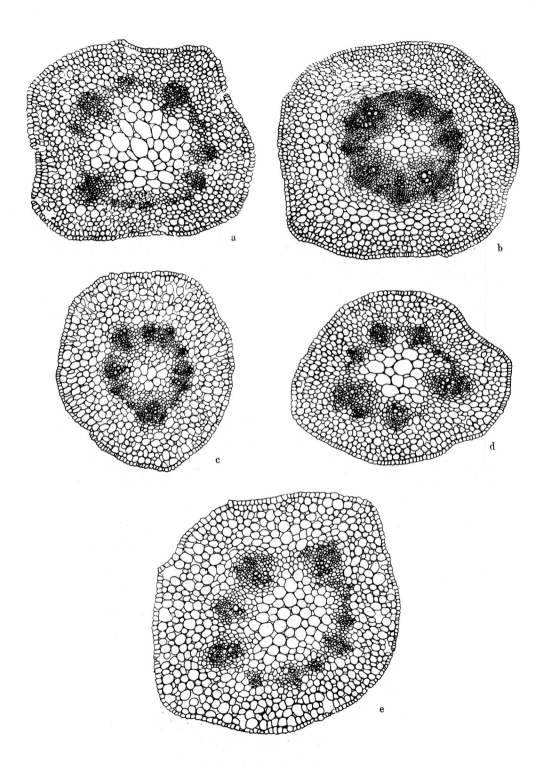

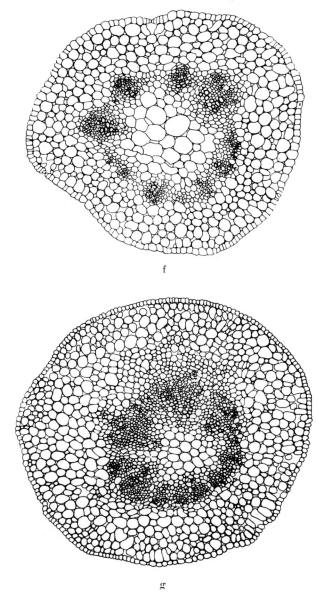

f

g

Fig. 21. Cross sections of the pedicels in *Raphanus sativus* (a), *Brassica oleracea* (b) and of their hybrids: diploid (c), triploid (d), tetraploid (e), pentaploid (f) and hypo-hexaploid (g). × 700

Table VIII.   Length of stomata in mm., while magnified 1350 times

|   | Plants in which the stomata were measured | n | M ± m |
|---|---|---|---|
| 1 | Cabbage . . . . . . . . | 100 | 7·09 ± 0·059 |
| 2 | Radish . . . . . . . . | 100 | 7·28 ± 0·058 |
| 3 | Diploid . . . . . . . . | 100 | 7·10 ± 0·086 |
| 4 | Triploid . . . . . . . . | 100 | 9·39 + 0·069 |
| 5 | Tetraploid . . . . . . . | 100 | 9·74 ± 0·060 |
| 6 | Pentaploid . . . . . . . | 110 | 10·60 ± 0·067 |
| 7 | Hypohexaploid . . . . . | 100 | 10·38 + 0·060 |

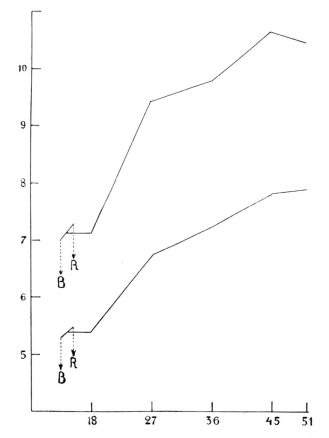

Fig. 22.  Increase of the length and width of the stomata in connection of the increase of the number of chromosomes in the hybrids.  On a vertical are shown the data obtained in measuring the length and width of the stomata; on a horizontal — the numbers of the chromosomes

From these curves, as well as from the drawings of the epidermis may be seen, that an especially great increase of cell dimensions takes place in a triploid, while later on, a somewhat diminishing of the increase process commences. As may be judged by the curves, the increase in length of the stomata is especially great in a triploid, evidently proceeding in parallel to the elongation of the petals; in the pentaploid the stomata are also long, longer even than in the hypohexaploid.

It is of interest to note, that notwithstanding the great polymorphism observed in diploids, triploids and tetraploids as to the size of the flower, the latter is quite definitely characterized anatomically; if, for instance, a large-flowered and a small-flowered tetraploid are investigated in regard to the size of flowers, the results will appear to be much the same. And so it is also for diploids and triploids.

The difference in the size of the flowers, observed within the groups with an equal number of chromosomes, depends chiefly on the number of cells, and not on their size. The petals of a large-flowered tetraploid consist of a greater number of cells, than those of a small-flowered one, the size of their cells being approximately the same.

Those tetraploids which in flower dimensions exceed our hypohexaploid and pentaploid, possess cells of smaller size than the latter plants, but in reward their cells are much more numerous. Evidently in general, the inconsiderable size of plants with many chromosomes, is due as may be stated, to the small number of cells they contain.

Most probably cells with many chromosomes divide less rapidly, or else the period of rest while dividing are of longer duration, and that is why in plants with many chromosomes, within one same space of time, occur much less cell divisions than in plants with few chromosomes.

From the drawings (fig. 21) may be perceived that in pedicels, an increase of cell dimensions, more right to say of their transversal section, is seen to take place in accordance to an increase of plant chromosome number, the same as was observed with epidermis cells, but the graduation of the increase is not so obviously manifested.

In hybrids with many radish chromosomes, in a triploid or a pentaploid, a more porous structure of the center may be clearly distinguished; in a hypohexaploid with much greater number of cabbage chromosomes than radish ones, the anatomical structure of its pedicel closely resembles that of the cabbage.

## 8. Meiosis in hybrids F₂

In studying the process of the formation of sexual cells in our hybrids F₂, the results which we are now proceeding to expose, we payed special attention to the following. First of all we investigated the metaphases of heterotypical division, as a stage in which the number of univalents, bivalents etc. in a hybrid may be readily ascertained; next were investigated the anaphases of first division, exhibiting the behaviour of chromosomes in splitting and their distributing to the poles; further on were examined the metaphases of second division and its anaphases in order to elucidate the character in which the distribution of chromosomes in this division took place and to determine how many chromosomes obtain cells of tetrads. All these observations were performed on pollen mother cells.

### a) Meiosis in triploids

Among triploid hybrids we investigated meiosis in several plants, differing in degree of fertility and in development of the two-valved part of the fruit. In all cases the results of the investigation were identical. In a metaphase of the first division 18 chromosomes were observed; whereas on most plates 9 chromosomes were large and of more or less globular form, while 9 chromosomes were smaller and of elongated form (fig. 23 a).

In the anaphases large chromosomes split normally and their halves distribute equally to the poles, while the small chromosomes distribute between the daughter nuclei irregularly; sometimes they do not split at all (fig. 23 b, c) and sometimes all or a part of them does (fig. 23 d); the latter occurs usually rather late, i. e. when the halves of large chromosomes are already reaching the poles. These observations prove without any doubt that in triploid hybrids we have in a first generation 9 bivalents and 9 univalents.

As has already been mentioned a somatic set of chromosomes in triploids consists most probably of 2 radish genoms and of 1 cabbage one. A study of the F₁ reduction division has shown that radish and cabbage chromosomes do not conjugate. It being so in a reduction division of triploids, we obviously are to meet with 9 bivalents and 9 univalents: chromosomes of radish genoms are homologous and therefore, conjugating with each other, form bivalents, while chromosomes of the cabbage genom having no homologues, remain univalents.

In metaphases of the second division in triploids the number of chromosomes in different cells varies from 27 to 36. If in a first division not a single univalent has split, we possess in the two plates of the homotypical metaphase total number 27 chromosomes (fig. 23 e); if all univalents have split, the number of chromosomes will be 36 (fig. 23 f). More often intermediate numbers occur (fig. 23 g).

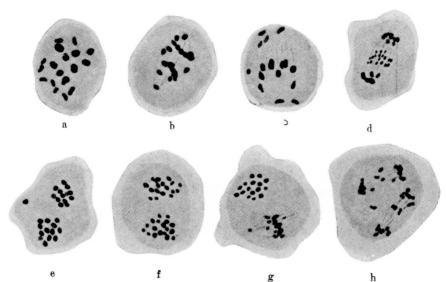

a            b            ɔ            d

e            f            g            h

Fig. 23. Division of the pollen mother cells in triploid hybrids. a. Metaphase of the 1st division with 9 bivalent and 9 univalent chromosomes. b—d. Anaphases of the 1st division. e—g. Metaphases of the 2nd division. h. Anaphase of the 2nd division.
× 1725

Those univalents which have not split in the first division do so in the second, while the univalents which have undergone splitting in the first division usually do not split in the second one (fig. 23 h).

In result of such a course of division in triploids, there arise tetrads of cells with different numbers of chromosomes from 9 to 18.

In several pollen mother cells of triploids one observes sometimes the same "omission" of the first division such as was described by us for the $F_1$. Several times we have observed one nuclear telophases of the first divison (fig. 24 a) and in a homotypical division metaphases and anaphases with an accumulation of all chromosomes in one spindle (fig. 24 b — e).

This deviation from a usual course of the division of pollen mother cells leads in triploids to the formation of dyads (fig. 24 f) consisting of cells containing about 27 chromosomes each.

In fig. 25 is given the scheme of a usual course of meiosis in triploids and the formation of sexual cells in them with somatic number of chromosomes.

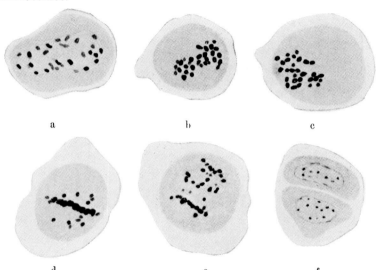

a                           b                           c

d                           e                           f

Fig. 24.  Division of the pollen mother cells in triploid hybrids.  a. Mononuclear telophase of the 1st division.  b—d. Metaphases of the 2nd division with one spindle. e. Anaphase of the same division.  f. Dyad.  × 1725

### b)  Meiosis in tetraploids

Just as in triploid hybrids we have investigated the meiosis in tetraploid hybrids on several plants differing as to the degree of fertility and development of the two-valved part of the pod.

With very few exceptions, which will be discussed further on, all the investigated tetraploids proved to have a quite normal meiosis. In metaphases of the 1st division they possess 18 bivalents (fig. 26 a, b, p. 46), the halves of which separate in anaphases quite regularly (fig. 26 c, p. 46), so that the metaphases of the 2nd division have again plates of 18 chromosomes (fig. 26 d, p. 46). Anaphases of the 2nd division are just as regular as the first (fig. 26 e, p. 46) and thus cells of tetrads in tetraploids receive a normally reduced, in regard to 36, number of chromosomes—18.

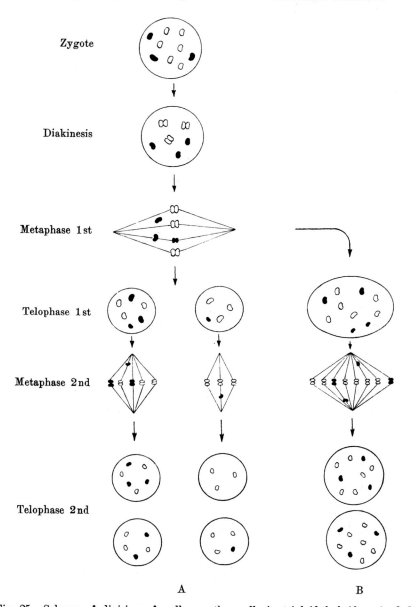

Zygote

Diakinesis

Metaphase 1st

Telophase 1st

Metaphase 2nd

Telophase 2nd

A                                                B

Fig. 25. Scheme of division of pollen mother cells in triploid hybrids. A. Ordinary course of division. B. Formation of dyad with somatic number of chromosomes in each cell. Each "genom" is represented by 3 chromosomes. The white chromosomes being radish, the black ones cabbage

The normality of meiosis in tetraploids evidently may be explained by the matter that in a somatic set of them we have two radish and two cabbage genoms, so that each chromosome is represented here by a duplicate number, each of them having a homologue.

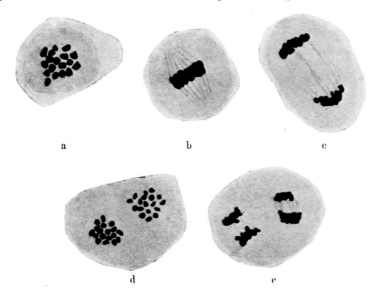

a                    b                    c

d                    e

Fig. 26. Division of the pollen mother cells in tetraploid hybrids. a. Metaphase of the 1st division with 18 bivalents. b. Metaphase of the 1st division, side view. c. Anaphase of the 1st division. d. Metaphase of the 2nd division. e. Anaphase of the 2nd division. × 1725

However sometimes in tetraploids as well a certain irregularity in distribution of chromosomes is to be observed: 1 or 2 chromosomes are left behind during distribution to the poles or else split earlier than the others; in rare cases in telophases of the first as well as the second division may be observed single chromosomes lying apart in the plasma, not being included in the daughter cells. But such deviations are inconsiderable; they may be observed as well in pure species.

Examining a division of pollen mother cells in a series of tetraploid plants by means of acetocarmine, we detected in them also the capability of producing gametes with a somatic number of chromosomes following the same process which was indicated by us for $F_1$ and triploids.

### c) Meiosis in the pentaploid

In a metaphase of the first division in our pentaploid plant on most of plates were reckoned 27 chromosomes being of different sizes and forms (fig. 27, a—b).

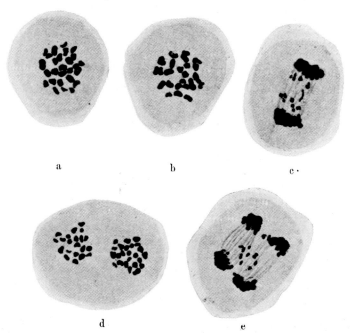

Fig. 27. Division of the pollen mother cells in a pentaploid hybrid. a. Metaphase of the 1st division with 27 chromosomes. b. Metaphase of the 1st division with 26-27 chromosomes. c. Anaphase of the 1st division. d. Metaphase of the 2nd division (in the plate on the left there are 21 chromosomes and in that on the right — 24). e. Anaphase of the 2nd division. × 1725

In anaphases of the first division part of chromosomes splits quite normally, part of them distributes between the poles without splitting, or else undergoes splitting after the remaining chromosomes reached the poles (fig. 27 c). Such behaviour of some chromosomes at first division in our pentaploid, as well as their size proves that here we are dealing with univalents.

We suppose the somatic set of chromosomes in a pentaploid to consist of 3 radish and 2 cabbage genoms. The chromosomes of the two radish genoms produce 9 bivalents, the same number of bivalents being provided by the two cabbage genoms, while the 9 chromosomes of the third radish genom finding no more partners, remain univalents. Such an assumption quite agrees with the views observed in divisions of pollen mother cells in a pentaploid, although to exactly determine by size and behaviour of chromosomes the number of bivalents and univalents in this plant, seems not to be possible.

In metaphases of the second division in a pentaploid the total number of chromosomes sometimes amounts to 45 on both plates (fig. 27 d), while in anaphase, as in the first division, a remaining behind of some chromosomes is observed (fig. 27 e).

### d) Meiosis in the hypohexaploid

In our hypohexaploid plant (2n = 51) in the metaphases of the first division the number of chromosomes varies, 27 being the most frequent (fig. 28 a), but in some metaphases we have counted also 25 chromosomes (fig. 28 c), while in others—about 31 (fig. 28 b). As to size, the chromosomes differ greatly. In anaphases of the first division, may be stated with certainty that some of them are univalents (fig. 29 b, c). On the other hand are observed also such chromosomes which must be considered as trivalents; when examining the metaphases from side view, usually these chromosomes appear very long (fig. 29 a) and sometimes, especially when they have dropped out of the spindle, and with a more acute differentiation of the preparate, it may be seen clearly that they consist of 3 small chromosomes united by their tips, or else of a large globular chromosome and a small elongated one (table III, 12—14). These trivalents break into their components at the very first division.

In the metaphases of the second division in hypohexaploid have been often reckoned more than 51 chromosomes (fig. 29 d) which doubtlessly results from the splitting of some univalents at the first division. In the anaphases of the second division may be observed a remaininig behind of some chromosomes and sometimes a splitting of them; these are obviously also univalents (fig. 29 e).

We assume that in a hypohexaploid there are 3 cabbage and 3 radish genoms, one of the latter lacking 3 chromosomes. It may be supposed that two cabbage and two radish genoms each produce 9 bivalents, in total—18 bivalents, and that the third cabbage genom supplies 9 univalent chromosomes, the 6 radish chromosomes remaining also univalent. With this assumption we ought to have in our hypohexa-

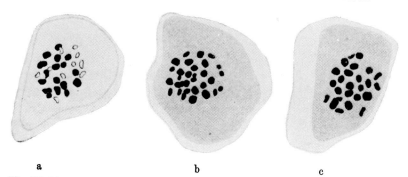

a     b     c

Fig. 28. Division of the pollen mother cells in a hypohexaploid hybrid.   a. Metaphase of the 1st division with 27 chromosomes (the non coloured chromosomes are supposed univalent).   b. Metaphase of the 1st division with 31 chromosomes.   c. Metaphase of the 1st division with 25 chromosomes.   × 1725

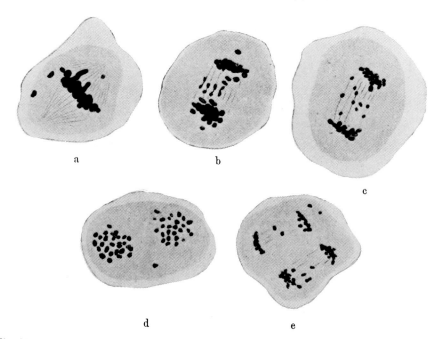

Fig. 29. Division of the pollen mother cells in a hypohexaploid hybrid.   a. Metaphase of the 1st division with trivalents, side view.   b. Anaphase of the 1st division with splitting univalents.   c. Anaphase of the 1st division with separating of split univalents. d. Metaphase of the 2nd division (in the plate on left are 28 chromosomes, in that on the right 26, apart in the plasma 1—their total number 55).   e. Anaphase of the 2nd division.   × 1725

ploid in metaphases of the first division 33 chromosomes, 18 of them
bivalents and 15—univalents. In reality as has been shown in the
greater part of metaphases of the first division, have been reckoned about
27 chromosomes. Evidently this decrease of the chromosome number is
due to the formation of trivalents. The possibility for trivalents
to form—the presence of three genoms—was possessed as well in a
pentaploid, but there they were not manifested. One must suggest
that the formation of trivalents in a hypohexaploid takes place among
cabbage chromosomes, while the radish chromosomes, as in a pentaploid,
produce only bivalents and univalents. The number of trivalents is not
constant, therefore the total number of chromosomes in metaphases of
the hypohexaploid varies[1]).

If in a metaphase we have 25 chromosomes, evidently here will
occur 8 trivalents, 10 bivalents and 7 univalents. If in a metaphase we
have 31 chromosomes then only 2 trivalents will be formed and 16 bivalents and 13 univalents. This calculation is easily found out on the
propositon that if in our hypohexaploids no trivalents had been produced,
we should always have in their metaphases of the first division 18 bivalents and 15 univalents, as has been indicated above. Always as many
trivalents are produced, as inferior to 33 is the number of chromosomes
in the metaphase—total sum of the just mentioned 18 bivalents and
15 univalents.

In the second division in a hypohexaploid we once observed the
formation of one large spindle containing all the chromosomes together
(plate III, 15), therefore in this plant, as in the preceding ones, sometimes
may arise gametes with a somatic complex of chromosomes.

### e) Meiosis in hypertriploid and hypopentaploid

As has been mentioned besides our hypohexaploid with 51 chromosomes, we possessed an inconsiderable number of other hypo- and hyperploids. Of them blossomed and were investigated as to meiosis one 29-
chromosome and one 41-chromosome plants.

In figure 30 are represented 4 metaphases of the first division in
a hypertriploid with 29 chromosomes. On the first three plates are
reckoned 19 chromosomes, and on the fourth—20.

---

[1]) The trivalents are in general apparently not stable ingredients (comp. LONGLEY,
1924, RANDOLPH and MCCLINTOCK, 1926, GAIRDNER, 1926).

It is more probable that this plant originated from the conjunction of a radish gamete of 9 chromosomes and a $F_1$ gamete of 20 chromosomes. If we assume that of the latter 20 chromosomes, 10 were $R$ and 10—$B$, it may be supposed on a basis of all the preceding data, that in the reduction division of this plant will be formed 10 bivalents, 9 of them—radish and 1—cabbage, while the remaining chromosomes,

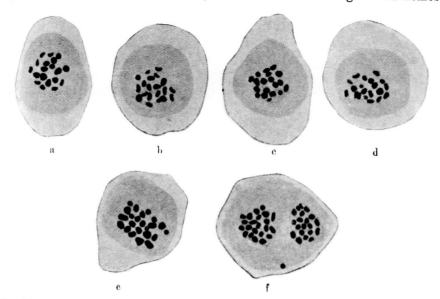

a                b                c                d

e                f

Fig. 30. Division of the pollen mother cells in hypertriploid (29 chrom.) and hypopentaploid (41 chrom.) hybrids. a—c. Metaphases of the 1st division in a 29 chromosome hybrid—plates with 19 chromosomes. d. The same—plate with 20 chromosomes. e. Metaphase of the 1st division in a 41 chromosome hybrid—plate with 23 chromosomes. f. Metaphase of the 2nd division in a 41 chromosome hybrid (22 chromosomes in the plate on the left, 20 in that on the right, 1 chromosome in the plasma separately, total number of chromosomes—43). × 1725

8 cabbage and 1 radish, having no partners, are to remain univalents. What we actually observe—the presence in the metaphases of the first division in this plant of 19 chromosomes of different size—agrees with the above assumption. Unfortunately to separate exactly here the bivalents from univalents is impossible.

The occurence of 20 chromosomes in single metaphases may be explained by a splitting of one of the produced bivalents—the cabbage one, for instance, it taking place earlier than that of all the rest. The

4*

character of the distribution of the chromosomes in the second division remains in the hypertriploid the same as in the triploids.

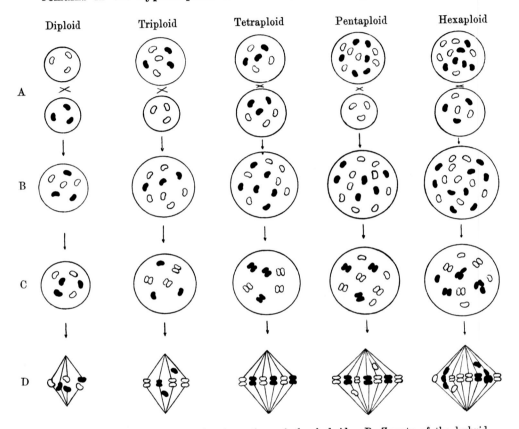

Diploid     Triploid     Tetraploid     Pentaploid     Hexaploid

A.   Gametes taking part in the formation of the hybrid.    B. Zygote of the hybrid. C. Conjugation of chromosomes in the hybrid.   D. Reduction of chromosomes in the hybrid.

Fig. 31.   Scheme of formation of diploid, triploid, tetraploid, pentaploid and hexaploid hybrids *Raphanus sativus* × *Brassica oleracea*, composition of their zygotes and chromosomes' behaviour in conjugation and reduction stages.   Each genom is represented by 3 chromosomes.   The white chromosomes being radish, the black ones — cabbage

The assumption, that our 29 chromosome plant has 19 radish and 10 cabbage chromosomes agrees also with the structure of the siliqua in this hybrid, its two-valved part being but inconsiderably longer than the two-valved part of the siliqua in the triploids.

In figure 30, e and f, are represented the metaphases of the first and second divisions in a 41 chromosome hybrid. The number of chromosomes in these metaphases as well as in the soma of this hybrid is smaller than in a pentaploid; the character of the meiosis remains the same as in a pentaploid but less univalents [1]).

### f) Conclusion

In conclusion of all we have said about the division of the pollen mother cells in our $F_2$ hybrids, we may establish that the study of this division leads to a final confirmation of the assumptions concerning the origin of the mentioned hybrids which were given by us on pp. 12—13.

Now we are able to form a clear conception of the gametes in result of the conjunction of which originate our polyploid hybrids, of the number of radish and cabbage chromosomes they possess, and of the chromosome behaviour in the conjugation and reduction stages in them (see scheme—fig. 31).

### 9. Tetraploid hybrids 7-13 and 7-150

Two tetraploid hybrids Ns. 7-13 and 7-150 require special consideration. Although both of them possess a somatic number of chromosomes amounting exactly to 36, yet the structure of their pistil and siliqua is not of „tetraploid"; in 7-13 is abnormally strongly developed the non dehiscent part of the fruit and in 7-150 the two-valved part (see the data for pistils in table V, p. 31, and the drawings of siliquas in fig. 32, p. 54, and 34, p. 56).

It seemed to us most important to elucidate if these hybrids really possessed the complex of chromosomes ordinary to tetraploids, if they were actually destroying the parallelism between chromosomes and pod structure of which we spoke on pp. 28—30.

We undertook the study of meiosis in hybrids 7-13 and 7-150 and thus were able to explain their morphological peculiarities.

In metaphases of the 1st division of the pollen mother cells in the hybrid 7-13, are reckoned 19-20 chromosomes (fig. 33 a, b) and in anaphases of the same division may be observed a remaining of chromosomes with late splitting (fig. 33 c); in metaphases of the second division

---

[1]) The siliqua in this hybrid approaches the pentaploid in structure.

are often counted 20 chromosomes or about so many on the plates (fig. 33 d), while in anaphases, some chromosomes just so as in the first division, remain backwards when distribution towards the poles takes place (fig. 33 e).

Evidently we have here several univalents, probably 4 of them, perhaps 2. It is clear that the hybrid 7-13 has not an equal number of radish and cabbage chromosomes. Probably it has more of the radish chromosomes since the non dehiscent part of its siliqua is more strongly developed than its two-valved part. But how could such a discrepancy arise if the zygote possesses exactly 36 chromosomes? Evidently here have been conjugatung not the gametes of the composition 9R + 9B but some other. If the hybrid 7-13 in the metaphase of the first division has 4 univalents and 16 bivalents, this plant may be supposed to have originated from a union of the gametes 11R + 9B and 9R + 7B or of the gametes 9R + 9B and 11R + 7B; then in the reduction division must be produced 9 radish and 7 cabbage bivalents and 2 radish and 2 cabbage univalents, consequently a total of 16 bivalents and 4 univalents. But making such a supposition we must admit in some cases the viability of hypodiploid gametes or of diploid, in which several cabbage chromosomes, owing to irregular distribution in meiosis, are replaced by radish chromosomes. We believe both cases to be some times possible. At any rate it is quite doubtless that the hybrid 7-13 proved a result of a fusion of some kind unusual gametes, being an exception among the viable sexual cells of $F_1$. It is important for us to note that the chromosome set of a tetraploid 7-13 is not ordinary, and because of this its siliqua structure is also particular.

20 R + 16 B

Fig. 32. Siliqua, somatic chromosome plate and formula for hybrid 7-13

If we assume for the hybrid 7-13 the formula 20 R + 16 B, its percentage of cabbage chromosomes would be 44·44 and the plant as to its chromosomes will occupy its place between a pentaploid (40 per cent cabbage chromosomes) and a tetraploid (50 per cent ·cabbage chromosomes); the same intermediate position between the pentaploid and a tetraploid occupies the hybrid 7-13 as to its pistil (see table V, p. 31, and fig. 18, p. 30—the dotted line).

Let us note, that in the hybrid 7-13, as in our other plants, was observed in the second division the formation of a spindle with all chromosomes assembled together and the production of dyads, with cells as to chromosome number close to the somatic ones instead of tetrads (fig. 33 f, g).

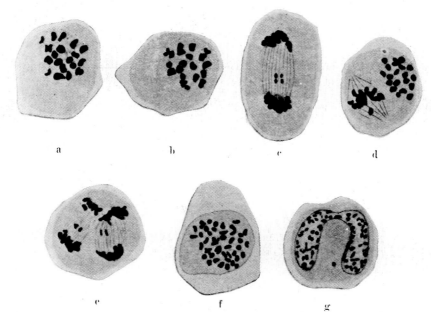

Fig. 33. Division of pollen mother cells in hybrid 7-13. a. Metaphase of the 1st division with 19—20 chromosomes. b. Metaphase of the 1st division with 20 chromosomes. c. Anaphase of the 1st division. d. Metaphase of the 2nd division. e. Anaphase of the 2nd division. f. Metaphase of the 2nd division with 45 chromosomes in one plate. g. Telophase of the 2nd division. × 1725

In the hybrid 7-150 in the metaphases of the 1st division the number of chromosomes is 19 (fig. 35 a, plate III—fig. 11); the metaphases being examined from side view there always might be discovered single chromosomes lying apart from others (fig. 35 b, c); in anaphases as of the first so of the second division, are observed a remaining and sometimes a splitting of small chromosomes (fig. 35 d, e). Apparently in the hybrid 7-150 we possess 17 bivalents and 2 univalents, and thus its somatic chromosome set is not 18R + 18B, but some other one. We suggest that the right formula for it would be 17R + 19B and that it arose either from a fusion of the gamete 9R + 9B and the gamete 8R + 10B or from a fusion of the gamete 8R + 9B and the gamete 9R + 10B. In both cases we are to obtain from radish chromosomes 8 bivalents and 1 univalent, and from cabbage chromosomes 9 bivalents and 1 univalent, accordingly in all 17 bivalents and 2 univalents.

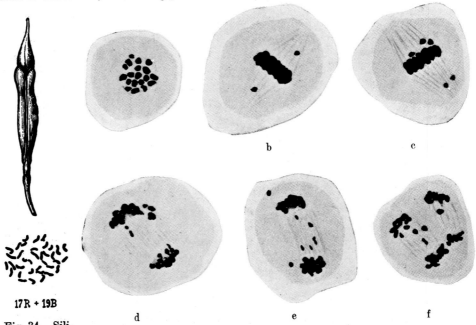

17R + 19B

Fig. 34. Siliqua, somatic plate and formula for hybrid 7-150

Fig. 35. Division of the pollen mother cells in hybrid 7-150. a. Metaphase of the 1st division — 19 chromosomes. b—c. Metaphases of the 1st division, side view. d—e. Anaphases of the 1st division. f. Anaphase of the 2nd division. ✕ 1725

But here, as in the hybrid 7-13 may be doubtlessly only stated, that the hybrid 7-150 has its chromosome set unusual to a tetraploid and proves a result of the fusion of some rare gametes, wherefore its siliqua has a structure of its own.

If the formula 17R + 19B for the hybrid 7-150 is right, the percentage of cabbage chromosomes in this plant will be 52·77 i. e. will be very close to the percentage of cabbage chromosomes in the hypohexaploid (52·94 per cent); it is interesting that this plant as to its pistil is also very close to the hypohexaploid (see table V, p. 31, and fig. 18, p. 30—dotted line).

## 10. The pollen in hybrids $F_1$ and $F_2$

We have already mentioned that in $F_1$ of our hybrids only a very inconsiderable number of pollen grains developed normally, while their majority degenerated in one or the other stage of development. When in $F_1$ mature pollen is being examined under microscope, only a small number of more or less normally developed pollen grains may be seen within the field of sight, while most of these represent small cells without any content, and often stick together in small groups (fig. 36 b).

In triploid plants the number of normal pollen grains is considerably greater than in $F_1$, but there are still many more of the small cells (fig. 36 c).

The tetraploid hybrids as to pollen are manifested as absolutely „pure" forms without any traces of hybrid origin (fig. 36 d). From their parental species, tetraploids as to pollen differ only in greater size of grains (comp. fig. 36 a and d).

In the pentaploid is observed a certain quantity of abortive pollen again; its normal pollen grains are larger than in tetraploids (fig. 36 e).

In the hypohexaploid the pollen is approximately the same as in the pentaploid (fig. 36 f).

On the whole, it is evident, that one or the other number of abortive pollen grains in our hybrids, is due to the character of division of their pollen mother cells and partly to the number of univalents and bivalents in the meiosis. The more irregularly proceeds the division, the greater the number of univalents in a hybrid, these disturbing the normal course of meiosis—the more abortive pollen it contains. The size of normal pollen grains depends on the chromosome number in the hybrid —the greater the number, the bigger the grains.

## 11. Fertility of hybrids $F_2$ and their crossings

Our hybrids $F_1$ as has already been said, were almost entireley sterile; out of 123 hybrids only 19 plants yielded seed and that in a very inconsiderable quantity.

Hybrids of the second generation differed in degree of fertility. The tetraploids in their majority proved quite fertile (see fig. 37). While freely blossoming, could be gathered in some of them up to 900—1000 seeds from 100 siliquas, while in *Brassica oleracea* 100 siliquas yielded 1200 till 1300 seeds, and in *Raphanus sativus* about 600 (see table IX, p. 60).

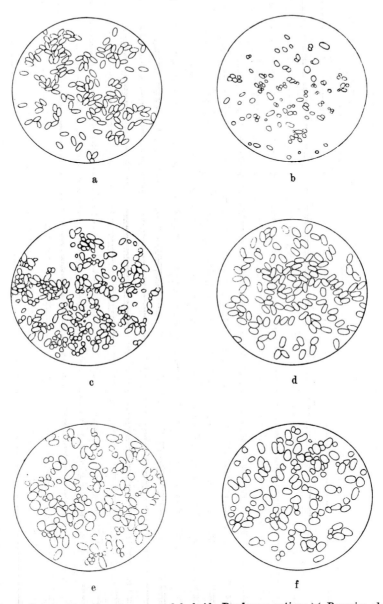

Fig. 36. Pollen in *Brassica oleracea* and hybrids *Raphanus sativus* × *Brassica oleracea*.
a. Pollen of *Brassica oleracea*. b. Pollen of diploid hybrid (F₁). c. Pollen of triploid
hybrid. d. Pollen of tetraploid hybrid. e. Pollen of pentaploid hybrid. f. Pollen of
hypohexaploid hybrid. × ca. 350

a                                        b

Fig. 37. Branches with pods in tetraploid (a) und diploid (b) hybrids

It is howewer interesting to note that a small part of tetraploids, notwithstanding regular meiosis and normal pollen, proved sterile or of reduced fertility.  Let us remark that the percentage of sterile tetraploids is higher in crosses where the paternal plant for the $F_1$ was *Brassica oleracea gemmifera*—brussels' sprouts and is lower in crosses with *Brassica oleracea capitata*—headed cabbage (table X).

Table IX.  Number of seeds selected from 100 siliquas free pollinated in *Raphanus*, *Brassica* and their tetraploid hybrids

| Plant | Number of seeds |
|---|---|
| *Raphanus sativus*  . . . . . . . | 627 |
| *Brassica oleracea*  . . . . . . . | 1200—1300[1]) |
| Tetraploid hybrid 7-93  . . . . . | 1067 |
| Tetraploid hybrid 7-229  . . . . . | 892 |

Table X.  Number of fertile and sterile tetraploids in $F_2$ from different crossings

| Crossing | Total number of tetraploids $F_2$ | Fertile | Reduced fertility | Sterile |
|---|---|---|---|---|
| *Raph. sativus* L. × *Br. oleracea capitata* L. . . . . . . | 44 | 29 (65·9 %) | 12 (27·3 %) | 3 (6·8 %) |
| *Raph. sativus* L. × *Br. oleracea gemmifera* DC. . . . | 31 | 16 (51·6 %) | 7 (22,6 %) | 8 (25·8 %) |

Sterile tetraploids, as has been shown by crossing, do not give any progeny neither as paternal nor as maternal plants.

Triploid hybrids also displayed different degree of fertility.  In most fertile triploids from 100 siliquas could be gathered up to 240 till 280 seeds.

A pentaploid and hypohexaploid showed considerably reduced fertility: the number of their siliquas was small and very few seeds were formed.

One hypopentaploid with 40—42 chromosomes (No. 17-9) proved absolutely sterile, while another with 41 chromosomes (No. 7-147) was fertile.  The tetraploids 7-13 and 7-150 showed reduced fertility.

Of great interest are the results of crosses of our hybrids with each other, with the parental species and other *Cruciferae*.  In table XI

---

[1]) According to KAKIZAKI (1925).

### Table XI.  Results of hybrid crossings

| ♀ | ♂ | Number of pollinated flowers | Number of seeds produced | Number of seeds in 100 pollinated flowers |
|---|---|---|---|---|
| Triploid | *Raphanus sativus* L. . . | **15** | **58** | **386·6** |
| ″ | *Brassica oleracea* L. . . | 98 | 16 | 16·3 |
| ″ | Triploid . . . . . . | **254** | **41** | **16·1** |
| ″ | Tetrapolid . . . . . | **150** | **0** | **0** |
| ″ | Pentaploid . . . . . | 57 | 2 | 3·5 |
| ″ | Hypohexaploid . . . . | 82 | 5 | 5·9 |
| ″ | *Raphanus raphanistrum* L. | 27 | 0 | 0 |
| ″ | *Brassica Napus* L. . . | 37 | 0 | 0 |
| ″ | *Brassica campestris* L. . | 19 | 0 | 0 |
| Tetraploid | *Raphanus sativus* L. . . | **62** | **1** | **1·6** |
| ″ | *Brassica oleracea* L. . . | **206** | **2** | **0·9** |
| ″ | Triploid . . . . . . | **125** | **0** | **0** |
| ″ | Tetraploid . . . . . | **580** | **247** | **42·4** |
| ″ | Pentaploid . . . . . | 97 | 7 | 7·4 |
| ″ | Hypohexaploid . . . . | 108 | 9 | 8·3 |
| ″ | *Raphanus raphanistrum* L. | 15 | 2 | 13·3 |
| ″ | *Brassica alboglabra* BAILEY. | 52 | 0 | 0 |
| ″ | *Brassica Napus* L. . . . | 57 | 0 | 0 |
| ″ | *Brassica campestris* L. . | 15 | 0 | 0 |
| ″ | *Brassica chinensis* L. . . | 45 | 0 | 0 |
| Pentaploid | *Raphanus sativus* L. . . | 28 | 0 | 0 |
| ″ | *Brassica oleracea* L. . . | 18 | 0 | 0 |
| ″ | Triploid . . . . . . | 70 | 0 | 0 |
| ″ | Tetraploid . . . . . | 48 | 2 | 4,2 |
| ″ | Pentaploid . . . . . | 15 | 0 | 0 |
| ″ | Hypohexaploid . . . . | 31 | 1 | 3,2 |
| ″ | *Brassica alboglabra* BAILEY. | 6 | 0 | 0 |
| ″ | *Brassica Napus* L. . . . | 19 | 0 | 0 |
| ″ | *Brassica chinensis* L. . . | 15 | 0 | 0 |
| Hypohexaploid 7-225 | *Raphanus sativus* L. . . | 14 | 0 | 0 |
| | *Brassica oleracea* L. . . | 18 | 1 | 5·5 |
| ″ | Triploid . . . . . . | 19 | 0 | 0 |
| ″ | Tetraploid . . . . . | 48 | 0 | 0 |
| ″ | Pentaploid . . . . . | 16 | 0 | 0 |
| ″ | *Brassica Napus* L. . . . | 12 | 0 | 0 |
| ″ | *Brassica chinensis* L. . . | 4 | 0 | 0 |
| *Brassica oleracea* L. | Triploid . . . . . . | 18 | 0 | 0 |
| | Tetraploid . . . . . | **74** | **0** | **0** |
| ″ | Pentaploid . . . . . | 34 | 0 | 0 |
| ″ | Hypohexaploid . . . . | 28 | 0 | 0 |
| *Brassica alboglabra* BAILEY | Triploid . . . . . . | 10 | 0 | 0 |
| | Tetraploid . . . . . | 25 | 0 | 0 |
| ″ | Pentaploid . . . . . | 10 | 0 | 0 |
| ″ | Hypohexaploid . . . . | 17 | 0 | 0 |

is given a reckoning of these results[1]). Particularly remarkable is it that tetraploids while crossing with each other, yield a sufficient quantity of seeds, but in crosses with *Raphanus sativus* and *Brassica oleracea*, almost no formation of seeds occurs, i. e. the tetraploid hybrids prove already singled out from the parental species as to sex[2]). It is also of interest that crosses of tetraploids and triploids and reciprocal crosses produce no results, while crosses of triploids with *Raphanus sativus* are most productive.

The inconsiderable number of crossings of our hybrids and other species of *Cruciferae*, for instance *Brassica alboglabra* Bailey possessing a diploid chromosome number 18, *Brassica Napus* L. with the somatic chromosome number 36, *Brassica campestris* L. and *Brassica chinensis* L. in which $2n = 20$, *Raphanus raphanistrum* L. in which $2n = 18$ till now have not been successful[3]).

## 12. The progeny of triploid, tetraploid and other hybrids

From triploid plants we collected a great number of seed which had formed under free pollinating conditions. Seeds were also obtained from triploids isolated for selfpollination and from intercrosses of triploids. Morphologically the progeny of the triploids obtained from free pollination as well as from selfpollination and from crosses of triploids inter se, proved heterogenous, but on the whole, approaching considerably *Raphanus sativus*.

In table XII are given the numbers of chromosomes found on examination of 28 plants having arisen from free pollination of one triploid. From the table may be seen, that the majority of plants in this case possessed 18 chromosomes i. e. that the triploids are apt to yield a diploid progeny. Plants obtained from intercrosses between triploids as far as can be judged by small number of investigated specimens,

---

[1]) In this table are given data only for 1926. The numerous crossings of 1927 (ca. 1250 crosses) presented the same picture. (Note incl. in the proof).

[2]) Our opinion on the difficulty of crosses of tetraploids with *Raphanus* we judge not only by experimental data but also by observing the progeny of tetraploids which grew side by side with *Raphanus*—it proved tetraploid.

[3]) Our data on chromosome numbers in species *Cruciferae* (KARPECHENKO, 1922) have already partly been confirmed in WINGE's (1924) and SHIMOTOMAI's (1925) papers.

also often possess in their soma a diploid, hyperdiploid, or hypotriploid number of chromosomes.

Table XII. Results of cytological investigation of triploid hybrid progeny

| Number of chromosomes (2n) . . | 18 | 19 | 20 | 21 | 23 | 24 |
|---|---|---|---|---|---|---|
| Number of plants . . . . . . | 13 | 8 | 4 | 1 | 1 | 1 |

From crossing a triploid and a *Raphanus sativus* we obtain in most cases plants morphologically much resembling the *Raphanus* and possessing the somatic chromosome number 18. The study of the reduction division in these plants, as well as the investigation of their fruit will enable us to tell, whether we have here a „pure radish" or plants containing cabbage chromosomes as well.

Neither hypodiploid plants nor triploid and hypertriploid plants were found in the progeny of triploids.

The progeny of tetraploids obtained through selfpollination proved exceedingly uniform (fig. 38). Great uniformity is exhibited also by plants obtained from tetraploids in free crossing and their intercrossing. Morphologically these plants exactly follow their parents; cytological investigation shows that their somatic number of chromosomes is 36 (at least such proved to be the 32 plants already examined as to chromosome number); on the whole there is no doubt that tetraploid hybrids prove constant.

We had a small number of seeds from the pentaploid and hypohexaploid and a fair quantity from the hypopentaploid (No. 7-147, with 41 chromosomes). Morphologically the progeny of these plants, as far as may be judged by the leaf rosettes, is not uniform. As yet 3 plants from a hypohexaploid have been submitted to cytological examination, all of which had about 40—43 chromosomes, and 3 plants from a hypopentaploid, in which 2n proved 39—41 chromosomes. Under examination has been also a plant—a hybrid obtained from the cross of a triploid and a hypohexaploid; this plant had attracted attention by its abnormal leaves (fig. 39). It proved to have 78 chromosomes (fig. 40) and consequently ought to be termed as hypoenneaploid.

Fig. 38. Progeny of

tetraploid hybrid 5

Probably this hypoenneaploid originated from the fusion of a 27 chromosome gamete from a triploid and a 51 chromosome gamete from

Fig. 39.   Hypoenneaploid hybrid obtained by crossing a triploid hybrid and a hypohexaploid

Fig. 40. Somatic chromosome plate in hypoennea-ploid hybrid. 78 chromosomes.  × 1725

the hypohexaploid or from close in chromosome number gametes; we have already spoken of the formation of such gametes in the above mentioned plants. It is of course possible that the hypoenneaploid may have originated from the duplication of the chromosome complex in a 39 chromosome zygote, having arisen from the fusion of the more common gametes of a triploid and the hypohexaploid.

This we hope might be established with precision when meiosis will have been studied and the structure of the siliqua in the hypoenneaploid will be known.

# Part II

## 1. Analysis of the phenomena of sterility in hybrids
### *Raphanus* $\times$ *Brassica*

On the basis of our study of meiosis in the F₁ hybrids *Raphanus* $\times$ *Brassica* and the determination of the chromosome number in the hybrids F₂, we made the supposition that in F₁ only those gametes were viable which had 18 or a larger number of chromosomes. The study of meiosis in the hybrids F₂ has lead us to the final conclusion, that gametes of F₁ were capable of producing progeny only when they had a minimum of 9 radish and 9 cabbage chromosomes. Only in most exceptional cases the one or two chromosomes of one genom could be absent or probably replaced by chromosomes of another genom (see pp. 54—56).

As in F₁ were produced very few of the indicated gametes, it is clear why the first generation had remained sterile to such a high degree.

The study of meiosis in triploid hybrids and the investigation as to chromosome number of their progeny as from selfpollination so from crosses with *Raphanus* and with each other showed, that in these hybrids these gametes were viable which possessed 9 or few more chromosomes.

This observation does not contradict what has just been said about the gametes F₁. The somatic set in F₁ consists of 9 radish and 9 cabbage chromosomes which do not conjugate with each other and distribute in the meiosis very irregularly so that when in F₁ haploid or hyperhaploid gametes are formed, they receive partly radish and partly cabbage chromosomes. In triploids are in the soma 18 radish and 9 cabbage chromosomes; in the meiosis radish chromosomes form bivalents and distribute to the poles normally by 9; only cabbage chromosomes distribute in the division unequally; thus haploid and hyperhaploid gametes in triploids must contain the radish genom fully.

This difference in the quality of the chromosomes of haploid and hyperhaploid gametes in diploid and triploid hybrids, is evidently what conditions the above indicated differences in their viability.

It is of interest that in the progeny of triploids, hypotriploids are seldom manifested and that not once we have discovered any triploids

5*

and hypertriploids while hypodiploid gametes which might produce them, arise very often in triploids. Just the same gametes doubtlessly arose also in $F_1$, but did not either realize in the progeny, although must be suggested they not seldom contained the genom of one of the parents fully. Hence we may conclude that the gametes possessing the genom of one of the parents plus a few chromosomes from the other parent, are viable, while the gametes, possessing a great surplus of the chromosomes of the latter, prove incapable of producing progeny. As soon as the gamete acquires both genoms from both parents its viability is restored.

Evidently besides the presence of definite chromosomes in the gametes, for their viability, is yet indispensable a definite „equilibrium" in their chromosome set. A lack of such an „equilibrium" is the probable cause of the nonviability of hypo- and hyperdiploid gametes, as of some gametes with still greater number of chromosomes.

We observe in hybrids much abortive pollen and therefore speak of nonviability of gametes; but of course there is much probability in the assumption of the nonviability of some zygotes in our plants as well. Cases are quite imaginable when fertilization proceeds undisturbed but a zygote not receiving the proper chromosome set is unable to develop, and perishes.

Such is the cytological explanation of sterility in our hybrids. But not all phenomena of sterility or reduced fertility observed in our crosses may be cytologically explained. We reported the facts showing that the fertility of our hybrids was not only determined by the presence of a definite chromosome set in their gametes. Tetraploid hybrids possessing quite normal sexual cells with 9 radish and 9 cabbage chromosomes prove sometimes sterile or slightly fertile. Triploids, similar in meiosis, also show different degrees of fertility. It is of interest to note, that in such cases sterility is manifested in both sexes of the plant and that the percentage of sterile plants in different crosses is different.

Here we are probably dealing already with a phenomenon of genetic order. As our triploids and tetraploids segregate according to size of flower, they evidently do so as well according to fertility.

Thus for the gamete to be able to take part in the production of progeny or for the zygotes to be able to develop, it must be not only of a definite cytological, but also of a genetic constitution.

Let us now dwell on the facts which are exhibited while crossing fertile tetraploids with fertile triploids and *Raphanus*. Tetraploids, when pollinated with triploid pollen do not form any seeds. Likewise ineffectual are the reciprocal crosses of triploids and tetraploids. The gametes of triploids and tetraploids are doubtlessly viable; the zygotes which should arise from a fusion of these gametes, the greater part hypertriploid and triploid, ought also to be viable, since we had 27 and 29 chromosome plants. Nevertheless we do not obtain any seeds from the mentioned crosses.

Tetraploid hybrids with great difficulty cross with *Raphanus*, while diploid (F₁) do readily. As to gametes taking part in crosses the diploid as well as the tetraploid hybrids are alike—they both have functioning 18 chromosomed sexual cells of the formula 9R + 9B. For an explanation of the results of these crosses we may suppose, that here is lacking the necessary agreement between pollen and stigma of the plants being crossed. Diploid hybrids have an 18 chromosome stigma, tetraploid ones—a 36 chromosome one; the 9 chromosome pollen of *Raphanus* may grow on the former, and may not be able to do so on the latter.

Then it is clear, that on the 36 chromosome stigma of a tetraploid, the gametes of a triploid will not as well germinate since they are in chromosome set much like the *Raphanus* gametes. On the other hand there may arise hindrances to the germination of the 18 chromosome pollen of tetraploids on the 27 chromosome stigma of the triploid.

Here naturally not the number of chromosomes is important, but those physiologic-chemical alterations which the stigma undergoes in accordance to the alteration of the chromosome complex. It may also not be the stigma playing here the rôle, but the leading tissues of the style or then the tissues of the ovules etc.

If the indicated phenomena really take place in the crossing of our hybrids they may be supposed to occur sometimes in selfpollination of hybrids producing various pollen.

But all these facts require further investigation. As yet we can but say that detailed study of the phenomenon of sterility in crosses *Raphanus* × *Brassica* plainly demonstrates the whole complexity of sterility causes. Here are undoubtedly playing a rôle the chromosome complex of gametes and zygotes, their genetic constitution and perhaps the interrelations occuring between style and pollen grains during the germination of the latter.

## 2. The cytological basis of constancy and segregation in hybrids of *Raphanus* × *Brassica*

We have reported that tetraploid hybrids in characters in which $F_1$ was more or less homogeneous, proved also homogeneous, while in characters in which $F_1$ was polymorphus the tetraploids were also polymorphous, though in a lower degree.

At the basis of this doubtlessly lies the fact, that in $F_1$ only gametes of a somatic chromosome set were viable and that the tetraploids originated as from crosspollination of plants $F_1$ between themselves, so from their selfpollination as well.

Let us imagine some plant of the first generation possessing, for instance, in one of the radish chromosomes the genes ABcD and another plant $F_1$ possessing in the respective chromosome the genes ABCd. In characters determined by the genes A and B, these plants are alike, while in characters determined by the genes C and D they are different, the former having a dominant D and recessive c, the latter on the contrary a dominant C and recessive d. If in a plant $F_1$ only those gametes are viable which contain the full somatic set, it is clear, that all gametes in our first plant will have the genes ABcD and all the gametes of the second plant—ABCd. In selfpollination of these plants we shall obtain from the first hybrid a tetraploid of the formula $\dfrac{ABcD}{ABcD}$ and from the second the tetraploid $\dfrac{ABCd}{ABCd}$; but if between the mentioned plants $F_1$ crosspollination also takes place, then among the progeny of both hybrids we shall find also tetraploids of the formula $\dfrac{ABcD}{ABCd}$. All these tetraploid hybrids will look phenotypically different just in the characters, in which differed also the plants $F_1$, and be homogeneous in the characters in which the $F_1$ plants were also homogeneous. These considerations may of course be applied to all hybrids $F_1$ and all their 18 chromosomes.

Thus is explained the uniformity as to some characters as well as the polymorphousness as to some other characters observed in our tetraploids. The greater homogeneousness of tetraploids in comparison with the plants $F_1$ is doubtlessly a result of the absorption of the recessive characters in crossings of hybrids of the first generation, and also that a part of the tetraploids had probably originated from selfpollination of plants $F_1$. In favour of the latter speaks the exceptional homogeneoussness of tetraploids obtained from some plants $F_1$, and their constancy in all characters.

If we have a tetraploid from the selfpollination of a hybrid $F_1$ then each pair of homologous chromosomes is represented in it by chromosomes absolutely identical and having one same set of genes. A tetraploid possessing two ABcD chromosomes will produce gametes, each of which will receive one of the mentioned chromosomes. This is right in regard to all other chromosomes. Thus all the gametes will be quite identical with each other and the progeny of this tetraploid (when self-pollinating of course or while crossing with quite the same tetraploid) will show no segregation whatever. This is what we actually observe on a great number of plants obtained from some tetraploids.

A tetraploid originated from crosspollination of plants $F_1$ between each other and possessing, for instance, the homologous chromosomes ABcD and ABCd will produce gametes, one half of which will receive the chromosome ABcD, and the other half—ABCd; because of this the progeny of such a tetraploid will segregate according to the characters determined by the genes C and D. We observe such cases in our cultures also.

In crosses of $F_1$ with *Raphanus* we obtain triploids. In the characters in which $F_1$ is heterogeneous they also show heterogeneousness. It is quite natural if triploids originate from different plants $F_1$. But sometimes inconsiderable differences are observed in triploids where the maternal plant has been one and the same hybrid $F_1$. Here we have doubtlessly to deal with heterozygousness of the paternal plant—*Raphanus*.

The progeny of the triploids is not homogeneous, a fact which is comprehensible as their gametes are of a different chromosome constitution, and with this genetically different as well.

The majority of plants obtained from crosses of a triploid with *Raphanus*, are morphologically closely related to the *Raphanus*; we know that the greater part of gametes in triploids as to their chromosome number also approach the gametes of *Raphanus*.

The heterogeneousness of the offspring of the hypopentaploid, pentaploid and hypohexaploid, which is observed in comparison to the progeny of tetraploids, may be doubtlessly explained by the fact, that the meiosis in the above mentioned plants is not quite normal, and that their gametes may differ in regard to their chromosome set.

Thus, in our crosses, we meet in tetraploids the phenomenon of constant-intermediate heredity.

SUTTON (1903) who the first expressed the opinion that the reduction division is a mechanism realizing the Mendelian segregation, supposed that constant-intermediate heredity is based on such a perfect conjugation of the parental chromosomes in the hybrid that their substance mixes and a restoration of the chromosomes in their initial form becomes already impossible.

FEDERLEY in his well-known work on interspecific hybrids of the butterfly *Pygaera* (FEDERLEY, 1913) has shown that the basis of constant-intermediate heredity is not a "perfect" conjugation of the parental chromosomes but on the contrary, the lack of conjugation, and as consequence, the formation of gametes with somatic number of chromosomes in hybrids. FEDERLEY was able to prove that the hybrids *Pygaera curtula* $\times$ *P. anachoreta*, forming gametes with a somatic number of chromosomes, exhibit no segregation in back crosses with *P. anachoreta*.

Our data confirm and supplement the investigations of FEDERLEY. Gametes with a somatic set of chromosomes in our hybrid $F_1$ also show no segregation and this is proved by back crosses of $F_1$ with one of the parents, as well as by way of obtaining the tetraploid $F_2$. We are also able to show the further constancy of tetraploids forming gametes the same as $F_1$.

It is perfectly clear, on the whole, that the reduction division is an indispensable condition of Mendelian segregation; if it drop out—we obtain a homogeneous progeny in the hybrid.

## 3. The results of the investigation and the theory of the hybrid origin of polyploid species

We commence the exposure of the present work with several instances of polyploid series in nature and the theory of the hybrid origin of polyploid species, our results being of exceptional interest for the problem of polyploidy. In result of the hybridization of the two species—*Raphanus sativus* and *Brassica oleracea*, having each 18 chromosomes and further crossings of their hybrids, we obtained forms with 27, 36, 45, 51—52, and even with 78 chromosomes. If our plants were to propagate apogamically, our "series" would then be perfectly secured in subsequent generations also. We would possess a row of apogamic polyploid and aneuploid[1]) forms herewith the most of them would manifest a "hybrid meiosis", similar to those apogamic species which one observes in nature (for instance, the species of *Rosa*, see TÄCK-HOLM, 1922).

But our plants are adapted only to sexual reproduction, and as in nature no triploid, pentaploid and similar forms occur among species propagating sexually, it is probable we should find no such forms in subsequent generations of our crosses. The triploids, as has been shown, will produce principally diploid progeny, the pentaploid will probably tend to give tetraploid offspring etc.

For the moment we may speak quite doubtlessly only of constancy in our tetraploids. The latter, as has been shown, are secured unchanged in $F_3$ and will, there is no doubt also keep constant in further generations. So let us now hang on to tetraploid hybrids more particularly.

With the present principles in systematics of *Cruciferae*, our tetraploids must be morphologically, according to their siliqua, singled out into an independent species or even more true into a genus. They distinguish themselves by quite normal development, by regular meiosis, they have normal pollen, and are most of them perfectly fertile; after several generations they will probably all be fertile, as progeny from fertile plants will increase much more rapidly than progeny from plants of reduced fertility, and will displace the latter.

While comparing it to the parental species *Raphanus sativus* and *Brassica oleracea*, tetraploids present a different chromosome number. They appear most difficult to cross with *Raphanus* and *Brassica*, i. e. we observe in them the occurrence of that sexual individualization to which BATESON (1914, 1922) attached special importance for determining "a specific nature".

This complex of characters for an independent "pure" species in our tetraploids is extremely interesting, and it seems to us that here more than ever we have been indeed approaching the experimental accomplishment of an eminent systematic unit.

Under our climatic conditions, tetraploid hybrids of *Raphanus* × *Brassica* can merely exist when cultivated, the same is it with species from the crosses of which they have originated. But in other climates, where *Raphanus sativus* and *Brassica oleracea* might grow without intervention of mankind culture, our tetraploids also would be apt to propagate with wild flora.

---

[1]) i. e., hypo- and hyperploids (TÄCKHOLM, 1922, p. 234).

For the moment our *"species nova"* is represented by rather uniform plants; but with time through mutations, and in some tetraploids through segregation as to small characters, we may expect races to occur with greatly differing rate of leaf blade dissection, with differing pubescence, with different size and colour of flowers, etc. We may obtain a whole system of forms by which a true Linnean species is usually manifested.

Thus we do seem to have exhibited an experimental verification of the theory of the hybrid origin of polyploid species.

But our tetraploids are not the first constant and fertile hybrids with a double number of chromosomes to have been obtained. Such hybrids have been already described and on these data we must now stop.

First of all must be mentioned the tetraploid *Primula kewensis*, known since 1912 (DIGBY, 1912). Two species—*Primula floribunda* and *Primula verticillata*, of 18 chromosomes each were crossed and produced the sterile hybrid—*Primula kewensis*, showing also 18 chromosomes. From this plant, by way of duplication of the somatic complex of chromosomes in some branches (see NEWTON and PELLEW, 1926) originated the fertile *Primula kewensis* with 36 chromosomes; each chromosome of the latter form is represented twice and only identical chromosomes conjugate in the reduction division, owing to which the tetraploid *Primula kewensis* proves constant.

In 1924 BLACKBURN and HARRISON described one fertile hexaploid hybrid in roses. *Rosa pimpinellifolia* is a "balanced" tetraploid species and forms male and female gametes with 14 chromosomes; *Rosa tomentosa* is a non balanced pentaploid species, and produced pollen with 7 chromosomes, while the ovules show 28 chromosomes. The hybrid *Rosa pimpinellifolia* × *Rosa tomentosa* ought to exhibit a somatic number of chromosomes = 21; but *Rosa Wilsoni*, which is with much probability considered to be such a hybrid, has 42 chromosomes and shows in the reduction division 21 bivalents. With that *Rosa Wilsoni* is a perfectly fertile plant. Evidently we possess here a hybrid with double chromosome complex.

The supposed hybrid from reciprocal cross—*Rosa tomentosa* × *Rosa pimpinellifolia*, the so called *Rosa Sobini* shows, as ought to be expected, the somatic chromosome number = 42, and in the reduction division 14 bivalents and 14 univalents.

CLAUSEN and GOODSPEED (1925) crossed *Nicotiana glutinosa* and *Nicotiana tabacum*. Of the two hybrids obtained from this cross one was sterile the other partly fertile. From the latter hybrid seeds were obtained through selfpollination and from them were raised 65 plants which proved similar to each other and to the plants $F_1$, only differing from them by somewhat larger dimensions of their organs and by their fertility. A cytological investigation manifested that *Nicotiana glutinosa* had $n = 12$ and *N. tabacum*—$n = 24$, while the hybrids of the second generation had $n = 36$, i. e. a duplicate chromosome number. The plant of the first generation that bore seeds, was not investigated, but the authors believe that the duplication of chromosome complex had taken place immediately or soon after fertilization, and therefore that this plant had already $n = 36$ chromosomes.

TSCHERMAK and BLEIER (1926) have published interesting data on the hybrids *Aegilops* $\times$ *Triticum*. Usually these hybrids are sterile, but in 1920 TSCHERMAK in crossing *Aegilops ovata* $\times$ *Triticum dicoccoides* obtained one fertile hybrid which was subsequently multiplied in pure lines for 6 generations and proved absolutely constant. In 1921, crossing *Aegilops ovata* $\times$ *Triticum durum*, TSCHERMAK obtained 4 fertile hybrids which also proved constant for several generations. BLEIER made a cytological investigation of the parental forms of these hybrids —*Aegilops ovata*, *Triticum dicoccoides* and *Triticum durum* and discovered $n = 14$ in all of them. Investigation of the $F_5$ and $F_6$ hybrids *Aegilops* $\times$ *Triticum* showed them to posses $n = 28$. As much hybrids $F_3$ from crosses *(Aeg. ovata $\times$ Trit. dicoccoides) $\times$ (Aeg. ovata $\times$ Tr. durum)* were likewise investigated by BLEIER and also showed $n = 28$[1]).

Such are the data found in literature on constant polyploid hybrids and they are ordinarily considered as an experimental verification of the theory of the hybrid origin of polyploid species. These papers are most interesting, but unfortunately they do not illustrate clearly, whether the crosses do really result here in polyploidy. In *Primula kewensis* the duplication of the chromosome complex is the result of a bud mutation, evidently not depending on hybridization. How the duplication of the chromosome complex in *Rosa Wilsoni* had occured, is quite unknown. The same must be said of the hybrids *Nicotiana* and *Aegilops* $\times$ *Triticum*. Perhaps the multiplication of the chromosome number appeared here as a result of hybridization and perhaps also it had been an independent mutation process, having incidentally coincided with hybri-

---

[1]) Here must be mentioned one more work having infortunately reached us with some delay. It is a paper of K. ICHIJIMA (Genetics, 1926, V. 11, pp. 590—604) in which the author reports his having obtained one tetraploid plant in $F_1$ from crossing *Fragaria bracteata* $\times$ *F. Helleri*. It does not alter our following discussion. (Note incl. in the proof).

zation. From the cross *Aegilops ovata* $\times$ *Tr. durum* TSCHERMAK obtained at a time 4 fertile plants $F_1$ consequently with probably already multiplied chromosome number; one would say that this puts aside the mutation question, but it must be said that here the very cytological data on hybrids do not appear very persuasive. BLEIER's drawings of the anaphase of the first division in hybrids[1]) speak as to our opinion in favour of the fact, that hybrids possess univalents as well as bivalents; it seems to us that further investigation is here required.

From genetic point of view, it is of course all the same, in what manner the duplication of chromosome complex in hybrids occured, be it by way of mutation or because of irregularities in the reduction division as a consequence of their hybrid nature—the results would be the same. But it is of importance to us to elucidate the true rôle of hybridization itself in the originating of polyploidy.

At any rate the radish-cabbage hybrids represent the most demonstrative and clear case of multiplication of the chromosome complex in result of hybridization. In the basis of this process is an absence of conjugation of parental chromosomes in plants $F_1$. The multiplication of chromosome number begins with the second generation.

The absence of parental chromosome conjugation has often been remarked as in artificially obtained hybrids, so in natural ones, and must be regarded as frequently occurring. We already mentioned FEDERLEY's hybrids *Pygaera curtula* $\times$ *P. anachoreta* (FEDERLEY, 1913) in which no conjugation takes place between the parental chromosomes. A same absence of conjugation between parental chromosomes is observed in the hybrids *Digitalis lutea* $\times$ *Digitalis purpurea* (HAASE-BESSEL, 1916), in the apogamous species *Hieracium laevigatum* and *H. lacerum* (ROSENBERG, 1917), in one of the races of *Saccharum officinarum* (BREMER, 1921), in the hybrid *Papaver atlanticum* $\times$ *P. dubium* (LJUNGDAHL, 1922), in the hybrid *Crepis setosa* $\times$ *Crepis capillaris* (COLLINS and MANN, 1923), in the hybrid *Populus Simoni Carrière* (MEURMANN, 1925), in the hybrid *Aegilops cylindrica* $\times$ *Triticum turgidum* (GAINES and AASE, 1926). In many of these hybrids in the process of formation of the sexual cells there are observed the same phenomena as the ones taking place in our plants $F_1$, and doubtlessly it is much possible that in them would arise dyads with a somatic number of chromosomes in their cells. Hence to obtain a second

---

[1]) In the original they are for some reason called telophases.

generation with a duplicate chromosome complex is possible. The constancy of the hybrids $F_2$ will be here secured since each chromosome will be represented in the somatic complex twice, only identical chromosomes will be able to conjugate with each other, and therefore all the gametes will receive one same chromosome set.

In this manner were probably born many of the polyploid species. We suppose that it is during the formation of the sexual cells in hybrids $F_1$ from remote crosses, that multiplication of the chromosome complex takes place in most cases. The thus resulting constant hybrids can more readily than any other polyploid forms by way of further genotypic alteration turn into an independent species.

The above said does not exclude the possibility for the chromosome complex to multiply in other way, and the originating of polyploid species without any participation of hybridization. We know from the phenomena of autosyndesis in *Crepis* (COLLINS and MANN, 1923) and *Papaver* (LJUNGDAHL, 1924) that certain polyploid species are manifested homogenomous. In such cases the duplication of the chromosome complex occured perhaps by way of mutation influenced either by internal or external causes, disturbing the normal course of the division of initial cells. Regarding the effect of external conditions on the course of meiosis in plants, we have already a whole row of experimental data.

NEMEÇ (1910) and SAKAMURA (1920) acted upon the stamens with narcotics and obtained pollen grains with various numbers of chromosomes. By a series of investigations it has been established that abnormal decrease and increase of temperature also strongly affects the course of meiosis, leading to a formation of gametes with various numbers of chromosomes sometimes reaching those of the diploid (SAKAMURA, 1920, BORGESTAM, 1922, BELLING, 1925, MICHAELIS, 1926, SAKAMURA and STOW, 1926, STOW, 1927). DE-MOL (1923) not only succeeded in inducing the hyacinth by means of external influences to produce diploid pollen grains, but even succeeded in obtaining triploid plants by pollinating with those grains.

Recently, based on the observation that in 3 tetraploid and 1 hexaploid species of *Salix*, were manifested by one pair of heterochromosomes, the opinion was expressed, that in the genus *Salix* and perhaps in a row of other genera, one cannot explain the occurrence of polyploid relations by the multiplication of the chromosome complex (HARRISON, 1926 a, b). We hold such a conclusion as not at all indispensable. There exists quite enough evident data about the multiplications of the chromosome complex in plants and animals, to doubt anyhow of the fact, that just this occurrence leads to polyploidy. The facts discovered by HARRISON cannot alter the accepted views on the origin of polyploidy; it seems to us they only set the question about the changes of chromosomes during the evolution process which certainly are to take place; the initial diploid species Salix could not possess any heterochromosomes whatever—the latter had perhaps appeared for the first time in polyploid species.

---

To learn to know in what way the multiplication of the chromosome complex in plants occurs in nature, and to conquer the process of multiplication of chromosomes experimentally, is not only of theoretical, but also of practical importance.

The multiplication of the chromosome complex in pure species doubtlessly already played a rôle in the breeding of new varieties. In the narcissus, for instance, according to DE-MOL (1922), at the beginning till 1885, were cultivated chiefly small diploid varieties, later on appeared in culture bigger triploid types until at last, about 1889, a first big tetraploid was obtained.

Applying the duplication of the chromosome complex to racial hybrids, we might change entirely the character of their segregation and obtain a great variety in combinations of characters and rate of expressiveness in their progeny (MULLER, 1914, BLAKESLEE, 1921, BLAKESLEE, BELLING and FARNHAM, 1923).

But in profiting of the duplication of the chromosome complex in interspecific or intergeneric hybridization, we may already, as follows from this investigation, dream of a creation of quite particular forms which in some cases might be of an exceptional practical interest.

The experimental conquest of the multiplication process of the chromosome complex in plants, ought to reveal to us new great possibilities in plant breeding work.

## 4. Literary data on the hybrids of *Raphanus* ✕ *Brassica*

The first hybrid of *Raphanus sativus* ✕ *Brassica oleracea* was obtained already a century ago by SAGERET (1826). The hybrid blossomed, but it produced few fruits, only two of which developed normally: one of them having, according to SAGERET's description, the shape of a cabbage siliqua, the other one—resembling the siliqua of a radish. From seeds of these siliquas, were obtained several plants which SAGERET did not investigate.

The second hybrid *Raph. sat.* × *Brass. ol.* described by GRAWATT (1914) was striking in its vigorous vegetative development. Having produced a strongly branched stalk, this hybrid reached the roof of the green-house in which it was being cultivated, herewith its branches passed through the ventilator and spread over the slopes of the roof. The hybrid had 15 per cent of flowers with 8 stamens. No siliquas developed on it and no seeds were obtained.

Further on, radish × cabbage hybrids were more than once obtained by BAUR (1922), who merely reports that the hybrids vegetated vigorously but were quite sterile.

Finally MOLDENHAWER (1924) has recently described several hybrids of radish and cabbage. He obtained one plant $F_1$ the pods of which „erinnerten an diejenigen des Kohls, da hier auch die falsche Scheidewand zu beobachten war". From selfpollination of the hybrid were obtained 4 plants $F_2$ of which three plants yielding seeds, gave pods containing two parts — „der untere Teil erinnerte an die *Brassica*-Schoten, der obere sah wie typische *Raphanus*-Früchte aus"; the fourth plant was sterile and had siliquas resembling those of $F_1$. From the seeds of one of the former three hybrids, plants $F_3$ were grown, which distinctly differed from each other in shape of leaves.

One must yet mention, that KAKIZAKI (1925) obtained through pollinating 86 flowers of *Raphanus sativus* with pollen of *Brassica oleracea* 56 seeds, but whether any hybrids were grown from the seeds, the author does not report.

KAKIZAKI has also mentioned in his work that the Hyôgo Agricultural Experiment Station also had seeds from a cross of *Raph. sat.* × *Brass. ol.*

Considering all these not numerous literary data on radish × cabbage hybrids from the point of view of our own results, we must note that all writers, just as we, succeeded in obtaining hybrids from crosses of *Raphanus sativus* × *Brassica oleracea*; hybrids from reciprocal crosses were never obtained up to date. As has been the case with us, $F_1$ in all works proved either quite sterile or slightly fertile. The structure of the pod in the hybrids, one must suggest, is always such as the one we described. When SAGERET said about one of his hybrid pods, that it resembled the radish, and of the other—that it was like the cabbage, he probably simply had to do with siliquas having in the first case a

more or less developed non dehiscent part, and in the second the
two-valved part. MOLDENHAWER describing the siliqua in $F_1$, considers
it reminding one of the *Brassica* pod because of its longitudinal partition; but such a partition is also always the case in the non dehiscent
radish part of the fruit, and we believe its development to take place
in MOLDENHAWER's hybrids $F_1$ as well as in his hybrids $F_2$. The
segregation of MOLDENHAWER's plants $F_3$ evidently finds its explanation
in the difference of the chromosome relations in his hybrids comparing
them to ours, but this requires further investigation.

As to the vigorous development of GRAWATT's hybrid, we are inclined to see its cause in the exceptionally favourable green-house
culture conditions of this plant. With us several hybrids $F_1$ also
attained considerable dimensions, but such exceptionally vigorous growth
as the one described by GRAWATT, we never observed.

Besides our hybrids had not as great a number of flowers with
8 stamens as was noted by GRAWATT in his hybrid. As we had reported, our hybrids did show 8 and even more stamens, but this only
as an exception.

A quite special genetic interest is due to the radish × cabbage
hybrids, only when studied on a large scale morphologically as well as
cytologically. Only such an investigation reveals facts of direct interest
for the resolution of fundamental problems of genetics.


# Summary

By way of a hybridisation — *Raphanus sativus* L. × *Brassica oleracea* L. possessing the diploid chromosome number = 18, hybrids $F_1$
were obtained, in which 2n was also equal to 18. In these hybrids was
not observed in the prophase of reduction division any conjugation of
parental chromosomes, the course of meiosis was very abnormal, through
which reason sexual cells occured having not only an approximate
haploid chromosome number, but a diploid and a tetraploid chromosome
number as well. From crossings of $F_1$ hybrids and *Raphanus*, triploid
hybrids and a pentaploid one of 27 and 45 chromosomes respectively, were
produced, and from crossings of the $F_1$ plants inter se and their selfpollination, tetraploid and hypohexaploid hybrids of 36 and 51—53
chromosomes. An inconsiderable number of hypertriploid, hypertetraploid
and hypopentaploid hybrids was also obtained.

Not a single hypotriploid or hypotetraploid plant occured in the $F_1$ progeny—evidently gametes of less than 18 chromosome number in $F_1$ were not viable. In all above mentioned $F_2$ hybrids, reduction division was being studied and the number of univalents, bivalents and trivalents that might have been expected knowing how the hybrids arose, was found out.

Morphological peculiarities of hybrids are perfectly determined by the correlation of their radish- and cabbage chromosome number and by the total number of chromosomes. A great accumulation of chromosomes has a desastrous effect on plant development.

In accordance to the increase of chromosome number in hybrids, increase the dimensions of their cells.

Gametes of hybrids $F_1$ possessing a somatic chromosome set, do not undergo segregation, thus—tetraploid and triploid hybrids prove uniform in specific characters. Tetraploid hybrids producing gametes quite identical to those of $F_1$, do not segregate as well—their progeny proves to be also uniform.

The remaining hybrids to a higher or lower extent do segregate. Triploids have a marked capability of yielding diploid progeny.

In connection to multiplication of chromosome complex increases the fertility in hybrids. Some tetraploids reach the same utmost fertility degree as do original pure species; often notwithstanding the full similarity in meiosis, the hybrids differ greatly in fertility degree.

In most polyploid hybrids one observed as in $F_1$ a formation of sexual cells possessing the somatic chromosome number. Because of this—from a single crossing of a triploid ($2n = 27$) and a hypohexaploid ($2n = 51$) has been successfully obtained a hybrid of 78 chromosomes.

The morphological peculiarities of tetraploid hybrids, their constancy, their normal meiosis, their fertility and as prove the experiments, their great difficulty in crossing with *Raphanus sativus* and *Brassica oleracea*, makes it possible to regard them as "species nova", obtained in result of hybridisation.

The present investigation was commenced in 1922 at the Plant Breeding Station of the Timiriasev Academy of Agriculture, Moscow, where it was being performed by me during the first three years with the kind aid of Prof. S. I. JEGALOV, Director of the Station, as well as of Miss I. N. SVESHNIKOVA and the late Miss A. G. NICOLAEVA.

In the autumn of 1925 the work being removed to the Section of Genetics of the Institute of Applied Botany, Detskoe Selo, the second generation of our hybrids was investigated here. Since that time Mrs. S. A. SHCHAVINSKAIA, Mr. A. N. LUTKOV, Miss O. N. SOROKINA and Mrs. E. P. GOGEYSEL participated greatly in the work, thanks to what it could be achieved on the appropriated scale.

Part of the cytological investigation concerning the formation of sexual cells in F₁ and in some other hybrids, was performed by me last year during my stay abroad in the laboratories of Prof. Dr. Ö. WINGE (Den Kgl. Veterinaer- og Landbohøjskole's Arvelighedslaboratorium, Copenhagen), Prof. Dr. E. BAUR (Institut für Vererbungsforschung, Berlin-Dahlem) and the late Dr. W. BATESON (John Innes Horticultural Institution, London-Merton), where all possible facility for work was being reserved to me.

To all above mentioned persons and Institutions having helped me in the achievement of this investigation I express my profound gratitude.

## Explanation of plates

### Plate I

1. Tetraploid hybrid in early stage of development.
2. Triploid hybrid in early stage of development.
3. Morphologically distinguished triploid plant, probably a hybrid F₁ with *Raphanus raphanistrum* L.
4. Root of tetraploid hybrid with outgrowths and leaf shoots.

### Plate II

5. Siliquas of a tetraploid.
6. Siliquas of a triploid.
7. Flowers taken from characteristic samples of a diploid, triploid, tetraploid, pentaploid and a hypohexaploid.
8. Flowers of different tetraploids.
9. Flowers of different triploids.

### Plate III

10. Somatic chromosome plate in a tetraploid.
11. Metaphase of the 1st division of a pollen mother cell in a tetraploid 7-150 (19 chromosomes).
12—14. Trivalents in metaphases of the 1st division of pollen mother cell in a hypohexaploid.
15. Metaphase of the 2nd division of pollen mother cell in a hypohexaploid—collection of all chromosomes into one spindle.

All microphotographs are taken with the kind assistance of Miss M. FLITNER; magn. 1150 times.

# Literature

1. BATESON, W., 1913. Problems of Genetics.
2. BATESON, W., 1922. Evolutionary Faith and Modern Doubts. Science, Vol. LV, No. 1412.
3. BAUR, E., 1922. Einführung in die experimentelle Vererbungslehre. 5. und 6. Auflage. Berlin.
4. BELLING, J., 1925. The Origin of Chromosomal Mutations in *Uvularia*. Journ. of Genetics, Vol. 15, No. 3, pp. 245—266.
5. BELLING, J., 1926. The Iron-Acetocarmine Method of Fixing and Staining Chromosomes. Biol. Bull., Vol. L, No. 2, pp. 160—163.
6. BLACKBURN, K. B. and HARRISON, J. W. H., 1924. Genetical and Cytological Studies in Hybrid Roses. 1. The Origin of a Fertile Hexaploid Form in the *Pimpinellifoliae-Villosae* Crosses. Brit. Journ. of Exp. Biology, Vol. I, No. 4, pp. 557—569.
7. BLAKESLEE, A., 1921. Types of Mutations and their Possible Significance in Evolution. Amer. Natur., Vol. 55, No. 638, pp. 254—267.
8. BLACKESLEE, A., BELLING, J. and FARNHAM, M., 1923. Inheritance in Tetraploid *Daturas*. Bot. Gaz., Vol. 76, pp. 329—373.
9. BORGESTAM, E., 1922. Zur Zytologie der Gattung *Syringa* nebst Erörterungen über den Einfluß äußerer Faktoren über die Kernteilungsvorgänge. Arkiv f. Bot., XVII, No. 15, pp. 1—26.
10. BREMER, G., 1921. Een cytologisch onderzoek van eenige soorten en soortsbastaarden van het geslacht *Saccharum*. Haag.
11. CLAUSEN, R. E. and GOODSPEED, T. H., 1925. Interspecific Hybridization in *Nicotiana*. II. A Tetraploid *glutinosa*-tabacum Hybrid, on Experimental Verification of Winge's Hypothesis. Genetics, Vol. 10, pp. 278—284.
12. COLLINS, J. L. and MANN, M. C., 1923. Interspecific Hybrids in *Crepis*. II. A Preliminary Report on the Results of Hybridizing *Crepis setosa* HALL. with *C. capillaris* (L.) WALLR. and with *C. biennis* L. Genetics, Vol. 8, pp. 212—232.
13. DIGBY, L., 1912. The Cytology of *Primula kewensis* and of Other Related *Primula* Hybrids. Ann. of Bot., Vol. 26, pp. 357—388.
14. FEDERLEY, H., 1913. Das Verhalten der Chromosomen bei der Spermatogenese der Schmetterlinge *Pygaera anachoreta*, *curtula* und *pigra* sowie einiger ihrer Bastarde. Zeitschr. f. ind. Abst.- u. Vererbgsl., IX, pp. 1—110.
15. FOCKE, W., 1881. Die Pflanzen-Mischlinge. Berlin.
16. GAINES, E. F. and AASE, H. C., 1926. A Haploid Wheat Plant. Amer. Journ. of Bot., Vol. XIII, pp. 383—385.
17. GAIRDNER, A. E., 1926. *Campanula persicifolia* and its Tetraploid Form "Telham Beauty". Journ. of Gen., Vol. XVI, pp. 341—351.
18. GRAWATT, F., 1914. A Radish-Cabbage Hybrid. Journ. of Heredity, Vol. 5, pp. 269—272.
19. HAASE-BESSEL, G., 1916. *Digitalis*-Studien I. Zeitschr. f. ind. Abst.- u. Vererbgsl., XVI, pp. 293—314.
20. HARRISON, H., 1926. a. Heterochromosomes and Polyploidy. Nature, January 9, p. 150.

6*

21. HARRISON, H., 1926. b. Polyploidy and Sex Chromosomes. Nature, February 20, p. 270.

22. HAYEK, A. V., 1911. Entwurf eines Cruciferensystems auf phylogenetischer Grundlage. Beitr. z. bot. Zentralbl., Bd. XXVII, p. 127.

23. KAJANUS, B., 1913. Über die Vererbungsweise gewisser Merkmale der *Beta*- und *Brassica*-Rüben, II. Zeitschr. f. Pflanzenzüchtung, Bd. I, p. 419.

24. KAJANUS, B., 1917. Über Bastardierungen zwischen *Brassica Napus* L. und *Brassica Rapa* L. Zeitschr. f. Pflanzenzüchtung, Bd. V, pp. 265—322.

25. KAKIZAKI, J., 1925. A Preliminary Report of Crossing Experiments with Cruciferous Plants, with Special Reference to Sexual Compatibility and Matroclinous Hybrids. Jap. Journ. of Genetics, Vol. III, pp. 49—82.

26. KARPECHENKO, G. D., 1922. The Number of Chromosomes and the Genetic Correlation of Cultivated *Cruciferae*. Bull. of Appl. Bot. and Plant-Breeding, XIII, pp. 1—14.

27. KARPECHENKO, G. D., 1924. Hybrids of *Raphanus sativus* L. $\times$ *Brassica oleracea* L. Journ. of Genetics, Vol. XIV, pp. 375—396.

28. KARPECHENKO, G. D., 1927. The Production of Polyploid Gametes in Hybrids. Hereditas, IX, pp. 349—368.

29. KIHARA, H. und ONO, T., 1926. Chromosomenzahlen und systematische Gruppierung der *Rumex*-Arten. Zeitschr. f. Zellforschung u. mikroskop. Anatomie, Bd. 16, H. 3, pp. 475—481.

30. LJUNGDAHL, H., 1922. Zur Zytologie der Gattung *Papaver*. Svensk. Bot. Tidskrift, Bd. 16, pp. 103—114.

31. LJUNGDAHL, H., 1924. Über die Herkunft der in der Meiosis konjugierenden Chromosomen bei *Papaver*-Hybriden. Svensk. Bot. Tidskrift, Bd. 18, pp. 279—291.

32. LONGLEY, A. E., 1924. Chromosomes in *Maize* and *Maize* Relatives. Journ. Agr. Research, Vol. XXVII, pp. 673—682.

33. LUND, F. og KIAERSKOU, 1885. Morphologisk-anatomisk Beskrivelse af *Brassica oleracea*, *Br. campestris* og *Br. Napus*. Bot. Tidskrift, Bd. XV.

34. MEURMANN, O., 1925. The Chromosome Behaviour of some Dioecious Plants and their Relatives with Special Reference to the Sex Chromosomes. Societas Scientiarum Fennica. Commentationes Biologicae, 11, 3, pp. 1—105.

35. MICHAELIS, P., 1926. Über den Einfluß der Kälte auf die Reduktionsteilung von *Epilobium*. Arch. f. wissenschaftl. Bot. (Planta), Bd. I, H. 5, pp. 569—582.

36. DE-MOL, W. E., 1923. The Disappearance of the Diploid and Triploid magnicoronata narcissi from the Larger Cultures and the Appearance in their Place of Tetraploid Forms. Proc. Koninklijke Akad. von Wetensch. te Amsterdam, XXV.

37. DE-MOL, W. E., 1923. Duplication of Generative Nuclei by Means of Physiological Stimuli and its Significance. Genetica, Deel V, pp. 225—272.

38. MOLDENHAWER, K., 1924. Über die Gattungskreuzungen *Raphanus* $\times$ *Brassica*. Mémoires de l'Institut de Génétique d'École supérieure d'Agriculture à Varsovie. Livraison 2, pp. 191—196.

39. MULLER, H., 1914. A New Mode of Segregation in Gregory's Tetraploid Primulas. Amer. Naturalist, Vol. 48, pp. 508—512.

40. NEMEÇ, B., 1910. Das Problem der Befruchtungsvorgänge und andere zytologische Fragen. Berlin.

41. NEWTON, W. C. F. and PELLEW, C., 1926. *Primula kewensis* and its Derivatives. 94th Report of the Brit. Ass-n. 1926 (Oxford), p. 407.

42. POMEL, A., 1883. Contribution à la classification méthodique des Crucifères. Thèse de Paris.

43. RANDOLPH, L. F. and MCCLINTOCK, B. 1926. Polyploidy in *Zea Mays* L. Amer. Naturalist, Vol. LX, No. 666, p. 99.

44. ROSENBERG, O., 1917. Die Reduktionsteilung und ihre Degeneration in *Hieracium*. Svensk Bot. Tidskrift, Bd. 11, pp. 145—206.

45. SAGERET, A., 1826. Considérations sur la production des hybrides, des variantes et des variétés en général, et sur celles de la famille de Cucurbitacées en particulier. Ann. des Sci. Nat., VIII, pp. 294—314.

46. SAKAMURA, T., 1920. Experimentelle Studien über die Zell- und Kernteilung etc. Journ. Coll. Scien., Tokyo Imper. Univ., Vol. 39, Art. 11, pp. 1—221.

47. SAKAMURA, T. and STOW, I., 1926. Über die experimentell veranlaßte Entstehung von keimfähigen Pollenkörnern mit abweichenden Chromosomenzahlen. Jap. Journ. of Botany, Vol. III, pp. 111—137.

48. SCHULZ, O. E., 1919. *Cruciferae-Brassicaceae*. Das Pflanzenreich von A. Engler. Heft 80 (IV, 105). Leipzig.

49. SHIMOTOMAI, N., 1925. A Karyological Study of *Brassica* L. Bot. Mag. Tokyo, Vol. XXXIV. pp. 122—127.

50. SHIMOTOMAI, N., 1927. Über Störungen der meiotischen Teilungen durch niedrige Temperatur. Bot. Mag. Tokyo, Vol. XLI, pp. 149—160.

51. STOW, I., 1926. A cytological Study on the Pollensterility in *Solanum tuberosum* L. Proc. Imp. Acad., Vol. 2, pp. 217—238.

52. SUTTON, W. S., 1903. The Chromosomes in Heredity. Biol. Bull., Vol. 4, pp. 237—251.

53. TÄCKHOLM, G., 1922. Zytologische Studien über die Gattung *Rosa*. Acta Horti Bergiani, Bd. 7, pp. 97—381.

54. TAHARA, M., 1921. Cytologische Studien an einigen Compositen. Journ. of the College of Science, Imp. Univ. of Tokyo, Vol. XLIII, pp. 1—53.

55. TROUARD RIOLLE, 1916. Hybrid entre une crucifere sauvage et une crucifere cultivée à racine tuberisée. Compt. rend. Acad. Paris, CLXII, pp. 511—513.

56. TSCHERMAK, E., und BLEIER, H., 1926. Über fruchtbare *Aegilops*-Weizenbastarde (Beispiele für die Entstehung neuer Arten durch Bastardierung). Ber. d. Deutschen Bot. Ges., Bd. XLIV, pp. 110—132.

57. WILSON, J., 1911. Experiments in Crossing Turnips. Trans. Highl. Agric. Soc. Scotland.

58. WINGE, Ö., 1917. The Chromosomes. Their Numbers and General Importance. Comptes-Rendus des Travaux de Laboratoire de Carlsberg, Vol. 13, pp. 131—275.

59. WINGE, Ö., 1924. Contributions to the Knowledge of Chromosome Numbers in Plants. La Cellule, Vol. XXXV, pp. 305—324.

60. WINKLER, H., 1916. Über die experimentelle Erzeugung von Pflanzen mit abweichenden Chromosomenzahlen. Zeitschr. f. Botanik, Bd. 8, pp. 417—531.

61. WINKLER, H., 1920. Verbreitung und Ursache der Partenogenesis im Pflanzen- und Tierreiche. Jena.

J. C. Walker and Rose Smith. 1930. Effect of environmental factors upon the resistance of cabbage to yellows. Journal of Agricultural Research 41(1):1–15.

J. C. Walker                                    Rose Smith

J. C. Walker, distinguished emeritus professor of the University of Wisconsin, was the premier pathologist of this century involved in disease-resistant genetics and breeding of vegetable crops. Walker forged a productive relationship between plant pathology and genetics in order to develop effective resistance screening, to distinguish between different types of resistance, to determine the inheritance of resistance, and to develop effective breeding strategies. He was also able to combine all the foregoing information, methods, and germplasm to produce useful disease-resistant cultivars and control methods which contributed to the survival and profitability of the vegetable industry in Wisconsin and other states and nations. Examples included cabbage (resistance to yellows, clubroot, and black leg disease-free seed), onion (control of smut by dipping seed in formaldehyde solution, resistance to smudge and pink root), peas (resistance to pea wilt), beet (application of boron to control black spot), and cucumber (resistance to cucumber mosaic virus and scab). The approaches and strategies, which he developed during the 1930s, have since provided the basis of present-day disease-resistant breeding programs.

R. H. Biffen, in 1905 in England, was the first to demonstrate that resistance to a particular disease in plants (yellow rust in wheat) was a simply inherited Mendelian trait. However, in the following decade the idea of the genetic control of resistance to a pathogen was not readily accepted by leading pathologists, who believed that resistance applied only to a particular cultivar in a particular locality (Butler 1905) and that hybrid plants acted as a bridging host between susceptible and resistant cultivars, leading to a breakdown of resistance (Pole-Evans 1911). It was not until E. C. Stakman and associates (Stakman and Levine 1922) established the basis of pathogenic variability that the arguments above could be discounted and that the variations in response of the host to the pathogen in different areas could be explained. However, even with these developments, breeding for disease resistance was not organized on a scientific basis. Reliable inoculation methods and testing under controlled environmental conditions were not practiced and well-designed genetic experiments were not conducted. It is in this area that Walker made his incisive experiments and discoveries.

Previously, Jones and Gilman (1915) had made cabbage selections for resistance to fusarium yellows [*Fusarium oxysporum* f. *conglutinans* (WR.) Snyder & Hansen] in an infested field in Wisconsin and released the resistant cultivar 'Wisconsin Hollander.' This resistance was useful but varied in its expression in the field. Later it was found that the expression was affected by temperature and all plants were susceptible under a temperature regime 27 to 33°C (Tims 1926). Walker (1930) had developed susceptible and resistant homozygous lines from selfing plants of 'All Head Early' and ''Glory of Enkhuizen' cabbage and determined that this resistance was controlled by a single dominant gene. In the experiment that forms the basis for this classic paper, Walker compared the expression of single-gene resistance with the resistance derived from mass selection of resistant plants in 'Wisconsin Hollander', 'Wisconsin All Seasons', and 'All Head Select', as well as the susceptible 'Copenhagen Market' and 'Danish Ballhead', to varying temperature regimes. The results of this straightforward experiment with its creative and remarkable interpretation had profound implications for plant breeding (genetics, plant pathology, and horticulture). The homozygous lines reacted differently to temperature than the mass-selected resistant progenies and the susceptible entries. The homozygous resistant lines expressed high and uniform resistance up to and at 24°C, but the mass-selected progenies and the susceptible lines were 100 percent susceptible at 24°C. 'Copenhagen Market' had 100 percent susceptible plants when grown in inoculated sterilized soil at 20 to 22°C, while the mass-selected resistant, 'Wisconsin Hollander', had 43 percent diseased and 14-29 percent dead plants under the same conditions. All types of genetic resistance became susceptible at 30°C. Walker was thus able to use temperature to separate-out two types of genetic resistance.

He hypothesized that the type of resistance in the mass-selected plants was controlled by modifying factors since various levels of resistance were detected in different progenies. Thi was later verified in genetic studies reported by Walker's student M. E. Anderson (1933), and also by L. M. Blank (1937). The expression of the ''modified factor'' (now known as ''multigenic resistance'') in cabbage (and also to fusarium wilt in tomato) was later found to be unstable in response to varying nutrient levels (Hoagland's solution) in contrast to the stable monogenic form of resistance (Walker and Hooker 1945; Walker and Foster 1946). The use of controlled temperature testing had led to reliable, effective, and predictable results saving both time and effort of breeders.

Currently, there is increasing interest in the identification and use of multigenic resistance since resistance due to major genes in the host can rapidly be overcome by virulence genes in the pathogen. Because it is difficult to separate monogenic from multigenic resistance under field conditions, it is necessary to use carefully controlled environment conditions to detect and to manipulate this type of resistance, as was originally shown by Walker. If breeders are to continue to investigate the value and use of this type of stable resistance, testing must be conducted under controlled conditions and finally evaluated in well-designed nurseries in the field. Our current direction and progress in this area rests securely on the basic discoveries and innovative approaches provided by Walker.

The highlights of J. C. Walker's career (1918–1969) are well documented in *Journal Phytopathology* (Vol. 60, No. 1, 1–5, 1970). Walker was a dedicated and inspiring teacher and mentor to 75 graduate students, many of whom became eminent in the United States and around the world. He authored over 300 technical papers and two widely used texts, *Plant Pathology* and *Diseases of Vegetable Crops*. He received numerous honors: among the most prestigious were election to the U.S. National Academy of Sciences (1945); Man of the Year (1954) from the Vegetable Growers of America; honorary Doctor of Science (1960), University of Gottingen, Germany; and fellow, Award of Distinction (1969), and president (1943) of the American Phytopathological Society. J. C. Walker has left a tremendous legacy to his profession and to the vegetable industry of Wisconsin and the nation.

Coauthor Rose Smith (1879–1972), a brilliant student of plant science, received her Ph.D. degree in 1926 at the University of Wisconsin, majoring in botany with her first minor in plant pathology. She was employed as a postdoctorate research associate from 1926 to 1930 at the University of Wisconsin, during which time she collaborated in this study. She spent most of her career as a professor of biology at the College of St. Teresa, Winona, Minnesota.

Dermot P. Coyne
Department of Horticulture
University of Nebraska
Lincoln, Nebraska 68583

## REFERENCES

ANDERSON, M. E. 1933. Fusarium resistance in 'Wisconsin Hollander' cabbage. J. Agr. Res. 47:639–661.

BIFFEN, R. H. 1905. Mendels' law of inheritance and wheat breeding. J. Agr. Sci. 1:4–48.

BLANK, L. M. 1937. Fusarium resistance in 'Wisconsin All Seasons' cabbage. J. Agr. Res. 55:497–510.

BUTLER, E. J. 1905. The bearing of Mendelism on the susceptibility of wheat to rust. J. Agr. Sci. 1:361–363.

JONES, L. R., and J. C. GILMAN. 1915. The control of cabbage yellows through disease resistance. Wis. Agr. Expt. Sta. Res. Bull. 38.

POLE-EVANS, I. B. 1911. South African cereal rusts, with observations on the problem of breeding rust-resistant wheats. J. Agr. Sci. 4:95–104.

STAKMAN, E. C., and M. N. LEVINE. 1922. The determination of biologic forms of *Puccinia graminis* on *Triticum* spp. Minn. Agr. Expt. Sta. Tech. Bull. 8.

TIMS, E. C. 1926. The influence of soil temperature and soil moisture on the development of yellows in cabbage seedlings. J. Agr. Res. 33:971–992.

WALKER, J. C. 1930. Inheritance of fusarium resistance in cabbage. J. Agr. Res. 40(8):721–745.

WALKER, J. C., and R. E. FOSTER. 1946. Plant nutrition in relation to disease development. III. Fusarium wilt of tomato. Am. J. Bot. 33:259–264.

WALKER, J. C., and W. J. HOOKER. 1945. Plant nutrition in relation to disease development. I. Cabbage yellows. Am. J. Bot. 32:314–320.

# EFFECT OF ENVIRONMENTAL FACTORS UPON THE RESISTANCE OF CABBAGE TO YELLOWS [1]

By J. C. WALKER, *Agent, Office of Horticultural Crops and Diseases, Bureau of Plant Industry, United States Department of Agriculture*, and *Professor of Plant Pathology, University of Wisconsin*, and ROSE SMITH, *Research Associate in Plant Pathology, University of Wisconsin*

## INTRODUCTION

In a previous paper (*8*) [2] the senior writer described the genetic behavior of resistance in cabbage to the yellows organism, *Fusarium conglutinans* Wr. It was shown that resistance and susceptibility are controlled by a single pair of genes (*Rr*), so far as could be determined from the reaction of the various progenies tested under field conditions in southern Wisconsin on soil naturally infested with the yellows organism.

Several years' work has yielded a number of strains of cabbage which are more constantly and uniformly resistant than any of the mass-selected varieties available to previous workers (*2, 4, 8, 9*). It was the purpose of the present investigation to study under controlled conditions their reaction to environmental factors, primarily temperature. Such inquiry is important for several reasons. From the practical point of view it is desirable to know to what extremes improved strains will maintain their resistance. The distinction between environmental and hereditary factors in the interpretation of the minor variations in severity of disease in susceptible plants is essential as a guide to further selection. A better understanding of this distinction was also germane to studies of the inherent differences in the relation of the parasite to resistant and susceptible host plants.

Considerable work has been previously done upon the relation of temperature to the development of yellows by Gilman (*1*), Tisdale (*7*), and Tims (*6*). Since these studies have been adequately summarized by Jones and associates (*3*), it is unnecessary to discuss them in detail here.

## METHODS OF EXPERIMENTATION

The naturally infested soil employed in these studies was taken directly from a thoroughly diseased field in southeastern Wisconsin used over a period of several years for testing the resistance of cabbage strains. For artificial inoculation of soil a single-spore isolation of the organism secured from a diseased plant collected in southeastern Wisconsin was grown for several weeks upon a mixture of sand and corn meal. The culture was then pulverized and mixed thoroughly with soil which had been previously sterilized at a pressure of 15 pounds for several hours. The inoculated soil was used in some experiments during the several weeks immediately following its preparation, and it was used again in other trials after storage for several months.

---

[1] Received for publication Aug. 20, 1929; issued June, 1930. This study has been supported jointly by the Department of Plant Pathology, University of Wisconsin, with the aid of a special grant from the general research fund of the university, and by the Office of Horticultural Crops and Diseases, Bureau of Plant Industry, U. S. Department of Agriculture.
[2] Reference is made by number (italic) to "Literature cited," p. 15.

Journal of Agricultural Research,
Washington, D. C.

105260—30—1

(1)

Vol. 41, No. I
July 1, 1930
Key No. G-710

Soil-temperature studies were carried out in the Wisconsin soil-temperature tanks described elsewhere (*3*). Plants were grown in infested soil in galvanized-iron cans which were inserted in tanks held with a reasonable degree of constancy at the desired temperatures. The seedlings were grown in soil free from the yellows organism and transplanted to the cans at the desired time. The cans were then kept at 15° C. or lower until the plants had recovered from transplanting and were then removed to the tanks. In records of these experiments the duration was calculated from the time the cans were placed in the tank, care having been taken to hold the soil temperature previous to that time to a point below 15°, the minimum for disease development. Unless otherwise noted, a common air temperature which fluctuated between 15° and 17° was maintained, except on sunny days, when it rose a few degrees higher.

In the study of air-temperature influences along with soil-temperature effects tanks were used which stood in greenhouses in which the air temperatures were regulated at different levels. Although the temperature of the air was not so constant as that of the soil, it was sufficiently constant to give reasonable contrast between the various levels used.

As indicated in the later discussion of experimental results, the typical appearance of yellows in the leaves was used as the chief criterion in recording the appearance and progress of the disease. In certain cases, however, this was insufficient, because of variation in symptoms under different environments. This was especially true at the higher soil temperatures in the case of resistant plants, where stunting of the plants and unnatural mottling of the leaves, not typical symptoms of the disease, were common as a result of exposure to the extreme environment. In these investigations the influence of soil moisture was not studied. For each series, however, the soil moisture was kept uniform throughout by frequent weighings and the addition of sufficient water to maintain the original weight of all pots. Following the evidence from Tisdale's work (*7*) that moderately dry soil gave the greatest quantity of disease, the moisture content was kept at about 60 per cent of the water-holding capacity.

## SOURCE OF HOST MATERIALS

As has already been explained in a previous paper (*8*), most commercial varieties of cabbage are extremely susceptible to yellows. In each variety, however, there are ordinarily a few plants that are resistant under field conditions. These are in practically all cases heterozygotes so far as resistance is concerned, and, resistance being dominant, they withstand the attack of the yellows organism. By pure-line selection from such individuals, lines were secured which are homozygous for resistance under field conditions, in Wisconsin at least and where tested in other localities. In like manner individuals selected from commercial varieties growing on noninfested soil in most instances yield progenies that are practically 100 per cent susceptible. Such lines are herein regarded as homozygous for the susceptible character.

A third type of material used in this connection is that referred to as mass-selected resistant. Examples of this are Wisconsin All Seasons and Wisconsin Hollander, resistant varieties now in general commercial use. These differ from the homozygous-resistant lines in that they were derived from commercial types, not by pure-line selection but by mass selection from the surviving plants on infested soil. They have thus remained to some degree heterozygous for resistance and in each generation continue to produce some purely susceptible individuals. As was pointed out earlier, resistant strains of this type were the only ones available to Tisdale (7) and to Tims (6). For purposes of more direct comparison the same varieties they used, Wisconsin Hollander and Wisconsin All Seasons, were included in these studies.

Thus the following two resistant and two susceptible types of material were used:

(1) The homozygous-resistant lines were progenies of individual plants, or mixtures of such progenies, which were proved to be homozygous by progeny tests in the field. Progenies 40–350s, 40–353s, 40–354s, 40–420s, 40–421s, 40–426s, 40–524s, 40–530s, and 40–535s were secured by self-pollination of individual plants selected from the All Head Early variety. 40–26–F was a mixture of remnants of selfed progenies from the same variety. 30–25–A was from the seed of several homozygous-resistant plants selected from the Copenhagen Market variety and allowed to be cross-pollinated by insects in an isolated location. 20–27–C was a mixture of remnants from selfed progenies of plants selected from the Jersey Wakefield variety. 20–29–A was the second generation from a homozygous resistant individual of the same variety.

(2) The mass-selected resistant varieties were secured from a reliable commercial source of these seeds. Wisconsin Hollander, Wisconsin All Seasons, and All Head Select were used.

(3) Homozygous-susceptible lines were from plants proved by progeny tests in the field to be homozygous for susceptibility. HG–5s was a selfed progeny from an $F_1$ hybrid plant resulting from a cross between an All Head Early and a Glory of Enkhuizen susceptible individual. HG–27–A was a mixture of sister progenies coming from the same original cross as HG–5s. Reciprocal crosses were made of two $F_1$ hybrid plants coming from a cross between an All Head Early and a Copenhagen Market susceptible individual; the flowers of these two plants (HC–19 and HC–3) were not emasculated, but a single camel's-hair brush was worked over the blossoms of each plant and the seed of each was saved separately. Seed from plant HC–3 was used and is designated as HC–3 × 19B. C–29–A was a homozygous-susceptible line from the Copenhagen Market Variety.

(4) Two commercial susceptible varieties, Copenhagen Market and Danish Ballhead, were used. Although these were not quite so homogeneous as the homozygous-susceptible lines, they were shown by field tests to contain a very small percentage of resistant individuals·

In addition to the above-mentioned lots, certain $F_1$ and $F_2$ progenies from crosses between resistant and susceptible lines were studied. These are described later in the text.

## EXPERIMENTAL RESULTS
### COMPARATIVE STUDY OF RESISTANT AND SUSCEPTIBLE LINES
#### INFLUENCE OF SOIL TEMPERATURE

The plants used in experiment 1 had grown in noninfested soil for about six weeks before they were transplanted to infested soil. Two homozygous-resistant lines (40–535s and 30–25–A) were compared with one commercial susceptible variety (Copenhagen Market). Table 1 shows that the severity of the disease in the susceptible variety increased with rise in soil temperature, reaching a maximum at 26° to 30° C. This is in accord with the findings of Tisdale (7) and Tims (6) for other susceptible varieties. In the case of the resistant strains, however, there was complete absence of disease at 20° and 22°. At 24° no external symptoms appeared, but upon careful examination of the vascular system there were found a few brownish streaks not unlike those present in greater abundance in susceptible plants at this and other temperatures. All attempts to isolate the fungus from such stems at 24° and 26° were unsuccessful. The significance of the browning of the veins in resistant lines at this temperature will be discussed later. In one resistant line (30–25–A) a few plants showed external evidence of disease at 28°, and both strains had a few such cases at 30°. It thus appeared that the homozygous-resistant lines were exhibiting uniform resistance in this case at a constant soil temperature as high as 26°. This was quite different from the results secured by Tisdale (7) and Tims (6) with the mass-selected resistant varieties, which showed considerable disease at 20° and comparatively little resistance above that point.

TABLE 1.—*Development of yellows in a commercial susceptible cabbage variety and two homozygous-resistant progenies when grown on naturally infested soil at various constant soil temperatures*

| Soil temperature (°C.) | Occurrence of yellows in— | | | | | |
| | A susceptible variety [a] | | A resistant variety [b] | | A resistant variety [c] | |
| | Plants | Diseased | Plants | Diseased | Plants | Diseased |
| | *Number* | *Per cent* | *Number* | *Per cent* | *Number* | *Per cent* |
| 20 | 10 | 60 | 20 | 0 | 23 | 0 |
| 22 | 10 | 70 | 24 | 0 | 22 | 0 |
| 24 | 9 | 89 | 24 | 0 | 23 | 0 |
| 26 | 9 | 100 | 24 | 0 | 23 | 0 |
| 28 | 8 | 100 | 24 | 0 | 23 | 17 |
| 30 | 9 | 100 | 24 | 25 | 23 | 13 |

[a] Copenhagen Market.
[b] Selected from All Head Early, 40–535s.
[c] Selected from Copenhagen Market, 30–25–A.

TABLE 2.—*Development of yellows in two homozygous-susceptible cabbage progenies and eight homozygous-resistant progenies when grown on naturally infested soil at various constant soil temperatures*

| Description of strain | Strain No. | Occurrence of yellows at soil temperature of— | | | | | |
| | | 24° C. | | 26° C. | | 28° C. | |
| | | Plants | Diseased | Plants | Diseased | Plants | Diseased |
| | | *Number* | *Per cent* | *Number* | *Per cent* | *Number* | *Per cent* |
| Homozygous-susceptible | HG–5s | 10 | 100 | 10 | 100 | 10 | 100 |
| | HC–3×19B | 7 | 100 | 7 | 100 | 6 | 100 |
| Homozygous-resistant: From All Head Early | 40–350s | 7 | 0 | 7 | 0 | 7 | 57 |
| | 40–353s | 10 | 0 | 9 | 11 | 10 | 0 |
| | 40–354s | 8 | 0 | 8 | 0 | 8 | 25 |
| | 40–420s | 9 | 0 | 9 | 0 | 9 | 22 |
| | 40–421s | 10 | 0 | 9 | 0 | 9 | 22 |
| | 40–426s | 10 | 0 | 10 | 0 | 10 | 10 |
| | 40–524s | 7 | 0 | 6 | 0 | 7 | 0 |
| From Jersey Wakefield | 20–27–C | 7 | 0 | 10 | 20 | 10 | 40 |

In experiment 2 (Table 2) the study was extended to two homozygous-susceptible lines (HG–5s, HC–3 × 19B), a number of homozygous-resistant progenies selected from All Head Early (40–350s, etc.), and a mixture of several homozygous-resistant progenies from Jersey Wakefield (20–27–C). In this instance the plants were much younger when exposed to infection, which exposure took place only 26 days from the time the seed was sown. The experiment was continued for 40 days. Only the critical temperatures 24°, 26°, and 28° C. were used. At all temperatures the susceptible strains succumbed completely. At 24° all the resistant lines remained free from disease symptoms. At 26° two progenies showed slight disease, and at 28° all but two progenies showed more or less disease. Thus in comparison with experiment 1, these lots were as resistant at 24°, but two of them were less stable at 26°, and most of them less so at 28°.

In experiment 3 a homozygous-susceptible line (HG–5s), two homozygous-resistant lines (40–420s and 20–27–C), and three mass-selected resistant varieties were used. (Table 3.) The plants in this case were 32 days old when exposed to the various soil temperatures (18°–33° C.), and the experiment was continued for 33 days. As noted earlier by Tisdale (7) and Tims (6), the disease developed very slowly at 18°, even in the susceptible strain, while its incidence also was much reduced at 33°. These extremes appear to limit the progress of the parasite materially, and this probably accounts in a large measure for the reduction in the amount of the disease. The behavior of the homozygous-susceptible and the homozygous-resistant lines otherwise coincides in general with that observed in experiments 1 and 2. As noted in the experiments of Tisdale and Tims, a small percentage of plants of Wisconsin Hollander and Wisconsin All Seasons became diseased at 21°, and the percentage increased with the rise in temperature up to 30°. The difference between the homozygous-resistant lines and mass-selected varieties appears at 21° and higher temperatures. It is evident that the latter, though quite resistant under Wisconsin field conditions, are inherently much more susceptible than the homozygous lines, while their exposure to constant temperatures above 21° reveals that they are unsuited to such conditions. The cans from this series were photographed near the close of the experiment, and some of them are shown in Figure 1.

TABLE 3.—*Development of yellows in susceptible and resistant cabbage progenies when grown on naturally infested soil at various constant soil temperatures*

[Ten plants of each strain at each temperature except 33°, where there were 20 plants]

| Description of strain or variety | Strain No. or variety | Percentage of plants yellowed when grown at soil temperature of— | | | | | |
|---|---|---|---|---|---|---|---|
| | | 18° C. | 21° C. | 24° C. | 27° C. | 30° C. | 33° C. |
| Homozygous-susceptible | HG–5s | 50 | 100 | 100 | 100 | 100 | 100 |
| Homozygous-resistant | 40–420s | 0 | 0 | 0 | 0 | 40 | 25 |
| | 20–27–C | 0 | 0 | 0 | 10 | 80 | 35 |
| Mass-selected resistant | Wisconsin Hollander | 20 | 30 | 50 | 80 | 90 | 25 |
| | Wisconsin All Seasons | 0 | 20 | 50 | 20 | 70 | 50 |
| | All Head Select | 0 | 20 | 20 | 60 | 80 | 60 |

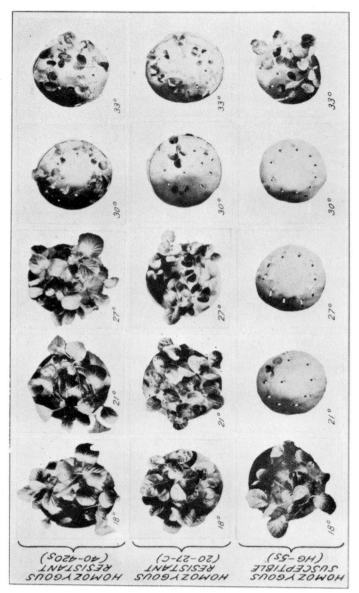

FIGURE 1.—Pots from experiment 3 (Table 3), photographed near the close of the experiment. The white markers indicate where the dead plants were removed. Homozygous-resistant progenies (40-42s and 20-27-C) and a homozygous-susceptible progeny (HG-5s) are shown. All plants of 40-42s remained healthy from 18° to 27° C., whereas at 30° the plants were stunted and some had died. They did not show typical yellows, however. The same is true of 20-27-C, except that one plant died at 27°. Some plants of the susceptible line (HG-5s) were still healthy at 18°, but at 21° and up to 30° all were diseased and all but one had died. At 33°, as in the resistant lines, the disease was less destructive than at 30°. Apparently the activity of the fungus was reduced at this temperature

In the three experiments just described certain new facts are brought out. In the first place, the homozygous-resistant strains, so called because of their perfect resistance under repeated field tests, are extremely resistant at quite high temperatures. Plants of the ages used were practically all stable up to and including constant temperatures of 26° to 27° C. Above this point the plants became stunted, showed a pale yellowing of the leaf parenchyma, and exhibited some browning of the vascular bundles. The external appearances of disease were not typical of those in the susceptible lines at these or lower temperatures. The browning of veins was very rarely accompanied by the presence of the organism.

It was important to determine whether this peculiar reaction of resistant plants at high temperature was due to the attack of the parasite or to the direct effect of the temperature upon the plants. The behavior of plants in infested and in noninfested soil over a range of temperature was therefore studied. (Experiment 4.) In order to make the infested and the noninfested soil as comparable as possible, a quantity of the latter was divided and to one portion inoculum of *Fusarium conglutinans* was added. A series of soil-temperature tanks ranging at 2-degree intervals from 18° to 28° C. were then set up so that in each tank there was one can each of homozygous susceptible plants (C–29–A) and one can each of homozygous resistant plants (20–29–A) in inoculated and in uninoculated soil. The plants were 43 days old at the beginning of the experiment, and the latter was run for 30 days. Six plants were placed in each can. In the inoculated soil the results were similar to those in the naturally infested soil in experiments 1, 2, and 3. The susceptible plants showed typical yellows, the severity increasing with the temperature. The resistant plants remained perfectly healthy up to 24°; they showed some atypical signs at 26°, and the atypical symptoms were severe at 28°. In the uninoculated soil no signs of disease whatever were noted, even at the highest temperature. This fact shows that this rather abrupt appearance of pathological symptoms in the resistant plants at 26° and above is the result of the yellows organism in the soil and not the effect of high temperature alone. The cans from the 24° and the 28° tanks, photographed on the thirteenth day, are shown in Figure 2.

Histological studies subsequently reported (5) indicate that there is invasion of the root tips of homozygous-resistant plants with little or no penetration of the vascular system. Furthermore, at 28° and above, the root systems of cabbage are at their low ebb of growth, while the causal organism is most active (7). The reduced rate of root growth, with invasion of the absorbing regions of the root system, is probably responsible for the gradual stunting and death of the resistant plants at 28° and above. Moreover, it is to be emphasized that this breaking down of resistance is of quite a different order in external appearance and in internal histology from the normal development of disease in susceptible plants at these or other temperatures.

The distinct difference between commercially successful mass-selected resistant varieties and pure-lined homozygous-resistant lines in their temperature reactions is the second matter of importance to be noted in these results. Many plants of the mass-selected strains behaved like susceptible ones under these controlled conditions, which is in accord with Tisdale's (7) and Tims's (6) results. While this may

105260—30——2

not appear to be in accord with the fact that these varieties are commercially successful upon infested soil, it should be pointed out that it is not uncommon to find a field of these mass-selected varieties with as high as 50 per cent of the plants showing slight symptoms of disease. As has been suggested before (8), it is possible that in the selection of these varieties mildly susceptible individuals were not eliminated and as a result they are apparently present in considerable numbers. Under average seasonal conditions where the crop is started during the cooler spring months so that the plants reach considerable size before infection occurs, these mildly susceptible individuals are not easily distinguished. The test under controlled environment is ap-

FIGURE 2.—Comparison of resistant and susceptible lines of cabbage upon inoculated and uninoculated soil at 24° and 28° C. Cans in left-to-right order in each row contain (1) susceptible plants in inoculated soil; (2) resistant plants in inoculated soil; (3) susceptible plants in uninoculated soil; (4) resistant plants in uninoculated soil. See further explanation in the text

parently more severe and the differences between them and homozygous-resistant plants are accentuated.

### INOCULATED STERILIZED SOIL

In the first three experiments naturally infested soil was used. While this method yielded results more comparable with those secured under field conditions, it did not afford a picture of the behavior of the plants exposed to the yellows organism when it was removed from competition with other soil organisms and thus allowed to accumulate in greater abundance. In experiment 4 the plants were 50 days old when exposed to various soil temperatures. A commercial susceptible variety, a mass-selected resistant variety, and a homozygous-resistant strain were used. There were four to seven plants in each can. The percentages of diseased and of dead plants were recorded after 30 days. The final results are shown in Table 4.

TABLE 4.—*Development of yellows in a susceptible cabbage variety, a mass-selected resistant variety, and a homozygous-resistant progeny when grown in naturally infested and inoculated sterilized soils at various constant soil temperatures*

[In each individual lot 7 plants were used, except of Wisconsin Hollander at 26°, where only 6 plants were used in naturally infested soil, and at 28°, where 5 and 6 plants were used, respectively, in naturally infested and inoculated sterilized soil]

| Soil temperature (° C.) | Occurrence of yellows in susceptible commercial (Copenhagen Market) variety grown in— | | | | Occurrence of yellows in mass-selected resistant (Wisconsin Hollander) variety grown in— | | | | Occurrence of yellows in homozygous-resistant (40–530s, selected from All Head Early) variety grown in— | | | |
|---|---|---|---|---|---|---|---|---|---|---|---|---|
| | Naturally infested soil | | Inoculated sterilized soil | | Naturally infested soil | | Inoculated sterilized soil | | Naturally infested soil | | Inoculated sterilized soil | |
| | Diseased plants | Dead | Diseased plants | Dead | Diseased plants | Dead | Diseased plants | Dead | Diseased plants | Dead | Diseased plants | Dead |
| | *P. ct.* | *P. ct.* | *P. ct.* | *P. ct.* | *P. ct.* | *P. ct.* | *P. ct.* | *P. ct.* | *P. ct.* | *P. ct.* | *P. ct.* | *P. ct.* |
| 20 | 86 | 57 | 100 | 100 | 14 | 0 | 43 | 14 | 0 | 0 | 0 | 0 |
| 22 | 100 | 86 | 100 | 100 | 57 | 14 | 43 | 29 | 0 | 0 | 0 | 0 |
| 24 | 100 | 100 | 100 | 100 | 100 | 57 | 100 | 86 | 0 | 0 | a 0 | 0 |
| 26 | 100 | 100 | 100 | 100 | 100 | 33 | 100 | 71 | 0 | 0 | a 100 | 0 |
| 28 | 100 | 100 | 100 | 100 | 100 | 80 | 100 | 100 | a 100 | a 14 | a 100 | a 43 |
| 30 | 100 | 100 | 100 | 100 | 100 | 100 | 100 | 100 | a 100 | a 100 | a 100 | a 100 |

a These plants did not show typical symptoms of yellows but were nevertheless stunted and showed yellowish spots on the leaves, dying back of the leaf margins, and more or less browning of the veins. Practically all attempts to isolate the organism from them were unsuccessful. Differences are illustrated in Figure 3.

It is clear that the increased potentiality of the inoculum which results from sterilization of the soil to remove the soil flora, temporarily at least, from competition with the yellows organism has an immediate effect upon the disease. The incubation period is somewhat shortened and the progress of the disease is hastened. Aside from these effects it is quite consistently noticeable that the total percentage of plants diseased in the susceptible and mass-selected resistant groups is increased in the inoculated soil. This is probably the result, in part at least, of a more uniform distribution of the parasite in the soil. The relative differences between susceptible and resistant strains are not changed; but the homozygous-resistant line showed, in inoculated sterilized soil, the pathological response characteristic of this type at high temperatures at a slightly lower point than in previous experiments wherein naturally infested soil was used.

### INFLUENCE OF AIR TEMPERATURE

In the first five experiments the air temperature during most of the time was at 15° to 17° C., although for a few hours during sunny days it rose a few degrees higher. This is somewhat lower than might often obtain under natural conditions during midsummer. Consideration was next given to the influence of the air temperature as well as of the soil temperature upon the expression of resistance.

In experiment 6 a homozygous-susceptible line (HG–27–A) and a homozygous-resistant line (40–26–F) were included. Plants 45 days old of each were used, and the experiment was run for 27 days in naturally infested soil. Soil temperatures of 20° and 24° were maintained at air temperatures of 15° to 17° and of 28°, respectively. In the susceptible variety the disease developed most rapidly at the 28°

air temperature in the two respective soil temperatures, which is in accord with the earlier findings of Tims (*6*). The final results (Table 5) show, however, that even at this air temperature the homozygous-resistant line remained healthy.

TABLE 5.—*Development of yellows in susceptible and resistant lines of cabbage at various air and soil temperatures*

| Temperature of— | | Occurrence of yellows in— | | | |
| | | Susceptible line | | Resistant line | |
| Air | Soil | Plants | Diseased | Plants | Diseased |
| ° C. | ° C. | *Number* | *Per cent* | *Number* | *Per cent* |
| 15 { | 20 | 5 | 100 | 10 | 0 |
| | 24 | 5 | 100 | 10 | 0 |
| 28 { | 20 | 5 | 100 | 10 | 0 |
| | 24 | 5 | 100 | 10 | 0 |

## EFFECT OF WOUNDING THE ROOT ON INFECTION AND RESISTANCE

It became of interest to determine whether or not severe pruning of the root system would permit the development of disease in resistant plants. Plants 10 weeks old of a homozygous-resistant strain (40–535s) and a commercial susceptible variety (Copenhagen Market) were transferred to a bench of naturally infested soil, a portion of each strain being handled as follows: (1) Transplanted with the least possible injury to the roots; (2) roots pruned moderately before resetting; and (3) all lateral roots pruned close to the taproot. The disease was recorded for each lot as it appeared, and the data are summarized in Table 6. It is significant that even the severest degree of root pruning did not bring about infection of the resistant strain. In the susceptible strain the appearance of the disease in slightly pruned plants was delayed several days as compared with the moderately and severely pruned plants, but at the end of the period the total infections in the three lots were not significantly different.

TABLE 6.—*Effect of varying degrees of root pruning at transplanting upon the occurrence of yellows in a resistant and a susceptible strain of cabbage*

| Strain or variety | Degree of root pruning | Total number of plants | Number of yellowed plants at end of— | | | | | | | |
| | | | 17th day | 19th day | 23d day | 25th day | 30th day | 33d day | 37th day | 44th day |
| 40–535s (resistant) | Slight | 19 | 0 | 0 | 0 | 0 | 0 | 0 | 0 | 0 |
| | Moderate | 20 | 0 | 0 | 0 | 0 | 0 | 0 | 0 | 0 |
| | Severe | 20 | 0 | 0 | 0 | 0 | 0 | 0 | 0 | 0 |
| Copenhagen Market (susceptible) | Slight | 10 | 0 | 0 | 4 | 4 | 7 | 9 | 9 | 9 |
| | Moderate | 16 | 3 | 4 | 6 | 7 | 7 | 13 | 13 | 13 |
| | Severe | 17 | 2 | 4 | 4 | 5 | 7 | 12 | 14 | 15 |

## GROWING CONDITIONS BEFORE EXPOSURE TO THE PARASITE IN RELATION TO SUSCEPTIBILITY AND RESISTANCE

Throughout the experiments just reported plants for transplanting were commonly grown at comparatively low temperatures—about 15° C.—and were then transferred to infested soil and placed at various soil and air temperatures. The question arose as to whether the conditions under which a plant is grown before exposure to the parasite under a given environment have any important bearing upon susceptibility or resistance. In fact Tims (6) suggested that plants grown at a comparatively high temperature became diseased less readily than those grown at lower temperatures more favorable for growth of the host. An experiment was therefore planned to yield some evidence on this point.

Seeds of a homozygous-resistant line (20–27–C) and of a homozygous-susceptible line (HC–27–A) were sown in cans of noninfested soil. They were placed in one set of soil-temperature tanks running at 20°, 24°, 28°, and 32°, at each of two air temperatures, 15°–17° and 28°. Thus each of these two varieties was growing under eight different combinations of air and soil temperature. After 32 days they were transplanted to naturally infested soil. After standing 3 days at about 20° for the plants to recover from transplanting, the cans were so divided that one half of the plants from each combination of temperatures was kept at a soil temperature of 24° and an air temperature of 28°, while the other half was held at 24° soil and 15°–17° air. The plants were thus held for 23 days. By that time every plant in the susceptible line was diseased and most of them were dead. No disease appeared in any of the resistant lines throughout the series. The rate of appearance of the disease and death of the susceptible plants, as shown in Table 7, does not indicate any marked predisposing influence of the growing conditions previous to infection upon disease development in plants of this age. It appears that the method employed of producing the plants at temperatures favorable to growth and later shifting them rather suddenly to various temperatures for exposure to the parasite does not introduce a significant experimental error.

TABLE 7.—*Relative development of yellows in susceptible cabbage plants* [a] *grown for about five weeks in noninfested soil at various soil and air temperatures and then transplanted to infested soil and exposed to a soil temperature of 24° and air temperatures of 15°–17° and 28° C.*

| Air temperature after transplanting (°C.) | Temperatures before transplanting from clean soil | | Total number of plants | Number of yellowed plants at the end of— | | | | Number of dead plants at the end of— | | | | | | |
|---|---|---|---|---|---|---|---|---|---|---|---|---|---|---|
| | Air | Soil | | 10 days | 12 days | 14 days | 16 days | 12 days | 14 days | 16 days | 18 days | 19 days | 22 days | 23 days |
| | °C. | °C. | | | | | | | | | | | | |
| 15–17 | 15–17 | 20 | 5 | 2 | 4 | 5 | ------ | 0 | 0 | 0 | 0 | 1 | 3 | 3 |
| | | 24 | 5 | 0 | 4 | 5 | ------ | 0 | 0 | 0 | 0 | 3 | 5 | ----- |
| | | 28 | 5 | 4 | 4 | 5 | ------ | 0 | 2 | 3 | 4 | 4 | 5 | ----- |
| | | 32 | 5 | 1 | 4 | 5 | ------ | 0 | 0 | 2 | 2 | 2 | 4 | 4 |
| | 28 | 20 | 5 | 0 | 5 | 5 | ------ | 0 | 0 | 0 | 3 | 3 | 5 | ----- |
| | | 24 | 5 | 0 | 0 | 1 | 5 | 0 | 0 | 0 | 0 | 3 | 3 | 3 |
| | | 28 | 5 | 0 | 1 | 3 | 5 | 0 | 0 | 1 | 1 | 3 | 4 | 5 |
| | | 32 | 4 | 2 | 3 | 4 | ------ | 0 | 1 | 1 | 1 | 3 | 4 | ----- |
| 28 | 15–17 | 20 | 5 | 4 | 4 | 5 | ------ | 0 | 0 | 2 | 3 | 3 | 4 | 4 |
| | | 24 | 5 | 0 | 5 | 5 | ------ | 0 | 0 | 3 | 4 | 4 | 4 | 4 |
| | | 28 | 4 | 0 | 3 | 3 | 4 | 1 | 1 | 3 | 4 | ----- | ----- | ----- |
| | | 32 | 4 | 0 | 2 | 3 | 4 | 0 | 0 | 1 | 2 | 3 | 3 | 4 |
| | 28 | 20 | 5 | 2 | 4 | 4 | 5 | 0 | 3 | 3 | 3 | 3 | 5 | ----- |
| | | 24 | 5 | 0 | 3 | 3 | 5 | 0 | 3 | 3 | 3 | 3 | 4 | 4 |
| | | 28 | 4 | 3 | 3 | 4 | ------ | 0 | 2 | 2 | 4 | ----- | ----- | ----- |
| | | 32 | 4 | 1 | 2 | 4 | ------ | 0 | 1 | 1 | 4 | ----- | ----- | ----- |

[a] HC–27–A.

## STUDY OF REACTION OF RESISTANT-SUSCEPTIBLE HYBRIDS

In the light of the evidence brought together in the foregoing pages, it is of interest to consider the reaction of resistant-susceptible hybrids as reported earlier (8).

### $F_1$ HYBRIDS

The $F_1$ hybrids from resistant and susceptible parents were first studied. Several progenies were grown in both inoculated sterilized and naturally infested soil at various temperatures. The hybrids, which were fully resistant in field tests, were found to be quite as resistant as homozygous-resistant plants except for the fact that in the case of four progenies they showed stunting and high-temperature symptoms at slightly lower temperatures. (Table 8.) In a fifth case (20–27–D) no pathological effects were noted at 26° C. in naturally infested soil. Certain of the lots are shown in Figure 3.

TABLE 8.—*Development of yellows in $F_1$ cabbage hybrids from resistant-susceptible crosses, and commercial susceptible and homozygous-resistant lines*

| Age of plant | Duration of experiment | Description of strain or variety | Strain No. or variety | Occurrence of yellows in plants grown on— | | | | | | | |
| | | | | Naturally infested soil at— | | | | Inoculated sterilized soil at— | | | |
| | | | | 24° C. | | 26° C. | | 24° C. | | 26° C. | |
| | | | | Plants | Diseased | Plants | Diseased | Plants | Diseased | Plants | Diseased |
| *Days* | *Days* | | | *Number* | *Per cent* | *Number* | *Per cent* | *Number* | *Per cent* | *Number* | *Per cent* |
| 50 | 30 | Susceptible | Copenhagen Market | 7 | 100 | 7 | 100 | 7 | 100 | 7 | 100 |
| | | Resistant | 40–530s | 7 | 0 | 7 | 0 | 7 | 0 | 7 | [a] 100 |
| | | $F_1$ hybrids | H–2×5–21 | 7 | 0 | 4 | [a] 100 | 4 | [a] 100 | 4 | [a] 100 |
| 82 | 28 | Susceptible | Copenhagen Market | 4 | 100 | 4 | 100 | 4 | 100 | 4 | 100 |
| | | Resistant | 40–530s | 4 | 0 | 4 | 0 | 4 | 0 | 4 | [a] 100 |
| | | $F_1$ hybrids | 5–32×C–1 | 4 | 0 | 4 | [a] 100 | | | | |
| | | do | 40–335×H–35 | 4 | 0 | 4 | [a] 100 | | | | |
| | | do | 5–32×H–35 | 4 | 0 | 4 | [a] 100 | 4 | [a] 75 | 4 | [a] 100 |
| 40 | 32 | do | 20–27–D | 10 | 0 | 10 | 0 | | | | |

[a] The symptoms in these cases were not those of typical yellows that occurred in the susceptible strain but were the atypical signs current in resistant strains at high soil temperature as described elsewhere in the text.

### $F_2$ HYBRIDS

From the evidence already presented it was to be expected that $F_2$ hybrid progenies would react according to the constitution of their segregants. Approximately 25 per cent, being homozygous-resistant, should survive as transplanted seedlings up to 26° C. Approximately 50 per cent, being heterozygotes, should survive at 24° and some become stunted at higher temperatures in naturally infested soil. The remaining one-fourth, being homozygous-susceptible, should succumb at 22° or above. Variation in the progress of the disease in the susceptible plants should also result, depending upon soil temperature.

The experiment reported in detail was conducted with a single $F_2$ progeny (5HIA) which in field trials showed 28.6 per cent diseased plants in 1926 and 22.2 per cent in 1927. Plants 37 days old were exposed to a range of soil temperatures in naturally infested soil. At

18° and 20° somewhat fewer than the expected number were affected; at 22° exactly one-fourth were diseased. (Table 9.) At 24° slightly above 25 per cent were diseased. At 26° the percentage was significantly higher, while at 28° only about one-fourth survived. At

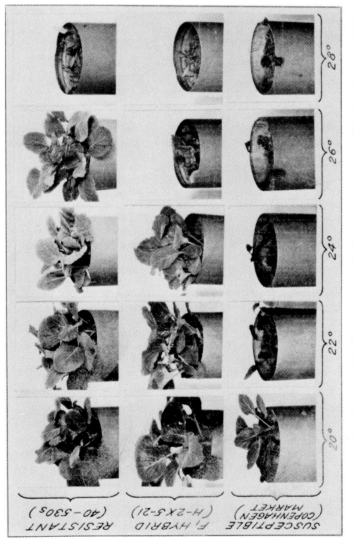

FIGURE 3.—Comparison of development of yellows in a homozygous-resistant progeny (40-530s), a susceptible variety (Copenhagen Market), and a hybrid progeny from a cross between a resistant and a susceptible plant (H–2×5–21). Plants grown on infested soil at various soil temperatures. (For details see Table 8.) The homozygous-resistant strain was healthy at 26° C. but decidedly stunted at 28°, whereas the hybrid plants were healthy at 24° but stunted at 26°. One plant of the susceptible strain was still healthy at 20°, but all plants were diseased or dead at 22° and above

26° part of the plants and at 28° all of those which survived showed distinct stunting of the type observed in heterozygous and homozygous-resistant plants at these temperatures. No attempt was made to differentiate the two types when the counts were recorded.

TABLE 9.—*Development of yellows in an F₂ hybrid progeny from a cross between a homozygous-susceptible and a homozygous-resistant parent*

| Soil temperature | Total number of plants | Yellowed plants | |
|---|---|---|---|
| | | Number | Per cent |
| °C. | | | |
| 18 | 58 | 12 | 20. 6 |
| 20 | 59 | 12 | 20. 3 |
| 22 | 60 | 15 | 25. 0 |
| 24 | 58 | 16 | 27. 5 |
| 26 | 60 | 19 | 31. 6 |
| 28 | 60 | 44 | 73. 3 |

## SUMMARY

The purpose of the study reported in this paper was to determine the effect of environmental factors upon the resistance or susceptibility in various strains of cabbage to yellows (*Fusarium conglutinans*).

The hosts tested consisted of homozygous-resistant progenies, homozygous-susceptible progenies, mass-selected resistant varieties in commercial use, and commercial susceptible varieties.

In the susceptible strains and mass-selected resistant varieties the typical disease symptoms appeared in increasing percentages with increase in soil temperature to about 28° C., and were retarded somewhat at 33°. In homozygous-resistant strains no evidence of disease was found up to 24° in naturally infested soil when young transplants were used. Slight evidence of disease was found in a few cases at 26° and to a greater degree at still higher temperatures. At this higher range the symptoms were not typical of yellows, and the fungus was rarely isolated from such plants. The homozygous-resistant lines, therefore, reacted in a distinctly different manner from the susceptible and mass-selected resistant types.

The increase of inoculum secured by steam sterilization of the soil and reinoculation with a pure culture of the yellows organism resulted in more uniform infection and more rapid disease development in susceptible and mass-selected resistant types, but had no effect upon the homozygous-resistant lines, except that the atypical high-temperature symptoms were evident at a slightly lower temperature.

Increase in air temperature up to 28° C. hastened the development of disease but did not alter the distinct difference in reaction between homozygous-resistant lines and the other types used.

Severe pruning of the root system during the process of transplanting served to shorten the incubation period in susceptible plants, but it did not facilitate the production of yellows in homozygous-resistant plants.

The temperature at which plants were grown prior to exposure to the parasite did not appear to affect the rate of disease development, nor did it have any influence upon the stability of resistance in homozygous lines.

Heterozygous plants, i. e., those resulting from crosses between resistant and susceptible parents, reacted as homozygous resistant except that they usually showed atypical high-temperature symptoms at a slightly lower tenperature.

At soil temperatures around 22°–24° C. transplants of F₂ segregating progenies from resistant-susceptible crosses showed approximately 25 per cent typical susceptible plants, as they had previously shown in the field. As the temperature increased above this point the atypical high-temperature symptoms appeared, as was expected, in the heterozygous and homozygous resistant members of the population. As the temperature fell below this point the full expression of typical disease symptoms in homozygous-susceptible plants decreased.

## LITERATURE CITED

(1) GILMAN, J. C.
    1916. CABBAGE YELLOWS AND THE RELATION OF TEMPERATURE TO ITS OCCURRENCE. Ann. Missouri Bot. Gard. 3: 25–84, illus.
(2) JONES, L. R., and GILMAN, J. C.
    1915. THE CONTROL OF CABBAGE YELLOWS THROUGH DISEASE RESISTANCE. Wis. Agr. Expt. Sta. Research Bul. 38, 70 p., illus.
(3) ——— JOHNSON, J., and DICKSON, J. G.
    1926. WISCONSIN STUDIES UPON THE RELATION OF SOIL TEMPERATURE TO PLANT DISEASE. Wis. Agr. Expt. Sta. Research Bul. 71, 144 p., illus.
(4) ——— WALKER, J. C., and TISDALE, W. B.
    1920. FUSARIUM RESISTANT CABBAGE. Wis. Agr. Expt. Sta. Research Bul. 48, 34 p., illus.
(5) SMITH, R., and WALKER, J. C.
    1930. A CYTOLOGICAL STUDY OF CABBAGE PLANTS IN STRAINS SUSCEPTIBLE OR RESISTANT TO YELLOWS. Jour. Agr. Research 41: 17–35, illus.
(6) TIMS, E. C.
    1926. THE INFLUENCE OF SOIL TEMPERATURE AND SOIL MOISTURE ON THE DEVELOPMENT OF YELLOWS IN CABBAGE SEEDLINGS. Jour. Agr. Research 33: 971–992, illus.
(7) TISDALE, W. B.
    1923. INFLUENCE OF SOIL TEMPERATURE AND SOIL MOISTURE UPON THE FUSARIUM DISEASE OF CABBAGE SEEDLINGS. Jour. Agr. Research 24: 55–86, illus.
(8) WALKER, J. C.
    1930. INHERITANCE OF FUSARIUM RESISTANCE IN CABBAGE. Jour. Agr. Research 40: 721–745, illus.
(9) ——— MONTEITH, J., JR., and WELLMAN, F. L.
    1927. DEVELOPMENT OF THREE MIDSEASON VARIETIES OF CABBAGE RESISTANT TO YELLOWS (FUSARIUM CONGLUTINANS WOLL.). Jour. Agr. Research 35: 785–809, illus.

O

H. A. Jones and A. E. Clarke. 1943. Inheritance of male sterility in the onion and the production of hybrid seed. Proceedings of the American Society for Horticultural Science 43:189–194.

H. A. Jones

A. E. Clarke

Hybrid vigor, as described in *Nicotiana* by Kolreuter in 1763 and later recognized by Darwin as having an adaptive value among wild species, has contributed significantly to the improvement of economic crops during the twentieth century. Both the vigor and uniformity of $F_1$ hybrids have been of great agricultural importance. Yield increases, uniformity of raw and processed products, and adaptation of crops to mechanization have resulted directly from the use of $F_1$ hybrid cultivars.

The most significant use of $F_1$ hybrids as commercial cultivars has occurred in maize. Maize is easily emasculated (detasseled). Wind conveys the pollen from male parent to female parent. Disomic inheritance without natural sexual barriers (e.g., incompatibility) permits the ready development of inbreds and useful hybrid cultivars. Few other species are so easily manipulated.

The inability to emasculate the seed parent has been the primary barrier to the agricultural application of heterosis in the many species of normally cross-pollinated crops. The solution to the problem of emasculation was found in research on onion by Henry A. Jones and Alfred E. Clark, who described the inheritance of male sterility and its potential use in hybrid seed production in the *Proceedings of the American Society for Horticultural Science* in 1943. Male sterility, controlled in concert by the cytoplasm and the nuclear genome, was to be the foundation upon which commercial $F_1$ hybrid seed production in onion, maize, sorghum, castor bean, beet, carrot, petunia, wheat, and rice developed.

The initial discovery of male sterility in onion in 1925 required 18 years of research before publication. The 18-year span reveals much about Henry Jones the scientist.

Henry Jones was born in Deer Park, Illinois in a log cabin in 1889. His early youth was spent on a family vegetable farm near Chicago, where Henry worked from a very early age, with his father. When Henry was 12 years old the Jones family moved to Seward, Nebraska, where they grew vegetables for local needs. His father died in 1905, at which time, as the oldest male in the family, Henry was forced to drop out of school and run the farm. At age 20, Henry Jones enrolled in the School of Agriculture, a secondary vocational school within the College of Agriculture at Lincoln, Nebraska. After graduation in 1912, he continued in the College of Agriculture for a baccalaureate degree (1916) and then proceeded to the University of Chicago for a Ph.D. degree (1918) in plant physiology. His background up to this point did not suggest either interest or training in genetics and/or plant breeding.

After short periods on the staffs of the Universities of West Virginia and Maryland, Henry Jones started (1922) a 14-year period of teaching, research and administration in the department of vegetable crops of the University of California, Davis. His research focused on onions and asparagus. During this period the landmark textbook *Truck Crop Plants* by Jones and Rosa was published (1928). From 1936 to 1957 he was on the USDA–ARS staff in Beltsville, Maryland.

The discovery of male sterility (1925) in the 'Italian Red' onion (Jones and Emsweller 1936a) was the start of a series of happy accidents that would have been lost by a lesser talent. The male sterile plant produced bulbils, or top sets, in the flowerhead. Initially, Jones reasoned that if the male sterility were stable, the male sterile seed parent could be increased clonally, via bulbils, and used as a seed parent of $F_1$ hybrids. Hybrid seed from the male sterile 'Italian Red' 13–53 sometimes gave progeny that were male fertile, sometimes the progeny segregated 1 male fertile : 1 male sterile, and some matings gave progeny all of which were male sterile. These last progeny (all male sterile) suggested that there might also be a sexual means of producing seed all of which were male sterile. If so, this trait could be used in the seed parent for hybrid seed production.

Henry Jones was unaware of any published reports of dominant genes for male sterility, and if dominant genes did exist, he wondered how an individual homozygous for the dominant gene could be synthesized. He was also aware of the classical papers in Germany by Michaelis and his coworkers on *Epilobiom* suggesting genetic-cytoplasmic intereactions controlling a number of traits, including pollen abortion. He was also aware of the thorough studies on cytoplasmic inheritance in maize at Cornell University by Marcus Rhoades. In Rhoades's papers (1931, 1933) the pollen sterility trait was either unstable or unpredictable. In any event, the matings by Jones and Clarke between male sterile and male fertile onions, reciprocal matings of fertiles with differing cytoplasms but similar nuclear genes, and the analysis of the respective progenies, gave repeatable results which demonstrated clearly that the interaction of a "sterile cytoplasm," maternally inherited, and the homozygous nuclear gene *ms* controlled pollen sterility. The male sterile trait was stable. The evidence reported by Michaelis (1933, 1935) supported the onion data.

In a paper only six small pages long the genetics of cytoplasmic–genic male sterility in onions, the method of developing male fertile and male sterile inbreds for hybrid seed production, the production of hybrid seed, and even the principle of fertility restoration, essential in crops such as maize, were clearly documented.

Immediately after the 1933 paper a series of papers were published by others describing cytoplasmic–genic male sterility in other crops and plant breeders pushed its application. Significantly, maize breeders reemphasized research on male sterility. As a student of D. F. Jones at the Connecticut Agricultural Experiment Station (New Haven), I was able to observe the profound effect the paper by H. A. Jones and A. E. Clarke had on D. F. Jones's research. Both S and T cytoplasm of maize were studied and subsequently, D. F. Jones (Connecticut) and John Rogers (Texas) showed that the stability of male sterility in maize with T cytoplasm approached the stability of male sterility in onion. D. F. Jones demonstrated that pollen sterility in maize could be used to produce hybrid seed without detasseling and that genetic restoration should be used to produce male fertile $F_1$ hybrid seed. All of these concepts were influenced significantly by the research results in onion. In fact, D. F. Jones's data did not add anything novel to Jones and Clark's understanding of cytoplasmic–genic control of pollen sterility and fertility restoration.

A. E. Clarke, a cytologist who coauthored the classic paper with H. A. Jones was a minor contributor for this research effort. Clarke was convinced, on the basis of a rather rare gene that restored plants to a "partial" level of fertility, that pollen sterility was unstable. In contrast, Henry A. Jones, fully aware that another locus may be involved, forged ahead with a successful research program on inbred development and hybrid seed production.

A. E. Clarke was Canadian by birth (1903). He was educated at the University of Alberta (B.S. degree in 1924, M.S. degree in 1927). He received his Ph.D. degree from the University of Wisconsin in 1931. His career with the U.S. Department of Agriculture (USDA) spanned a period starting in 1936 until his death in 1952. As a cytologist he worked closely in the potato research program with F. J. Stevenson and in the onion research program with H. A. Jones. He was a quiet, methodical, competent, laboratory-oriented scientist.

Henry Jones also pioneered in the production of hybrid spinach using female phenotypes as seed parents of hybrid seed. Amid a swirl of controversy over the genetic control of sex (dioecism versus monoecism) in spinach by others, Henry Jones quietly developed a practical method by which male segregates were rogued from dioecious populations subjected to inbreeding. The "female" plants eventually produce pollen permitting the development of "female" inbred lines. This method has been eminently practical. Again, Henry Jones had reduced a complex phenomena to simplicity and manageability. The development of commercial $F_1$ hybrid spinach soon followed.

The examples stated above for both onion and spinach tell us much about Henry Jones as a scientist. His research was logical, precise, and published only after variables were understood. His concern for application of basic research is borne out in both the development of hybrid onions and hybrid spinach.

At the age of 43, Henry Jones was elected president of the American Society for Horticultural Science. The University of Nebraska bestowed on him an honorary D.S. in 1952. The following year the U.S. Department of Agriculture (USDA) awarded him the Distinguished Service Award. He also received other notable recognition, but I wonder what level of recognition might have been his if his research had been on a major food crop of the world.

After retiring from the USDA in 1957, Henry Jones joined the Desert Seed Co., El Centro, California, where for 22 years he enjoyed his first love—breeding hybrid onions. He died in 1981; his career as a plant breeder and horticulturist was one of the most distinguished in this century.

W. H. Gabelman
Department of Horticulture
University of Wisconsin
Madison, Wisconsin 53706

## REFERENCES

JONES, D. F., and P. C. MANGELSDORF. 1951. The production of hybrid corn seed without detasseling. Conn. Agr. Expt. Sta. Bull. 550.

JONES, H. A., and A. E. CLARKE. 1942. A natural amphidiploid from an onion species hybrid. J. Hered. 33:25–32.

JONES, H. A., and A. E. CLARKE. 1947. The story of hybrid onions. p. 320–326. In: Science and farming. Yearbook of agriculture. U.S. Department of Agriculture, Washington, D.C.

JONES, H. A., and S. L. EMSWELLER. 1931. The vegetable industry. McGraw-Hill, New York.

JONES, H. A. and S. L. EMSWELLER. 1936a. A male-sterile onion. Proc. Am. Soc. Hort. Sci. 34:582–585.

JONES, H. A. and S. L. EMSWELLER. 1936b. Development of the flower and macrogametophyte of *Allium cepa*. Hilgardia 10:415–428.

JONES, H. A., and L. K. MANN. 1963. Onions and their allies: Botany, cultivation, and utilization. Leonard Hill, London.

JONES, H. A., D. M. McLEAN, and B. A. PERRY. 1956. Breeding hybrid spinach resistant to mosaic and downy mildew. Proc. Am. Soc. Hort. Sci. 68:304–308.

JONES, H. A. and J. T. ROSA. 1928. Truck crop plants. McGraw-Hill, New York.

JONES, H. A., D. M. McLEAN, and B. A. PERRY. 1956. Breeding hybrid spinach resistant to mosaic and downy mildew. Proc. Am. Soc. Hort. Sci. 68:304–308.

MICHAELIS, P. 1933. Entwicklungsgeschichtlichgenetische Untersuchungen an Epilobium. II. Die Bedeutung des Plasmas für die Pollenfertilitat des *Epilobium luteum-hirsutum*-Bastardes. Z. Indukt. Abstamm. Vererbungsl. 65:1–71.

MICHAELIS, P., and M. V. DELLINGSHAUSEN. 1935. Entwicklungsgeschichtlichgenetische Untersuchungen an *Epilobium*. VII. Experimentelle Untersuchungen über die Beeinflussungen der Pollenfertilitat unter besonderer Berucksichtigung der Plasmawirkung. Jahrb. Wiss. Bot. 82:45–64.

RHOADES, M. M. 1931. Cytoplasmic inheritance of male sterility in *Zea mays*. Science 76:340–341.

RHOADES, M. M. 1933. The cytoplasmic inheritance of male sterility in *Zea mays*. J. Genet. 27:71–93.

# Inheritance of Male Sterility in the Onion and the Production of Hybrid Seed

By Henry A. Jones and Alfred E. Clarke, *U. S. Department of Agriculture, Beltsville, Md.*

THE utilization of hybrid vigor in plants had its inception and greatest stimulus in the production of hybrid corn. The use of hybrid seed to increase yields of squash, tomatoes, and muskmelons has been advocated by a number of investigators. Jones and Emsweller (7) suggested that hybrid onion seed might be produced on a commercial scale by using a male-sterile onion found in the variety Italian Red as the female parent. A knowledge of the genetics of the male-sterile character is desirable in order to determine the best breeding procedure for developing male-sterile lines. This paper describes the mode of inheritance of male sterility in the onion, which has been found to be partly cytoplasmic and partly nuclear in nature, and also discusses the manner in which this character may be utilized in a breeding program.

## Review of Literature

Male sterility may result from any one of several causes, including unfavorable environmental conditions, chromosomal irregularities during meiosis, or the presence of Mendelian genes for sterility. In a male-sterile line of maize Rhoades (10, 11) showed that the pollen sterility was inherited through the maternal cytoplasm.

East (4) found that certain $S$ factors from *Nicotiana sanderae* produced male sterility in the presence of cytoplasm derived from *N. langsdorfii* but produced viable pollen in the presence of *N. sanderae* cytoplasm.

In a cross between procumbent and tall flax Bateson and Gairdner (1) obtained the ratio 3 normal : 1 male-sterile in the $F_2$ generation. Male-steriles were obtained only when a plant of the procumbent strain was pollinated by a tall plant, not when a tall was pollinated by a procumbent. Later Chittenden and Pellew (3) and Chittenden (2) explained these results as the interaction of a Mendelian recessive in the cytoplasm of the procumbent strain.

Owen (9), working with the sugar beet, found that several pairs of Mendelian factors may influence pollen development when carried by plants with sterile cytoplasm, but the same factors have no effect when carried by plants with normal cytoplasm.

Jones and Davis (5) have found that by combining suitable inbred lines it is possible to obtain very productive onion hybrids. Consequently, if the seed could be produced at reasonable cost the utilization of these hybrids should be practicable on a commercial scale.

The male-sterile onion described by Jones and Emsweller (7) produced bulbils and has been perpetuated as a clonal line under the number 13–53. Monosmith (8) found no irregularities in chromosome behavior of 13–53 during the reduction divisions, but later the microspores degenerated, so that no viable pollen was produced.

189

## MATERIALS AND METHODS

To obtain data on the inheritance of male sterility crosses were made between the male-sterile line 13–53 and a large number of commercial varieties. While these male-sterile plants produce no viable pollen they set seed readily when hybridized with pollen from normal male-fertile plants.

To facilitate crossing flies were used as pollinating agents, as described by Jones and Emsweller (6). Data were obtained from $F_1$, $F_2$, $F_3$, first and second generation backcross populations grown in the greenhouses at the Bureau of Plant Industry Station, Beltsville, Maryland.

## EXPERIMENTAL RESULTS

When male-sterile plants of the line 13–53 were crossed with various male-fertile plants, three types of breeding behavior were observed in the $F_1$, some progenies being entirely male-fertile, others entirely male-sterile, whereas still others produced both male-sterile and male-fertile plants in a 1:1 ratio (Table I).

### TABLE I—INHERITANCE OF MALE STERILITY IN CROSSES BETWEEN MALE-STERILE AND NORMAL LINES OF ONIONS

| Parent or Cross | Generation | Parental Genotypes | Expected Ratio | | Number Plants Obtained | | Deviation/ Standard Error |
|---|---|---|---|---|---|---|---|
| | | | Male-Sterile | Male-Fertile | Male-Fertile | Male-Sterile | |
| Male-sterile 13-53 ×male-fertile | $F_1$ | $S\,ms\,ms \times N\,Ms\,Ms$ | 1 | 0 | 62 | 0 | —— |
| Male-sterile 13–53 ×male-fertile | $F_1$ | $S\,ms\,ms \times N\,ms\,ms$ | 0 | 1 | 0 | 225 | —— |
| Male-sterile 13-53 ×male-fertile | $F_1$ | $S\,ms\,ms \times N\,Ms\,ms$ | 1 | 1 | 98 | 86 | Less than 1 |
| $F_1$ male-fertile | $F_2$ | $S\,Ms\,ms$ selfed | 3 | 1 | 375 | 155 | 2.3 |
| $F_2$ male-fertile | $F_3$ | $S\,Ms\,ms$ selfed | 3 | 1 | 96 | 41 | 1.3 |
| Male-sterile 13-53 ×$F_1$ male-fertile | Backcross selfed | $S\,Ms\,ms$ selfed | 3 | 1 | 98 | 49 | 2.4 |
| $F_1$ male-sterile ×male-fertile | Backcross selfed | $S\,Ms\,ms$ selfed | 3 | 1 | 104 | 24 | 1.6 |
| $F_1$ male-sterile ×male-fertile | Backcross | $S\,ms\,ms \times N\,Ms\,Ms$ | 1 | 0 | 20 | 0 | —— |
| $F_1$ male-sterile ×male-fertile | Backcross | $S\,ms\,ms \times N\,ms\,ms$ | 0 | 1 | 0 | 458 | —— |
| $F_1$ male-sterile ×male-fertile | Backcross | $S\,ms\,ms \times N\,Ms\,ms$ | 1 | 1 | 157 | 163 | 0.3 |
| Male-sterile 13-53 ×$F_1$ male-fertile | Backcross | $S\,ms\,ms \times S\,Ms\,ms$ | 1 | 1 | 200 | 246 | 2.2 |
| $F_1$ male-fertile ×male-fertile | Backcross | $S\,Ms\,ms \times N\,ms\,ms$ | 1 | 1 | 33 | 32 | Less than 1 |
| Male-fertile ×$F_1$ male-fertile | Backcross | $N\,ms\,ms \times S\,Ms\,ms$ | 1 | 0 | 53 | 0 | —— |
| Male-sterile ×male-fertile | Second backcross | $S\,ms\,ms \times N\,ms\,ms$ | 0 | 1 | 0 | 275 | —— |
| Male-sterile ×male-fertile | Second backcross | $S\,ms\,ms \times N\,Ms\,ms$ | 1 | 1 | 125 | 142 | 1.0 |

When a self-fertile $F_1$ plant is selfed, the $F_2$ approximates the ratio 3 normal:1 male-sterile. The total for all of the $F_2$ populations is shown in Table I. It will be noted that the fit to a 3:1 ratio is rather poor, the deviation being 2.3 times the standard error. An examination of the table will show that there is a slight excess of male-steriles in almost every case where the populations are assumed to be segregating for 3:1 or 1:1 ratios. Since high temperatures in the greenhouse during the flowering period reduced the percentage of viable pollen produced by male-fertile plants, it is probable that occasionally a heterozygous plant, while genotypically male-fertile, was classified as male-sterile.

Segregating $F_3$ and selfed backcross populations were also found to approximate 3 : 1 ratios.

When a male-sterile $F_1$ is backcrossed to the male-fertile parent three types of segregation are obtained, as in the $F_1$ (Table I). When the 13–53 male-sterile parent is backcrossed with an $F_1$ male-fertile plant, a 1 : 1 segregation is obtained.

In crosses between certain $F_1$ male-fertile plants used as the female parent and the male-fertile parent line a ratio of 1 male-fertile to 1 male-sterile is obtained. But in the reciprocal backcross, when the $F_1$ male-fertile plant is used as the pollen parent, all of the progeny are male-fertile (Table I).

Some second backcrosses gave a 1 : 1 segregation; other second backcrosses gave all male-steriles (Table I).

## DISCUSSION

### EXPLANATION OF EXPERIMENTAL RESULTS

These results may be accounted for by assuming that the male-sterile condition results from an interaction between a recessive nuclear gene and a non-nuclear or cytoplasmic factor. On this hypothesis it is assumed that there are two types of cytoplasm. All plants with normal cytoplasm (designated for convenience as N) produce viable pollen. All male-sterile plants possess the sterile type of cytoplasm ($S$). The experimental results throw no light on the nature of the non-nuclear or cytoplasmic factor which differs in the two types. A gene for male sterility ($ms$) also influences pollen development when carried by plants with $S$ cytoplasm but has no effect when carried by plants with $N$ cytoplasm. Consequently, the 13–53 male-sterile plants belong to the genotype $S\ ms\ ms$. Plants with $N$ cytoplasm are male-fertile always and may belong to the genotypes $N\ Ms\ Ms$, $N\ Ms\ ms$, or $N\ ms\ ms$, since the ms gene has no effect in the $N$ cytoplasm. Apparently there is no natural selection for or against the recessive ms gene, when it is present in $N$ cytoplasm. Plants with the genetic constitution $S\ Ms\ Ms$ and $S\ Ms\ ms$ will also be male-fertile, in spite of the $S$ cytoplasm, because they carry the dominant gene $Ms$.

The non-nuclear or cytoplasmic factor is inherited only through the egg (maternal transmission) and not through the male parent, presumably owing to the very small amount of cytoplasm present in the male gamete. From the cross $S\ ms\ ms$ x $N\ Ms\ Ms$ all $F_1$ plants will be $S\ Ms\ ms$ and, in spite of the $S$ cytoplasm, are male-fertile because they carry the gene $Ms$. $S\ ms\ ms$ x $N\ ms\ ms$ gives all male-sterile, and $S\ ms\ ms$ x $N\ Ms\ ms$ gives 1 male-sterile : 1 male-fertile. When a self-fertile $F_1$ plant ($S\ Ms\ ms$) is selfed, the expected $F_2$ ratio is 3 normal : 1 male-sterile.

All male-sterile $F_1$ plants belong to the genotype $S\ ms\ ms$. $S\ ms\ ms$ x $N\ Ms\ Ms$ produces all male-fertile. $S\ ms\ ms$ x $N\ ms\ ms$ produces all male-sterile and $S$ $ms$ $ms$ x $N$ $Ms$ $ms$ produces 1 male-fertile : 1 male-sterile. When the 13–53 male-sterile parent is backcrossed with an $F_1$ male-fertile plant, $S$ $ms$ $ms$ x $S$ $Ms$ $ms$, a 1 : 1 segregation is expected.

When an $F_1$ male-fertile plant, $S$ $Ms$ $ms$, is used as the female parent and backcrossed to $N$ $ms$ $ms$ a ratio of 1 male-fertile to 1 male-sterile is obtained. But in the reciprocal backcross $N$ $ms$ $ms$ x $S$ $Ms$ $ms$ all of the progeny are male-fertile, since all carry $N$ cytoplasm. The unlike behavior of these reciprocal backcrosses is additional evidence in support of the validity of this hypothesis.

Second backcrosses of the type $S$ $ms$ $ms$ x $N$ $Ms$ $ms$ gave a 1 : 1 segregation. Second backcrosses of the type $S$ $ms$ $ms$ x $N$ $ms$ $ms$ gave all male steriles. This confirms the expectation that 100 per cent male-sterile progenies can be obtained in repeated backcrosses to a stock with the genetic constitution $N$ $ms$ $ms$. As will be shown later this is of great practical importance in developing a breeding program.

### APPLICATION OF GENETIC RESULTS TO A BREEDING PROGRAM

The production of hybrid onion seed of all types and in quantity is now possible. To perpetuate the pure male-sterile line two lines must be carried along: a male-sterile line of the genotype S ms ms and a fertile line of the genotype N ms ms. All of the progeny of this cross will be male-sterile. Fig. 1 illustrates the method of perpetuating the male-sterile line through the seed.

As the male-sterile plants cannot be selfed, seed is secured by continually backcrossing to the normal or male-fertile line. Backcrossing continues as long as the particular male-sterile line is to be perpetuated. After a few backcrossings the male-sterile line should be practically identical with the male-fertile except for the sterility factor of the cytoplasm. This backcross seed makes it possible to perpetuate the male-sterile line, as well as produce the male-sterile female parents used in the production of hybrid seed.

The next step is to make crosses between the male-sterile line and other selected lines to determine which combination produces the best commercial hybrid (Fig. 2). The constitution of the male parent that enters into the cross for the production of commercial hybrid seed may be $N$ $ms$ $ms$, $N$ $Ms$ $ms$, or $N$ $Ms$ $Ms$, the particular one selected being based on progeny tests. The behavior of the commercial hybrids as to fertility is not important, because the commercial crop must be grown from hybrid seed each year. It is important, however, to get a favorable combination of growth factors.

As stated above, the cytoplasmic factor for male sterility (S) was found in the Italian Red 13–53 variety. No doubt this character may also be present in other varieties. The genotype $N$ $ms$ $ms$ that produces all male-sterile plants when crossed with $S$ $ms$ $ms$ has been found in Brigham Yellow Globe, Early Yellow Globe, Sweet Spanish, Southport White Globe, Crystal Wax, Creole, and Stockton G 36, so it is possible to get male-sterile cytoplasm into these varieties by continuous backcrossing. As many as three backcrosses have been made to a number of them. Male-sterile lines could be produced in an *N Ms Ms* variety by alternating backcrossing and selfing.

The best method of interplanting male-sterile and male-fertile lines in the field to secure the highest return of hybrid seed has not been worked out as yet.

The type of breeding program just outlined should be applicable to any crop plant in which male sterility is inherited in this way, provided that the plant, like the onion, is naturally cross-pollinated and displays hybrid vigor in the $F_1$.

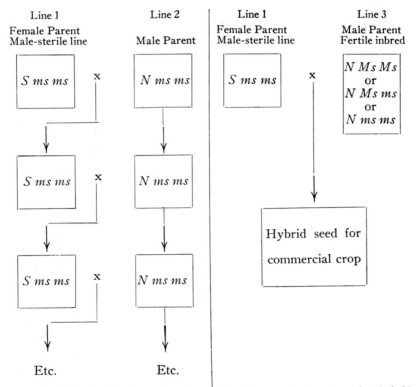

FIG. 1. Method of perpetuating the male-sterile lines.

FIG. 2. Method of perpetuating hybrid seed for the commercial crop.

LITERATURE CITED

1. BATESON, W., and GAIRDNER, A. E.  Male sterility in flax, subject to two types of segregation.  *Jour. Genetics* 11: 269–275.  1921.
2. CHITTENDEN, R. J.  Cytoplasmic inheritance in flax.  *Jour. Hered.* 18: 336–343.  1927.
3. ————— and PELLEW, C.  A suggested interpretation of certain cases of anisogeny.  *Nature* 119: 10–12.  1927.
4. EAST, E. M.  Studies on self-sterility. IX. The behaviour of crosses between self-sterile and self-fertile plants.  *Genetics* 17: 175–202.  1932.

5. JONES, H. A., and DAVIS, G. N.   Inbreeding, heterosis, and the development of new varieties of onions.   (In press).

6. ———— and EMSWELLER, S. L.   The use of flies as onion pollinators. *Proc. Amer. Soc. Hort. Sci.* 31 : 160–164.   1934.

7. ———— A male-sterile onion.   *Proc. Amer. Soc. Hort. Sci.* 34 : 582–585. 1937.

8. MONOSMITH, H.   Male Sterility in *Allium cepa* L.   Thesis, library, University of California.

9. OWEN, F. V.   Male sterility in sugar beets produced by complementary effects of cytoplasmic and Mendelian inheritance.   *Amer. Jour. Bot.* 29 : 692.   (Abstract.)   1942.

10. RHOADES, M.   Cytoplasmic inheritance of male sterility in *Zea mays. Science* 73 : 340–341.   1931.

11. ———— The cytoplasmic inheritance of male steriilty in *Zea mays. Jour. Genetics* 27 : 71–93.   1933.

H. Kihara. 1951. Triploid watermelons. Proceedings of the American Society for Horticultural Science 58:217–230.

H. Kihara

The seedless triploid hybrid watermelon represented a novel horticultural creation when it first appeared. Many research pathways led to its development, including the polyploidization effects of colchicine, hybrid breeding, and studies on the physiology of parthenocarpy—the phenomenon of seedless fruit. That the seedless watermelon became a reality is due to the vision of Hitoshi Kihara, a world-renowned Japanese plant geneticist and breeder. The story of its creation and its subsequent publication in English is a bit of horticultural history that I am in a unique position to relate.

Three independent steps are required to produce seedless watermelons. The first is to induce tetraploidy, a doubling of the chromosome number; the second is to cross the tetraploid with a normal diploid to produce triploid seed; and the third step is to pollinate the triploid with pollen of a diploid to induce parthenocarpy. Successful coordination of these steps to produce a seedless watermelon represents a fine application of the breeder's art and the horticulturist's skill.

The use of colchicine to induce tetraploidy represents a special niche in the history of genetics. *Colchicum autumnale,* the autumn crocus, has been known since antiquity (it is probably referred to in the Ebers papyrus, 1550 B.C.) as a plant with both medicinal and poisonous properties. It has been specifically prescribed as a cure for gout and, indeed, is still used for this purpose. The active ingredient, now known as colchicine, has unique action on mitosis. As early as 1889, B. Pernice in Palermo, Sicily published evidence that colchicine causes mitotic arrests in epithelial and endothelial cells in the stomach and intestines of dogs. A number of independent studies in the 1930s demonstrated that colchicine had remarkable effects on mitosis, and in February 1937 I saw from my experiments that colchicine was very specific in creating polyploid restitution nuclei in *Allium* root tip cells, opening a new era for the induction of polyploidy in plants.

The induction of parthenocarpy by auxin application was a landmark discovery made by the late Felix G. Gustafson at the University of Michigan (1936), but his attempts to produce seedlessness in watermelons was unsuccessful with indolebutyric acid treatment. Various studies with pollen extracts expanded our knowledge of parthenocarpy and subsequent attempts were made to induce parthenocarpy in watermelon in both the United States and Japan. C. Y. Wong, a visiting Chinese scholar working at Michigan State College, induced parthenocarpy in watermelons with auxin application (NAA). During his studies, Wong included colchicine in his experimental materials and obtained completely seedless fruit on a colchicine-treated plant treated with NAA. A photograph of this fruit appears in his 1941 paper published in the *Botanical Gazette*, but Wong was probably unaware that the plant was a tetraploid. His studies were evidently discontinued.

At about the same time, two Japanese scientists, J. Terada and K. Masuda of Kyoto University, independently studied the induction of seedlessness in watermelon with growth regulators. Following their 1938 paper, Kihara, also at Kyoto, conceived the concept of exploiting triploid sterility to produce seedlessness. He initiated experiments to induce tetraploidy by colchicine in 1939, and his first paper was published in Japanese with an English summary in 1947.

Kihara reported on these studies at the 8th International Genetics Congress in Stockholm in 1948. On his way back to Japan from Sweden he stopped at my laboratory at Northwestern University and provided photographs, diagrams, and data which were later included in my book *Colchicine* (Eigsti and Dustin 1955). I suggested that he write up his work in English. A copy of his manuscript arrived in the summer of 1950 and was taken in person to Henry Munger, then editor of the *Proceedings of the American Society for Horticultural Science*. I learned that a copy of the manuscript had already been received and rejected! I urged reconsideration. To cover publication costs, a number of seed companies were solicited, and $15 was requested from each to make up the $75 required. The seed companies who contributed are listed in a footnote in the original paper. The merits of the paper were quite obvious and it won the Vaughan award in 1952 for the outstanding paper in vegetable crops provided by the Vaughan Seed Company. Ironically, the Vaughan Seed Company had been solicited but declined to contribute to the publication fund.

The paper received worldwide attention. Many horticulturists attempted to repeat creation of seedless watermelons but failures were many and successes few. Although seedless watermelons never became very popular because of high seed costs and problems in germination, they are offered in seed catalogs worldwide. In the United States about 4000 pounds of triploid seed (Tri X hybrids) were sold in 1986 by the American Seedless Watermelon Seed Corporation, a company I established to further this technology. Seedless watermelons are now popular in Taiwan and Israel, and I believe we are entering an era of expanded usage with more improved hybrid combinations.

Kihara died in 1986 at the age of 92, still active in genetic and cytogenetic investigations. Kihara had a brilliant career in science. He was on the staff of Kyoto University from 1920 to 1956 and spent two years (1925–1927) with Carl Correns, one of the "rediscoverers" of Mendelism. He later directed the institute that bears his name, the Kihara Institute of Biological Research, and was a leader of an expedition that led to the identification of *Aegilops squarrosa* as an ancestor of the bread wheats. An expert skier, he became president of the Ski Association of Japan and led the Japanese team at two Winter Olympics. His classic paper on seedless watermelons is an ingenious example of a genetic solution to a practical problem and remains a remarkable achievement in horticultural science.

O. J. Eigsti
American Seedless Watermelon Seed Corp.
Goshen, Indiana 46526

## REFERENCES

EIGSTI, O. J. 1938. A cytological study of colchicine effects in the induction of polyploidy in plants. Proc. Natl. Acad. Sci. (USA) 24:56–63.

EIGSTI, O. J., and P. DUSTIN, JR. 1955. Colchicine—in agriculture, medicine, biology and chemistry. Iowa State College Press, Ames, Iowa.

GUSTAFSON, F. G. 1936. Induction of fruit development by growth promoting chemicals. Proc. Natl. Acad. Sci. (USA) 22:628–636.

KIHARA, H. 1979. Seedless fruits. Seiken Jiho 27–28:37–44. Report of the Kihara Institute for Biological Research, Yokohama, Japan.

KIHARA, H., et al. 1947. An application of sterility of autotriploids to the breeding of seedless watermelons (in Japanese with English summary). Seiken Jiho 3.

TERADA, J., and K. MASUDA. 1938. Seedless watermelons by means of artificially induced parthenocarpy (in Japanese). Kyoto Engei 36.

WONG, C. Y. 1941. Chemically induced parthenocarpy in certain horticultural plants with special reference to the watermelon. Bot. Gaz. 103:64–86.

YASUDA, S., et al. 1930. Parthenocarpy caused by the stimulation of pollination in some plants of Solanaceae. Agr. Hort. 5:287–294.

# Triploid Watermelons[1]

By H. Kihara, *Kyoto University, Japan*

There are numerous cases of edible fruits, completely or practically seedless, i.e. banana, pineapple, citrus, and others. These may be classified under most conditions into three groups, on the basis of causes for failures in seed production: (a) functionless ovules and pollen grains, (b) fertilization failures caused by internal or external condition even though both male and female gametes are functional, (c) functional ovules but non viable pollen.

Illustrative of the first group is the banana which is triploid and highly sterile, hence seedless.

Pineapple varieties produce abundant seeds when cross pollination takes place, but owing to self-incompatibility seeds are not produced under usual growing conditions. This type characterizes the second group.

Satsuma orange of the third group has functional ovules that produce seeds by artificial or open pollination if good pollen of the correct citrus species is within reach; however the anthers are shrivelled and no pollen grains are shed when plants are cultivated under conditions of the Japanese climate.

Generally, pollination has at least two important effects: 1. The entrance of pollen tubes into the embryo sacs, resulting in the union of the male with female gametes. 2. The supply of growth hormones which, while stimulating the ovaries to further development, induces the formation of fruits.

Sufficient amounts of growth hormones apparently are present in the ovaries to initiate ovulary growth without pollination among the plants mentioned above that regularly produce seedless fruits. Fruit formation can be artificially induced by treatment with growth hormones. In other words, growth hormones supplied by pollen grains for the development of the ovaries and formation of fruits can be substituted by chemicals acting as such. This was demonstrated in the watermelon by Terada and Masuda (9).

Consequently, if female sterility could be achieved, seedless fruits should be obtained even though normal pollen grains are used for pollination. This reasoning led the author to experiments designed to make the production of triploid watermelons in which fruit for the most part would be seedless or parthenocarpic.

Accordingly, this report covers the collective investigations relative to seedless watermelon carried out in various parts of Japan since 1939.

## Production of Triploids

The common watermelon, *Citrullus vulgare,* Shrad, is diploid, with somatic (2*x*) and gametic chromosome (*x*) numbers of 22 and 11,

[1] Financial assistance in publishing this paper was generously provided by the following seed companies: W. Atlee Burpee Co., Philadelphia, Pa., Joseph Harris Co., Rochester, N. Y., Earl E. May Co., Shenandoah, Iowa, and Lawrence Robinson and Sons, Modesto, Calif.

Received for publication April 20, 1950.

217

respectively. After tetraploids are at hand, triploids are produced by crossing tetraploids with diploids ($x = 22$ x $x = 11$). Young diploid seedlings are treated with an aqueous solution of colchicine (0.2 or 0.4 per cent, applied daily for four successive days. One drop of colchicine solution, each day, on the growing point of young seedlings, constitutes a successful schedule to produce tetraploids.

Pollen grain and stomatal sizes are good criteria for the identification of tetraploid shoots, but the offspring from such branches are by no means all tetraploid. As a matter of fact, Kihara and Nishiyama found the size and shape of the seeds in the first generation after treatment more helpful in distinguishing the tetraploid from the diploid seeds (2).

Once tetraploids are obtained, they breed true. Thereafter, triploids can be produced season after season by crossing tetraploids ($x = 22$) with diploids ($x = 11$). However, only those crosses using the diploid pollinator and tetraploid seed parent produce viable triploid seed, the reciprocal cross, ($x = 11$ x $x = 22$) is unsuccessful as illustrated in Fig. 1.

### MORPHOLOGY OF $2x$, $3x$, AND $4x$ PLANTS

The leaves of $3x$ and $4x$ individuals, are thicker and darker green than the $2x$ plants. Accordingly surface hairs of tetraploid and triploid, are coarser and longer, than corresponding diploids. This characteristic feature is most distinct in $4x$ plants.

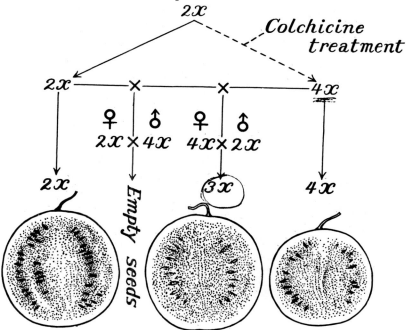

FIG. 1. Procedure of formation of triploids.

The size of flowers of 2*x*, 3*x* and 4*x* individuals appears to increase in proportion to the chromosome number. Stomatal guard cells, and pollen grains (Fig. 6) are larger in 4*x* (Fig. 6C) than in 2*x* plants (Fig. 6A) as usual for a polyploid series (see measurements in Table 1.). However, the triploids produce pollen grains that are very irregular, being mostly empty, and generally the pollen tetrads do not separate (Fig. 6B).

*Table 1.* Diameter of pollen grains of 2*x*, 3*x* and 4*x* plants.

| Chromosome numbers | Mean[a] |
|---|---|
| 2*x* | 19.1 ± 0.15 |
| 3*x* (smaller grains) | 20.7 ± 0.01 |
| 3*x* (larger grains) | 33.5 ± 0.48 |
| 4*x* | 22.5 ± 0.16 |

[a]1 unit = 3 micra    (Suemoto 1948)

In the first stage of development the growth of 2*x* plants at an early stage is most rapid, while 3*x* plants are the slowest, with 4*x* plants intermediate. However, after about one month and a half, the triploids rapidly overtake both the tetraploid and the diploids. The tetraploids show the longest period of growth, persisting until the end of August, even when the 2*x* plants are fully grown at this time (Kihara 1951).

## Karyological Observations

Cytological studies (4, 6, 11) reporting, $x = 11$ chromosomes, for watermelon were confirmed for five Japanese varieties (Yamato, Asahi-Yamato, Shin-Yamato, Kurobe Snake and Kanro) (Fig. 2).

Diploids present no unusual divisions of the PMC's, whereas tetraploids usually form tetravalent chromosomes at the first meiotic division, i.e. $11_{IV}$, as in Figs. 3A and 3B which show $11_{IV}$ and $10_{IV} + 2_{II}$, respectively. Tetravalents tend to separate into 4 homologues at anaphase which, as a rule, are evenly distributed to each pole. Fig. 3C confirms such distribution, since 22 chromosomes are easily detected in each group of the anaphase plate.

Among triploids, the maximum trivalent conjugation, $11_{III}$, obtains frequently (Fig. 4A). Occasionally 10 trivalents, 1 bivalent and 1 univalent are found (Fig. 4B). Disjunction to the poles on a 2 and 1 basis of the trivalent is at random. Consequently, random disjunction and segregation of the trivalents sister plates, in 17 and 16, or 18 and 15, are most usually observed. Fig. 4C shows the side view of an anaphase in which a few chromosomes are omitted, as they were out of focus. Abnormalities in the first division are shown in Fig. 4D, where one lagging chromosome can be seen.

Tetrads are formed among diploids after the completion of the second division. In the triploids, however, the formation of the cell wall is sometimes disturbed (Fig. 4E and 4F), thus leading to the occurrence of quadripartite and giant pollen grains (Fig. 5). Comparative photo micrographs of pollen grains of 2*x*, 3*x* and 4*x* individuals are shown in Fig. 6A, B, C, respectively.

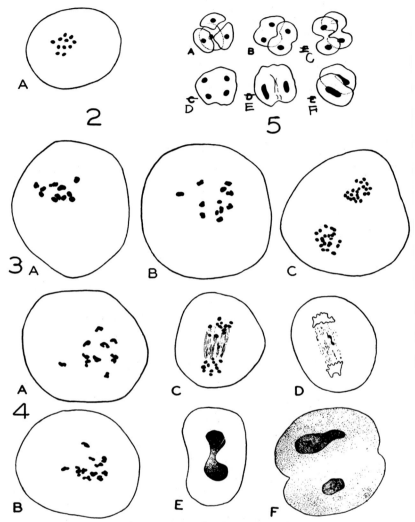

Fig. 2. Chromosomes of water-melon. A PMC with $11_{II}$.

Fig. 3. Chromosomes of tetraploids. A. $11_{IV}$. B. $10_{IV}$. + $2_{II}$. C. anaphase showing 22 chromosomes in the daughter plates.

Fig. 4. Chromosomes of triploids. A. $11_{III}$. B. $10_{III}$ + $1_{II}$ + $1_{I}$. C. Anaphase showing normal segregation of trivalents. Some chromosomes are omitted in this side view. D. one chromosome is lagging in the telophase. E. a rare case of restitution nucleus. F. interkinesis.

Fig. 5. Young microspores of triploids after the second meiotic division. A. normal pollen-tetrad. B. C. and D. abnormalities showing incomplete cell wall formation. E. and F. microspores with unreduced number of chromosomes corresponding to Fig. 5, 3, B. C. and D will result in the formation of giant pollen grains.

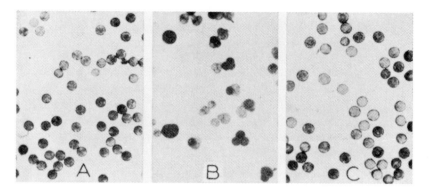

FIG. 6. Pollen grains of $2x$, $3x$ and $4x$. A. $2x$ B. $3x$ C. $4x$.

### CHARACTERISTIC FEATURES HELPFUL IN THE IDENTIFICATION OF THE TRIPLOIDS

*Seeds:*—Distinct morphological differences exist between the $2x$ and the $4x$ seeds. For example, seeds of Yamato, a cross-bred variety from the Nara Experiment Station, are oblong, whereas the auto-tetraploids vary from oblong to almost round. Tetraploid seeds in addition to increased thickness and weight have coarser appearance than diploid seeds. Moreover, the seed-coat of tetraploids has one, two, or more distinct fissures, clearly distinguishable along the longitudinal axis (Fig. 8).

Seeds obtained from a cross $4x \times 2x$, carry triploid embryos and for convenience's sake, shall be called the "$3x$ seeds". Their morphological characters are almost the same as the $4x$ seeds, except for their slightly thinner dimension, as shown by the frequency curve in Fig. 9. This remarkable feature has a great practical value in the sorting out of "$3x$-" from "$4x$-" seeds in cases where $4x$-plants were left to open pollination and pollen from tetraploids and diploids was available in the same field.

According to Nishida (1951) the seeds produced by $4x$ individuals, which contain either $3x$ or $4x$ embryos, have a considerably thicker seed-coat than corresponding diploid seeds. A single layer of enlarged and elongated palisade cells forms the epidermis hence the fissures on seed coats are caused by the "cracking up" of the dried palisade layer (Fig. 8).

In longitudinal section of seeds one may note that where $2x$ seeds show a completely filled cavity (Fig. 10A), the $3x$ (Fig. 10B, C) and $4x$ (Fig. 10 D) embryos are filling the space by only 82.5 and 90.1 per cent, respectively. This weak development of the embryo may account for poor germination encountered among some of the $3x$ seeds (Table 4).

*Seedlings:*—Morphological differences between the $2x$ and $4x$ seedlings may be mentioned, notably, increased size and thickness and deeper green color of the $4x$ cotyledons. Moreover, abnormal seedlings

Fig. 7. Fruits of 2x, 3x, and 4x. fruits of Shin-Yamato (2x), Yamato (4x) and their hybrid (3x). dark green stripe of Shin-Yamato is dominant. longitudinal sections of the above.

occur sometimes in 3x and 4x offspring. This abnormality is conspicuous only early in development following germination, and does not appear to be associated with aneuploidy or an extra chromosome. Abnormal seedlings, in some instances, emerge from the seed-coat with great difficulty.

Seedlings raised from self-pollinated tetraploids gave about 8 per cent abnormals, while approximately 30.6 per cent 3x seedlings from the cross 4x × 2x were abnormal (Table 2). Among open pollination of the tetraploids, 4x and 3x offspring are to be expected, because the 4x pistils are pollinated by pollen grains from 4x or 2x plants.[2]

In such cases of open pollination, the seeds thus obtained gave rise to 37.6 per cent abnormal seedlings (Table 2) in one case, or 27.1 per cent (Table 3) in another instance. It was found that about one half of these abnormal seedlings were triploid. On the other hand, the 4x seedlings were mostly normal.

It is interesting to note the high viability among abnormal seedlings, the percentage being 70.4 per cent (Table 4).

---

[2] 2x and 4x pollen grains means pollen grains from 2x and 4x plants, respectively.

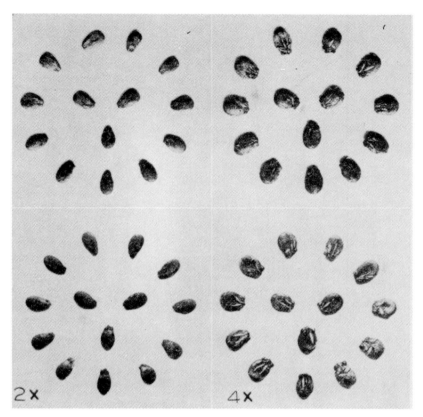

FIG. 8. Seeds of watermelons (left, $2x$; right, $4x$). A. Yamato.
B. Shin-Yamato.

*Pollen Grains:*—Another distinctive feature of $3x$ individuals is the appearance of pollen grains, which are mostly shrivelled, empty or may be of giant proportion abnormally quadripartite. This latter feature is a conspicuous character and distinguishes the triploids from the diploids and tetraploids (Fig. 6B).

*Genetic Markers:*—To distinguish the triploid fruits from $4x$ and $2x$ fruits, genetic markers are useful. The simplest one, is a certain gene, controlling parallel dark green stripes (Fig. 7A). Therefore triploids are marked like the pollinator parent and tetraploids without this pollinator gene are not striped (Fig. 7).

## FRUITS OF TRIPLOIDS AND TETRAPLOIDS

Self- and open-pollinations were studied in relation to fruit setting in the tetraploid and diploid. Table 5 shows that tetraploids may have

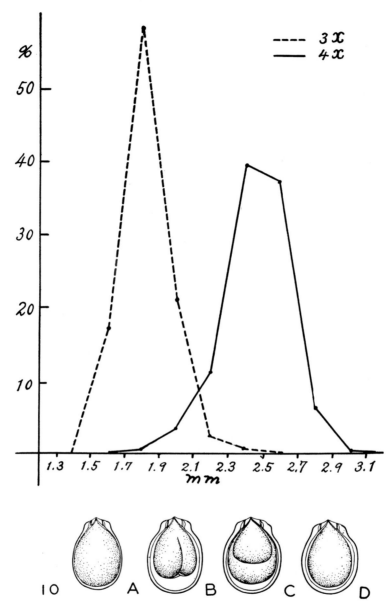

FIG. 9. (above) Frequency curves showing the difference in the thickness of 3x- and 4x-seeds.

FIG. 10. (below) Longitudinal sections of 2x (A), 3x (B, C), and 4x (D) seeds.

*Table 2.* Normal and abnormal seedlings among *3x* and *4x* individuals.

| Exp. stations[a] | | Normal | Abnormal number | Per cent | |
|---|---|---|---|---|---|
| *4x × 3x* | | | | | |
| Gifu (1949).......................... | *3x* | 84 | 54 | 39.1 | |
| Nara (1949).......................... | *3x* | 404 | 142 | 26.0 | 30.6 |
| T. A. C. (1949)..................... | *3x* | 101 | 64 | 38.7 | |
| *4x × 4x* | | | | | |
| Gifu (1949).......................... | *4x* | 109 | 14 | 11.4 | |
| T. A. C. (1949)..................... | *4x* | 97 | 4 | 3.9 | 8.0 |
| *4x (Open-pollinated)* | | | | | |
| Gifu (1949).......................... | *4x + 3x* | 89 | 26 | 22.6 | |
| K. I. B. (1947)..................... | *4x + 3x* | 65 | 58 | 47.1 | |
| K. I. B. (1948)..................... | *4x + 3x* | 124 | 71 | 36.4 | 37.6 |
| T. A. C. (1949)..................... | *4x + 3x* | 402 | 101 | 20.1 | |

[a]Abbreviations. Gifu = Agricultural Experiment Station of Gifu Prefecture.
Nara = Agr. Exp. Sta. of Nara Prefecture.
T. A. C. = Tokyo Agr. College.
K. I. B. = Kihara Institute for Biological Research.

*Table 3.* Chromosome numbers and seedling-types in the offspring obtained by open pollination of *4x* plants.

| | Chromo-some number | Normal | Abnormal | |
|---|---|---|---|---|
| | | | Number | Per cent |
| K. I. B........................... | *4x* | 81 | 5 | 5.8 |
| | *3x* | 38 | 45 | 55.0 |
| T. A. C........................... | *4x* | 247 | 19 | 7.1 |
| | *3x* | 66 | 52 | 44.0 |
| Total........................ | | 432 | 161 | 27.1 |

*Table 4.* Viability of normal and abnormal seedlings grown from seeds obtained from open pollination of *4x* plants.

| Number of seeds sown | Number of germinated seeds | Number of seedlings | Chromosome number | | Viability (per cent) |
|---|---|---|---|---|---|
| | | | *4x* | *3x* | |
| 1081 | 772 (71.3 per cent) | Normal 600 | 324 | 185 | 84.8 |
| | | Abnormal 169 | 23 | 97 | 70.4 |

(T. A. C. 1949)

*Table 5.* Fruit set of *2x*, *3x* and *4x* plants.

| | Mean percentage | Variation percentage |
|---|---|---|
| *2x* (14 varieties)............................................. | 26.0 | 21–30 |
| *3x* (3 combinations)......................................... | 61.5 | 50–68 |
| *4x* (22 varieties)............................................. | 28.5 | 5–60 |

(Gifu 1949)

a better fruit set than the diploids. However, in other experiments where selfing was practiced the tetraploids had the poorest fruit set. The amount of fruits set by the tetraploids shows a wide seasonal variation from 14.4 per cent early in the season to 3.4 per cent at a later date (Table 6). Fruit set becomes somewhat higher if tetraploids are pollinated by diploids as compared with selfing. These observations

*Table 6.* Fruit set in 4x plants.

| Date of pollination | Selfing | | | | Intervarietal mating (× 2x) | | | |
|---|---|---|---|---|---|---|---|---|
| | Pistils | Fruits set | | | Pistils | Fruits set | | |
| | | No. | Per cent | | | No. | Per cent | |
| June 14–July 15.......................... | 187 | 27 | 14.4 | | 13 | 12 | 92.3 | |
| July 16–Aug. 4........................... | 269 | 9 | 3.4 | | 136 | 34 | 25.0 | |
| August................................ | 0 | 0 | — | | 75 | 8 | 10.7 | |
| Totals........................... | 456 | 36 | 7.9 | | 224 | 54 | 24.1 | |

(T. A. C. 1949)

along with higher sterility of tetraploids, indicates difficulties in preservation of the 4x strains.

The relationship between the setting of fruits and type of pollen used, was studied in triploids, showing that pollen of diploid plants is superior to tetraploids. An experiment with a mixture of pollen from diploid and tetraploid plants showed that the former is favored over the latter.

*Fruit Shape:*—The fruit shape of most Japanese watermelon varieties is slightly elongated spherical. The tetraploid fruits are pronouncedly spherical and smaller. But good care given to the tetraploids grown under optimum conditions gives fruits not smaller than the corresponding diploids.

The triploid fruits have shape and size similar to the diploid fruits. However, a cross section of triploid fruit may be triangular or tetrangular according to the number of placentas. This feature is common in fruits obtained by artificially induced parthenocarpy.

*Inner Parts of the Fruits:*—The rind of 4x fruits is somewhat thicker than corresponding 2x fruits (Fig. 7 B), but no difference in the texture of flesh is observed, even though tetraploid fruits among certain other species is known to be coarser. Triploids compare favorably with diploids for texture (Fig. 7 B).

The sugar content, by hand refractometer, shows that triploid fruits are not inferior in sweetness to 2x or 4x fruits.

*Sterility and Empty Seeds:*—Triploids are highly sterile and produced viable seed from 3x x 2x crosses only on rare occasions. These gave diploids, no plants with intermediate chromosome numbers being observed. So far, only one triploid plant has been found from open pollination of 3x-plants. As a rule, the 3x plants have no true seeds but they contain small, white, more or less rudimentary seeds which can be eaten like unripe cucumber seeds. But, sometimes, empty seeds, that is seeds of normal size and colour, are found in triploid fruits. Their number varies widely, ranging from a few to over 100 per fruit.

Though the percentage of 3x fruits with many coloured, empty seeds does not exceed 10 per cent throughout the season, their presence, nevertheless, renders the name "seedless" more or less inadequate. Usually fruits with abundant empty seeds are found among the first ones. Then, they seem to disappear all of a sudden, to reappear again, sporadically, as the season advances. This was one of the reasons for hesitation in our laboratory before distributing the seeds of the triploid water-melon. However, as the production was increased they were entrusted to experiment stations and ordinary farmers for experimental cultivation. As a result, since 1949, a small quantity of triploid fruits have made their first appearance in the market. The public opinion has not been unanimously favorable, as, in general, the expectations for complete seedlessness have been too high.

The number of empty seeds is closely correlated with the varieties used, and varies also according to the combination of varieties which take part in the production of the triploids. Moreover, the same combination sometimes showed varying results in different localities. For instance, according to the Nara Experiment Station, many empty seeds were found when Miyako was used as one of the parents; however, similar combinations gave good fruits in Kanto (a district near Tokyo).

In order to elucidate the external agents affecting the size and colour of empty seeds, many investigations have been carried out. Furusato et al. (1951) could not find any difference in the amount of empty seeds following the use of $n$ and $2n$ pollen. The combination of the $2x$- and $4x$-strains used in the production of the triploids played the more important role. Furusato (1951) believes that grafting of $2x$, $3x$ and $4x$ scions on *Lagenaria* is beneficial as far as fruit set and yield are concerned. But he can not tell yet, how the development of empty seeds is influenced by the grafting.

According to Koyama (1951), parthenocarpic fruits induced by plant hormones had fewer empty seeds, coloured and non-coloured, than those obtained by pollination with $x$-pollen. The hormone-induced parthenocarpic $3x$-fruits had the drawback of being conspicuously smaller in size. But they increased in size, when later treated pistils were pollinated by normal pollen grains. The size was inversely proportional to the interval of time between the hormone treatment and pollination.

The tetraploids are highly sterile and contain many undeveloped white seeds. The number of good seeds per fruit varies roughly from 50 to 100, sometimes more.

The number of seeds per fruit in $2x$, $3x$ and $4x$ plants of the same variety and triploid hybrids between two varieties can be seen from Table 8. Figure 7 shows the fruits of $2x$ and $4x$ plants together with that of their $3x$ hybrid.

*Yield:*—It can be said that, in general, the yield of the triploids is high. However, owing to the combination of the parental varieties or, possibly, to a delayed sowing, the development of the triploids has been sometimes retarded and negative results, though rare, have been recorded in a few instances.

In comparing the yield of various $2x$ varieties and $3x$ strains, Mr. Kondo of T. A. C. found that the yield of $3x$ was better in most cases, namely 1.3–1.97 times higher than that of the diploid parents. This difference in favour of $3x$ was most conspicuous when Kaho — a variety with small fruits (1.7 kg. per fruit) but high fruit set — was used as the diploid parent. The results obtained by the Gifu Experiment Station are summarized in Table 7.

*Table 7.* The yield of triploids as compared with their diploid
and tetraploid parents.

| Varieties | Yield per unit area[a] | | Comparison in weight | |
|---|---|---|---|---|
| | No. of fruits | Weight (kg) | | |
| Shin-Yamato 2x........................... | 165 | 352.1 | 138 | 100 |
| Shin-Yamato 4x........................... | 115• | 155.6 | 61 | —— |
| ╳ - - - - - - 3x........................... | 215• | 596.2 | 235 | 169 |
| Otome    2x........................... | 150 | 253.7 | 100 | —— |

(Gifu 1949)

[a]20 *tsubo* (1 *tsubo* is approximately 6 × 6 ft.).

*Comparison between the triploids and their parents:*—Preliminary
investigations have been undertaken in order to compare the triploids
with their parents. The results obtained, though tentative, are shown
in Table 8. There are indications that the triploids mature more quickly than their respective diploid and tetraploid parents. However, according to an expert in watermelon culture in Nara Prefecture, the
maturation of the first fruit of the 3x plants is about a week later than
in the diploids.

*Table 8.* Comparison[a] between 3x-plants and their parents with regard
to various characteristic features.

| Varieties (or hybrids) | | Date of first flow-ering ♀ | Date of harvest (1st fruit) | Days to maturity | Thickness of rind (cm) | No. of seeds | Sugar content (per cent) | Fruit set (per cent) |
|---|---|---|---|---|---|---|---|---|
| Shin-Yamato | 2x | July 2 | Aug. 28 | 32 | 1.3 | 290.0 | 9.0 | 26 |
| | 3x | July 2 | Aug. 23 | 30 | 1.5 | 1.0 | 11.0 | 67 |
| | 4x | July 13 | Aug. 7 | 36 | 1.8 | 92.7 | 8.3 | 21 |
| Asahi | 2x | July 12 | Aug. 15 | 33 | 1.1 | 292.0 | 10.0 | 26 |
| | 3x | July 1 | Aug. 11 | 30 | 1.4 | 2.0 | 11.5 | 50 |
| | 4x | July 18 | Aug. 26 | 31 | 1.1 | 53.6 | 7.4 | 41 |
| Otome | 2x | July 2 | Aug. 5 | 26 | 1.1 | 223.0 | 8.0 | 28 |
| ╳ - - - - - | 3x | July 6 | Aug. 7 | 27 | 1.3 | 1.5 | 12.2 | 68 |
| Shin-Yamato | 4x | July 13 | Aug. 23 | 36 | 1.8 | 92.7 | 8.3 | 21 |

(Gifu 1949)

[a]Mean or mode obtained from 10 individuals cultivated in a unit area (20 *tsubo*).

Worthy of notice is the fact that tetraploid as well as triploid plants
seem to be more resistant to wilt (*Fusarium niveum* Smith), one of
the most devastating diseases of watermelon in Japan. On account of
this disease farmers can cultivate watermelon in the same field only
once in eight years.

## WHY IS THE TRIPLOID WATER-MELON SEEDLESS?

The most important factor responsible for the sterility of the triploid water-melon and the seedlessness of its fruits is the sterility of
the gametes due to the incompleteness of their chromosome comple-
ments. In general, viable, normal gametes are produced only when
they possess one or more complete chromosome sets (genomes). As

each of the 11 trivalents in the first maturation division of the triploid plants divides into three chromosomes, one of which goes to one pole, while the remaining two move to the opposite pole, the segregation occurs independently. As a consequence, gametes with 11–22 chromosomes are formed. Their frequency can be calculated from the expansion of $(1 + 1)^{11}$.

|  |  | Gametes with |  |
|---|---|---|---|
| Fertile | $\left\{\begin{array}{l} 1/(2)^{11} \\ 1/(2)^{11} \end{array}\right.$ | 11 chromosomes $\left.\begin{array}{l} \\ \end{array}\right\}$ 22 chromosomes | 0.1% |
| Sterile | $1 - 1/(2)^{10}$ | Intermediate chromosome numbers | 99.9% |

On the basis of the above calculation practically complete seedlessness should be expected.

Thus, parthenocarpy in triploid water-melon is induced by growth hormones provided by insect-carried n pollen grains. When pollinated by such pollen, the pistil develops quite normally and fruit formation ensues. If pollination is prevented by bagging the flowers, the ovaries do not develop.

## A Few Suggestions Concerning the Cultivation of Triploid Watermelon

Although the cultivation of the triploid water-melon does not differ fundamentally from that of the diploid, there are a few important points the grower has to bear in mind, concerning germination and fruit setting. The following hints are for the use of Japanese farmers and may be found helpful outside Japan.

1. The "3x" seeds originating from tetraploid ovularies and ovules have a hard thick seed coat which interfers with germination. Consequently, to facilitate germination, part of the seed-coat is removed, exercising care to prevent injury to embryo. Later experiments showed that sometimes germination between treated and untreated seeds was comparable.

The soil in the seed-bed and the seeds, should be disinfected. Sterilizing soil for two or three days, prior to use, either with chemicals or by steam is desirable.

2. The seed-bed should be kept at a temperature of about 30 degrees C and at an optimum degree of humidity. When the seeds begin to germinate they should be transplanted into pots 1 or 2 inches wide and placed in a greenhouse. The soil in the pots should be made up of straw manure and turf grass, and covered with glass or cellophane to protect from direct sunshine. However, after a week or ten days, the young plants need all available sunshine.

During the seedling stage the plants should not be overwatered. When the seedlings develop two or three leaves, and when they are nicely rooted in the small pots, permanent transplantation may be made.

3. In order to insure a sufficient supply of pollen, it is advisable to intermingle triploid with diploid plants. For this purpose a proportion

of 1 diploid to 4–5 triploids will suffice. This procedure permits a confusion of diploid with triploid fruits unless satisfactory diploid genetically marked varieties are used, that differ from triploids (See Fig. 7A).

Pollination by insects in large scale cultivation is satisfactory. However, artificial pollination is recommended if possible.

## General Considerations

Though there are still many difficulties to overcome, the breeding of the triploid water-melon seems to be very promising. At present, improvements should be attempted along the following lines:

a. Elimination of empty seeds, especially in the earliest fruits.

b. Improving upon the irregular fruit shape.

c. Maintenance of $4x$ strains.

d. Search for combinations most efficient in promoting heterosis and, if possible, combine it with the least possible production of empty seeds.

## Literature Cited

1. Kanda, T.  How to grow water-melon.  (Japanese. Publ. Takii Seeds Co.) 1950.
2. Kihara, H., and Nishiyama, I.  An application of sterility of autotriploids to the breeding of seedless watermelons.  *Seiken Ziho* No. 3.  (Japanese with English résumé).  1947.
3. Kondo, N.  On the triploid (seedless) watermelons.  *Agri. and Hort.* 5: No. 2.  (Japanese).  1950.
4. Kozhukhow, S. A.  Karyotypishce Eigentümlighkeiten der kultivierten *Cucurbitaceae*.  *Bul. Appl. Bot. and Pl. Breed.* 14, No. 4.  1925.
5. Matsui, K.  The culture of seedless watermelons.  (Japanese).  1950.
6. Passmore, S. F.  Microsporogenesis in the *Cucurbitaceae*.  *Bot. Gaz.* 90. 1930.
7. Takasugi, K.  Breeding of watermelons.  Report of the Gifu Experiment Station.  (Japanese).  1949.
8. Suemoto, H.  Investigations on polyploid watermelons.  (A thesis, unpublished).  1949.
9. Terada, J., and Masuda, K.  Seedless watermelon by means of artificially induced parthenocarpy.  *Kyoto Engei* 36.  1938.  (Japanese).
10. ———— Parthenocarpy of triploid watermelons.  *Agri. and Hort.* No. 18.  (Japanese).  1943.
11. Whitaker, T. V.  Chromosome numbers in cultivated cucurbits.  *Amer. Jour. Bot.* 17: 1033–1040.  1930.
12. Yanagisawa, S., Hosono, M., and Yasui, A.  Breeding of watermelons. Report of the Nara Experiment Station.  1949.  (Japanese).
Note. Papers mentioned in text and not given above will be published in Seiken Ziho.  No. 5 (1951).

2

Peter S. Carlson, Harold H. Smith, and Rosemarie D. Dearing. 1972. Parasexual interspecific plant hybridization. Proceedings of the National Academy of Sciences. 69:2292–2294.

Peter S. Carlson, Rosemarie D. Dearing, and Harold H. Smith

In the late 1960s and early 1970s, there was great interest in the potential application of cellular genetic approaches to modify plants. Haploid plants were first produced in 1964, and the genetic implications of this technique were widely recognized. Several European laboratories expressed interest in publications and at scientific meetings on the prospects of creating new hybrid plants by protoplast fusion. This early work culminated with the publication by Takebe and coworkers (1971) of plant regeneration from cultured protoplasts.

Despite the well-publicized effort in several European laboratories to obtain somatic hybrid plants, the first recovery of plants from protoplast fusion was in the United States by Carlson and coworkers in 1972. The beauty of this classic paper is that it presents in simple eloquent fashion the recovery and characterization of the first somatic hybrid plants. This paper reported a process that to some degree was limited to the two species used in this fusion combination, and hence to some degree may have been guilty of oversimplification. However, the paper stimulated a great deal of interest in the application of protoplast fusion to crop improvement. Moreover, this paper helped initiate an era in which many corporations and universities have directed research at applying tools such as protoplast fusion for crop improvement.

This paper was the result of a collaboration among Peter S. Carlson, Harold H. Smith, and Rosemarie D. Dearing, when Peter was an associate biologist and Rosemarie was research technician at Brookhaven

National Laboratory in 1971. Harold Smith had published several papers on genetics of the genus *Nicotiana* (see the review in Smith 1968). Although Smith had some exposure to plant cell culture, he had only a few coauthored papers in this area during his scientific career. However, he was the world authority on genetics and interspecific hybridization in *Nicotiana*. Smith had authored numerous papers on genetic tumors in interspecific hybrids. Occurrence of spontaneous tumors was first reported by Kostoff (1930). These tumors were similar to, but generally more differentiated than, the crown gall tumors caused by *Agrobacterium*. Of more than 300 different interspecies hybrids in *Nicotiana*, only 34 produced genetic tumors spontaneously. One interspecies hybrid (GGLL) derived from *N. glauca* (GG) × *N. langsdorffii* (LL), with prolific tumor growth, was characterized in great detail by a series of studies in the 1950s and 1960s. It had been determined by a series of crosses that tumor formation is under genetic control. In all genome combinations tested, tumors developed as long as at least one genome of each species (e.g., GLLL or GGGL) was present. Smith also completed a great deal of work to determine if tumor formation was the result of somatic mutation or developmental regulation. It was demonstrated that exposure to X-irradiation at dosage levels that resulted in a 100-fold increase in mutation frequency failed to increase tumor initiation (Smith 1957). Hence early studies had demonstrated that tumor formation, common with this interspecies hybrid, was under genetic control and developmental regulation.

Braun (1958) demonstrated that crown gall (*Agrobacterium*) tumors are capable of rapid growth on a minimal culture medium, whereas normal cells require auxin and cytokinin for growth. Schaeffer and Smith (1963) demonstrated the same response for genetic tumors. Whereas cells of the parent species, *N. glauca* or *N. langsdorffii*, require auxin and cytokinin for growth, the interspecies allotetraploid sexual hybrid (GGLL) is capable of profuse growth in the absence of hormones. This difference in response to growth regulators permitted the development of the cell selection system used for isolation and recovery of somatic hybrid plants in this paper.

Peter Carlson received his B.A. degree from Kenyon College in 1966 and his Ph.D. degree from Yale University in 1971. After brief stints as an assistant professor at Wesleyan University and a visiting scientist at the Max-Planck Institute in Cologne, Germany, Peter spent 1971–1974 as associate biologist at Brookhaven National Laboratory. In his early career, he developed a strong background in genetics and has even published papers on *Drosophila* genetics (e.g., Carlson 1971). Carlson worked for several years on developing and demonstrating the feasibility of somatic cell genetics in plants. This paper on protoplast fusion was one of a series of papers that also included mutant isolation, mitotic crossing over, and haploid selection in plants. Each paper received a great deal of discussion in scientific meetings in the period of 1972–1976, and most important was instrumental in getting new, young scientists (including myself) attracted to this area of research.

The experiment reported in this paper was simple and straightforward. Protoplasts were isolated from leaves of *N. glauca* and *N. langsdorffii* using a standard mixture of commercial cellulase and macerozyme. Mixed populations of protoplasts in 1 : 1 ratio were suspended in NaNO$_3$

to stimulate fusion, since this work was done prior to the discovery of PEG as a fusogen. Callus pieces, when visible, were transferred to culture medium containing no growth regulators. As only fusion products are capable of growth in hormone-free media, all regenerated shoots were presumed to be somatic hybrids. The remainder of the publication is aimed at characterizing the regenerated plants. As the sexual hybrid between these two species had been identified, characterization of the somatic hybrid was unambiguous. Hybrids were distinguished based on leaf shape, trichome density, tumor formation, chromosome number, and flower and seed capsule morphology.

There was a great deal of criticism leveled at this publication in scientific meetings. In one case this criticism surfaced in a publication (Melchers and Labib 1974). In this paper Melchers reported the production on intraspecific hybrids of tobacco, essentially a step backward in gene transfer from the Carlson paper. Melchers seemed weak on technical criticisms and strong on emotional criticisms, as his primary complaint seemed to rest on a remark by Carlson that the somatic hybrids were formed by "good fortune." However, it was clear that the primary reason for Carlson's success was the use of a very tight selection system to distinguish the somatic hybrids from their parents.

This paper opened the door for a large number of publications in protoplast fusion from laboratories throughout the world. It is a testimony to the uniqueness of this work that the next interspecific somatic hybrids were not reported for four years (Power et al. 1976). The key technical development that opened the door for large-scale fusion was the discovery by Kao et al. (1974) that polyethylene glycol (PEG) could be used to induce protoplast agglutination and fusion at high frequency. In fact, Smith et al. (1976) reconfirmed the original experiment using PEG. In this reconfirmation, somatic hybrid plants of *N. glauca* + *N. langsdorffii* were obtained in high frequency, thereby laying to rest all the earlier criticisms. A completely independent confirmation was published two years later from a French laboratory (Chupeau et al. 1978). In each case fertile somatic hybrid plants were obtained. Several reviews of progress in somatic hybridization have been published (see Evans, 1984).

Peter Carlson continued to work on a wide range of applications of plant cell genetics, both at Brookhaven and in his subsequent position as distinguished professor at Michigan State University. In the years between 1970 and 1978, he published several papers in this area of research. His overall goal was to demonstrate that plants were amenable to the same type of genetic analysis as fungi. He isolated and characterized the first auxotrophic mutants in higher plants (Carlson 1970), developed a method to select for haploid cells versus diploid cells (Gupta and Carlson 1972), demonstrated uptake of chloroplasts into protoplasts (Carlson 1973a), isolated amino acid analog-resistant mutants (Carlson 1973b), and demonstrated mitotic crossing over in higher plants (Carlson 1974). As an extension of the work on mitotic crossing over, he developed methods to selectively isolate genetically altered cells *in situ*. By examining a leaf after mutagen treatment, he could detect, then rescue, genetically altered cells. By using this approach, he isolated virus-resistant plants (Carlson and Murakishi 1978) and herbicide-resistant plants (Radin and Carlson 1978). By 1980 he had succeeded in founding a new science of plant somatic cell genetics.

Peter approached his work with unparalleled enthusiasm. At many lectures in the 1970s, he kept his audience completely spellbound. By combining the attributes of good scientific training and childlike enthusiasm, he had the unique ability to reduce complicated scientific principles to simple scientific experiments. In 1981, Peter become founder of a company with the aim of commercializing the opportunities in plant biotechnology. Crop Genetics International (Hanover, Maryland) has prospered and grown.

The application of these techniques, and in particular protoplast fusion, in horticulture has not yet been fully realized. Although ornamentals of *Petunia* and *Nicotiana* have been synthesized, as well as new breeding lines of tomato, potato, carrot, and other horticultural crops, no new cultivars have yet been produced by this technique. However, the enthusiasm for this area of research in ornamental horticulture is perhaps best exemplified by the establishment of the Eric E. Young Somatic Hybrid Orchid Prize, in which a reward of $50,000 has been offered for the first successful somatic hybrid orchid plant (Fiske 1982). As we look forward into the next decade, it is clear that the leadership provided by this classic paper will lead to many new developments in horticulture.

David A. Evans
DNA Plant Technology Corporation
2611 Branch Pike
Cinnaminson, New Jersey 08077

## REFERENCES

BRAUN, A. C. 1958. A physiological basis for autonomous growth of the crown-gall tumor cell. Proc. Natl. Acad. Sci. (USA) 44:344–349.

CARLSON, P. S. 1970. Induction and isolation of auxotrophic mutants in somatic cell cultures of *Nicotiana tabacum*. Science 168:487–489.

CARLSON, P. S. 1971. A genetic analysis of the rudimentary locus of *Drosophila melanogaster*. Genet. Res. Camb. 17:53–81.

CARLSON, P. S. 1973a. The use of protoplasts for genetic research. Proc. Natl. Acad. Sci. (USA) 70:598–602.

CARLSON, P. S. 1973b. Methionine sulfoximine-resistance mutants of tobacco. Science 180:1366–1368.

CARLSON, P. S. 1974. Mitotic crossing over in a higher plant. Genet. Res. Camb. 24:109–112.

CARLSON, P. S., and H. H. MURAKISHI. 1978. Evidence on the clonal versus nonclonal origin of dark-green islands in virus infected tobacco leaves. Plant Sci. Lett. 13:377–381.

CHUPEAU, Y., C. MISSONIER, M. C. HOMMEL, and J. GOUJAUD. 1978. Somatic hybrids of plants by fusion of protoplasts. Observations on the model system "*Nicotiana glauca–Nicotiana langsdorffii*." Molec. Gen. Genet. 165:239–245.

EVANS, D. A. 1984. Protoplast fusion. p. 291–321. In: D. A. Evans et al. (eds.). Handbook of plant cell culture, Vol. 1. Macmillan, New York.

FISKE, M. D. 1982. The Eric E. Young somatic hybrid orchid prize. Am. Orchid Soc. Bull. 51:43–46.

GUPTA, P. K., and P. S. CARLSON. 1972. Preferential growth of haploid plant cells in vitro. Nature (London) New Biol. 239:86.

KAO, K. N., F. CONSTABEL, M. R. MICHAYLUK, and O. L. GAMBORG. 1974. Plant protoplast fusion and growth of intergeneric hybrid cells. Planta 120:215–227.

KOSTOFF, D. 1930. Tumors and other malformations on certain *Nicotiana* hybrids. Zentralbl. Bakteriol. Parasitenkd. Infektionskr. Hyg. Abt. 81:244–260.

MELCHERS, G., and G. LABIB. 1974. Somatic hybridization of plants by fusion of protoplasts I. Selection of light resistant hybrids of "haploid" light sensitive varieties of tobacco. Molec. Gen. Genet. 135:277–294.

POWER, J. B., C. HAYWARD, D. GEORGE, P. K. EVANS, S. F. BERRY, and E. C. COCKING. 1976. Somatic hybridization of *Petunia hybrida* and *P. parodii*. Nature 263:500–502.

RADIN, D. M., and P. S. CARLSON. 1978. Herbicide-tolerant tobacco mutants selected *in situ* and recovered via regeneraton from cell culture. Genet. Res. Camb. 32:85–89.

SCHAEFFER, G. W., and H. H. SMITH. 1963. Auxin–kinetin interaction in tissue cultures of *Nicotiana* species and tumor-conditioned hybrids. Plant Physiol. 38:291–297.

SMITH, H. H. 1957. Genetic plant tumors in *Nicotiana*. Ann. New York Acad. Sci. 71:1163–1178.

SMITH, H. H. 1968. Recent cytogenetic studies in the genus *Nicotiana*. Adv. Genet. 14:1–54.

SMITH, H. H., K. N. KAO, and N. C. COMBATTI. 1976. Interspecific hybridization by protoplast fusion in *Nicotiana*. J. Heredity 67:123–128.

TAKEBE, I., G. LABIB, and G. MELCHERS. 1971.Regeneration of whole plants from isolated mesophyll protoplasts of tobacco. Naturwissenschaften 58:318–321.

Reprinted from

*Proc. Nat. Acad. Sci. USA*
Vol. 69, No. 8, pp. 2292–2294, August 1972

# Parasexual Interspecific Plant Hybridization

## (*Nicotiana*/leaf mesophyll/plant tissue culture/genetics/selective media)

PETER S. CARLSON, HAROLD H. SMITH, AND ROSEMARIE D. DEARING

Department of Biology, Brookhaven National Laboratory, Upton, New York 11973

*Communicated by A. M. Srb June 13, 1972*

**ABSTRACT**     Interspecific plant hybrids have been produced by parasexual procedures. Protoplasts of *Nicotiana glauca* and *N. langsdorffii* were isolated, fused, and induced to regenerate into plants. The somatic hybrids were recovered from a mixed population of parental and fused protoplasts by a selective screening method that relies on differential growth of the hybrid on defined culture media. The biochemical and morphological characteristics of the somatically produced hybrid were identical to those of the sexually produced amphiploid.

Recent advances in plant cell culture have demonstrated that protoplasts isolated from leaf mesophyll cells can be induced to regenerate into entire plants (1, 2), and that they can be stimulated to fuse by defined experimental manipulations (3). A combination of these two techniques should permit the fusion of protoplasts isolated from two different species and the regeneration of a somatically produced hybrid plant without having to involve a normal sexual cycle. This paper reports the successful parasexual production of a hybrid between two different species of *Nicotiana*, *N. glauca* Grah. and *N. langsdorffii* Weinm. The amphiploid hybrid between these two species has been produced by sexual means, and the characteristics of the hybrid plant have been thoroughly studied (4–7). Known biological differences between the hybrid and its parental species have been used in a selective screen to recover preferentially regenerated fused hybrid protoplasts from a mixed population of protoplasts. We have also used the distinctive characteristics of the hybrid tissue to verify that parasexual hybridization was achieved.

## MATERIALS AND METHODS

The species used were *Nicotiana glauca* ($2n = 24$), *N. langsdorffii* ($2n = 18$), and the amphiploid ($2n = 42$) of the tumorous hybrid of these species.

Protoplasts were isolated from leaf mesophyll cells by stripping the lower epidermis from sterilized, young, expanding leaves. Stripped leaf pieces were placed in an enzyme solution consisting of 4% cellulase (Onozuka SS, All Japan Biochemicals Co. Ltd.), 0.4% macreozyme (All Japan Biochemicals Co. Ltd.), and 0.6 M of sucrose at pH 5.7. Flasks containing the leaf pieces in the enzyme solution were evacuated briefly, then returned to standard atmospheric pressure to facilitate penetration of the enzyme solution into the intercellular spaces. These flasks were incubated for 4–6 hr at 37°, after which the protoplasts were harvested by low-speed centrifugation ($<100 \times g$).

Experimental conditions and regeneration medium used for protoplast culture were exactly those described by Nagata and Takebe (2). Protoplast density was always greater than $5 \times 10^3$ protoplasts per ml. In the Nagata and Takebe medium, proroplasts of *N. glauca* and *N. langsdorffii* will re-

generate a cell wall and occasionally go through one division cycle. Protoplasts of these two species were never observed to regenerate into a callus. Protoplasts of the amphiploid hybrid react similarly; however, about 0.01% of the protoplasts will continue to divide and give rise to a callus mass of cells. The different growth characteristics of protoplasts from the two parental species and from the amphiploid hybrid on Nagata and Takebe medium constitutes a selection method with which to recover preferentially hybrid individuals from a mixed population of protoplasts. Since protoplasts from both parental species are unable to regenerate on the Nagata and Takebe medium, the only protoplasts capable of forming viable colonies will be those with a hybrid genetic constitution.

Mixed populations of protoplasts of *N. glauca* and *N. langsdorffii* in an approximate 1:1 ratio were stimulated to fuse by their suspension in 0.25 M of $NaNO_3$ for 30 min, and then pelleted by low-speed centrifugation. This pellet was then resuspended in the regeneration medium and plated in petri dishes. After the fusion procedure, the population consisted of protoplasts of both parental types and fused clumps of protoplasts involving various numbers of cells. About 25% of the protoplasts were involved in a fusion event. On the regeneration medium, only the cells containing the genetic information of both parental species were able to regenerate into calli. Regenerated calli were removed from the regeneration medium and placed on the medium of Linsmaier and Skoog (8), which was solidified with agar and contained no hormones. This constitutes a further selective step, since tissue from neither parental species is able to grow on a medium without added hormones, while the amphiploid hybrid grows vigorously without exogenous hormones present (6).

Recovered calli formed rudimentary shoots and leaves in culture, but failed to form roots. In order to obtain further differentiation of presumed hybrid tissue, the regenerated shoots were grafted onto the freshly cut stem surface of young plants of *N. glauca*. The grafts were wrapped with parafilm and were kept under high moisture conditions in the mist propagation section of a greenhouse until the graft had taken and a few leaves had developed.

The chromosomes were prepared from young leaves by the method of Burns (9). Electrophoresis and staining for peroxidase isozymes was as described by Smith *et al.* (10, 11).

## RESULTS

More than $10^7$ protoplasts of *N. glauca* and $10^7$ protoplasts of *N. langsdorffii* were taken through the fusion procedure and plated on a regeneration medium that permits the growth of only cells containing the genetic information of both parental species. 33 Regenerated calli were recovered after 6 weeks, and placed on a medium containing no added hor-

FIG. 1. Leaves typical of (*left* to *right*): *N. glauca*, amphiploid *N. glauca* × *N. langsdorffii*, somatic hybrid *N. glauca–N. langsdorffii*, and *N. langsdorffii*.

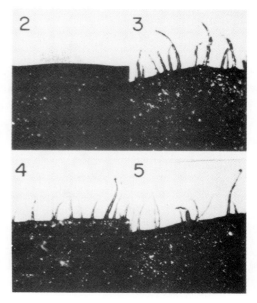

FIGS. 2–5. Glabrous leaf surface of *N. glauca* (2), dense trichomes on *N. langsdorffii* leaf (3), intermediate trichrome condition on leaves of amphiploid (4), and somatic hybrid (5).

mones. All 33 isolates grew vigorously with no exogenous hormone source. This observation provided circumstantial evidence that the recovered calli had a hybrid genetic composition. Several of the recovered calli that were presumed parasexual hybrids were chosen for further tests to confirm their hybrid genetic composition.

The following analysis was completed with three isolates. Photographs and figures are presented for only one, although the described characteristics were common to all three.

*Morphology of the Tissue in Culture.* The characteristic tissue morphology and growth requirements of the somatically produced hybrid are identical to those of the sexually produced hybrid. Tissues from both sources grow vigorously in culture in the absence of exogenous hormones. On an agar medium containing no hormones, both tissues form a semifriable callus. In a liquid medium containing no hormones, both tissues will regenerate shoots and leaves. Tissue from either parental species is not capable of growth and differentiation in medium lacking added hormones.

*Morphology of the Leaf.* The morphology of leaves regenerated on the somatically produced hybrid is identical to that of the sexually produced hybrid, and distinct from either parental type (Fig. 1). The leaves of *N. langsdorffii* are sessile, those of *N. glauca* are petiolate, and the hybrid has a leaf of intermediate morphology (12).

*Trichome Characteristics.* The leaves of *N. langsdorffii* are densely covered with trichomes, while leaves of *N. glauca* are glabrous without trichomes (12). On leaves of both the somatically and sexually produced hybrid trichomes are present, but in a much lower density (Figs. 2–5).

*Tumor Formation.* The somatically produced hybrid spontaneously forms tumorous outgrowths on the stem (Fig. 6). Spontaneous tumor formation is a genetically determined trait that is characteristic of the F₁ hybrid and amphiploid, but is not found in either parent species, and is not transmitted across a graft union (7).

*Chromosome Numbers.* A somatic chromosome number of 42 (Figs. 7 and 8) was determined for the somatically produced hybrid. This is a summation of the diploid somatic numbers of the parental species (24 + 18), and is distinct from a whole ploidy change in either parental type. The sexually produced amphiploid has been shown to contain a chromosome number of 42 (12). Although the somatically produced hybrids all

FIG. 6. Tumor formation on scion of somatic hybrid, *N. glauca–N. lagsdorffii*, grafted onto stock of *N. glauca*.

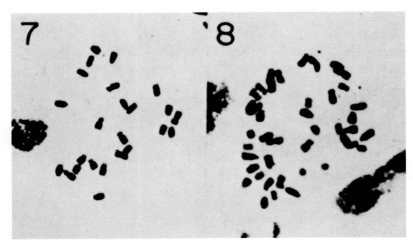

Figs. 7–8.    Metaphase chromosomes in cell of young leaf of *N. glauca* 2n = 24 (Fig. 7), and of somatic hybrid (Fig. 8) with additional 18 *N. langsdorffii* chromosomes, giving a total of 42 chromosomes.

demonstrated a chromosome number of 42, deviations from simple addition of the parental chromosome numbers might be expected to occur due to the complexity of the fusion event and divisions after fusion.

*Peroxidase Isozymes.* The leaf peroxidase isozymes in the somatically produced hybrid are identical to those of the amphiploid. The isozyme bands of the hybrid are a summation of those found in the parental species.

The characteristics of the somatic hybrid are not due to a chimerial association of cells. Single cells derived from calli of the somatically produced hybrid were regenerated into calli, and their characteristics were analyzed. In every case, the regenerated callus displayed characteristics of hybrid tissue, and was distinct from either parantal type. Hence, the characteristics of the somatically produced hybrid are not due to a chimerial association of cells of the parental species. All cells of the somatically produced hybrid contained only one nucleus. The possibility that the somatically produced hybrid is due to contamination by sexually produced amphiploid cells is ruled out by the experimental procedure used.

Summation of this evidence leaves no doubt that the calli and plants recovered from fused cells are of a hybrid genetic constitution corresponding to the sexually produced amphiploid.

## DISCUSSION

Each of the individual steps in the procedure of protoplast isolation, fusion, and regeneration has, as noted above, already been performed. The successful recovery and analysis of a parasexually produced hybrid, as reported here, has depended primarily on the availability of a selective technique to permit preferential recovery of fused hybrid cells, and recognition of known distinctive characteristics of the hybrid amphiploid. Further attempts to produce a somatic interspecific hybrid and hybrids between more distantly related species in our laboratory have been hampered by a lack of familiarity with the kind of characteristics the tissue will display. Preliminary attempts to recover preferentially intra- and interspecific and intergeneric hybrids with the available auxotrophic mutants of *N. tabacum* (13) have been incon-

clusive. Many of the auxotrophic protoplasts will grow at a reduced rate in minimal medium when mixed with other auxotrophic mutants or with wild-type protoplasts of other species, presumably due to the effect of crossfeeding between the cell types. We are investigating one further method for preferentially recovering parasexually produced hybrids. Protoplasts containing potentially complementing recessive nuclear albino mutations from different species are isolated and fused. Only calli that have regenerated from fused cells should appear as green colonies on the petri plate.

In general, the potential offered by somatic hybridization may be expected to exceed the limitations imposed by sexual processes, and extend the possibilities of combining widely divergent genotypes of plants.

### Note added in proof

The somatic hybird has produced flowers and fertile seed capsules that are identical with the *N. glauca* × *N. langsdorffii* amphiploid.

This work was performed at Brookhaven National Laboratory, which operates under the auspices of the U.S. Atomic Energy Commission. The research was supported in part by USPHS Grant GM 18537 to P.S.C. We thank Prof. Ian Sussex and Dr. Gerhard Wenzel for their helpful suggestions, and Mr. Nicholas Combatti for cytological determinations.

1.  Takebe, I., Labib, G. & Melchers, G. (1971) *Naturwissenschaften* 6, 318–321.
2.  Nagata, T. & Takebe, I. (1971) *Planta* 99, 12–20.
3.  Power, J. B., Cummings, S. E. & Cocking, E. C. (1970) *Nature* 225, 1016–1018.
4.  Kostoff, D. (1939) *J. Genet.* 37, 129–209.
5.  White, P. R. (1939) *Amer. J. Bot.* 26, 59–64.
6.  Schaeffer, G. W. & Smith, H. H. (1963) *Plant Physiol.* 38, 291–297.
7.  Smith, H. H. (1972) *Prog. Exp. Tumor Res.* 15, 138–164.
8.  Linsmaier, E. M. & Skoog, F. (1965) *Physiol. Plant.* 18, 100–127.
9.  Burns, J. A. (1964) *Tobacco Sci.* 8, 22–23.
10. Bhatia, C. R. Buiatti, M. & Smith, H. H. (1967) *Amer. J. Bot.* 54, 1237–1241.
11. Smith, H. H., Hamill, D. E., Weaver, E. A. & Thompson, K. H. (1970) *J. Hered.* 61, 203–212.
12. Goodspeed, T. H. (1954) *The Genus Nicotiana* (Chromica Botanica, Waltham, Mass.).
13. Carlson, P. S. (1970) *Science* 168, 487–489.

Robert B. Horsch, Robert T. Fraley, Stephen G. Rogers, Patricia R. Sanders, Alan Lloyd, and Nancy Hoffmann. 1984. Inheritance of functional foreign genes in plants. Science 223:496–498.

Robert B. Horsch

Robert T. Fraley

Stephen G. Rogers

Patricia R. Sanders

Alan Lloyd

Nancy Hoffmann

Conventional plant breeding is based on the identification of useful characteristics that are heritable and the incorporation of these traits into a suitable genetic background to produce improved cultivars. Superior individuals, lines, or populations are produced by exploiting the sexual process to recombine favorable genes. The results of such breeding programs have been, in many cases, spectacular. There are, however, limitations to this traditional approach, including the inability of the sexual system to incorporate variation from nonrelated species, the reliance on naturally occurring or randomly induced mutations, the inability to obtain specific genetic changes, and finally, the difficulty in detecting infrequent recombinants. Furthermore, the availability of genetic variation does not guarantee rapid progress in developing an improved cultivar, as the selection and propagation of superior recombinants can be a lengthy process.

The application of new methods of biotechnology may overcome some of the limitations of conventional plant breeding. These techniques include cell culture, cell selection, protoplast fusion, and recombinant DNA. Recombinant DNA technology, although dependent on cell culture techniques, is particularly powerful because it allows genes that are responsible for a particular phenotype to be identified, modified, and transferred between nonrelated species. The paper by Horsch et al. describes the principles behind this approach and demonstrates the successful incorporation of new genetic material in *Nicotiana plumbaginifolia*, a diploid tobacco species. This model system has provided the basis for both practical and experimental applications of this technique to plant science research. Although it may be presumptious to describe a paper that was published so recently as a "classic," this technology is likely to have a significant impact on the future of plant breeding and horticulture.

The development of methods for genetic transformation of plant cells has rested almost exclusively on an understanding of the biology of the pathogen *Agrobacterium tumefaciens*. This soilborne bacterium is the causative organism of crown gall disease on various dicotyledonous plants (Bevan and Chilton 1982, Chilton 1983). In the late 1970s it was demonstrated that the molecular basis of this disease was the transfer of DNA from the bacterium to the plant host (Chilton et al. 1977). The transferred DNA (T-DNA) is carried on the tumor-inducing (Ti) plasmid in the bacterium. During infection, the T-DNA is transferred from the bacterium and integrated within the nuclear genome of the plant cell. One of the fascinating aspects of this pathogen is that the T-DNA, although normally present in a prokaryote, carries genes that are designed to function in a eukaryote. Three of these genes encode proteins involved in the synthesis of plant growth regulators. The activity of these enzymes alters the levels of auxins and cytokinins and results in the disorganized proliferaton of infected cells—the formation of a tumor. Another protein encoded by the T-DNA is responsible for the synthesis of a compound that *A. tumefaciens* is able to use as its sole carbon and nitrogen source. The tumor thus provides a niche for the bacterium. This unique transformation system, involving the natural transfer of genetic material from a prokaryote to a eukaryote, has been studied intensively during the last 10 years. Horticultural species such as tobacco and petunia have been used as the plant hosts for this research because *A. tumefaciens* does not normally infect monocotyledonous plants, such as maize, rice, and wheat. Thus it is likely that some of the first applications of this technology will be within the horticulture industry, using crops such as potato (Shahin and Simpson 1986) and tomato (Shahin et al. 1986).

The paper by Horsch et al. describes the use of *A. tumefaciens* to introduce DNA into protoplasts of *N. plumbaginifolia*, the selection of transformed cells and their regeneration into plants, and the sexual transmission of the newly introduced DNA. The success of this work depended on the application of two related developments. Earlier experiments using *A. tumefaciens* to introduce DNA into plants had relied on the production of tumors to identify transformants. However, it is not possible to regenerate plants from tumors because of the high levels of phytohormones. To overcome this obstacle it was necessary to develop alternative methods for identifying cells that had received the T-DNA and to modify the T-DNA so that it no longer carried the genes involved in the production of auxins and cytokinins. Genes encoding resistance to antibiotics had been isolated from a number of bacteria. To function in a plant cell, the DNA encoding a bacterial protein for resistance to an antibiotic must be flanked by DNA sequences that allow this gene to be expressed in a plant cell. A number of research groups had shown that chimeric antibiotic resistance genes were functional in tumors (Bevan et al. 1983, Fraley et al. 1983, Herrera-Estrella et al. 1983). In an elegant series of experiments, Garfinkel et al. (1981) showed that the T-DNA contained a number of genetic loci that affected tumor morphology. These loci have since been shown to correspond to the T-DNA genes involved in phytohormone synthesis. These genes are not required for the transfer of T-DNA from bacterium to plant cell, so it was postulated that if these genes were deleted from the T-DNA, transformation could still occur without the formation of a tumor. In the absence of a tumor to identify plant cells that had received the T-DNA, transformants could be selected on the basis of their resistance to the antibiotic. In this paper the tumor-producing genes were not deleted from the T-DNA; instead, the T-DNA was modified so that one of two possible transfer events could occur, both involving the transfer of the antibiotic resistance gene. The "long transfer" would result in the formation of a tumor, whereas the "short transfer" would carry none of the genes required for tumorigenesis and produce transformants that could be regenerated. Protoplasts of *N. plumbaginifolia* were infected with the strain of *A. tumefaciens* carrying the modified T-DNA constructed by Horsch et al. The antibiotic-resistant plant cells that were obtained after infection could be divided into two categories, as predicted. Those that where able to regenerate into plants contained the "short transfer" T-DNA. Progeny of these plants inherited the T-DNA in a simple Mendelian manner and were morphologically normal.

In the three years since this publication, further progress has been rapid. Methods are now available for the routine introduction of virtually any piece of DNA into the genome of a variety of plant species that are susceptible to infection by *A. tumefaciens* (Bevan 1984, Fraley et al. 1985). Regeneration of transformed plants is also possible for some of these species. Many of the tools developed here have also been of use in designing transformation systems for species that are not readily infected by *A. tumefaciens*. For example, maize cells have been transformed using electroporation to introduce DNA into protoplasts (Fromm et al. 1985). The problem of regeneration from single transformed cells remains for many species, notably the cereals. However, recent observations suggest that this may be irrelevant as transformation of rye has been obtained by applying DNA directly to developing floral tillers (de la Pena et al. 1987). The rapid progress in this area of plant molecular biology has surpassed most expectations, and it is likely that this will be maintained in the future.

This paper comes from a research group at Monsanto Company and is a reflection of the commercial interest in developing the necessary skills for applying recombinant DNA technology to plants. To date, no cultivars have been released as a result of this research. There are clearly many problems to be addressed in identifying genetic traits that are amenable to this sort of manipulation. Nevertheless, this paper has laid the groundwork for future development.

Peter B. Goldsbrough
Department of Horticulture
Purdue University
West Lafayette, Indiana 47907

## REFERENCES

BEVAN, M. 1984. Binary *Agrobacterium* vectors for plant transformation. Nucleic Acids Res. 12:8711–8721.

BEVAN, M. W., and M. D. CHILTON. 1982. T-DNA of the *Agrobacterium* Ti and Ri plasmids. Annu. Rev. Genet. 16:357–384.

BEVAN, M. W., R. B. FLAVELL, and M. D. CHILTON. 1983. A chimeric antibiotic resistance gene as a selectable marker for plant cell transformation. Nature 304:184–187.

CHILTON, M. D. 1983. A vector for introducing new genes into plants. Sci. Am. 248(6):50–59.

CHILTON, M. D., M. H. DRUMMOND, D. J. MERLO, D. SCIAKY, A. L. MONTOYA, M. P. GORDON, and E. W. NESTER. 1977. Stable incorporation of plasmid DNA into higher plant cells: The molecular basis of crown gall tumorigenesis. Cell 11:263–271.

DE LA PENA, A., H. LORZ, and J. SCHELL. 1987. Transgenic rye plants obtained by injecting DNA into young floral tillers. Nature 325:274–276.

FRALEY, R. T., ET AL. 1983. Expression of bacterial genes in plants. Proc. Natl. Acad. Sci. (USA) 80:4803–4807.

FRALEY, R. T., ET AL. 1985. The SEV system: A new disarmed Ti plasmid vector system for plant transformation. Bio/Technology 3:629–635.

FROMM, M., L. TAYLOR, and V. WALBOT. 1985. Expression of genes introduced into monocot and dicot plant cells by electroporation. Proc. Natl. Acad. Sci. (USA) 82:5824–5828.

GARFINKEL, D. J., R. B. SIMPSON, L. W. REAM, F. F. WHITE, M. P. GORDON, and E. W. NESTER. 1981. Genetic analysis of crown gall: Fine structure map of the T-DNA by site-directed mutagenesis. Cell 17:143–153.

HERRERA-ESTRELLA, L., A. DEPICKER, M. VAN MONTAGU, and J. SCHELL. 1983. Expression of chimeric genes transferred into plant cells using a Ti-plasmid-derived vector. Nature 303:209–213.

SHAHIN, E. A., and R. B. SIMPSON. 1986. Gene transfer system for potato. HortScience 21:1199–1201.

SHAHIN, E. A., K. SUKHAPINDA, R. B. SIMPSON, and R. SPIVEY. 1986. Transformation of cultivated tomato by a binary vector in *Agrobacterium rhizogenes:* Transgenic plants in normal phenotypes harbor binary vector T-DNA, but no Ri plasmid T-DNA. Theor. Appl. Genet. 72:770–777.

# Inheritance of Functional Foreign Genes in Plants

Abstract. *Morphologically normal plants were regenerated from* Nicotiana plumbaginifolia *cells transformed with an* Agrobacterium tumefaciens *strain containing a tumor-inducing plasmid with a chimeric gene for kanamycin resistance. The presence of the chimeric gene in regenerated plants was demonstrated by Southern hybridization analysis, and its expression in plant tissues was confirmed by the ability of leaf segments to form callus on media containing kanamycin at concentrations that were normally inhibitory. Progeny derived from several transformed plants inherited the foreign gene in a Mendelian manner.*

*Agrobacterium tumefaciens*, the causative agent of crown gall disease, is capable of transferring a DNA segment (designated T-DNA), located between specific border sequences, from its tumor-inducing plasmid (Ti plasmid) into the nuclear DNA of infected plant cells (*1*). Expression of T-DNA-encoded tumor genes in the transformed cell provides a selectable trait for recognition of those cells in culture; namely, the ability to grow on medium without added phytohormones. Unfortunately, this trait interferes with regeneration of normal fertile transformed plants (*2*).

Recently, we (*3*) and others (*4*) have constructed chimeric genes that function as dominant selectable markers in plant cells, thus making the tumor genes unnecessary for identification of transformants. Our chimeric gene contains the coding sequence of the bacterial gene for neomycin phosphotransferase II (NPTII) joined to the 5′ and 3′ regulatory regions of the nopaline synthase (NOS) gene, which is expressed constitutively in higher plant cells (*5*). We have shown that petunia and tobacco cells transformed with this chimeric NOS/NPTII/NOS gene are readily selected and are highly resistant to kanamycin (*3*).

We now report that kanamycin-resistant plant cells obtained with our vector system regenerate to morphologically normal plants. These plants carry a functional kanamycin-resistance gene and produce viable seeds. Analysis of progeny shows that the chimeric kanamycin-resistance gene is inherited and is expressed as a dominant Mendelian trait.

We used the previously described pMON120 intermediate vector to introduce the chimeric NOS/NPTII/NOS gene into the *A. tumefaciens* Ti plasmid. The pMON120 plasmid also contains an intact NOS gene as a scorable transformation marker. This NOS fragment includes the nopaline-type T-DNA right border sequence (*6*). Because this additional border sequence is initially carried on a separate plasmid, we refer to our system as the split end vector (SEV) system. Figure 1 shows how this system is used. The pMON128 plasmid [pMON120 containing the chimeric kanamycin-resistance gene (Fig. 1B)] is introduced by conjugation into *A. tumefaciens* cells, where homologous recombination with a resident octopine-type Ti plasmid [pTiB6S3 (Fig. 1A)] occurs. The resultant cointegrate plasmid pTiB6S3::pMON128 (Fig. 1C) contains a hybrid T-DNA in which the nopaline-type right border sequence is positioned between the kanamycin-resistance gene and the tumor genes of the resident Ti plasmid. Use of the nopaline T-DNA border sequence during infection results in the transfer of a short T-DNA segment (Fig. 1E) which contains the kanamycin-resistance gene and an intact NOS gene but does not contain genes for tumor formation or octopine synthase (*7*). The short-transfer transformants can be regenerated to give intact plants as described below.

Transformation of *Nicotiana plumbaginifolia* cells was carried out with the engineered *A. tumefaciens* train containing the chimeric NOS/NPTII/NOS gene by the method of cocultivation (*3*).

Fig. 1. Steps in the SEV system for plant cell transformation. The arrows represent the T-DNA border sequences. *LIH* is a region of homologous DNA for recombination. The tumor genes are represented by *tms* and *tmr* (*11*); *OCS* and *NOS* are octopine and nopaline synthase genes, respectively. The chimeric kanamycin-resistance gene is designated as *kan^r*. The bacterial spectinomcyin - streptomycin resistance determinant for selection of cointegrates is designated *spc/str^r*. Reciprocal recombination of (A) a resident Ti plasmid (pTiB6S3) and (B) pMON120 derivative (pMON128) yields (C) the cointegrate, pTiB6S3::pMON128. After cocultivation and selection for kanamycin-resistant plant cells either (D) the entire hybrid T-DNA or (E) a truncated T-DNA without tumor genes is transferred into the plant genome.

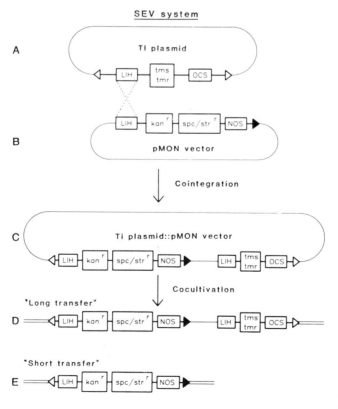

SEV system

Ti plasmid

pMON vector

Cointegration

Ti plasmid::pMON vector

Cocultivation

"Long transfer"

"Short transfer"

496

Leaf mesophyll protoplasts of *N. plumbaginifolia* were obtained and cultured as described by Pollock *et al.* (*8*). Two-day-old cultures of protoplast-derived cells were inoculated with live *A. tumefaciens* cells ($10^7$ cells per milliliter). After 48 hours of coculture, the plant cells were collected by centrifugation (100*g* for 5 minutes), washed twice, and replated in medium containing carbenicillin (500 μg/ml) to kill the remaining bacteria. On the sixth day after isolation of protoplasts, the procedure described for petunia cells (*3*) was used to transfer the cultures from liquid to feeder plates (*9*) of the same medium. After 7 to 10 days on the feeder plates, the colonies were transferred to fresh medium (without feeder cells). After another 7 days, colonies were transferred to a different medium [MS salts (Gibco), B5 vitamins, 3 percent sucrose, carbenicillin (500 μg/ml), kanamycin (100 μg/ml), benzyl adenine (1.0 μg/ml), and naphthalene acetic acid (0.1 μg/ml); *p*H 5.7]. After 7 days, the most promising colonies (largest and greenest) were picked from the filter paper substrate and transferred to the same medium at low density [10 colonies per petri plate (100 by 15 mm)].

In our first experiment, four kanamycin-resistant colonies were recovered from approximately $8 \times 10^4$ protoplasts cocultivated with pTiB6S3::pMON128. No resistant colonies were found among the same number of untreated control colonies. Southern blot hybridization with an NPTII specific probe showed that all four resistant colonies contained the chimeric gene (data not presented). One of the four colonies was morphogenic and produced a kanamycin-resistant plant, NPK3.

Leaf segments from NPK3 were able to form callus and proliferate new shoots on medium containing kanamycin (100 μg/ml) (Fig. 2A). In contrast, leaf segments from wild-type plants were completely inhibited. Analysis with Southern blot hybridization showed that the 1.5-kilobase (kb) Eco RI fragment containing the chimeric NOS/NPTII/NOS gene was present in the leaves of NPK3 but not in tissue from wild-type plants (lanes c and d in Fig. 2B).

Twenty-one first-generation progeny plants ($S_1$) from the self-fertilized transformed parent NPK3 were grown to maturity and tested for kanamycin resistance in the leaf callus assay. Fifteen of the 21 were able to form callus on medium with kanamycin (100 μg/ml). Another 80 seedlings (germinated under sterile conditions) were transferred to medium containing kanamycin (100 μg/ml). Of

these, 62 grew several times larger and formed callus, whereas 18 ceased growth and did not form callus. Thus the trait was inherited in a Mendelian manner with a 3:1 ratio. The final proof of the

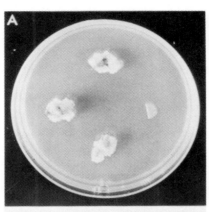

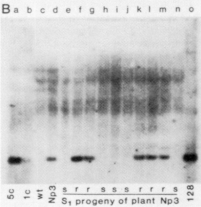

Fig. 2. Analysis of transformed progeny. (A) Leaf callus assay. Surface-sterilized segments of leaves were placed on medium [MS salts (Gibco), B5 vitamins, 3 percent (weight to volume) sucrose, benzyl adenine (1 μg/ml), and naphthalene acetic acid (0.1 μg/ml), *p*H 5.7] containing kanamycin (100 μg/ml). Explants from wild-type plants were unable to grow on this medium, whereas explants from our transformed plants callused and generated shoots within 3 weeks (data not shown). The explants shown here are from four separate $S_1$ progeny of NPK3. One of the progeny plants is clearly sensitive to kanamycin, whereas the other three are resistant. DNA blot hybridization analysis. (B) Total plant DNA was extracted, purified by CsCl gradient centrifugation, and digested (10 μg) with the restriction enzyme Eco RI as described (*3*). After transfer of the DNA to nitrocellulose, a nick-translated DNA probe specific for transposon Tn*5* was used to identify a fragment containing the chimeric NOS/NPTII/NOS gene (*3*). (Lanes a and b) Five-copy (*5c*) and one-copy (*1c*) reconstruction experiments; (lane c) DNA from wild-type (*wt*) control plants; (lane d) DNA from parental NPK3 plant (*Np3*); (lanes e to n) DNA from $S_1$ progeny of NPK3 plant; and (lane o) digested pMON128 plasmid showing the position of the 1.5-kb fragment of the chimeric gene. The letters *r* and *s* denote kanamycin resistance and sensitivity, respectively, in the leaf callus assay.

correspondence between the presence of the chimeric gene and the antibiotic-resistant phenotype was established by the perfect correlation between inheritance of the chimeric gene and the kanamycin-resistant phenotype in $S_1$ plants (lanes e to n in Fig. 2B).

Three subsequent cocultivation experiments gave high frequencies of transformation, averaging 6 percent of the total colonies or 1.2 percent of the total initial protoplasts plated. The control populations consistently failed to yield any kanamycin-resistant colonies. Most (about 90 percent) of the kanamycin-resistant colonies produced both octopine and nopaline and were nonmorphogenic, as expected for transformants arising when the octopine T-DNA right border was utilized (Fig. 1D). About 10 percent of the colonies were morphogenic, producing shoots that could be excised, rooted, and grown in soil. Of 22 plants examined, 9 were escapes or revertants that showed none of the markers of transformation. The other 13 plants produced nopaline, but not octopine, and were resistant to kanamycin as measured by a leaf callus–induction assay.

The $S_1$ progeny from three of the independently isolated, nopaline-producing and kanamycin-resistant plants (NPK7, NPK9, NPK10) were scored for nopaline content. In each case, the progeny showed normal Mendelian inheritance and expression of the inserted DNA segment: 71 of 105 progeny of NPK7, 39 of 48 progeny of NPK9, and 34 of 44 progeny of NPK10 produced nopaline. In addition, axenically grown seedlings from each of the transformants showed similar segregation for ability to form callus on medium containing kanamycin (100 μg/ml). For example, 37 of 51 progeny of NPK10 formed callus in the presence of kanamycin, and all 37 resistant progeny produced nopaline.

We have shown that (i) the chimeric NOS/NPTII/NOS gene is expressed in *N. plumbaginifolia*, (ii) the regenerated transformed plants are phenotypically normal and fertile, and (iii) normal Mendelian inheritance of an engineered gene can occur in the progeny of transformed plants. The genetic transmission of chimeric antibiotic-resistance genes has now been confirmed for the $S_2$ progeny of the NPK3 plant.

Normal Mendelian inheritance of the chimeric gene has also been demonstrated for petunia plants transformed with pMON120-type vectors (*10*). The availability of dominant selectable markers and transformation vectors that permit the regeneration of phenotypically normal plants will greatly facilitate studies of gene expression and regulation in plants.

ROBERT B. HORSCH
ROBERT T. FRALEY
STEPHEN G. ROGERS
PATRICIA R. SANDERS
ALAN LLOYD
NANCY HOFFMANN

*Monsanto Company,*
*St. Louis, Missouri 63167*

### References and Notes

1. M.-D. Chilton, M. H. Drummond, D. J. Merlo, D. Sciaky, A. L. Montoya, M. Gordon, E. Nester, *Cell* 11, 263 (1977); N. Van Larebeke, G. Engler, M. Holsters, S. Van der Elsacker, I. Zaenen, R. Schilperoort, J. Schell, *Nature (London)* 252, 169 (1974); M.-D. Chilton, R. Saiki, N. Yadav, M. Gordon, F. Quetier, *Proc. Natl. Acad. Sci. U.S.A.* 77, 4060 (1980); N. Yadav, K. Postle, R. Saiki, M. Thomashow, M.-D. Chilton, *Nature (London)* 287, 458 (1980); L. Will-mitzer, M. Debeuckeleer, M. Lemmers, M. Van Montague, J. Schell, *ibid.*, p. 359.
2. A. Braun, *Cancer Res.* 16, 53 (1956); _____ and H. Wood *Proc. Natl. Acad. Sci. U.S.A.* 73, 496 (1976); F. Yang *et al., Mol. Gen. Genet.* 177, 707 (1980).
3. R. T. Fraley *et al., Proc. Natl. Acad. Sci. U.S.A.* 80, 4803 (1983).
4. L. Herrera-Estrella, M. DeBlock, E. Messens, J.-P. Hernalsteens, M. Van Montague, J. Schell, *EMBO J.* 2, 987 (1983).
5. J. Tempe and A. Goldmann, in *Molecular Biology of Plant Tumors*, G. Kahl and J. Schell, Eds. (Academic Press, New York, 1982), pp. 427–449.
6. P. Zambryski, A. Depicker, K. Kruger, H. Goodman, *J. Mol. Appl. Genet.* 1, 361 (1982); N. Yadav, J. Vanderleyden, D. Bennet, W. Barnes, M.-D. Chilton, *Proc. Natl. Acad. Sci. U.S.A.* 79, 6322 (1982).
7. Evidence that the nopaline border sequence functions in this manner will be presented (R. B. Horsch, in preparation).
8. K. Pollock, D. G. Barfield, R. Shields, *Plant Cell Rep.* 2, 36 (1983).
9. R. B. Horsch and G. E. Jones, *In Vitro* 16, 103 (1980).
10. R. B. Horsch, unpublished results.
11. D. Garfinkel, R. Simpson, R. Ream, F. White, M. Gordon, E. Nester, *Cell* 27, 143 (1981).

1 November 1983; accepted 30 November 1983